Functional Thinking for Value Creation

Jürgen Hesselbach • Christoph Herrmann
Editors

Functional Thinking for Value Creation

Proceedings of the 3rd CIRP International Conference
on Industrial Product Service Systems,
Technische Universität Braunschweig,
Braunschweig, Germany, May 5th - 6th, 2011

Springer

Editors
Prof. Dr.-Ing. Dr. h.c. Jürgen Hesselbach
Technische Universität Braunschweig
Institut für Werkzeugmaschinen
und Fertigungstechnik (IWF)
Langer Kamp 19B
38106 Braunschweig
Germany

PD Dr.-Ing. Christoph Herrmann
Technische Universität Braunschweig
Institut für Werkzeugmaschinen
und Fertigungstechnik (IWF)
Langer Kamp 19B
38106 Braunschweig
Germany

ISBN 978-3-642-19688-1 e-ISBN 978-3-642-19689-8
DOI 10.1007/978-3-642-19689-8
Springer Heidelberg Dordrecht London New York

Library of Congress Control Number: 2011924878

Cover design: eStudio Calamar S.L.

Printed on acid-free paper

Springer is part of Springer Science+Business Media (www.springer.com)

Preface

„Die rasch fortschreitende Spezialisierung auf allen Gebieten der Naturforschung erschwert es dem Einzelnen, sich auch nur auf seinen Nachbargebieten zu orientieren. Geradezu unmöglich wird ihm die Orientierung aber auf den ferner liegenden Gebieten. Andererseits muß sich jedem das geistige Bedürfnis um so fühlbarer machen, den Zusammenhang mit dem Ganzen nicht zu verlieren, je mehr er gezwungen ist, das Feld der eigenen Arbeit einzuengen – ist er doch meistens auf die Hilfe von anderen Zweigen der Naturwissenschaft angewiesen."[1]

[1] *"The rapidly proceeding specialisation in all areas of natural sciences makes it difficult for the individual to find orientation even in his neighbouring fields. It is further almost impossible to find orientation in the more distant fields. But then again, the individual must feel the intellectual need to understand the context of the whole, and even more so when he is forced to localize his own area of research – as he is mostly dependent on the assistance of other branches of natural sciences"*

Arnold Berliner. Die Naturwissenschaften (The Science of Nature). Vol.1/1, Springer, Berlin (1913).

Keeping track of the whole context in spite of further specialisation, the chance and the necessity to cooperate with other disciplines, to learn from them and to get inspiration from them is not only true for the natural sciences but also for engineering sciences. For the introduction of this year's conference let us take a look into ecology. A well-known phenomenon in ecology is the edge effect. The edge- or transition zone between two habitats enables the exchange of energy, organisms and organic material. Those edge- or transition zones are often characterised by a high structural diversity. Large edge- or transition zones are also named ecotone. Ecotone is a combination of the word eco(logy) and tone, where tone comes from the Greek word tonos (engl. tension), expressing a place where ecologies are in tension. The contact zones in ecotones may take very different shapes, starting from simple ecotones with equal and homogeneous surfaces to more complex ones.

Analogous to ecotones, tangible product(s) and intangible service(s) meet in Product-Service Systems (PSS). They are interlocked or tensed (braced to one another). Service(s) can be defined as a series of activities designed to enhance the level of customer satisfaction. Product and service are equally important for the function fulfilment, but the relation between physical product and service varies, similar to that of ecotones, from simple combination of one product and one service to more complex models. Whereas PSS as a first priority include consumer products in the range of services, Industrial Product-Service Systems (IPS^2) focus on industrial customers and on their associated demands. Thus, IPS^2 is defined as "an integrated industrial product and service offering that delivers value in use". The customers expect comprehensive solutions, which are adapted to their individual needs. IPS^2 offers the possibility to stand out from competition and for long-term customer loyalty. Particularly in times of economic crisis it becomes apparent which producing companies understand to satisfy the needs and requirements of their customers. Or in terms of ecology: The success of IPS^2 is limited by its range of tolerance for local environmental factors or is rather dependent on how fast and reliable the system can be adapted to changing factors.

After Cranfield in the year 2009 and Linköping in 2010, this year's International CIRP Conference on Industrial Product-Service Systems takes place in Braunschweig. The topics of this year's conference are amongst others planning and design of IPS^2, planning- and design methods, IPS^2 business models as well as case studies from different industries. Particularly in this relatively new domain IPS^2 it will be important to keep track of the whole context and to seek cooperation with other research fields and disciplines.

Jürgen Hesselbach

Christoph Herrmann

Table of Contents

IPS² - From Concept to Application

IPS² - Integration and Process Management

IPS² - Knowledge & Information Management

IPS² - Evaluation Methods

IPS² - Business Models and Case Studies

Organization

CHAIRMEN

Prof. J. Hesselbach
PD Dr.-Ing. Christoph Herrmann

ORGANIZING COMMITTEE

Chief Organizers

Dipl.-Wirtsch.-Ing. Tim Heinemann
Dipl.-Wirtsch.-Ing. Mark Mennenga

Organizing Committee

Hannah Jule Schäfer, M.A.
Dipl.-Ing. (FH) Stefan Andrew
Dr.-Ing. Ralf Bock
Dipl.-Ing. Gerrit Bogdanski
Dr.-Ing. Tina Dettmer
Dipl.-Wirtsch.-Ing. Patricia Egede
Dipl.-Wirtsch.-Ing. Philipp Halubek
Dipl.-Ing. Mohamad Jamal Kayasa
Dipl.-Ing. Michael Krause
Dipl.-Wirtsch.-Ing. Katrin Kuntzky
Dr.-Ing. Tobias Luger
Dipl.-Chem. Gerlind Öhlschläger
Anne-Marie Schlake, M.A.
Dipl.-Wirtsch.-Ing. Tim Spiering
Dipl.-Wirtsch.-Ing. Julian Stehr
Dipl.-Wirtsch.-Ing. Sebastian Thiede
Dipl.-Wirtsch.-Ing. Marius Winter
Dipl.-Wirtsch.-Ing. André Zein

INTERNATIONAL SCIENTIFIC COMMITTEE

Prof. T. Arai / JP
Prof. J. Aurich / DE
Prof. M. Björkman / SW
Prof. C. Boks / NO
Prof. D. Brissaud / FR
Prof. J. L. Chen / TW
Prof. J. Duflou / BE
Dr. T. Hara / JP
Prof. S. Kara / AU
Prof. J. M. Leimeister / DE
Prof. H. Meier / DE
Prof. N. Morelli / DK
Prof. G. Ölundh Sandström / SW
Prof. S. Robra-Bissantz / DE
Prof. R. Roy / UK
Prof. G. Schuh / DE
Prof. G. Seliger / DE
Prof. Y. Shimomura / JP
Prof. D. Spath / DE
Prof. T. Spengler / DE
Prof. S. Takata / JP
Prof. T. Tomiyama / JP
Prof. E. Uhlmann / DE

Service Cost Estimation Challenges in Industrial Product-Service Systems

Rajkumar Roy[1], John Ahmet Erkoyuncu[1]

[1] Manufacturing Department, Cranfield University, Cranfield, Bedfordshire, MK43 0AL, UK

Abstract

The aim of this keynote paper is to demonstrate that with emerging challenges experienced in delivering integrated product and service solutions the adopted techniques for service cost estimation is evolving. Whilst representing the current practice and existing challenges in service cost estimation, through interaction with four major organisations in the defence industry, the paper focuses on the implementation of agent based modelling (ABM) as a possible approach to be adopted more prevalently in the future. Subsequent to presenting a sample agent based model, the paper presents three architectures of ABM that would potentially be considered for service cost estimation, including representations of interaction across organisations, equipments and humans.

Keywords:

Cost Estimation; Industrial Product-Service Systems; Service; Uncertainty; Agent Based Modelling

1 INTRODUCTION

Industrial Product-Service Systems (IPS^2) apart from having a physical product core tends to be high in net worth and result in business to business relationships [1]. Three business models have commonly been suggested for the delivery of IPS^2 including product, use and result oriented, whereby the transformation from a traditional business model varies across each of these [2]. The product oriented model operates on the pay on order principle. Payment in the use oriented model concentrates on the delivered availability level, and the approach has found much application in the defence industry in the UK through Contracting for Availability (CfA). This involves a commercial process which seeks to sustain an equipment/system/part at an agreed level of readiness, over a period of time (e.g. equipment operational life - 30 to 40 years) by building a partnering arrangement [3]. These performance driven contracts require enhanced interaction across the supply chain as typically a large proportion of parts or support is brought in by the solution provider [4]. The range of service offerings varies across CfA, including maintenance, repair, asset and operation management, supply chain management and training. In the result oriented model the payment is based on delivered capability [5].

Co-creation of value has often been considered in association to the delivery of IPS^2 (e.g. [6]). Along these lines, the collaborated nature of processes across the supply network enables to achieve the targeted value delivery. IPS^2 delivery has progressively increased in scale and complexity. Examples range across major infrastructure projects (e.g. Private Finance Initiative hospitals) and large defence projects (e.g. complete sea, air or land platforms). A common example of CfA includes Total Care Products, which integrates hardware and service support systems [7].

In CfA, the sales value of each transaction with respect to costs incurred, which may be negotiated case-by-case, determines the supplier's profitability whilst transaction throughput determines the affordability for the customer. As opposed to CfA, under traditional contract arrangements, suppliers are typically paid according to the throughput of 'spares and repairs', or other offerings such as mitigations for obsolescence, maintenance, asset management, query resolution etc [8]. Under CfA the mean time to repair (MTTR) or other performance criterion is made the essence of the contract. At the time of bidding, the supplier typically offers a fixed price to the customer whilst assuming responsibility for estimating the cumulative number of transactions needed to sustain the MTTR.

There is a breadth of literature that has discussed the challenges experienced in transforming in to service driven organisation from a product context (e.g. [9, 10, 11, 12]). Though, on the other hand, some of the main advantages of adopting CfA include, better risk management, guaranteed innovation which enables technological advancement, through life consideration of equipment, lean response to customer's service requirements [7]. For the Original Equipment Manufacturer (OEM), it enables a steady cash flow, long term relationship enabling sharing of equipment related data, sustainability in design and competitive advantage [6]. For the customer, a major potential benefit includes lower servicing costs through aligned objectives.

The financial implications of CfA necessitate adequate consideration of whole life cost estimation. However, the emergence of IPS^2 has further enhanced the influence of uncertainty in service cost estimation, which undermines the ability to forecast. Uncertainty is the stochastic behaviour of any physical phenomenon that causes the indefiniteness of outcomes meaning the expected and actual outcomes are never the same. The uncertainty is largely driven by changes in the responsibilities of the solution provider and the customer. As a result, a gradual evolution is experienced in the adopted techniques for service cost estimation. Along these lines, this paper aims to demonstrate the perspectives from literature and industry in order to represent the current and future challenges for service cost estimation. The following section presents the methodology followed.

2 RESEARCH METHODOLOGY

In order to achieve the objectives for this study an iterative process was followed including literature review and industry interaction.

J. Hesselbach and C. Herrmann (eds.), *Functional Thinking for Value Creation: Proceedings of the 3rd CIRP International Conference on Industrial Product Service Systems, Technische Universität Braunschweig, Braunschweig, Germany, May 5th - 6th, 2011*,
DOI 10.1007/978-3-642-19689-8_1,

2.1 Literature review

The dominant interests within this research included service delivery systems, uncertainty, uncertainty management, cost estimation, uncertainty modelling, supply chain, and IPS[2]. Some of the main journals contributing to this research included Harvard Business Review (for service oriented research) and AIAA (for uncertainty in cost estimation). Other contributors included: Production and Operations Management, International Journal of Production Economics and Management Science.

2.2 Industrial interaction

In depth literature review facilitated preparation for interaction with industry through semi-structured interviews, case studies, and workshops. The participants were limited to the service based cost estimation context in the defence industry. The industry interaction for the presented material in this paper totalled over 100 hours. The interaction aimed to gauge various aspects such as service cost estimation techniques, role of uncertainty and experienced challenges. The analysis involved transcription of the interviews, developing mind maps and writing reports to illustrate the learning. The background of the respondents ranged across cost engineers, project managers, support managers, engineering managers, and functional experts (e.g. in risk and uncertainty). Their experience ranged between 5 and 30 years. The outcomes from industrial interaction are presented in section 3. The industry participants included the UK defence customer and three major organisations in the defence industry. An overview of the interacted organisations is provided below:

- Organisation A: Their range of products and services accounts for air, land and naval forces, as well as advanced electronics, security, information technology solutions and customer support
- Organisation B: They are a world-leading provider of jet engines, components and integrated systems for commercial and military aircraft
- Organisation C: They are a leader in systems integration working on major programs spanning the aerospace, defence, civil and commercial sectors
- Organisation D: They are the defence customer in the UK government department responsible for implementing government defence policy and is the headquarters of the British Armed Forces

3 CURRENT PRACTICE

3.1 Service Cost Estimation

According to the Association for the Advancement of Cost Engineering (AACE), Cost Engineering is defined as "the area of engineering practice where engineering judgment and experience are used in the application of scientific principles and techniques to problems of cost estimating, cost control, business planning and management science, profitability analysis, project management, and planning and scheduling" [13]. It is a methodology being used for predicting/forecasting/estimating the cost of a work's activity or output [14]. The application of CfA has further promoted the concept of life cycle costs (LCC), which is the total cost over a product's life cycle span [15]. The industrial business processes have moved from delivering spares and parts to total care packages through the whole lifetime of a product [16]. These require integrated analysis of design cost, manufacturing cost, operating cost, and disposal cost. Other terminologies for LCC are WLCC (Whole Life Cycle Cost) and TLC (Through-Life Cost).

The overall growth in adoption of WLCC has been relatively slow. This can be attributed to a number of factors. For instance, estimators may experience too much uncertainty across the phases of the life cycle. The in-service phase sets the context to this study. The segmentation of the life cycle enables a cost breakdown structure (CBS) to be produced for use in allocating budgets to individual cost centres and recording actual spend [17]. The approach taken in developing the CBS may be driven by various factors such as a legacy structure, customer requirement, the product breakdown, project organisation, or functional organisation.

Service cost estimating considers various activities that enable or enhance the operational life of given equipment [18]. Hansen and Mowen [19] emphasise that traditional costing has focused on companies manufacturing physical products and virtually ignored costing of services. Cost assessment of CfA offerings remains a challenge and has not been addressed in literature [20]. Some of whole life cycle cost literature focuses on assessing the maintenance or in-service costs [18, 21, 22]. But, most of this literature has an emphasis on costing the service associated with stand-alone products. Furthermore, the two main areas of interest in relation to the service phase relates to the prediction of the service life and performance modelling, and cost estimation approaches vary being either deterministic or stochastic methods [21]. This process covers a number of areas including the prediction of the remaining service life of the facility components and the prediction of the rate of their deterioration [23]. Some of the main objectives that need to be identified and assessed at this stage include survey condition of the existing facility components, assumptions about the remaining service lives of components, updating budget requirement, priority of components updating-critical components, quality of maintenance and replacement components, assumption about time-lag replacement or maintenance delays, the effect of delayed maintenance on budget and deterioration of facility, evaluating the economic viability with a view to disposal [24]. Furthermore, with CfA, industry has become responsible of service and support activities such as bidding, acquisition, product repair and overhaul, retrofitting and upgrades, training, obsolescence, and disposal, in addition to planned and unplanned maintenance in order to provide an agreed level of performance (i.e. availability level).

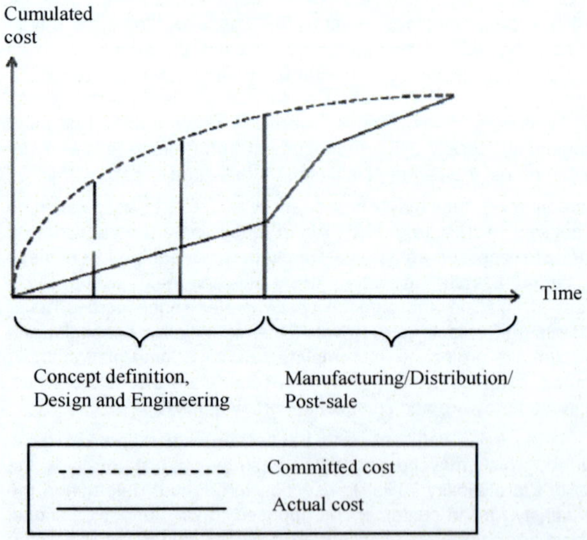

Figure 1: Committed versus actual cost.

As illustrated in Figure 1, up to 75 to 80 percent of the WLCC is often committed before contract award, or shortly after, because the early design activity must scope the product solution and service solution concurrently owing to interdependency [18, 25]. In fact this behaviour is often driven by requirements in the customer's request for proposal that ask for early visibility of the WLCC predictions [26]. The percentage uncertainty in cost estimates for the product solution (including development, manufacturing, testing, integration, certification, and acceptance) is often much less than that of the support solution [27]. Furthermore, by the time cost outcomes become evident in the first year or two of the support phase it is difficult to modify the support solution from either the technical or contractual viewpoint (e.g. in the event that it becomes necessary to pre-empt predicted cost overruns). Some availability contracts mitigate this problem by means of the 'evergreen renewal' principle that permits re-negotiation at intervals, e.g. every five years for a thirty year contract, but this is not seen as an ideal solution. For these reasons it is a priority to address uncertainty in the cost estimating of the support phase of CfA, and to prepare initial estimates at an earlier phase of the lifecycle than under traditional arrangements where the product solution and support solution are decoupled under separate, sequential contracts.

In service cost estimation, historical data provides guidance in terms of costs and priority of required maintenance, rehabilitation and replacement. However, it is necessary for the data based cost estimates to be supplemented with expert opinions in order to perform whole life cycle analysis and risk assessment [27]. In terms of service cost estimation, interest has been growing recently. A commonly adopted methodology to estimate service costs includes 5 major steps [28]:

- Define decision problem and objectives
- Detailed description of the service(s)
- Identify and classify resource items and units of resources to deliver service
- Measure resource consumption
- Place monetary value on each resource item (goods, activities) and calculating the unit costs of a particular service

In order to achieve accurate cost estimates it is necessary to have the required inputs, techniques, processes as well as tools in place [29]. The inputs include various sources such as a work breakdown structure, assumptions list embedded in a Master Data Assumptions List (MDAL), customer requirements, historical data and expert judgement [30]. Among all inputs data commonly is referred to as a major source of challenge. Some of the main reasons causing data issues relate to variability in data, reliability, completeness, relevance, representativeness, accessibility, usability, and repeatability in retrieving data [31]. The availability of historic data from previous projects could improve the accuracy of cost estimates.

The stages of the lifecycle for the IPS^2, for example as in the Concept, Assessment, Demonstration, In-Service and Disposal (CADMID) cycle, are widely varied in scope and scale and require a variety of methods to most effectively predict the respective costs. Three well-recognised methods are mentioned in [32] and a fourth has been identified from industrial interaction:

- **estimating by analogy:** reads-across the cost outcomes from past projects having similarities with the one being estimated, traditionally used for the non-recurring engineering costs during development and unit cost during manufacture.
- **activity based costing**: identifies activities in an organisation and allocates the cost of each (e.g. in terms of man-hours, facilities, and materials) to products and services according to their actual consumption.
- **the parametric method:** derives cost estimating relationships (CERs) that can predict cost as a function of the basic attribute(s) of an item (e.g. weight, volume, complexity). Models in this group include regression analysis, fuzzy logic, and neural networks [33].
- **extrapolation:** particularly for operational and support phases, experience from prior contracts (e.g. spares and repairs) can inform projections of future costs when the incumbent supplier now needs to estimate for follow-on availability contracts.

Selection of the method largely depends on the available data. Relatively new methods such as fuzzy logic and evidence theory [33, 34, 35] are available but have not been widely adopted in industry because the uncertainty in the estimates is too large to assure the supplier of profitability and the customer of affordability. Ideally, more than one method is applied to a given estimating challenge, and this in itself can reduce uncertainty in the prediction of cost.

3.2 The Role of Uncertainty and Service Cost Estimation

There has been an increasing emphasis on understanding the link between uncertainty and cost since the Second World War, which has resulted in numerous approaches to assess cost uncertainty. The initial focal point was on defining the sources, scope, and types of uncertainties that impact the cost of future systems [36]. The main focus of the literature was particularly on understanding costs of military systems with long range decisions. However, the cost estimates were often lower than the actual cost estimate or an estimate developed at a later phase, where the cause was referred to as uncertainty [37].

CfA to a large extent has increased the level of uncertainty that the manufacturer faces [38]. In CfA delivery of a holistic support package necessitates a new form of service provision, while the interaction across the service network has drastically changed compared to the traditional form of contracting, which centres on ad-hoc provision of support. At the heart of this change is the performance driven nature of CfA.

To capture differences across projects, it is necessary to identify and define the scope of variables that contribute to uncertainty through ground rules and assumptions [39]. For instance, the focus of assumptions may include, mission scenario, deployment plans, system quantity, system availability, operating environment, and selection of cost estimating methods [17]. This also explains why the modelling of support costs is typically more challenging than the estimation for design and manufacturing. Furthermore, from a IPS^2 context some of the main challenges in integrating uncertainty to cost estimation include, inability of existing tools/methods to estimate technical scope, change in customer request, modelling the integrated product service delivery, lack of data to reflect the through life perspective. There are also more general concerns that affect the uncertainty modelling process. These include issues in classifying and prioritising risks and uncertainties. Furthermore, there are also issues in the modelling process (e.g. interdependencies, selecting modelling approach), and in defining the cost estimation boundaries.

3.3 Uncertainty Based Service Cost Estimation

Integration of uncertainty follows the cost estimation procedure [40]. In developing a framework of uncertainties to integrate into cost estimation, the work breakdown structure tends to be considered. The uncertainty level in a work breakdown structure will be influenced by the toughness of the required operation. Furthermore, prioritization of uncertainties depends on the cost impact and/or influence over schedule and performance. From a more concentrated view, each service offering derives its own specific uncertainties. Areas of interest for the commonly offered set of services can be found below:

- Spares services: The focus of the assessment is the frequency of support and the range of requirements.

- Repair services: The trade off between the costs of new spare parts and repair is considered. These offerings also take obsolescence into account. This may be in the case of an unrepairable component due to obsolescence. The uncertainty relates to how many times failures occur. Additionally, there is much uncertainty in determining effort.
- Defect response: Based on the likelihood of occurrence of defects in components, appropriate management procedures are developed. These responses depend on the priority of the defect.
- On-call service: Related to support that is provided on unplanned bases, 24/7. Some requirements may not be included in the contract.
- Health checks: Involve regular checks on equipment, which follows a planned process.
- Performance assessment: Focuses on measuring the performance of a contract. This may involve assessing the availability of support.

The integration of uncertainty to cost estimation follows a step by step procedure [41]. In this process initially all cost drivers affected by uncertainties need to be identified. In parallel a single point estimate needs to be calculated in order to add the influence of uncertainty on this value, which is typically represented through a probability distribution devised for each cost driver. For this purpose it is necessary to select the type of distribution (e.g. commonly normal, log-normal, and triangular) to apply. Subsequently it is necessary to estimate the distributions' parameters such as maximum, minimum and most-likely values, which refer to a three point estimate. These values reflect the possible outcomes, which yield numerical values that can be used in quantitative risk analysis to support decision making. Setting these values involves a degree of subjectivity as it relies on expert opinion; however it is commonly acknowledged that defining a cost range enables to capture the world more realistically compared to a single point estimate [42]. Definitions for the minimum, most likely and maximum cost estimates are defined as follows [43]:

- Minimum – Reflects the optimistic scenario where the cost outcomes are most in favour of the estimator.
- Most Likely – Represents the cost level which is right more often than any other (i.e. the mode, in statistical language).
- Maximum – Covers the pessimistic view of cost outcomes, where the worst scenario is assumed, but excluding the very unexpected - e.g. "Acts of God".

In order to reflect the influence of uncertainty on cost typically three-point estimates are devised, which, typically, are inputs to relatively complex analysis through Monte Carlo analysis [40]. With the repetition over a large number of times, by randomly combining different input values, results from Monte Carlo analysis are represented through probability distributions and the impact of uncertainty on cost drivers is realised [23]. This enables to replace single points with probability distributions of possible values based on the three point estimates.

Data collection is a major requisite for uncertainty analysis and cost estimation. In order to achieve the desired level of quality in cost estimates, good data whether based on historical analogies or engineering understanding is necessary [30]. Once the key objectives of the program (e.g. requirements, scope, and schedule) have been established, then it is necessary to specify the data requirements to build cost estimates. For instance, this involves identifying all potential variables in the cost model affected by risk and uncertainty and identifying the potential data sources for estimating risk and uncertainty for each of these variables (e.g. specifying probability distributions).

Figure 2 presents the way in which the cost of software development is driven by the Source Lines of Code (SLOC) and the evolution of the range estimates. Furthermore, through industrial interaction the authors did not come across a unique estimating methodology for CfA.

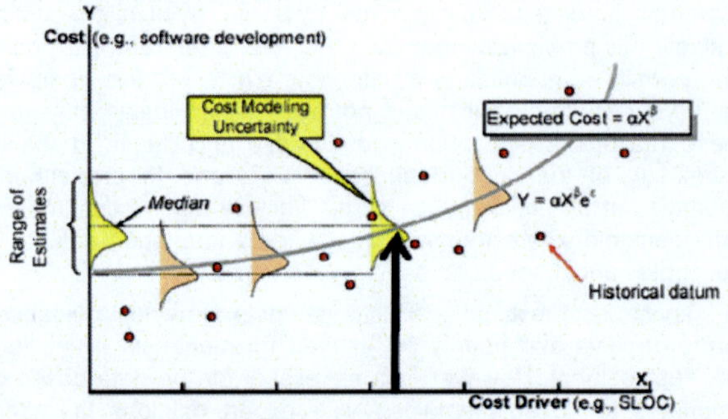

Figure 2: Baseline cost estimate [41].

3.4 Tools Used for Service Cost Estimation

There are several commercial software packages available that attempt to quantify cost uncertainty using the simulation method. Some of which provide stand-alone models (such as @RISK), whilst others incorporate the simulation option into the cost model (such as PRICE). These programs use a statistical sampling technique, such as Monte Carlo or Latin Hypercube, to generate the cumulative probability distribution curve for the total project cost [23]. Commercial models tend to be used for validation purposes, as the provided data may not fit into requirements. This has been the case because commercial software packages have been considered to be limited in adapting applications to the be-spoke nature of projects. A study conducted by NATO showed that most countries tend to use in-house tools to estimate costs across the life cycle [41]. On the other hand, commercial models tend to be used for validation purposes, as the provided data may not fit into requirements. These models have been considered to be limited in adapting applications to the be-spoke nature of projects:

- Predict (schedule only): Is a decision support tool that aims to enhance the management of uncertainty in predicting cost.
- @risk: Is the most widely used uncertainty analysis tool that utilizes Excel to perform Monte Carlo simulation.
- Crystal Ball: Is a spreadsheet based software suit used to predict software and service costs.

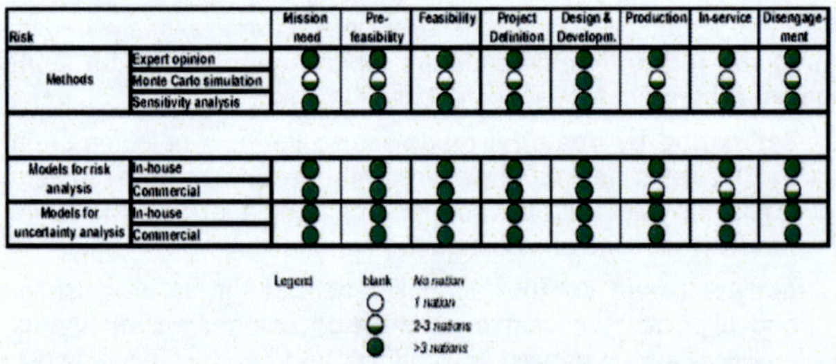

Figure 3: Cost uncertainty modelling across countries [41].

Figure 3 illustrates survey results from a NATO study, which indicates that when the uncertainty analysis is conducted the two most commonly used techniques seem to be expert opinion and sensitivity analysis [41]. Detailed analysis, using e.g. Monte Carlo simulation seems to be undertaken with varying degree of frequency by only two to three countries, using typically in-house built models.

3.5 Comparative Analysis of Industrial Findings

This section presents a comparison across the four interviewed organisations by focusing on the current practice in uncertainty based cost estimation. The similarities, differences and the unique aspects associated to each organisation concerning the uncertainty considerations is discussed below.

Similarities across the organisations

- Three-point estimating and application of Monte Carlo simulation
- Dependency on experience for identification, prioritisation, and defining degree of uncertainty in cost drivers. This also causes difficulties in standardising terminologies and associated processes
- Application of in-house developed tools for cost uncertainty analysis, whilst using commercial tools for validation and verification purposes. Additionally, the Monte Carlo simulation is typically conducted using commercial packages

Differences across the organisations

- Definitions of uncertainty vary, whilst all organisation concurred with the concept that uncertainty causes a difference between actual and targeted levels
- Variation in defining the Cost Breakdown Structure (CBS) and Work Breakdown Structure (WBS), which is driven by the equipment that follows be-spoke characteristics
- Terminologies for the types of uncertainty and in defining the types of uncertainty in the risk registers

Unique characteristics

Each organisation was observed to have a different level of experience in CfA. This was recognised to cause a variation in the perception of uncertainties. A comparison of the key uncertainties that were specified is illustrated in Table 2.

	Key types of uncertainty
Organisation A	Obsolescence, supply chain, failure rate
Organisation B	Pricing, flying hours, obsolescence, failure rate
Organisation C	Reliability, whole life cycle cost, lines of maintenance, equipment utilisation, training
Organisation D	Pattern of equipment usage, equipment utilisation, failure rate

Table 2: Comparison of key uncertainties across the organisations.

Organisation A and D appeared to have a higher degree of experience in CfA, and focused on “failure rate” as a key issue affecting the service delivery. On the other hand, Organisation B and C appeared to have lesser experience, and focused on uncertainties associated to pricing and the process of whole life cycle cost estimation.

3.6 Current Challenges in Service Cost Estimation

Through interaction with the specified organisations a number of challenges were recognised. At a high level, it was realised that with CfA the role of data flow from the customer to the OEM grows in service cost estimation driven by variation in equipment usage levels and conditions of usage. Furthermore, it was emphasised that an alignment between the customer and supplier is necessary in order to unify the cost and uncertainty considerations. From a service delivery point of view the reliance on human performance was also highlighted as a key area of challenge. An overview of the key challenges observed across the organisations includes:

- Sustaining the reliability level is very important in CfA, though, measuring this metric is difficult. All interviewed organisations agreed with this idea and suggested that improvements should be made to understand the concept of reliability. This will necessitate better understanding of equipment failure over time and in association obstacles such as technology maturity or obsolescence that may arise in the future
- All organisations that were interviewed highlighted the large reliance on the service network for be-spoke parts. Furthermore, issues related to quality of parts and timeliness of the supply chain was commonly highlighted. There was also a commonality among interviewees in suggesting the need for better understanding uncertainty that derives from the supply chain; especially from the perspective of integrating uncertainties into cost estimation
- Another area of common agreement involved challenges in predicting support costs. Three out of the four organisations that were interviewed on this topic, suggested that services lack standardization and need an approach to reduce non-recurring costs, which may originate from obsolescence mitigation strategies and/or failure costs
- Two out of the four organisations mentioned difficulties in prioritising uncertainties at the early stages of the CADMID cycle. This means among uncertainties that are considered their influence over potential costs may not easily be captured.
- Growth in service offerings has brought about new issues (e.g. who does what?). This is an issue of scope and was highlighted by all four organisations that were interviewed; this also implies that agreeing CfA is still a challenging task.

As an overview of the experienced challenges it is recognised that there are two particular areas that require additional attention. Firstly, the delivery of service promotes dynamism in cost, driven by various uncertainties, which in particular equipment failure rate can have a triggering affect. Along these lines, the dynamism refers to the evolution of a particular uncertainty and the cost driver. For instance, the number of failure events are not static over time (e.g. 10 events) and face a different level of likelihood of occurrence over time (e.g. -10/+25%). Secondly, with enhanced interaction, the behaviour of supply chain can affect the whole life cycle costs. For instance, if the actual cost of the spares supplier is below the x% profit level it may trigger a risk sharing or incentive mechanism to allocate costs across the manufacturer and supplier.

4 POTENTIAL TECHNIQUES TO TACKLE SPECIFIED CHALLENGES

There is a wide selection of cost estimation techniques that are suitable for the CfA context. Approaches vary depending on two aspects including service life stage (e.g. design, delivery, and adaptation) and level of information available (e.g. low, medium, and high) [20]. These aspects also vary along the bidding stage, where early on there is lower information available and the service to be delivered may be vaguely defined. The types of information that are considered in a cost assessment of a CfA include user requirements, supplier data, industry standards, historical data and expert opinion. These are used to define assumptions, risks and uncertainties, procurement, deployment and support. In the case where low information exists and at the design stage of service delivery, expert opinion becomes the key determinant of cost assessments. Datta and Roy [20] also define various simulation approaches including systems dynamics, discrete event, and agent based modelling suitable for this context. At the early stages, a top-down view of the given context tends to be taken of costs. The cost estimates have low reliability at this stage. On the other hand, as the amount of information grows and considerations for service are more mature the use of techniques such as detailed estimates (e.g. activity based costing) or simulation becomes used more commonly. Such approaches yield improved accuracy and detailed information regarding costs.

For the medium to high level information available approaches such as analogy and parametric estimation have been commonly used. The advantages of the parametric approach centre on the repeatability and less information requirements. Alternatively, with the analogy approach it is necessary to establish an analogous case that represents the given context. Furthermore, the approach serves as a good means to reach rough order magnitude estimates in absence of data. Figure 4 illustrates the use of the cost estimating approaches across possible CfA contexts.

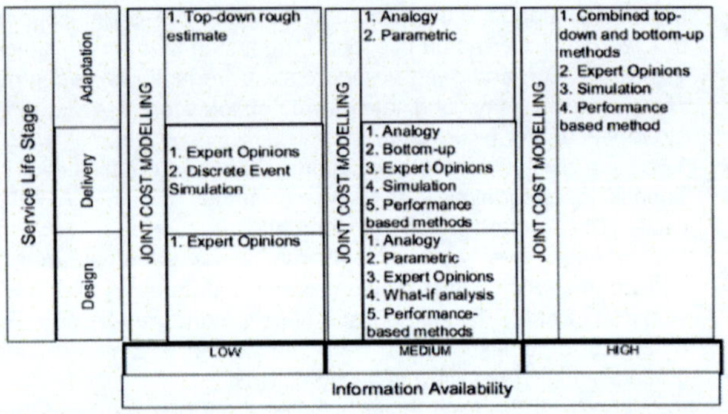

Figure 4: Cost assessment for Contracting for Availability [20].

4.1 Techniques aiming to handle data issues

In order to tackle various issues that arise from data a wide range of approaches have been proposed including possibility theory, fuzzy set theory, neural networks (NNs) and evidence theory. Possibility theory and fuzzy set theory are forms of artificial intelligence, which can be considered to be extensions to probability theory [44]. These approaches are capable of representing uncertainty with much weaker statements of knowledge and more diverse types of uncertainty [33, 45]. Fuzzy set theory has, in many occasions, been proposed to substitute the traditional probability theory [14]. The advantage of using this method lies in its capability to assign probabilities to ambiguous events or vague knowledge, which suits whole life cycle applications [24]. Fuzzy techniques have the ability to exploit the tolerance for imprecision and uncertainty while representing the real world.

4.2 Techniques aiming to handle the stochastic nature of services

Traditional models in maintenance cost uncertainty modelling take a static perspective by assuming that the system operates in a certain fixed time instant (e.g. Monte Carlo simulation). On the other hand, stochastic models use random variables to reproduce or visualise the possible occurrence of events or disturbances that are unknown a priori. Thus, such models define a representation of stochastic phenomena, which is typically achieved through a set of probability distributions and/or a set of relevant statistical parameters to generate suitable values for the random variables over time. The supply chain literature has commonly applied stochastic techniques to represent dynamism in systems. Furthermore, three simulation approaches have typically been applied: discrete event simulation (DES), system dynamics (SD) and agent based modelling (ABM), as represented in Figure 5.

Their applications have varied depending on the problem at hand whether it be at a strategic, operational or planning level [46]. Strategic refers to issues such as deciding the structure of the supply chain over many years or modes of transport to be used. Planning, for instance involves consideration of which markets will be supplied from which locations. Finally, operational problems can be daily and the focus is on the supply chain configuration (e.g. allocation of inventory). In literature, SD and ABM have been used equally to address strategic and planning problems. On the other hand, the use of DES heavily focuses on planning problems, while it has also been used for the operational context [46].

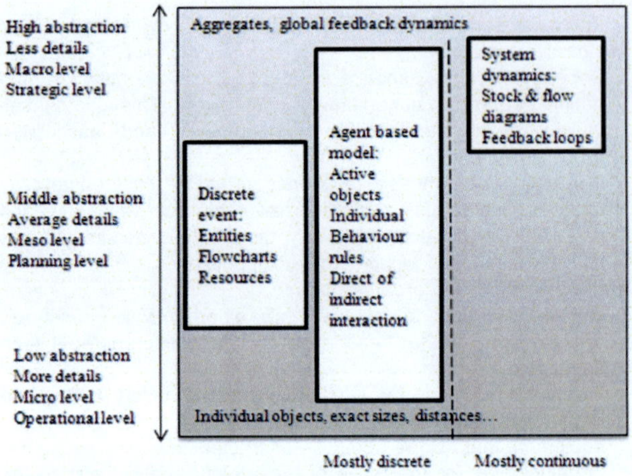

Figure 5: Simulation approaches across problems [47].

The application of ABM is widely growing due to a number of reasons. An important reason is associated to the growth in complexity across activities, which is driven by the increased interdependencies [48]. Whilst this phenomenon is not new, traditional approaches have in some cases built in potentially unrealistic assumptions for modelling purposes. For instance, in the case of modelling economic markets notions of perfect markets have been relied upon, whereby assumptions such as homogeneous agents, and long-run equilibrium have made the problems analytically and computationally tractable. To the contrary, ABM provides the possibility of modelling processes out of equilibrium. Furthermore, the approach enables to model more fluid-turbulent social conditions when modelled agents and their identities are not fixed or given, but susceptible to changes that may include birth or death of individual agents as well as adaptation of their behaviour [49]. Although, there are a number of potential benefits to be realised from the approach, its application in cost estimation has been limited.

The case of service delivery, as experienced in IPS^2, requires better understanding of a number of aspects in order to achieve performance targets. As discussed in [50], these aspects include demand management, capacity and resource management, customer relationship management, supplier relationship management, service delivery management, service quality management, service recovery management, and information flow and technology management. Each of these aspects has cost implications and in section 6 potential agent based architectures to address some of these issues are proposed. Furthermore, in order to tackle these challenges the authors believe that driven by the benefits proposed in literature for applying ABM, its application would be suitable within the IPS^2 context, whilst yielding beneficial information regarding the outcomes of interactions across various scenarios (e.g. [51]).

4.3 Agent Based Modelling

In literature ABM has mostly been used to define the engagement across a supply chain, but the approach has wide application including domains such as economics and manufacturing [52]. In the supply chain literature there tends to be a set of fixed relationships and the effects of different patterns of decision making on overall stock levels is explored [53]. The main theme has been to capture the interaction with the customer. In defining an agent four key properties have been referred to including autonomy (e.g. function without user intervention), proactive (e.g. independently working towards a goal), reactive (e.g. respond to environment) and social (e.g. interact with other agents) [54].

Across literature there is no general agreement on the term agent, whilst various definitions focus on different aspects such as software entity with thread of control over chosen operations [53] or a self contained problem solving agent [55]. The fact that agents can react to changes, adapt and re-plan if a better approach is realised, based on information sharing between agents, makes it a dynamic system, which is key to representing a continuously changing world (e.g. CfA). In ABM the focus is on agents and their relationships with other agents or entities [52].

ABM serves explanatory rather than predictive purposes, which is particularly suitable for the context of the early stages of the bidding phase, where issues originating from data are influential. Other reasons for the growth of the application of ABM is associated to increased amount of data at lower levels of granularity and the enhanced computational power, which enables to conduct more detailed analysis. The key difference of ABM from other simulation approaches was highlighted in [55] as:

> "... it is the flexibility and high-level nature of these interactions (cooperation, coordination, negotiation) which distinguishes multi-agent systems from other forms of software and which provides the underlying power of the paradigm"

Based on literature review it was recognised that existing network approaches to cost estimation take a deterministic perspective, whilst in some cases taking the abstractions and assumptions limit representing reality. Furthermore, the implications of costs across the supply chain were understood in an isolated manner. The factors contributing to the complexity of supply chain uncertainty include scale, technological novelty, quantity of sub-systems, and the number of feedback loops [56]. In order to manage these aspects organisations need to enhance the information system, configuration and organisational structure and the inherent processes related to supply and demand.

4.4 Existing Research in Agent Based Modelling for Service Cost Estimation

This section presents an overview of existing research in ABM that is applied to service cost estimation. Based on literature review the authors came across two papers addressing ABM in cost estimation. Firstly, [57] proposes a model, where each agent represents one kind of cost estimation method.

Secondly, [51] presents an agent based model that focuses on the variation in the behaviour of a supply network driven by a dynamic uncertainty (including failure rate), an incentive mechanism and various risk sharing options. The model focuses on CfA, where formulation of incentives is an important driver. There are different ways to construct the incentive mechanism, including contractual, relational, and performance based approaches. Furthermore, the performance driven context has received more interest due to the stability in price, and financial motivation through a share-out mechanism across partners.

The agents in the presented model illustrate a service supply network, which consists of a customer, solution provider, a spares and resource supplier. The customer agent pre-dominantly aims to acquire the desired level of equipment usage, whilst paying the solution provider on pre-determined basis the cost originating from the varying equipment usage. On the other hand, the solution provider agent aims to respond to the customer requirements that are triggered by the dynamic evolution of the failure rate/event over time. The solution provider interacts with the customer on the basis of closing the gap between the actual and target cost level through the incentive mechanism. Additionally, the solution provider interacts with the spares supplier on the basis of meeting the spares requirements, originating from the failure rate, which is outside the expected level of spares requirements. In the model, the associated additional cost is apportioned between the spares supplier and the solution provider based on three major risk sharing scenarios. The considerations for the spares and resource suppliers are fairly similar in that a multiplier is used to translate a target cost level into an actual cost figure. An overview of this architecture is demonstrated in Figure 6.

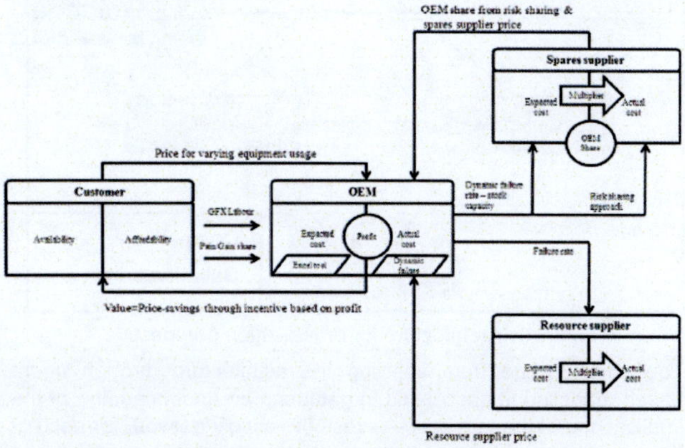

Figure 6: Architecture of a sample agent based model.

As for outputs from the model, the user is able to recognise information about customer affordability, equipment availability, maintenance costs for relevant cost drivers, and the influence of uncertainty. In the presented sample agent based model the agents are defined as real-life organisations identified in the context of interest, characterised with varying degrees of autonomy (i.e. execution ability and self-control), and characteristics based on policies, behaviours, states and constraints. Though, in other contexts such as in the manufacturing and logistics perspective an agent might represent a machine.

5 FUTURE RESEARCH: BUILDING AN UNDERSTANDING OF THE APPLICATION OF AGENT BASED MODELS

The architectures proposed in this section aim to demonstrate how agent based models could conceptually be developed within the context of service cost estimation for IPS[2]. In the subsequent sub-headings three architectures are proposed: interaction across the supply chain, the evolution of material/part requirements of the equipment and the role of humans in service provision.

5.1 Architecture 1: Organisational perspective

The organizational perspective concentrates on the interaction between organisations. This architecture aims to answer the question such as what is the cost impact of collaboration and competition across the supply network for service delivery. Along these lines, each of the agents has a different form of interaction with each other, which may be either in a collaborative or a competitive form. Each organisation is considered as an agent and represents different parts of a typical supply network, as represented in Figure 7. The collaboration refers to the service delivery, which takes into account the co-creation of value (e.g. [6]). From the "Customer" perspective this refers to adequate flow of information regarding equipment utilisation to the "Company". From the "Company" perspective collaboration refers to the quality and timely provision of service. Competition affects the cost of materials.

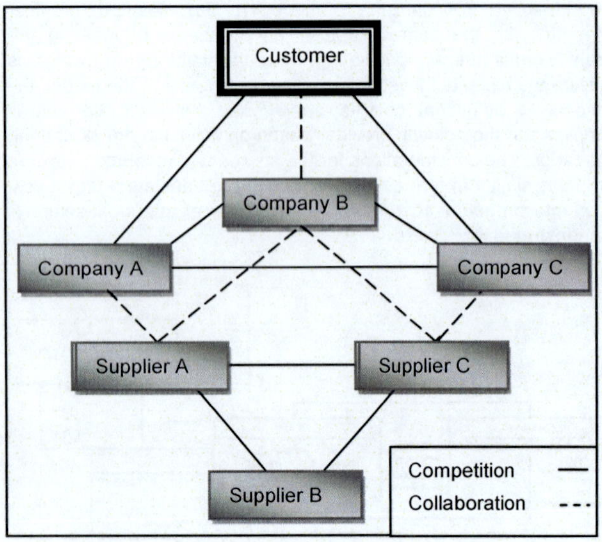

Figure 7: Architecture for organisation perspective.

The added value from applying the architecture through agent based modeling is associated to gathering an understanding of the implications of the interaction across the supply network for each of the associated organizations and also reduction of overall uncertainty coming from the supply chain. Furthermore, of the challenges specified in Section 3.6, it enables to reflect the pros and cons of the increasing reliance on the service network by considering aspects such as quality and timely delivery of parts.

5.2 Architecture 2: Equipment perspective

The architecture for the agent based model in the equipment perspective aims to assess the performance of the equipment across a large set of equipments that a solution provider needs to support. Within the architecture each equipment is defined as an agent, whilst driven by the specified granularity (e.g. component or system) the characteristics of the agent is formed. As a result, a different set of material and part requirements emerge for each agent and cause the equipment performance to vary. Furthermore, the interaction across agents is considered to be driven by the material and part requirements, whilst these requirements are sourced from suppliers, as represented in Figure 8. Though, additionally each equipment may also be able to support one another. The cost of maintaining each equipment varies depending on the characteristics of the agents.

The benefit of developing such an architecture is particularly from a materials and parts perspective, where the decision making regarding equipment performance can be visualised with respect to a portfolio of equipments, which represents reality.

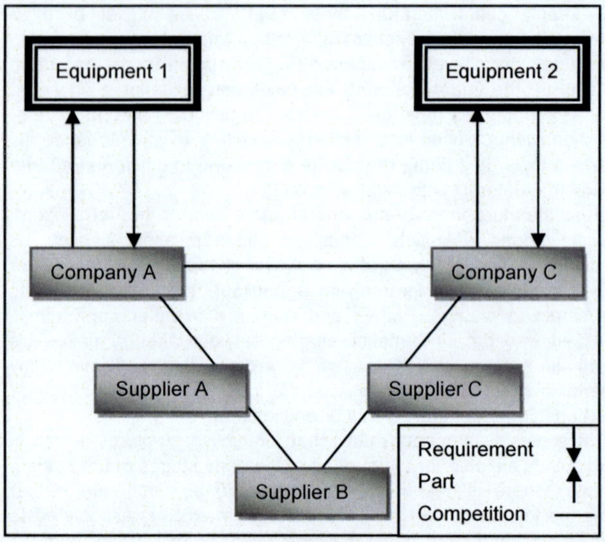

Figure 8: Architecture for equipment perspective.

5.3 Architecture 3: Human perspective

The human perspective focuses on assessing the human performance in service delivery with respect to its impact on the cost estimates. The approach takes into account a departmental/role breakdown of an organisation, based on the relevant stakeholders for a given project. Individuals represent agents and the characteristics of each agent are built by considering aspects such as skill level, motivation, contribution to the customer value and experience. In Figure 9 the architecture for the human perspective is illustrated.

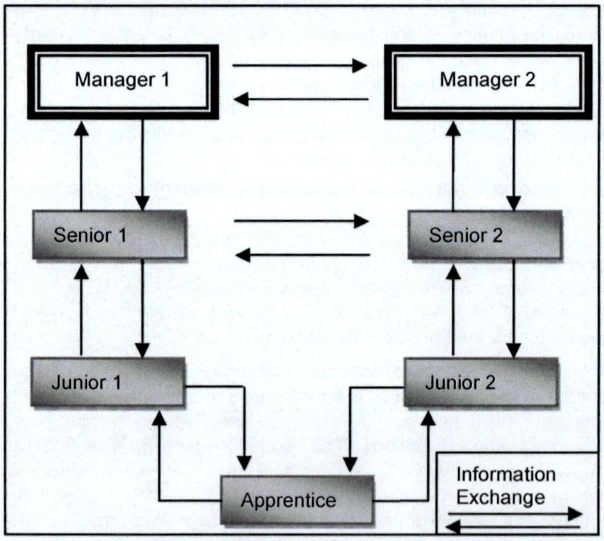

Figure 9: Architecture for human perspective.

Driven by the performance the cost of the given task varies. The aim of this perspective is to represent a resource model given these aspects and how they affect the interaction between individuals and ultimately the total service delivery in terms of the cost and value outcome. The interaction reflects the departmental associations and varies over time based on the service requirements. There are a number of benefits in adopting this perspective, Most importantly, it provides an ability to demonstrate the variation in human performance across departments and how it impacts on the overall service delivery.

6 CONCLUDING REMARKS

This keynote identifies the importance of service cost estimating within the IPS^2 context. The research presented focuses on the CfA type projects within the defence industry. CfA projects invite additional uncertainties, especially at the service delivery stage. There is a lack of research in incorporating the uncertainties in the service cost estimating in a systematic manner. The estimating at the early stage of a CfA contract, for example at the bidding stage, is even more challenging as the uncertainties change over time and the behaviour and performance of the supply chain change over time based on the incentives and their emerging relationships. Initial attempts to assess the impact of the dynamic aspects on cost uses an agent based modelling approach. Various established techniques are mature for particular contexts, whilst there is an increasing interest to use simulation techniques because of the perceived benefits in visualising the dynamic environments and their impacts on cost. The paper discusses the potential of agent based modelling for service cost estimation and proposes three perspectives for the future research.

7 ACKNOWLEDGMENTS

Authors are grateful for the kind support from the Cranfield IMRC for funding this research and the industrial partners. Authors are also grateful to the members of the Decision Engineering Centre and Cranfield PSS Community for their support and contribution.

8 REFERENCES

[1] Meier, H., Roy, R., Seliger, G. (2010): Industrial Product-Service Systems - IPS², in: CIRP Annals Manufacturing Technology, Vol. 59, No. 2, pp. 607-627.

[2] Cook M.B., Bhamra T.A., Lemon M., (2006): The transfer and application of Product Service Systems: from academia to UK manufacturing firms, in: Journal of Cleaner Production, Vol. 14, No. 17, pp. 1455-1465.

[3] Roy, R. Cheruvu, K.S. (2009): A competitive framework for industrial product service systems, in: International Journal of Internet Manufacturing and Services, Vol. 2, No. 1, pp.4–29.

[4] Kim, S.-H., Cohen, M.A. Netessine, S., (2006): Performance Contracting in After-sales Service Supply Chains, in: The Wharton School, Philadelphia, PA.

[5] Baines, T.S., et al. (2007): State-of-the-art in product-service systems, in: Proceedings of the Institution of Mechanical Engineers Part B – Journal of Engineering Manufacture, Vol. 221, No. 10, pp. 1543-52.

[6] Ng, I., Williams, J., Neely, A. (2009): Outcome- based contracting: Changing the boundaries of B2B customer relationships, in: AIM Executive Briefing

[7] Alonso-Rasgado, T., Thompson, G., Elfstrom, B. O. (2004): The design of functional (Total Care) products, in: Journal of Engineering Design, Vol. 15, No. 6, pp. 515-540.

[8] Erkoyuncu, J. A.; Roy, R.; Shehab, E.; Wardle, P. (2009) Uncertainty challenges in service cost estimation for product-service systems in the aerospace and defence industries, in: Proceedings of the 1st CIRP Conference on Industrial Product Service Systems, Cranfield Univ., pp. 200-207, 2009.

[9] Oliva, R., Kallenberg, R. (2003): Managing the transition from products to services, in: International Journal of Service Industry Management, Vol. 14, No. 2, pp. 160–172.

[10] Brax, S. (2005): A manufacturer becoming service provider–challenges and a paradox, in: Managing Service Quality, Vol. 15, No. 2, pp. 142–155.

[11] Shehab, E.; Evans, S.; Baines, T.; Lightfoot, H.; Tiwari, A.; Johnson, M.; Pepperd, J. (2008): Challenges of product-service systems: a real life case study, in: Proceedings 6th Int. Conference on Manufacturing Research, Brunel, UK, pp. 557-563.

[12] Goffin, K., New, C. (2001): Customer support and new product development: an exploratory study, in: International Journal of Operations and Production Management, Vol. 21, No. 3, pp. 275–301.

[13] The Association of the Advancement of Cost Engineering recommended practice – AACE International Recommended Practice No. 17R-97, (2003): Cost Estimation Classification System. 7.3 – Cost Estimating and Budgeting.

[14] Stewart, R., Wyskida, R., M., Johannes, J., D., (1995): Cost estimator's reference manual, John Wiley & Sons, New York.

[15] Xu, Y., et al., (2010): Current and Future Research in Cost Engineering, in: International Journal of Computer Integrated Manufacturing, Accepted, November 2010.

[16] Roy, R.; Datta, P.; Romero, F.J.; Erkoyuncu, J.A, (2009): Cost of industrial product service systems (IPS^2), in: Keynote, Digital Proceedings of the 16th CIRP International Conference on Life Cycle Engineering, Cairo, Egypt.

[17] Garvey, P. R. (1996): Modeling Cost and Schedule Uncertainties - A Work Breakdown Structure Perspective. In: Military Operations Research, Vol. 2, No. 1, pp. 37-43.

[18] Asiedu, Y., Gu, P. (1998): Product life cycle cost analysis: state of the art review, in: International Journal of Production Research, Vol. 36, No. 4, pp. 883–908.

[19] Hansen, D. R., Mowen, M. (2006): Cost management: accounting and control, Fifth Edition, Thomson, South Western, Toronto.

[20] Datta, P., Roy, R., (2010): Cost modelling techniques for availability type service support contracts: a literature review and empirical study, in: CIRP Journal of Manufacturing Science and Technology, In Press.

[21] Kirkham,R.J. (2005): Re-engineering the whole life cycle costing process, in: Construction Management and Economics, Vol. 23, pp. 9–14.

[22] United States Government Accountability Office-GAO, (2009): Cost Estimating and Assessment Guide, Best Practices for Developing and Managing Capital Program Costs, March, [Online] Accessed from: **http://legislative.nasa.gov/hearings/3-5-09%20CostEstimatingAssessGuide.pdf** Access date: 02/09/09.

[23] NASA, (2004) Cost Estimating Handbook, [Online] Accessed from; **http://www.ceh.nasa.gov/webhelpfiles/Cost_Estimating_Handbook_NASA_2004.htm**

[24] Boussabaine, A., Kirkham, R. (2004): Whole Life-cycle Costing: Risk and Risk Responses, 1st edition. Blackwell Publishing, London.

[25] Roy R (2003): Cost engineering: Why, what and how? Decision Engineering Report Series, Cranfield University.

[26] Griffin, J.J. (1988): Whole life cost studies: A defence management perspective, in: Engineering Costs and Production Economics, Vol. 14, pp. 107–115.

[27] Erkoyuncu, J. A., Roy, R, Shehab, E.,Cheruvu, K. (2010): Understanding service uncertainties in industrial product-service system cost estimation, Journal of Advanced Manufacturing Technology, In-press.

[28] Brouwer, W., Rutten, F., Koopmanschap M. (2001): Costing in economic evaluations, in Drummond M, McGuire A, (Eds.) Economic evaluation in health care: Merging theory with practice. Oxford University Press.

[29] Flanagan, R., Norman, G., Meadows, J. and Robinson, G. (1989): Life cycle costing-Theory and practice', London: Blackwell Scientific Publication-BSP Professional Books.

[30] Ministry of Defence (MoD), (2009): Cost estimation process, [Online] Available via: **http://www.aof.mod.uk/aofcontent/tactical/engineering/content/fe_costestprocess.htm** Access date: 11/12/2009.

[31] Durugbo, C., Erkoyuncu, J. A., Tiwari, A., Alcock, J. R., Roy, R., Shehab, E. (2010): Data uncertainty assessment and information flow analysis for Product-Service Systems in a library case study. International Journal of Services Operations and Informatics, In press.

[32] Curran, R., Raghunathan, S., Price, M., (2004): Review of aerospace engineering cost modelling: the genetic causal approach, in: Progress in Aerospace Sciences, Vol. 40, pp. 487–534.

[33] Oberkampf, W., Helton, J., Sentz, K., (2001): Mathematical representation of uncertainty, AIAA Paper, pp. 2001-1645.

[34] Kishk M. (2004): Combining various facets of uncertainty in whole-life cost modeling, in: Construction Management and Economics, Vol. 22, No. 4, pp 429-435.

[35] Harding A., Lowe D., Emsley M., Hickson A. and Duff R. (1999): The role of neural networks in early stage cost estimation in the 21st century, COBRA 1999: [Online] **http://www.rics.org/Practiceareas/Management/Businessmanagement/Finance/role_of_neural_networks_19990101.html**, Access date: 12/01/2008

[36] Garvey, P. R. (2000): Probability Methods for Cost Uncertainty Analysis, *Marcel Dekker*, New York.

[37] Fisher, G. H. (1971): Cost considerations in Systems Analysis, New York: American Elsevier Publishing Company.

[38] Erkoyuncu, J. A., Roy, R., Shehab, E., Cheruvu, K., (2010): Impact of uncertainty on Industrial Product Service System delivery, in: *Proceedings of the 2nd CIRP Conference on Industrial Product Service Systems*, Linkoping University, Sweden, pp. 481-489.

[39] DeLaurentis, D., Mavris, D. (2000): Uncertainty modeling and management in multidisciplinary analysis and synthesis, AIAA Paper 2000-0422, January. Accessed from: **http://smartech.gatech.edu/bitstream/1853/6396/1/AIAA-2000-0422.pdf** Access date: 28/04/08.

[40] Arena, M. V., et al. (2006): Impossible Certainty: Cost Risk Analysis for Air Force Systems, Santa Monica, Calif.: RAND Corporation, MG-415-AF, 2006. As of September 26, 2006: **http://www.rand.org/pubs/monographs/MG415/**.

[41] NATO, (2007): *RTO Technical Report*, [Online]. NATO, Annex D, Pg D1. Available from: **http://www.rta.nato.int/pubs/rdp.asp?RDP=RTO-TR-SAS-054** [Accessed 8th July 2009].

[42] Curran, M. W. (1989): Range Estimating: Measuring uncertainty and reasoning with risk, in: Cost Engineering, Vol. 31, No. 3 March, pp. 18-26.

[43] Ministry of Defence (MoD), (2007) Three point estimates and quantitative risk analysis- A process guide for risk practitioners, Unclassified, [Online] **http://www.aof.mod.uk/aofcontent/tactical/risk/downloads/3pepracgude.pdf** Access date: 03/06/2010.

[44] Emblemsvag, J., (2003): Life-cycle costing: Using Activity-Based Costing and Monte Carlo Methods to Manage Future Costs and Risks. *John Wiley and Sons,* New York, NY.

[45] Cavalieri S., Maccarrone P., Pinto R., (2004): Parametric vs. Neural network models for the estimation of production costs: A case study in the automotive industry, in: International Journal of Production Economics, Vol. 91, No. 2, pp. 165-177.

[46] Owen C., Albores P., Greasley A. (2010): Simulation in the supply chain context: matching the simulation tool to the problem, in: Proceedings of the Operational Research Society Simulation Workshop.

[47] Buxton, D., (2010): Anylogic training manual.

[48] Macal, C. M., North, M. J. (2006): Tutorial on agent based modeling and simulation Part 2: How to model with agents, in: Proceedings of the 2006, Simulation Conference, pp. 73-83.

[49] Srbljinovic, A., Skunca, O. (2003): An introduction to agent based modelling and simulation of social processes, in: Interdisciplinary Description of Complex Systems, Vol. 1, No. 1-1, pp.1-8.

[50] Cohen, M., Agrawal, N., Agrawal, V. (2006): Achieving breakthrough service delivery through dynamic asset deployment strategies, in: Interfaces, Vol.36,No 3,pp.259-271.

[51] Erkoyuncu, J. A.: Roy, R.; Shehab, E.; Cheruvu, K. (2010): Service cost estimation with uncertainty using agent based simulation, in: Joint ISPA/SCEA Conference Proceedings, 08-10 June, San Diego, US.

[52] Nilsson, F, Darley, D. (2006): On complex adaptive systems and agent-based modelling for improving decision-making in manufacturing and logistics settings: experiences from a packaging company, in: International Journal of Operations & Production Management, Vol. 26, No. 12, pp. 1351-1373.

[53] Allwood, J. M., Lee, J. H. (2005): The design of an agent for modelling supply chain network dynamics, in: International Journal of Production Research,Vol.43, No.22, pp. 4875-4898.

[54] Wooldridge, M., Jennings, N. R. (1995): Intelligent agents: Theory and Practice, in: The Knowledge Engineering Review, Vol. 10, No: 2, pp.115-152.

[55] Jennings, N. R., Sycara, K., Wooldridge, M. (1998): A Roadmap of Agent Research and Development, in: International Journal of Autonomous Agents and Multi-Agent Systems, Vol. 1, No. 1, pp. 7-38.

[56] Datta, P.P., Christopher, M., Allen, P. (2007): Agent-based modelling of complex production/distribution systems to improve resilience, in: International Journal of Logistics Research and Applications, Vol. 10, No. 3, pp. 187–203.

[57] Ping, L., Yongtong, H., Bode, J., Shouju, R, (1996): Multi-agent system for cost estimation, in: Computers Industrial Engineering, Vol. 3, No. 3/4, pp. 731-735.

The Volkswagen combined Heat and Power Unit - Automotive Know-How for Intelligent Energy Production

Falko Rudolph[1], Holger Manz[1]

[1] Volkswagen AG, Wolfsburg, Germany

Abstract

As the climate change can be especially traced back to CO_2-emissions, it is a worldwide aim to reduce those CO_2-emissions. Therefore, it is necessary to make use of regenerative energy sources and highly efficient technologies. Besides the reduction of emissions within automotive mobility significant attention is paid to heat- and electricity generation in Germany. The reason for this is that more than half of the consumed energy in Germany is used for heat generation. Besides, large parts of the heating plants are not state-of-the-art. This shows that there is a considerable savings potential in this field.

In 2007, German government decided that the percentage of combined heat and power (CHP) in electricity generation is supposed to be increased to 25 % until the year 2020. In order to reach this goal, an annual additional construction of CHP-units with a power output of 700 MW becomes necessary.

A contribution to this aim can be achieved by small CHP-units, so-called combined heat and power plants. A combined heat and power plant is a unit in which a combustion engine generates electricity by means of a generator in a highly efficient manner. At the same time, the developing waste heat of the engine is used for heat generation. The joint electricity- and heat generation leads to an overall efficiency which is considerably superior to any conventional heat- and electricity generation. In direct comparison, the primary energy input is up to 40% lower. Compared to a coal-fired power plant the CO_2-emissions are even 60% lower in a CHP powered with natural gas.

Efficient energy conversion and intelligent control technology are only exemplary requirements, which are important issues in the CHP as well as in the automotive industry and which are mastered by Volkswagen. The Volkswagen combined heat and power unit "EcoBlue 2.0" stands out for a modular and compact design, and shows many similarities to the front part of a vehicle. Power unit, generator, engine control, heat exchanger and exhaust system are only a few examples for those similarities. Additional measures, such as modified valve springs, an optimised camshaft as well as a supplementary oil tank enable a long life-time in stationary operation and generate a product which stands out in competition.

Sales as well as the subsequent control of the equipment are effected by the LichtBlick AG. For this purpose, the LichtBlick AG has developed the so-called fluctuating power concept, which implies a control center which can turn the combined heat and power plants on and off at the customer by remote control. The aim is the network connection of thousands of CHPs to a virtual power plant, which is able to close the gap between power requirement and -capacity within a short time and thus is a major advantage compared to the inert large power stations.

The combination of the Volkswagen "EcoBlue 2.0" and the innovative fluctuating power concept of the LichtBlick AG is a totally new business model, which enables the introduction of high quantities on the energy market and thus strongly contributes to the reduction of CO_2-emissions.

Keywords:

Combined Heat and Power; Decentralized Energy Generation

J. Hesselbach and C. Herrmann (eds.), *Functional Thinking for Value Creation: Proceedings of the 3rd CIRP International Conference on Industrial Product Service Systems, Technische Universität Braunschweig, Braunschweig, Germany, May 5th - 6th, 2011*,
DOI 10.1007/978-3-642-19689-8_2,

Service Knowledge Management for Product Service Systems – a systems design approach –

Andy Harrison[1]

[1]Rolls-Royce PLC, Derby, England

Abstract

Knowledge management represents a critical success factor for complex product service systems. Do we know how our product impacts service delivery? Do we understand the interactions between the product design, service design and actual customer utilisation of the two? Can we design a product aligned to our service vision? Relative to original equipment manufacture, the introduction of services brings a significant increase in the range of conceptual concepts and functional specialties that have to be integrated and considered within the design phase. This presentation explores the fundamentals of closed loop service knowledge management in the context of a product service system (PSS) design. What is knowledge in this context? What are the capability building blocks? How do they work together to deliver visible service delivery value? In particular it describes a system engineering approach to the deconstruction of the knowledge management problem.

Keywords:

Industrial Product Service Systems; Closed Loop Knowledge Management; Aviation and Aerospace

J. Hesselbach and C. Herrmann (eds.), *Functional Thinking for Value Creation: Proceedings of the 3rd CIRP International Conference on Industrial Product Service Systems, Technische Universität Braunschweig, Braunschweig, Germany, May 5th - 6th, 2011*,
DOI 10.1007/978-3-642-19689-8_3,

Consumable and Expendable Material Support as an Element of Modular Product Service Systems in the Aviation MRO Industry

Markus Mansour[1]

[1] Lufthansa Technik Logistik, Hamburg, Germany

Abstract

The aviation maintenance, repair and overhaul (MRO) industry has a long history of developing and applying product service systems and similar service packages. Aircraft operator value the idea of shifting part of the operating risk to service providers by having them guarantee aircraft availability. The revenue model for such services is often based on operated flight hours and thereby closely related to the revenue model of an aircraft operator. Nevertheless, true total care packages for aircraft have not yet prevailed on the market since aircraft operators prefer modular packages that offer them higher cost transparency and more flexibility in deciding what activities to outsource.

Consumable and Expendable Material in the aviation industry is used during all MRO activities especially during airframe maintenance and overhaul. The material scope ranges from standard hardware like nuts, bolts and rivets to liquid consumables such as oils, greases and hydraulic fluids to electrical and electronic parts and small components. All of this material is not suitable for repair processes either because of technological or economical reasons.

Many operators decide to contract the supply of consumables and expendables separately from the actual maintenance activity, breaking up an important link in a normal product oriented service systems. However, the general expectations towards the service providers in terms of availability guarantees and pricing model persist. The provisioning of consumables and expendables in this context is no longer part of product service system of a capital good. It becomes a stand alone service product that is subject to similar expectations as a product service system for a capital good.

This presentation analyses this industry trend and discusses advantages and challenges for such a service products on an overall level as well as for each player in the service system.

Keywords:

Aviation MRO Industry; Consumable Material; Supply Service Contracts

J. Hesselbach and C. Herrmann (eds.), *Functional Thinking for Value Creation: Proceedings of the 3rd CIRP International Conference on Industrial Product Service Systems, Technische Universität Braunschweig, Braunschweig, Germany, May 5th - 6th, 2011*,
DOI 10.1007/978-3-642-19689-8_4, © Springer-Verlag Berlin Heidelberg 2011

Fan Out of Japanese Service Engineering
- the State of the Art -

Yoshiki Shimomura[1], Kentaro Watanabe[1], Fumiya Akasaka[1], Koji Kimita[1]
[1] Department of System Design, Tokyo Metropolitan University, Japan

Abstract
Recently, we have seen a rapid rise in expectations that engineering and scientific approaches will bring dramatic improvements in the design and production of services. There is a critical need to establish a method to measure customer satisfaction with services that guarantees certain objectivity and use this information to create services that are better in quality. However, this cannot be achieved by systematically exploiting information or more intense labor. To tackle this problem, we have been focusing our efforts on service engineering since 2001. This paper provides the state of the art in service engineering and the tools for it that we developed.

Keywords:
Service Engineering; Computer Aided Design; PSS; Customer Value

1 INTRODUCTION

Recently, we have seen a rapid rise in expectations that engineering and scientific approaches will bring dramatic improvements in the design and production of services (e.g., [1,2]). Indeed, now that the service industry accounts for some 70% of the workforce and Gross Domestic Product (GDP) in Japan, it is only obvious that the service industry must see dramatic improvement in productivity. In other words, there is a critical need to establish a method to measure customer satisfaction with services (fairly ambiguous up to now) that guarantees certain objectivity and then use this information to create services that are better in quality than anything we have had before. However, this cannot be achieved by systematically exploiting information or more intense labor. Rather, it is only by introducing scientific analysis and intelligent design tools that we will be able to achieve sustained growth for the first time.

We have been focusing our efforts on service engineering [3] since 2001. Here, we define a service as "providing a means of integrating a tangible object (a physical product) with an intangible object (an action product) that realizes some required value for a customer," and the goal of our work is to develop a way of measuring customer satisfaction as a change in the state of the recipient of the service. More specifically, we came up with procedures for modeling and analyzing services and developed a service design computer-aided design software (service CAD) system using those procedures. Scientific design methodology and service engineering that develops tools based on the methodology investigate "what values are required and what societal and artificial systems are needed to achieve those values." This should help curb excessive production of artificial objects (products in the narrow sense), contribute to the formation of a sustainable society in harmony with the environment, and provide a fundamental principle and engineering basis for design, production, and education regarding services in the years ahead. This paper will provide an overview of service engineering and its methodology, which support various activities for service design, development, and operation. In addition, we introduce several projects to apply our study to actual service activities, and describe our future involvement in research.

2 CIRCUMSTANCES OF SERVICE RESEARCH IN JAPAN

The phenomenally rapid growth of Japan's manufacturing sector in the 1960s and 1970s can be largely attributed to the country's advanced knowledge, technology, and skills in designing and manufacturing products and the advanced system that the country put in place to achieve these goals. Product design has retained its position as a subject of academic interest in the engineering field for the past three decades; thus, not surprisingly, the theoretical framework relating to product design is currently well established. Then, more recently, as computers came on the scene, computer-aided design (CAD) systems permitting design work to be done on computers spread rapidly through the manufacturing sector, greatly contributing to the high productivity of manufacturing in Japan. For the manufacturing of products, product design procedures have already been broken down into detailed step-by-step design phases: concept design, embodiment design, production design, and so on. However, services are different. Given the four attributes of services (intangibility, inseparability, transience, and heterogeneity [4]), engineering has never really dealt with services design, which has primarily only been discussed in terms of management and marketing in the management area.

In this section, we will review our proposed service modeling approach, which we developed out of our work on service engineering.

2.1 Definition of Services in Service Engineering

In previous discussions of service engineering [5,6,7], services have been defined as "actions initiated by a service provider in exchange for compensation that produces a state change that is desirable to a recipient." In other words, a recipient receives something tangible or intangible from the provider, and this induces a state change in the recipient. According to this definition, all actions are defined as services. This approximates the definition of service in the classic theater approach model, namely, services in

J. Hesselbach and C. Herrmann (eds.), *Functional Thinking for Value Creation: Proceedings of the 3rd CIRP International Conference on Industrial Product Service Systems, Technische Universität Braunschweig, Braunschweig, Germany, May 5th - 6th, 2011*,
DOI 10.1007/978-3-642-19689-8_5,

the broad sense as actions that create value. In other words, if the state change obtained through use of a product and the state change caused by receipt of a service are based on this definition, then both can be defined as the results of services being provided. Conventional acts of manufacturing and selling products can then be defined as a form of service in the sense that they are actions creating value.

Now, let us incorporate the idea of the Receiver State Parameter (RSP) into the above definition to represent the state of the recipient. The RSP is a measure of customer satisfaction such that "when the RSP moves in the desired direction, the recipient is satisfied." Recipients will all make their own subjective evaluations of services; thus, RSPs will differ from one recipient to the next. This subjectivity represents a major problem when incorporating the point of view of recipients. To overcome this problem, we introduce the persona concept [8] and a scenario model (as will be described later).

2.2 Fundamental Models in Service Engineering

In configuring a model to represent services, it is essential that the model include information regarding view and perspective: "who will be viewing the service and what is the scope of the service." Service engineering thus introduces three sub-models--view, scope, and flow--for representing services [5].

- View Model

The View Model focuses on one RSP for the service of interest and reveals the functions, attributes, and linkages between functions and attributes that can effect changes. It is configured as a function and attribute graph with the RSP at the top. In other words, service engineering uses functions and attributes that induce customer state changes to represent the content of services. Services are shaped by a combination of physical products and facilities on the one hand and service activities on the other, but using both together permits us to consolidate functions by type and capture services.

Figure 1 illustrates the View Model for the RSP Comfortable Environment for a coffee shop service. Each function is represented by a function parameter (FP) including a lexical representation and target parameter, and the RSP Comfortable environment is here linked to the FPs Volume of environmental sound and Seating comfort. These FPs are then linked to other functionally related FPs. Finally, the bottom tier of FPs is associated with Attribute Parameters (APs) of the structure implementing the service.

- Scope Model

Actual services form complex structures involving multiple providers and receivers; therefore, the scope of the service must be clearly specified when representing the service. The Scope Model represents the scope of a service as the set of RSPs described in the View Model.

- Flow Model

Services generally involve the participation of many interested parties. For this reason, service engineering represents the overall structure of services as a multi-agent system. This is the Flow Model. In other words, even if we look at services by just focusing on the relationship between the provider and the receiver, there are numerous intermediate agents (relay agents) between providers and receivers. Service evaluation is done from the perspective of the receiver's RSP. The relay agents themselves are also service receivers who evaluate provided value and implementing structures and providers of equivalent or new services. Service engineering represents this chain of relay agents using the Flow Model.

3 RESEARCH OVERVIEW OF SERVICE ENGINEERING

Based on the fundamental study on services, various methods have been proposed in service engineering. These methods were developed to solve the problems in various phases of service design, development, and operation. To overcome these problems, knowledge and methodologies in various research fields have been introduced in addition to the conventional design engineering approach.

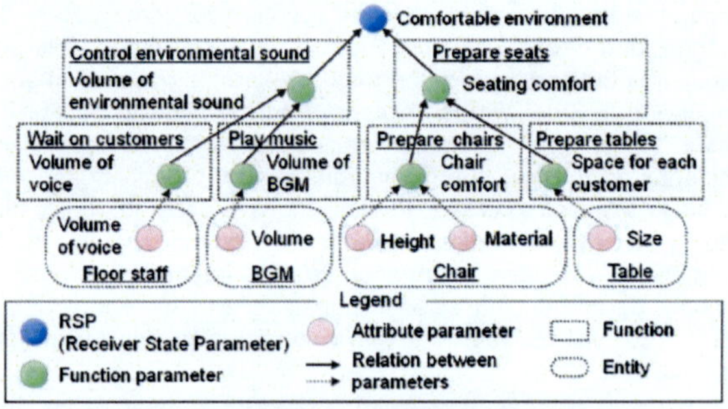

Figure 1: View model.

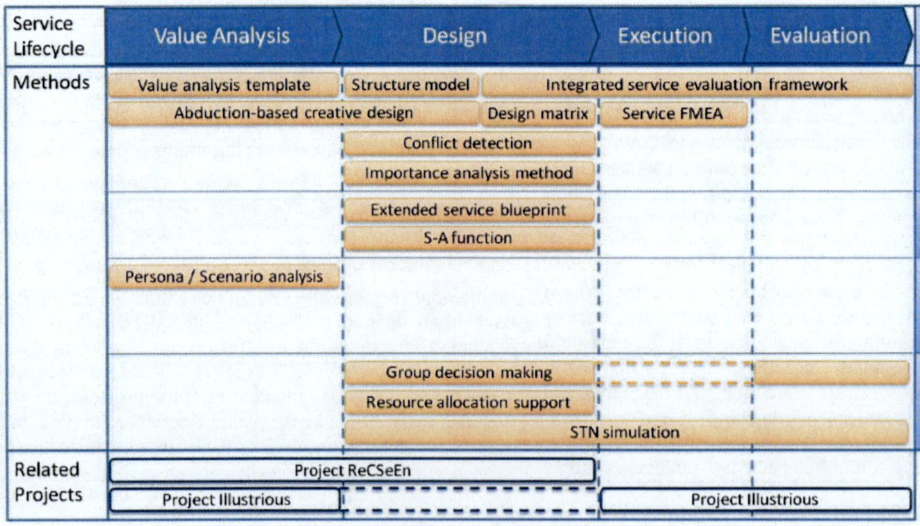

Figure 2: Research map of service engineering.

Figure 2 shows the existing methods of service engineering and the related projects. The arrows on the top describe how a service is designed, developed, and offered, which we define as the service lifecycle [9]. The methods described in Figure 2 are used in the related phase of the service lifecycle. In addition, the column on the right side of the figure describes the related research fields, namely, service marketing, systems engineering / requirement engineering, and operations research, which have affected the development of the methods in the same row. In this chapter, we explain each phase of the service lifecycle and related methods as follows.

3.1 Value Analysis

Generally, as discussed in the Service Marketing field, the quality of a service is always determined by service receivers [10]. Service designers, therefore, need to focus more on service receivers to understand their requirements to the service. The value analysis phase is to extract requirements of service receivers, i.e., RSPs.

- Persona / Scenario analysis

For the extraction of the RSPs of service receivers, a persona is described for each agent that works as a receiver in the service. The persona is a tool used mainly for software interface design to give a simplified description of a customer and works as a compass in a design process [8]. According to this persona, subsequently, a scenario is developed to clarify the context in which the service is received. The scenario is described in the form of a graph representing a scene transition. For each described scene, the receiver's state is represented as a set of parameters called state parameters (SPs). From the SPs, RSPs, which correspond to target parameters in the service design, are extracted.

- Value analysis template

While the persona model and scenario model are scientific methods for setting RSPs, it uses a form of representation based on natural language. It would, thus, require a certain amount of training, given the diversity of vocabulary selection and the difficulty of determining synonymy. Therefore, to describe these models more simply, we developed templates as a convenient means of preparing data corresponding to each stage of model building. To address these issues, we developed a framework consisting of five templates, shown in Table 1, that assist in the tasks of (1) describing personas, (2) describing scenarios, and (3) setting RSPs. We were, thus, able to implement a system that supports the comprehension and analysis of personas, the comprehension and analysis of scenarios, and the understanding and analysis of provided values.

- Value analysis template for B2B service design

Persona and scenario concepts are effective to extract the customer requirements in a B2C (Business to Consumer) service. However, there are difficulties in extracting requirements in B2B (Business to Business) services. In brief, a client organization is generally comprised of several stakeholders, who play different roles, even in the same business. Their requirements for a service, therefore, include various kinds of concerns. This situation makes it difficult for service designers to define the target scope of requirement extraction in service design. Furthermore, the developed service should not only meet the requirements of the service itself but also be in line with the objectives of their business tasks.

To solve these problems, a method to identify requirements in a B2B service has been proposed [11]. In the method, the concept of the business-activity cycle, which is an extension of the concept of customer-activity cycle [12], was introduced to visualize clients' business activities as a B2B service receiving context and to define a scope for requirements extraction. In addition, the concept of goals [13] was adopted to extract customers' requirements in consideration of the objectives of their business tasks.

No.	Template	Summary
1	Persona template	A template to define a persona that represents a hypothetical client as an "individual." A persona is defined by describing demographic data, e.g., name, gender, and carrier, as well as psychological data, e.g., character and lifestyle.
2	Character /Intent template	A template to configure the character/intents of the defined persona using a prepared vocabulary list.
3	Script template	A template to describe detailed behavior (including service-receiving context) that the persona, defined in the persona template, takes to accomplish his/her specific goals or objectives.
4	Keyword template	A template to convert the described script into unified lexical expressions by arranging keywords from the viewpoints of 6 phases in the "Phases of Service Encounter[4]" and 4W1H (What, What like, How, Where, When).
5	RSP extraction template	A template to associate required items/qualities and quality elements with the keywords identified in the keyword template.

Table 1: Structure of the value analysis template.

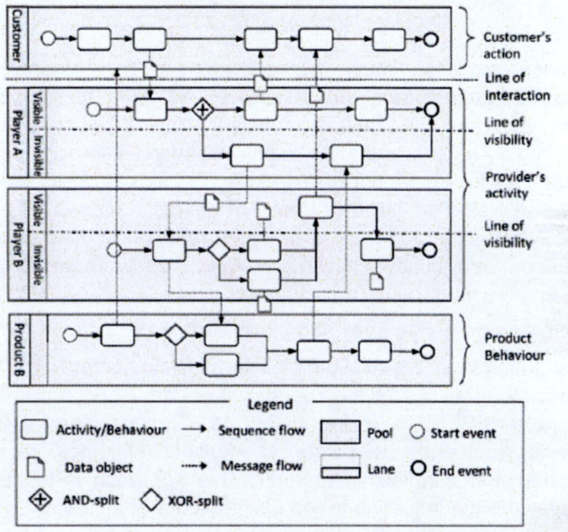

Figure 3: Extended service blueprint.

Furthermore, a template has been developed to help service designers execute this method by extending the aforementioned value analysis template.

3.2 Design

In order to fulfill the values which are extracted by the value analysis, the specifications of a service and its realization structure should be determined in the design phase. The methods for service design can be categorized into three types: model, evaluation, and design support. These methods are implemented in the computer-aided design tool for a service design called service CAD [5].

Model

A service is quite complicated as a design object. The aforementioned models of service engineering, namely, the view model, scope model, and flow model, describe the complicated structure of a service. In addition to these structure models, the model to describe a service process has been proposed.

- Extended service blueprint

The extended service blueprint is a modeling method to describe a service process which consists of service activities and product behaviors [14, 15]. The origin of its concept is the service blueprint, which is commonly used in the service marketing field to describe service activities undertaken by a customer and a front-line and a support team in the performance of a service [16]. However, a service blueprint is not available to describe the role of products in

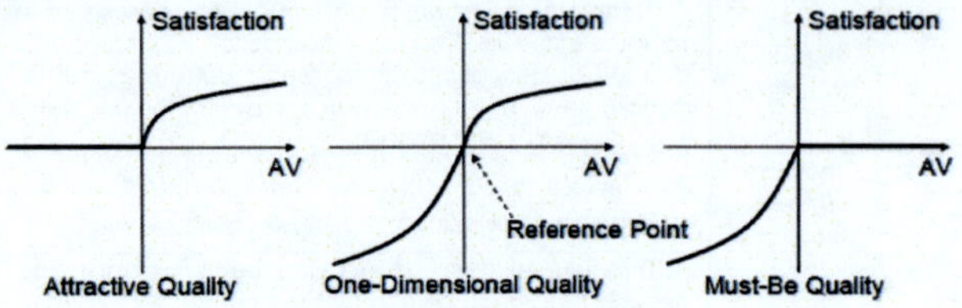

Figure 4: S-A functions [21]

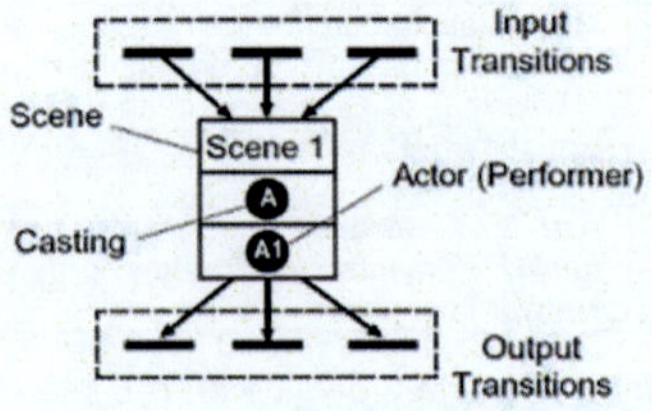

Figure 5: Components of STN. [24]

a service process. Thus, the extended service blueprint was developed to describe both service activities performed by humanware and its related software and product behaviors performed by hardware and its related software. The notation of an extended service blueprint is based on Business Process Modeling Notation (BPMN) for the description of business processes [17]. In Figure 3, the extended service blueprint represents the activities of employees and customers and product behaviors in paralleled lanes and the interactions among them.

- Integrated Service Evaluation Framework

The simultaneous satisfaction of various stakeholders is required for feasible service development. Integrated Service Evaluation Framework (ISEF) is a framework to evaluate service design solutions from the viewpoints of multiple stakeholders in an integrated and quantitative manner. This approach is to describe and evaluate the influence to the stakeholders of a service, who are not only service receivers but also service providers and other related agents in a broader context.

In ISEF, the viewpoints of stakeholders are taken into account as evaluation standards of a service. They are described as state parameters. In addition, the influence to them is described in service process model. The service process model covers the entire service process, which includes pre-activities before the main service activity, such as manufacturing, distribution of products, hiring, and training of employees, and post-activities, such as maintenance, disposal, and after-sales care. The service process model was developed on the basis of IDEF0 [18], a well-known functional modeling method in system modeling to describe the influence among various stakeholders. ISEF is intended to perform the computational evaluation before the service operation. Furthermore, it is utilized as a fundamental model to analyze and improve the operation of the target service in the evaluation phase.

Evaluation

To improve design solutions or to determine the operation of a certain design solution as an actual business, they should be evaluated from various viewpoints. Therefore, evaluation methods to analyze a service and its influence on service receivers have been proposed.

- Importance analysis method

By means of the aforementioned structure models of services, their detailed specifications can be described. Based on the structure models, a method to analyze the importance of design parameters by means of QFD and Analytical Hierarchy Process

	Attribute 1	Attribute 2	Attribute 3	Attribute 4	Attribute 5		
Function 1	X					M1*	Component 1
Function 2		X				M2	Component 2
Function 3		X	X			M3	
Function 4				X	X	M4	Component 3
Function 5				X	X	M5	

*M: Module

Table 2: A simple example of Design matrix. [26]

(AHP) [19] has been proposed to improve a service structure. Quality Function Deployment (QFD) is a systematic analysis method to translate customer needs into requirements and specifications of a design object [20]. Meanwhile, the AHP method numerically computes the importance weights of RSPs according to bilateral comparisons among parameters. After the RSP importance is obtained, it is converted into the FP importance by using the matrix of the QFD [7]. The matrix is created on the basis of the specifications described in view models. According to the matrix, the improvement point of the service structure can be specified.

- S-A function

Differently from the importance analysis method, which evaluates the relative importance of the requirements, the function of S-A (Satisfaction-Attribute) is to evaluate the satisfaction of service receivers directly [21]. The S-A function describes the condition of service receivers' satisfaction from the aspects of their expectations toward the service and the type of the service function.

The receivers' expectations toward a service are used as comparison standards to determine whether the receiver is satisfied or dissatisfied by the provided quality. Namely, the function is divided into a gain side and a loss side at the expectation value. The type of the service function is originally proposed by Kano [22], who classifies functions of products or services into attractive, one-dimensional, or must-be. Figure 4 shows the difference of the shapes of functions. The shape of an S-A function is determined by the analysis of the service function type. By means of S-A functions, the numerical evaluation of a service receiver's satisfaction can be performed.

- STN simulation

The evaluation should be done in both a static and a dynamic manner. We adopt the STN as a simulation model of a service process [23]. STN is a graphic modeling method for discrete-continuous hybrid systems [24]. Figure 5 shows the components of STN. STN can describe the discrete state change with tokens called Actor and the continuous state change described with differential equations which is triggered by the movement of actors. STN is suitable for a service process simulation, since it can describe both the process transition and the state change of each stakeholder simultaneously.

Design support

To determine the specifications of a service, various design support methods have been developed on the basis of conventional product design research and operations research.

- Design matrix

The design matrix method is to optimize the composition of service modules from the aspect of Suh's axiomatic design [25, 26]. In service engineering, a module is determined in terms of the (function/attribute) relationship. Table 2 is the design matrix for service design which shows the (function/attribute) relationship with the character 'X.' For the determination of service components, service designers first determine modules from the (function/attribute) relationship. A module uncoupled from the others is defined as a single component which is a minimum subset of decoupled modules to satisfy the Independence Axiom [25]. By means of this matrix, service designers can configure a flexible and reliable service structure.

- Abduction-based creative design

Abduction is a reasoning method that generates the hypothetical knowledge needed to explain a certain fact and is realized by analogy in many cases. Abduction by analogy has mainly been studied in the research on artificial intelligence and applied to product design. We have applied abduction by analogy to the creative design of services by integrating knowledge from different domains. By means of this approach, innovative ideas can be obtained by the integration of different design domain knowledge. It is implemented as a computer design support system named Universal Abduction Studio [27].

- Conflict detection

In the designed service models, some kinds of conflicts can be found. For example, two functions, "increase staff" and "decrease staff," may exist in separated service models of the same service. To avoid this, the conflict-detection method has been proposed by analyzing objects and predicates of functional expressions with the lexical expression database [6]. In addition, the method to solve the extracted conflicts with TRIZ [28] has also been proposed [6].

- Group decision making

This method attempts to derive the group opinion and to minimize customer dissatisfaction by using grading scores [29]. The grading score is the weight of each customer that would influence the total opinion of the group. The dissatisfaction of all the customers is defined as the sum of the distances between individual opinions and the group opinion, which is affected by the grading score. This method is used to derive the requirement values for a target service by minimizing the dissatisfaction of all the customers and is particularly effective for the design of public services.

- Resource allocation support

The resource allocation problem is usually not the focus of the conventional product design. In service engineering, two issues on service resource allocation have been discussed.

One of them is how to distribute resources to improve a certain service efficiently [30]. The resource distribution is not easily designed because the search space is large and the problem to be solved is complicated. Therefore, a method to determine the optimal value of the resource distribution by means of Genetic Algorithm (GA) [31] has been proposed.

The other issue is how to distribute resources to correlated services for the optimal revenue. To determine the optimization problem of the total revenue of correlated services efficiently, a support method to develop it from the service model has been proposed [32]. We utilize the basic framework of the revenue management network, which is the general representation of revenue models [33], as a blueprint for the development of revenue models. By converting the service model to a revenue model, the resource allocation problem can be determined more easily.

3.3 Execution

Based on the design result, a service is operated for a business or public purpose. The execution phase consists of three sub-steps: transition, operation, and maintenance. Especially in the maintenance step, we consider that some of the quality management methods in product design can be applied. The following idea is one of the research topics.

- Service FMEA

Failure Mode and Effect Analysis (FMEA) is a method to analyze the failure mode of a product which may cause malfunction [34]. Since the structure of a service is complicated, various reasons for malfunctioning can be considered. Service FMEA attempts to analyze the possibilities of such malfunctioning and prevent it.

3.4 Evaluation

The evaluation phase is to evaluate the actual result of a service and compare it with the ideal result, which is obtained in the design phase. The aforementioned ISEF is suitable to analyze it. A simulation method such as STN can also be performed to improve the service operation.

4 RELATED PROJECTS

The proposed methods have been applied to real service cases, such as projects funded by academic organizations or government offices. The following are representative examples.

- Project ReCSeEn

Project ReCSeEn (Research Consortium on Service Engineering) was carried out by the Research Consothium on Service Engineering, which is comprised of 4 companies and 3 universities in Japan, from Sep. 2007 to Feb. 2008. The objectives of this project were to (a) exemplify engineering methods to represent and analyze services, (b) develop a practical tool that makes customer requirements analysis simpler and more general, and (c) execute 4 case studies and improve these services. The value analysis template reported in 3.1 was developed in this project.

- Project Illustrious

Project Illustrious (Integrated loop-line universe of strategic innovation for outasite services) was implemented from Jul. 2009 to Feb. 2010 by 1 company and 3 universities in Japan for the temporary staffing service field. The objective of this project was to improve the quality of human resource matching by realizing (a) correct identification of client company's requirements and (b) selection of temporary staff based on the requirements of a client company. To achieve the objectives, two tools were newly developed: a tool to analyze the required staff and a tool to automatically select the best staff for a client from the viewpoint of customer satisfaction. Especially, in the latter tool, a neural network is utilized to predict customer satisfaction quantitatively. The developed tools were applied to a temporary staffing service, and it was confirmed that these tools could present temporary personnel who matched the client's requirements for service designers and would be useful to improve the quality of their service.

5 CONCLUSION

We have overviewed the contributions of service engineering in Japan and proposed methods which tackle a variety of problems throughout the service lifecycle by introducing the knowledge from various research fields. In addition, we have explained projects which apply service engineering methodology to actual business cases. We will promote more projects to reflect the contribution of research to the industry.

6 ACKNOWLEDGEMENT

This research was partially supported by the Ministry of Education, Science, Sports, and Culture through a Grant-in-Aid for Scientific Research (A), 21240012, 2009.

7 REFERENCES

[1] Tukker, A. and Tischner, U. (2006): Product-services as a research filed: Past, present and future. Reflections from a decade of research, in: Journal of Engineering Design, Vol. 14, No. 17, pp. 1552-1556.

[2] Spohrer, J. and Maglio, P. (2006): The Emergence of Service Science: Toward systematic service innovations to accelerate co-creation of value, in: Production and Operations management, Vol. 15, pp. 329-343.

[3] Yoshikawa, H. (2008): Introduction to service engineering, in: Synthesiology – English edition, Vol. 1, No. 2, pp. 103-113.

[4] Fisk, R.P., John, J., and Grove, S.J. (2000): Interactive Services Marketing, Houghton Mifflin College.

[5] Shimomura, Y. and Tomiyama, T. (2005): Service Modeling for Service Engineering, in: IFIP International Federation for Information Processing, 167, (ISSN 1571-5736 (Paper) 1861-2288 (Online)), pp. 31-38, Springer Boston.

[6] Shimomura, Y. and Hara, T. (2010): Method for Supporting Conflict Resolution for Efficient PSS Development., in: CIRP Annals -Manufacturing Technology, Vol. 59, No.1, (ISSN 1660-2773), pp. 191-194.

[7] Shimomura, Y., Hara, T. and Arai, T. (2008): A Service Evaluation Method using Mathematical Methodologies., in: CIRP Annals - Manufacturing Technology, Vol. 57, No.1, (ISSN 1660-2773), pp. 437-440.

[8] Cooper, A. (1999): The Inmates Are Running the Asylum: Why High Tech Products Drive Us Crazy and How to Restore the Sanity, Sams.

[9] Kimita, K., Akasaka, F., Hosono, S., and Shimomura, Y. (2010): Design Method for Life Cycle Oriented Product-Service Systems Development, in: Proceedings of the 43rd CIRP Conference on Manufacturing Systems, CIRP, CD-ROM, Vienna, Austria.

[10] Gronroos, C., (2000): Service Management and Marketing: A Customer Relationship Management Approach, John Wiley & Sons.

[11] Akasaka, F., Hosono, S., Yamamura, K., Kimita, K., Shimomura, Y., and Arai, T., (2009): A Method for Analyzing Customer Requirements in a B2B Service, in: Proceedings of the 9th Japan Korea Workshop on CAD/CAM -Design Engineering Workshop-, pp. 7-12, Okinawa, Japan.

[12] Vandermerwe, S. (2000): How increasing value to customers improves business results, in: *Sloan Management Review*, Vol. 42 (fall 2000), pp. 27-37.

[13] Lamsweerde, A. V. (2001): Goal-Oriented Requirements Engineering: A Guided Tour, in: Proceedings of the 5th IEEE International Symposium on Requirements Engineering (RE'01), pp. 249-263.

[14] Hara, T., Arai, T., Shimomura, Y., and Sakao, T. (2009): Service CAD System to Integrate Product and Human Activity for Total Value, in: CIPR Journal of Manufacturing Science & Technology, CIRP, Vol. 1, No. 4, pp. 262-271.

[15] Shimomura, Y., Hara, T. and Arai, T. (2009): A unified representation scheme for effective PSS development, in: CIRP Annals -Manufacturing Technology, Vol. 58, No.1, (ISSN 1660-2773), pp. 379-382.

[16] Shostack, G.L. (1982): How to Design a Service, in: European Journal of Marketing, Vol. 16, No. 1, pp.49-63.

[17] Havey, M. (2005): Essential Business Process Modeling. O'Reilly Media, Inc.

[18] FIPS PUB 183 (1984): Integration definition for Function Modeling (IDEF0) Federal Information Processing Standards, United States National Institute of Standards and Technology (NIST), Computer Systems laboratory, Gaithersburg.

[19] Saaty, L. (1980): The Analytic Hierarchy Process, McGraw-Hill.

[20] Akao, Y. (1990): Quality Function Deployment, Productivity Press.

[21] Kimita, K., Shimomura, Y., and Arai, T. (2009): Evaluation of Customer Satisfaction for PSS Design, in: Journal of Manufacturing Technology Management, Special issue on Transferring Best Practices between the Manufacturing and Service Sectors, Vol. 20, No. 5, pp. 654-673.

[22] Kano, N., Seraku, N., Takahashi, F. and Tsuji, S. (1984): Attractive Quality and Must-Be Quality, in: Quality, Vol.14, No.2, pp. 39-48, the Japan Society of Quality Control.

[23] Tateyama, T., Shimomura, Y. and Kawata, S. (2009): Development of Scene Transition Nets (STN) GUI Simulator for Service Flow Simulation, in: Proceedings of the 17th International Conference on Engineering Design -ICED09-, pp. 337-346, Stanford, USA.

[24] Kawada, S., Kawata, S. and Watanabe, A. (1994): A Manufacturing System Modeling Using a Scene Transition Net, New Direction in Simulation for Manufacturing and Communication, pp.86-92.

[25] Suh, N.P. (1998): Axiomatic Design Theory for Systems, in: Research in Engineering Design, Vol. 10, No.4, pp. 189-209.

[26] Kimita, K., Akasaka, F., Hosono, S., and Shimomura, Y. (2010): Design Method for Concurrent PSS Development, in: Proceedings of CIRP IPS2 Conference 2010, pp. 283-290, CIRP, Linkoeping, Sweden.

[27] Takeda, H., Sakai, H., Nomaguchi, Y., Yoshioka, M., Shimomura, Y., and Tomiyama, T. (2003): Universal Abduction Studio -Proposal of A Design Support Environment For Creative Thinking In Design-, in: Proceedings of the 14th International Conference on Engineering Design -ICED03-, CD-ROM, Stockholm, Sweden.

[28] Mann D. (2002): Hands On: Systematic Innovation. CREAX Press Ieper, Belgium.

[29] Yamamoto, S., Kimita, K. and Shimomura, Y. (2010): An Importance Decision Method of Customer's Demands for Highly Public Service, in: Proceedings of the 15th Design for Manufacturing and the Lifecycle Conference -DFMLC2010-, CD-ROM, The American Society for Mechanical Engineering (ASME), Montreal, Canada.

[30] Chiba, R., Akasaka, F., Shimomura, Y., Tateyama, T., and Arai, T. (2010): Contents Quality Design Considering Corporate Resource with Service Engineering, in: Proceeding of the 11th International Design Conference, pp. 173-182, Dubrovnik, Croatia.

[31] Holland, J. H. (1975): Adaptation in Natural and Artificial Systems. Univ. of Michigan Press.

[32] Watanabe, K., Yamagishi, M., Akasaka, F., and Shimomura, Y. (2010): Proposal of the unified methodology of revenue management and service design, in: The Journal of Revenue and Pricing Management, Special issue on: Pricing and Revenue Models in the New Service Economy, Vol. 9, Issue 3, pp. 249-259.

[33] Bitran, G. and Caldentey, R. (2003): An overview of pricing models for revenue management, in: Manufacturing & Service Operations Management, Vol. 3, No. 3, pp. 203-229.

[34] Stamatis, D. H. (2003): Failure Mode and Effect Analysis: FMEA from Theory to Execution, Amer Society for Quality.

Elements of Change in the Transformation towards Product Service Systems

Doroteya Vladimirova[1], Stephen Evans[1], Veronica Martinez[2], Jennifer Kingston[3]
[1] School of Applied Sciences, [2] School of Management, [3] School of Engineering,
Cranfield University, Cranfield, United Kingdom

Abstract

This paper presents a set of challenges in the transformation of a traditional manufacturer into a product service system provider. These challenges constitute the main elements of change in the transformation. The paper forms part of the development of a transformation model for servitization and advances prior conceptual research on the shift from a product-centric mode to a servitized one. This is qualitative research based on data collected from 41 industry and academic experts. Nine elements of change were identified, which will allow industry practitioners, who take on the journey of servitization to understand, analyse and plan their organisation's transformation.

Keywords:
Organisational Transformation; Product Service Systems; Servitization

1 INTRODUCTION

Many manufacturers of both industrial and consumer goods have sought growth in revenue and profits through increased sales of services, as the manufacturing sector forms a decreasing share of Western economies. Margins are being squeezed by global competition which forces manufacturers in the western world to develop new and innovative ways to compete with producers from lower cost economies. One strategy is to create value through the addition of services to traditional product offerings [1].

However, the challenge is to know how to transform a business from being production-oriented into servitized. Servitization is defined as the journey or transformation process whereby an organisation enables its product-service offerings [1]. Organisations such as IBM, General Electric, Xerox, Rolls-Royce, and Ericsson have had a significant share of revenues and profits from services since the middle of 1990s, which has been attributed to a shift from product to service perspective [2].

Academic literature often argues that during this transformation to a combined product-service offering, organisations are likely to change their strategies, operations and value chains, technologies, people expertise and system integration capabilities [3] [4] [5]. However, the transformation from a product-centric mode to a servitized one is still poorly understood and remains a new and complex concept [1].

This paper forms part of a wider research programme into the transformations towards servitized organisational forms. It advances prior conceptual work undertaken for developing a transformation model for servitization [6] [7]. The purpose of this paper is to increase the understanding of the challenges faced by manufacturers when transforming towards product service systems (PSS), based on empirical evidence. These challenges are then interpreted as the main elements of change during a manufacturer's transformation towards a product-service mode and built into a transformation model. The scope of each element of change is also defined, based on the concepts emerging from the empirical data.

2 BACKGROUND

2.1 Literature Review

Vandermerwe and Rada (1988) first introduced the notion of servitization in their study of companies which bundle products and services to add value to their business offering [8]. They also recognized the fact that servitization poses its own special challenges for top management: how to blend services into the overall strategies of the company. Since then, there has been a steady discussion in the product service systems literature on the challenges which traditional manufacturers face when undergoing servitization [1] [3] [9] [10] [11] [12] [13] [14]. Several authors went on to categorize the challenges stated below.

Brax (2005) proposed a set of six challenges as follows [9]. The *marketing challenge* category consists of problems that are rooted in the manufacturer's prevailing marketing practices. It is critical that both the provider and the customer understand the concept of service. The manufacturer faces a *production challenge* when they need precise documentation of their products and installed base, as well as operation and maintenance data from the customer and effective information management is required. *The delivery challenge* involves cultural problems and challenges based on bad timing. Services require support and change in thinking throughout the company. Better scheduling of new equipment installation and follow-up services, e.g. training customers how to use the new equipment, are also needed, to avoid hectic and stressful handovers. The *product-design challenge* comes from the unsophisticated design of the offering which creates problems in its use. Services need to be designed to support the customers' business goals and practice. Manufacturers face *communication challenges* when they have no established processes to obtain and utilize feedback from their customers. As services are processes, communication with the customers is needed throughout the service relationship. The *relationship challenge* consists of the problems arising from the nature of the relationship between the provider and its customers. The implicit transaction-oriented business philosophy of the manufacturer does not support service offerings. The relationship factors, product support and the

J. Hesselbach and C. Herrmann (eds.), *Functional Thinking for Value Creation: Proceedings of the 3rd CIRP International Conference on Industrial Product Service Systems, Technische Universität Braunschweig, Braunschweig, Germany, May 5th - 6th, 2011*,
DOI 10.1007/978-3-642-19689-8_6, © Springer-Verlag Berlin Heidelberg 2011

surrounding business context are important in the customer's decision making [9].

Neely (2008) identified three main sets of challenges to servitization: *shifting mindsets; timescale;* and *business model and customer offering* [14]. In the process of servitization, a shift of the mindsets of marketing (from transactional to relational), sales (from selling multimillion dollar products to selling service contracts and capability), and customers (from wanting to own the product to being happy with the service) is required [14]. Timescales change for manufacturers when they have to manage and deliver multi-year partnerships, model and understand the cost and profitability implications of long term partnerships, and control long term risk and exposure. Traditional manufacturers have to understand what value means to customers and consumers. They have to develop the capability to design and deliver services rather than products, and alongside that, develop a service culture. Their main challenge is to embed all of the above into a service organisation [14].

Martinez *et al.* (2010) developed a set of five categories of challenges that a company has to consider when moving from being a product oriented organisation to a product-service one. These five main categories of challenges are as follows [1]. *Embedded product-service culture* that traditional manufacturers need to embrace and develop into a passion for service, in order to meet customer expectations. *Delivery of integrated offering* which implies multiple touch-points on the interface between provider and customer. With the integrated offering, the product-centric orientation has to be replaced with service-centric. Lack of organisational responsiveness can inhibit provision of an integrated offering. *Internal processes and capabilities,* i.e. acquisition of new capabilities that enable the organisation to compete in new service spaces. Alignment of product and service design processes is required, as well as performance metrics which measure the delivery of the integrated offering. *Strategic alignment*, or the alignment of mindset and understanding towards service provision, e.g. internal cooperation and common language. *Supplier relationships* call for a greater degree of co-operation between a provider and its supporting network. Changes in the relationships between the product-service provider and its customers have to be reflected in the relationships with the provider's suppliers [1].

2.2 Conceptual Transformation Model for Servitization

The authors have previously developed a Conceptual Transformation Model for Servitization [6] [7]. The Model has been derived as a foundation for further empirical work, with the aim to build new theory on the transformation towards product service systems. The purpose of the Conceptual Model is to give an holistic view and an overall picture of all aspects of the change that occurs when an organisation takes on the journey of servitization. The Conceptual Model is based on the authors' classification of the transformation challenges. This classification was built on the existing sets of challenges discussed the product service systems literature, with the specific aim to identify the various aspects of change during the servitization process. The following categories of challenges and areas of change were found [6] [7]:

Organisational challenges include new structures in the organisation (e.g. front office – back office), new organisational principles, a new leadership role for management, accepting new timescales. Another major group of organisational challenges is related to a change of culture and shifting the mindsets of ownership [6] [7]. Developing new capabilities to design and deliver services, developing service culture and managing the change itself are also some of the issues that organisations face. Understanding the value of the service in the organisation, as well as raising the servitization awareness in the organisation and improving the internal communication are also considered critical in the transformation towards product service systems [6] [7].

Customer challenges during the transformation process refer to developing a new relationship with the customer, appreciating the customer, and managing and delivering a multi-year partnership with the customer. This can be achieved through improved communication with the customer, obtaining new customer knowledge, connecting the company capabilities with the new customer opportunities, changing the customer mindset of ownership and developing a new contract. The manufacturer is viewed to be taking new responsibilities and in some cases even taking over customer operations [6] [7].

Business process challenges refer to a new business model, new competencies, new performance measures, devising a new service strategy, adopting new marketing practices and developing the new integrated offering. Often new production processes, new product design, a service delivery system and information management are also required [6] [7].

Financial challenges of the transition towards product service systems include increased investment into the integrated offering and redistribution of resources within the organisation. The company often changes its risk profile and it has to understand the costs of undertaking PSS. The transforming organisation is in a position where it has to manage and control long-term risk and exposure, new goals, metrics and incentives. It often faces difficulties with quantifying the savings from PSS in economic and environmental terms [6] [7].

Network challenges relate to a new alignment in the supply chain. The transforming organisation and its network of suppliers and distributors may have to shift responsibilities as to who will best deliver a function. New actors may appear in the network, e.g. end-of-life partners. In this process, new competitive dynamics are established and sometimes a shift of power in the network is created [6] [7].

Market challenges refer to the elements of the market which play a role in the servitization process. These could be existing service providers already in the market; the market demand for services and solutions or the lack of market demand for some PSS offerings. The transforming organisation needs to be aware of the threat of substitute products and new entrants. It also has to consider developing a service offering which is not easily replicated by competitors. Public policies or legislation can also be a factor in the servitization process [6] [7].

Our interpretation of the challenges as elements of change is based on the understanding that these challenges represent areas within the organisation, its processes and/or its network, which undergo specific, or complete, transformations. The challenges can also represent new elements which emerge in the process of enabling the delivery of a product-service offering.

The Conceptual Transformation Model for Servitization [6] [7] defined three dimensions of change: Organisation, Network, and Processes. Each dimension consists of a set of elements of change. The *Organisation* dimension includes the following elements of change: Product-Service; Leadership; Vision; Communcation; Strategy; People; Culture; Structure; Behaviour; and Competencies. The *Network* dimension is comprised of Customer; Suppliers; Competitors; Contract; Partnering; and Informal contract. The elements in the *Processes* dimension are New offering; Value; Risk; Cost; Standards; and Measures. The model also incorporates enablers and barriers representing the factors which affect the successful transformation towards product service systems [6] [7].

The current study builds on the Conceptual Transformation Model in various ways:

1. We identify a set of nine challenges based on the analysis of collected data.
2. We map the nine challenges, as elements of change, onto the Conceptual Transformation Model.
3. This study presents our first attempt at empirical validation of previous conceptual propositions.

3 METHODOLOGY

The paper is based on qualitative research where data was collected from 41 industry and academic experts. A one-day workshop on the subject of transformations towards servitized organisational forms was organised which was attended by industrialists and academics from the UK, Ireland, Italy, France and the USA. In terms of the participants' competence and bias, it is pertinent to note that the invited academics were active researchers in the field of Product Service Systems. The industrialists represented companies which, self-reportedly, had achieved a certain level of servitization or were considering a shift towards product service systems in their organisations. The companies that took part in the workshop were from the following industries: Defence, Transport, Steelmaking, Water and Cable industry, therefore, creating a bias towards high-value manufacturing representation. A small number of business services and consultancy providers in the area of servitization also took part in the workshop.

During the workshop, the participants were organised into four groups; each group had the same ratio of academic vs. practitioner representation to minimize influence and bias. The participants were tasked to answer the question "What are the challenges in transforming an organisation from a product-oriented into a servitized one?". First, each participant individually had to note down the transformation challenges, which they had experienced in their professional life, or challenges they believed would be stumbling blocks on the road to servitization. This was followed by a discussion and analysis of the individual lists of challenges within the group. Each group then had to agree on a "short list" of the top four challenges critical for the successful transformation towards product service systems, and provide the rationale behind their selection to the remaining groups.

The workshop provided us with a dataset containing an extensive list of challenges, based on those of each individual; and a short list of the most critical transformation challenges, selected based on the group discussions and analysis. The results of the workshop are presented, and analysed in the next section.

4 FINDINGS

4.1 Analysis and results

Data were analysed through an inductive approach. The workshop yielded a long list of 147 challenges and a short list of 15 top challenges, which included duplicate concepts in the groups' results. Analysis of the workshop data showed that collectively the four groups ranked the following nine challenges as the most critical to the successful shift to a product-service mode of business: *Culture, Customer, Contract, Value, Supply chain, New offering, Service delivery system, Measurements, and Risk management.* The *Culture* category appeared in all four groups' short lists of top challenges and the *Customer* challenge appeared in three out of the four groups' short lists of most critical factors. The remaining seven challenges were only found once in the short lists. Therefore, we can argue that the participants have weighted *Culture* and *Customer* as being the two most critical factors in the transformation towards product service systems. The rest of the top challenges appear to be of a somewhat lesser importance to the transformation process, albeit, still making it into the collective top challenges list.

We have presented below the original workshop data (Figure 1 - Figure 9). Analysis of the original participants' lists of challenges was conducted, in order to define the scope and develop a description for each of the top challenges.

Culture

Traditional manufacturers taking on the journey of servitization face the challenge of changing mindsets. This shift of culture takes place within the provider but also involves its customer and supply network. The shift of ownership is challenging to the customer as they move from owning the product to using the service. The provider is often put in a position when they have to educate their customer and the market. The shift of organisational culture from technology-led to service-oriented and from features-lead to value-based is a major challenge in the process of transformation of a traditional manufacturer into a more servitized organisational form. The product legacy inhibits service thinking. Products have to be designed along the customer perspective and in view of the future product support and total cost of ownership. The change of mindsets can be facilitated by strong commitment of organisational leadership; training and developing new skills; and improved and continuous communication. The mentality change can also be encouraged by financial incentives and reward motivation. The emerging service culture has to be then protected and supported throughout the entire transformation process.

- *Mindset change.*
- *Culture: Technology vs. Service.*
- *Shift of ownership.*
- *Incentives in the organisation.*
- *Clear strategy and leadership from the top.*
- *Educate the market and the customer.*
- *Contract based model.*
- *Organisational leadership, mindset change.*
- *Educate sales.*
- *Product legacy inhibits service thinking.*
- *Engineers generally not interested in support.*
- *Lack of communication.*
- *Lack of commercial imagination.*
- *Mentality change – reward motivation, incentives.*
- *Develop new skills.*
- *Protect the emerging service culture.*
- *Integration/communication of two cultures.*

Figure 1: Excerpts from original dataset on "Culture".

Customer

The customer poses another set of challenges during the transformation towards product service systems. The servitizing manufacturer has to build a new type of relationship with its customers; a relationship which is a dynamic partnership. The new relationship involves continuous collaboration, improved communication, and better understanding of customer's requirements. The provider has to create an 'account management' structure as the customer facing function of the business. Lack of knowledge about customer processes and product lifecycle at customer locations, as well as the traditional slow speed of the provider's response in the provision of services can be challenging to resolve. The provider has to recognise the diversity in the customer needs which the service offering can potentially meet. The manufacturer has to realise that the product is the start, not the completion of the offering. It is important to demonstrate to the customer the value of the integrated product-service offering and translate that into the contract.

- *Customer collaboration.*
- *Different relationship – dynamic.*
- *'Product' is start not completion.*
- *Understanding customer requirements (not just selling your product): build new type of relationship with your customer.*
- *Create an 'Account Management' structure.*
- *Contracting limitations.*
- *Educate the customer.*
- *Lack of knowledge about customer processes and product lifecycle at customer's locations.*
- *Affordability for the customer.*
- *OEM responsiveness slow – service fast.*
- *More understanding of how service product and user client interrelate to form part of a system that is less product-centric.*

Figure 2: Excerpts from original dataset on "Customer".

Contract

The servitizing manufacturer has to design a new contract which has to be agreed with the customer for the provision of the new combined product-service offering. The existing product contract is inappropriate for the service delivery: it normally contains product incentives and penalties related to the volume of product sales, and does not contain sufficient information on new sharing of responsibilities. It is important that the provider understands the performance measures and the risks prior to signing a service contract. The service also requires a closer relationship with the customer which often enables the establishment of trust between the provider and the customer.

- *Contract: never right.*
- *Product incentives/penalties – contract inappropriate for service.*
- *Services require/ enable much closer customer intimacy. Really understand risks and performance prior to signing service contracts.*
- *Establish contracts which deliver/enable end service received.*

Figure 3: Excerpts from original dataset on "Contract".

Value

Understanding, assessing and communicating the value of the integrated product-service offering internally within the organisation, and externally across to the customer, are critical for the successful transformation towards product service systems. It is difficult to identify and quantify the value from a potential service offering. This makes it challenging to define the business case internally in the organisation. The provider has to develop the capability to understand the value chain and co-create value with the stakeholders.

- *Develop a capability to understand the value chain and co-create value with the stakeholders.*
- *Identify and quantify value from a potential service offering.*
- *Value management.*
- *Making the business case internally.*
- *Measuring the value of the service.*
- *Understanding offering's value.*
- *Manage the offering rather than the product.*
- *Develop the capability to adapt over time regarding service value provision.*

Figure 4: Excerpts from original dataset on "Value".

Supply chain

During the transformation towards product service systems, new mechanisms of service network are established. The supply chain has to be restructured and re-aligned for the delivery of the combined product-service offering. This increases the complexity of the relationships in the supply network and new governance mechanisms, incentives, performance measurement systems have to be put in place.

- *Complexity of relationships.*
- *Supply chain alignment.*
- *Service-chain perspective – governance mechanisms, e.g. outsourcing, mechanisms of service network.*
- *Structuring the supply chain.*

Figure 5: Excerpts from original dataset on "Supply chain".

New offering

Defining the business case and designing the new integrated product-service offering can be challenging for the manufacturer. It requires gathering information, obtaining new knowledge and analysis of data. It is also critical to link the new offering to the existing processes within the organisation.

- *Define business case.*
- *Information, knowledge, data and analysis key to service.*
- *Link offering to process and organisation.*
- *Define overall level of service.*

Figure 6: Excerpts from original dataset on "New offering".

Service delivery system

The traditional manufacturer needs to set up an infrastructure or a service delivery system, in order to deliver the product-service offering.

- *Structuring the service network.*
- *Organisational structure.*
- *Rigidity due to infrastructure, processes, artifacts.*
- *Cross functional integration key to service optimization.*
- *Scale up the service delivery system.*

Figure 7: Excerpts from original dataset on "Service delivery system".

This involves creating new organisational structures, as well as cross functional integration within the company. The challenge is to overcome the rigidity of the existing infrastructure and traditional processes.

Measurements

Another major challenge in the servitization process is to measure and quantify the economic impact of the product service system. New key performance indicators have to be established in the organisation, along with new metrics, pay structure and incentives for the sales force. It is challenging to shift from 'product' to the new and different 'service' measures of success.

- *Process and business measurement.*
- *Put appropriate business measurements KPIs (revenue/head, margin/head, utilization).*
- *Financial tracking and management – new metrics, new incentives, new costings.*
- *Shift of focus from product measures to service measures. Pay structure changes.*
- *Different measures of success.*
- *Identify and quantify a profitable service opportunity.*

Figure 8: Excerpts from original dataset on "Measurements".

Risk management

Moving into services business can impose certain risks to the traditional manufacturing company. The company first has to establish an understanding of the risks involved in the transformation towards product service systems. The financial impact on the organisation's performance has to be assessed and measured. The new integrated product-service offering requires an investment that has to be quantified and sourced within the organisation. In the long term, the manufacturer faces the challenge of losing revenue from its product sales. From the market perspective, the transforming manufacturer can face competition from established service providers or in some cases start competing with its own customers.

- *Understanding the risk first.*
- *Assess the financial impact.*
- *Develop new capabilities / measures of risk.*
- *Risk of competing with our own customers.*
- *Pricing risk.*
- *Assess the investment into PSS transformation.*
- *Loss of revenue from product sales.*

Figure 9: Excerpts from original dataset on "Risk management".

4.2 Transformation Model for Servitization

We have identified a list of nine challenges based on the analysis of the data collected. These challenges are then mapped as elements of change onto the Transformation Model for Servitization as shown in Figure 10. These are the elements that have to be put in place in order for a company to complete the shift to a product service system. Based on our findings, the Organisation dimension now includes the elements Culture and Service delivery system. The Processes dimension in the model is now comprised of Value, New offering, Measurements and Risk management. The Network dimension now includes Customer, Contract and Supply chain.

The empirical data in this study provided us with evidence for the existence of several of the originally proposed elements of change in the Conceptual Transformation Model for Servitization, discussed

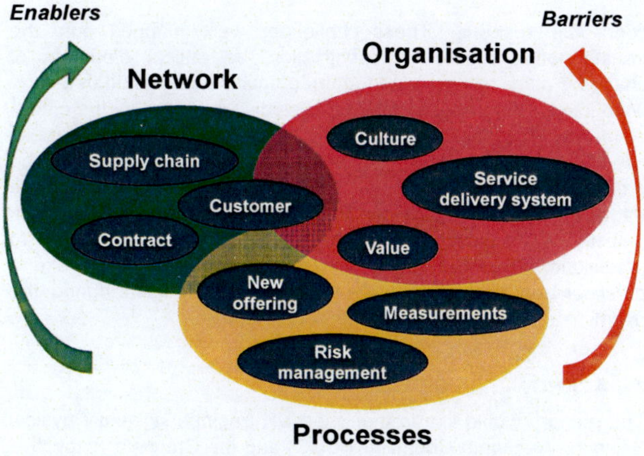

Figure 10: Transformation Model for Servitization.

earlier in Chapter 2.2, while a set of elements remained unconfirmed by the data. The unconfirmed elements are: Product-Service, Leadership, Vision, Communcation, Strategy, People, Behaviour and Competencies in the Organisation dimension of the conceptual model; Competitors, Partnering and Informal contract in the Network dimension of the conceptual model; and Cost and Standards in the Processes dimension of the conceptual model. We can, however, argue that the elements of change emergent from the empirical data largely incorporate these concepts, as evident in the lists of challenges collected during the workshop. The empirical validation of the conceptual model, therefore, has achieved two goals: 1) It helped us refine and structure the model as shown in Figure 10. 2) It also provided us with content for defining the scope and clarifying the concepts for each element of change in the transformation model for servitization.

This is our first attempt to validate the model empirically.

5 CONTRIBUTION AND RESEARCH LIMITATIONS

This study forms part of the ongoing validation of the conceptual transformation model for servitization. A list of nine elements of change was developed, which were mapped onto the transformation model for servitization. The purpose of this model is to facilitate the process of transformation towards a servitized organisation. Practitioners who take on the journey of servitization can use the model as a knowledge tool and for signposting, to be used when developing their change management plans in order to reduce the cost, the time and the risk of implementing the change. This article increases the comprehensiveness of the wider study into the transformations towards servitized organisational forms, by providing qualitatively derived richness, in order to achieve a more complete understanding of the phenomenon of servitization. The results of this work will also be used in multiple methods triangulation against findings from future case studies, in order to off-set biases and to validate our findings.

A workshop data collection method is not without its limitations. The bias of the academic and the industry representatives has been acknowledged. We also recognise that the workshop data collection is just one of our multiple methods of data collection in conducting this research.

6 CONCLUSIONS AND FUTURE RESEARCH

We have identified a set of nine challenges based on analysis of the empirical evidence in this study, which advances our prior conceptual research. These challenges were mapped onto the transformation model for servitization, as critical elements of change. Future research in the form of multiple case studies will be conducted. The purpose of the case studies is to identify the critical elements of change during the transformation towards product service systems, and the enablers and barriers affecting the process of servitization. These future findings will be built into the transformation model for servitization for use and application in industry. This research addresses issues of importance to practitioners and the model will be further refined to make it practical and user-friendly, so it can easily be used during the planning and the execution of the transformation.

7 ACKNOWLEDGMENTS

The authors would like to thank the UK Engineering and Physical Sciences Research Council (EPSRC) and the Cranfield Innovative Manufacturing Research Centre (IMRC) for funding this project, under grant number [IMRC 154], as part of the Product - Service Systems research programme at Cranfield University (UK).

8 REFERENCES

[1] Martinez, V., Bastl, M., Kingston, J., Evans, S. (2010): Challenges in transforming manufacturing organisations into product-service providers, in: Journal of Manufacturing Technology Management, Vol. 2, No. 4, pp. 449-469.

[2] Quinn, J.B., Doorley, T.L. and Paquette, P.C. (1990): Beyond products: service-based strategies, in: Harvard Business Review, Special report, March-April, 1990, pp. 58-67.

[3] Oliva, R., Kallenberg, R. (2003): Managing the transition from products to services, in: International Journal of Service Industry Management, Vol. 14, No. 2, pp. 160-172.

[4] Davies, A., Brady, T., Hobday, M. (2006): Charting a path toward integrated solutions, in: MIT Sloan Management Review, Vol. 47, No. 3, pp. 39-48.

[5] Araujo, L, Spring, M. (2006): Services, products, and the institutional structure of production, in: Industrial Marketing Management, Vol. 35, pp. 797-805.

[6] Vladimirova, D., Evans, S., Martinez, V. (2010): Transforming the Manufacturer into a Servitized Organisation, presented at: EurOMA's 3rd International Service Operations Management Forum, University of Bath School of Management, 20-21 September 2010, Bath, United Kingdom.

[7] Vladimirova, D., Evans, S., Martinez, V. (2010): Conceptual Transformation Framework for Servitization, presented at: International Conference on Advances in Production Management Systems (APMS) 2010, Competitive and Sustainable Manufacturing, Products and Services, 11-13 October 2010, Cernobbio, Como, Italy.

[8] Vandermerwe, S., Rada, J. (1988): Servitization of business: adding value by adding services, in: European Management Journal, Vol. 6, No. 4, pp. 314-324.

[9] Brax, S. (2005): A manufacturer becoming service provider - challenges and a paradox, in: Managing Service Quality, Vol. 15, No. 2, pp. 142-155.

[10] Baines, T., Lightfoot, H., Evans, S., Neely, A., Greenough, R., Peppard, J., Roy, R., Shehab, E., Braganza, A., Tiwari, A., Alcock, J., Angus, J., Bastl, M., Cousens, A., Irving, P., Johnson, M., Kingston, J., Lockett, H., Martinez, V., Michele, P., Tranfield, D., Walton, I., Wilson, H. (2007): State-of-the-art in product-service systems, in: Proceedings of the Institution of Mechanical Engineers, Part B: Journal of Engineering Manufacture, Vol. 221, pp. 1543-1552.

[11] Baines, T., Lightfoot, H., Kay, J. (2009): The servitization of manufacturing: a review of literature and reflection on future challenges, in: Journal of Manufacturing Technology Management, Vol. 20, No. 5, 2009, pp. 547-567.

[12] Baines, T., Lightfoot, H., Kay, J. (2009): Servitized manufacture: practical challenges of delivering integrated products and services, in: Proceedings of the Institution of Mechanical Engineers, Part B: Journal of Engineering Manufacture, Vol. 223, pp. 1-9.

[13] Spring, M., Araujo, L. (2009): Service, services and products: rethinking operations strategy, in: International Journal of Operations & Production Management, Vol. 29, No. 5, pp. 444-467.

[14] Neely, A.(2009): Exploring the financial consequences of the servitization of manufacturing, in: Operations Management Research, Vol. 1, No. 2, pp.103-118.

[15] Antonacopoulou, E., Konstantinou, E. (2008): The New Service Model: a review, a critique and a way forward, in: The Service Industries Journal, Vol. 28, No. 6, pp. 845-860.

Research on Services in the Manufacturing Industry based on a Holistic Viewpoint and Interdisciplinary Approach

Tomohiko Sakao, Christian Berggren, Mats Björkman, Christian Kowalkowski, Mattias Lindahl, Jan Olhager, Jörgen Sandin, Erik Sundin, Ou Tang, Patrik Thollander, Lars Witell

Department of Management and Engineering, Linköping University, Sweden

Abstract

This paper begins by consolidating industrial challenges and research issues concerning Product/Service Systems obtained through various activities by the authors. Based on this, it points out the importance of the holistic view in further research in this area so that PSS providers do not fall into local optimization. The intent of this contribution to our research community includes shedding light on interesting issues that thus far have been relatively invisible and with narrower scope.

Keywords:

Research Priority; ManuServ; PSS

1 INTRODUCTION

Given the significant interest in services in the manufacturing industry, a research group at Linköping University, named Centre for Service Research in Manufacturing Industry (ManuServ), has aimed at advancing service knowledge acquisition and exploitation, thereby increasing the competitiveness of manufacturing firms. Academics participating in ManuServ in collaboration cover a wide variety of disciplines, such as industrial marketing, R&D, product development, service innovation, quality management, production system development, and production economics.

The below summarizes various activities in ManuServ for more than a decade in the area of Product/Service Systems (PSS). Its aim is sharing with the reader our experiences and competencies that helped us arrive at our recognition of important industrial challenges and research issues, which are described later in Sections 2 and 3. The research has been carried out in close cooperation with leading Swedish and international companies and research institutions. Examples of projects, some of which are ongoing, are:

- Adapting Manufacturing Strategy to Industrial After-Sales Service Operations
- Energy Services in Industry
- Expanding Engineers' Opportunities: from Ecodesign to Integrated Product Service Engineering
- Flexible Business Models for Sustainable Competitive Advantage – Swedish Industrial Companies' Transition towards Service Orientation
- Integrated Product Service Engineering and Integrated Product Service Offerings
- Integrated Product Service Offerings
- Integrated Product Service Offerings of the Railway Infrastructure System
- Integration of Developing Products and Services towards Customer-oriented Solution Provider
- Managing Innovation Processes for a Business-Driven Collaborative Network
- Mapping of Integrated Product Service Offerings in Sweden
- Routinized Activities for Balancing Products and Services
- Solvent-Free Industrial Cleaning
- Technology - Business - Design for successful functional sales

These projects cover various disciplines and different aspects, including environment and energy[1]. Figure 1 depicts the processes of interest (IPSO, standing for Integrated Product Service Offering, is used to emphasize integration but as an equivalent term to PSS).

Based on ManuServ's wide range of research experiences by all the authors in various industries, this paper first consolidates industrial challenges and research issues in Sections 2 and 3, respectively. Then, Section 4 presents a discussion, mainly from a holistic viewpoint, which is crucial in research and practice in this area. Indeed, the intent of this initiative to the research group includes, by presenting a holistic viewpoint, shedding light on interesting issues that have thus far been invisible given their narrower scopes. Section 5 concludes the paper.

2 INDUSTRIAL CHALLENGES

2.1 Marketing and sales

In order to develop, sell, and deliver PSS, several marketing and sales practices need to be revised, and many new capabilities are needed in goods-centric firms. For example, many challenges relate to the manufacturing and R&D-orientated corporate culture, lack of top management commitment to service strategies, the sales organisation's focus on equipment sales, and incentive systems and key performance indicators (KPIs) which disfavour service operations. In a survey of more than 300 manufacturing companies, [1] identified pricing as a major challenge facing manufacturing

[1] For more details of the projects above and their associated publications, see manuserv.iei.liu.se.

J. Hesselbach and C. Herrmann (eds.), *Functional Thinking for Value Creation: Proceedings of the 3rd CIRP International Conference on Industrial Product Service Systems, Technische Universität Braunschweig, Braunschweig, Germany, May 5th - 6th, 2011*,
DOI 10.1007/978-3-642-19689-8_7,

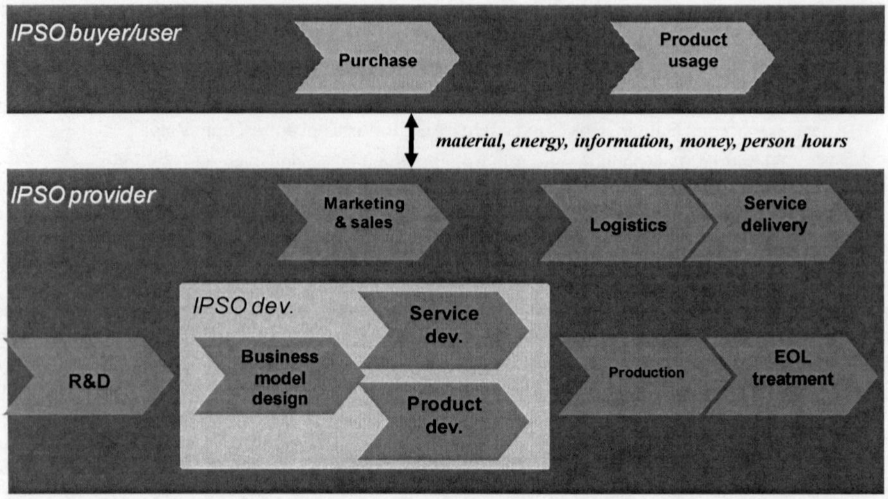

Figure 1: Processes of interest in ManuServ.

firms. Thus, pricing is an important challenge among others. In terms of pricing, many find it easier to price their goods than their services. In particular, they have difficulties in estimating the value of the "intangibles" associated with services. In addition, many manufacturing firms have traditionally given away service in order to make a sale, i.e. services are free. Sales forces typically find it more difficult to sell services than products, and the financial incentive systems are often not designed to reward this more challenging behaviour [2]. For PSS, the firm's charges are often based more on the customer's value-in-use than on the monetary exchange value [3, 4]. Specifically, the process-orientation of solutions implies that pricing policy is linked, to varying degrees, to outcomes delivered to the customer. The price can be fixed, such as when a fixed fee is charged for a given level of availability in outcome-based agreements, or dynamic, as in a "gain-sharing" arrangement with payments directly linked to the customer's business performance [5].

However, pricing-related challenges do not only concern services. What is challenging at a higher level is pricing different services and products of a PSS along the entire product lifetime. Our observations of a manufacturer providing long-term products has found them charging relatively lower prices on products at instalment but higher prices on services (incl. spare parts), which has led to low customer satisfaction. This is at present recognized as a problem for increasing customer satisfaction in that company. This issue needs investigation with a holistic view on another dimension.

A related challenge that firms need to manage is to have different revenue mechanisms (e.g. transactional and value-based pricing approaches) in parallel. Firms offering PSS have to manage dual revenue models, i.e. both the traditional one where customers pay for service hours and parts used and one where the customer pays a fixed cost per time period, regardless of any fluctuations in the provider's service activity costs [2].

New industrial services and PSS also require companies to have the ability to visualise service-based and relationship-based values, i.e. to communicate and demonstrate the values both internally and externally. Managers need be able to make use of several different visualisation strategies, depending on the actual content of the offering and its position in the development process. Value visualisation strategies have traditionally focused on external sales activities. However, companies need to have a broader approach to visualisation, being able to visualise the (potential) value during the service development process, during sales, and during service provision, as well as including different visualisation techniques. Value visualisation has become vital for winning new contracts and retaining existing ones. It is therefore a strategic resource that managers need to pay attention to, and continuously develop, in order to compete with PSS [6, 7].

2.2 Business model development

Manufacturing firms have typically exhibited a product and technology orientation, and many of them are relatively new to service logic and to service innovation. Service development often becomes more complex in manufacturing firms than it does in pure service firms, since manufacturing firms have a core product that has traditionally carried most of the value and which is heavily embedded in the company culture [8, 9]. Thus, identifying and implementing service-based business models in goods-centric companies is often a very long-term process [10]. Firms need to focus on all areas of their business models in a holistic fashion, and not just change isolated elements. Among other challenges, companies must develop their abilities to build relationships with customers and to advance a dynamic service offering portfolio that is adaptive to changing customer needs [9]. From a development perspective, many firms provide services in an unstructured manner and lack the capability to sense market opportunities. Furthermore, many firms have problems getting investment decisions and commitment and they lack the necessary service infrastructure and technology [8]. Therefore, firms need to balance exploration and exploitation, and local and central innovation and learning. In contrast to goods innovation, many new services are developed locally, often in an ad hoc manner in close interaction with customers [2, 11]. Firms should not only blueprint complex product development models but also consider service-specific aspects, such as more iterations and interactions with lead customers. This also relates to marketing and organizational issues, such as developing measures and sales tools, and changing existing mindset, norms, and values [8].

2.3 PSS development and R&D

Solutions in order to improve service (including maintenance and remanufacturing) are often ignored when designing new products. In other words, there is lack of feedback from the field. Regarding service design, our observations found that service is often added on already existing products (e.g. [12, 8]). Integration between product design and service design is weak. In these cases, there is obviously a risk of sub optimisation (and even one of ending up with unsuccessful solutions). For this problem, developing physical products (artefacts) in parallel to adjacent services instead of having a traditional sequential development needs to be established (also argued in [13, 14]). This needs an integrated approach across different divisions, especially in large firms.

In particular, developing models and methods that facilitate the integration of the new requirements that PSS put on the development process is also a challenge (see e.g. [15]). Another challenge in integration within a firm is integrating PSS design methods, if any, and customer value management methods. Some companies have customer value management systems, but face difficulty in making use of it to conduct PSS design effectively.

New requirements on development, products, offerings and operations might occur and the need for crosstalk between different functions increases. The majority of the functions in the organization need to work in an integrated manner to a greater extent. Processes need to be parallel, flexible, frontloaded and to support learning. Moreover, contemporizing and connecting support tools and systems to the processes might be another challenge. System and infrastructure need to support reversed logistics, not only with physical material, but also with information, both upstream and downstream.

Among others, a great challenge for industrial manufacturing companies moving towards PSS is that the structure, including processes, support tools and systems, are not yet completely sufficient for PSS. A need for new features might occur and there is most often a demand for an increased holistic view that attends to the complete product life-cycle. Examples on new features required are fleet management, product monitoring and return systems. Furthermore, content based visualising system displaying flow of value and also material, energy, information and man hours may be critical.

Looking at the service design part, effective and efficient service development in goods-centric companies is an issue at present, since that was not so important before.

Looking at an earlier phase, e.g. R&D, evaluating and improving R&D projects towards increasing customer value is also an issue. Some companies find the need to modify the procedure of engineering activities prior to product design. One reason is the emerging need to focus on customer value even in R&D, which used to be often oriented toward technology push. For instance, new technology can support reducing the system downtime and thereby adding customer value. The challenge is to be able to do this at the same time as the costs for Maintenance, Repair and Overhaul (MRO) is minimized (see e.g. [16]).

2.4 Production

An ideal product to sell is the product you don't need to produce, and PSS opens the door for this opportunity if you as a PSS provider can take back used products and with minimal effort can reuse them again for existing or new customers. However, in order to do so in an efficient and effective manner, the products must be engineered for take-back and remanufacturing (see previous section). Previous research [12] has shown that this is not often the case.

PSS, when including having take-back systems, change the way of looking at production. The focus in production moves from producing as many products (based on virgin components) as possible into, as long as it is economically and technically possible, producing products based on used products and components with as little use of new components as possible.

PSS put new requirements on the products and the production systems throughout the products' technical and economic lifecycles [15]. For instance, when using return items, it creates some challenges in production and planning, due to the uncertainty of return time, quantity, and quality. In addition, returns can be substantially diversified, thus creating difficulties in applying "standard" production process. At the same time, those challenges, if handled in the development process, can be prevented or eased and turned into advantages because of the cost reduction reuse of products can give. It could mean that a supplier sets up a system for reverse logistics and remanufacturing in order to benefit from the economic and environmental potential found in the PSS business model [17]. Toyota Material Handling is an example of a company that has done just that. They provide material handling through forklift truck rental agreements, and have set up a fleet management centres including product monitoring and remanufacturing activities throughout Europe [17].

To summarize, based on ManuServ's previous research, PSS business models elucidate the potential of developing models for setting up e.g. preventive maintenance, product monitoring, product return systems / reverse supply chains, part exchange systems, remanufacturing / recycling and fleet management centres.

2.5 Supply chain management

In case the business involves return, the reverse logistics could be a major problem. It is important to identify the return network, its connection with the forward network, and acquisition/sorting issues. It is also strategically important to balance the returns and the demand of returns (considering the production life cycle issue). In case the business does not involve returns, the integration of cost, market and knowledge should be a concern in defining a supply chain structure.

2.6 Service delivery

The lack of a service delivery organization and the risk of building up one – i.e. entering a new field – is for many traditionally product-focused companies a huge challenge since they need new competencies, management, etc. Added to this is the economic challenge in building up the service delivery infrastructure, and the concern of what will happen if the demand for the service disappears. A more detailed but important challenge is to maintain the quality of the delivered offering. For products it is easier to optimize based on fraction of a percent, but when optimizing service delivery, i.e. working with people and not machines, it is trickier to optimize and the fluctuation of the results increases.

For companies with an existing external service delivery organization designed for traditional sales of products, a move into PSS offerings might be a big challenge if they cannot control their service organization or cannot build up a parallel one suited for PSS. Traditionally, service delivery organisations are often built up on earning money from service and spare part sales. The problem is that when selling PSS offerings, e.g. get paid for the function, spare parts and service become costs instead of incomes for the PSS provider. This might imply tensions and problems in the existing service delivery organisation. Hence, it can be a great challenge to implement a new mindset from the very top of the organization, all the way down to the development and the operations.

When designing PSS offerings, managing internal data about ongoing service delivery with their contracts (including the planned and actual costs) has been identified to be a big challenge [4]. This data contributes to more precise projection of uncertain events during a contract period, and assists the provider in making decisions on uncertainty allocation. However, this needs to be tackled from a company-wide perspective, because this kind of data will be related to existing systems such as the ERP (enterprise resource planning) system.

2.7 Managing energy and material consumption – application-oriented challenges

Increasing raw material and energy prices have become a problematic issue for both producers and users. In line with this, some larger companies have also realized that if they want to increase their market, e.g. in China and India, they need to change their way of offering into more material and energy efficient and effective solutions. If not, they foresee that their own increased need of material and energy will have a notable negative impact on the price levels and will be an obstacle for their ambitions. At the same time, society's focus on environmentally-related issues has forced many companies to rethink their way of offering goods and services. Since environmental problems are tightly linked to energy and material use, they need to find new solutions that reduce this use.

This has led to an increasing focus and interest in reducing the use of material and energy in order to cut down their costs, and PSS has in this context become an interesting solution.

Looking at energy consumption, great potential for energy efficiency exists in industry, but the adoption of energy efficiency measures is often inhibited by various barriers. Different means to overcome these barriers and promote energy efficiency in industry exist, one of the most promising being energy services. Earlier research shows that while many barriers could be overcome by energy services, the industry's ranking and adoption of energy services is very low. Therefore, developing methods on how to integrate/promote energy services in manufacturing companies is an interesting issue.

Regarding raw material consumption, prior research indicates the large potential for more efficient and effective use. However, to do so, products used in the offerings need to be designed to fit into the PSS business model (see previous section about product development and production).

2.8 Organizational management

One major industrial challenge facing manufacturing companies trying to increase their emphasis on services is that the organization is not ready to work with services [1] [2]. Many manufacturers have embarked on services without any clear strategy. These firms have viewed their service transition as another growth initiative, and have failed to understand the organizational change required. Most of a manufacturing company's tacit knowledge is with its products, rather than existing or potential service offerings, and many companies had attempted to market new services in the same way as they had always sold new products—that is, by adding new responsibilities to existing managers' roles while maintaining current organizational structures, practices and incentives.

As the division between goods and services becomes ever more blurred, there is an increasing need for cooperation between the service and the product organizations [18]. Applying a service logic means that the traditional division between goods sales and after-sales services becomes outdated. Instead, the customer relationship becomes the centre of the offering regardless of its combination of services and goods. Thus, a transition to service logic implies much more than an increased emphasis on services; it implies a reframing of the purpose of the firm and its role in value co-creation. For instance, it means that research and development, sales, service, finance, human resources, and other local and central organizational functions must work together to a greater extent than is usually the case in manufacturing firms [4]. What has been found among large firms delivering PSS is that the product development department often has a long tradition and is much larger than the service development department, which makes it harder for the "service people" to have their voice heard [19]. Further, the infusion of service in manufacturing firms means that more service processes and interfaces have to be managed simultaneously [20].

Very few manufacturing firms have all the physical and service components and all the competencies needed to provide PSS. Therefore, firms are to various extents dependent on partner firms and other actors in the business network. From a managerial point of view, firms need to determine which services and service-related capabilities to provide themselves and which can, or should, be externalized. Furthermore, other firms, such as consultants, may have significant influence on the customers' specifications and inquiries, and, thus, these network relationships must be managed in accordance with the firm's market and service strategies [2]. This is particularly challenging if firms operate in antagonistic service networks with limited trust and communication among actors, and a considerable amount of power play. In such contexts, actors strive to maximize their own value from a zero-sum game perspective, and there is no room for "real" partnerships [21].

3 RESEARCH ISSUES

3.1 Towards a value-oriented provider

There are several research avenues of interest that relate to service infusion and value creation and capture through new industrial services and PSS. Manufacturing firms need to focus on these issues in a holistic and interdisciplinary fashion, due to the systemic nature of value creation and customer value management. Research in this area is scarce, however, and there are many calls for further research [22, 23].

For example, managing a cultural shift from maximizing unit sales to selling performance and facilitating customer value co-creation is yet an under-researched area, including issues such as shifting from a goods-centric to a service-oriented sales force, coordinating and determining the degree of integration between local and central, and between goods, services, and solutions units, and aligning goods, services, and solutions strategies.

Furthermore, service and business model innovation should be a research priority [24]. It includes issues such as identifying the role of local and central units in the innovation process, identifying how and when to involve customers and partners, and generating, evaluating, and realizing innovation ideas. More research is called for regarding developing a competitive offering portfolio and PSS value propositions. For example, research should facilitate a deeper understanding of what perceived benefits and sacrifices should be emphasized, and the relative emphasis on value-in-exchange and value-in-use in value propositions [25] . How should firms balance requirements on cost-efficient operations and standardized processes with customization and flexibility, how can service modularity be utilized, and how can we better understand customers' value-creating processes and which customers want services and PSS?

Finally, research should focus on how firms can derive equitable value capture. That is, determining how to share gains and losses across a service system. Researching service systems and business networks implies a holistic approach, and other areas of interest include: optimizing service networks and value constellations, and improving intra- and inter-organizational collaboration locally and globally.

3.2 Basic research

Looking at fundamental insights from conventional engineering, revisiting the well-known relationship between the cost and the influence of each lifecycle phase (or stage) can be a research issue. E.g. a rule, "the task clarification stage of product design accounts for only 10-15% but determines 50-70%" of the total cost is well known. But, on the other hand, a PSS provider has an opportunity to reduce 30-40% of the total lifecycle cost of a customer through provided services. More related pieces of information are expected to be collected from different cases. Apparently, a substantial change to such rules in comparison to those in a product sales business is foreseen. It would be valuable for companies to generalize such knowledge in PSS, even though it depends on the types of products.

3.3 Application-oriented research

The potential for energy services in the non-trading sector of the company is stated to be vast. Related to the need for mitigation of global climate change, Energy Services is stated to play a key role. However, research in the area is scarce, both in relation to theoretical ideas and also in terms of bottom-up studies of how Energy Services are actually implemented in manufacturing companies. This may need an interdisciplinary approach due to its different types of barriers (see [26]).

4 DISCUSSION

In our research we have found that services in manufacturing have received greater attention lately due to several reasons. One reason is that there is more for the manufacturers to make profit on besides the pure physical products. Changing the manner which the manufacturer makes profit needs a new mindset in the internal organization, e.g. the service department needs to get more attention regarding maintenance schemes and systems of reverse logistics and remanufacturing, where suitable. In addition, these people's needs also must be considered more in the development of business models and product design. All this should be conducted with the aim of gaining a holistic view of what the manufacturers are offering their customers.

The description of the industrial challenges and research issues in Sections 2 and 3 highlight the importance of the holistic view in this research theme. Therefore, ManuServ aims in particular at tackling industrial challenges with a holistic viewpoint to avoid local optimization within a firm or a group of companies within the same value chain. In order to deal with one industry challenge, several other interdependent ones frequently need to be coped with. For example, organizational and managerial challenges often need to be overcome in order to succeed with new marketing and sales practices, supply chain management, PSS and business model development, etc. Likewise, PSS development and design processes and routines, marketing and sales metrics, etc. may need to be revised in order to successfully implement organizational changes.

The ManuServ temple in Figure 2 describes the basic idea behind the ManuServ initiative – a temple needs several pillars to become robust. A PSS company cannot simply focus on one area to be successful – they need to balance and modify parameters from all the described areas. The idea is to link together different PSS research areas and their researchers in order to achieve cross communication and initiate new interdisciplinary research that take advantage of the opportunities to study the challenges from different perspectives and evaluate different ways of doing things. An example could be how the design of a product and service could affect the service delivery and business model and the interactions between those three areas.

When aiming for an improved way of working, adjusted to PSS, those types of changes are often a long term process and there is a great need for being assiduous. There is also a challenge of comprehending which capabilities that is realistic to develop and where to depend on partners.

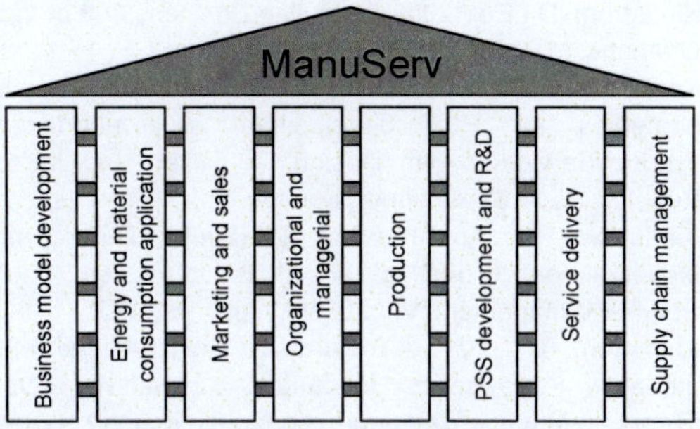

Figure 2: The ManuServ temple.

Note: The grey lines between the research areas illustrate linkages and communication between the research areas.

As described in Section 2, previous research has generated new knowledge and also identified numerous relevant research issues for each described research area. However, the researchers behind the ManuServ initiative believe it is a problem that the links between those research areas today are so weak and undeveloped. They believe that PSS research would be more successful with a more integrated and holistic approach. At the same time, it is a bit paradoxical that the two basic ideas behind the PSS concept are the lifecycle perspective and integrated approach. It is natural to assume that the very different research areas, as described in Section 2, as are linked to one another. In other words, changes in one will affect the other, and it ought to be possible to balance different parameters from different areas between them. This is also partly highlighted in Section 2.

5 CONCLUSION

This contribution presented the authors' recognition of important industrial challenges and research issues about Product/Service Systems obtained through various activities by researchers in ManuServ. Based on this, it pointed out the importance of the holistic view in researching this area. This was shown to be crucial from the different problems and challenges exemplified in this contribution.

6 ACKNOWLEDGMENTS

We extend our sincere gratitude to the Swedish and international firms who provided us with invaluable experiences in this area.

7 REFERENCES

[1] Brown, S. W., Gustafsson, A., Witell, L. (2009): Strategy beyond Products, The Wall Street Journal.

[2] Kowalkowski, C. (2008): Managing the Industrial Service Function. Linköping, Linköping University.

[3] Nordin, F., Kowalkowski, C. (2010): Solutions offerings: a critical review and reconceptualisation, Journal of Service Management. Vol.21, No.4, pp. 441-459.

[4] Kowalkowski, C. (2011): What does a service-dominant logic really mean for manufacturing firms?, CIRP Journal of Manufacturing Science and Technology, in print.

[5] Kowalkowski, C., Brehmer, P.-O. Kindström, D. (2009): Managing industrial service offerings: requirements on content and processes, International Journal of Services Technology and Management. Vol.11, No.1, pp. 42-63.

[6] Kindström, D., Ed. (2009): Creating Business Out of Industrial Offerings. MTC, Stockholm.

[7] Kowalkowski, C., Kindström, D. (2009). Value visualization strategies for PSS development, in: Introduction to Product/Service-System Design. T. Sakao and M. Lindahl Eds., pp. 159-182, Springer, London.

[8] Kindström, D., Kowalkowski, C. (2009): Development of industrial service offerings: a process framework, Journal of Service Management. Vol.20, No.2, pp. 156-172.

[9] Kindström, D. (2010): Towards a service-based business model – Key aspects for future competitive advantage, European Management Journal. Vol.28, No.6, pp. 479-490.

[10] Gebauer, H., Fleisch, E. (2007): An investigation of the relationship between behavioral processes, motivation, investments in the service business and service revenue, Industrial Marketing Management. Vol.36, No.3, pp. 337-348.

[11] Gremyr, I., Löfberg, N., Witell, L. (2010): Service innovations in manufacturing firms, Managing Service Quality. Vol.20, No.2, pp. 161-175.

[12] Sakao, T., Napolitano, N., Tronci, M., Sundin, E., Lindahl, M. (2008): How Are Product-Service Combined Offers Provided in Germany and Italy? – Analysis with Company Sizes and Countries -, Journal of Systems Science and Systems Engineering. Vol.17, No.3, pp. 367–381.

[13] Sakao, T., Shimomura, Y. (2007): Service Engineering: A Novel Engineering Discipline for Producers to Increase Value Combining Service and Product, Journal of Cleaner Production. Vol.15, No.6, pp. 590-604.

[14] Sakao, T., Shimomura, Y., Sundin, E., Comstock, M. (2009): Modeling Design Objects in CAD System for Service/Product Engineering, Computer-Aided Design. Vol.41, No.3, pp. 197-213.

[15] Sundin, E., Lindahl, M., Ijomah, W. (2009): Product design for product/service systems - design experiences from Swedish industry, Journal of Manufacturing Technology Management. Vol.20, No.5, pp. 723-753.

[16] Candell, O., Karim, R., Söderholm, P. (2009): eMaintenance-Information logistics for maintenance support, Robotics and Computer-Integrated Manufacturing. Vol.25, No.6, pp. 937-944.

[17] Sundin, E., Bras, B. (2005): Making functional sales environmentally and economically beneficial through product remanufacturing, Journal of Cleaner Production Vol.13, No.9, pp. 913-925.

[18] Gebauer, H., Fleisch, E., Friedli, T. (2005): Overcoming the Service Paradox in Manufacturing Firms, European Management Journal. Vol.23, No.1, pp. 14-26.

[19] Sundin, E., Sandström, G. Ö., Lindahl, M., Rönnbäck, A. Ö., Sakao, T., Larsson, T. C. (2009): Industrial Challenges for Product/Service Systems: Experiences from a large company network in Sweden, in: CIRP IPS^2 Conference, pp. 298 - 304, Cranfield.

[20] Kowalkowski, C., Brehmer, P.-O. (2008): Technology as a driver for changing customer-provider interfaces: Evidence from industrial service production, Management Research News. Vol.31, No.10, pp. 746-757.

[21] Matthyssens, P., Vandenbempt, K. (2008): Moving from basic offerings to value-added solutions: Strategies, barriers and alignment, Industrial Marketing Management. Vol.37, No.3, pp. 316-328.

[22] Lindgreen, A., Wynstra, F. (2005): Value in business markets: What do we know? Where are we going?, Industrial Marketing Management. Vol.34, No.7, pp. 732-748.

[23] Ulaga, W., Eggert, A. (2006): Value-Based Differentiation in Business Relationships: Gaining and Sustaining Key Supplier Status, Journal of Marketing. Vol.70, No.1, pp. 119-136.

[24] Ostrom, A. L., Bitner, M. J., Brown, S. W., Burkhard, K. A., Goul, M., Smith-Daniels, V., Demirkan, H. Rabinovich, E. (2010): Moving Forward and Making a Difference: Research Priorities for the Science of Service, Journal of Service Research. Vol.13, No.1, pp. 4-36.

[25] Kowalkowski, C. (2011): Dynamics of value propositions: insights from service-dominant logic, European Journal of Marketing. Vol.46, No.1/2, in print.

[26] Thollander, P., Palm, J., Sakao, T. (2010): Increasing industrial energy efficiency through energy services – an interdisciplinary approach with engineering- and social science aspects, in: 2nd CIRP International IPS^2 Conference, pp. 51-56, Linköping.

Product Adaptation for Industrial Product-Service Systems – Characteristics, Motives and Challenges

Sabine Biege[1], Marcus Schröter[1], Marion Weissenberger-Eibl[1]

[1] Fraunhofer Institute for Systems and Innovation Research ISI, Karlsruhe, Germany

Abstract

Industrial product-service systems have become an increasingly researched topic during the last decade. Although product and service components of these integrated systems strongly interact, literature on adaptation of physical goods to the service needs of delivering these product-service bundles is rare. The paper contributes to closing this gap by analyzing expert interviews on industrial product-service systems. Based on this empirical basis, to begin with the diffusion of product adaptation in the capital goods producing industries is discussed. Furthermore, characteristics of product adaptation processes are identified. Motives behind doing so are revealed and challenges for companies are substantiated.

Keywords:

Capital Goods Industry; Product Adaptations; Qualitative Research

1 INTRODUCTION

In many manufacturing industries a shift from traditional product-oriented towards service-oriented business models could be witnessed during the last years. Consequently, industrial services and product-service systems have become increasingly researched topics in the last decade and a multitude of aspects of this phenomenon have been discussed intensively.

Yet, research on an integrated development of product and service constituents of industrial product service systems seems to be in its initial phase. Up to now, service requirements are rarely considered in the development of new products and existing products are usually not adapted for product-service systems [1]. Neither design principles for physical goods nor for service products are appropriate for products used in these product-service concepts. It was pointed out in the literature that development of product and service components of product-service systems needs to be done in an integrated way [2,3,4], yet only some authors have made an attempt to relate engineering guidelines to servitization issues [1,3,5,6,7,8].

Product-service systems pose new challenges for the design of products, as new and complex dimensions need to be considered during the product development process [9]. Yet, „design research on service has been minimal compared to design research relevant to manufacturing" [10]. Product-service system development has been explored only from a management perspective, although developing product-service systems entails primarily technical innovations [11]. The physical and the service component, however, need to be developed in an integrated way to ensure the proper functionality of the product-service system [11,12,13]. Considering service requirements throughout the development of the physical component contributes to increasing the efficiency of service delivery. Furthermore, the costs for the service activities can be decreased so that the overall competitiveness of the offer rises [14].

As a conclusion, the following research question tackled in the paper raises: In which way do the physical assets used in product-service systems need to be adapted to the requirements of servitization so that delivery of product-service systems is facilitated and risks are reduced? The aim of the research described in the paper is to firstly determine to what extent manufacturing companies already do adapt their physical products to the product-service systems they deliver to their customers. Secondly, the research intention is to frame design guidelines for manufacturing companies to enable them to deliver product-service systems in an efficient way. The results of interviews with experts from research and industry are presented to explain characteristics of product adaptations for product-service systems, to learn about the motives behind adaptation processes and to identify challenges for manufacturing companies which implement product adaptation strategies.

The remainder of this paper is structured as follows. In the second chapter an excerpt of the available literature dealing with integrating product and service development in product-service systems is given. In the third section of this paper, the methodology applied is described. The results of the expert interviews can be found in the fourth section of this paper. After a short summary, need for further research is outlined in section five of the paper.

2 LITERATURE REVIEW

The combination of physical products and immaterial services implicates that product development and service development depend on each other and can no longer be carried out separately. When developing product-service systems, elements of product design specificities as well as elements of service design need to be implemented.

In the current literature dealing with product-service systems and industrial product-service systems, several publications can be found that address the subject of developing products and services for these integrated offers. Many authors state that the development of services and products needs to be executed in an integrated way. Yet only a few authors have concretely discussed the implications for the design of products that arise by the infusion of services into manufacturing industries. A short overview is given below.

J. Hesselbach and C. Herrmann (eds.), *Functional Thinking for Value Creation: Proceedings of the 3rd CIRP International Conference on Industrial Product Service Systems, Technische Universität Braunschweig, Braunschweig, Germany, May 5th - 6th, 2011*,
DOI 10.1007/978-3-642-19689-8_8, © Springer-Verlag Berlin Heidelberg 2011

In [15] two kinds of processes are defined according to the relationship of product and service innovation processes in capital goods producers. If no links exist between service development and product innovation processes, the authors call this way of innovating separated innovation. Services are being developed for products which have already been introduced into the market. In an integrated development process, service innovation processes are integrated into the development of new products and both, new product and new service, are introduced into the market simultaneously.

In [1] and [2] a process model for product-services development is presented which assigns functions of the concept to product or services modules. The authors assume that product-services systems on the one hand but also their development on the other hand can be carried on more effectively and efficiently if the development processes are integrated.

Spath and Demuß present in [8] an ideal development methodology for developing product and service components of product-service systems. By doing so, the authors attempt to bridge a gap they identified. According to Spath and Demuß, the requirements accruing from industrial services are currently considered by implementing design for service, i. e. design for maintenance, and design for recycling methods. Methods of regarding requirements beyond maintenance and recycling do currently not exist. Consequently, they propose a method of developing hybrid products consisting of physical product and service components which is based on integrated product development methods and service engineering methodologies.

Although several process models for an integrated development of products and service components of product-service systems exist, integrated development can only be found in producers of capital goods to a limited degree. Sturm and Bading surveyed 99 German producers of capital goods and found out that 21 % of these developed their goods and services in an integrated way whilst 16 % of the surveyed companies stated that new products often had their origin in new service offers [4]. Furthermore, the companies participating in the study were asked how important they consider the integrated development of products and services. Despite the low share of those surveyed who stated that product and service development are actually done in an integrated way, 90 % of the study participants regarded it as important or even very important.

Berkovich et al. did a study on requirements engineering for hybrid products [16]. 15 company representatives were surveyed via qualitative interviews to gain insights into the state of practice into requirements engineering for hybrid products. The findings indicate that requirements engineering for products and for services is actually done, but in parallel and without coordination. The single elements of product-service systems are not integrated until the very end of the separate development processes. Hence this study also shows that an integrated development of product and service components of product-service systems is very limited in practice.

As a consequence from the theoretical and empirical findings described above, the following questions can be derived. If an integrated development of product-service systems cannot or only be found to a limited degree in practice, how common are product adaptation strategies to the service needs of industrial product service systems? To be more precise, which adaptations are needed and how do the characteristics of industrial product-service systems influence these adaptation strategies?

3 METHODOLOGY

A qualitative approach was chosen to tackle the research questions lined out in the first two paragraphs. Sixteen experts from several industries were interviewed. They were identified on the basis of a literature review on industrial product-service systems. Persons were selected who had contributed to the international debate on industrial product-service systems during the last years. The experts came either from research, from industry associations or were employed in companies. Furthermore, the interviewees came from several European countries: experts from Germany, Sweden, Denmark, Austria, Belgium and Italy were surveyed. Five interviewees were experts for the manufacturing sector in general. A total of three experts were interviewed to gain insights into the chemical industry, three more experts were spoken to from the compressed air sector. Four experts from the machine tool sector were surveyed, whilst one expert gave insights into the assembly industry. An overview about sectors and professional background of the experts is given in Table 1.

To achieve comparability between the interview results, an interview guideline was elaborated which contained several blocks of questions. The first group of questions dealt with the diffusion of industrial product-service systems in the industry and their future potentials. The next three groups of questions were focused on economical, ecological and social effects of industrial product-service systems. The last set of questions dealt with barriers, promoters and the legal framework for industrial product-service systems in the experts' industry.

Overall 16 interviews with 19 experts coming from 16 organizations were conducted. In one interview two experts, in another interview three experts were surveyed together. The interviews took place between April 2009 and May 2010. They took between 45 and 105 minutes and were held by two interviewers either on the telephone or as a face-to-face interview. As findings from qualitative research cannot fully be generalized, selection of experts needs to focus on building types instead of their representativeness [17]. Hence, the number of cases seemed adequate for covering the researched topic. Based on notes taken during the interview, minutes of the survey were prepared afterwards. They were sent back to the interviewees and the content was corrected by them. These revised records were analyzed by means of a content analysis [18] using the qualitative data analysis software Atlas.ti.

The content analysis technique used structured the statements given in the interviews aiming at identifying linkages and filtering structures from the texts [18]. The coding scheme was derived from the research questions tackled. Hence, all kinds of product adaptations were coded as well as characteristics of industrial product-service systems, which were deduced from the morphologic box proposed in [19]. These describing elements of "service-based business models" are

- ownership of the physical good during the use phase
- ownership of the physical good after the use phase
- employer of the operating personnel
- employer of the maintenance personnel
- location of operation
- payment modus in the product-service system and
- mode of operation (for a single or for multiple customers) [19].

The results of the analysis of the expert interviews can be found in the following section of this paper.

4 RESULTS FROM EXPERT INTERVIEWS

4.1 Product adaptations in the capital goods industry

The experts interviewed were asked whether products are being specifically adapted to the services offered within the scope of industrial product-service systems. Seven experts gave affirmative answers to this question: two without specific sectoral affiliation, two chemical leasing experts, two experts from mechanical and plant engineering and one expert from the field of compressed air. The statements of the experts resemble the findings of [4]. In their survey, Sturm and Bading found out that in 16 % of surveyed manufacturers new products had their origin in new service offers.

While expert N points out that product adaptations always depend on the conditions of the new business model, expert F emphasizes that there has to be a given critical mass of products in industrial product service systems before design modifications should be being considered. Expert E puts it like this: "Every innovation creates a new market in which old products no longer fit and have to be adapted". Expert B argues in more general terms that it is the perspective from which products are regarded which changes. Products no longer represent a value in themselves; in these business models they have become "value carriers".

Five of those questioned stated, in contrast, that products are not currently being adapted to services; two of these experts were from mechanical and plant engineering and one each from compressed air and chemical leasing. One expert, who shared the view that products are not being adapted so far, was without sectoral affiliation. Three of those questioned who gave a negative answer to the question added, however, that this kind of adaptation would be desirable in the future. For instance, expert A argues that adapting products to the leasing models being offered "would be fundamental for successful implementation [...], because the old business model's target system is very different to that of the leasing model". Expert O has a similar approach to this question: "There are not that many new business [models], which is why products have not yet been adapted. But if you want to be successful with these business [models], you have to adapt the products to the service". Expert P links product adaptations with reducing costs: "Currently, companies do not adapt their products to [product-service systems] and doing so is not always necessary. Sometimes, it is necessary to lower costs, but the companies rarely do it". These statements are as well in line with the quantitative findings of [4]. In this study, 90 % of the surveyed company representatives stated that they consider an integrated development of products and services as important or very important, although only a small share of the companies participating in this study actually do develop products and services in an integrated way.

Finally, expert C points out that product adaptations are actually the wrong approach because the customer should be the starting point and "a bundle of services [should be] created which includes the product and the service and simultaneously solves the customer's problem".

The experts also talked about obstacles to implementing product adaptations. For instance, expert P said that it is not possible for companies which also provide services for machines and plants, i.e. utilities, which they have not produced themselves, to carry out modifications on these. They could only adapt self-produced equipment to the demands resulting from service-based business models. The expert also points out that it may be small and medium-sized enterprises which are unable to access the necessary technologies.

Expert O pointed out that data security issues may be a barrier to applying the relevant technologies to products when implementing systems for the remote monitoring of machines and plants: "the customer does not want this because he is worried about the confidentiality of his data and is therefore reluctant to expose them to third parties".

Tight restrictions on the part of the customer with regard to designing products to be used in service-based business models were mentioned by several experts as a barrier to implementing the necessary adaptations. Expert O argued in favor of business models with guaranteed life cycle costs: "Because via the concept of life cycle costs, the provider is responsible and he would utilize those components he feels make sense and not those the customer wants".

4.2 Types of product adaptations

In the following section of the paper, design principles mentioned by the experts interviewed, are clustered.

Condition monitoring and remote control systems

Implementing condition monitoring and remote control systems was mentioned by three experts. Interviewees can imagine these being used in different business models; different motives were listed by the experts for integrating these condition monitoring systems. Expert A offered the following thoughts on leasing models:

"Because [...] the system is returned once the contract period is completed, it is in the producer's own interests to monitor its condition during the contract period. This is why it is essential to develop or adapt the existing construction structure to process monitoring technologies".

Expert P also considered condition monitoring systems and remote control units suitable for being able to guarantee the availability of capital goods. Expert O saw the chance to compensate for information asymmetries as a motive for integrating condition monitoring systems in the machines and systems being used when offering guaranteed life cycle costs:

"The core problem is the fact that it is difficult for the provider to determine the life cycle costs of their plant since he does not know how carefully the user will treat the plant. [...] Manufacturers are turning to condition monitoring systems to check how their

Sector	Research	Industry Associations	Industry	Total
Manufacturing sector (general)	3 (A, B, E)	2 (C, D)		**5**
Chemical industry	2 (F, H)	1 (G)		**3**
Compressed Air		1 (K)	2 (I,J)	**3**
Machine tool industry	2 (L, P)	2 (M, N)		**4**
Assembly			1 (O)	**1**
Total	**7**	**6**	**3**	**16**

Table 1: Overview about the interviewees (expert identifiers in parentheses).

machines are being used and to equip machines and systems with the sensors necessary in order to see what the customer is doing to them [...] if the manufacturer is responsible for the system's condition within the scope of guaranteed life cycle costs, he also has to be allowed to see how the customer is treating it".

According to expert L, another point in favor of implementing these monitoring systems from the perspective of the service-based business model provider is that he has the chance to obtain information about the operation phase of the machine he has provided with the help of these systems, on the one hand, and, on the other hand, also through on-site inspections – for example as part of maintenance and service activities. This knowledge can then be fed back into the construction of machines and systems: in availability-oriented business models, "obtaining information is the most important factor for the manufacturer to generate product improvements and innovations". He also made it clear that he considers the chance to access possibilities for exerting influence a crucial factor for success when implementing business models with guaranteed availability of machines and systems:

"If the provider is paid for the level of availability, the client must give him the chance to keep this level as high as possible. Alongside technical equipment, this also includes the possibility to measure the machine's performance during operation and regular checks".

Standardization

Standardization of components and modules was cited by three experts as a product feature for products used in service-based business models. Expert A referred here to leasing models in which the provider has the responsibility for maintaining and servicing the leased systems. As a reason for the increased used of standardized components, he compared the target systems in traditional transaction-oriented business models with leasing models: "Whereas the old model focuses on maximizing profits, the leasing model targets a maximization of the residual value which can be realized by a higher level of standardization." He also highlighted that generating added value with this business model depends on the corresponding technical design: "In order to generate the added value, however, the technology provider has to invest greater efforts in the standardization of his components". Expert O linked the increased standardization of modules with lowering the risks associated with service-based business models: "The greater the degree of standardization, the more modules are completely designed and the lower the risks". The standardization expert D makes similar arguments, especially for companies actively involved in standardization activities. Using standardized components accords investment security: "standardization activities tend to move in the same direction as the firm's own research; this accords security since too large risks are not taken". He also added that participating in the development of standards as well as the use of standardized components results in "lowering the transaction costs, e.g. by making purchasing procedures easier" – not only in service-based business models - rationalization, client confidence and quality assurance.

Flexibility and adaptability

The greater flexibility or adaptability of machines and systems was mentioned by two experts. In this context, expert A referred to the products used in leasing models and emphasized that the standardization of components has to be balanced here with the flexible design of the used products in order to be able to generate added value in the business model; expert N mentioned the adaptability of material assets to various application locations. When leasing machines and systems it is important that these are able to be easily adapted to the respective production facilities.

Mutability

The mutability of machines and systems is also cited by one expert as an important feature for products used in service-based business models. Flexibility means that a production system can be adapted quickly and at low cost to changed framework conditions but that this has to be managed within certain limits. Mutability, on the other hand, means that a production system is able to overcome these limits depending on demand with the relevant technical and organizational measures [20]. The technical principles of universality, mobility, scalability, modularity and compatibility represent the so-called enablers of mutability.

Modularity

Modularity was the only one of these enablers of mutability specifically mentioned by two experts as a particularly relevant product feature in connection with implementing service-based business models. Expert A linked the feasibility of a service-based business model with a modular structure of the capital goods used: "Companies offering the leasing model rely on modularly designed systems; otherwise the business model could not be realized".

Maintainability and serviceability

The maintenance-friendly design of machines and systems was mentioned by one expert as one way in which products are adapted to service-based business models which can already be observed. Expert L referred to TCO guarantees in this context, which focus on lowering the total cost of ownership by "increasing the technical availability via reducing downtimes" among other things. In concrete terms, he meant that these business concepts "[promote] the development of maintenance-friendly and longer-lived products through the changed incentive structure compared to the traditional business model".

Longevity

Two experts considered the use of longer-lived products to be meaningful. Both referred to the changed incentive structure of the service-based business models which is accompanied by a new remuneration mode for manufacturers. Thus expert O voiced the following opinion: "If the seller were the lessor of these systems, then he would build them differently. More long-lived components [...] would be utilized, as each exchange means a loss for the provider, and not a profit, as in earlier times". In the experts' opinion the use of longer-lasting products contributes to lowering the downtimes of the goods employed, thus reducing the loss and replacement costs incurred by the capital goods manufacturers.

High-quality materials

One expert mentioned, as a relevant product feature in business models based on guaranteed life cycle costs, the use of high quality materials and named as motivation for using them not only extending the lifetime of the capital goods employed, but also removing the pressure to realize as low a selling price as possible for the capital good and reducing the additionally assumed risks: "Abandoning the purchasing price as THE criterion for buying favors the utilization of higher-quality materials, in addition to the priority of this factor in the context of risk-taking" (expert L).

Ease of assembly and disassembly

One expert cited new requirements, i.e. the ability to easily assemble and disassemble machines and systems as a result of their use in service-based business models. Expert A argues that a "revision of the assembly and disassembly concept will take place as a result of the new demands of the leasing concept." Expert O was of the opinion that it should be made easier to exchange wearing parts.

Industrial product-service systems	Recommended product adaptations
Leasing/renting models with full service contracts	Implementing condition monitoring systems Standardizing components Flexible system design Modular system design Use of more long-lived products Ease of assembly and disassembly
Guaranteed life cycle costs / guaranteed total cost of ownership	Implementing condition monitoring and remote control systems Maintenance-friendly Use of more durable products Use of higher quality materials
Availability guarantee	Implementing condition monitoring and remote control systems
Industrial product-service systems in general	No need for protection against incorrect operation
Hybrid value added concepts in general	Standardizing components

Table 2: Industrial product-service systems and recommended product adaptations.

Convertibility

Convertibility and easy reprogrammability of the machines and systems utilized were noted as important product characteristics with one mention each, because if the employee who was originally responsible for programming the software of a capital good leaves the manufacturer, "having another employee familiarize himself with the original programming would take twice the time" (expert O).

Protection of machinery against incorrect handling and manipulation

Protecting capital goods against incorrect operation receives one mention. One expert referred to the training of the staff involved. As one example of adapting products to new service concepts, expert O states that "for example less effort [would have] to be made to safeguard the systems components against incorrect operation [if their] construction were foolproof". The interviewee implicitly assumes that the client's personnel is less knowledgeable about the operation and/ or maintenance of machines and equipment than staff members who are entrusted by capital goods manufacturers with maintenance in use-oriented business models and with the operation of physical goods in result-oriented business models. The expert then concludes that if the goods are appropriately designed, there is less likelihood of incorrect operation. This statement could refer for example to aspects like user-friendliness or intuitive user interfaces.

4.3 Product adaptations and service-based business models

In the previous section the product adaptations mentioned by the experts were summarized according to the named principles. In Table 2, the observations of the interviewees are structured in a different manner. The product adaptations cited by the experts are listed in the following table in the context of the business models associated with the constructive changes.

It is striking that the experts have given statements concerning hybrid value added concepts and service-based business models in general and have commented on various user-oriented business models. None of the experts however quoted a result-oriented business model, e. g. a pay on production model, as an exemplary concept.

5 SUMMARY AND NEED FOR FURTHER RESEARCH

Based on sixteen interviews with experts from several industries, it could be shown that the diffusion of product adaptation to service needs of industrial product-service systems is not very wide spread in the capital goods industry. Yet, the experts see a need for these adaptation processes and name a multitude of design principles which should be applied in an integrated design of product and service components of these business models. The most frequently mentioned principles are the implementation of condition monitoring systems, the use of standardized components and modules and flexibility and adaptability of machinery to changing requirements.

In practice, as also shown by [4], the need for considering service requirements in product development via integrated innovation processes has been realized. However, there is a huge lack in its actual implementation.

In the literature, some attempts have been made to show how product and service components of industrial product-service systems can be designed in an integrated way. In this paper, this question was tackled by using a qualitative approach. This kind of research is aimed at exploring situations that are not completely understood and does not provide results which are generalizable. This limitation of the study presented above implicates a need for further research on this issue. To identify the actual diffusion of product adaptation to industrial product-service systems, quantitative data is needed. A broad data set can contribute to finding generalizable design guidelines for companies from the capital goods sectors which intent to turn their business into a service business.

6 ACKNOWLEDGMENTS

The research results presented in this paper come from the project "Hywert – New hybrid value added concepts as opportunities for sustainable development" funded by the German Federal Ministry of Education and Research within its program "Innovationspolitische Handlungsfelder für die nachhaltige Entwicklung" (Innovation policy actions for sustainable development). The funding source was not involved in study design, the collection, analysis and interpretation of data, the writing of the report, or in the decision to submit the paper for publication.

7 REFERENCES

[1] Müller, P., Blessing, L. (2007): Development of product-service systems - Comparison of product and service devel-

opment process models, Proceedings of the International Conference on Engineering Design, ICED'07, 28-31 August 2007, Cité des Sciences et de L'Industrie, pp. 1-12.

[2] Welp, E. G., Sadek, T., Müller, P., Blessing, L. (2008): Integrated modelling of products and services – The conceptual design phase in an integrated IPS^2 development process, Design Synthesis, 2008, Lab. Design, Production and Manufacturing, University Twente.

[3] Doultsinou, A., Roy, R., Baxter, D., Gao, J., Mann, A. (2009): Developing a service knowledge reuse framework for engineering design, in: Journal of Engineering Design, Vol. 20, No. 4, pp. 389-411.

[4] Sturm,F., Bading, A. (2008): Investitionsgüterhersteller als Anbieter industrieller Lösungen - Bestandsaufnahme des Wandels anhand einer Umfrage, in: Wirtschaftsinformatik, Vol. 50, No. 3, pp. 174-186.

[5] Weissenberger-Eibl, M.A., Biege, S. (2010): Design for Industrial Product-Service Combinations - A Literature Review, in: Journal of Applied Management and Entrepreneurship, Vol. 15, No. 3, pp. 34-49.

[6] Sundin, E., Lindahl, M., Ijomah, W. (2009): Product design for product/service systems: Design experiences from Swedish industry, in: Journal of Manufacturing Technology Management, Vol. 20, No. 5, pp. 723-753.

[7] Sundin, E., Lindahl, M. (2008): Rethinking product design for remanufacturing to facilitate integrated product service offerings, in: IEEE International Symposium on Electronics and the Environment, San Francisco.

[8] Spath, D., Demuß, L. (2006): Entwicklung hybrider Produkte - Gestaltung materieller und immaterieller Leistungsbündel, in: Bullinger, in: H.-J., Scheer, A.-W.: Service Engineering. Entwicklung und Gestaltung innovativer Dienstleistungen, 2. edition, Springer, Berlin, pp. 463-502.

[9] Brad, S. (2009): Perspectives on high-tech product design for better supporting product-service systems, in: Rossetti, M. D., Hill, R. R., Johansson, B., Dunkin, A., Ingalls, R. G.: Proceedings of the 2009 Winter Simulation Conference, pp. 3036-3046.

[10] Kim, S.K., Ishii, K., Beiter, K. A., Uchihira, N., Kyoya, Y. (2010): Design for service innovation: A methodology for designing service as a business for manufacturing companies, in: International Journal of Services Technology and Management, Vol. 13, No. 1/2, pp. 40-62.

[11] Morelli, N. (2003): Product-service systems, a perspective shift for designers: A case study: The design of a telecentre, in: Design Studies, Vol. 24, No. 1, pp. 73-99.

[12] Kindström, D., Kowalkowski, C. (2009): Development of industrial service offerings: a process framework, in: Journal of Service Management, Vol. 20, No. 2, pp. 156-172.

[13] Martin, C.R.Jr., Horne, D. A. (1992): Restructuring towards a service orientation: The strategic challenges, in: International Journal of Service Industry Management, Vol. 3, No. 1, pp 25-38.

[14] Goffin, K. (2000): Design for supportability: Essential component of new product development, in: Research Technology Management, Vol. 43, No. 2, pp. 40-47.

[15] Gebauer,H., Krempl, R., Fleisch, E. Friedli, T. (2008): Innovation of product-related services, in: Managing Service Quality, Vol. 18, No. 4, pp. 387-404.

[16] Berkovich, M., Leimeister, J. M, Krcmar, H. (2009): An empirical exploration of requirements engineering of hybrid products, in: 17th European Conference on Information Systems (ECIS), Verona, Italy.

[17] Lamnek, S. (2005): Qualitative Sozialforschung, 4. edition, Weinheim: Beltz Verlag.

[18] Mayring, P. (2003): Qualitative Inhaltsanalyse. Grundlagen und Techniken, 8. edition, Weinheim: UTB.

[19] Lay,G., Schröter, M. and Biege, S. (2009): Service-based business concepts: A typology for business-to-business markets, in: European Management Journal, Vol. 27, No. 6, pp. 442-455.

[20] Nyhuis, P., Kolakowski, M., Heinen, T. (2007): Adequate and economic factory transformability - Results of a benchmarking, in: 2nd International Conference on Changeable, Agile, Reconfigurable and Virtual Production (CARV 2007). Toronto, Canada, 22.07. - 24.07.2007.

Mechanisms to conduct Life Cycles of Extended Products

Marcus Seifert, Klaus-Dieter Thoben, Jens Eschenbächer
Bremer Institut für Produktion und Logistik GmbH – BIBA, Bremen, Germany

Abstract
Product life cycle (LC) models are mainly based on the traditional product understanding where the phases from ideation to recycling are sequential steps [1]. Today, products have to be understood as a combination of physical products and related services, so called "extended products" (EP) [2]. A famous example of an extended product is Apples iPhone, a cellular phone as physical core extended by iTunes as access point to software offerings to customize the core with services [3]. LC models addressing the characteristics of the EP concept are rarely discussed in literature [4]. Within this paper, the relationship between product and service LC is discussed to get a better understanding of the extended product LC. It will be shown that there can be both the product as well as the service the incubator for an EP in any phase of their life cycle.

Keywords:
Extended Product; Life Cycle; Product Service System

1 INTRODUCTION

Value creation on emerging markets is strongly connected to customer satisfaction. The customer is only willing to pay for an offer if it is able to serve his specific needs. Industrial companies learned that customer satisfaction is not only guaranteed by providing the right functionalities of a physical product but also by providing appropriate services in the different phases of the product life cycle. An example comes from the automotive industry: Today, the manufacturer does not sell cars anymore; he guarantees mobility to his customer [5]. This example shows that a market offer can be described based on its materialization degree which is the contribution of the physical product on the one hand and service on the other hand to the value proposition. Figure 1 illustrates the materialization degree and three different typical product concepts.

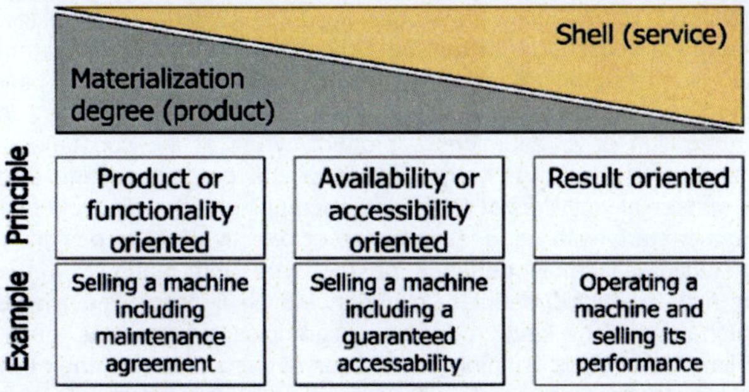

Figure 1: Materialization degree and value proposition of products [6].

Actual studies in the manufacturing industry state that in average the contribution of services to customer satisfaction became higher during the last years while the relevance of the physical product decreased [5]. In consequence, the ability to enrich the original offer by providing adequate product service combinations to the market becomes an essential asset for companies. Any combination of a physical product with a service can be understood as an Extended Product [7].

1.1 Problem statement

Actual studies prognosticate that the turnover share in industry generated by services will grow up to 40% of the total turnover while the contribution generated by selling a physical product will decrease. In the same time, 90%-95% of the engineering efforts are linked to the physical product which means that tomorrows' customers are not willing to pay at least for the costs of a physical product any more [8]. This means that the role of services to generate the needed margins in any phase of the product life cycle will become an essential aspect in upcoming future industries.

As value proposition takes place along the entire life cycle of a market offer, it is essential to describe the life cycle of a specific market offer. For traditional physical products, life cycle models from different scientific disciplines such as business management and engineering are available in literature. The same can be stated for the discipline of service engineering which proposes several service life cycle models.

Actually, there have only a few works been done which are related to model the life cycle of an extended product. The understanding of the extended product life cycle would be a prerequisite to identify and structure value creation opportunities.

1.2 Purpose of this paper and approach

The objective of this paper is to approach the understanding of the generic life cycle of an extended product. For this purpose, it is analyzed how the physical product life cycle can be linked with the service life cycle to derive an integrated concept supporting the extended product idea. The discussion will be based on examples from emerging markets such as the mobile phone market. In such branches, innovative ideas to integrate service proposition with product commercialization have revolutionized the traditional business models. Such case studies could be an incubator to transfer these approaches towards traditional industries.

Chapter 2 of this paper introduces the concept of the extended product as any combination of products and services to generate value for the customer. Based on this definition, the state of the art on life cycle models for both products and services is introduced.

J. Hesselbach and C. Herrmann (eds.), *Functional Thinking for Value Creation: Proceedings of the 3rd CIRP International Conference on Industrial Product Service Systems, Technische Universität Braunschweig, Braunschweig, Germany, May 5th - 6th, 2011*,
DOI 10.1007/978-3-642-19689-8_9,

Chapter 3 discusses potential incubators for the generation of an extended product which determine the starting phase of the extended product life cycle. These incubators can be product or service based and they can be located in any phase of their life cycles. This discussion will be based on industrial examples. Based on the discussion in chapter 3, chapter 4 introduces a generic extended product life cycle model and discusses possible instantiations of the model.

2 STATE OF THE ART ON LIFE CYCLE MODELS FOR PRODUCTS AND SERVICES

The first section introduces the concept of the extended product. In the following, life cycle models for both physical products and services are introduced. Literature distinguishes between three basic life cycle types which are the technological life cycle, the business management life cycle and the intrinsic life cycle. The intrinsic life cycle describes the different phases, a product or a service runs from ideation to end of life. Within this chapter, the intrinsic life cycles for physical products and for services will be used.

2.1 Concept of the Extended Product: Kernel and Shell

The literature introduces the extended product as any combination of products and services to generate value to the customer [9]. Some authors call the combination of products and services "hybrid products" [10] or "product service systems [11], [12] and the services "accompanying services" [13]. Figure 2 introduces the anatomy of the extended product which consists of the product kernel, in many cases a physical product, and the product shell, in many cases the related services.

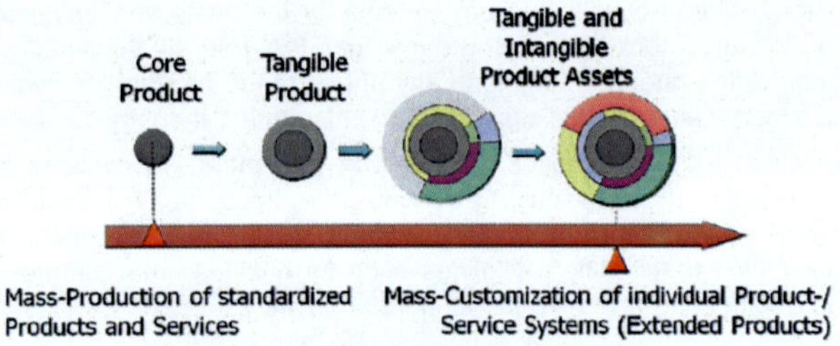

Figure 2: The Extended Product concept [4].

This concept shows that the original understanding of the extended product is to extend an existing product kernel, the physical product, with value adding services. In consequence, the prerequisite for the design of an extended product is the availability of the kernel which is complemented with a shell. This shell can be designed and developed together with the kernel in an integrated process or separated in any other time.

A formal criterion to describe an extended product is the materialization degree which reflects the contribution of both physical product and service to the extended product. This contribution can be understood and measured in different ways: On the one hand, their contribution to the value creation in terms of the realize profit can be an indicator for the materialization degree. On the other hand, the related cost share can be an indicator.

In the following, life cycle models from literature to describe both physical products and services are introduced which is the later basis to identify and to describe potential opportunities to generate value.

2.2 Intrinsic Life Cycle Models for products and services

Literature differentiates between three different types of product life cycle models. Pahl et al introduce business management oriented models, technological models and intrinsic life cycle models [14]. While business management oriented models distinguish phases in relation to the generated turnover during the on-market phase (Five phases models), technological models analyze the maturity level of different product generations over time (S-curves). From an engineering perspective, intrinsic life cycle models are a typical approach to analyze the life cycle of a physical product.

The intrinsic life cycle describes all phases a product runs from ideation to end of life. These phases are normally understood as sequential steps. There is no standardized model for generic products which means that the phases differ from reference to reference. Figure 3 shows the Product life cycle after [15]. It has to be defined if the life cycle model should describe the life cycle of a specific entity (e.g. a concrete, specific car) or the life cycle of a product type (e.g. a general line of a car).

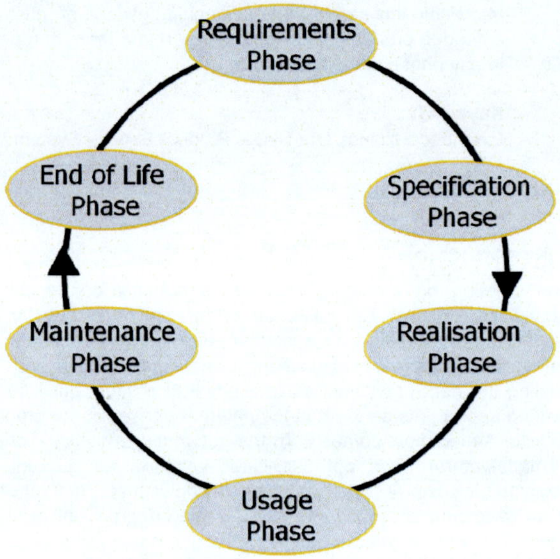

Figure 3: Product life cycle model [2].

Independent of a concrete model, the three phases begin of life, mid of life and end of life can be distinguished. While the "begin of life" summarizes all phases from the product idea until the consumer starts to use the product, the "mid of life" is an aggregation of all situations a product cycles at the consumer's. The end of life starts with the end of the products usage and comprises all activities of recycling or scrapping. If the life cycle of a product starts with an ideation phase or directly with the design or production/assembly depends on the type and maturity of the product: Products where less experience on its life cycle is available (such as one of a kind products or customized designs), the "begin of life phase" plays a major role. In case of mass products, the early product life cycle phases are fulfilled based on the experiences the producer/designer made with former product generations.

For services, the main difference on life cycle models is that provision and consumption of the service are made in the same time. Services can't be stored and they need an interaction between provider and consumer [16]. In some cases, this interaction is substituted by a so called "external factor" [17]. An example for an external factor is the maintenance service for a machine. Service provider and customer do not have to interact directly; carrier of the interaction is the machine where the service is linked to. Traditional services such as a customer helpdesk require a direct contact between service provider and consumer to fulfill the service. For extended products, both kinds of services are possible.

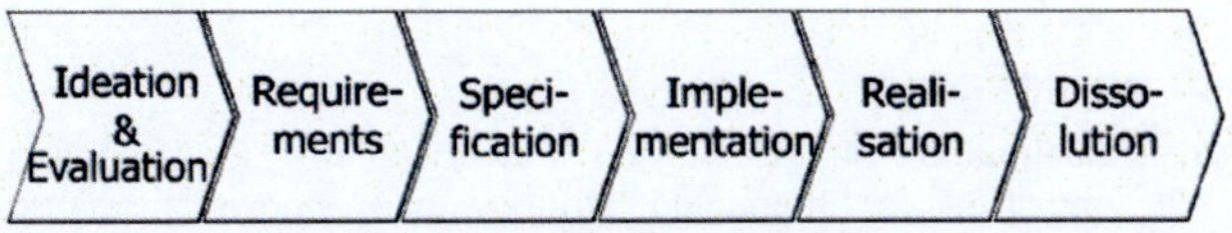

Figure 4: Service life cycle model [17].

Figure 4 introduces a service life cycle model after DIN 1998 [18]. An overview on available service life cycle models can be found in [19]. Similar to the product life cycle, the early phases comprise activities such as ideation and specification. The implementation phase ensures the availability of the service for potential customers, the realization the consumption of the service by the customer. The end of life can be defined as the dissolution of the service.

An approach to describe the life cycle of product service systems or extended products as an extension of the business management model has been introduced by Blinn et al. [20]. While the traditional five phases model focuses on the commercialization phase of a product, Blinn also considers the product planning and the after sales phase. Within this context, he analyses additional value creation opportunities during the after sales phases based on service provision.

3 INCUBATORS TO INITIATE EXTENDED PRODUCT LIFE CYCLES

Within this chapter, principal approaches to initiate a life cycle of an extended product are introduced. Due to the fact that an extended product bases on its kernel (product) and the shell (service), the initiation of the life cycle is analyzed from these two perspectives. Section 3.1 discusses examples there the incubator for the extended product life cycle is the kernel while section 3.2 discusses examples where the incubator is the shell. Chapter 4 is a proposal derived from this discussion to integrate these different life cycle types into a generic life cycle model for extended products.

3.1 Kernel driven Extended Products

Within this paper, kernel driven extended products are understood as offers where the starting point of the extended product design has been the physical product (the kernel). Services are built up around this pre-existing kernel in a way that they are accompanying certain phases of the kernel. This is the traditional understanding of "accompanying services" as part of an extended product. The objective is to enrich the value of a physical product for the customer by providing useful services in different phases of the physical product life cycle. Two main streams of extended products incubated by the kernel can be identified. They are highlighted in Figure 5.

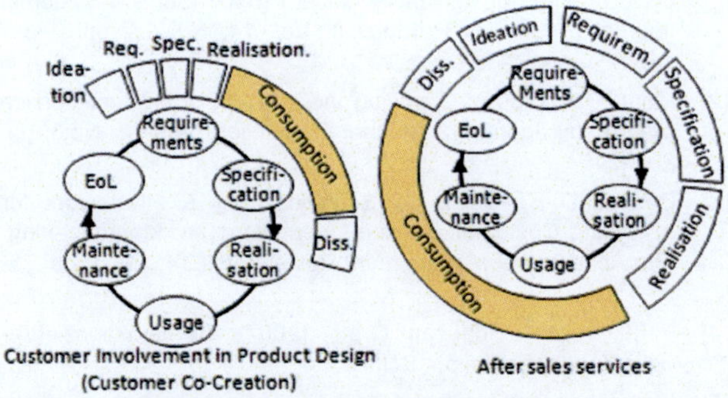

Figure 5: Kernel based Extended Products.

The inner cycle represents the product life cycle of the kernel, the outer one the service life cycle as introduced in chapter 2.

One opportunity to derive services based on a pre-existing kernel is shown on the left hand side: The service is provided and consumed during the early life cycle phases of the physical product. The objective could be a customer involvement in the product design to realize customer specific products. An area where this principle can be found is the mass customization: Nike e.g. provides a service allowing the customer to configure his individual sports shoe via the internet which is realized in short term by the OEM. The approach to design the extended product is "inner-out" (from the kernel to the shell). Starting from the core competence to realize sportswear, the OEM extends his offer by providing services allowing the customer to participate the product design (Customer Co-Creation).

The other option to come up with an extended product based on the kernel is illustrated on the right hand side of Figure 5: The service is provided and consumed during the late life cycle phases of the physical product. The objective is to support the user of a product during the usage phase by applicable services. These services might just provide special maintenance or repair conditions. Other approaches are to guarantee the accessibility or availability of a product. Examples are mobility guarantees in the automotive industry. The OEM not only sells a car- he understands himself as a mobility provider which is a car plus the guarantee of a reliable availability of its functionality. The approach is similar to the first example: The extended product is designed "inner out". Starting from an existing physical product, this offer is extended to provide convenience or reliability to the customer.

3.2 Shell driven Extended Products

Shell driven extended products are understood as market offers where the service is the starting point for the product design. The physical product is designed according to the needs of the service and might be replaced by other substitutes. This approach is mainly recognized in areas where the materialization degree of the extended product is low. Figure 6 illustrates how the product and the service life cycle can be linked together.

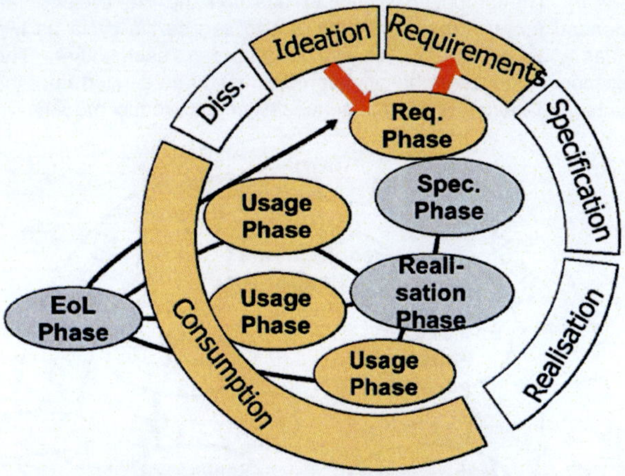

Figure 6: Shell based Extended Products.

This kind of extended product bases on a pre-existing shell and designs the market offer in an "outer-in" approach (from the shell to the kernel). The provision and consumption of the service is related to the usage of the accompanying product. A typical example for this type of extended product comes from the mobile phone industry: Pre-existing infrastructure and services for mobile phones (SMS and MMS services, mobile internet, localization services, iTunes as access to many software services etc.) require a certain technological specification of the hardware (kernel). Especially in this branch, the life cycle of the kernel is extremely short with the consequence that several generations of kernels need to support the shell. For the user, the kernel is not the competitive advantage anymore- it is just the enabler to access an attractive service.

Figure 6 links the ideation/requirements phase of the service life cycle with the requirements phase of the product life cycle with arrows. This option is not obligate- it just considers examples where a new hardware enables a first time access to a new service (e.g. the first mobile phones with GPS navigation). In this case, the requirements have to be identified and collected in an integrated approach.

4 CONCLUSION: LIFE CYCLE MODEL FOR EXTENDED PRODUCTS

Chapter 3 introduced two different approaches and related examples to design extended products: "inner-out" (from the kernel to the shell) and "outer-in" (from the shell to the kernel). It has been shown that a life cycle of an extended product requires the integration of the product with the service life cycle. It has been shown that there are several options how these life cycles interact with each other. In consequence, there is not one single extended product life cycle model- the instantiation of the model depends on the evolution of the extended product. For this reason, it is important to identify the incubator as trigger for the extended product design (kernel or shell and which life cycle phase). Current industrial examples from emerging markets such as Apples iPhone demonstrate that engineering approaches become more and more complex. In consequence, the traditional understanding of a life cycle is not applicable anymore to describe the underlying engineering principles. Kernel and shell do not have to be one-to-one relationships- iTunes as a shell can be accessed via multiple kernels (iPod, iPad, iPhone). This example shows the diversity of actual extended products: On the one hand, customers are willing to pay for the ownership of an attractive physical product, on the other hand they want access to an attractive App-store (consume a service). The proposed model to describe the life cycle of an extended product considers product and service life cycle as two cycles with a flexible relationship between each other. The dependency between the single phases has to be derived from the incubator. Figure 7 shows the generic extended product model.

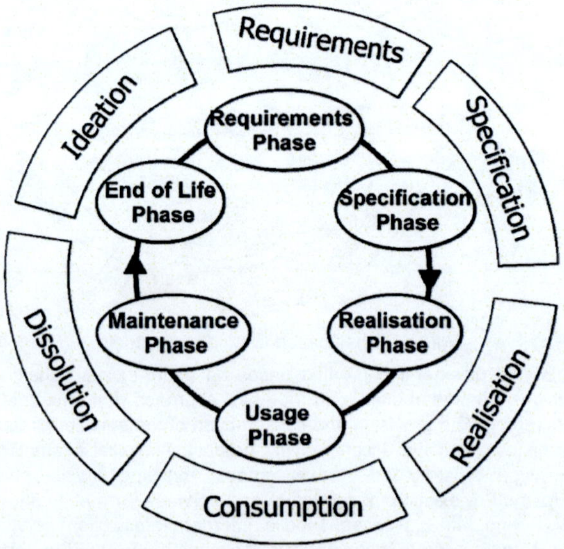

Figure 7: Generic Extended Product Life Cycle Model.

5 REFERENCES

[1] Cusumano, M. A: Staying Power (2010): Six enduring principles for managing strategy and innovation in an uncertain world. Oxford University Press, New York.

[2] Thoben, K.-D., Jagdev, H., Eschenbaecher, J. (2002): Emerging concepts in E-business and Extended Products. in: Gasos, J., Thoben, K.-D. (Eds.): E-Business Applications - Technologies for Tomorrow's Solutions; Advanced Information Processing Series, Springer, New York.

[3] Voida, A.; Grinter, R. E. ; Ducheneaut, N. ; Edwards, W. K. ; Newman, M. Listening (2005): Practices surrounding iTunes music sharing. Proceedings of ACM Conference on Human Factors in Computing Systems (CHI 2005); 2005 April 2-7; Portland; OR; USA. NY: ACM; pp.191-200.

[4] Thoben, K.-D., Eschenbächer, J., Jagdev, H.: Extended Products (2001): Evolving Traditional Product Concepts; Proceedings of the 7th. International Conference on Concurrent Enterprising (ICE 2001), Bremen, Germany, pp. 429-439.

[5] Johnston, R., Clark, G. (2008) Service operations management: improving service delivery. Financial Times Prentice Hall, Essex.

[6] Burianek, F., Ihl, C., Bonnemeier, S. und Reichwald, R. (2007): Typologisierung hybrider Produkte: Ein Ansatz basierend auf der Komplexität der Leistungserbringung. Arbeitsbericht Nr. 01 / 2007 des Lehrstuhls für Betriebswirtschaftslehre – Information, Organisation und Management der Technischen Universität München.

[7] Thoben, K.-D., Eschenbaecher, J. (2003) Die Erweiterung des Produktbegriffs: Konzept und Praxisbeispiele. In: Industrie Management Vol. 19, No. 4, pp 48-51.

[8] Sihn, W., Proksch, R., Lehmann, F. (2000) Produktbegleitende Dienstleistungen unter der Lupe – Was Kunden wirklich wollen. In: Service Today 14 (2000), No.6 pp. 38-40.

[9] Pine II, J. (2000): Mass Customization: The New Frontier in Business Competition. Harvard Business School, Boston, Mass.

[10] Chao, P. (1993): Partitioning Country of Origin Effects: Consumer Evaluations of a Hybrid Product. in: Journal of International Business Studies, Vol. 24, No. 2, pp. 291-306.

[11] Aurich, J.C., Fuchsa, C. and Wagenknecht, C. (2006): Life cycle oriented design of technical Product-Service Systems. Journal of Cleaner Production, Vol. 14, No. 17, pp. 1480-1494.

[12] Mont, O.K. (2002): Clarifying the concept of product–service system. in: Journal of Cleaner Production, Vol. 10, No.3, pp. 237–245.

[13] Bowen, D,-E., Siehl, C., Schneider, B.: A Framework for Analyzing Customer Service Orientations, in: Manufacturing. The Academy of Management Review, Vol. 14, No. 1, pp. 75-95.

[14] Pahl, Beitz, Feldhusen Grote (2007): Konstruktionslehre-Grundlagen, Berlin, pp. 97ff.

[15] Eschenbaecher, J., Hirsch, B.-E., Thoben, K.-D., Schumacher, J., Jansson, K., Ollus, M. Karvonen, I. (2002): Extended Products: Results of the first roadmapping study. In: Brian Standford-Smth, Enrica Chiopzza and Mireille Edin (Ed.): Challenges and Achievements in E-business and E-Work (Venice), pp. 677-684.

[16] Fitzsimmons J A, Fitzsimmons M. J. (2004): Service Management: Operations, Strategy, and Information Technology, McGraw Hill, Houston, TX, U.S.A.

[17] Bitner, M-J. (1990): Evaluating Service Encounters: The Effects of Physical Surroundings and Employee Responses. in: Journal of Marketing, Vol. 54, No. 2, pp. 69-82.

[18] DIN Deutsches Institut für Normung e. V. (1998.): DIN-Fachbericht 75, Entwicklungsbegleitende Normung (EBN) für Dienstleistungen, Berlin et al.

[19] Klostermann, T (2007): Optimierung kooperativer Dienstleistungen im Technischen Kundendienst des Maschinenbaus, Gabler, Stuttgart, pp. 35.

[20] Blinn, N et al. (2010): Lebenszyklusmodelle hybrider Wertschöpfung: Modellimplikationen und Fallstudie an einem Beispiel des Maschinen- und Anlagenbaus, in: Hybride Wertschöpfung, Springer Berlin Heidelberg 2010, pp.130-143.

Towards a Framework for developing Product/Service Systems

Ola Isaksson[1, 2], Tobias C. Larsson[2], Pär Johansson[1, 2]

[1] Volvo Aero Corporation, 468 81 Trollhättan, Sweden

[2] Luleå University of Technology, 971 87 Luleå, Sweden

Abstract

Increasingly, manufacturers become service providers rather than product providers. In this work, a framework based on industrial studies is suggested for developing product service system. The framework is represented by six life cycle phases where three actor categories views are represented. The framework is described using examples from different industries/businesses. The intention is that the framework can be used for both B2C and B2B type of relations. One significant characteristic is that the "traditional" sales phase is replaced by a sales occasion that occur within different life cycle phases depending on which business context/model is used.

Keywords:

Product/Service Systems; Framework; Development; Life-Cycle

1 INTRODUCTION

One effect of introducing the notation of Product/Service Systems (PSS) solution is the change in responsibility and roles the various stakeholders have and there is a need to facilitate this shift in working. The aim in this paper is to present a framework that facilitates the common understanding of PSS development for both the providers and the users of these solutions.

There is a direct link between products (what is used) and services (how it is used). The increased interest to combine product and service development is driven by both manufacturers business development (increase value of proposal) and requirements (legislation enforcing life cycle responsibilities). From a consumer point of view, the focus tends to shift from ownership of products to availability to its function.

There are several aspects that make co-development of products and services difficult. One aspect is how products are developed and manufactured. The prevailing paradigm states that the product is developed based on a trade-off between function (including appearance etc.) and the cost of realization. Although life cycle cost is referred to as a mechanism to integrate products and services – user perspective is not explicitly included. A second aspect is how supply chains are integrated. A product is seen as a physical artefact where hardware, and software, is build from components and sub-systems to work together as a system. Products are seen as systems built of components, sub-systems or simply by material ordered from suppliers. This view works for hardware, but is less suitable for services due to its intangible nature. A third aspect is the relation between the manufacturer and the user and customer. Service literature emphasize that the discrete device is one of several elements in strategic relationships [1] hence putting the customer relationship before "requirements" that more relate to the product paradigm. This is because services are activities in collaboration with customers. Product development literature points out how the product is first settled and then services take form to complement that thing [2]. In engineering industry the word "aftermarket" is commonly used to describe the life of a product after it has been designed and developed, while also indicating that a service perspective comes in second place for engineering and manufacturing firms [3]. It can be argued that the value carrier for the customer is *either* the product or the service provided. Customers however tend to view their purchase from a more holistic perspective [4]. This difference in perception is one reason why in the last decade, the vision for manufacturing industry is to provide their customers with functional offers, meaning that what is sold is the function or the use of the product [5]. Manufacturers are becoming service providers, and as a first step the focus has been put on service provision solely, and not on the combination of services and products.

In this paper we accept the integral view of a Product Service System and focus on the question of *how an integral solution can be developed*. The potential if the product, and its accompanying services, can be co-developed is significant.

2 METHOD FOR DATA GENERATION

The research builds on empirical data from, primarily, aerospace manufacturing companies, and secondly from manufacturing industry. The data has been generated by studying internal documents, deliverables in projects and from experiences in industrial projects. Data has also been generated by applying a participative action research approach [6]. The empirical data set has been analysed in reference to literature studies on product/service systems, and its likes.

3 THEORETICAL FRAMEWORK

The theoretical framework builds on three parts, namely some principles of product development, service development, and product/service systems.

3.1 Product development

Product Development is typically developed in a process, finding solutions to requirements that have been defined based on understanding customers/market needs. In general, a product development process composes of a number of sequences, going from

J. Hesselbach and C. Herrmann (eds.), *Functional Thinking for Value Creation: Proceedings of the 3rd CIRP International Conference on Industrial Product Service Systems, Technische Universität Braunschweig, Braunschweig, Germany, May 5th - 6th, 2011*,
DOI 10.1007/978-3-642-19689-8_10,

early design phases to the launch of the discrete device [2]. Today, most models emphasize an integrated, parallel or concurrent approach – integrating functions within the organisation [7]..

Today, commonly, a product lifecycle perspective is important to take environmental design aspects into consideration. Thus, a circle-shaped form can symbolize, for example, that taking care of and/or reduce the waste is vital. In such models, the phases recycle, delivery of the product, its maintenance and its use are visualized as additional aspects to consider (see Figure 1).

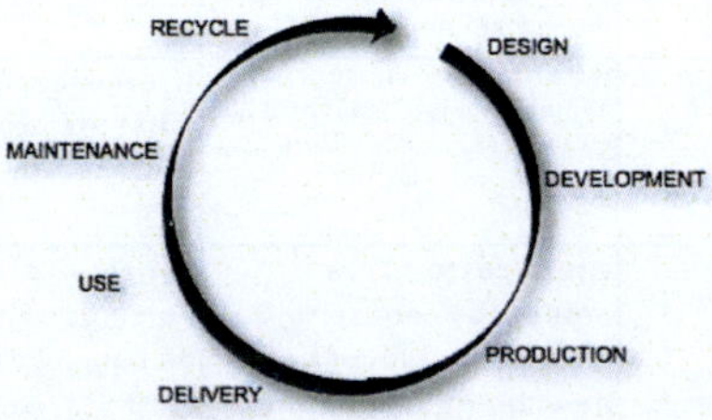

Figure 1: A product life-cycle.

One firm's capabilities and competences are not enough to take care of all lifecycle aspects of a product. For example, taking responsibilities for the products impact on environment, e.g., chose material and production methods with respect to their eco-friendliness, can be done within the manufacturing company. And, for example, another company provides recycling. So, at the same time, the picture of several companies working together cross-company is starting to evolve in the models, hence the development of "spiral models" that are increasingly inspiring "traditional" product development models [9].

3.2 Service development

For service development, visualization of process models becomes problematic since services are partly produced by the customers and regarded as activities [1]. In this sense, services are partly intangible and unfold in relationships, i.e., people interacting with each other to achieve a goal. Due to this interaction the users are co-producers in services, thus also have an effect on the outcome. Therefore, the customer's own processes are part of the service concept and affect the quality of the service. So, the goal of service development is to make up the conditions for the right customer outcome [10].

A service offer is built upon three main development components, namely, a service concept, a service process and a service system. The service concept describes the customer needs, and links these to how the service should fulfil them. Since partners and customers are co-producers in a service process, they are included to some extent in the model (see Figure 2).

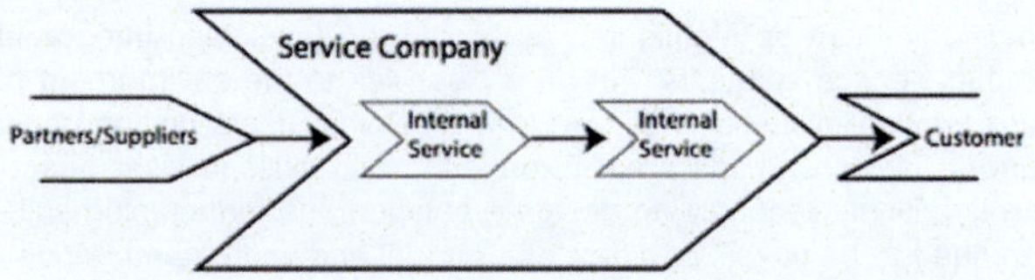

Figure 2: Service process, after Edvardsson [10].

Within the service company the internal services (middle of Figure 2) shows that a service perspective is used throughout the whole process. The internal organizational functions should also be seen as a supplier-customer relationship. A service is always seen from a customer point of view [1], meaning that suppliers should take the perspective of their customer. In service literature [10,11] the customer is a central actor in service provision.

A manufacturer, adopting the service paradigm must, first together with customer's, change their way of communicating needs from specifications to communicating effect of its use. Second, the organization has to adapt a service perspective. This highlights that, for example, having a product structure might be a barrier. A product structure can prompt people to regard the company as a provider of those specific things. Third, the culture has to change into a service culture. A simple interpretation of this is that all connections and relationships should be seen from a customer point of view, no matter if they are internal or external. A culture can be described as the result of different actions over a long period of time, thus cannot be instilled over night nor fully managed [12]. For a firm focusing on producing excellent goods and provide additional services to complement those devices, the cultural change will be a really challenging issue. This is particularly true, if the development processes are firmly focusing on the commodity as the main carrier for customer value and services are, more or less, something developed haphazard. The physical resources (middle row at right in Figure 3) are from a service perspective focusing on the equipment, premises, technical systems etc. [11]. From a service perspective this means that all actors' resources are included in the production process.

3.3 Product/Service Systems

A PSS business firm is enforced to understand services, use and performance in a more holistic way than in a classical product situation. As denoted in the PSS concept, the products and services have the companionship of 'systems'. Already in 1968 Ludwig von Bertalanffy [13] stated that the concept systems had become popular in all fields of science and made its way into popular thinking, jargon and mass media. The word system can be used to label all sorts of systems, a software application can be a system, the roads and traffic signs are systems, and the laws and regulations in a society are yet another system. These different systems cannot build up the same system, therefore the interfaces between systems, that is how they are linked and how they have an effect on each other is an important issue for understanding wholes [14]. In a system view, treating products and services jointly, we get what is commonly called a Product/Service-System offering, a special case of servitization [15]. Fransson [5] has developed a model for how the service degree increases in engineering firms, calling such offerings "functional offers". A functional offer starts from a traditional product perspective where the discrete device is supported by services (at bottom left in Figure 3). In the next degree of functional offerings, the service part is extended with services that are not usually offered, e.g., customized offerings of expertise from production processes. Then, there are two middle degrees of offerings which put forward that the shift towards a service perspective change how the customer is invoiced, i.e., the engineering firm is paid on the basis of the performance of the device including some services, and at the next degree, all services are included. These kinds of business models are suggested to incorporate an organizational structure of a virtual enterprise, i.e., where the collaboration takes place cross several company boundaries [16].

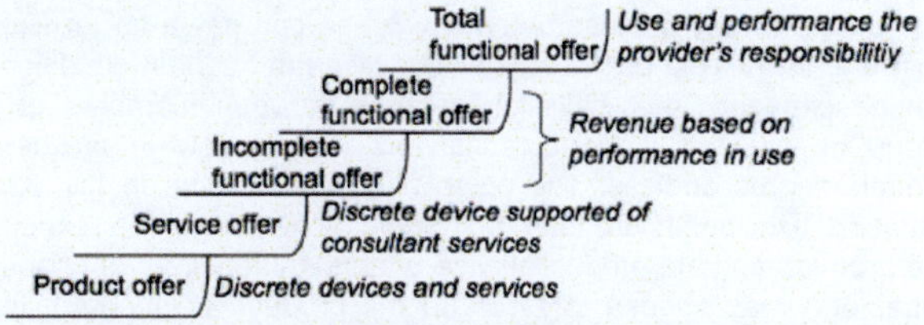

Figure 3: A progress model for functional offerings, after Fransson [5, p.128].

There are challenges to realize PSS and provide successful functional offerings, for example, to understand the voice of the customer more profoundly than merely obtaining the requirements [17] and, to implement an innovation approach there is both internal and external issues to tackle [18]. These examples illuminate that additional capabilities besides the classical engineering ones are part and parcel of developing products for a successful PSS offering. Yet, if the challenges can be handled the advantages abound. For instance, the collaboration in PSS development are expected to reduce some competition through making the business relationships more stable [19], also, PSS triggers changed use patterns that reduces the waste, i.e., provides for ecological sustainability [20,21].

Today, engineering firms state that they continuously on an everyday basis deal with innovation due to developing products; however such innovation is focusing on, e.g., new features on a known end-product, here called incremental innovation. As discussed above, classical product development seems delimited when it comes to manage and progress innovation in view of PSS. That is such innovation that ends in breakthrough products, here called radical innovation. A radical innovation situation has similarities with wicked problems [22], where several aspects are vague and not fully understood, e.g., What is going to be designed? What should it do? Who is going to use it? And, in what circumstances? [23]. To deal with these questions the design task needs to be addressed from different point of views, i.e., multiple perspectives

For various development projects, distinct process models give the team guidance for how to begin and proceed, as well as pointers to what is needed to fulfil their missions. These processes, whatever they intend to finally produce, are vital. Also, there is a link between industry and applied research, to develop better products and to become better designers; the processes must be continually improved [24]. This motivates engineering design researchers to study them. Process models are not blueprints of reality; rather they are representations of the design world to deal with the relational complexity in the processes. In view of this, the models depend on how the team's actors interpret and perceive them [25].

4 INDUSTRIAL CONTEXT

Increasingly manufacturers integrate services close to their product offer and this requires the involvement of customers and actual users of the intended service. The drivers to combine products and services are several, four of these listed in Table 1, where the consequence from a manufacturers point of view is highlighted. Another industrial context that is of importance to define as we continue is the similarity and difference between Business to Business (B2B) and Business to Consumer (B2C) markets. In the Business to Consumer situation, the role of the user (consumer) is apparent. The user pays for a service, as he/she enters a contract. A relevant example is mobile phones, where the user can choose between a large variety of service offers. He/she may choose to buy a phone, and pay per use. Alternatively, he can pay a larger fixed cost and lower cost per consumer phase. Under certain conditions he/she can get "free" calls for instance between members of a group or within the operators network. The variety in business models seems endless. The point made here is that in the B2C situation, user behaviour is evident and directly possible to relate to the product, and the product/service offer. Manufacturers of phones (enablers) may choose to focus on the product, or integrate the product with service functionality within the platform. In the B2B case there are several reasons why the relations between the user/customer and manufacturer differ from the B2C case.

Drivers towards PSS solutions	Description	Consequence for manufacturers
Regulations	Regulations that enforces produces through life/end-of-life responsibility of products.	Responsibility after delivery increases.
Competition	Services that tied to the product and based on product knowledge, can be seen as a differentiator on the marketplace.	Opportunity to add value for customers.
Life Cycle Cost	Increasingly – the cost of ownership, total cost and life cycle cost are considered by the user.	Userincreasingly value what the products do, rather that what they are
Business Cycle Management	Users/custerers as a group may choose to either maintain or buy new depending on the	Dynamics of sales volumes of products may be balanced through life cycle services.

Table 1: Drivers towards PSS.

The actual end user is not always the same as the one who pays for the classical product. An example is that the passenger (user) that buys a ticket from a bus-company or an airline. He/she pays for the service. The transportation company (airline, bus company etc.) buys the bus, or aircraft from a manufacturer. In this case the B2C relation between Passenger and User relates to the service whereas the Bus company buys the product, or the combined product-service system from a manufacturer. These have a B2B relation. Also, in B2B relations the actual person, or group of persons involved in the negotiation about requirement on the product are specialists and seldom themselves users of the resulting product.

For services – user involvement is decisive. There is most often several steps between the user and the manufacturer. If the manufacturer now increasingly offers services – who is then the user? If user behaviour is tightly coupled with value and service consumption the identification of user, and the involvement of users, need special consideration. In the B2C case the understanding and involvement of users is more evident, since the customer and the user are the same. Involving the customer is the same as involving the user. In the B2B case, involving the customer is quite relevant since the agreement of what has to be developed has to be agreed between the business partners. From a service perspective - the end user may use the service.

Based on the observations and arguments presented, we argue that *there exists an imbalance between the drivers to provide integrated product-service system solutions and the coordination of capabilities to develop such solutions.* Primarily the way services and products are developed are not as integrated as needed.

Four topics can be highlighted as critical for developing integrated product-service solutions. First – a clear link to the environment of consumption of the service, i.e. how and what users do, and how current user conditions area. Experience and skills in User operation is critical. Secondly, to design a solution, the forthcoming solution need to be possible to describe, model and understand. Modelling the service attributes is another area. Third, since products are developed in one way, and services typically in other ways, their integrated PSS development procedure is another key topic. A fourth topic is the understanding of PSS as such..

An assessment of basic capabilities, current best practices and future state of the art capabilities is summarized in Table 2.

Topic facet	Basic Capability	Current best in class capability	Future state of the art capability
Experience from Use/Operate	Feedback, Continous improvement, Lessons Learnt	Continous update of best practice /Instruction/Processes	Methods/Tools that include service restraints/experiences into design methods/tools
Modeling Service Attributes	Business Modeling,Cost modeling, ABC	Account for operation conditions for design (MTBF etc). LCC based design	Possible to "Design" Service attributes
Integrated PSS development	"Concurrent Engineering" – domain and life cycle roles involved early on	Customer/user involvement in start/spec and Evaluation of solutions – Open Innovation oriented work	Enterprise development that includes the all roles user, provider and enablers
Understanding of PSS	Sell products together with services	Design products so these can be used for services	New competence mix required.

Table 2: Basic, Best Practice, and Future capabilities.

5 A FRAMEWORK FOR DEVELOPMENT OF PSS

The main purpose of the framework is to enable a common tool for co-develop product service systems, facilitate communication and understanding of the concepts emerging.

5.1 Framework roles

First we focus three complementary roles useful when forming the framework for developing product service systems. We choose to use only three roles, the User, the Provider and the Enabler, see Figure 4.

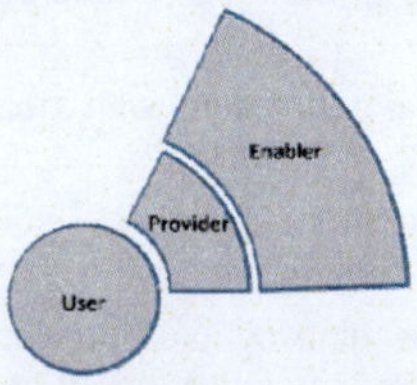

Figure 4: Framework roles.

The roles presented have all different views on the solution.

The User is a consumer of the PSS. Example: Passenger in an aircraft, user of cell phone or a professional user such as an airline (service provider).

The Provider is the organization responsible to offer the integrated PSS solution. The provider has direct relation to the user.

The Enabler is a collective term for organizations providing technologies or services necessary for the provider to provide solutions.

5.2 The PSS solution

The notation used for the relation between Need, Solution and the Solution components is seen in Figure 5.

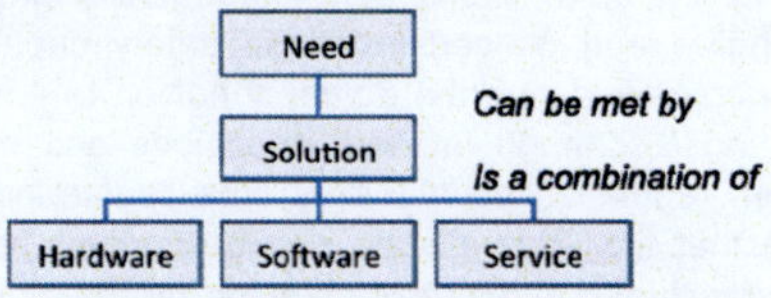

Figure 5: Relation between Need, Solution and Solution Components.

5.3 Life-cycle phase model

Secondly, we introduce a simple, generic life cycle model that can be applied to products, services and its combination.

The life cycle phases are generic in the sense that both typical product development processes and service development processes can be mapped onto the process. Each phase is described below.

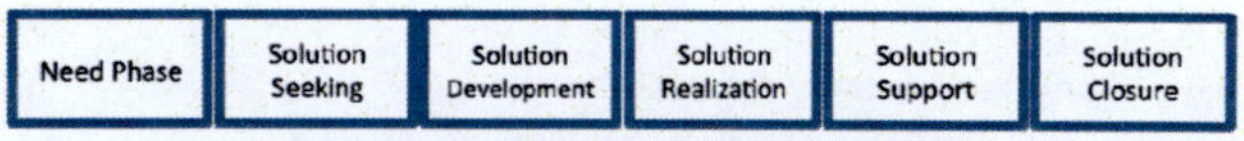

Figure 6: A six-phase life-cycle model.

Need Phase: The need phase is where there a user has a need. The phase is also applicable where there is an idea of a solution to a need that few users are aware of. In the latter situation there is a marketing work to "Create" a need.

Input to the need phase can be a decision of a strategic move for a company to offer new, or improved functionality in a market, or simply a solution that has some limitations or restraints that omit the user to perform as desired.

Output from the need phase is an identified need, opportunity or idea, that is defined to the extent that a solution can be sought after.

Solution Seeking: The solution seeking phase is characterized by search for existing solutions, alternative solutions, sub-solutions etc. that may be possible to use. This phase is equivalent to Pre-development, Conceptual development or technology development in traditional product development processes.

Input to the Solution seeking phase is the identified need, opportunity of idea, generated in the Need Phase.

Output from the Solution phase is the conceptual solution. The solution mix is identified, where a part of the solution may require physical products and software, other parts of the solution may be suitable to be met by a service.

Solution Development: The solution development phase is the coordinated development of PSS components. The solution development phases may include a "traditional" product development process for developing the physical part of the solution. Coordination with service development is needed. The important aspect is that the service is developed in parallel with the product, allowing enabling technologies/solutions necessary for the intended service can be supported by features in the product.

Input to the Solution Development phase is solutions identified in the Solution Seeking phase.

Output from the Solution Development phase are solutions that offered to customers.

Solution Realization: This is the most intriguing phase. Solution realization in traditional hardware terms is the manufacturing and delivery of the product. The realization of services is the actual use event, i.e. a repair activity, a training activity or the bus driver, driving the bus.

Input to the Solution Realization phase is the defined Product-Service offer.

Output from the Solution Realization phase is the actual provision of the product-service system. Notably, the deliverable of a physical product to be used in a product-service system is a partial delivery of the offer. The consumption of a training activity (example) is a part of the realization phase.

Solution Support: The solution support phase represent the change in state of a product-service offer. New technologies such as upgrades of existing, or offer of new, software is a part of the Solution support phase.

Input to the Solution Support phase is that there exists a provision of a product-service system.

Output from the Solution Support phase is a changed state of the product-service system.

Solution Closure: The solution closure phase covers the end of use, and possible re-cycling state. Input is theproduct and services in use. Output is the closed PSS system.

5.4 Combined framework model

Following one of the principles behind the Zachman Framework [26] for architecting information systems, the views of each role are introduced into each life cycle phase. The viewpoint of each role in relation to the evolvement of the PSS solution can be expressed.

Notably, there is no explicit «market» phase or «sales» phase. Nor is there an aftermarket phase. This is a necessity since the «sales» occasion differ depending on what business, or what solution strategy is being developed. In the model, the sales occasion is represented as a marker – that is positioned in different phases, see Figure 7.

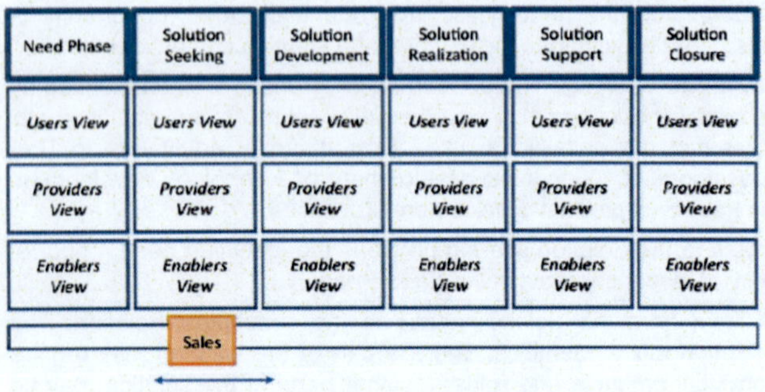

Figure 7: The sales occasion is variable.

6 DEMONSTRATION OF FRAMEWORK USAGE

In this section we introduce the proposed framework by an example (Figure 8). The example used is an illustrative situation labelled a *Product Availability optimization offer* in the aerospace business.

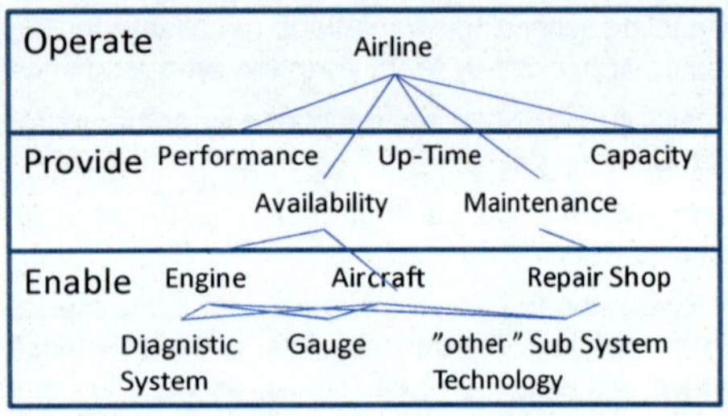

Figure 8: Framework example case.

The case is described as follows:

The professional user in this case is an airline, providing travel services to end *end-users* (passengers). *The provider* is a product-service system provider in the aeronautical sector, which is providing an availability service together with its core product as an OEM. As we shall see the PSS provider enterprise actually forms as a part of the development process.

In the case we make the assumption that the «always on time» strategy has been analysed initially by the airline which concluded, amongst others, that availability and turnaround time for clarifying the engine was one critical factor for success.

Enablers are several such as the hardware enabler (gauges) and a service enabler (repair) and a transport logistics provider.

A walk through of the development using the framework follows as;

Need Phase: Passengers value timeliness and reliability for the travel service. An airline has chosen to use «always on time» as the differentiator on the market. The work starts with framing the consequences and expressing needs. One of the identified needs is the timely availability of carriers, and a pre-defined way of dealing with unexpected incidents with inspections and unplanned maintenance.

Limiting the example to availability of ready-to-use engines, the airline, together with the engine manufacturer and engine overhaul services identifies the need for a solution that covers several areas, including condition based maintenance, training of clarification personnel and repair methods. In addition, the solution must also include solutions to compensate for potential delay situations – yet ensuring schedules for passenger to be on time. The user value for the passenger is timeliness, and for the airline availability of equipment and a service to overcome unintended problems.

Solution seeking phase

The Solution seeking phase identifies that a mix of services and product features are needed. Several actors in the solution chain need to collaborate to find a solution.

As a result from a workshop between representatives from the airline together with the aircraft and engine representatives, ideas are identified for how to provide solutions that can beet the challenge to ensure «always on time». A virtual company is seen as an option where several different actors collaborate to offer the product/service system solution.

There are several solution strategies identified, amongst them

- A travel logistics service, requiring more accurate engine condition data
- A condition monitoring tool, capable of providing data to the logistics service
- Data logging and analysis tools to feed the condition monitoring system
- Data measuring and acquisition, with more accurate data provision
- Engine control features, enabling « safe operation » modes despite some, non-flight worthiness critical, incidents
- A number of less engine coupled services ensuring timeliness for travellers requiring the involvement of ATM (Air Traffic Management) actors etc.

All together the entire solution requires the joint effort of a number of actors.

In the **Solution realization phase** an enterprise is formed including the actors who share risk and revenue of providing the service to the airline.

Development of software, service and hardware is co-ordinated within the enterprise. The engine is equipped with dedicated integrated sensors, enabling real-time diagnostics required for the high performance condition monitoring system. In turn the condition monitoring system enables the desired combination of condition-based maintenance with the logistic planning service that need to be available in real time. The engine manufacturer has knowledge about the engine and its certified flight safety capabilities. This knowledge is combined with the actual condition to – in real time - analyse the most common incident conditions and make engine control system regulation solutions as time constrained solutions. The result is that the engine state can be instantly and correctly analysed, and included in dynamic planning for the airline. Simultaneously the airline and the transport logistic services take advantage of the extended condition monitoring system that together increase availability of the engine service, despite disturbances. Also the flight logistics service provision has developed agreements with other parties to complement the more technical achievements in the solution, such as a software company who develops and supports the tool for airline personnel.

Finally, in the **solution support phase**, there is a contractually agreed solutions improvement program where adjustments and improvements to all constituent PSS components are introduced.

Out of Scope improvements are offered as new functionalities, or up-graded functionalities, whereas incremental solutions, of win-win- type are introduced continuously. The life cycle is completed with **solution closure** services to terminate and possible re-cycle the products and transfer to new PSS solutions.

7 DISCUSSION

It can be argued that this scenario is already happening in real life. All services and technologies in the story may exist, but the integrated perspective is rarely seen. Why?

At first, the manufacturer in the scenario had to use solutions mainly intended to be used for internal purposes. The algorithms needed to offer the advanced condition monitoring service to a customer, was based on internal know how. The externalization of such capabilities require another dimension – the packaging and provision of such algorithms in a commercially used end user IT system. Internally, it was used within the engineering organization and never intended for commercial use. In the case – a liaison with a professional IT vendor was a must.

Secondly, the scenario might have been the result of an "open innovation" effort, i.e. the solution was not apparent within a single organization. Despite the power of open innovation, the full penetration is not yet seen in established organizations.

Third, there is the problem of timing. All parties in the scenario must be ready, technically, mentally and business wise to address risk and revenue aspects.

Finally, what is essentially addressed here, is that there is a need for a communication and development tool, that enables service development and product/technology development to co-happen; the proposed framework.

For further work the suggested framework will be further defined and validated with case studies from different industries. The framework can also be used as a reference model to associate specific tools, such as enterprise modelling and value modelling and simulation tools.

8 ACKNOWLEDGMENTS

The financial support from SSF (Swedish Foundation for Strategic Research) through ProViking, and from VINNOVA through the Faste Laboratory is gratefully acknowledged.

9 REFERENCES

[1] Grönroos, C. (2000): Service management and marketing: a customer relationship management approach. Wiley, Chichester.

[2] Ulrich, K.T., Eppinger, S.D. (2008): Product Design and Development. McGraw-Hill, Inc.

[3] Normann, R. (2001): Reframing business – when the map changes the landscape. Chichester: John Wiley & Sons Ltd.

[4] Mello, S. (2002): Customer-centric product definition: the key to great product development. Boston: USA: AMACOM.

[5] Fransson, M. (2004): Verkstadsindustrins tjänstefiering – en explorativ resa i det nya affärslandskapet. Licentiate thesis. CTF, Karlstads Universitet, Sweden, No. 2004:24.

[6] Stringer, E.T. (1999): Action Research, 2nd ed. Thousands Oaks, Sage Publications Inc, CA.

[7] Andreasen, M.M. and Hein, L. (1987): Integrated product development. Berlin, New York: Springer-Verlag.

[8] Ullman, D.G. (2003): The mechanical design process. 3rd ed. New York: McGraw-Hill.

[9] Unger, D., Eppinger, S. (2010): Improving product development process design: a method for managing information flows, risks, and iterations, Journal of Engineering Design, pp 1–11.

[10] Edvardsson, B. and Modell, S., eds. (1996): Service management. Interdisciplinary perspectives. Stockholm: Nerenius & Santérus Förlag.

[11] Edvardsson, B., Johnson, M.D., Gustafsson, A., Strandvik, T. (2000): The effects of satisfaction and loyalty on profits and growth: products versus services. Total Quality Management, 11, vol. 7.

[12] Grönroos, C (1996): Service culture, perceived service quality and their marketing impact. In Edvarsson, B. and Modell, S. Editors. (1996). Service management. Nerenius & Snatérus Förlag AB, Sweden, 113-122.

[13] von Bertalanffy, L. (1968): General System Theory: Foundations, Development, Applications. George Braziller, NY, USA.

[14] Checkland, P., Holwell, S. (1998): Information, Systems and Information Systems: making sense of the field, Wiley, Chichester.

[15] Baines, T.S., Lightfoot, H.W., Evans, S. et al. (2007): State-of-the-art in product-service systems, in: Proceedings of the Institution of Mechanical Engineers, Part B: Journal of Engineering Manufacture. Vol. 221, No. 10.

[16] Johansson, C., Larsson, A., Larsson, T., Isaksson, O. (2008): Gated Maturity Assessment: Supporting Gate Review Decisions with Knowledge Maturity Assessment. CIRP Design Conference 2008. Twente, Netherlands.

[17] Ericson, Å. (2007): A need-based approach to product development. Thesis (PhD). Luleå University of Technology, No. 2008:01, ISSN 1402-1757.

[18] Parida, V. (2008): Challenges in Open Innovation Practices for Industries. Nordic Innovation Research – NIR 2008, nr 3. December 8-9 2008, Oulo, Finland.

[19] Alonso-Rasgado, T., Thompson, G., Elfström, B-O. (2004): The design of a functional (total care) product. Journal of Engineering Design, 15 (6): 515-540.

[20] Mont, O.K. (2002): Clarifying the concept of product-service system. Journal of cleaner production 10: 237-245.

[21] Mont, O., Dalhammar, C., Jacobsson, N. (2006): A new business model for baby prams based on leasing and product remanufacturing. Journal of Cleaner Production, 14: 1509-1518.

[22] Rittel, H. and Webber, M. (1973): Dilemmas in general theory of planning. Policy Sciences, 4, 155–169.

[23] Randall, D., Harper, R., and Rouncefield, M. (2007): Fieldwork for design – theory and practice. London: Springer-Verlag.

[24] Dubberly, H. (2009): How do you design? A Compendium of Models. Working paper. Dubberly Design Office, San Francisco, CA, USA.

[25] Engwall, M., Kling, R., Werr, A. (2005): Models in action: how management models are interpreted in new product development. R&D Management, 35(4), 427-439.

[26] Zachman, J. (1987): A framework for information systems architecture. IBM Systems Journal, Vol 26. No 3.

Communicating the Value of PSS Design Alternatives using Color-Coded CAD Models

Alessandro Bertoni[1], Marco Bertoni[1], Ola Isaksson[12]

[1] Division of Functional Product Development, Luleå University of Technology, Luleå, Sweden

[2] Volvo Aero Corporation, Trollhättan, Sweden

Abstract

The paper proposes an approach to increase the decision makers' awareness at the gate, when evaluating PSS design alternatives from a lifecycle and value oriented perspective. The paper illustrates a lightweight value visualization tool, running on top of existing CAD systems, supporting value visualization of a given part or assembly, thanks to color coding. Information from value assessment is in this way translated into visual features of the CAD 3D model. Despite the approach is still on a start-up phase, an early mock-up of the tool has been developed and applied to a real aircraft engine component, in order to verify the feasibility of the approach.

Keywords:

Product/Service Systems; Color Coding Approach; Value Visualization

1 INTRODUCTION

In the aerospace business, a stage-gate approach [1] is commonly used to facilitate projects from idea conception to product launch. The key components of the stage-gate process are the stages, a set of information-gathering activities, and the gates, where information is assessed and decisions are made, The role of the gate is to evaluate what has been done in the previous stage and to decide the way forward, what should be done and what resources should be allocated for the next stage [2].

Reviewers evaluate the information developed during the stage and matched against a number of criteria to make a decision both external, such as environment and commercial pressure, and internal, e.g., established processes, organizational structures or incentive policy, that are often determined by the company's overall strategy [3]. A range of strategies aims at supporting decision making at the gate, such as Technology Readiness Level (TRL) [4].

Engineers, however, are no longer solving the problems they used to solve. In the aerospace industry, for instance, the design of a new aircraft engine cannot merely be reduced to a pure technical activity, such as the stress calculation on the blades or on the intermediate case. Engineers are no longer dealing with "tame" problems only; rather have to pay increasingly attention to "wicked" problems [5] as well, such as developing a "passenger-friendly" airplane [6].

Additionally, it can be observed a move towards extending traditional product-based offers to incorporate more intangible assets, i.e. software and services, taking on life cycle responsibilities to secure the aftermarket and to satisfy increasingly sophisticated customer needs. Initiatives such as Functional Products [7], Product Service Systems (PSS) [8] and Integrated Product Service Engineering [9] highlight the opportunity to add value to users and stakeholders by providing a "function" instead of merely selling hardware.

The complexity of functional product implies that companies need not only to assess the readiness of technological components, but also need to develop a shared understanding from a wide set of disciplinary concerns and perspectives, and finally assess how each piece of information or knowledge contributes to the knowledge base used to make decisions. Early on in the aircraft and engine development process it is necessary to reason upon how to improve hardware, software and service to provide a more comfortable, timely, and entertaining flight experience. Although the decisions made at this stage should always add value to the solution space, it is a great challenge to effectively understand the impact on the overall system "value" of changing design variables at the micro-level. Determining the most value-adding alternative is not a straightforward process, thus there is a need for better guidance for design choices, a guidance that translates the desires of customers and business developers into terms that are immediately meaningful to PSS engineers.

The evaluation needs to be based on a different set of criteria, related to the different stakeholders and life cycle phases. Early in the development process, it should be possible to capture, model and communicate the value contribution of different technologies, and their impact on the super-system, in a clear and easily understandable way. Methods and tools are needed to gather, merge and present in a coherent and relevant way all the information coming from different contexts and environments, including application fields far from the direct knowledge of the design team background [7].

2 MOTIVATION AND OBJECTIVES

The research framework in which this work has been conducted aims at contributing to the achievement of the ACARE [10] targets by changing the way product development is initiated in the aerospace industry, through the development of innovative mechanisms to 1) capture, model and understand customers' and stakeholders' needs and expectations, 2) to incorporate the value dimension into preliminary design and 3) to identify criteria affecting customer perceived value to be used in preliminary design studies evaluation.

The paper summarizes the outcomes of an explorative study

J. Hesselbach and C. Herrmann (eds.), *Functional Thinking for Value Creation: Proceedings of the 3rd CIRP International Conference on Industrial Product Service Systems, Technische Universität Braunschweig, Braunschweig, Germany, May 5th - 6th, 2011*,
DOI 10.1007/978-3-642-19689-8_11,

conducted in the frame of a European research project in the aerospace domain. It proposes an approach to support decision making at the gate, by increasing the decision makers' awareness about the life cycle value of a set of PSS alternatives. The paper illustrates a lightweight value visualization tool, running on top of existing CAD systems, to support the visualization of the life cycle value contribution of a given part or assembly.

An early mock-up of the tool has been developed for demonstration purpose and applied to a real aircraft engine component to verify the feasibility of the approach.

3 RESEARCH APPROACH

The study has been performed in the frame of a European Commission's research project within the FP7 programme. The research project has provided access to several aerospace companies (i.e. major aircraft, engine and sub-systems manufacturers and other companies with experience in aerospace development projects), which have contributed to the problem domain definition with their empirical data and expertise.

The research approach adopted can be described as inductive [11], qualitative [12] and participatory [13]. The findings have emerged form the understanding of a real industrial problem that involves all the stakeholders of a product/service system life cycle. The work has been conducted through physical meetings and workshops involving both academic and industrial partners. Semi structured interviews, virtual workshops and company site visits represent the main data gathering activities.

The data gathered have been further analyzed in view of theory; improvements and implications have been proposed using as a reference the scenario created together with industrial partners.

4 EARLY STAGE DECISION MAKING IN PSS DESIGN

In an ideal world, design teams would be able to deliver the best possible design calculating costs and benefits of everything they do beforehand. In the real world, however, rational decision making is a rare occurrence [14]. The design process paradox [15] highlights in fact that the more the team learns, the less freedom it has to use what it knows. Designers are in the unfortunate situation of having limited knowledge in the early stages, when they have good opportunities to influence the design, whereas when they have established a more developed knowledge base in the later stages, major decisions have already been made, capital has already been committed, and it is therefore more costly and time-consuming to make changes.

This is particularly evident when developing products that target life cycle commitments. Approaching the development of functions, it is virtually impossible to know everything about all alternatives, and even to talk about what "best" means:

"there are no true or false answers. [...] solutions are expressed as 'good' or 'bad' or, more likely, as 'better or worse' or 'satisfying' or 'good enough.'" [5].

Gut-feelings and intuition play an important role in the decision making process [16] of PSS. Making a decision is often about settling for what is good enough rather than waiting for the optimal solution to emerge. One of the reasons is that the downstream implications of a PSS alternative is not very visible or well communicated at the gate. Especially when taking decisions about the hardware, life cycle issues tend to be neglected in favor of technical performances of the solution, i.e. better the physical characteristics of the product are, more likely the hardware will be considered for further design.

4.1 Value assessment at the gate

The novelty introduced by a PSS solution causes a radical change in the way the engineering team approaches the decision-making problem. Traditional 'hardware' development, in fact, takes the move from a 'static' and well-defined product usage and from a clear understanding of the required capability. Such context is established by the customer and not generally subjected to significant evolution over the programme life. As opposite, the design of a PSS needs to consider a more service-oriented solution where companies "do not have the luxury of rigidity within their requirement definition" [3]. The initial requirements are subjected to changes in the use/operation phase in response of the changing environment.

Reasoning from a life cycle perspective, merging hardware, software and services, engineers have to deal with plenty of secondary impacts on design decisions that are difficult, if not impossible, to foresee and balance beforehand. Sometimes it is preferable to sacrifice some sub-system performances to optimize the overall system. For components that are going to be sold as PSS, and supported for 20, 30 or even 50 years, the decision making task at the gate should not merely focus on the readiness assessment of technological components. It should be oriented towards understanding the value contribution of a given solution in a much wider perspective, estimating the value contribution at super system level, as outlined by one of our informants in the aircraft engine components manufacturing business:

"Nowadays you can easily tell why a solution is the optimal one in terms of performances, however it is not straightforward to see if it is optimal also from a value perspective. Hence, we have to look at people, tools, processes for developing the optimal solution both from a business as well as customer viewpoint"

Nowadays the overall goodness of a solution is mainly expressed by technical performance figures and cost. There is a need to assess design solutions across a wider range of value-related dimensions, such as monetary value, perceived value, confidence on the product, and requirement fulfillment levels; those need to be represented in a more holistic value model. A main challenge, especially when working on cross-functional and cross-disciplinary teams, is to aggregate and communicate this value information in an easy and understandable way to enable quicker and more agreed design decisions.

4.2 Maturity assessment at the gate

In PSS development, there are many environmental and contextual variables, with varying degrees of controllability. While it would not be feasible to create innovative PSS and, at the same time, completely eliminate uncertainty and ambiguity, companies would most likely benefit from learning more about what exactly those uncertainties and ambiguities involve.

Decision makers at the gate require a deeper understanding of the status of the knowledge base, to deal with uncertainty and ambiguity and to make more confident decisions. A crucial challenge is how to assess knowledge sources and knowledge assets with respect to their fitness-for-purpose. What is the level of readiness of the presented information? Is it reflecting assumptions or verified facts? Is there anything missing? Is the information current or out of date? Are there specific knowledge assets that would need further development to contribute more clearly to the objective? Is there a need to prioritize refinement of some aspects over others?

A key concept in the PSS decision making process is maturity, which is defined by Grebici et al. [17] as a compromise between the target uncertainty and the expected uncertainty, denoting maturity as the distance between the level of completeness relative to what should be the level of completeness, i.e. as-is status versus to-be status. Knowledge maturity builds on the assertion that everything cannot be known – some things are known, while other things are not, i.e. some variables are exogenous and uncontrollable, and others can be controlled. It measures the state of the development of a piece of information with respect to achieving a purpose, which implies that a piece of information may be mature for one purpose and immature for another. For example, a preliminary analysis result, concerning the heat tolerance of an aero engine component, may be good enough in the feasibility stage of component design, whereas the same numbers may be too inaccurate to be valuable in the detailed design stage.

The knowledge maturity assessment may therefore represent a valuable approach for companies to take more conscious decisions at the gates, making sure that implicit assumptions, ingrained views and provisional results are not mistaken for verified facts. In these situations, there is a degree of uncertainty that needs to be handled, perhaps not by directly focusing on reducing the uncertainty, rather by assisting the decision makers in achieving a better understanding of what those uncertainties, ambiguities, and assumptions actually involve. Creating an increased awareness on the status of information and knowledge assets could allow companies to move forward with known risk, rather than making decisions based on unclear rationale.

5 THE INFLUENCE OF COLOR-CODED INPUTS ON DESIGN DECISIONS

The ground hypothesis for this study is that the use of multiple cues to encode information can enhance life cycle information processing at the gate, taking advantage of associative processing.

Associative or peripheral processing, in fact, provides information quickly and automatically, decreasing time and effort needed to complete a task [18]. The theory of cue-summation proposes that multiple cues presented both across and within media/channels can improve information processing, mainly because these cues provide more opportunities for the learner to discern the new information being presented [19].

In particular, the authors believe that color-coded 3D CAD representation might enable the combination and communication of heterogeneous information, when deciding upon the most suitable configuration of the hardware for the functional product (e.g. displaying the value of a solution across the different life cycle phases, communicating the impact of the solution at super-system level and showing the maturity of the knowledge used to compute the value model)

Color emerged as one of the key cues for this study for several reasons. The influence of color and graphical information presentation in a managerial decision environment is an argument discussed since the 70s. Colors have been found to be the most effective coding technique for aiding visual search [20], and properly used colors has been found to improve the usefulness of an information display system [21].

Further, Chute [22] stated that color's ability to delineate figure-ground relationships, to show interrelatedness, and to make discrimination, underscores color's effectiveness in learning. A later experiment conducted by Benbasat [22] showed that color has several beneficial effects for decision making. Subjects with color-coded reports managed to obtain a significantly higher average profit over the first 10 trials and completed the task using fewer trials. The processing of color, in fact precedes the processing of other attributes [24]. Further studies have highlighted the importance of colors as supplementary information cue in interfaces design to encourage associative processing [25].

In the specific area of Computer Aided Design, color-coded 3D models support several analyses such as cost calculation, mechanical and electromechanical simulation, tooling and fixture design and engineering process management. A multilayer color coding visualization, able to access the input data from the system and subsystem models and summarize them in a clear, understandable and useful way, can therefore represent a relevant instrument during design and re-design phases.

The idea of the color visualization aims to help the choice of the designers thanks to the creation of a structure that allows the automatic visualization of the value generated by different design alternatives, from different life cycle points of view, in a unique representation in a CAD environment, trough a constant link to different product models.

6 LIVERY: THE LIGHTWEIGHT VALUE VISUALIZATOR

LIVERY (LIghtweight ValuE visualizatoR) is intended as a practical decision support for PSS development to increase decision makers' awareness of the life cycle value associated to a PSS design solution. LIVERY is organized in a multilayer visualization structure where at different level of detail, from component level to whole aircraft level, a different value visualization is associated. The color-coding visualization consists of two layers of representation, the first concerning the concept value, the second representing the knowledge maturity of the concept.

Value visualization consists in the association of a scale of colors, from green to red to each component/product. The value is related not only to the physical characteristics of the provided hardware, instead it aims to embed the value of the underlying service for a given hardware solution. The input data are qualitative or quantitative; each color is linked to a particular numerical value or range of values, which are the results of the aggregation of weighted values of underlying models. The different levels perspective allows the visualization of value intensive phases along the life cycle.

A main issue in the preliminary design phase of a PSS is that decisions have to be based on very qualitative data, assumptions and forecasts, since an historical series of data as reference is missing. A transparency layer is over-imposed to the 3D model to provide a visual feedback about the maturity of the knowledge used in the representation. The underlying value models can be generated from either qualitative or quantitative studies. Values can be the results of deep empirical analysis, easy measurable and defined, as well as they can be based on designers´ assumptions, panel of experts´ suggestions, or market forecasts, implying therefore either a qualitative or statistical nature. Hence, LIVERY aims to display the maturity readiness of the knowledge assets used to compute the value model, in order to take more conscious decisions at the gates, making sure that implicit assumptions, ingrained views and provisional results are not mistaken for verified facts.

7 APPLICATION EXAMPLE

The LIVERY mock-up has been applied to visualize the value contribution of two alternative intermediate compressor cases (IMC) for turbojet engines. The intermediate case structure includes a core structure, an outer ring, a mechanical connection between the two parts and, occasionally, integrated structural fan outlet guide vanes. Figure 1 shows where the IMC is located in the engine. The intermediate case supports the high-pressure bearing structure and a low-pressure bearing support structure that includes the thrust bearings. The forces that act on the bearing generally occurs as a result of aircraft manoeuvres and forces are typically g-forces, gyro momentum and other flight induced loads.

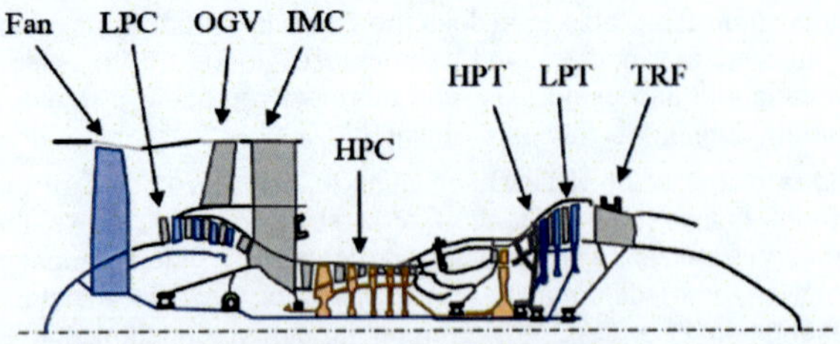

Figure 1: Intermediate case (IMC) position in a jet engine.

The IMC is a so-called "cold structure" since the operating temperature is considered to be fairly low compared to the parts downstream of the combustion chamber. This is also one of the reasons why the IMC is usually made of materials like titanium and aluminum. In the past the fan outlet guide vanes were mounted as separate parts between the fan rotor and the intermediate case. The current trend is however to integrate the vanes into the load carrying IMC-structure. Aerodynamically shaped guide vanes are used to redirect the flow from approximately a 35-degree swirl angle to axial flow.

From a product development perspective, several trends are currently affecting the design of this component. First, aircraft engines are getting larger in diameter to increase the bypass ratio and so needs the intercase. Secondly, aerodynamic functions are increasingly integrated with the intercase. Thirdly, the intercase needs to withstand higher temperatures and heavier loads, meanwhile reducing its weight. Eventually, the intercase is mounted on increasingly electrically powered aircrafts, so needs to provide room for attaching electrical equipments. In the example, two different intercase alternatives have been considered: 1) a casted titanium intercase and 2) a fabricated titanium+aluminum intercase with foam core hollow vanes.

The LIVERY mock-up has been realized using Siemens UGS NX® software. The results of a simple value analysis, performed in MS Office environment, have been exported in NX and automatically hued after converting the input file a common .txt file.
The data used for the value computation are entirely fictitious and the example has only a demonstration purpose.

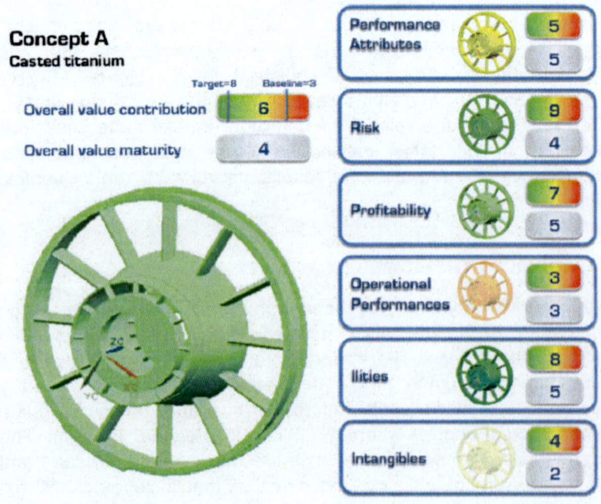

Figure 2: Concept A value contribution.

Figure 2 represents in a schematic way the visualization approach applied to concept A. The approach aims to visualize the value contribution of the design alternative against six main value dimensions, namely: Performance attributes, Risk, Profitability, Operational Performances, Ilities and Intangibles [26].

The overall value contribution is computed in form of a single scalar on a scale from 1 to 9, conceptually derived from TRL, where level 3 represent the baseline and level 8 the target for each value dimension. Different colors helps designers to understand which are the value intensive parts and where improvement should focus. Engineers can, therefore, compare two design alternatives (e.g. concept B – Figure 3) and visualize their value contribution across the given parameters.

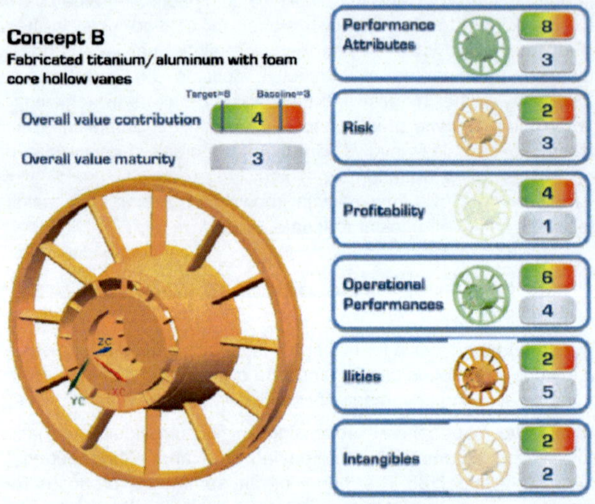

Figure 3: Concept B value contribution.

Each of the 6 main dimensions has to be considered the result of a merging of lower level parameters, which detail value gains and losses for each alternative design. The value for each sub-parameter is computed by accessing a heterogeneous set of models.

In LIVERY, knowledge maturity assessment is seen as a complementary activity to the value assessment. The value models built in a preliminary design phase suffer from poor maturity and reliability of the data used. It is particularly important, therefore, to have pointers that can help indicate at which level people may trust the material entering in the value assessment activity. It is highly beneficial to understand the expected level of quality of the final output, i.e. correctness of the decision taken regarding the design solutions proposed at the gate. In order to calculate the maturity, LIVERY adopts the maturity calculation approach proposed by Johansson et al. [27], who have developed a knowledge maturity model to assess the state of readiness of a knowledge asset using a narrative scale. Knowledge maturity is computed over three dimensions: input, method (tool), and expertise (experience) on a scale from 1 to 5. Johansson [27] rank as 5 an Excellent knowledge maturity, meaning that content and rationale have been tested and proven and reflect a known confidence, that the procedure to produce the content and rationale reflects an approach where tried out methods are used, where workers continually reflect and improve and where lessons learned are recorded. Level 4 is defined as Good and level 3 as Acceptable. Knowledge maturity is Acceptable when content and rationale are more standardize, when there is a greater extent of detailing and definition and the procedure to produce the content and rationale is stable with an element of standardization and repeatability. Level 2 is ranked Dubious and level 1 as Inferior. A knowledge maturity level ranked 1 means that the content and rationale is characterized by instability (e.g. poor/no understanding of knowledge base), the procedure to produce the content and rationale is dependant on individuals and formalized methods are non-existent.

This enables an assessment of input data both of the tools to refine or develop the input into an output, and of the individuals contributing to the work. Furthermore it allows worldwide-located teams share an artifact around which they can identify and discuss issues of concern, visualize the current status of the knowledge base, and negotiate a shared understanding of the advantages and drawbacks with the available knowledge base. In a system-of-systems perspective, especially in business-to-business situation, it is also important how a given design alternative impacts on the value scale of the different levels of customers. Figure 4 shows a more detailed representation, visualizing the impact of Concept A and Concept B on engine and aircraft level.

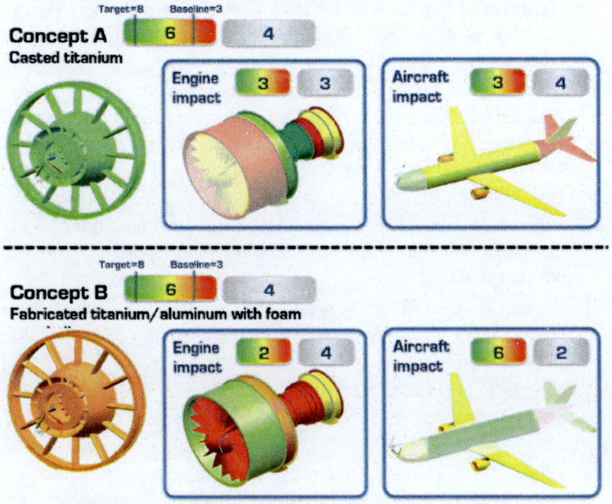

Figure 4: Concept A vs. B system level impact.

A new intercase configuration could have a major value because it allows the engine to be lighter, or more efficient, or requires less maintainability during the life. This could generate consequences on the whole aircraft design allowing a different mounting system or a different wing shape. Alternative intercase concepts could have radical or minor impacts on the engine and on the aircraft structure. The lightweight visualization aims to raise the awareness about system level impacts and supports a more intuitive comparison of different product/service solution alternatives.

Eventually, concept B would be preferable merely looking at performance attributes, but the evaluation changes when taking the all picture in consideration. Concept A, in fact, shows better characteristics in terms of profitability (cash flows), in terms of possibility to upgrade the solution to face changing environmental conditions, as well as in terms of Intangibles. Moreover, the value computation is based on knowledge that is more mature.

8 DISCUSSION

LIVERY is an approach that presents some relevant issues to be discussed. It provides the possibility to make a more intuitive visualization of life cycle value of a sub-system/component to be included in a functional offer.

LIVERY is currently under implementation in a CAD environment, thus no formal evaluation has been performed yet. However the approach has been presented and discussed with several partners from the aerospace industry and has received positive feedbacks.

The approach can help in communicating where the value is "hidden" in the product, increasing the awareness of decision makers about life cycle issues. The design team could visualize the impact in term of value of a decision with different system perspectives, choosing the activity layer of interest. Browsing through the different visualizations, the decision making team could understand sweet spots and weak spots of the alternatives. The development of visualization technologies needs to consider the problem of information overload. The target is to condense information to a minimum in order not to dump engineers and designers with too many inputs. It may also link service issues with the physical product and immediately relate the value associated to the service part to the hardware.

The definition of the value metric can be ambiguous when revenue is not the main measure of the product. It might not be straightforward in many situations to summarize everything to a number representing the overall value associated to a given choice. An important aspect is also to define which criteria to be used for the value calculation, in the example the criteria proposed by Bertoni [28] have been considered, however different criteria could be related to different products in different application fields. Furthermore to grant the results consistency it is necessary to previously establish a baseline value level and a target value level; how baselines and targets should be set, related to the product considered, strongly depends from customers and management needs and requirements. On the same line is the definition of parameter weights, allowing prioritizing some variables instead of others during the value assessment phase.

This approach, even if discussed with different industrial partners, needs to be deeper analyzed and validated together with designers in a real product development process, in order to obtain more constructive feedbacks to improve the overall functionality.

Two additional fields of study related to the topic, but outside the paper intent, are the analysis of how the process at the gate would really change after the adoption of LIVERY, and how and where the information treated can be retrieved.

9 CONCLUSIONS

In this paper an approach is proposed for communicating the value of different PSS design alternatives.

In order to face with the multi attributes decision making problem, considering all the aspects related to value generation in a product/service system, the authors have developed and used a methodology aiming to consider both value, tangible or intangible, and knowledge maturity.

The work led to the formulation of a coloring approach reflecting the value embedded in a PSS design alternative, linking this information to the concept knowledge maturity.

To verify the feasibility of the idea, the functionality of the approach has then been implemented and tested in a CAD/PLM environment.

Although the approach is still in a start-up phase, we believe that in the future this methodology could contribute to the design of a PSS solution, and we consider it enough advanced to be included in a value drive design assessment approach.

Since the objective of the paper was to propose an approach, despite the possibility to access some real models and parameters, the example described uses fuzzy data in order to avoid private data secrecy problems.

Further developments will focus on correlated fields: First the refinement and testing of the mock-up in a laboratory environment with a selected set of engineers/designers actively involved in gate decisions activities; second, the extension of the system-impact layer to be able to visualize the alternatives' impact at higher customers levels (i.e. airlines and the passengers), and third the evaluations of the implications for the product development process performances, i.e. measuring the gains in term of lead time reduction and quality improvement of the functional offer.

10 ACKNOWLEDGMENTS

The research leading to these results has received funding from the European Union Seventh Framework Programme (FP7/2007-2013) under grant agreement n° 234344 (**www.crescendo-fp7.eu/**).

11 REFERENCES

[1] Cooper, R. G. (2001): Winning at New Products: Accelerating the Process from Idea to Launch, 3rd ed. Reading, MA: Perseus Books.

[2] Cooper, R. G. (2008): Perspective: The Stage-Gate® Idea-to-Launch Process-Update, What's New, and NexGen Systems*, in: Journal of Product Innovation Management, Vol. 25, pp. 213-232.

[3] Molloy, E.; Siemieniuch, C.; Sinclair, M. (2009): Decision-making systems and the product-to-service shift, in: Journal of Manufacturing Technology Management, Vol 20, No. 5, pp. 606-625.

[4] Mankins, J. C. (1995): Technology Readiness Levels, NASA Advanced Concepts Office, Washington, DC, USA.

[5] Rittel, H.; Webber, M. (1973): Dilemmas in a General Theory of Planning, in: Policy Sciences. Vol. 4, pp.155-169.

[6] Boeing (2006): Technology Redefines Joy of Flight, [online] available at: **http://www.boeingcapital.com/p2p/archive/06.2006/techredefinesjoy.htm** [Accessed 13 November 2010]

[7] Isaksson, O.; Larsson, T.; Rönnbäck, A.Ö. (2009): Development of product-service systems: challenges and opportunities for the manufacturing firm, Journal of Engineering Design, Vol. 20, No. 4, pp.329-348.

[8] Baines, T.S.; Lightfoot, H.W.; Evans, S.; Neely, A.; Greenough, R.; Peppard, J.; Roy, R.; Shehab, E.; Braganza, A.; Tiwari, A.; Alcock, J.R.; Angus, J.P.; Bastl, M.; Cousens, A.; Irving, P.; Johnson, M.; Kingston, J.; Lockett, H.; Martinez, V.; Michele, P.; Tranfield, D.; Walton, I.M.; Wilson, H. (2007): State-of-the-art in product-service systems, in: Proceedings of the Institution of Mechanical Engineers, Part B: Journal of Engineering Manufacture. Vol. 221, No. 10.

[9] Lindahl, M.; E. Sundin, A.; Rönnbäck, A.Ö.; Ölundh, G.; Östlin, J. (2006): Integrated Product and Service Engineering – the IPSE project, in Proceedings of Changes to Sustainable Consumption, Workshop of the Sustainable Consumption Research Exchange (SCORE!) Network, Copenhagen, Denmark . April 20-21.

[10] ACARE, (2002) Strategic Research Agenda Volume 1. Advisory Council for Aeronautics Research in Europe, ACARE. October 2002. URL: **www.acare4europe.org/docs/es-volume1-2/volume1.pdf**. [accessed 13 November 2010]

[11] Marshall, C.; Rossman, G.B. (2006): Designing qualitative research, fourth edition, Sage Pubblication, Inc.

[12] Michalski, R.S. (1983): A theory and methodology of inductive learning, in: Artificial intelligence, Vol. 20, pp. 111-161.

[13] Cornwall, A.; Jewkes, R. (1995): What is participatory research?, in: Social Science & Medicine, Vol. 41, No. 12, pp. 1667-1676.

[14] Simon, H. A.; (1979): Rational Decision Making in Business Organizations, in: The American Economic Review, Vol. 69, No. 4, pp. 493-513.

[15] Ullman, D.G.; (1992): The mechanical design process, McGraw-Hill, New York.

[16] Ericson, Å.; Bergström, M.; Johansson, C.; Larsson. T. (2007): On the way to knowledge awareness in early design, CIRP 2007, Design Seminar on "The Future of Product Development", Berlin, Germany, March 26-28.

[17] Grebici, K.; Yee Mey Goh; Sunny Zhao; Blanco, E.; McMahon, C. (2007): Information Maturity Approach for the Handling of Uncertainty within a Collaborative Design Team, in: 11th International Conference on Computer Supported Cooperative Work in Design.

[18] Bargh, J.A. (1994): The four horsemen of automaticity: Awareness, intention, efficiency, and control in social cognition, in: R.S. Wyer and T.K. Srull (Eds.) Handbook of social cognition, Lawrence Erlbaum Associates, Inc., Hillsdale, NJ, pp.1-40.

[19] Severin, W. (1967): Another look at cue summation, in: Educational Technology Research and Development, Vol. 15, No. 3, pp. 233-245.

[20] Christ, R. E. (1975): Review and analysis of color-coding research for visual displays, in: Human Factors, Vol. 17, No. 6, pp. 542-570.

[21] Murch, G.M. (1984): Physiological principles for the effective use of color, in: IEEE Computer Graphics and Applications, Vol. 4, No. 11, pp. 49-54.

[22] Chute, A. G. (1979): Analysis of the instructional functions of color and monochrome cuing in media presentations, in: Educational Communication and Technology Journal, Vol. 27, No. 4, pp. 251-263.

[23] Benbasat, B. (1986): An experimental program investigating color-enhanced and graphical information presentation: an integration of the findings, in: Communications of the ACM archive, Vol. 29, No. 11, pp. 1094 – 1105.

[24] Karayanidis, F.; Michie, P.T. (1997): Evidence of visual processing negativity with attention to orientation and color in central space, in: Electroencephalography and Clinical Neurophysiology, Vol. 103, No. 2, pp. 282-297.

[25] McNab A.L.; Hess T.J.; Valacich, J.S. (2009): Designing Interfaces for Faster Information Processing: Examination of the Effectiveness of Using Multiple Information Cues, in: Proceedings of AMCIS 2009, Paper 699.

[26] Bertoni, M.; Eres, H.; Isaksson, O. (2011): Criteria for assessing the value of Product Service System design alternatives: an aerospace investigation. Submitted to CIRP IPS² Conference, May 5th – 6th, Braunschweig, Germany.

[27] Johansson, C.; Hicks, B.; Larsson, A.; Bertoni, M. (2010): Knowledge maturity as a means to support decision making during product-service systems development projects in the aerospace sector, Project Management Journal, in press.

Influences of the IPS² Business Model on the Development of a Micro Milling Spindle

Eckart Uhlmann, Christian Gabriel, Niels Raue, Christian Stelzer

Institute for Machine Tools and Factory Management Technische Universität Berlin, Berlin, Germany

Abstract

In the IPS² planning phase, defined customer and provider needs have to be considered for the realization of product and service shares. In contrast to a traditional product with add-on services, the service shares are an integral part of an IPS². This paper presents the influence of the IPS² business model on the development of a micro milling spindle. The IPS² micro milling spindle is a core component for a milling process. An overview of possible solutions of product and service shares for a micro milling spindle under consideration of the IPS² business model and suitable development methods are presented.

Keywords:

Industrial Product Service Systems; Micro Milling Spindle; Design for IPS²

1 INTRODUCTION

Detection and utilization of customer needs are important for individualized industrial solutions and lead to competitive advantages. These can be used by machine tool manufacturers, if they consider the integrated approach of Industrial Product Service Systems (IPS²) already during the planning phase of a product. Here, different IPS² business models are possible.

The business model can be function (product) oriented, availability (use) oriented, or result oriented [1], [2], [3]. The descriptions for these three exemplary IPS² business models are mainly the same. For the description of the different IPS² business models in this paper, the classification of [2] is used.

Within the business models a change of the responsible partners is to observe. The function oriented model shows a clear responsibility of the customer. The availability oriented model is characterized by a greater responsibility of the IPS² provider. This also leads to a greater dependency of the provider on the equipment reliability. This tendency ranges up to the responsibility for the whole process in the product oriented business model. For the assurance of the reliability, e. g. of a machine tool, the critical components have to be identified. It is also necessary to determine measures for the component monitoring. The monitoring process must be able to detect all data for the valuation of the equipment's condition during the whole life cycle. This can be realized by an automated agent-based communication that notifies the IPS² provider in case of an incident or any other occurrence with potential impact on the equipment's reliability.

The business models are linked to different process design requirements. These requirements are pointed out in this paper by the example of a micro milling spindle. Here, the challenges of an IPS² development are demonstrated that occur under the restrictions induced by the business model.

2 INDUSTRIAL PRODUCT SERVICE SYSTEMS (IPS²)

2.1 Definition of IPS²

An Industrial Product-Service System is characterized by a combination of tangible product and intangible service shares including the immanent software via the complete lifecycle to fulfill the customer and also the IPS² provider requirements [3], [4], [5], [6]. The IPS² provider offers a customized solution in terms of an IPS² business model to the customer [1], [2].

2.2 IPS² lifecycle

The main phases of an IPS² lifecycle are:

1. IPS² creation;
2. IPS² realization;
3. IPS² operation and;
4. end of IPS² (termination);

The IPS² creation is separated into two sequential process steps, which are the IPS² planning and IPS² development.

The IPS² planning phase starts with the first customer contact. In a second step of the IPS² planning the customer requirements are identified by the IPS² provider [2]. The IPS² planning phase ends with an offer for a specific IPS². This is based on a requirements list and is connected to a customized IPS² business model [2].

A functional and conceptual model of the IPS² is generated in the IPS² development phase. The needed IPS² function can be realized by service, product shares or a combination of both shares. Three different developments, methodically based on conventional design process, can be distinguished [2]:

- new development;
- variation development;
- configuration;

These three developments need different process configurations and are based on special knowledge on side of the IPS² provider [2].

J. Hesselbach and C. Herrmann (eds.), *Functional Thinking for Value Creation: Proceedings of the 3rd CIRP International Conference on Industrial Product Service Systems, Technische Universität Braunschweig, Braunschweig, Germany, May 5th - 6th, 2011*,
DOI 10.1007/978-3-642-19689-8_12,

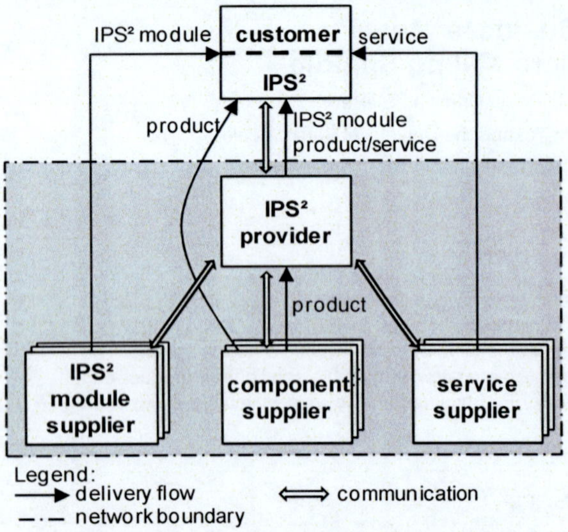

Figure 1: Typology IPS² network organization [4].

In the IPS² realization phase, which follows the development, the IPS² with all product and service shares is built up.

The main characteristic of the subsequent IPS² operation phase is the parallelism of the IPS² product share's use and the delivery of the IPS² service shares [2]. In this phase, the IPS² creates value and the processes of the IPS² service shares are supplied.

2.3 IPS² network

To deliver IPS² in an effective and efficient way during the IPS² life-cycle, a network of different partners is needed. Therefore the definition of the different roles must be given, which are also characterized by flows of information and products (Figure 1).

In addition to the customer of the IPS², four different roles are defined for an IPS² network [4]. The IPS² provider coordinates the network of IPS² module supplier, components supplier and service supplier.

2.4 IPS² business models

The term business model raised its popularity in the last decade of the 20th century due to the use in papers concerning business and computer science. The business model idea was associated with the increase of commercial activities on the Internet and the thereby created market [7]. Companies who belong to the so called "New Economy" created a room for maneuver and became attractive for investors by describing their business strategy with an innovative business model. Due to the dynamic advancement the term was used widely but had never been defined. This caused a commingling between the terms business model and business strategy [8]. In the primary understanding a business model demonstrates an abstract description of the business operations and is based on figures of business units, transformation processes, transfer flows, influencing factors, resources and their interaction [4], [9]. Information and communication technology supports the process and data modeling. As an example, processes, tasks and communication relationships had been modeled in an existing business to support strategic decisions and to get a clear description of the organization units [8].

To consider the customer individual characteristic of an IPS² the exemplary business models can be defined as follows:

Function oriented IPS² business model

While developing an IPS² in a function oriented business model, e. g. a micro milling spindle as a component of a machine tool, the IPS² provider aims to guarantee its functionality for a defined operating time. The design considers the earlier identified requirements [10]. The responsibility for the supply of service shares, e. g. maintenance or training, can vary between the IPS² provider and customer (see 2.3), depending on the negotiation of all participants [10]. In contradiction to different IPS² business models the customer mainly creates value in the use phase and becomes owner of the IPS².

This IPS² business model is similar to a conventional product sell of machine tool manufacturers, where the machine tool with an appropriate functionality is sold (see also [3]).

Availability oriented IPS² business model

The availability oriented IPS² business model is determined by the term availability. As a service share on the industrial market the IPS² provider guarantees a defined availability of the IPS². The micro milling spindle represents a critical component for the use of a micro milling machine tool. A failure of this component leads to a downtime of the micro milling machine tool and therefore the availability can not be guaranteed any more.

In this business model, the use of the IPS² is defined by the availability. This availability is much more complex to calculate [11] than to determine the function in a function oriented IPS² business model. Therefore, special monitoring solutions can be used.

To prevent unscheduled downtimes and their consequential financial losses, the IPS² provider has to take over responsibilities of industrial services. Maintenance and condition monitoring are services that should be delivered in the use phase of the micro milling spindle [10], [12]. By delivering service shares, the IPS² provider creates value and divides the responsibilities for the use of the IPS² between the network partners (see 2.3). That approach requires an intensive coordination in the IPS² network for delivering industrial services to secure the availability of the spindle [4].

Result oriented IPS² business model

A result oriented business model is determined by the fact that the IPS² provider remains owner of the machine tool and will be paid by the customer for manufactured parts. The requirements concerning quality and units per period have to be met [10]. Therefore the IPS² provider assumes the responsibility for all operations necessary for manufacturing a part while using an IPS². To avoid or at least reduce financial losses due to a breakdown of an IPS², it is useful to create alternative options for manufacturing parts. A suitable IPS² network partner could handle the production in this case [4].

3 DEVELOPMENT OF A MICRO MILLING SPINDLE

The following chapter shows by example the development of a micro milling spindle. The effects of different IPS² business models on the development are analyzed and summarized into a list of requirements. However, this step will not be presented in this paper.

Afterward the function structure for micro milling spindles is the basis for the development of different solution principles according to the method of systematic development [13].

3.1 Description of the scenario

The micro milling machine tool (Figure 2) consists of an aluminum frame and three orthogonal feed axes. The spindle is fixed to the vertical axis via a mounting device with cooling capability and realizes a maximum rotational speed of 50.000 rpm.

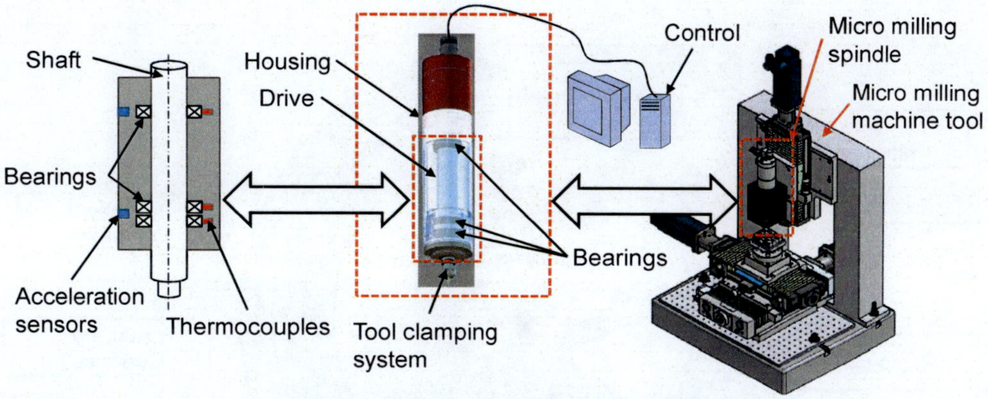

Figure 2: Components of the micro milling spindle.

A pneumatic work piece clamping unit fits four watch baseplates, which are held in place by a locking screw. The clamping unit can be fixed and released by applying compressed air.

3.2 Identification of relevant spindle components

By the analysis of the manufacturing process for a watch baseplate, the relevant process steps executed by the micro milling spindle can be identified. Subsequently, they can be assigned to the corresponding components of the micro milling machine tool.

The following components of the micro milling machine tool are considered as relevant for the discussion of designs in different IPS² business models:

- control of the micro milling spindle;
- drive of the micro milling spindle;
- bearing of the micro milling spindle;
- housing of the micro milling spindle;
- tool clamping system of the micro milling spindle;

The drive and the bearings are located in the housing (Figure 3), whereas the control is an external device. Below, these components are discussed in further detail.

Control

The control of the micro milling spindle enables the machine tool user to set cutting parameters such as spindle rotational speed and feed rate and to execute NC programs. Depending on the type of drive, different control units are needed.

Drive

The options for a high speed spindle drive are either a synchronous or an asynchronous motor. A synchronous motor consists of a permanent magnet as a rotor and a stator winding with a three phase current flow. With the current changing its amplitude and polarity, the magnetic field of the stator winding rotates and forces the rotor to move synchronously. For an asynchronous motor on the other hand, the permanent magnet in the rotor is replaced by a second winding. This winding, which transports current induced by the stator winding, establishes a second magnetic field which interacts with the stator's magnetic field. This way, a rotational motion is generated.

Bearings

For a micro milling process the requirements for the bearings are characterized by high rotational speeds, which can be as high as 200.000 rpm. A conventional machine part for this task is a roller bearing. However, high velocities lead to a great centrifugal force on the rolling elements as well as high friction between bearing rings and rolling elements. Strategies to counteract these problems are high velocity bearings with smaller rolling elements that decrease the effect of the resulting centrifugal force. Hybrid bearings feature ceramic rolling elements and thus have a positive impact on the friction at high rotational speeds.

Alternatives to roller bearings are electromagnetic as well as aerostatic bearings, which are more cost intensive because of the needed extra equipment.

Housing

The choice of the housing largely depends on the necessary supply lines for the other components. The housing diameter is mainly determined by the stator diameter of the spindle drive (see above). The sealing of the spindle is realized by a labyrinth sealing. To avoid the intrusion of debris and foreign particles, sealing air is applied to the gap between housing and shaft. Furthermore, the maintenance-friendly and tamper-proof design of housing openings is an important aspect of an IPS²-compatible design.

Tool clamping system

Lastly, the tool clamping system is a component considered relevant for the IPS²-based design. Connected to the central compressed air supply of the machine tool, the clamping system allows a quick and automated tool change.

The standardized interface geometry between tool holder and clamping system can either be a steep angle taper or a hollow shank taper lock. The advantages of a hollow shank taper lock are an increased possible rotational speed due to its modified geometry, an improved concentricity and radial as well as axial stiffness. A smaller weight and decreased height facilitates an automated tool change.

Common methods for the fixation of the tool shank into the holder are the use of a collet chuck, a hydraulic expansion chuck or a shrink chuck. Collet chucks are characterized by a relatively low clamping force and decreased accuracy. However, its capability to mount tools of varying shaft diameter provides a high degree of flexibility. Hydraulic expansion chucks and shrink chucks do not feature this flexibility and are more expensive, therefore their accuracy is superior.

3.3 Introduction to the development of relevant spindle components and industrial services

The development of IPS² aims to raise the customer benefit by integrating product and service shares along the whole IPS² lifecycle. Therefore it is necessary to define the framework of proper service shares, which the IPS² should deliver.

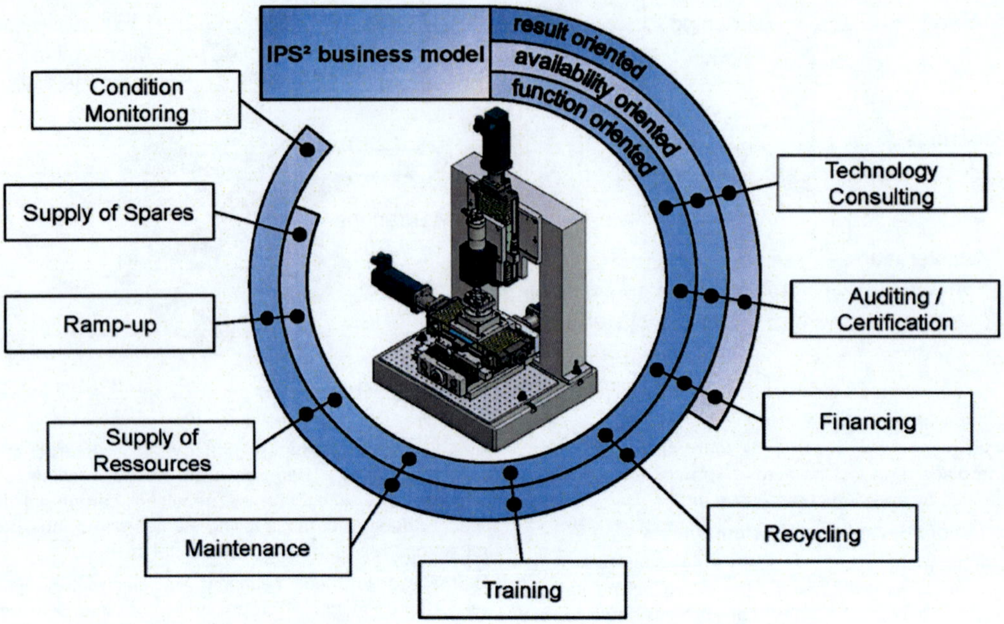

Figure 3: Applicable industrial services for IPS² business models.

A general overview of industrial services is illustrated in Figure 3 along with their applicability for a micro milling spindle in the defined IPS² business model. The figured industrial service shares vary in the way they impact the structuring of the micro milling spindle.

In a function oriented IPS² business model the spare part service is able to generate a higher customer benefit and should be provided as an industrial service. In other IPS² business models, the IPS² provider is already responsible for a stable supply of spare parts (Figure 3). A customer who is responsible for the use of an IPS², e. g. in the function or availability oriented business model, increases the machine efficiency if the employees are continuously educated and skilled. Therefore the IPS² provider should create and offer technology as well as user training, e. g. for service technicians or machine tool users. A machine tool user who is technologically skilled will manufacture parts with a high quality standard at low production time by choosing the optimal cutting speed and milling strategy. In addition staff trainings lead to a minimization of machine breakdowns as a result of less user failures.

A further service is a special financing model for the customer in a function and availability oriented IPS² business model to divide the costs in a defined contract period. Holdback payment enables customers the acquirement of machine tools with a high technological standard and increase the number of possible customers for the IPS² provider. The risk for possible customer's insolvency demands a well chosen financing model and could be accomplished by the cooperation with a finance institute.

All service shares have a more or less direct relation to the constructive characteristics of the relevant components mentioned above. These impacts will be discussed below for different IPS² business models.

3.4 Spindle components in a function oriented IPS² business model

In a function oriented IPS² business model, the IPS² provider guarantees the functionality of the micro milling spindle. In this IPS² business model, the IPS² customer owns and operates the micro milling machine tool. Beyond that, the IPS² provider can develop and provide industrial services on customer request.

The design of the micro milling spindle has to meet the minimum requirements which are necessary to fulfill its function and the defined warranty. An extended warranty as part of an agreement between IPS² provider and customer would be a possible way to strengthen their relationship. Such an agreement would imply the customer's obligation to obtain maintenance services only from the IPS² provider.

Control

Services regarding the spindle control in a function oriented business model include the machine ramp-up on the site of the customer and a training to prevent machine failure induced by operating errors.

Drive

The drive has to be designed in a way that it can fulfill the function demanded by the customer. Mainly, these requirements are given by the achievable rotational speed and torque. This determination limits the customer's flexibility in handling unforeseen production tasks. However, since the value creation process on IPS² provider side ends nearly completely at the point of sale, considerations about future changes of production parameters on customer side are not of interest. The lower price of the asynchronous drive at similar operation characteristics makes this solution an economic choice for the micro milling spindle in a function oriented business model.

Bearings

The use of roller bearings in a function oriented business model implies the advantage of using standardized and therefore cheaper parts (see 2.2), which can be designed based upon reliable empirical values. Commonly, a fixed-floating bearing arrangement is chosen, whereas the fixed mounting is located on the tool side of the shank. This ensures that thermal expansion of the shank has a minimal influence on the axial accuracy of the tool center point.

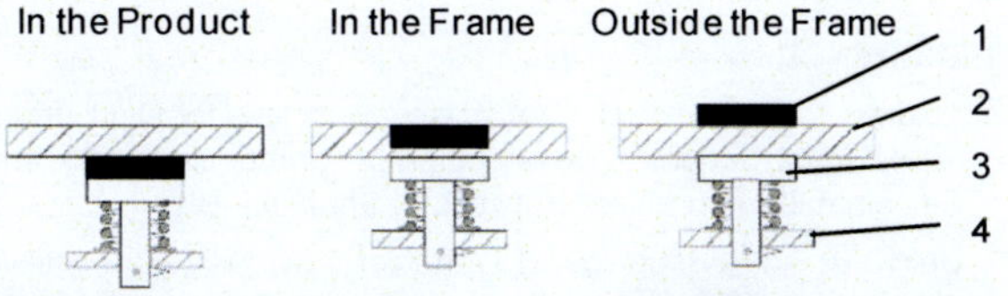

1 – Magnet
2 – Frame
3 – Connection Element
4 – Frame of Product

Figure 4: Arrangements of magnet and connecting element [14].

Housing

The housing should be designed in a tamper-proof way, so that damages caused by improper handling or intrusion during the warranty period can be verified by the IPS² provider [14]. A simple method of restricting the access to critical components is the use of slotted nuts, which can only be released by a special screwdriver. Instead of conventional connecting elements like screws, electromechanical locks can be installed.

During a maintenance process it ensures that the connected components can only be separated by authorized personnel like a service technician. The IPS² provider has to authorize the person. The login and access procedure to unlock this mechanism can be realized by an internet connection, so that the provider is able to control who is maintaining the machine (Figure 4) [14].

This measure avoids manipulations by the customer. Furthermore, a stable flow of properly treated sealing air has to be ensured for providing a safe operation of the spindle. The IPS² provider could ensure the monitoring of sealing air by an integrated sensor guard upon request.

Tool clamping system

The collet chuck tool holder is recommended in a function oriented IPS² business model, since the collet chuck enables the customer to use different tool shank diameter and thus remain flexible.

3.5 Spindle components in an availability oriented IPS² business model

In this type of IPS² business model, the provider assures a determined degree of production equipment availability to the customer, who owns and operates the micro milling machine tool in his facilities. This implies two general approaches for the micro milling spindle:

1.) Optimization of critical components to increase their service life.

2.) Condition monitoring of critical micro milling spindle components by means of sensor integration into the relevant components.

Via condition monitoring, the IPS² provider can observe or even predict any incidents that endanger the availability of the micro milling spindle and counteract these disturbances by prompt service actions. In the following, the five critical spindle components (see 3.2) will be designed according to these approaches.

Control

To draw conclusions about the operation conditions of the micro milling spindle, the spindle control should feature the possibility of monitoring and recording operation parameters such as drive current. Component failures due to operating errors, e. g. too high rotational speeds, can be detected this way. In case of a complex error or complete failure of the micro milling spindle, an exchange of the micro milling spindle will become necessary. To ensure a quick and correct exchange by a service technician, the control should provide a tamper-proof and prompt connection of control and supply lines. This can be realized by a coupling unit that unifies all needed connections in one plug.

Drive

The drive of the micro milling spindle in an availability oriented IPS² business model should be realized by a synchronous motor. The heat generated in the stator coils is lead away by a cooling device based on a stream of compressed air. The air constantly flows through grooves located in the spindle housing. To monitor the temperature in the drive, which is mainly depending on the applied torque, thermocouples are used. To detect critical operating conditions, the temperature has to be monitored, while the milling process parameters are taken into account. A steep rise of temperature under rough cutting condition can be normal, whereas a slowly increasing temperature plot during idling can be a sign of critical operating conditions.

Bearings

A high rotor temperature of the micro milling spindle leads to an increased stress on the roller bearings. In milling spindles based on roller bearings, 85 % of spindle failures are due to bearing defects [15]. To ensure availability, it is important for the IPS² provider to enable the monitoring of the bearings. By integrating piezo based acceleration sensors and thermocouples, vibration and temperature values can be tracked. By comparing the data plots to characteristic damage progress plots, conclusions about critical tendencies can be drawn. Another method is the use of hybrid bearings, featuring wear-resistant ceramic rolling elements.

Housing

The housing in an availability oriented IPS² business model is characterized by simple, but tamper-proof accessibility to critical components. The service delivery can be supported by the use of electromechanical locks. This can be similar to the solution in the function oriented IPS² business model. They ensure a quick and easy access to the critical components inside the housing, like the bearings of the micro milling spindle.

Also, the housing should allow for ways of monitoring the flow and temperature of compressed air. The air ensures the cooling of the micro milling spindle and the tool change.

Tool clamping system

The tool clamping system can support the availability assurance by additional sensor functionality. Through integrated and wirelessly communicating sensors, the process forces can be measured and taken into account when the bearing condition is determined.

3.6 Spindle components in a result oriented IPS² business model

The IPS² provider is paid for the result of the IPS² in a result oriented IPS² business model. In this business model, the provider's interest in the machine's proper function goes beyond the sole assurance of availability. The productivity and manufacturing quality are directly influencing the payment at the IPS² provider. Therefore the process reliability has to be guaranteed. On the other hand, a failure of the micro milling spindle does not imply an immediate financial loss for the IPS² provider. In such a situation, he could reschedule the production of the batch at an IPS² network partner (see 2.3) that can fulfill the same technological requirements. In this IPS² business model, the machine is owned and operated by the provider at his facilities.

Control

The control of the micro milling spindle should feature possibilities for the monitoring of energy and auxiliary supplies. The consumption of these items directly influences the provider's efficiency in this IPS² business model. This also implies that the frequency converter itself should possess a high degree of efficiency.

Drive

This thought of efficiency can be extended to the drive of the micro milling spindle as well. Synchronous motors have a smaller dissipation loss and thus, a higher degree of efficiency compared to asynchronous motors. Also the better dynamical behavior of the synchronous motor reduces non-productive times and thus allows for an increased productivity.

Bearings

The spindle bearings influence surface quality and form accuracy parameters of the produced work pieces and are therefore of special interest for the IPS² provider. Since aerostatic and electromagnetic bearings do not require solid contact for their function, they positively influence accuracy parameters such as concentricity error and are especially suited for the use in this IPS² business model.

Housing

The housing of the micro milling spindle in the result-oriented IPS² business model should be have a high degree of flexibility. This can be realized by an interchangeability of the micro milling spindle with other machine tools at the IPS² provider's production facilities.

Tool clamping system

The mentioned productivity and quality considerations, which are important in the result oriented IPS² business model, can as well be applied to the tool clamping system. Key parameters influencing the accuracy are the balancing quality and the clamping length. The choice of the clamping system geometry also influences the productivity. The use of hollow shank taper locks leads to decreased non-productive time, since they allow for a quicker tool holder change, e. g. between roughing and finishing steps.

4 SUMMARY

In order to demonstrate the importance of the influences of different IPS² business models on the development process of a micro milling spindle, three exemplary IPS² business models were introduced. Subsequently, the critical components of a micro milling spindle were identified and possible design solutions were presented. These solutions were then applied to the three IPS² business models defined above. This paper proposes a business model specific IPS² development.

This approach clarifies the necessity of considering IPS²-specific needs in the development phase. Through recognizing the requirements induced by different IPS² business models as early as possible, the successful development and provision of service shares during the IPS² use phase can be facilitated. In a next step, this procedure could be extended to other components and ultimately, the complete machine tool.

5 ACKNOWLEDGMENTS

We express our sincere thanks to the Deutsche Forschungsgemeinschaft (DFG) for funding this research within the Collaborative Research Project SFB/TR 29 on Industrial Product-Service Systems – dynamic interdependency of products and services in the production area.

6 REFERENCES

[1] Tukker, A., Tischner, U. (2005): New Business for Old Europe – Product-Service Development, Competiveness and Sustainability, Greenleaf Publishing, Sheffield, UK.

[2] Uhlmann, E., Bochnig, H., Stelzer, C. (2010): Partially automated IT support for planning processes in IPS² development, In: Proceedings of the 2nd CIRP IPS² Conference (CIRP Industrial Product-Service Systems), pp. 399 - 405.

[3] Baines, T. S., Lightfoot, H. W., Evans, S., Neely, A., Greenough, R., Peppard, J., Roy, R., Shehab, E., Braganza, A., Tiwari, A., Alcock, J. R., Angus, J. P., Bastl, M., Cousens, A., Irving, P., Johnson, M., Kingston, J., Lockett, H., Martinez, V., Michele, P., Tranfield, D., Walton, I. M., Wilson, H. (2007): State-of-the-art in product-service systems, Proceedings of the Institution of Mechanical Engineers, Part B: Journal of Engineering Manufacture, Professional Engineering Publishing, Volume 221, Number 10 / 2007, pp. 1543 – 1552.

[4] Meier, H., Uhlmann, E., Krug, C. M., Völker, O., Geisert, C., Stelzer, C. (2009): Dynamic IPS² Networks and Operation Based on Software Agents, In: Proceedings of the 1st CIRP IPS² Conference (CIRP Industrial Product-Service Systems), 1-2 April 2009, pp. 305 – 310.

[5] Tan, A. R., McAloone, T. C. (2006): Characteristics of strategies in Product/Service-System Development, International Design Conference – Design, May 15-18, Dubrovnik, Croatia.

[6] Matzen, D., Tan A. R., Andreasen, M. (2005): Product/Service-Systems: Proposal for Models and Terminology, 16. Symposium "Design For X", October 13-15.

[7] Stähler, P. (2002): Geschäftsmodelle in der digitalen Ökonomie, Josef Eul Verlag, Lohmar, Germany.

[8] Bach, N., Buchholz, W., Eichler, B. (2003): Geschäftsmodell für Wertschöpfungsnetzwerke, Gabler Verlag, Wiesbaden, Germany.

[9] Scheer, C., Deelmann, T., Loos, P. (2003): Geschäftsmodelle und internetbasierte Geschäftsmodelle – Begriffsbestimmung und Teilnehmermodelle, In: Working Paper of the Research Group Information Systems & Management (Paper 12), Johannes Gutenberg-Universität, Mainz, Germany.

[10] Müller, P., Stelzer, C., Geisert, C., Uhlmann, E., Knothe, T. (2008): Kernprozesse hybrider Leistungsbündel, wt Werkstattstechnik online, Jahrgang 2008 (7-8/2008), pp. 581 – 586.

[11] VDI-Guideline 3422 (2002): Technical availability of machines and production lines, Beuth Verlag GmbH, Berlin, Germany.

[12] Gegusch, R., Geisert, C., Höge, B., Stelzer, C., Rötting, M., Seliger, G., Uhlmann, E. (2009): Multimodal User Support in IPS² Business Model, In: Proceedings of the 1st CIRP IPS² Conference (CIRP Industrial Product-Service Systems), 1-2 April 2009, pp. 125 – 131.

[13] VDI-Guideline 2221 (1993): Systematic approach to the development and design of technical systems and products, Beuth Verlag GmbH, Berlin, Germany.

[14] Grudzien, W. (2002): Beitrag zur Steigerung der Nutzenproduktivität von Ressourcen durch eine Life Cycle Unit, Fraunhofer IRB Verlag, Stuttgart, Germany.

[15] Jacob, E., Manner, P. (2009): Motorspindelreparatur ist preiswert, erhöht aber die Stillstandzeit, MM Das IndustrieMagazin, Vogel Business Media GmbH & Co. KG, Würzburg, Germany.

Eco-Innovative Design of Product Service Systems by using the Substance-Field Analysis Method

Jahau Lewis Chen[1], Shih-Che Huang[1]

[1] Department of Mechanical Engineering, National Cheng Kung University, Tainan, Taiwan

Abstract

This paper presents an eco-innovative design methodology to support designers in developing product service systems (PSS) by using functional analysis and the TRIZ substance-field model. This paper collects existing PSS cases and identifies the relation between products and services. The defects existing in current PSS models within the relation between products and services are discussed. This paper also analyzes the characteristics of products, services and the environment. By using this method, the designer is able to identify the problems step by step and get some innovative suggestions.

Keywords:

Product Service Systems (PSS); Substance-field; TRIZ

1 INTRODUCTION

The development of technology plays a crucial role in modern economic growth, but is also the key factor of environmental crisis. It is usually emphasizing the novelty and economic usefulness of an innovation product but neglects its environmental impact. Currently, many eco-design methods have been combined with eco-innovation. However, although there is an eco-design method to design products, it's hard to say that all the eco-products are used by an eco-efficiency concept. Therefore a new concept of product service system (PSS) that defined a system of products, services, and supporting infrastructure to fulfil a customer need with less environmental impact than traditional ones is proposed [1-3]. Though there are many PSS examples that can be followed, it still needs a method to provide product designers with some inspirations.

The use of the TRIZ method in PSS development methodology is proposed by Abdalla [4-5]. Lin and Su [6] apply the TRIZ inventive principles to create new services. Kondoh and his co-authors [7] have proposed some rules for planning new eco-business or PSS. Chen and his co-authors [8] have proposed TRIZ ideality and trend of evolution in eco-business. Chen and Huang [9-10] have presented an eco-innovation method to support designers to develop product service systems (PSS). This method is based on TRIZ 40 inventive principles and contradiction matrix tool. Current PSS cases are collected and divided into five categories. A relationship table between TRIZ inventive principles and PSS cases is established in this paper to present a new design guideline for PSS. Some green evolution rules are observed and identified by investigating patterns and lines of technological evolution of different PSS. Chen and Li [11] collected and analyzed current PSS cases to identify consumers' use habits in the PSS cases. Find out the relations between product and service in those cases. Several design tables are developed and used as tools for PSS eco-innovation.

In this paper, an eco-innovative design methodology to support designers in developing product service systems (PSS) by using functional analysis and the TRIZ substance-field model is proposed. PSS innovation example was used to illustrate the effectiveness of the proposed methods.

2 SUBSTANCE-FIELD ANALYSIS FOR PSS

2.1 Substance-field analysis

Substance-field analysis (Su-field analysis) is a modeling approach in TRIZ for the analysis and innovation of physical phenomena in product systems. This method is performed by building a substance -field model for design problems and analyze the modification possibility of this model to innovate new solutions for design problems. Substance is any object or tool within a product system. Field refers to the energy required for the interaction between two substances. Two substances (objective and tool) and the energy of their interaction are the minimal requirements to build a minimal Su-field model for performing one function, as shown in Figure 1. The notation of three situations in Su-field model is illustrated in Table 2. Several standard modification rules can help the designer to modify the Su-field model to eliminate harmful effects [12-15].

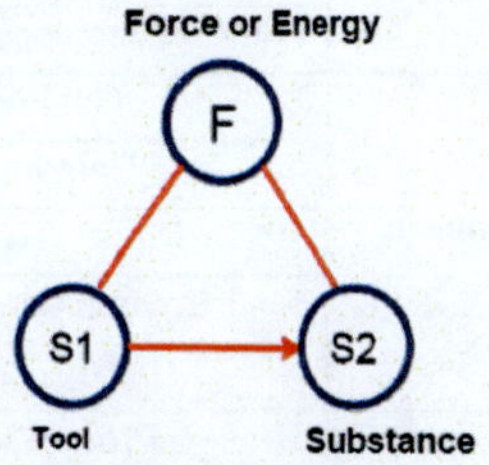

Figure 1: A complete Su-field triangle.

Symbol	Meaning
——→	Desired Action
- - - →	Inadequate Useful Action
∿∿→	Harmful Action

Figure 2: Notation of three situations in Su-field model.

J. Hesselbach and C. Herrmann (eds.), *Functional Thinking for Value Creation: Proceedings of the 3rd CIRP International Conference on Industrial Product Service Systems, Technische Universität Braunschweig, Braunschweig, Germany, May 5th - 6th, 2011*,
DOI 10.1007/978-3-642-19689-8_13, © Springer-Verlag Berlin Heidelberg 2011

2.2 Substance-field analysis in PSS

For using the Su-field analysis technique with PSS problems, we redefine a complete PSS Su-field triangle model as shown in Figure 3. The meanings of substance and field in PSS are redefined as illustrated in Table 1 and Table 2, respectively.

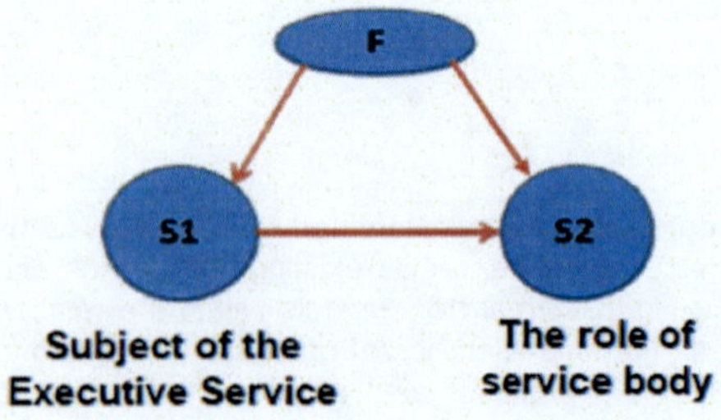

Figure 3: A complete PSS Su-field triangle.

Type of Substance	Examples
Product	Product or Function
System Network	Producers, consumers, the company role or service
Capital construction	Roads, waste recovery system

Table 1: Substances used in PSS problems.

Field	Service Function
Supplier	Planning and Setup
	Maintenance and Repair
	Maintenance (Update, Warranty)
Customer	Leasing
	Sharing
	Pooling
Technology	New Service
	Replace
	Multi-function Service
	New Technology (Web, e-service)
Product Potential	Get Information
	Remanufacturing
	Reuse
	Environmental Protection
	Resale
Professional Competitiveness	Provide Information
	Human Power Service
	Technical Support
	Peripheral Service
	Personalized
	Easy to Obtain

Table 2: Field used in PSS problems.

The process of a Su-field analysis in PSS begins by defining the subject of the executive service (S1) and the role of the service body (S2). Next, the field (F) acting in substance S1 and S2 is identified based on the definition of Table 3. Then, the situation (desired action, inadequate useful action, or harmful action) between the field and the substances is identified. After building the Su-field model, one can obtain a solution from TRIZ standard solutions [12-15].

3 TOOLS FOR PSS ECO-INNOVATION

3.1 Function analysis

The function analysis [16] is proposed in this study to identify the relationship between different parts of a system. Function analysis is performed before doing the Su-field analysis. Functions are divided into useful functions (UF) and harmful functions (HF). As shown in Figure 4, parts A and B of a system are UF. Part C is HF. B is required for achieving function A. However, B will produce harmful function C simultaneously. Therefore, harmful function C will also produce harmful influences to function A. The result of influences can be the information to identify a Su-field model.

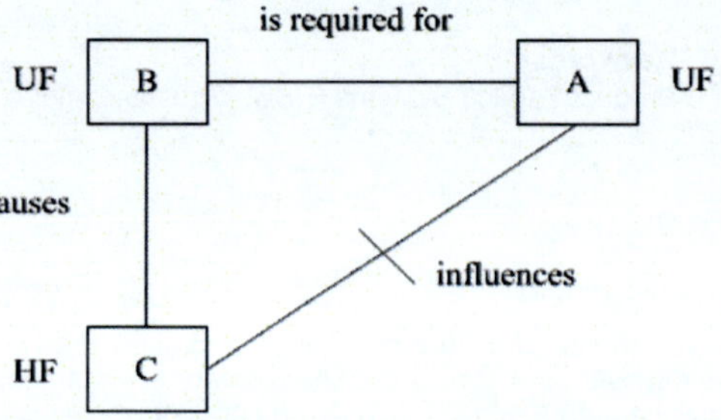

Figure 4: An example of function analysis.

3.2 Product characteristic factor level table of PSS

After having collected 109 PSS cases, an index system to exploit similarities among PSS cases is proposed for identifying most similar PSS cases in a database as eco-innovation direction for new PSS problems. An index system provides all the important characteristics required to classify different PSS groups. Each group is identified by a five-digit alphanumeric code. Each digit is a notable feature and represented by a different number for each different PSS according to its characteristics. The significance of each digit is represented in detail later and also shown in Table 3.

Five product characteristic factors for PSS are product property, product life, price, idle-using time, and using period. The level of each factor and the rank of each level are all shown in Table 3.

3.3 PSS case database

Each PSS case is recorded in the database with information about this PSS, such as the case number, the name of PSS case, the five product characteristic factors, the eco-efficiency and the service effectiveness categories, the main product, and the product design guide, as illustrated in Table 4. In Table 4, only a partial list of a PSS case database is shown. A detailed list of all 109 PSS case databases can be found in reference 17.

3.4 Eco-efficiency vs. service effectiveness table

The World Business Council for Sustainable Development (WBCSD) has pointed out seven major elements for a company in considering the eco-efficiency of developing environmentally friendly products or processes in order to reduce environmental impacts.

A. Reduce the material intensity of its goods and services

B. Reduce the energy intensity of its goods and services

C. Reduce the dispersion of any toxic materials

D. Enhance the recyclability of its materials

E. Maximize the sustainable use of renewable resources

F. Extend the durability of its products

G. Increase the service intensity of its goods and services

The seven elements of eco-efficiency mentioned above can be used as the design goal for PSS.

New service model categories for PSS based on effectiveness of service are proposed in this study. PSS cases are classified into six big service mode categories, such as environmental protection, saving of time and convenience, promotion competitive power, meets more needs, more specialized, and reducing cost, respectively. A position table for identifying the relationship between eco-efficiency and service effectiveness for all 109 PSS cases is shown in Table 5. The number in Table 5 is the PSS case number in the PSS case database. Detailed information about the case number shown in Table 6 can be obtained from Reference 17.

A. Product Property		B. Product Life		C. Price (NT)		D. Idle-Using Ratio		E. Using Period	
Level	Rank	Level	Rank	Level	Rank	Level	Rank	Level	Rank
Fixed/Solid, Waste Energy	**1**	1Week ↓	**1**	<100	**1**	<1	**1**	1 hr ↓	**1**
Fixed/Solid, Save Energy	**2**	1Week ~1Month	**2**	100~1000	**2**	1~2	**2**	1 hr~ 1 day	**2**
Movable, Waste Energy	**3**	1Month ~1Year	**3**	1000~10000	**3**	>2	**3**	1 day ~ 1 week	**3**
Movable, Save Energy	**4**	1Year ~10Year	**4**	10000~ 100000	**4**			1 week ~ 1 month	**4**
Produce Chemical Reaction During Use	**5**	10Year ↑	**5**	100000~ 10000000	**5**			1 month ↑	**5**
Service	**6**			>10000000	**6**				

Table 3: Product characteristic factor level table of PSS problems.

Case No.	PSS case	Characteristic factor level					Eco-efficiency vs. Service Effectiveness		Main product	Product design guide
		A	B	C	D	E	Eco-efficiency	Service Effectiveness		
7	Car surface treatment	3	4	3	3	1	G	5	Car surface	Df Repair
50	Complex Office	2	4	3	2	2	A	4	Office furniture	Df Disassembly, Modularity
61	GPS positioning	3	4	3	3	2	E	3	GPS	Reduce energy
62	Baggage transportation	4	4	3	3	3	G	2	Baggage	
98	Warm blanket by human being	2	4	3	2	2	G	4	Warm blanket	Change material

Table 4: Partial list of PSS case database table.

Service Effectiveness / Eco-efficiency	1.Environmental Protection	2.Saving Time	3.Promotion Competitive Power	4.Meet More Needs	5.More Specialized	6.Reducing Cost
A. Reducing Intensity of Material	13.70.84.	93.95.	21.27.58.93	39.41.50.57 .	13.18.52.	2.23.24.26. 29..30.40. 42.
B. Reducing Intensity of Energy	101.	94.	63.	11.	10.16.48.90 .91.108.	3.
C. Reducing Toxcity	5.13.64.86.87. 101.	20.	56.	74.	6.12.13.45. 47.	64.
D. Recycling of Material	1.14.32.33.38. 64.67.68.69.75 .84.85.	66.		41.	80.	1.38.64.
E. Sustainable Use of Resource	14.38.67.68.69 .85.86.87.92.	66.93.100.	61.83.93.	11.	4.15.80.	16.24.28.38 .42.49.65. 73.102.105.
F. Long Life of Product	14.72.92.	53.54.		46.55.74.76 .	37.45.59.91 .	23.40.
G. Increasing Intensity of Service	72.	8.20.43.44. 53.54.62.77 .81.94.95. 96.99.104. 106.107.	19.21.56.58. 71.82.	39.41.43.46 .52.54.57. 76.88.89.97 .98.109.	7.9.12.15. 16.22.25.34 .35.36.51. 78.79.103.	49.60.65.

Table 5: Relationship between eco-efficiency and service effectiveness.

4 ECO-INNOVATION OF PSS

4.1 Design process for PSS eco-innovation

The tools, such as Figure 1-4 and Table 1-5, the Su-field model, function analysis, categories, PSS cases, and eco-efficiency vs. service effectiveness table, for PSS eco-innovation developed in section 2 and 3 are organized as a design process for PSS innovation, as shown in Figure 5. As illustrated in Figure 5, the design procedure can be recognized as a three stages process. The significations of these three parts are explained in detail as follows.

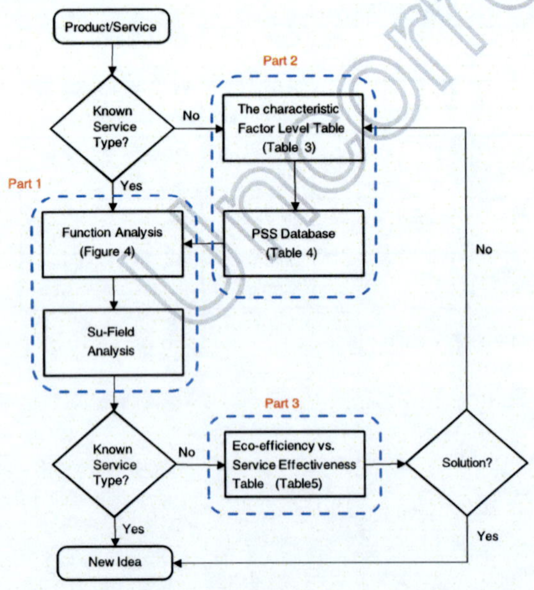

Figure 5: Flowchart of the PSS eco-innovation process.

4.2 First part: function analysis and Su-field analysis for PSS eco-innovation

This part is the major process in PSS eco-innovation. First, one can use the function analysis tool to realize one's problem, as presented in section 3.1. Next, by building the PSS Su-field model by using information from function analysis results, one can obtain innovative ideas from TRIZ standard solutions.

4.3 Second part: find similar PSS cases

If the service type of product is unknown, then the designer can first find similar PSS cases based on Table 3 and Table 4. The first step is to identify the level value of five product's characteristic factor from the definition of Table 4 for target product. Next, the designer can search the most similar PSS cases from the PSS case database of Table 5 with target product by comparing the level value of five product characteristic factors. The PSS cases with a low degree of difference will be the most similar cases. The process of searching the most similar PSS cases is divided into two stages. The first stage is the comparison of the first three characteristic factors (product property, product life, and price) which are related to product characteristics.. The degree of difference is calculated by the following equation.

$$\text{Product Degree of difference} = \sum_{i=1}^{3} | M_i - m_i | \qquad (1)$$

The second stage is the comparison of the last two characteristic factors (idle-using time, and using period) which are related to consumer habit characteristics.. The degree of difference is calculated by the following equation.

$$\text{Consumer Degree of difference} = \sum_{j=1}^{2} | N_j - n_j | \qquad (2)$$

After finding the most similar PSS cases, the designer can find suitable service types for his problem and process into the next step.

4.4 Third part: find eco-innovation idea by eco-efficiency vs. service effectiveness table

As the designer cannot obtain innovative ideas from the standard solution of Su-field analysis in part one, the designer can search the related cases from the eco-efficiency vs. service effectiveness table (Table 5). Eco-efficiency concept and service effectiveness concept can be obtained from function analysis, C and A, as illustrated in Figure 3. Therefore, the designer can try to obtain new idea for PSS eco-innovation from the other PSS cases with the same motivation of service category of the most similar PSS cases.

5 EXAMPLE

5.1 Problem description

Recently, the e-book is becoming a popular portable electronic device. It can replace traditional paper books and is an environmentally friendly eco-product. How to innovate PSS idea for the e-book to increasing its eco-efficiency and service effectiveness is the object of this e-book eco-design case. The effectiveness of the proposed PSS innovation method will be shown in this section.

5.2 Find similar PSS cases

By following the steps in Figure 5, one can first find the level value of characteristic factor for the e-book from the definition of Table 3, as shown in Table 6. Next, the designer can search the most similar PSS cases from the database of Table 4 with e-book by comparing the level value of five product characteristic factors between Table 6 and each case in Table 4.

Product characteristic	Product property	Product life	Price	Idle-Using ratio	Using period
Index value	3	4	3	3	2

Table 6: The characteristic factor level of an e-book.

The most similar cases in this problem with degree of difference equals 0 and 1 from two search stages are shown in Table 7. From the information in Table 7, the case 61 (GPS positioning) can be identified as most similar case with e-book for degree of difference equal 0 in two search stages.

1st stage / 2nd stage	Degree of difference =0	Degree of difference =1
Degree of difference =0	61. GPS positioning (New technology)	
Degree of difference =1	7. Car surface treatment (Technology support)	50. Complex office (Multi-function) 62. Baggage transportation (Human resource support) 98. Warm blanket by human being (Human resource support)

Table 7: The most similar PSS cases for e-book.

5.3 Function analysis and Su-field analysis

The function analysis of an e-book is illustrated in Figure 6. The harmful functions of an e-book are battery pollution and energy consumption. The useful functions of an e-book are the replacement of traditional paper, reducing tree cutting, and increasing environmental protection. However, battery pollution will influence environmental protection.

Based on the function analysis of Figure 5 and the PSS substance and field definition in Table 1 and Table 2, a Su-field model of an e-book is built, as shown in Figure 6. “Produce” and “new technology” are the available fields. The e-book producer, e-book, and consumer are identified as substances. This S-field model contains two fields and three substances. Since battery pollution and energy consumption cause environmental pollution, the “new technology” will produce harmful action from the e-book to the consumer, as shown in Figure 7.

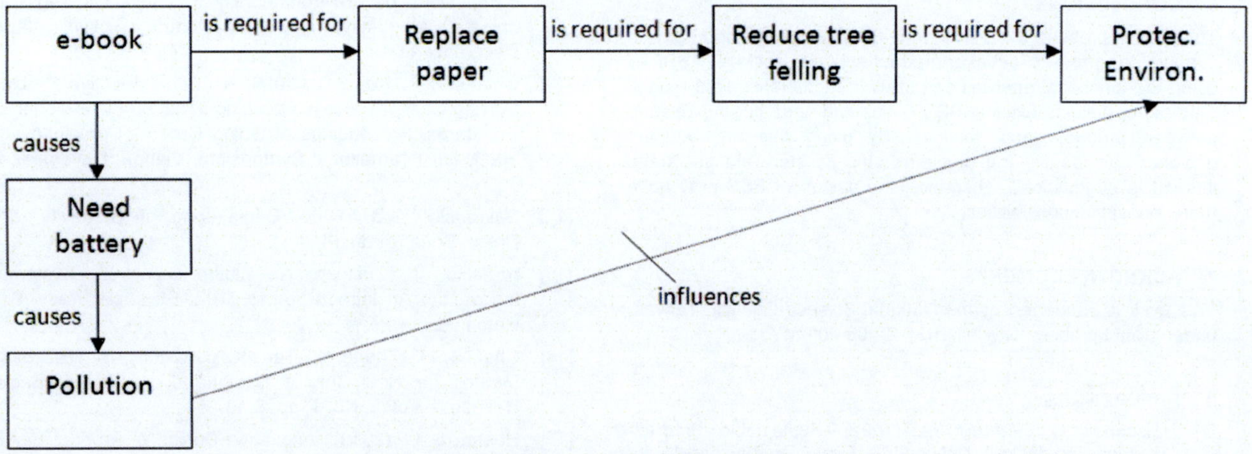

Figure 6: Function analysis of e-book problem.

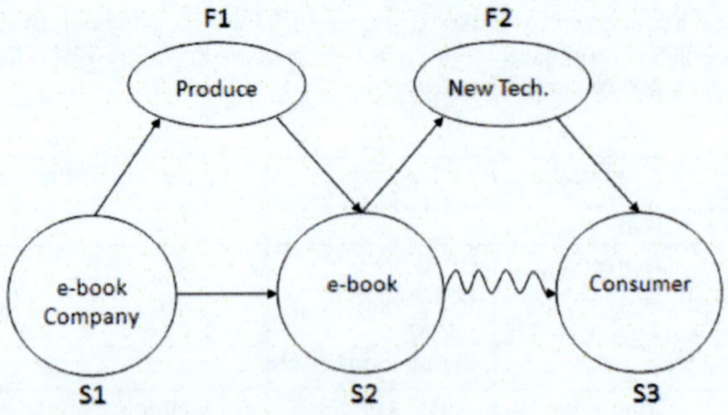

Figure 7: Su-field model of e-book.

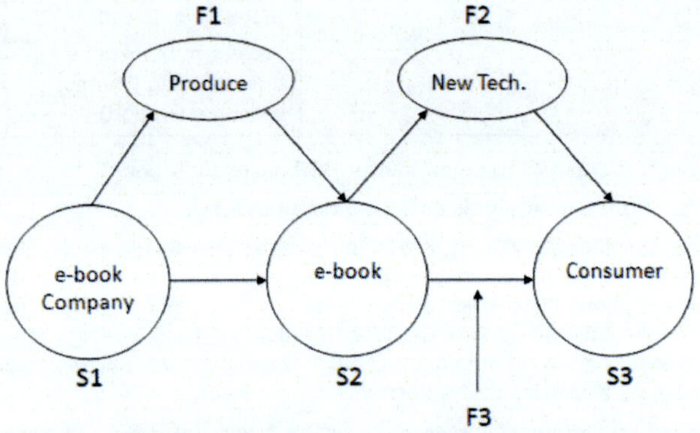

Figure 8: Improved Su-field model of e-book.

5.4 Eco-innovation by TRIZ standard solution

One of the TRIZ standard solutions [12-15] for Figure 7 is "increasing another field F3 to balance the field F2 which produces harmful action", as shown in Figure 8. An innovative idea is using another new technology – a solar battery (F3) instead of a traditional battery in order to reduce pollution.

6 CONCLUSIONS

This paper presented an eco-innovative design methodology to support designers to develop product service systems (PSS) by using the functional analysis and the TRIZ substance-field model. This paper collects existing PSS cases and finds out the relation between products and services. By using this method, the designer can identify the problems step by step and get some innovative suggestions. Therefore, the design of PSS may have more innovative possibilities.

7 ACKNOWLEDGMENTS

This work is supported by the National Science Council, Taiwan, under grant numbers: NSC96-2621-Z-006-001-MY3.

8 REFERENCES

[1] Tischner, U., Verkuijl, M. (2002): Sustainable Product-Service co-design Network - First Draft Report, In: SusProNet, Web: www.suspronet.org.

[2] Goedkoop, M. J., van Halen, C. J.G., Riele, H. R., Rommens, P.J.M. (1999): Product Service systems, cological and Economic Basics, In: Report for the Dutch ministries of Environment (VROM) and Economic Affairs (EZ), pp. 17-19,March.

[3] Mont, O. (2002): Clarifying the concept of product service system, In: Journal of Cleaner Production, Vol. 10, pp. 237-245, June.

[4] Abdalla, A., Bitzer B., Morton, D. (2004): Benchmarking TRIZ in the field of Product Service systems "PSS, In: Proceedings of the 4th ETRIA TRIZ future Conference, Florence, Italy, 3-5 Nov.

[5] Abdalla, A. (2006): TRIZ Innovation Management Approach for Problem Definition and Product Service Systems, Ihaker Verlag, Inc., Aachen, April.

[6] Lin, C. S., Su, C.T. (2007): An Innovative Way to Create New Services: Applying the TRIZ Methodology, In: Journal of the Chinese Institute of Industrial Engineers, Vol. 24, No. 2, pp. 142-152.

[7] Kondoh, S., Kurakwa, K., Kato, S., Umeda, Y., Takata, S. (2006): Analysis of Key Success Factors for Eco-Business through Case Studies in Japan, In: Proceedings of the 13th CIRP International Conference on Life Cycle Engineering, Leuven , Belgium . May 31st to June 2nd.

[8] Chen, J. L., Masui, K., Kondoh, S., Mishima, N. (2006): Ideality and Trends of Evolution in Minimal Manufacturing and Eco-business, In: Proceedings of EcoDesign 2006 Asia Pacific Symposium Program, Tokyo, Japan.

[9] Chen, J. L., Huang, C.-P. (2009): A TRIZ Based Eco-Innovation Method for PSS, In: Proceedings of the 16th CIRP International Conference on Life Cycle Engineering, Cairo, Egypt, May 4-6.

[10] Huang, C.-P. (2008): The Innovation of Product Service System Based on TRIZ Method, Master Thesis, Department of Mechanical Engineering, National Cheng Kung University, June 2008 (in Chinese).

[11] Chen, J. L., Li, H.-C. (2010): Innovative Design Method of Product Service System by Using Case Study and TRIZ Method, In: Proceedings of the second CIRP International Conference on Industrial Product Service Systems (IPS² 2010), April 14-15, Linkoping, Sweden.

[12] Chang, H.T., 2005, The Study of Integrating Su-Field Analysis Modeling with Eco-Innovative Concept for Product Design, In: Proceedings of the Fourth International Symposium on Environmentally Conscious Design and Inverse Manufacturing, Ecodesign05, Tokyo, Japan, December 11-14.

[13] Chang, H.T., Ko, Y.C., 2006, An Eco-Innovative Problem-Solving Design Process Combing TRIZ Su-Field Model and Standards, Proceedings of Going Green - Care Innovation 2006, 6th International Symposium, Vienna, November 13-1.

[14] Savransky, S.D., 2000, Engineering of Creativity, CRC Press, Boca Raton, FL.

[15] Terninko, J., Zusman, A., Zlotin, B., 1998, Systematic Innovation: An Introduction to TRIZ, St. Lucie Press, Boca Raton, FL.

[16] Chai, K. H., Zhang, J., Tan, K. C. (2005) : A TRIZ-Based Method for New Service Design, In: Journal of Service Research, Vol. 8, No. 1, pp. 8-48.

[17] Huang, S.-C. (2010): Innovative Design of Product Service Systems by Using Substance-Field Analysis Method, Master Thesis, Department of Mechanical Engineering, National Cheng Kung University, June (in Chinese).

Case Based Idea Generation Method for Sustainable Businesses

Shinsuke Kondoh[1], Nozomu Mishima[1]

[1] National Institute of Advanced Industrial Science and Technology (AIST), 1-2-1 Namiki, Tsukuba, Ibaraki Japan

Abstract

Sustainable business design has gained growing interest recently. However, it is often difficult to clarify which business activities actually contribute to (or harm) the sustainability of the earth due to high complexity of environmental and economical system of our society. In order to help a business to generate sustainable business ideas handling this complexity in a systematic and comprehensive manner, this paper proposes a method to formulate a wide variety of causalities in our society into a cause-effect pattern library, based on which sustainable business case base is constructed. An idea generation method is proposed using the case base.

Keywords:

Sustainable Business Design; Industrial Product Service Systems; Cause-Effect Pattern Library

1 INTRODUCTION

Sustainable development (SD) has become a critical issue due to growing concern about environmental problems in recent years and firms are obliged to justify their presence in society more than ever before [1]. From the point of view of sustainability as well as an economic perspective, business planning, monitoring, evaluation, and communication with customers and the wide variety of other stakeholders are becoming more and more important. In this context, a growing number of methods and tools for sustainable product and business design ([2]-[9]) are being proposed.

However, it is often difficult to determine which business activities actually contribute to (or harm) the sustainability of the earth due to the highly complex nature of the environmental and economical systems of our society. The rebound effect is a typical example that shows such complexity. One can think that the development of energy-saving technology such as new vehicles with better mileage may contribute to a reduction of energy consumption in our society. However, such developments sometimes result in increased energy consumption. One explanation for this is that the diffusion of energy efficient technology encourages greater usage of services that consume energy. Thus, consideration of the indirect (delayed) influence as well as direct (immediate) influence of energy efficient technology are both indispensable for realizing sustainable development in this case.

In order to handle such complexity in a systematic and comprehensive manner, businesses need to think upstream in cause-effect chains and consider indirect (delayed) influences as well as direct (immediate) ones in their decision making and communication processes. To support such upstream thinking, collecting a wide variety of cause-effect chains based on existing sustainable business cases holds considerable promise.

The objective of this study is to propose a case-based idea generation method for sustainable business. Firstly, cause-effect patterns in various successful sustainable businesses, including Industrial Product Service Systems [3], are identified and formulated into cause-effect pattern library. Then, using information contained in the library, businesses can generate practical sustainable business ideas handling the complexity in a systematic and comprehensive manner.

To this end, this paper first discusses the prerequisites for sustainable businesses and characterizes its ideal state using a causal loop diagram (CLD) [10]. Then, we classify the gaps between ideal sustainable business (ISB) and conventional ones into three categories, and explain how to structure a cause-effect pattern library focusing on the gaps. Finally, we propose an idea generation method for sustainable businesses using the library, and discuss future needed developments.

2 IDEA GENERATION METHOD FOR SUSTAIANBLE BUSINESSES

2.1 Causal-loop diagram (CLD)

In order to represent complex cause-effect chains in economic and environmental systems, we introduce a causal-loop diagram (CLD) in this study. A CLD is a kind of system dynamics (SD) [10] tool. The basic idea behind SD is that every action triggers a reaction. This reaction is called feedback, and the behavior of a system is expressed by a set of circles of causality. This set of circles is called a causal-loop diagram (CLD). CLDs contain two types of feedback; namely, (i) reinforcing feedback (R) and (ii) balancing feedback (B). Reinforcing feedback accelerates the existing trend of a process. In contrast, balancing feedback counteracts and opposes the change.

Figure 1 shows examples of CLDs. The arrows indicate causal relationships. The + or - sign at the arrowhead indicates whether the effect relates positively or negatively to the cause, i.e. whether an increase in the cause leads an increase or decrease in the result. Dotted arrows represent delays in the processes. As shown on the left in Figure 1, if a firm lowers its price to gain market share, its competitors also lower their prices and force the firm to lower its price again. Thus, the price becomes lower and lower. This is an example of reinforcing feedback. As shown on the right in Figure 1, the higher the price of a commodity becomes, the lower the demand becomes and the more the stock increases. This leads to

J. Hesselbach and C. Herrmann (eds.), *Functional Thinking for Value Creation: Proceedings of the 3rd CIRP International Conference on Industrial Product Service Systems, Technische Universität Braunschweig, Braunschweig, Germany, May 5th - 6th, 2011*,
DOI 10.1007/978-3-642-19689-8_14, © Springer-Verlag Berlin Heidelberg 2011

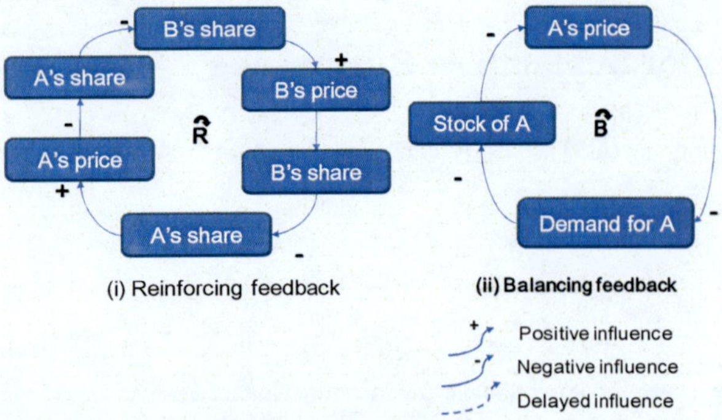

Figure 1: Examples of CLD.

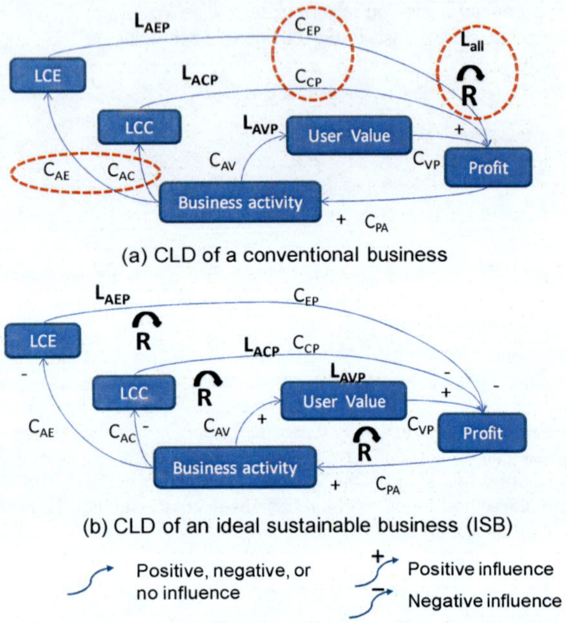

Figure 2: CLD for a sustainable business.

pressure to lower the price of the commodity and the price will fall. This is an example of balancing feedback. In some cases, a feedback or reaction does not occur immediately, so delays in processes should also be identified.

2.2 Representation of and prerequisites for sustainable business

CLD of sustainable businesses.

Since all businesses are driven by profit, causality among life cycle environmental load (LCE), life cycle cost (LCC), user value (UV), profit, and business activities should be taken into account in the process of designing sustainable businesses. Thus, the CLD of a sustainable business should include seven causal links; C_{AE}, between business activity and LCE; C_{EP}, between LCE and profit; C_{AC}, between business activity and LCC; C_{CP}, between LCC and profit; C_{AV}, between business activity and user value; C_{VP}, between user value and profit; and C_{PA}, between profit and business activity as shown in the upper part of Figure 2. These links can form three feedback loops; L_{AEP}, consisting of C_{AE}, C_{EP}, and C_{PA}; L_{ACP}, consisting of C_{AC}, C_{CP}, and C_{PA}; L_{AVP}, consisting of C_{AV}, C_{VP}, and C_{PA}. Each can be either a reinforcing or a balancing (immediate or delayed) feedback loop.

The analysis of 130 Japanese sustainable businesses [11] have identified four kinds of UV ; (i) improvement of service quality, (ii) reduction of user's cost, (iii) avoidance of user's risks, and (iv) improvement in user's eco-image from society and market. All types of UVs should be taken into account.

Requirements and ideal causality configuration of sustainable businesses

Theoretically, there exist 3^7 causality configuration patterns because there are three possibilities, no-influence, positive influence, and negative influence, for each causal link. In practice, however, the numbers of configuration patterns describing conventional and sustainable businesses are much smaller.

In order to accelerate a certain business activity, the overall causality trend L_{all}, which is the resultant of L_{AEP}, L_{ACP}, and L_{AVP}, should form a reinforcing feedback loop. In addition, profit and investment in business activity generally increases as the user value it provides increases so C_{VP} and C_{PA} are positive influences. These are prerequisites for success in conventional businesses.

Since sustainable businesses can be defined as businesses that, directly or indirectly, lead to a reduction in environmental load and its cost to society through their business activities, their causality configuration patterns are more limited than those of conventional businesses. In addition to the prerequisites for conventional businesses, C_{AE} and C_{AC} should have negative influences in sustainable businesses. The ideal sustainable business (ISB) is regarded as one for which the L_{AEP}, L_{ACP}, and L_{AVP} loops are reinforcing loops as shown in the lower part of Figure 2. This model shows that an ISB increases user value and decreases both societal environmental load and cost simultaneously, and all of these effects increase profit, which may promote investment in the business's activities. Causality configuration patterns for conventional businesses, sustainable businesses, and ISB are summarized as shown in Table 1. Characters +, -, 0, R, B in each cell of the table denote positive influence, negative influence, no influence, reinforcing feedback loop, and balanced feedback loop, respectively.

2.3 Structuring sustainable business case base

As mentioned in the introduction section, a cause-effect pattern library should be structured focusing on how each successful case overcomes the gaps between the conditions for ISB and conventional businesses. The gaps are classified into three types as shown in Figure 2;

(i) Insufficient or no C_{AE} or C_{AC}

Since every business pursues profit, its activity does not necessarily reduce society's LCE and LCC. Therefore, the first gaps to be overcome are C_{AE} and C_{AC}, which should be negative influences.

(ii) Insufficient or no C_{EP} or C_{CP}

Even if a business reduces the LCE and LCC of its products, the reduction does not necessarily increase the firm's profit. Green consumers might accept environmental conscious products, but their market segment is not generally large enough to insure success. Therefore, how to relate societal LCE and LCC reduction to profit is an important issue for sustainable business design.

(iii) Insufficient drivers

As in conventional businesses, in order to drive sustainable businesses, the overall causality trend L_{all} should form a reinforcing loop even if the businesses incur additional costs to reduce their societal LCE through actions such as providing cleaner end-of-life treatment for their products.

We constructed a cause-effect pattern library based on an analysis of 130 sustainable business cases in Japan [11]. Table 2 summarizes typical examples. Taken from these examples, Figure 3 depicts a common problem pattern where adoption of energy efficient but expensive technology causes a delay in C_{AC}. This can be categorized as a gap of the 1st type, insufficient C_{AE} or C_{AC}. Introduction of more energy efficient technology like photovoltaic electricity and hybrid vehicles promises to reduce both users' and society's LCE and LCC. However, energy-efficient products are sometimes more expensive than conventional ones which hinders their diffusion. This problem is represented in the CLD as shown in Figure 3 and its causality pattern is summarized as shown in the 5th row in Table 1; adoption of energy-efficient technology may increase the initial cost (price) for users although it may reduce the LCE and LCC for society. Thus, UV in the case is cost reduction and the delay in C_{AC} , which causes the delay in C_{AV}, is the gap to be overcome. One solution pattern for this problem is found in ESCO businesses [12]. In ESCO business, the business provides

No.	Business case		User	UV	L_{AEP}		L_{ACP}		L_{AVP}		C_{PA}	L_{all}
					C_{AE}	C_{EP}	C_{AC}	C_{CP}	C_{AV}	C_{VP}		Overall trend
1	Situations in conventional business		[redacted]		+,–, or 0	+,–, or 0	+,–, or 0	+,–, or 0	+,–, or 0	+	+	R
3	Requirements for susitainable business				–	+,–, or 0	–	+,–, or 0	+,–, or 0	+	+	R
4	Condtions of Ideal Sustainalbe Business (ISB)				–	–	–	–	+	+	+	R
5	Insufficient (no) C_{AE} or C_{AC}	Selling energy–efficient but expensive technologies	Factory, office building	Reduction in energy cost (result in a reduction in user's LCC in long run)	–	0	+ at the beginning, – in the end (delay)	0	– at the beginning, + in the end	+	+	B
6		ESCO business	Factory, office building	Reduction in energy cost without initial investment (result in a reduction in user's LCC in long run)	–	0	+ at the beginning, – in the end (delay)	0	+	+	+	R
7	Insufficient (no) C_{EP} or C_{CP}	Selling green electricity	Direct electricigy user	Improvement of user's eco–image from society with higher electricity price	–	– (insufficient)	+	–	–	+	+	B
8		Green electricity certificate business	Direct green electricity user	Improvement of user's eco–image from society with reasonable electricity price	–	–	+	–	+	+	+	R
9			Indirect green electricity users who buy green electricity certificate	Improvement of user's eco–image from society with reasonable electricity price	–							

Table 1: Causality configuration patterns of sustaianble and conventional businesses.

		Sustainable business case	Explanation of solution pattern
Gaps between ISB conditions and conventional business	Insufficient (no) C_{AE} or C_{AC}	Car sharing, household rental for students [14], etc.	Product sharing (lease, rental, product pooling) can reduce the production volume over the whole society by improving the utilization rate of under–utilized products.
		Proper treatment of EOL products that contain hazardous materials (e.g., fluorescent lamps [15])	Proper management and control of product life cycle can reduce societal LCE and LCC.
		ESCO [12], air conditioner with heat pump etc.	Improvement of environmental performance of products and services can reduce societal LCE and LCC.
		Reuse of toner cartridges [16], etc.	Reutilization of waste materials and energy can reduce societal LCE and LCC.
		Eco–drive training service for automobiles [17], etc.	Consultation and receipt of advice on proper product usage can reduce societal LCE and LCC.
	Insufficient (no) C_{EP} or C_{CP}	Zero emission factory, etc.	Cooperation among different business sectors can sometimes contribute to societal reduction of LCE and LCC.
		Carbon tax, etc.	Proper legislation, taxation and subsidy can increase the profit from the reduction in societal LCE and LCC.
		Green power certification business [13]	Enlargement of the green market size is effective for strengthening C_{EP}.
	Insufficient driver	Carbon offset products, selling automobiles with maintenance contracts [18], etc.	Proper combination of various user value (improvement of service quality, cost reduction, improvement of image, and avoidance of risks) can make the overall causalities of the business into a reinforcing loop.

Table 2: Examples from a cause-effect pattern library.

energy saving service based on a long-term performance contract so that the user need not invest in necessary equipment at the beginning, thus overcoming the delay in C_{AC} and C_{AV} as summarized in the 6th row in Table 1.

Figure 4 depicts the green electricity certification business [13]. The upper part of the figure and the 8th row of Table 1 show the problem of this case, which can be categorized as a gap of the 2nd type, insufficient C_{EP} or C_{CP}.

Green electricity generated by hydro and wind turbine has a potential to reduce LCE of society. However, its power generation cost is more expensive than those of conventional ones and this hinders the diffusion of green electricity (L_{ACP} is balanced loop). Although some users prefer green electricity and are willing to pay more money for it, the volume of the users is insufficient to compensate the increase in the power generation cost because only limited users who live near from the plant can buy the green electricity. Thus, the insufficient C_{EP} is the gap to be overcome in the case. The solution idea of the business is depicted in the lower part of the Figure 4 and its causality pattern is summarized as shown in the 9th row in Table 1. In this business, green electricity certificates organization, which is a different organization from the power company, calculates the LCE reduction potential, divides LCE reduction certificate into small pieces and sells them to the customers who cannot directly use the electricity. The revenue is used to compensate the cost for green electricity. By that way, C_{EP} is strengthened in this example.

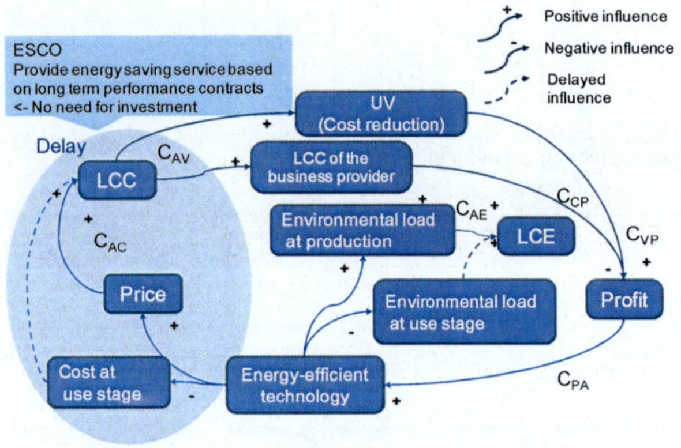

Figure 3: Introduction of energy efficient but expensive technology (delay in C_{AC}).

2.4 Sustainable business design and communication procedure

The procedure by which the cause-effect pattern library can be used to design sustainable businesses (SB) and communicate the design with pertinent stakeholders is summarized as follows;

Step 1: Identification of current business environment

First, the business designer identifies a set of key parameters that describes the given business environment in terms of UV, LCE, LCC, and profit, with causality among these parameters and describes the situation using a CLD, so that he/she can decide the types of causality among the seven causal links and identify the gaps between the conditions necessary for an ISB and his/her business environment.

Step 2: Idea generation for a sustainable business

The next step is to generate the ideas that overcome the gaps identified in step 1. By referring to the cause-effect pattern library, which is structured based on the gaps, as discussed in section 3.1, a designer can easily find adequate solution patterns for his/her problems.

Step 3: Validation of CLD through communication with stakeholders

Since the business model developed in step 2 may contain a variety of technological, economical, and social causalities, its validity and feasibility should be evaluated through discussion with experts in various fields. In addition, in order to implement the business model, negotiation among all the business's stakeholders is indispensable. Thus the model should be simplified for this purpose after evaluating its validity and feasibility.

Step 4: Evaluation of business ideas

After evaluating the CLD of the business model, the performance of the business should finally be evaluated. In order to calculate the LCE, LCC, and profit of the business, the CLD of the business model should be interpreted as a life cycle model (LCM) for its products, defining all material and energy flows related to them. Based on the LCM, the LCE and LCC of the products are calculated by using conventional life cycle assessment (LCA) and life cycle costing (LCC) methods.

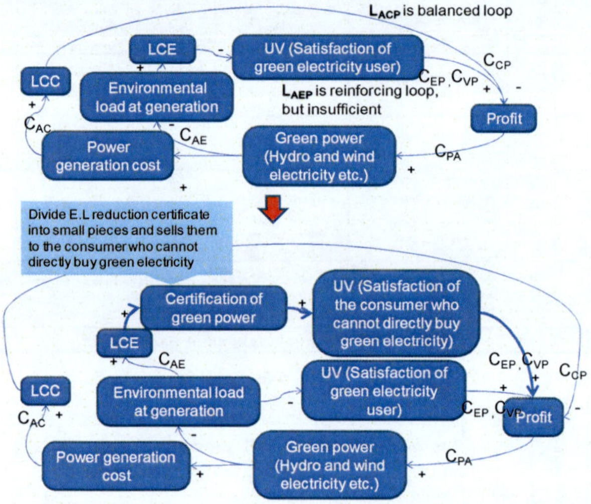

Figure 4: Green power certification business (Insufficient C_{EP}).

3 CASE STUDY

In order to illustrate how to use the library to come up with sustainable business ideas (step 1 and 2 in section 3.2), we present a case study of an energy-efficient vehicles business (EcoFleet) in this section. First, the designer identifies his/her customers, their user value, and causal links among business activities, UV, LCE, LCC, and profit as shown in the upper part of Figure 5. Although energy-efficient vehicles such as electric vehicles and hybrids have the possibility to reduce LCE and LCC in many cases, they are more expensive than conventional vehicles. In addition, environmental reduction through replacement of vehicles with more energy efficient ones has limited benefit to the firm's customers. Thus, the gaps to be overcome in this case are *insufficient* C_{AC} and *insufficient* C_{EP}. Referring to the cause-effect pattern library, the designer finds two cases (ESCO businesses and a green power certification business) that overcome these respective gaps. The business idea they arrive at by applying the solution patterns from the successful businesses to this case is summarized as shown in the lower part of Figure 5:

The 'EcoFleet' business provides a gasoline saving service to carriers based on long-term performance contracts similar to those used by ESCO businesses. First, EcoFleet determines the best mix of multiple types of vehicles based on the customer's fleet usage pattern, and replaces some of them with electric vehicles and plug-in hybrid vehicles to reduce the fleet's gasoline consumption. The investment required for replacement is covered by reduced gasoline consumption and there is no need for initial investment by the customer. In addition, to enhance the profit from the reduction of environmental load, green electricity such as solar or wind power is used to charge the batteries of electric and plug-in hybrid vehicles. Selling green electricity certificates to electricity users who cannot use green electricity directly increases the profit that can be earned from the environmental impact reduction.

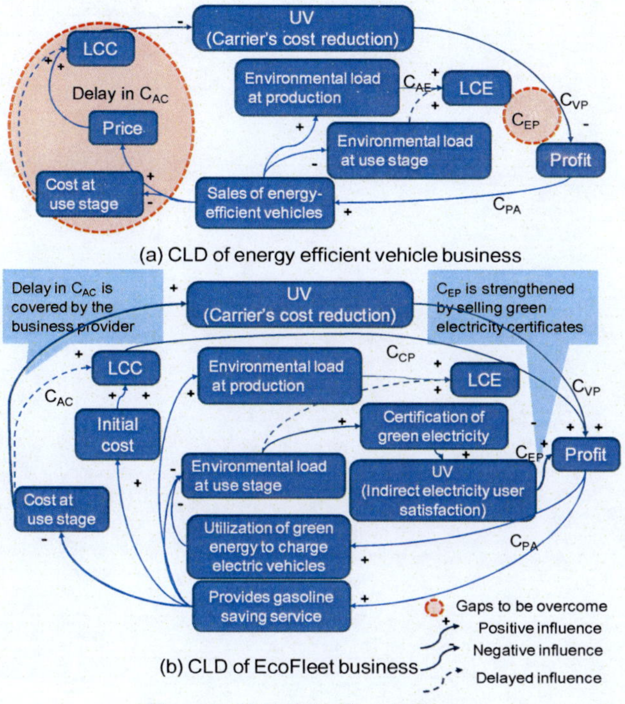

Figure 5: CLD of EcoFleet business.

4 ARCHITECUTURE OF THE SUSTAIANBLE BUSINESS DESIGN SUPPORT SYSTEM

To support a case based design for sustainable businesses, its implementation as software tool is quite effective. Figure 6 shows the architecture of the system. The system consists of three sub modules; namely, CLD editor (CLDE), life cycle modeler (LCM), and sustainable business evaluator (SBE). CLDE is the module that directly supports the design flow described in section 2.4. CLDE provides a designer with a CLD template of sustainable business, which includes the seven causal links among business activity, LCE, LCC, and profit, and a designer edit the template to represent his/her given business environment. After representing a given business environment on CLDE, it identifies the gaps between ISB and the business environment and provides a designer with a set of sustainable business cases that overcomes the gaps. Referring the provided cases, a designer generates an idea of sustainable business and reedit the CLD of his/her business to represent the idea.

LCM helps a designer to interpret CLD of the business to a set of material and energy flow throughout the whole product life cycle. Based on the life cycle model, SBE helps a designer to evaluate the environmental and economic performance of the business. To this end, SBE includes three different tools, total performance analyzer (TPA) [19], discrete event simulator (DES), and multi agent simulator (MAS). TPA, which has been developed by authors group, is a simple environmental and economical performance evaluation tool throughout whole product life cycle. Since the business model has dynamic nature, DES is used for detail analysis of its performance. In addition, to represent the complex combination of behaviors of multiple stakeholders, MAS is also utilized [for example, 20]. SBE is also connected to a set of database including global statistics (e.g., Gloss Domestic Product, population, etc.), life cycle inventory of energy and materials, and the cost.

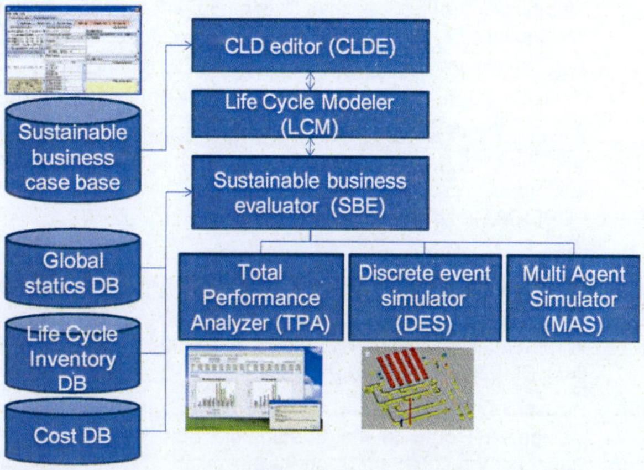

Figure 6: Architecture of the case based design support system for sustainable business.

5 CONCLUSION

This paper proposed a case based design method for sustainable business. We first explained how to describe various sustainable businesses by using CLD, and structured the sustainable business case base focusing on the gaps between the conditions of ISB and the conventional ones. Then, we explained idea generation procedure for sustainable business by using the case base and discussed the system architecture of its software system. Future work includes the following topics:

- Collection and analysis of existing cases:

  Collection and analysis of various exiting cases is required to build useful case base. Environmental (or sustainability) report of the businesses in various industrial sectors can be also a source for the library. In addition to the business cases, macro level causality patterns (e.g., market reaction to the environmental legislation or taxation, causal description of rebound effect etc.) are also indispensable for the method.

- Implementation of the system:

  Implementation of the system is also included in our future work. In addition to, TPA, sustainable business case base, and MAS, which have already been partially implemented, CDE and LCM should be implemented and integrated into the overall framework of the proposed system.

6 ACKNOWLEDGMENTS

This research is partially financially supported by Grant-in-Aid for Scientific Research (No. 20246130), JSPS, Japan.

7 REFERENCES

[1] Karl-Henrik, R. (2000): Tools and concepts for sustainable development, how do they relate to a general framework for sustainable development, and to each other?, in: Journal of Cleaner Production, Vol. 8 , pp. 243-254.

[2] Umeda, Y.; Nishiyama, T.; Yamasaki, Y.; Kishita, Y.; and Fukushige, S. (2009): Proposal of sustainable society scenario simulator, in: CIRP Journal of Manufacturing Science and Technology, Volume 1, Issue 4, pp. 272-278

[3] Meier, H.; Roy, R.; Seliger, G. (2010): Industrial Product-Service Systems - IPS2, in: CIRP Annals - Manufacturing Technology 59, pp. 607-627.

[4] Kobayashi, K.; Kumazawa, T. (2005): A Simulation-based Decision Support Methodology for Reuse Business, in: Proc. of Eco-Design 2005, 3A-1-2F (CD-ROM).

[5] UNEP, (1997): ECODESIGN: a promising approach to sustainable product and consumption, UNITED NATIONS PUBLICATION.

[6] Umeda, Y.; Nonomura, A.; Tomiyama, T. (2000): Study on life-cycle design for the post mass production paradigm, in: AIEDAM, Vol. 14, No. 2, pp.149-161.

[7] Wimmer, W.; Zust, R.; Lee, K. (2004): ECODESIGN IMPLEMENTATUION, Springer

[8] Hauschild, M.Z.; Wenzel, H.; Alting, L. (1999): Life Cycle Design. A Route to the Sustainable Industrial Culture? In: Annals of the CIRP 48(1), pp.93-96.

[9] Lindahl, M.; Olundh, G. (2001): The Meaning of Functional Sales, in: Proc. of the 8th CIRP Int. Seminar on Life Cycle Engineering, pp.211-220.

[10] Sterman, J. D. (2000):, Business Dynamics, McGraw-Hill.

[11] Kondoh, S.; Kurakawa, K.; Kato, S.; Umeda, Y.; Takata, S. (2006): Analysis of Key Success Factors for Eco-Business through Case Studies in Japan, in: Proceedings of 13th CIRP International conference on Life Cycle Engineering, pp. 275-280, Tokyo.

[12] Association for ESCO Business Introduction in Japan, (1998): Committee report for ESCO business introduction in Japan, Energy Conservation Centre Japan.

[13] Press release of Green Power Certification Council, http://eneken.ieej.or.jp/greenpower/eng/temp/04/01.pdf

[14] TOSHIBA SERVICE & ENGINEERING Co.,Ltd., http://www.toshiba.co.jp/tcn/pack/single/index_j.htm (in Japanese) (accessd on 12th, Jan. 2011).

[15] Matsushita Electric Works, Ltd., http://biz.national.jp/Ebox/akarianshin/index.html (in Japanese) (accessed 9th, July, 2008).

[16] Nikkei Business Publications, Inc., http://eco.nikkeibp.co.jp/article/report/20100614/104022/ (in Japanese) (accessed on 12th, Jan, 2011).

[17] Volkswagen Japan, http://www.volkswagen.co.jp/brand/news/2007/2007-1207.html (in Japanese) (accessed on 12th, Jan, 2011).

[18] Toyota Motor Corporation, http://toyota.jp/after_service/tenken/maintepack/index.html (in Japanese) (accessed on 12th, Jan, 2011).

[19] Kondoh, S.; Masui, K.; Hattori, M.; Mishima, N. (2007): Total performance analysis of product life cycle considering the deterioration and obsolescence of product value, in: Int. J. Product Development, Vol. 6, Nos. 3/4, pp.334.352.

[20] Kondoh, S.; Tateno, T.; Matsumoto, M. (2009): Multi-agent Simulation of Component Reuse Focusing on Variations in User Preference, in: CIRP Journal of Manufacturing Science and Technology, 1, pp. 287-293.

Ontology-Based Approach for Supporting Creativity in a PSS Design Methodology

Mircea Fulea[1], Stelian Brad[1]

[1] Research Lab of Competitive Engineering in Design and Development, Technical University of Cluj-Napoca, Romania

Abstract

In recent years serious efforts were directed towards building methodologies for PSS design. Regardless of these efforts, the degree of acceptance in industry for these methodologies cannot be clearly determined. Moreover, any PSS design methodology requires ideative processes and relies on creativity and tacit knowledge of the design team; this cannot be easily formalized within a traditional methodology. This paper approaches these issues by enhancing an already-existing methodology (based on acknowledged tools such as TRIZ and QFD) by adding a support ontology to it. The entire concept is then exemplified on a software-PSS project related to supporting innovation in SMEs.

Keywords:

PSS Innovative Design; New Software Development; Ontologies

1 INTRODUCTION

As sustainability tends to become the ultimate goal in almost every strategy, product-service systems start representing a new approach in many domains. This is also the case in software - the domain of this paper - thus allowing developers to better react to the increasing customer need of personalized software products. Software systems are flexible, complex [1] and imply a lot of planning and continuously overpassing a multitude of challenges [2]. Making such systems customizable by turning them into PSSs further complicates things.

There is a statement, originating from the 60s and nowadays still valid, saying that people do not buy products, but rather the expectation of benefits [3]. Meeting user expectations is the key factor that differentiates successful software projects from unsuccessful ones [4] [5] [6]. Judging after the alarming figures of software development project failure mentioned in [7] or [8], we might conclude that building successful software-PSSs is not at all an easy job.

As PSS design became an important concern for the scientific community, serious efforts have been directed in recent years towards building methodologies and setting up frameworks to face the challenge of building successful PSSs.

While using the latest software technologies, adopting user-centred design philosophies, standards or proven software development methodologies might seem a valuable option, achieving sustainability implies relying on the tacit knowledge of the development team and on the creativity and innovation potential of its members [9]. In simple words, one could gain a strategic advantage over the competitors by benefiting of a well-trained, innovative and creative human resource.

Yet, effectively employing such a human resource while designing PSSs needs a systematic support. According to [9], a design methodology to qualify as such a support would need to be collaborative, to facilitate a common vocabulary for all stakeholders (as highlighted in [2]), to be flexible, to be integrated within the development process and assist it's every phase, to support creativity and innovation and - most important - to offer to it's users the information that is relevant to a specific context.

2 THE PROBLEM

A review of existing engineering methods and tools for PSSs is done in [10]. Basically, they rely on service engineering, use cases and scenarios or life-cycle oriented design. The review carried out in [10] reveals also that these methods mainly deal with general service and PSS design rather than serve as a practical implementation; it is noted that they cannot yet readily assist developers in implementing sustainable PSSs [10], [11]. In an adjacent perspective, in [12] it is stated that, up to date, an insufficient number of studies have been conducted on methods to support the PSS design process.

The importance of the design process, when dealing with PSSs, is stressed out in [13], while also mentioning the importance of specific methodologies that should handle critical aspects of the design process. The same paper states that an accurate design process is mandatory for maximizing the innovation potential of a PSS. A design activity, based on scenarios, is also proposed. Another PSS design methodology, also scenario-based, is introduced in [14].

Market surveys and direct customer contacts are the starting point for the design methodology introduced in [15]. This data is further transformed, basically by ideative processes, into the final PSS concept. Creativity tool usage is encouraged for developing this concept, but no other note towards supporting creativity is made.

Another methodology to support a PSS design process is introduced in [12]. It deals with identifying existing conflicts in design solutions and with developing the basic strategies required for resolving them computationally. It makes use of lexical expressions of functions, ranges of design parameters and TRIZ. Although innovative and flexible, it still relies on the designer's tacit knowledge (actually on his experience and creativity), but without supporting it.

J. Hesselbach and C. Herrmann (eds.), *Functional Thinking for Value Creation: Proceedings of the 3rd CIRP International Conference on Industrial Product Service Systems, Technische Universität Braunschweig, Braunschweig, Germany, May 5th - 6th, 2011*,
DOI 10.1007/978-3-642-19689-8_15,

A methodology for concurrent planning and design of software-PSSs against multi-target functions, developed by the authors of this paper, is presented in [2]. The designer is guided by the methodology through 11 steps in order to obtain an overall solution to address the market needs. The methodology relies on well-known tools like AHP, QFD and TRIZ for data transformation and problem solving, supported by several ideative processes.

In [16] the application of specific methods in service design is shown, as it resulted from a survey within companies that provide services as core or add-on products, with various sizes and from different industry sectors. Surprisingly, one of the "classic" methods among the research community, QFD, is used by only 19% of the respondents, of which almost a half use it rarely. Other methods, like service blueprinting, FMEA or gap-analysis, seem to have a similar degree of acceptance within the industry. With this data in mind, it might be appropriate to conclude that, in the industry (even in the more dynamic ITC sector), the degree of acceptance of PSS design methodologies is even lower.

According to [12], in PSS design, ad-hoc processes - basically trial and error methods - are still dominant, and the quality of a design solution largely depends on a designer's experience and intuition.

All the methodologies mentioned above systematize and guide the design process, but lack an effective way of exploiting the tacit knowledge of the design team. Also, its creative potential is not supported and there is no emphasize on collaboration among its members. Yet, the real challenge is to identify **those** target functions, metrics, market needs, functionalities or whatsoever that get a unique and distinctive flavour for the PSS to be developed. Still, up to now, all this information is required by methodologies as a simple input data.

3 CREATIVITY IN THE DESIGN PROCESS

Creativity transcends human nature, academic disciplines and cultural boundaries; there is no accepted approach or methodology for studying, teaching or assessing creativity or creative processes [17] [18] [19]. Creativity does not necessarily consist of novel or new ideas, but also of a combination of what is already known - an unusual juxtaposition of current but formerly unrelated facts and ideas.

A creative process consists, according to [18], of four phases. The first one - the preparation phase - involves considerable effort in understanding the problem and searching for solutions. The second one - the incubation phase - implies much less conscious effort; an individual rather filters information from conscious awareness to the subconscious, to be used for creative insight. Creative insight comes in the third phase, called the illumination phase. Its outcome (the possible solutions) is then passed through the verification phase of testing, elaborating and developing (the fourth phase). It is worth noting that the four phases are not a linear path, but rather iterative [18].

Considering the ideative processes that accompany any design methodology, we can assert that only the first phase of a creative process might be supported by such a methodology; this could be done basically by finding a way of properly disseminating suitable information.

In line with the idea that collaboration is a must in the PSS design process, [20] states that although society often thinks of creative individuals as working in isolation, intelligence and creativity result mainly from interaction and collaboration with other individuals. The same paper asserts that much human creativity is social, arising from activities carried out in a context in which interaction with other people and artefacts that embody collective knowledge are essential contributors.

As highlighted in [2], in the equation of a PSS's success all the stakeholders must be included, because of their different perspective on what success represents. In this respect, a consortium that plans to develop a PSS could be associated to a community of interest, as described in [20]. One of the strengths of such a community is its potential for creativity, as different backgrounds and different perspectives can lead to new insights. Yet, there is also a barrier: its members usually fail to create a common ground and a shared understanding [20].

Another issue that needs to be taken into account in the equation of PSS design is that, in general, a successful project often starts as a result of individual creativity [20]. This starting point could be in the form of a requirement set, prototype, blueprint or feature list. When shared with other individuals (designers, users), they will be able to interact with this initial version and extend or use it creatively in more situations than the original designer had intended.

Although the importance of creativity in requirement engineering has been acknowledged, research into it is still at an early stage of theoretical development [19].

This paper tries to fill this gap by introducing an approach to support creativity, collaboration and tacit knowledge exploitation while designing successful software-PSSs.

4 THE SOLUTION

The aim of the PSS design methodology introduced in this paper (represented by the acronym *inSight*) is to be as accessible as possible, in order to increase the chances of being accepted by PSS practitioners. In addition to this, *inSight* should overpass the previously highlighted limits regarding creativity, collaboration and benefiting of tacit knowledge.

4.1 The basic idea

A natural choice was to enhance an already existing methodology, introduced by the authors of this paper and presented in reference [2]. It is straightforward, it can be easily integrated in any main software development methodology, and it makes use of "standard" tools like AHP, QFD or TRIZ.

The most important idea that *inSight* relies on is that a market need (that has to be satisfied) does exist at the ground of every commercial product or service. There are several options to formulate a solution for meeting a targeted market need. In addition to that, such a solution could be a pure service, a pure product or a mix of product and service. The methodology helps a designer to find the most suitable product and service combination, within some given constraints.

The methodology is built around an ontology that describes (through a series of concepts) the target domain of the PSS to be developed, the market needs, the stakeholders and the user particularities. The ontology also includes a data dictionary, in order to make sure that all the parties implied in the PSS design, development and exploitation processes "speak the same language".

The users of *inSight* are all the people involved in the design process of a PSS, but mainly the designers and the project manager.

The ideative processes required by *inSight* (which are marked with an "idea bulb" in Figure 1) rely on the experience and creative potential of the designer. They cannot be automated, but the methodology *supports* them through the ontology. This is done by offering relevant and specific information to the designer, thus supporting the first step of his creative processes.

4.2 Software ontologies

An ontology, in the software field, is a formal representation of knowledge as a set of concepts within a specific domain, along with relationships and attributes between such concepts [21] [9].

Ontologies are used to get and / or share a common understanding of the structure of information among people or software agents, to allow the reuse of a domain knowledge, to effectively analyze it and to make its assumptions explicit [21]. Thus, an ontology could perfectly support a data dictionary (used as an information repository), which is the ground for any effective blueprint [22].

Ontologies are reusable - therefore existing knowledge can be easily integrated into a starting project. Previous experience from older projects can also be integrated in this way, allowing a PSS-design support ontology to "learn" and evolve from one project to another. Another advantage of ontologies is that they are more flexible and more natural to a PSS designer than a classical database, because they are usually derived from a human language. There is also a drawback - ontologies are not yet very popular among software practitioners.

An ontology is a perfect mean of supporting the first phase of a creative process, because it allows reasoning and exploring information in a natural, human way.

4.3 The *inSight* methodology

A very important task of *inSight*, which is also the first to be carried out, is to gather as much data as possible for the concepts mentioned above. For example, if building a PSS related to organizational excellence, understanding the target domain would mean not only to gather information about what organizational excellence is, but also about how it is assessed, about models of organizational excellence that exist or about generating specific improvement projects based on the balance between business key components.

The data dictionary is more than a glossary of terms. It clearly includes definitions or assertions, but in addition it contains data structure descriptions or specific data transformations.

This task of the methodology should be repeated any time in the PSS lifetime, because of the dynamic character of the information it handles.

We should note that every piece of information used and implied by *inSight* is stored in the support ontology as an instantiation of a specific concept, represented by a class, and by object relations and properties.

Although the support ontology provides an initial set of concepts, extensively using it will certainly imply adding new concepts and adjusting relevance levels or relations between them.

To benefit from both individual and group creativity, the PSS design process starts by selecting a key designer, which should determine an initial PSS feature list, or an initial overview of the forthcoming PSS - the seed solution. The key designer should be, if possible, the most creative person involved in the design process.

To actually determine the PSS feature list, *inSight* uses the following concurrent algorithm (presented in Figure 1): for each market need, a target function is determined, along with its key requirements and key metrics. Further, through a series of data transformation, a local solution set is determined for each target function. A local solution set consists of local PSS features, each as a mix of product and service. The local solution sets are finally aggregated into an overall PSS solution, which is actually an overview of the PSS functionalities.

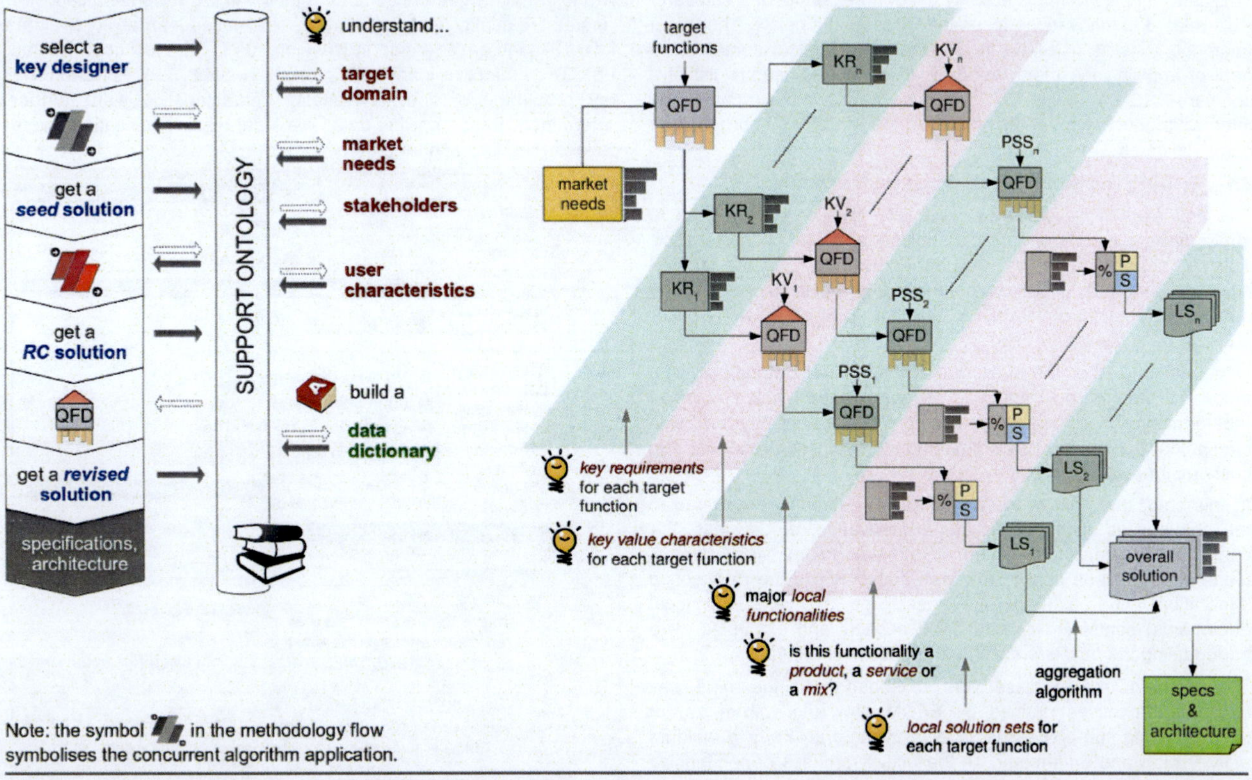

Figure 1: The *inSight* PSS design methodology: the left part represents the workflow; the right part represents the concurrent algorithm.

Each step of the algorithm that implies an ideative process is supported by examining all the relevant concepts from the support ontology and by reasoning, if suitable. All the information that the key designer produces throughout this algorithm is stored in the ontology (both output data of each step but also additional information related to the ontology concepts that the key designer considers useful).

The PSS design process continues by passing again through the concurrent algorithm, but this time in a group session and having the seed solution as a reference. The output is the release candidate (RC) solution, which is as well under the form of a set of PSS functionalities (whether as a product, service or as a mix).

Further on, the PSS functionalities are deployed against the stakeholder requirements. The goal of this task is to effectively check that all the requirements were properly considered. The QFD graphical support can be used to complete this step. A revised solution is developed and - based on it - the detailed architecture and the final requirement specifications are produced. All this data forms an effective blueprint that can adequately support developers actually architecting and implementing the PSS.

As a final note, the tasks described above are not linear. If the design team experiences a breakdown at some point, it should return to an earlier step and resume the methodology.

While *inSight* might be well used "on paper", an important aim of the authors is to automate its use through a software application. Therefore, it was built accordingly and resources will be invested in the near future to build this software application.

5 APPLICATION EXAMPLE

To illustrate how *inSight* is actually used, an example of a software-PSS related to innovation processes is presented below. The target audience consists of SMEs in the electromechanical domain. The goal of applying the methodology is to obtain the feature list of a software-PSS that helps SMEs to systematize and structure their internal product innovation processes, in order to improve their competitiveness.

5.1 Getting and sharing a common understanding

The first step of *inSight* is to instantiate the concepts related to target domain, market needs, stakeholders and user characteristics. This is actually done by creating individuals, in the support ontology, and adding adequate annotations, object properties and data properties. To effectively do this, the Protégé ontology editor [23] can be successfully used.

Understanding the target domain means understanding what innovation means, how it should be approached, what innovation methodologies and tools exist and how SMEs currently deal with innovation. Additionally, understanding the business processes and costs might also be necessary.

In this step, the most relevant innovation methods were added to the ontology as instantiations of the *InnovationMethod* class. For example, an individual named *TRIZ-Methodology* was created. General information about TRIZ was added (using the *Description* annotation) and external references (on the web or local documents) containing specific TRIZ aspects and examples were added using the *hasReference* data property.

The *DomainConcept* class was extended with the nine key business processes defined by EFQM [24] and with a set of corresponding sub-processes, in order to help an SME to build its activity map and understand all the activity/process dependencies that might exist.

Understanding the business processes of the SMEs not only helps themselves to innovate; it is a vital information for the PSS designer when deciding **which** system feature and **how much of it** will be provided as a service (in the final steps of the concurrent algorithm, see figure 1).

In a similar way, the market information was added using concepts like *LegalEntity*, *MarketActivities*, *Product*, object properties like *hasCompetitor*, *hasCustomer*, *hasFocusOn* and data properties like *hasFeature*, *hasGrowthRate* or *hasMarketShare*. The stakeholders and users were described in the same spirit.

The market needs for our software-PSS and their relative importance (determined using AHP) were identified as follows: RQ_1 - training on innovation (importance: 0.7), RQ_2 - systematizing organizational processes (importance: 2.1), RQ_3 - identifying product or process problems (importance: 2.3), RQ_4 - improving products or processes (importance: 3.9), and RQ_5 - systematizing the company knowledge (importance: 1.0).

Figure 2 shows how RQ_1 is effectively represented in the support ontology by using the Protégé ontology editor.

Actually, RQ_2 was identified due to exploring and detailing the ontology concepts of the domain, in the very first step of the methodology. As a side note, while initially stating the vision on the software PSS, nobody from the design team considered taking organizational processes into account.

5.2 Getting the seed solution

A key member of the design team, which has a strong collaboration with a SME in the electromechanical field, was nominated to produce the seed solution, applying the concurrent algorithm and using the information from the support ontology.

The following target functions (and their weight) were identified: TF_1 - increased quality (weight: 2.3), TF_2 - increased learnability (weight: 1.4), TF_3 - increased ergonomy (weight: 0.6), TF_4 - low cost (weight: 1.8), TF_5 - effective innovation projects (weight: 3.9). For each TF, key requirements and key quality characteristics were further determined. Based on this data, five local PSS specifications were defined (one for each target function).

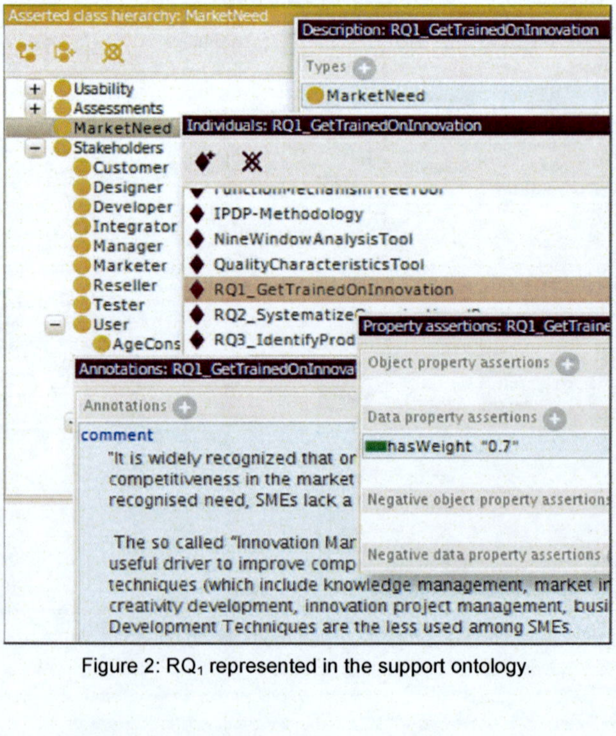

Figure 2: RQ_1 represented in the support ontology.

TF_1: effective innovation projects			$PSS_{5.1}$	$PSS_{5.2}$	$PSS_{5.3}$	$PSS_{5.4}$	$PSS_{5.5}$	$PSS_{5.6}$
Adaptability	2.3	p	3	1	9	3	9	1
		s	9	3	3	9	0	9
Development time	0.9	p	3	1	3	3	1	3
		s	0	9	0	3	0	3
Development cost	1.6	p	3	1	1	9	3	1
		s	1	9	3	3	0	3
Implementation time	0.6	p	0	0	1	3	3	3
		s	1	3	3	1	0	3
Implementation cost	2.0	p	1	0	3	3	3	1
		s	3	9	3	1	0	9
Quality of deliverables	2.6	p	1	1	9	9	9	1
		s	9	9	0	9	0	9
Product (p)	Mark		19,0	7,4	55,0	55,2	57,6	13,0
	Weight [%]		27%	9%	**74%**	**51%**	**100%**	15%
Service (s)	Mark		52,3	72,6	19,5	54,2	0,0	71,4
	Weight [%]		**73%**	**91%**	26%	49%	0%	**85%**

Table 1: Product and service distribution for TF_5.

The following criteria were defined (and ranked) for analysing the opportunity of implementing major PSS specifications as a product or as a service: adaptability (weight: 2.3), development time (weight: 0.9), development cost (weight: 1.6), implementation time (weight: 0.6), implementation cost (weight: 2.0), and quality of deliverables (weight: 2.6). Such an analysis, carried out for TF_5 - *effective innovation projects*, which is the most important target function, is summarized in Table 1.

The local PSS specifications for TF_5 are: $PSS_{5.1}$ - a business process feature, $PSS_{5.2}$ - an adaptive innovation roadmap feature, $PSS_{5.3}$ - an innovation toolbox, $PSS_{5.4}$ - a knowledge base and search centre, $PSS_{5.5}$ - an innovation project centre, and $PSS_{5.6}$ - a documentation centre (regarding both the system and innovation methodologies and tools).

While assessing the opportunity of implementing each local PSS specification as a product, service or mix, the key designer made use of the information in the support ontology regarding user particularities. As an example, implementing $PSS_{5.1}$ as a product wouldn't be the best option if we consider the organizational culture of an average SME (described by the *OrganizationalCulture* class in the support ontology); providing it as a service would be much more effective because the customer would benefit of adequate expertise regarding business processes and organizational issues.

The key designer then formulated local solutions, for each of the target functions, and aggregated them using the method from [2] into a global solution (the seed solution), based on nine major modules (briefly listed below, because of space constraints):

- $PSS_{f.1}$ - an installation and customization module, suited for the customer's environment (different languages and environments should be supported)
- $PSS_{f.2}$ - an integrated help module (covering system usage, innovation, business processes etc.)
- $PSS_{f.3}$ - an organizational excellence module, to understand and assess the customer's business processes
- $PSS_{f.4}$ - an innovation project management module, to choose the adequate innovation roadmap, project description and goal, user access rights
- $PSS_{f.5}$ - an innovation toolbox (implementation of tools like House of Quality, TRIZ contradiction matrix, Nine Window Analysis, S-Curve, SuField, Ideal Final Result, Laws of Evolution)
- $PSS_{f.6}$ - an information repository, structured by marketing and technical (electromechanical) concepts, where each piece of information is a [problem - solution] pair
- $PSS_{f.7}$ - an open innovation module, based on $PSS_{f.6}$, which allows problem solving by (external) experts (the problems are the ones from $PSS_{f.6}$)
- $PSS_{f.8}$ - a search module, to extend the information repository (search should be carried out in the "outside" world - on the web and in dedicated databases such as patent repositories)
- $PSS_{f.9}$ - an expert module to suggest innovation projects based on key business process assessments (it should assess the gaps in providing robust connections among the key business processes and suggest / prioritize innovation projects for minimizing these gaps)

All the information produced within the *inSight* steps above was stored in the ontology, in the subclasses of the *FunctionalPerspective* concept.

5.3 Getting the revised solution

In the next step, a **release candidate** solution was produced, this time by the entire design team. The information about the target domain, market and user characteristics was further refined and extended by detailing the specific classes (mainly *DomainConcept* and *Stakeholders*) in the support ontology. Regarding the target domain, information was added about types of costs that a SME deals with.

While applying the concurrent algorithm for the second time, the local PSS specifications for TF_5 - *effective innovation projects* were extended, based on the information in the support ontology under the *Cost* class, by considering an approach in which innovation projects could be driven by reducing overall quality costs without affecting product quality.

Based on the information in the support ontology regarding market and user characteristics, a new PSS specification was added to target function TF_3 - *increased ergonomy*: $PSS_{3.5}$ - a trend monitoring service (as a 64% product and 36% service mix).

The overall solution was completed with three more major modules:

- $PSS_{f.10}$ - a quality cost monitoring and assessment centre, based on business processes, sub-processes and activities
- $PSS_{f.11}$ - an expert module to correlate innovation projects with quality cost reduction goals
- $PSS_{f.12}$ - a trend monitoring centre, to estimate future changes in customer behavior or competitor strategies

It is worth noting that the features regarding business processes and innovation projects driven by quality cost reduction objectives were considered only after applying *inSight*. During the initial talks regarding the vision about the software PSS to be developed, none of these features were even thought of.

The last phase of *inSight* allows the design team to make sure that all stakeholder requirements were properly taken into account (*Stakeholder* / *Requirement* class instances are used to store these requirements in the support ontology). The QFD graphical support was used to complete this task (due to space constraints the House of Quality will not be listed here).

Within this analysis, the "requirement" entries were the stakeholder requirements and the "characteristics" were the PSS feature list in the RC solution. As on each row and column of the House of

Quality at least one strong correlation did exist, the design team concluded that all stakeholder requirements were properly addressed. Thus, the 12 major modules ($PSS_{f.1}$ to $PSS_{f.12}$) could be considered the final output of the *inSight* methodology. Based on it, the final software requirements were established, the possible implementation technologies were assessed and the architecture was done. All this information was eventually compiled into the software-PSS blueprint.

6 CONCLUSIONS AND FUTURE WORK

Within a complex environment, in which the perceived user needs guide a company's strategy, product-service systems can be the key in attaining sustainability. Yet effectively designing such PSSs is a real challenge. To properly address it, there is a clear need of a methodology that encapsulates all those fine aspects of design that finally lead into a unique product: effective tools, creativity support, flexibility and a shared understanding for all the stakeholders.

The ontology-based methodology introduced in this paper systematically addresses all these aspects, while trying to remain simple and straightforward enough so that PSS practitioners might effectively use it. PSS designers are *guided* by this methodology throughout the entire process of shaping a unique product service system, starting from the ground point of any successful product: the market needs.

An application example is presented, highlighting the benefits of the support ontology in terms of understanding problems, getting a shared view and supporting ideative processes.

Finally, the methodology aims to support PSS practitioners in developing systems that will eventually turn them from simple suppliers into strategic partners for their customers. Moreover, the methodology emphasizes the most valuable asset that a PSS practitioner has - its human resource, by supporting and better exploiting their creativity. Thus, it could be an important aid in the equation of sustainable development.

To maximize the potential this methodology has, a software application will be developed to automate it. For the design process, the methodology itself will be used and we are looking forward to build this software application under the form of a product-service system.

Further research will also be done towards refining the ontology classes and relations among them, thus allowing designers to get a more comprehensive view over the target domain, users, stakeholders or market needs.

7 ACKNOWLEDGMENTS

Financial support from the European Commission within the FP7 research project TECH-IT-EASY/232410 is acknowledged with gratitude.

8 REFERENCES

[1] Futrell, R., Shafer, D., Shafer, L. (2001): Quality Software Project Management, Prentice Hall.

[2] Brad S., Fulea M., Brad E. (2010): A PSS Approach in Software Development, in: Proceedings of the CIRP IPS² Conference 2010, pp. 291-298, Linköping, Sweeden.

[3] Levitt, T. (1969): The Marketing Mode: Pathways to Corporate Growth, McGraw-Hill, New York.

[4] Petter, S. (2008): Managing User Expectations on Software Projects: Lessons from the Trenches, in: International Journal of Project Management, Vol. 26, No. 7, pp. 700-712.

[5] Juristo, N., Moreno, A.M. and Sanchez-Segura, M-I. (2007): Analyzing the Impact of Usability on Software Design, in: The Journal of Systems and Software, Vol. 80, pp. 1506-1516.

[6] Mantere, T., Alander, J.T. (2005): Evolutionary Software Engineering, a Review, in: Applied Soft Computing, Vol. 5, No. 3, pp. 315-331.

[7] Stamelos, I. (2010): Software Project Management Anti-Patterns, in: Journal of Systems and Software, Vol. 83, No. 1, pp. 52-59.

[8] Han, W-M., Huang, S-J. (2007): An Empirical Analysis of Risk Components and Performance on Software Projects, in: Journal of Systems and Software, Vol. 80, No. 1, pp. 42-50.

[9] Fulea, M. (2010): Contributions to industrial software usability design and assessment, Technical Report RTH022010, Technical University of Cluj-Napoca, Romania.

[10] Yang X., Moore P., Pu J.-S., Wong C.-B. (2009): A practical methodology for realizing product service systems for consumer products, in: Computers & Industrial Engineering, Vol. 56, pp. 224-235.

[11] Wong, M. (2003): Product Service Systems in Consumer Goods Industry, Ph.D. Thesis, Cambridge University, UK.

[12] Shimomura, Y., Hara, T. (2010): Method for supporting conflict resolution for efficient PSS development, in: CIRP Annals - Manufacturing Technology, Vol. 59, pp. 191-194.

[13] Morelli, N. (2006): Developing new product service systems (PSS): methodologies and operational tools, in: Journal of Cleaner Production, Vol. 14, pp. 1495-1501.

[14] Maussang, N., Zwolinski, P., Brissaud, D. (2009): Product-service system design methodology: from the PSS architecture design to the products specifications, in: Journal of Engineering Design, Vol. 20, No. 4, pp. 349-366.

[15] Aurich, J.C., Fuchs, C., Wagenknecht, C. (2006): Life cycle oriented design of technical Product-Service Systems, in: Journal of Cleaner Production, Vol. 14, pp.1480-1494.

[16] Aurich, J.C., Mannweiler, C., Schweitzer, E. (2010): How to design and offer services successfully, in: CIRP Journal of Manufacturing Science and Technology, Vol. 2, pp. 136-143.

[17] Gumusluoglu, L., Ilsev, A. (2009): Transformational leadership, creativity, and organizational innovation, in: Journal of Business Research, Vol. 62, pp. 461-473.

[18] Johnson, H., Carruthers, L. (2006): Supporting creative and reflective processes, in: International Journal of Human-Computer Studies, Vol. 64, pp. 998-1030.

[19] Nguyen, L., Shanks, G. (2009): A framework for understanding creativity in requirements engineering, in: Information and Software Technology, Vol. 51, pp. 655-662.

[20] Fischer, G., Giaccardi, E., Eden, H., Sugimoto, M., Ye, Y. (2005): Beyond binary choices: Integrating individual and social creativity, in: International Journal of Human-Computer Studies, Vol. 63, pp. 482-512.

[21] Noy, N.F., McGuinness, D.L. (2001): Ontology development 101: a guide to creating your first ontology, Stanford Knowledge Systems Laboratory Technical Report KSL-01-05.

[22] Gill, T. (2002): Planning Smarter: Creating Blueprint-Quality Software Specifications, Prentice Hall PTR, 2002.

[23] Protégé (2010): Protégé <**http://protege.stanford.edu/**>, accessed 12.09.2010.

[24] EFQM (2010): The European Foundation for Quality Management < **http://www.efqm.org/**>, accessed 4.01.2011.

Customer oriented Configuration of Product-Service Systems

Carsten Mannweiler[1], Jan C. Aurich[1]

[1] Institute of Manufacturing Technology and Production Systems, University of Kaiserslautern, Kaiserslautern, Germany

Abstract

Customer orientation is one of the key aspects to obtain competitive advantages. Especially in the capital goods industry, the offered Product-Service Systems (PSS) have to fulfill all customer requirements to ensure the performance expected. Today, the development of the product and service portfolio is mostly separate. Thus, the manufacturers are faced with the challenge to configure the "right" solution out of these separate portfolios. Therefore, it becomes necessary to support manufacturers with a systemized configuration, regarding the specification of configuration and the characteristics of PSS. This paper illustrates a framework for a customer oriented configuration of Product-Service Systems.

Keywords:

Product-Service Systems; Service; Configuration

1 INTRODUCTION

The following scenario describes a change most of the capital goods manufacturers have recognized during the last years. In former times, customers wanted to buy (physical) products for their specific needs. For example, a road building company wanted to buy road milling machines. Today, customers require holistic solutions for their specific needs. So, the road building company wants to have a solution for milling a road. These solutions consist of (physical) products (e.g. road milling machine) and (non-physical) services (e.g. technical, qualifying or logistical services) – so called Product-Service Systems (PSS) [1, 2, 3]. The purpose of the PSS thereby is to ensure or to enhance the product performance expected by the customer throughout the whole customer lifecycle. [4]. Therein, the customer lifecycle includes all tasks and activities that the customer has to fulfill regarding his specific needs.

The described scenario results from different changes in the manufactures' environment. On the one hand, the merging of international markets and therewith, the increasing number of competitor forces the manufacturers to obtain their competitiveness [5, 6]. Offering PSS helps to differentiate and to diversify from competitors and results in obtaining the market position [7, 8]. At the best, the market position could be improved. On the other hand, the markets changed from a seller's market to a buyer's market. Therein, the customer expects far more than the simple product functionalities. He expects that his specific needs will be fulfilled with individual solutions. As mentioned before, these solutions have to consist of customized physical and non-physical components [9, 10]. Thereby, the focus has to be placed on how to create customer specific values [11].

These changes force capital goods manufacturers to gradually transform from manufacturing companies to PSS-providers [12]. This transition constitutes major managerial challenges regarding organizational as well as operational aspects [13], e.g. an integrated PSS-development process and new mindsets along with PSS-design [14]. Obviously, this transition will take time, but the customers require their individual solutions now. Hence, there are two key questions capital goods manufacturers have to answer: Are we able to offer such specific solutions to our customers? Are we able to offer these solutions in an adequate time? For answering these questions, the manufacturers' underlying circumstances have to be considered.

2 MANUFACTURERS' UNDERLYING CIRCUMSTANCES

Up to now, in manufacturing firms, services are often thought of as add-ons, and initial services (installation, commissioning, etc.) are frequently "given away" during the negotiations to sell the product [13]. In Addition, companies are still using a "hit-and-miss" approach to handle their services [10]. There are no standardized and enterprise-wide service development processes available. The offered services have traditionally grown in different parts of the organization and a very few companies do not even have specific service units [13]. Although different authors state that the development of products and services have to be integrated [15, 16, 17] the offered products and services are still separately developed in industry. Further on, during the development process the existing interdependencies are neglected, even though the combined solutions have to fulfill customer needs and create customer value [18].

For configuring and offering customer specific PSS a systematical gathering of customer's requirements is needed. Nevertheless, in

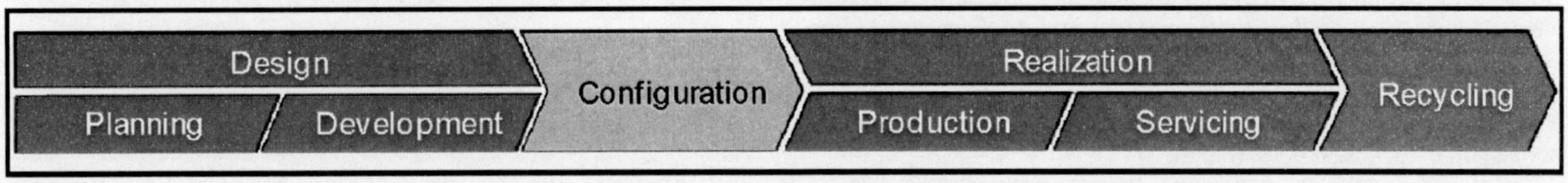

Figure 1: Manufacturer's PSS lifecycle [according to 19].

J. Hesselbach and C. Herrmann (eds.), *Functional Thinking for Value Creation: Proceedings of the 3rd CIRP International Conference on Industrial Product Service Systems, Technische Universität Braunschweig, Braunschweig, Germany, May 5th - 6th, 2011*,
DOI 10.1007/978-3-642-19689-8_16, © Springer-Verlag Berlin Heidelberg 2011

industry there is still a lack of understanding how the customer will use the PSS and how to gather the corresponding requirements systematically [20]. There are different approaches dealing with requirements gathering. These approaches are product [21, 22, 23] or service oriented [24, 25, 26]. But regarding PSS, an entire and systematical requirements gathering is missing, especially in industry. Therefore, a systematical approach has to be implemented. The manufacturers' underlying circumstances could be summarized as following:

- separate developed product and service portfolios
- unknown interdependencies between product and service components
- unknown configurability aspects concerning product and service components
- non-systematical gathering of customer PSS-requirements

Regarding these circumstances, the questions above have to be answered with "no". Consequently, for not losing customers and therewith, competitive position during the transition, the capital goods manufacturers have to be provided with means to offer customer oriented PSS. Therefore, this paper illustrates a framework for a customer oriented configuration of PSS.

3 TOWARDS PSS-CONFIGURATION

Configuration is defined as the compilation of a product by selecting and combining pre-defined modules/components within given rules [27]. The components are described by a set of properties and the connections between the components [28]. A further definition describes the configuration as the process to transfer customer's wishes into a concrete product specification, while the solution space is set up at the company level [29]. For guaranteeing the consistence of the configured product with the products structure and the customer requirements a systematical procedure supported by software tools is helpful [30]. The configuration process is usually a selective event at the end of the design phase. At this point the customer's requirements are gathered by the manufacturer. Based on these requirements an adequate product is configured and offered [31]. With respect to this, a successful configuration could only be conducted fulfilling the following prerequisites:

- systematical gathering of customer requirements
- knowledge of interdependencies and connections of the components
- methodical configuration process.

Traditional capital goods manufacturer fulfill these prerequisites mostly, when they have to configure pure products. Therefore, the configuration of a customer oriented product will not be a challenge. But with respect to the aspects above– changed manufactures' environment and the manufacturers' underlying circumstances – the configuration of customer oriented PSS is not possible.

4 FRAMEWORK FOR PSS-CONFIGURATION

Because of the importance of the configuration, the former product lifecycle from the manufacturers' point of view – configuration phase was seen as part of the design phase – has to be extended. Therein, the configuration will be seen as an own phase (Figure 1). This phase provides a new framework for a PSS-configuration, consisting of two main parts (Figure 2).

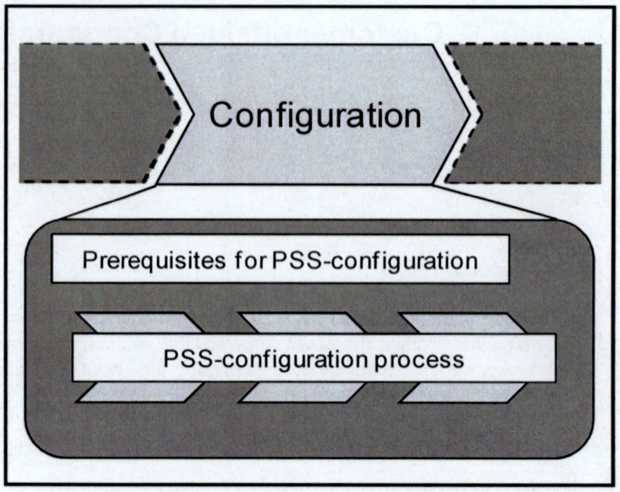

Figure 2: Parts of the framework.

The first part deals with the prerequisites that provide capital goods manufacturers with means for configuring adequate customer oriented PSS. For example, these means are tools that help to analyze and to structure the existing product and service portfolio. The second part deals with the configuration process itself. The framework provides a configuration process with three sub phases – the requirements gathering phase, the configuration process phase and the purchasing phase. By implementing this framework the following goals could be achieved by the capital goods manufacturers:

- systematical gathering of PSS-requirements
- increase of product and service portfolio performance
- methodical PSS-configuration process
- support of customer's investment decisions by specifying the degree of PSS-fulfillment
- identification of portfolio gaps
- detailed resource planning for PSS-realization

Further on, the presented framework and the tools are practically oriented and include existing company standards. Additionally, the implementation could be managed without spending lots of effort in it and is adaptable to company's specifications. The following sections describe the parts in detail.

4.1 Prerequisites for PSS-configuration

Before capital goods manufacturer are able to configure customer oriented PSS, three main prerequisites have to be fulfilled. These prerequisites are:

- systematical gathering of customer's PSS-requirements
- analysis of the existing product and service portfolio
- standardized PSS-offer forms.

For fulfilling the prerequisite adequate tools are necessary. Figure 3 shows the prerequisites and the corresponding tools.

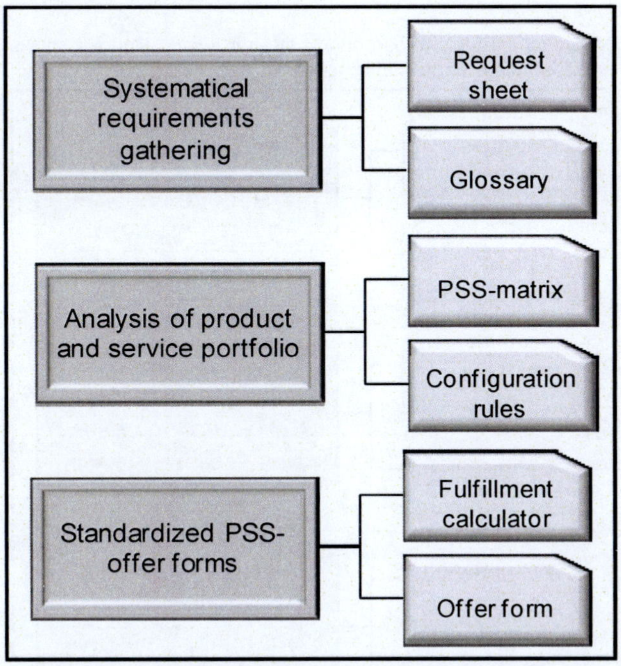

Figure 3: Prerequisites for PSS-configuration and corresponding tools.

For configuring adequate PSS, a comprehensive gathering of customer's requirements is necessary. Thereby, the difficulty lies in gathering requirements for a PSS-solution instead of for products and services separately. Further on, the tasks in the customer's lifecycle have to be considered, because they affect the requirements strongly. For example, the availability of a machine for a road builder in northern Europe is different than the availability for a road builder in southern parts. Hence, in this framework, the request sheet is based on the customer lifecycle with its sub-phases. Table 1 presents the schematic view of the request sheet. Another challenge while gathering customer's requirements, is the transformation of the "customer speech" into a manufacturer specific one. Requirements are often expressed in a native way, while the manufacturer needs a technical one. In this context, "technical" contains product specific as well as service specific properties that are necessary for configuring PSS. Further on, to avoid misunderstandings, the manufacturers have to develop a glossary. This glossary contains all terms that a relevant for gathering the requirements. For example, the availability of a service hotline is specified with a 24/7 availability. To ensure a comprehensive and precise requirement gathering, an appropriate request sheet has to be developed. This sheet has to fulfill the just mentioned criteria and to support manufacturers' employees during their dialog with the customer (see requirements gathering phase). Different approaches (compare [32, 33]) are dealing with an integrated gathering and provide checklists for a comprehensive gathering.

Based on the request sheet, the configuration of the customer oriented PSS is conducted. Therefore, the transformation of the customer requirements into technical properties is necessary. To ensure an entire gathering, an overview about the product and service portfolio is needed. Further on, the existing products and services have to be described in a systematical and standardized way. Based on this description, the checklist for the requirement gathering has to be derived and adapted. Regarding the products, traditional manufacturer possess holistic descriptions of their products and the products' structure. For example, these structures are function or assembly oriented [22]. They represent the product in a more or less configurable manner. Comparing the products, a detailed knowledge as well as a systematical and standardized description about the services and their structures is mostly missing. This results from the ad-hoc and non-systematical development of the services. Hence, tools supporting such a description for services are necessary. Promising approaches are presented in [34] and [35]. The described tools for products and services have to be implemented as a corporate standard and ensure the configuration of PSS. Therefore, the requirements of this standard, e.g. level of detail for description, have to be derived. The derivation depends on the company's specifications and standards. By means of the standard, all products and services are analyzed before the PSS-configuration is possible.

The configuration is represented by a set of configuration rules [27]. Therefore, all existing interdependencies between products and services as well as the configurability aspects must be identified. Hence, it is necessary to describe product and service components by a set of decisive dependencies on other components. Therefore, a method for identifying and describing the dependences between product and service components is essential. The resulting interdependencies and configurability aspects also have to be summarized in a description standard. Appropriate sheets, concerning the definition and derivation of the configuration rules, have to be developed. Further on, the rules have to be implemented in a software tool that will support the PSS-configuration.

For the standardized analysis and description of products and services as well as for the identification and description of the interdependencies and configurability, interdisciplinary knowledge is necessary. Therefore members of different units (e.g. design, service, marketing, production) have to be involved.

Regarding the separate developed portfolios, configuring PSS that fulfill customer's requirements at 100% is not always possible. There are two possibilities dealing with this circumstance. On the one hand, the manufacturer could offer PSS that do not fulfill the

Lifecycle phase	Sub-phase	Customer requirement (native)	Weight (D1, D2, W)	Technical requirement
Decision	Consulting			
	...			
Purchasing	Financing			
	...			

Table 1: Schematic view of the request sheet.

customer's requirements without any other explanation. This would lead to customer dissatisfaction and consequently to customer losing. On the other hand, the manufacturer could offer under fulfilled PSS with a specification of the degree of fulfillment. This degree transparently shows the customer how the PSS supports him during the lifecycle. Regarding the different kinds of requirements (demands and wishes) [22], unfulfilled wishes could result in a lower degree of fulfillment. Nevertheless, the customer could accept this and might be satisfied. Therefore, it is essential to transparently represent PSS-offers. For calculating the degree of fulfillment, a calculating software tool has to be developed. This tool must consider the specified requirements for the products and services as well as the importance of the requirements – specified during the requirement gathering by the customer. The difficulty by developing such a calculator lies in quantifying all requirements and evaluating the variance.

The realization of services throughout the usage of a PSS needs several resources. For ensuring the requested services, it is necessary to know, which resources are needed and which resources are available. Therefore, a standardized resource plan should be used (Table 2). This resource plan is based on a systematical analysis and description of all resources that could be needed regarding all offered services. These resources are e.g. spare parts, tools or personal. For this, suitable sheets for analyzing and structuring the resources are necessary. Especially with respect to services, there is still a lack of tools, also because of the non-systematic treatment of services. The counterpart of the resource plan is the parts list of the products for assembling the configured product components. In contrast to resource plans, the parts lists are mostly available and standardized in traditional manufacturing companies. Herein, the parts lists should be adapted to the new PSS-configuration conditions if necessary. The resource plans and the parts lists are automatically generated by the software tool after configuring the PSS. Finally, new PSS-offer forms for supporting the customer's investment decision are needed. These offer forms include a brief description of the PSS-components, the degree of PSS-requirements fulfillment as well as the calculated lifecycle costs. A method for calculating the PSS-lifecycle cost is presented in [36].

All presented tools have to be developed and implemented before a PSS-configuration is possible. Nevertheless, these tools have to be dynamically adaptable. The linkage between the prerequisites and the PSS-configuration process is presented in the following section.

4.2 PSS-configuration process

The configuration process presented in this framework consists of three sub phases:

- requirements gathering phase
- configuration process phase
- purchasing phase.

Regarding the tools of part one of the framework. Figure 4 shows their assignment to the sub phases.

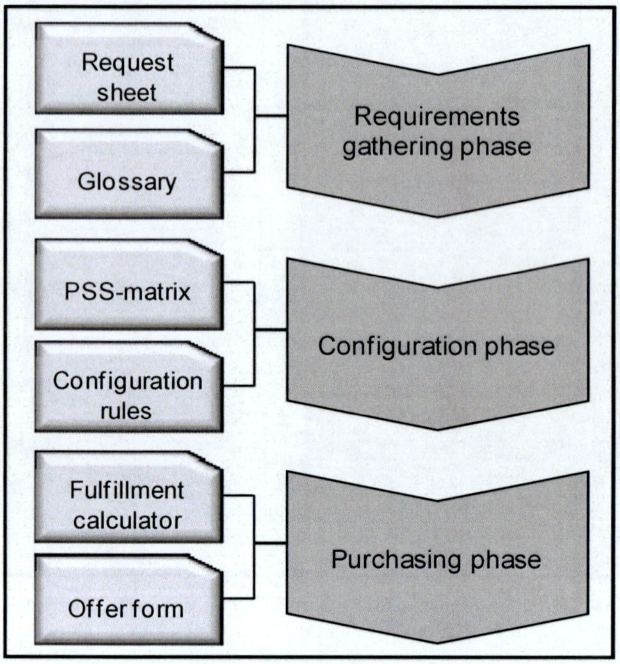

Figure 4: PSS-configuration process.

Requirements gathering phase

After the standardized description of the products and services as well as their interdependences and configurability aspects within the interdisciplinary teams the configuration is possible. The configuration process starts with the gathering of the customer's requirements. The requirement gathering consists of three steps and is supported by the developed request sheet. The gathering starts with a systematic customer interview (step 1). Therein, the customer's requirements are gathered, based on the specific customer lifecycle. The requirements are listed in the "customer speech" and evaluated as demands or wishes. At the beginning of the meeting, the developed glossary is presented, so that the probability of misunderstandings is reduced. Based on the colloquial gathered requirements, the manufacturer has to transform the requirements into technical properties (step 2). For transforming the requirements the knowledge of different units (e.g. design, service or production) could be necessary. Depending on the customer's knowledge, most of the requirements could be expressed as technical properties, so that the transformation is reduced. Step 3 deals with a meeting wherein the transformed

Resource type	Resource specification	Needed resources	Available resources	Purchase or build up	Comment
Spare parts	Filter	4	4	yes **no**	---------------
	Valve	6	6	yes **no**	---------------
Service technicians	Foreman	2	1	**yes** no	Recruitment

Table 2: Resource plan.

requirements are presented to the customer. If there are some misinterpretations the corresponding requirements will be adapted. Based on these results the PSS-configuration could be conducted.

Configuration process phase

The configuration process phase consists of the configuration itself. For configuring a PSS, an approach with three steps seems to be a promising starting point. Therein a product and service configuration has to be conducted primary (step 1 and 2). The third step contains the PSS-configuration, i.e. the configuration of product and service parts. Regarding these steps and the support by a software tool, the above presented tools are necessary. How the PSS specific configuration-software, the configuration rules as well as the detailed procedures have to look like, have to be part of further research work, including the review of existing approaches (e.g. configuration and compatibility matrix [37]).

Purchasing phase

After the configuration, the degree of PSS fulfillment is calculated. As mentioned before, the calculation is based on the evaluated requirements – done by the customer during the requirements gathering – and the comparison of the demanded and achieved requirements. A corresponding calculator has to be developed in further research work, considering existing approaches.

Mostly, there are different PSS-variants available that fulfill the specific customer tasks. Therefore, for each PSS-variant a PSS-offer form has to be filled in. These forms should support the customer's investment decision and have to contain all relevant information. This information is at least a brief PSS-description, the degree of PSS-fulfillment and the corresponding PSS-lifecycle costs. Based on these forms the customer's investment decisions are supported and simplified.

Further on, for supporting the customer with the required service throughout the whole lifecycle, a resource adjustment is necessary. Based on a resource plan of the selected PSS-variant, the existing resources are compared with the needed resources. Therefore, a form presented in Table 2 could be used. By comparing the resources, missing resources are detected. These resources have to be purchased (e.g. tools for repairing) or build up (e.g. qualification of service technicians). Further on, the product parts, based on the created parts list, have to be assembled. These steps finish the PSS-configuration and the PSS-realization follows.

5 SUMMARY

In the capital goods industry, a well-structured PSS-configuration enables traditional product oriented manufacturer to fulfill the customer requirements during the gradual change to a PSS-provider. The paper presents a PSS-configuration framework that provides manufacturers with means to offer customer oriented PSS during this time. For that reason, necessary prerequisites and systematical process for PSS-configuration are needed. Hence, adequate tools have to be developed and implemented. For providing these tools, there are several approaches that could be used as promising starting points. But there is still a lot of research work to do, especially when considering the different characteristics of products and services with respect to interdependencies and configurability aspects or the calculation of PSS-fulfillment.

6 REFERENCES

[1] Aurich, J.C., Fuchs, C., Wagenknecht, C. (2006): Life Cycle Oriented design of Technical Product-Service Systems, in: Journal of Cleaner Production, 14/7, pp. 1480-1494.

[2] Baines, T.S., et al. (2007): State-of-the-art in product-service systems, in: Journal of Engineering Manufacture, 221 (B), pp. 1543-1552.

[3] Schuh, G., Klotzbach, C., Gaus, F. (2008): Service provision as a sub-model of modern business models, Production Engineering, 2/1, pp. 79-84.

[4] Aurich, J.C., Wolf, N., Siener, M., Schweitzer, E. (2009): Configuration of product-service systems, in: Journal of Manufacturing Technology Management, 20/5, pp. 591-605.

[5] Meier, H., Roy, R., Seliger, G. (2010): Industrial Product-Service System - IPS², in: Annals of the CIRP – Manufacturing Technology, doi: 10.1016/j.cipr.2010.05.004.

[6] Fähnrich, K.-P., Meiren, T. (2007): Service Engineering: State of the Art and Future Trends, in: Spath, D., Fähnrich, K.-P.: Advances in Services Innovations, Springer, Berlin, pp. 3-16.

[7] Johnson, M., Mena, C. (2008): Supply chain management for servitised products: a multi-industry case study, in: International Journal of Production Economics, 114/1, pp. 27-39.

[8] Randall, T., Terwiesch, C., Ulrich, K.T. (2007): User Design of Customized Products, in: Marketing Science, 26/2, pp. 268-280.

[9] Fueglistaller, U. (2007): From Service Management towards Service Competence – An entrepreneurial approach, in: Spath, D., Fähnrich, K.-P.: Advances in Services Innovations, Springer, Berlin, pp. 113-129.

[10] Legnani, E., Cavalieri, S., Ierace, S. (2009): A framework for the configuration of after-sales service processes, in: Production Planning and Control, 20/2, pp. 113-124.

[11] Sundin, E. (2009): Life-Cycle Perspectives of Product/Service-Systems: In Design Theory, in: Sakao, T., Lindahl, M.: Introduction to Product/Service-System Design, Springer, London, pp.31-46.

[12] Schuh, G., Friedli, T., Gebauer, H. (2004): Fit for Service: Industrie als Dienstleister, Carl Hanser Verlag, München.

[13] Oliva, R., Kallenberg, R. (2003): Managing the transition form products to services, in: International Journal of Service Industry Management, 14/2, pp. 160-172.

[14] Sakao, T., Lindahl, M.: Introduction to Product/Service-System Design, Springer, London.

[15] Aurich, J.C., Fuchs, C. (2004): An approach to life cycle oriented technical service design, in: Annals of the CIRP, 53/1, pp. 151-154.

[16] Manzini, E., Vezolli, C. (2003): A strategy design approach to develop sustainable product service systems: examples taken from the 'environmentally friendly innovation' Italian prize, in: Journal of Cleaner Production, 11/8, pp. 237-245.

[17] Kindström, D., Kowalkowski, C. (2009): Development of industrial service offerings: a process framework, in: Journal of Services Management, 20/2, pp. 156-172.

[18] Lindahl, M., Ölundh, G. (2001): The Meaning of Functional Sales, in: Proceedings of Life-cycle Engineering: Challenges and Opportunities, Varna, Bulgaria.

[19] Fuchs, C. (2007): Life Cycle Management investiver Produkt-Service Systeme - Konzept zur lebenszyklusorientierten Gestaltung und Realisierung, Dissertation, TU Kaiserslautern.

[20] Deutsche Gesellschaft für Qualität (2001): QFD, Quality Function Deployment, Beuth Verlag, Berlin.

[21] VDI Richtlinie 2221 (1993): Methodik zum Entwickeln und Konstruieren technischer Systeme und Produkte, Beuth Verlag, Berlin.

[22] Pahl, G., Beitz, W. (2007): Engineering Design - A Systematic Approach, Springer, London.

[23] Ahrens, G. (2007): Das Erfassen und Handhaben von Produktanforderungen - Methodische Voraussetzungen und Anwendungen in der Praxis. Dissertation, TU Berlin.

[24] van Husen, C. (2007): Anforderungsanalyse für produktbegleitende Dienstleistungen, Jost-Jetter, Heimsheim.

[25] Schmitz, G. (2000): Die Ermittlung der Kundenanforderungen an industrielle Dienstleistungen, in: Zeitschrift für Planung, 11/2, S. 195-215.

[26] Kepper, G. (1999): Methoden der Qualitativen Marktforschung, in: Herrmann, A., Homburg, C.: Marktforschung – Methoden, Anwendungen, Praxisbeispiele, Gabler, Wiesbaden.

[27] Scheer, C. (2006): Kundenorientierter Produktkonfigurator: Erweiterung des Produktkonfiguratorkonzeptes zur Vermeidung kundeninitiierter Prozessabbrüche bei Präferenzlosigkeit und Sonderwünschen in der Produktspezifikation, Dissertation, Berlin.

[28] Mittal, S., Frayman, F. (1998): Towards a Generic Model of Configuration Tasks, in: Proceedings of the 11th International Joint Conference on Artificial Intelligence (IJCAI), Detroit, USA, pp. 1395-1401.

[29] Tseng, M.M, Piller, F. (2003): The Customer Centric Enterprise - Advances in Mass Customization and Personalization, Springer, Berlin.

[30] Aldanondo, M., Hadj-Hamou, K., Moynard, G., Lamothe, J. (2003): Mass customization and configuration: Requirement analysis and constraint based modeling propositions, in: Integrated Computer-Aided Engineering 10, pp. 177–189.

[31] Aurich, J.C., Schweitzer, E., Mannweiler, C. (2008), Integrated Design of Industrial Product-Service Systems, in: Mitsuishi, M., Ueda, K., Kimura, F: Manufacturing Systems and Technologies for the New Frontier, Springer, London, pp. 543-546.

[32] Steinbach, M. (2005): Systematische Gestaltung von Product-Service Systems - Integrierte Entwicklung von Product-Service Systems auf Basis der Lehre von Merkmalen und Eigenschaften, Dissertation, Saarbrücken.

[33] Müller, P., Schulz, F., Stark, R. (2010): Guideline to elicit requirements on industrial product-service systems, in: Proceedings of the 2nd CIRP IPS² Conference (IPS² 2010), pp. 109-116, Linköping, Schweden.

[34] Aurich, J.C., Clement, M.H. (2010): Produkt-Service Systeme – Gestaltung und Realisierung, Springer, Berlin.

[35] Hermsen, M. (2000): Ein Modell zur kundenindividuellen Konfiguration produktnaher Dienstleistungen, Shaker, Aachen.

[36] Mannweiler, C., Siener, M., Aurich, J.C. (2010): Lifecycle Cost oriented Evaluation and Selection of Product-Service System Variants, in: Proceedings of the 2nd CIRP IPS² Conference (IPS² 2010), pp. 21-26, Linköping, Schweden.

[37] Bongulielmi, L. (2003): Die Konfigurations- und Verträglichkeitsmatrix als Beitrag zur Darstellung konfigurationsrelevanter Aspekte im Produktentstehungsprozess, Fortschritt-Berichte VDI Reihe 16 Nr. 151, VDI-Verlag, Düsseldorf.

An Engineering Method for Supporting Customer-Oriented Service Improvement

Fumiya Akasaka,[1] Ryosuke Chiba,[1] and Yoshiki Shimomura[1]

[1] Department of System Design, Tokyo Metropolitan University, Tokyo, Japan

Abstract

Offering customized solutions is one of the important characteristics of the integrated provision of a product and a service. Generally, customer requirements for a service might change depending on the environments which customers are facing. In order to meet such changeable requirements and realize a long-term relationship, the key is continual customer-oriented service improvement. This paper proposes an engineering method to support the planning of such customer-oriented service improvement. To be more precise, an optimum resource allocation method using a Genetic Algorithm is adopted to prioritize problems to be solved for the purpose of maximizing the degree of customer satisfaction.

Keywords:

Service Engineering; Continual Service Improvement; Planning

1 INTRODUCTION

Our society's limitations of natural resources, energy supplies, and capability to accept industrial waste have led to serious problems related to producing and disposing of artifacts [1]. We have to immediately search for solutions or business models to realize a sustainable society. In this context, services are becoming increasingly important in the manufacturing industry, since a longer life or more added value of a product can be achieved by offering services (e.g., a maintenance service) combined with a product (e.g., a car). As a result, new concepts such as the Product-Service System (PSS) [2] and Industrial Product Service System (IPS^2) [3] have been proposed and are attracting much attention, especially in developed countries.

1.1 Japanese Service Engineering

The authors have carried out fundamental research on Service Engineering (SE) in Japan, which aims at providing design methodology of services from an engineering viewpoint [4]. In SE, a service is defined as *an activity between a service receiver and a service provider to change the state of the receiver* [4]. This definition includes a broader sense than the typical definitions in a service marketing field, which are used to clarify the difference between services and products (e.g., [5]). In this definition, we regard a service as a combination of service activities and physical products [4]. The term service used in this study, therefore, corresponds to PSS.

1.2 Continual service improvement

Offering a customized solution for each customer is one of the important characteristics of integrated provision of a product and service [6]. Understanding and fulfilling customer requirements is, therefore, a key issue in designing services [7].

Generally, customer requirements for a service might change depending on the environment (i.e., trends or situation) which customers are facing [8]. In order to meet such changeable customer requirements and realize a long-term relationship with a customer, the key is continual service improvement adjusted to customer requirements. In this study, such a type of service improvement is called *customer-oriented service improvement.*

2 CUSTOMER-ORIENTED SERVICE IMPROVEMENT

2.1 Service lifecycle

Service lifecycle can be divided into three phases: service design, transition, and operation (e.g., [9]), as shown in Figure 1.

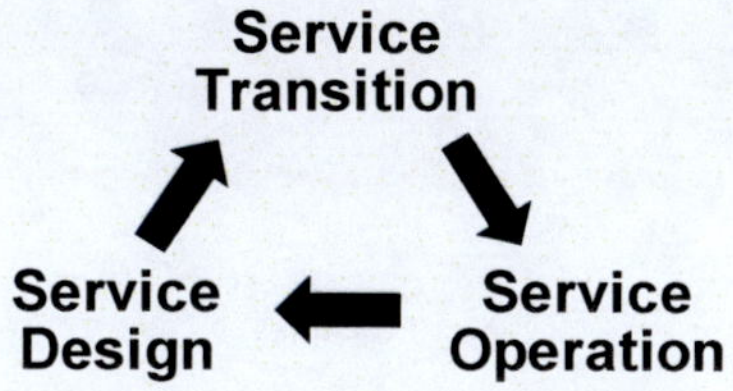

Figure 1: A model of service lifecycle.

Service design

The service design phase is to generate a conceptual model of a service based on required specifications. Entities, including both physical products and human resources, to construct a service are specified as well.

Service transition

In the transition phase, both providers and customers make preparations for the operation phase. These preparations include not only manufacturing and installing physical products but also obtaining knowledge and skills required for service operation.

Service operation

This phase is to execute the designed service in the real world. During this phase, maintenance is carried out as necessary for products as well as human resources, such as employees.

2.2 Customer-oriented service improvement

As reported in 1.2, for the purpose of building a long-term relationship with a customer, a service provider should maintain customer satisfaction by improving his service and adjusting it to

J. Hesselbach and C. Herrmann (eds.), *Functional Thinking for Value Creation: Proceedings of the 3rd CIRP International Conference on Industrial Product Service Systems, Technische Universität Braunschweig, Braunschweig, Germany, May 5th - 6th, 2011,*
DOI 10.1007/978-3-642-19689-8_17, © Springer-Verlag Berlin Heidelberg 2011

the customer requirements, namely, by implementing customer-oriented service improvement continually. In parallel, the improvement has to be implemented under the constraints of resources that the service provider possesses.

Figure 2 depicts an overview of customer-oriented service improvement. This indicates that Continual Service Improvement (CSI) should be done through an overall service lifecycle and it is important for a service provider to improve his service from two perspectives: (1) customer requirements and (2) constraints of internal resources.

In this framework, service improvement is regarded as *problem solving*. The offered service is evaluated from the two perspectives aforementioned; then, the provider's problems in service offering would become apparent (e.g., a lack of user support). In the service design phase, solutions for the identified problems that make customer satisfaction improve under the constraints of the provider's internal resources are then explored. Subsequently, a new service is conceptually designed (e.g., development of a device for user support). Finally, a procedure to introduce the designed service to the real world is prepared, and, after the introduction, customers can receive the improved service.

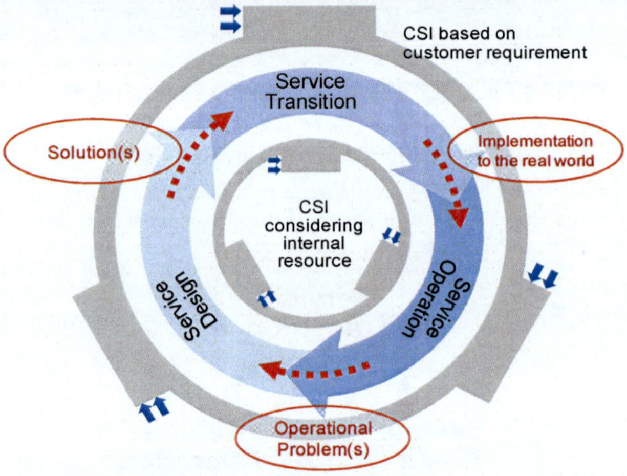

Figure 2: An overview of customer-oriented service improvement.

2.3 Scope of this research

In this paper, an early stage of the service design phase in customer-oriented service improvement is focused on; namely, the scope of this paper is the planning stage of service improvement.

Customer-oriented service improvement should be planned to realize the maximum customer satisfaction under the constraints of resources that a service provider possesses. Namely, it is important for a service provider to specify the problem whose solution would lead to high customer satisfaction.

This paper proposes an engineering method for planning the most effective service improvement. To be more precise, in this paper, a model to quantify customer satisfaction is proposed. In addition, a resource allocation method using a Genetic Algorithm (GA) [10] is adopted to prioritize problems to be solved for the purpose of maximizing customer satisfaction.

3 MODELING CUSTOMER SATISFACTION

3.1 Means-end theory and customer satisfaction

Consumers' behavior in choosing to consume or use products or services is often described as purposeful and goal-oriented [11]. The means-end theory [12] suggests that a consumer forms a conception of the suitability of a product or service attributes in question (means) for fulfilling specific values (end) and this theory can be modeled on three elements: attributes, utility expectations, and set of values (Figure 3). Here, attributes are elements that define the quality of a product/service, which corresponds to "quality elements" in Quality Function Deployment (QFD) [13]. Utility expectations refer to customer requirements or needs. Value, which is defined as an "intervening variable that leads to action when activated [14]," indicates the goal or aim of an action. Based on the means-end chain model, a customer perceives satisfaction for a product or service when not only his requirements but also his values are fulfilled (Figure 3).

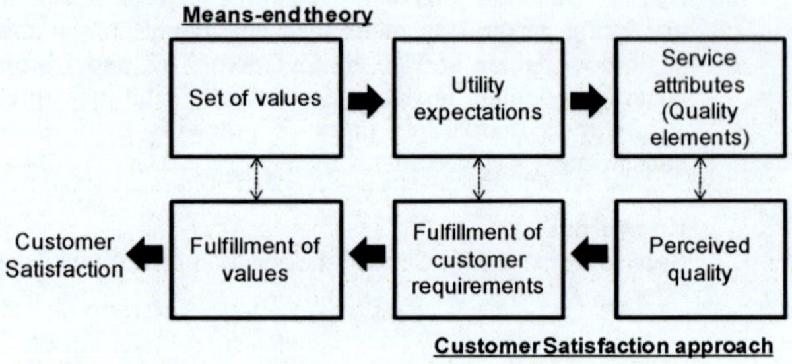

Figure 3: Customer satisfaction constructs [11].

3.2 Modeling and quantification of customer satisfaction

Based on the theory reported in 3.1, in this study, a customer's perception of satisfaction is hierarchically structured through three layers of (required) service quality elements, customer requirements/needs, and values/goals (Figure 4). This model illustrates the relationship between the service qualities and the customer satisfaction. Each element is expressed as a node that possesses its value to represent the fulfillment. Each node is associated with other nodes in its upper layer with arcs. The weight on each arc represents the degree of influence of a lower node on an upper node. These nodes, arcs, and weights are determined from the viewpoint of customers by conducting, e.g., interviews or questionnaires. Especially, the laddering method, which is an in-depth interviewing technique to develop an understanding of a customer's cognitive structure to a product/service [15], is useful to decide linkages (arcs) between elements (nodes).

In this model, the degree of fulfillment of a customer requirement (CR_l) is calculated from the value of related service qualities (SQ_j) and its influences (R_{ij}^{SQ}) (Equation (1)). In the same manner, the degree of fulfillment of a value/goal (V_k) is determined by CR_l and R_{kl}^{CR}, which represent the weight between a value/goal and customer requirement (Equation (2)). Customer satisfaction is quantified by the calculation with Equation (3).

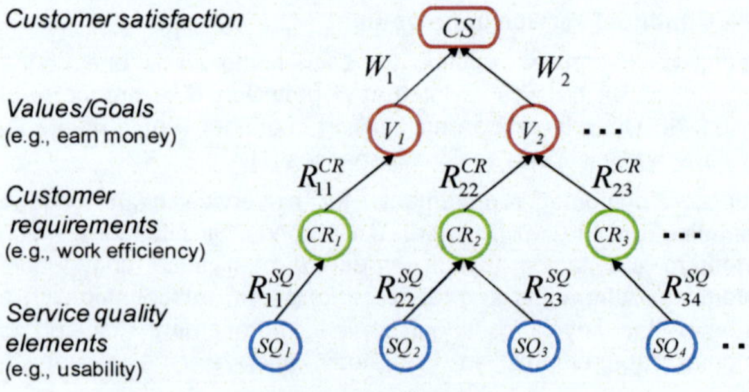

Figure 4: A model of customer satisfaction.

$$CR_i = \sum_{j=1}^{n} SQ_j \cdot R_{ij}^{SQ} \qquad (1)$$

, *where* n *is the number of service qualities.*

$$V_k = \sum_{l=1}^{o} CR_l \cdot R_{kl}^{CR} \qquad (2)$$

, *where* o *is the number of customer requirements.*

$$CS = \sum_{m=1}^{p} V_m \cdot W_m \qquad (3)$$

, *where* W_m *indicates a weight of a value/goal on customer satisfaction and* p *is the number of values/goals.*

4 A METHOD FOR SERVICE IMPROVEMENT PLANNING

4.1 Overview

Solving problems that result in the degradation of service quality leads to an improvement of service quality. According to the model of customer satisfaction in Figure 4, the improvement of service quality triggers further fulfillment of customer requirements and, eventually, an improvement of customer satisfaction (Figure 5 (A)).

Meanwhile, a service provider possesses internal resources. Here, the term "resources" includes, e.g., human resources, products, facilities, information, money, and time, where money and time are primitive resources because they are essential and consumed by all of the activities in service improvement. In general, time and money that can be utilized for service improvement are limited (Figure 5 (B)). Hence, at the planning stage of service improvement, a service provider has to prioritize problems to be solved to realize the most effective service improvement under the constraints of time and monetary resources.

In this study, an approach to prioritize problems by adopting an optimal resource allocation method is proposed for the planning of service improvement. Namely, a service provider's limited resources (time and money) are optimally allocated to each problem for the purpose of maximizing customer satisfaction. Then, based on the allocation result, problems can be quantitatively prioritized.

In the method, customer satisfaction is calculated according to Equations (1)-(3) in 3.2. The resource allocation is optimized by GA to maximize customer satisfaction, in which the relationship between a problem and a resource is represented as the sigmoid function.

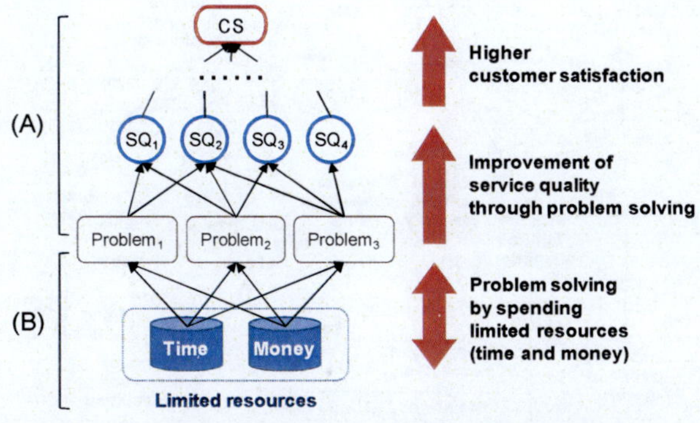

Figure 5: Service improvement through problem solving.

4.2 Relationship between problem solving and resources

Generally, a high use of resources, i.e., time and monetary resources, makes the achievement of problem solving more likely. Meanwhile, its effect depends on the type of resources and problems. For instance, in a maintenance service, solving the problem "lack of special skills of staff" would require a significant amount of time as well as a certain amount of monetary resources. On the other hand, a shorter time and a smaller amount of money would be needed to solve the problem "lack of tools for maintenance."

The sigmoid function, which is a kind of the monotone-increasing function, is appropriate to represent such a relationship. It models several relationships between two variables with two parameters, center and gain value. Figure 6 shows the examples. Here, a high gain means a quick effect on problem solving, and a low gain means a slow-acting effect on it. A lower center value indicates that a smaller amount of resources is needed to begin to exert the effect. For instance, a problem that the effect of resource consumption appears in binary can be represented with low center and gain (Figure 6 (b)).

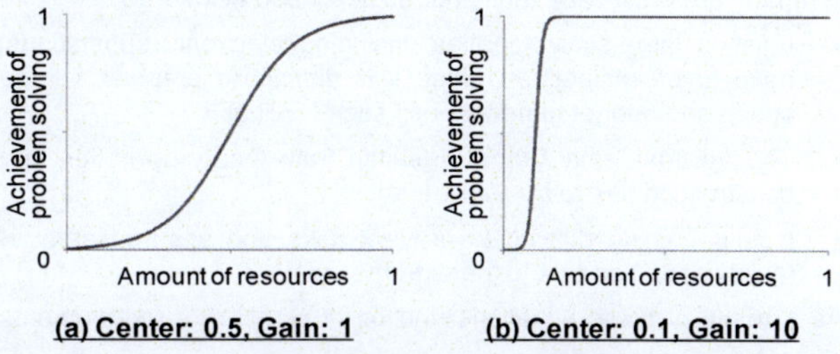

Figure 6: Examples of the sigmoid function.

4.3 Service improvement planning with Genetic Algorithm

Optimal resource allocation for maximizing customer satisfaction is adopted as a method for supporting the planning stage of a service improvement. Namely, the amount of resources to be consumed for the improvement of each problem when the value of customer satisfaction is the maximum is decided. The calculated amount of resources can be clear criteria to prioritize problems to be solved in service improvement.

In this optimal resource allocation problem, the search space must be very large, and the search model could be complicated, since the proposed structure contains several sigmoid functions as well as many nodes and arcs. The GA is a heuristic search method that mimics the process of natural evolution. Heuristic searching is routinely used to generate a solution for optimization. GA is effective to solve a search problem with local minimum and multimodal search space. GA is therefore suitable to the optimization problem addressed in this study, since it contains the locality due to the fact that it is regarded that a strategy of intensive investment is generally allowed in a resource allocation problem.

In this optimization, a gene, which indicates a design parameter, is set as the ratio of resources to be allocated. The detailed process of the proposed method is as follows:

1. The genes of individuals are randomly set from 0 to 1 as the temporary value representing the amount of resources to all of the provider's problems.
2. The temporary resource values are normalized to fit the resource summation.
3. For each problem, the achievement of problem solving is calculated according to the sigmoid functions.
4. The customer satisfaction is calculated through the structure depicted in Figure 4 according to Equations (1)-(3).
5. Each individual is evaluated on the basis of the calculated value of customer satisfaction, which is regarded as the fitness value.

6. Each individual evolves based on the fitness value, i.e., customer satisfaction.
7. When the number of generations is achieved to a certain number, i.e., the set number of generation, the calculation is terminated. Otherwise, go back to step 2.

5 APPLICATION

The proposed method was applied to a real service, a facility construction and maintenance service for an electronic substation, in which a utility company was the service receiver. In this case, the utility company offers a social infrastructure service and maintains lifelines for the public. The utility company thus has to operate its facilities safely. A facility construction/maintenance company constructs and maintains the facilities or equipments (Figure 7).

Nowadays, the situation and environment surrounding the utility company are drastically changing, as described below:

- Existing large-scale facilities, including electronic substations, have to be maintained, since it is difficult to prepare a large space and budget to newly build such facilities.
- The environmental burden resulted from facility operation and construction has to be considered.
- It is becoming difficult to prevent risks and ensure safety in facility operation due to the extreme weather.

As a result of these situations, the utility company's requirements

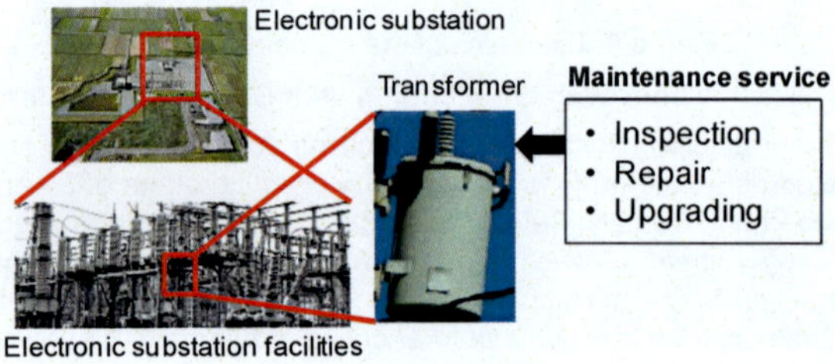

Figure 7: Case example.

for maintenance service were changing. Therefore, the maintenance service required improvement. To plan this improvement, in this application, a structure of the customer satisfaction was firstly constructed. Subsequently, for the planning of service improvement, the provider's problems were prioritized by applying the optimal resource allocation method.

5.1 Structure of customer satisfaction

A structure of customer satisfaction could be that shown in Figure 8. This figure illustrates relationships among required service quality elements, customer requirements, and customer's values/goals. These three elements were extracted by utilizing a tool for the extraction [16] as well as conducting interviews.

The importance of customer's values/goals was obtained with the Analytic Hierarchy Process (AHP) [17], which quantifies the relative importance of elements on a subjective scale by conducting a paired comparison. In addition, to determine the weights on the other arcs, two binary tables were prepared. In each table, the degree of association between elements was comparatively evaluated using figures 1, 3, and 5, where 5 meant a strong association. These evaluations were conducted by service customers.

5.2 Provider's problems and internal resources

It was revealed, on the basis of some discussions and data collection, that the service provider had problems in service operation, such as "insufficient technical tradition" and "lack of manuals on regular maintenance activities," as shown in the topmost row of Table 1. The influence of the identified problems on the service qualities was obtained using a binary table, namely, in the same manner as reported in 5.1. Table 1 shows the result.

The relationships between problem solving and the consumption of resources are represented with sigmoid functions. The center and gain values of each sigmoid function were set as shown in Table 2.

The achievement of problem solving was calculated as the product of values that are determined by sigmoid functions. Namely, for instance, when the amount of time resources invested to solve a problem is 0, the achievement of problem solving will be 0, even if a

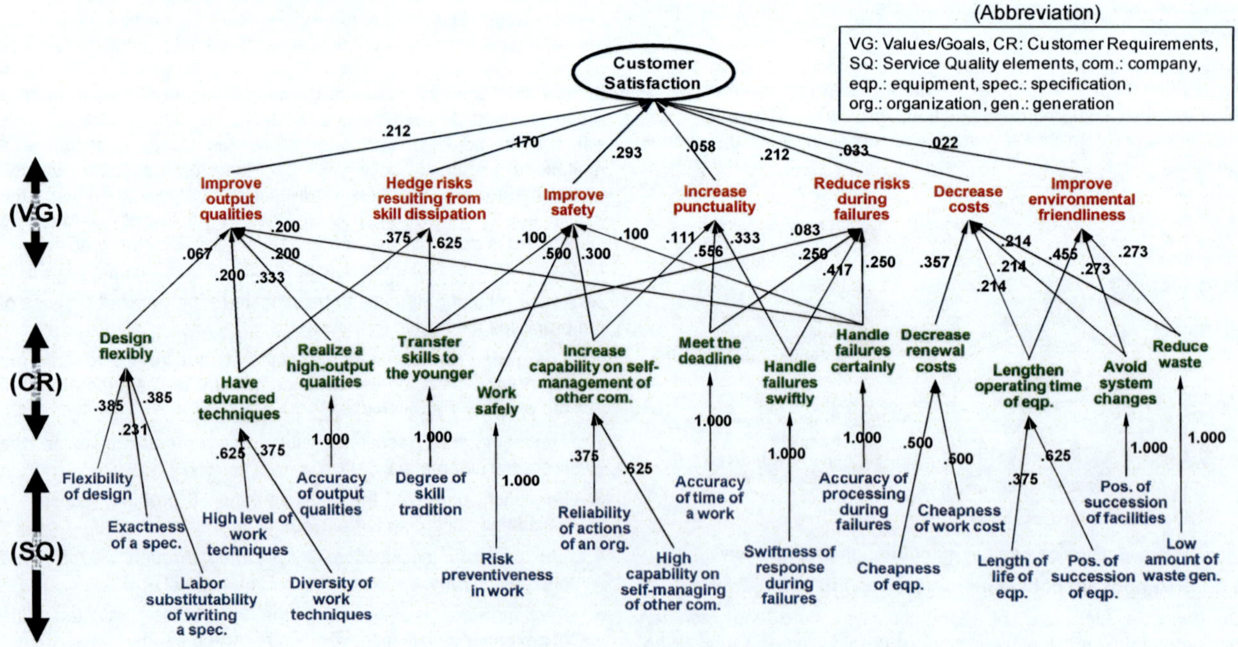

Figure 8: Structure of customer satisfaction of the utility company.

(Abbreviation)
spec.: specification, eqp.: equipment, org.: organization, com.: company

Service quality elements / Provider's problems	Insufficient technical tradition	Waste of time to prepare materials/equipment	Lack of information on work preparation	Uncertain preparation of materials/equipment	Lack of manuals on regular maintenance activities	Lack of rules for emergent situations	Insufficient general education	Old tools	Lack of cooperation with the client/partners	Increasing of older engineers	Few cross-trained workers	Time consumption in preparing ISO ducuments	Lack of management of work progress information	Lack of information sharing among companies	Disposal of usable equipment	Lack of managers	Degradation of practical competence
Flexibility of design	0.15	0.00	0.00	0.00	0.00	0.00	0.05	0.00	0.25	0.00	0.15	0.00	0.00	0.15	0.00	0.00	0.25
Exactness of a spec.	0.00	0.00	0.00	0.00	0.00	0.00	0.00	0.00	0.42	0.00	0.25	0.00	0.08	0.00	0.00	0.00	0.25
Labor substitutability of writing a spec.	0.23	0.00	0.00	0.00	0.00	0.00	0.00	0.00	0.08	0.00	0.23	0.00	0.23	0.00	0.00	0.00	0.23
High level of work techniques	0.28	0.00	0.00	0.00	0.00	0.00	0.06	0.00	0.00	0.06	0.17	0.00	0.00	0.17	0.00	0.00	0.28
Diversity of work techniques	0.27	0.00	0.00	0.00	0.00	0.00	0.00	0.00	0.00	0.00	0.45	0.00	0.00	0.00	0.00	0.00	0.27
Accuracy of output qualities	0.14	0.00	0.00	0.09	0.14	0.09	0.00	0.09	0.00	0.03	0.09	0.00	0.09	0.09	0.00	0.03	0.14
Degree of skill tradition	0.29	0.00	0.00	0.00	0.18	0.18	0.00	0.00	0.00	0.06	0.06	0.00	0.00	0.18	0.00	0.06	0.00
Accuracy of time of a work	0.03	0.10	0.10	0.00	0.17	0.00	0.00	0.00	0.00	0.03	0.00	0.03	0.17	0.17	0.00	0.10	0.10
Swiftness of response during failures	0.14	0.05	0.00	0.05	0.00	0.23	0.00	0.00	0.14	0.14	0.14	0.00	0.00	0.00	0.00	0.00	0.14
Accuracy of processing during failures	0.00	0.00	0.00	0.00	0.19	0.19	0.12	0.00	0.00	0.00	0.04	0.19	0.00	0.04	0.00	0.12	0.12
Cheapness of an eqp.	0.00	0.00	0.00	0.00	0.00	0.00	0.00	0.00	0.25	0.00	0.00	0.00	0.00	0.25	0.42	0.00	0.08
Cheapness of work cost	0.00	0.05	0.05	0.14	0.14	0.14	0.05	0.00	0.00	0.05	0.00	0.00	0.00	0.14	0.23	0.00	0.05
Length of life of an eqp.	0.00	0.00	0.00	0.00	0.00	0.00	0.00	0.00	0.45	0.00	0.00	0.00	0.00	0.00	0.27	0.00	0.27
Possibility of succesion of eqp.	0.00	0.00	0.00	0.00	0.00	0.00	0.00	0.00	0.00	0.00	0.00	0.00	0.00	0.00	0.63	0.00	0.38
Possibility of succession of facilities	0.00	0.00	0.00	0.00	0.00	0.00	0.00	0.00	0.00	0.00	0.00	0.00	0.00	0.00	0.63	0.00	0.38
Low amount of waste generation	0.00	0.12	0.19	0.00	0.04	0.00	0.04	0.00	0.19	0.00	0.00	0.00	0.12	0.12	0.19	0.00	0.00
Risk preventiveness in work	0.12	0.00	0.00	0.00	0.12	0.12	0.04	0.12	0.00	0.12	0.00	0.00	0.12	0.04	0.00	0.04	0.19
Reliability of actions of an org.	0.10	0.00	0.00	0.00	0.17	0.17	0.10	0.03	0.00	0.03	0.03	0.00	0.03	0.10	0.00	0.10	0.10
High capability on self-managing of other com.	0.03	0.00	0.00	0.00	0.09	0.09	0.15	0.00	0.00	0.09	0.00	0.00	0.15	0.15	0.00	0.15	0.09

Table 1: The influence of provider's problems on service quality elements.

Provider's problems	Monetary		Time	
	center	gain	center	gain
1. Insufficient technical tradition	0.2	2	0.6	1
2. Waste of time to prepare materials/ equipment	0.3	7	0.3	9
3. Lack of information on work preparation	0.3	7	0.4	1
4. Uncertain preparation of materials/ equipment	0.7	1	0.6	2
5. Lack of manuals on regular maintenance activities	0.1	9	0.1	10
6. Lack of rules for emergent situations	0.1	10	0.1	10
7. Insufficient general education	0.3	9	0.8	1
8. Old tools	0.2	10	0.1	10
9. Lack of cooperation with the client/partners	0.7	10	0.5	1
10. Increasing number of older engineers	0.3	10	0.6	9
11. Few cross-trained workers	0.4	4	0.4	10
12. Time consumption in preparing ISO documents	0.3	9	0.2	8
13. Lack of management of work progress information	0.3	6	0.4	2
14. Lack of information sharing among companies	0.4	6	0.4	2
15. Disposal of usable equipment	0.3	9	0.5	1
16. Lack of managers	0.5	2	0.6	1
17. Degradation of practical competence	0.2	2	0.6	1

Table 2: Parameters in the sigmoid function.

much higher amount of monetary resources were consumed.

5.3 Other settings

We considered two cases. In the first case, the amount of time and monetary resources to be consumed in the improvement was respectively set as 1.0 and 3.0. The first case meant service improvement with a small amount of monetary and time resources consumed, and the second one meant that the provider tried to improve his service with a large amount of them. In this application, the total amount of resources was represented as relative values. Moreover, the settings in GA were as follows:

- Fitness value: Customer satisfaction,
- Possibility of crossover: 0.7, Possibility of mutation: 0.01,
- Number of individuals: 200, Number of generations: 15000,
- One-point crossover and elite selection were applied.

5.4 Results

Figures 9 and 10 illustrate the result of optimal resource allocation, with Figure 9 showing the results when the total amount of resources was 1.0 and Figure 10 showing those when the total was 3.0 for each. The horizontal axis corresponds to the list of provider's problems to be solved in the improvement, and the vertical axis corresponds to the amount of resources, where the left bar denotes the monetary resource and the right indicates time.

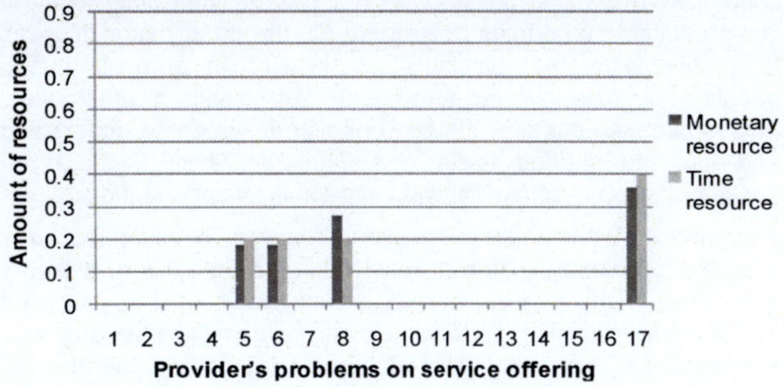

Figure 9: Results of resource allocation (amount = 1.0).

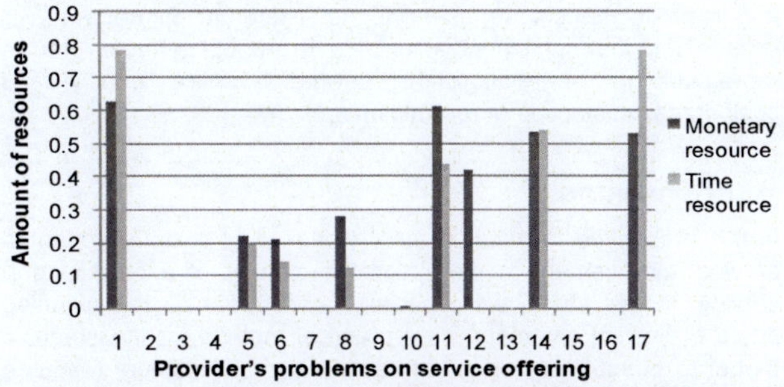

Figure 10: Results of resource allocation (amount = 3.0).

6 DISCUSSION

The result illustrated in Figure 9 shows that, in smaller-scale service improvement, a large amount of resources were allocated to provider's problems such as "Lack of manuals on regular maintenance activities (no. 5)," "Lack of rules for emergent situation (no. 6)," "Old tools (no. 8)," and "Degradation of practical competence (no. 17)." This indicates that these problems should be preferentially addressed in service improvement, since this allocation was optimized to maximize the value of customer satisfaction under the constraints of the provider's limited resources. Namely, the allocation could be regarded as the prioritization of problems that should be solved from the viewpoint of customer satisfaction. Therefore, it can be said that the proposed method was useful to support designer's decision-making on the planning of customer-oriented service improvement.

Meanwhile, the result shown in Figure 10 indicates that the provider's problems "Insufficient technical tradition (no. 1)," "Few cross-trained workers (no. 11)," and "Lack of information sharing among companies (no. 14)" as well as four problems should be solved in the smaller-scale improvement (no. 5, 6, 8, and 17) were important to be solved in the larger-scale service improvement.

The amounts of resources allocated to the problems "Insufficient technical tradition" and "Degradation of practical competence" in Figure 10 were much larger than those in Figure 9. This means that these problems should be solved preferentially and intensively to improve customer satisfaction in the longer-term improvement. It is considered that this result was fairly reasonable, since these were educational problems, which generally require much time to be solved and, furthermore, had a strong influence on a variety of service quality elements, as shown in Table 1.

In this study, the relationship between problem solving and resource consumption is represented as a sigmoid function. This kind of simplification would not be suitable for a detailed evaluation conducted in the final phase of service design, where the concrete design solution is already developed. On the other hand, this was fairly suitable for the planning stage, in which a large number of people should be involved and the design solution is ambiguous. This is because the simplification enabled designers to understand and use the method easily and intuitively, even if there are designers who are unfamiliar with mathematical presentations.

The model of customer satisfaction proposed in this paper was effective to quantify customer satisfaction on the basis of a theory developed in the service marketing field. However, the proposed model was simplified in terms of the relationships among four elements, i.e., service quality elements, customer requirements, customer's values/goals, and customer satisfaction. In the model, all relationships were represented as linearly increasing, although they are generally more complicated, e.g., negative relationships or non-linear relationships must be found. Therefore, future work will include the introduction of a function to enable a more precise representation of relationships between service quality and customer's satisfaction or recognition.

7 CONCLUSION

In this article, we focused on the importance of customer-oriented service improvement in the integrated offering of a product and service. In this study, we especially concentrate on the planning stage of the improvement. As a practical method for the stage, a model to quantify customer satisfaction and an optimum resource allocation method using GA are proposed to prioritize problems to be solved from the viewpoint of maximizing customer satisfaction. The application result showed that the proposed method was useful to support designer's decision-making on the planning of customer-oriented service improvement.

8 ACKNOWLEDGEMENT

This research was partially supported by the Ministry of Education, Science, Sports, and Culture through a Grant-in-Aid for Scientific Research (A), 21240012, 2009.

9 REFERENCES

[1] Tomiyama, T. (1997) A Manufacturing Paradigm Toward the 21st Century, Integrated Computer Aided Engineering, Vol. 4, pp. 159-178.

[2] Tukker, A. and Tischner, U. (2006) Product-services as a research field: Past, present and future. Reflections from a decade of research, J. of Engineering Design, Vol. 14, No. 17, pp. 1552-1556.

[3] Meier, H., Roy, R., and Seliger, G. (2010) Industrial Product-Service Systems-IPS², CIRP Annals - Manufacturing Technology, Vol. 59, No. 2, pp. 607-627.

[4] Shimomura, Y. and Tomiyama, T. (2005) Service Modeling for Service Engineering. In IFIP Int. Federation for Information Processing, 167, pp. 31-38, Springer Boston.

[5] Fitzsimmons, J.A. and Fitzsimmons, M.J. (2001) Service management: Operations, Strategy, and Information Technology, 3rd Edition, McGraw-Hill.

[6] McAloone, T.C. and Andreason, M.M. (2004) Design for Utility, Sustainability and Social Virtues, Developing Product Service Systems, In Proc. of the 8th Int. Design Conf., pp. 1545-1552.

[7] Uhlmann, E., Bochnig, H., and Stelzer, C. (2009) Characterization of Customer Requirements in IPS² creation, In Proc. of CIRP IPS2 Conf. 2010, CIRP, pp. 399-405.

[8] Gronroos, C. (2000) Service Management and Marketing: A Customer Relationship Management Approach, John Wiley & Sons.

[9] Office of Government Commerce (2007) Service Design – ITIL Version 3, The Stationery Office.

[10] Holland, J.H. (1975) Adaptation in Natural and Artificial Systems, Univ. of Michigan Press.

[11] Herrmann, A., Huber, F., and Braunstein, C. (2000) Market-driven product and service design: Bridging the gap between customer needs, quality management, and customer satisfaction, Int. J. of Production Economics, Vol. 66, pp. 77-96.

[12] Gutman, J. (1982) A means-end chain model based on consumer categorization processes, J. of Marketing, Vol. 46, No. 1, pp. 60-72.

[13] Akao, Y. (1990) Quality Function Deployment, Productivity Press.

[14] Rokeach, M. (1973) The Nature of Human Values, Free Press.

[15] Reynolds, T.J. and Gutman, J. (1988) Laddering Theory, Method, Analysis, and Interpretation, J. of Advertising Research, February/March, pp. 11-31.

[16] Akasaka, F., Hosono, S., Yamamura, K., Shimomura, Y., and Arai, T. (2009) A Method for Analyzing Customer Requirements in a B2B Service, In Proc. of the 9th Japan Korea Workshop on CAD/CAM, pp. 7-12.

[17] Satty, L. (1980) The Analytic Hierarchy Process, McGraw-Hill.

Customer Value Optimization in Product Service System Design

Yuanling Tan, Xuening Chu, Zaifang Zhang, Xiuli Geng
School of Mechanical Engineering, Shanghai Jiao Tong University
Shanghai, 200240 PR China

Abstract

Product service systems are the holistic design of products and services. In product design, traditional customer needs are confined to specific needs of tangible products, which are not appropriate for product service system design that intends to provide more comprehensive and higher level solutions. Thus, a notion of customer value-oriented product service system design was proposed, in which customer value is divided into economic value, functional value, and emotional value. An optimization model dealing with discrete and continuous variables was given to determine appropriate values of functional characters. The approach was applied in a real-world example for fork truck systems.

Keywords
Product Service System; Customer Value; Function Character; Optimization Model

1 INTRODUCTION

With the increasing competition in manufacturing industry, product service systems (PSS) were presented as a new manufacturing model to gain competitive edge. In PSS, what the enterprise provides shifts from purchasing just a physical product to a result or a functionality which means products combining with corresponding services [1][2][3]. Compared with the focus on customer specific needs of physical level in product design, PSS intend to provide more comprehensive and higher level solutions, including various monetary and non-monetary cost and emotional appeal. For example, in the design of dining-room PSS, customers do not need to tell that the color of the tables they want is red or blue, they just request that it should beharmonious with the environment. PSS design doesn't need customers to know details of products or services, but on a more abstract level to embrace more information. Accordingly, traditional expressions of customer needs are not suitable for the PSS design. A new driven target of customer value for PSS design is proposed in this article. There are three evident advantages: Firstly, customer value is the key factor deciding if the customer will purchase or not. Customers may weigh the actual costs (including monetary costs and non-monetary costs incurred) against the actual benefits and are inclined to make a buying decision if the latter outweigh the former. The more the customers benefit the more they tend to buy. [4] provided evidence supporting an influential role of value in the consumers' purchase decision making. Secondly, customer value plays a mediating role between trust and loyalty [5], which is a critical factor for enterprises to establish long-term relationships with customers. If the quality is good and the price is high, customers will probable continue searching for better alternatives for the same product or service. This easily leads to the loss of the customers. Thirdly, customer value brings what customer focus on together and provides an integrative balance between benefits and costs. Customer value-oriented PSS design can provide the best cost-effective choice for customers.

In this study, customer value is used as the starting point. The importance of customer value was already recognized by many scholars. [6] argued that customer value plays a decisive role in business performance. What the customer purchases is the value rather than the product, and it is the responsibility of the enterprise is to provide customers with more value. [7] said the next major source for competitive advantage likely will come from the more outward orientation towards customers, as indicated by the many calls for organizations to compete on superior customer value delivery. [8] believed that the perceptions of value directly impact customers' satisfaction with suppliers, which influences customer retention. [9] created a customer value-based theory of the firm. [10] indicated that customer value is a strategic weapon in attracting and retaining customers and has become one of the most significant factors in the success of both manufacturing businesses and service providers. In the PSS research field, [11] tried to bring customer value to PSS design. He proposed a customer value model for sustainable service design, in which the change of receiver state parameters was regarded as customer value. [12] optimized the functional solution from a life cycle perspective, with reference to the customer value,.

In this study, customer value is used as the starting point of the PSS design to describe customer expectation information. A customer value index system consisting of economical value, functional value, and emotional value is proposed to quantize customer value. After customer value expectation is obtained, what we care about is how to convert customer information into design information (functional characters). Designing customer value translation tools for converting customer value into FCs is also an urgent problem to be solved. A customer value optimization model which can deal with discrete and continuous FCs is proposed to determine values of FCs.

The paper is organized as follows: The second section offers an index system to measure customer value and gives a framework of customer value-based PSS design. The third section presents an optimization model to determine the parameters of the function,

J. Hesselbach and C. Herrmann (eds.), *Functional Thinking for Value Creation: Proceedings of the 3rd CIRP International Conference on Industrial Product Service Systems, Technische Universität Braunschweig, Braunschweig, Germany, May 5th - 6th, 2011*,
DOI 10.1007/978-3-642-19689-8_18, © Springer-Verlag Berlin Heidelberg 2011

maximizing customer value of PSS. The forth section applies the proposed method to the forklift truck's PSS design, to demonstrate the efficiency and feasibility of the method. Finally, the last section of the article summarizes advantages of customer value-oriented PSS design, and discusses implications for future customer value-related research.

2 CUSTOMER VALUE-BASED PSS DESIGN

2.1 Customer value index system

How to describe customers' expectation of value is *the question of questions to be solved.* Building a customer value index is efficient to determine which variables about customers relate most directly to sales and use of PSS. The development of a customer value measure index system should begin with a deep understanding of the concept itself. [13] deemed customer value as the quality that customers can afford. [4] argued that value was the consumer's overall assessment of the utility of a product based on perceptions of what is received and what is given. [14] conceptualized perceived value as a tradeoff between perceived quality and perceived psychological as well as monetary sacrifice. [7] laid out that customer value is a hierarchically structured construct at levels of consumption goals, consequences, and attributes. These perceptions typically involve a trade-off between what the customer receives (e.g. quality, benefits, worth, utilities) and what he or she gives up to acquire and use a product (e.g., price, sacrifices). In the past research, most researchers focused on customer perceived value, which is widely used for evaluation in the after-purchase stage. In fact, in the post-purchase stage, customers may imagine what value they want in the product/service. Defining customer expectations of value can be used to guide the following PSS design. It has a future-oriented significance for enterprises to sustain customers. Thus, a customer value index system was built to quantize and identify what customers expect in this study.

A customer value index should comprise what customers focus on and what influences their purchase decision. [15] proposed that consumers' decisions are influenced by five consumption value: functional value, social value, emotional value, epistemic value and conditional value. In PSS, enterprises try to satisfy customers by offering functions, considering costs and emotional factors. Thus, we divide customer value into three parts: the economic, the functional, and the emotional part. Economic value denotes costs for quantification and measurement, such as purchase cost, time cost, etc. Functional value denotes benefits for quantification and measurement, such as the quantity level, functional specifications, etc. Correspondingly, emotional value particular puts emphasis on describing customer psychological feelings, such as riding comfort, disappointment arising from appearance, etc. ; which is fuzzy and hard to measure accurately. Note that emotional value may be benefits, but also costs, basically depending on the specific definition of index variables. Index variables at the next lower level should be determined according to the specific circumstances. They may be changed depending on different PSS features and the actual situation in application. Because for different kinds of PSS, factors that influence customer purchase decisions are different. Determining index variables of PSS often requires research to quantify information already available about customers, which can be done by surveying a portion of the potential customers. An important part of building a successful customer value index is that it should be as simple and understandable as possible. By keeping each index variable simple, customers can better describe expectation, while managers can better understand and use the index system.

3 A FRAMEWORK FOR CUSTOMER VALUE BASED PSS DESIGN

[16] introduced the four domains framework into product design. In his opinion, the world of design is made up of four domains: the customer domain, the functional domain, the physical domain and the process domain. The domain on the left relative to the domain on the right represents 'what we want to achieve', whereas the domain on the right represents the design solution for 'how we propose to satisfy the requirements specified in the left domain'. In this article, to illustrate the design process of customer value-based PSS, an integrated framework is proposed as shown in Figure 1. We use the four domains framework of Axiomatic Design Theory for reference, proposing a framework of mapping processes among value domain, function domain and concept domain. Different from traditional design processes, customer value is used as the starting point of the PSS design to describe customer expectation information. Moreover, the physical domain and the process domain are merged Into the concept domain to reduce the loss of valid information in the multi-layer mapping.

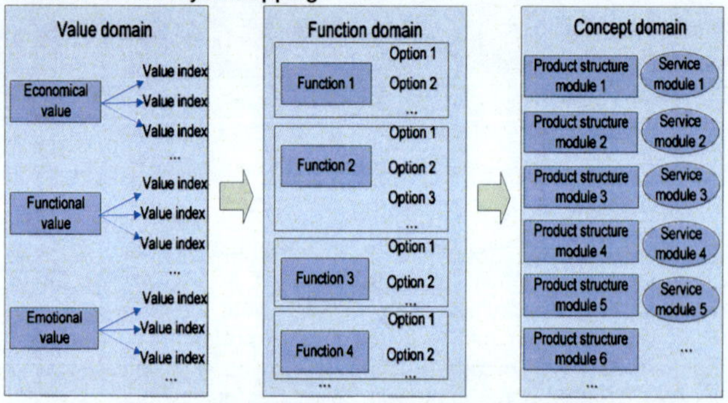

Figure 1: A framework for customer value-based PSS design.

The three domains are described as follows:

Value domain: this domain mainly describes the needs information from the angle of customers. Customer value can across-the-board describe the customer expectation information from a macroeconomic perspective. It describes customer expectations from three dimensions, which are economic value, functional value, and emotional value.

Functional domain: this domain needs to determine the contents of two levels, first, to determine what features need to achieve, and second, to determine the corresponding parameters of the function needs to achieve. What is described in this domain is design information and the mapping from customer domain to the function domain translates customer expectation information into concrete design parameters. Note that this feature is optional, not all features must be achieved, as customers will not pay for functional modules they don't need. For example, there are two models of laptops with basically the same configuration, the only difference is that model A has a built-in camera. Model A is 20 $ more expensive than B. Many customers may choose model B, because they do not need the camera. Here, there is no clear distinction between product function and service function. For each function, it may be achieved by the product or by services, or even may be the combination of product and services.

Concept domain: this domain is composed of product and services together. Different from service design which lags behind product design in traditional product design, they are generated together in PSS design. Product components and services can complement and substitute each other to form final solutions - the most cost-effective one for customers. The main problem in this domain is how to deal with the mutual constraints between product components, as well as between product components and services.

4 SETTING UP YOUR CUSTOMER VALUE ORIENTED OPTIMIZATION MODEL

This paper focuses on transforming customer domain into function domain. After the customer value index system is established, the problem is how to determine value of FCs to get the maximum customer value. Quality function deployment (QFD) is a cross-functional planning tool widely used to help designers to determine to what extent certain characteristics of products are to be met for the sake of gaining a higher level of overall customer satisfaction. Most of the existing approaches and models for QFD planning seldom consider the different types of FCs: continuous type and discrete type. [17] proposed a mixed optimization model, which can deal with continuous and discrete FCs. As a basis, a mixed optimization model is advanced to realize the calculation of FCs. The major difference is the target function is not maximizing customer satisfaction, but maximizing customer value. By analyzing relations between customer value index and FCs and considering cost and technical constraints, the appropriate value for FCs are determined. The optimization model can be described as follows:

$$\max V(\Delta v_{11},\Delta v_{12},\ldots,\Delta v_{1a},\Delta v_{21},\Delta v_{22},\ldots,\Delta v_{2b},\Delta v_{31},\Delta v_{32},\ldots,\Delta v_{3c},) \quad (1)$$

$$s.t.\ \Delta v_{1i} = f_{1i}(\Delta x_1,\Delta x_2,\ldots,\Delta x_n)\ \ i=1,2,\ldots,a$$

$$\Delta v_{2i} = f_{2i}(\Delta x_1,\Delta x_2,\ldots,\Delta x_n)\ \ i=1,2,\ldots,b$$

$$\Delta v_{3i} = f_{3i}(\Delta x_1,\Delta x_2,\ldots,\Delta x_n)\ \ i=1,2,\ldots,c$$

$$\Delta x_j = g_j(\Delta x_1,\Delta x_2,\ldots,\Delta x_{j-1},\Delta x_{j+1},\ldots,\Delta x_n)\ \ j=1,2,\ldots n$$

$$x_j=\sum_{k=1}^{J_j} z_{jk}x_{jk}\ \ j=1,2,\ldots l$$

$$\sum_{k=1}^{J_j} z_{jk}=1\ \ j=1,2,\ldots l$$

$$z_j\Delta x_{jL}\le \Delta x_j \le z_j\Delta x_{jH}\ \ j=l+1,l+2,\ldots n$$

$$\sum_{j=1}^{l}\sum_{k=1}^{J_j} z_{jk}c_{pjk}+\sum_{j=l+1}^{n}(D_j z_j + c_{pj}\Delta x_j)\le B$$

$$z_j\in\{0,1\}\ \ j=l+1,l+2,\ldots n$$

$$z_{jk}\in\{0,1\}\ \ j=1,2,\ldots l\ \ k=1,2,\ldots J_j$$

$$\Delta x_j=\frac{x_j-x_{j0}}{x_{jH}-x_{jL}}$$

Where a is the number of economic value indexes, b is the number of functional value indexes, c is the number of emotional value indexes, n is the number of FCs, l is the number of discrete variables . Δv_{1i} is a variation of the economic value index, similarly, Δv_{2i} is a variation of the functional value index, Δv_{3i} is a variation of the emotional value index. x_j is the value of FCs, while Δx_j is the *variation rate* of x_j, in which, x_{j0} is the initial value of x_j, x_{jH} is the upper bound of x_j, x_{jL} is the lower bound of x_j. z_{jk} and z_j are Boolean variables. The objective function is a variation of the customer value, which is simply the sum of contributions from each member of customer value. Constraints (7) ensures that total costs do not exceed the given budget limit.

5 CASE STUDY

Forklift trucks are powered industrial trucks used to lift and transport materials. They are most commonly used to move loads stored on pallets. Company A is a professional forklift truck manufacturer, which continues fulfilling customer's expectations by providing high-quality cost-effective products and superior technical services. In recent years, in order to obtain competitive advantage, company A constantly increases the proportion of services, provides comprehensive services and solutions, as well as logistics design and consultancy, etc. How to design a reasonable product configuration and the corresponding services is an urgent problem to resolve for company A. This case presents the mapping process from value domain to function domain of forklift truck PSS systems.

According to the expectation for customer value in forklift industry, the customer value index system is built as Figure 2. In the customer value system, customers are only concerned about their target value needs to be met, without caring about how it is achieved by products or services.

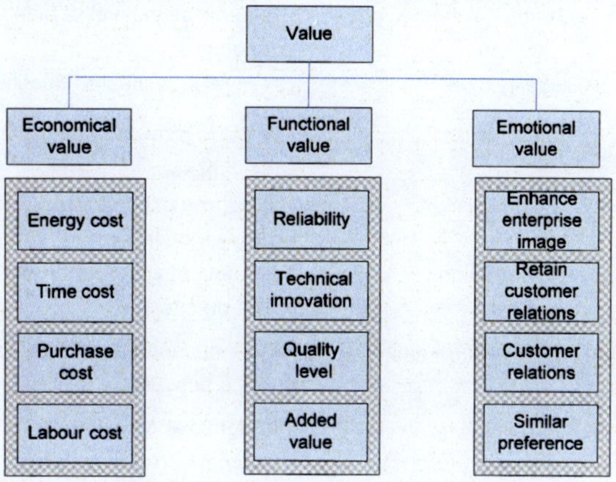

Figure 2: Customer value index system for forklift trucks.

Detail of customer value indexes is represented in Table 1.

5.1 Construct the HoQ

Figure 3 shows the HoQ of the translation from the required customer value to FCs. Information in the HoQ is gathered from customers and designers.

The calculating process of HoQ has the following steps [17].

Step 1: Determine the importance weights of each customer value index. The data are obtained from interviewing customers.

Step 2: Identify FCs and their correlation matrix. Function specifications parameter options are represented in Table 2. Among FCs, FC1, FC2, FC3, FC4, FC7are continuous variables, while FC5, FC6, FC8, FC9 are discrete variables. The correlation matrix is given by designers according to their experience.

No.	Customer Value Index	Explanation
v_1	Energy cost	The consumption of energy in use
v_2	Time cost	Including the time spent in use, maintenance and repair process, etc
v_3	Purchase cost	One-time purchase expenses
v_4	Labour cost	The physical consumption for participants
v_5	Reliability	The ability or possibility for system to realize designated function in certain period, under certain conditions.
v_6	Technical innovation	The application level of high technology
v_7	Quality level	The level in respect to quality
v_8	Added value	The value provided by additional functions
v_9	Riding comfort	The degree of comfortable feeling for operators
v_{10}	Presentation	The effect of appearance and image on customer
v_{11}	Customer relations	Length of cooperation relationship between enterprise and customer
v_{12}	Similar preference	Customers preference of similar products

Table 1: Customer value indexes detailed instruction.

No.	Function specifications	Parameter Options
FC 1	Maximum running rate	16-50 (km/h)
FC 2	Rated capacity	3000-12000 (kg)
FC 3	Lift height	3-12 (m)
FC 4	Lifting speed	1.2- 2.5 (km/h)
FC 5	Safety protection level	I , II ,III
FC 6	Engineer rated power	26, 50, 162 (kw)
FC 7	Biggest climbing power (full load)	35- 50(°)
FC 8	Maintain level	I , II ,III
FC 9	Curing cycle	100,500(h)

Table 2: Functional characters parameter options list.

Step 3: Elaborate relationships matrix between customer value indexes and FCs. Each value in the matrix means the variation of value index arising from the variation of FC. Data are obtained from interviewing experts and customers.

Step 4: Identify the affection weights of FCs on customer value. Two influence coefficients are defined in this paper, which are affection weight of FC on benefits and affection weight of FC on costs. The affection weight of FCs on customer value is the difference between affection weight of FC on benefits and affection weight of FC on costs. We use the difference instead of ratio here to avoid two negatives make a positive. It doesn't affect calculation results.

5.2 Optimization calculation

Objective function:

$$-0.0927\Delta x_1 - 0.017\Delta x_2 - 0.0077\Delta x_3 - 0.0211\Delta x_4 + 0.0549\Delta x_5 - 0.007\Delta x_6 + 0.0634\Delta x_7 + 0.07\Delta x_8 + 0.000814\Delta x_9$$

Disperse FCs constraint condition to the scope of the technical characteristics:

$$x_5 = 1\times z_{51} + 2\times z_{52} + 3\times z_{53}$$

$$x_6 = 26\times z_{61} + 50\times z_{62} + 162\times z_{63} \quad x_8 = 1\times z_{81} + 2\times z_{82} + 3\times z_{83}$$

$$x_9 = 100\times z_{91} + 500\times z_{92}$$

$$z_5 = z_{52} + z_{53};\ z_6 = z_{62} + z_{63};\ z_8 = z_{82} + z_{83};\ z_9 = z_{92};$$

$$z_{51} + z_{52} + z_{53} = 1;\ z_{61} + z_{62} + z_{63} = 1;\ z_{81} + z_{82} + z_{83} = 1;$$

$$z_{91} + z_{92} = 1.$$

Constant FCs constraint condition according to the scope of the technical characteristics:

$$\Delta x_2 = (x_2 - 3000)/9000;\ \Delta x_3 = (x_3 - 3)/9;$$

$$\Delta x_4 = (x_4 - 1.2)/1.3;\ \Delta x_7 = (x_7 - 35)/15;\ 0 \le \Delta x_1 \le z_1;$$

$$0 \le \Delta x_2 \le z_2;\ 0 \le \Delta x_3 \le z_3;\ 0 \le \Delta x_4 \le z_4;\ 0 \le \Delta x_7 \le z_7.$$

Considering the interaction relationship between FCs:

$$\Delta x_1 + \Delta x_2 - \Delta x_4 < 0.1;\ \Delta x_7 - \Delta x_4 \le 0.2;\ \Delta x_2 - 2\Delta x_3 > 0.2.$$

Cost constraint condition:

FCs	FC 1	FC 2	FC 3	FC 4	FC 7
Fixed cost coefficient	0.65	0.4	0.82	0.7	0.54
Variable cost coefficient	0.07	0.054	0.048	0.05	0.069

Table 3: Continuous FCs unit change relation coefficient with cost.

FC5	I	II	III
corresponding cost	1.65	2.23	2.36
FC6	26	50	162
corresponding cost	1.54	2.36	2.7
FC8	I	II	III
corresponding cost	0.8	1.63	2.75
FC9	100	500	
corresponding cost	2.4	1.32	

Table 4: Disperse FCs for each value of the corresponding cost.

$$C_1 = 0.65z_1 + 0.07\Delta x_1 + 0.4z_2 + 0.054\Delta x_2 + 0.82z_3 + 0.07\Delta x_3 + 0.7z_4 + 0.05\Delta x_4 + 0.54z_7 + 0.0069\Delta x_7$$

$$C_2 = 1.65z_{51} + 2.23z_{52} + 2.36z_{53} + 1.54z_{61} + 2.36z_{62} + 2.7z_{63} + 0.8z_{81} + 1.63z_{82} + 2.75z_{83} + 2.4z_{91} + 1.32z_{92} \quad C_1 + C_2 \le 12$$

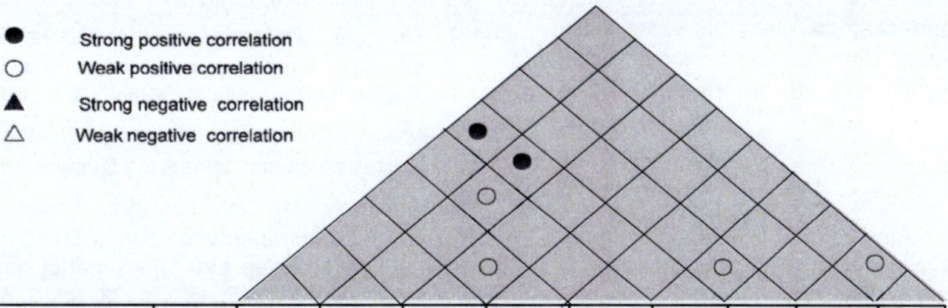

Customer index	RW	FC1	FC2	FC3	FC4	FC5	FC6	FC7	FC8	FC9
v_1	0.02	0.2		0.01	0.02		0.35			
v_2	0.07	-0.1			0.1				-0.1	0.2
v_3	0.21	0.48	0.1	0.12		0.3			0.4	-0.22
v_4	0.14			-0.15				-0.18		
v_5	0.12			-0.06	-0.24	0.2			0.37	0.5
v_6	0.01	0.01	0.02	0.01	0.16			0.1		
v_7	0.15	0.02	0.02	0.02	0.01	0.26		0.2	0.3	0.2
v_8	0.08					0.3			0.1	-0.08
v_9	0.05					0.1				
v_{10}	0.03					0.05				
v_{11}	0.04	0.01	0.01		0.1	0.15		0.08	0.52	-0.2
v_{12}	0.08	0.02		0.01	0.1	0.23		0.05	0.36	-0.33
FC unit		km/h	kg	m	km/h	rank	kw	°	rank	h
FC initial value		16	3000	3	1.2	I	26	35	I	100
FC min value		16	3000	3	1.2	I	26	35	I	100
FC max value		50	12000	12	2.5	III	162	50	III	500
Affection weight of FC on benefits		0.0051	0.0036	-0.0033	-0.0137	0.1179		0.0382	0.147	0.0492
Affection weight of FC on costs		0.0978	0.021	0.0044	0.0074	0.063	0.007	-0.0252	0.077	-0.0322
Affection weight of FC on customer value		-0.0927	-0.017	-0.0077	-0.0211	0.0549	-0.007	0.0634	0.07	0.0814

Figure 3: HoQ for forklift truck.

The computational results

$z_{51} = 0$; $z_{53} = 1$; $z_{62} = 0$; $z_{81} = 0$; $z_{83} = 0$; $z_{91} = 0$; $z_{92} = 1$; $\Delta x_1 = 0$; $\Delta x_2 = 0.2$; $\Delta x_3 = 0$; $\Delta x_4 = 0.8$; $\Delta x_7 = 1$.

The values of FCs are given in Table 5:

No.	Parameter Options
FC 1	16 (km/h)
FC 2	4800 (kg)
FC 3	3 (m)
FC 4	2.24 (km/h)
FC 5	III
FC 6	26 (kw)
FC 7	50(°)
FC 8	II
FC 9	500(h)

Table 5: Final values of FCs

In order to simplify, we just list scope constraints and cost constraints here. In fact, for the actual problem, constraint conditions are quite complex, including time constraints, design constraints, etc.

6 SUMMARY

In PSS, product design and service design is carried out at the same time to provide integrated solutions to customers, where the traditional expression of customer needs does not fit. Accordingly, a customer value-oriented PSS design framework is proposed in this paper. Customer value has a very important significance for both customers and enterprises by offering the most cost-effective choice for customers, helping enterprises to establish a long-term cooperative relationship with customers to gain greater economic benefits. Focusing primarily on customers' post-purchase decision-making process, the customer value index system is composed of economic value, functional value, and emotional value in this paper, which brings factors that influence customers purchase decision together. Besides, a customer value optimization model is put forward to identify appropriate values of FCs through calculating the change of customer value caused by the change of FCs. To simplify, in this study only scope constraints and cost constraints are taken into account.

Further work would consider perfecting the customer value index system to make it more standard. And for the customer value optimization model, constraint conditions are not enough, further complement is needed.

7 ACKNOWLEDGMENTS

The project was supported by National Natural Science Foundation, China (No.51075261), Shanghai Science and Technology Innovation Action Plan (No.09dz1124600), No.10dz1121600), Shanghai Jiao Tong University Innovation Fund For Postgraduates. The authors would also like to express their grateful appreciation to the anonymous referees for their helpful comments to improve the quality of this paper.

8 REFERENCES

[1] Lifset, R. (2000): Moving from products to services, in: Journal of Industrial Ecology, Vol. 4, No. 1, pp. 1-2.

[2] Mont, O. (2002): Clarifying the concept of product-service system, in: Journal of Cleaner Production, Vol.10, pp. 237-245.

[3] Maussang, N.; Zwolinski, P.; Brissaud, D. (2005): Design of product-service systems, in: The 10th ERSCP, Antwerp, Belgium.

[4] Zeithaml; Valarie A. (1988): Consumer Perceptions of Price Quality, and Value: A Means-End Model and Synthesis of Evidence. In: Journal of Marketing, Vol. 52, pp. 2-22.

[5] Sirdeshmukh; Deepak; Jagdip S., et al. (2002): Consumer Trust, Value, and Loyalty in Relational Exchanges, in: Journal of Marketing, Vol.66 (January), pp. 15-37.

[6] Drucker, P. F. (1954): The Practice of Management. New York: Harper & Row.

[7] Woodruff; Robert B. (1997): Customer Value: The Next Source for Competitive Advantage. In: Journal of the Academy of Marketing Sciences, Vol. 25, No. 2, pp. 139-153.

[8] Flint, D. J.; Woodruff, R. B.; Gardial, S. F. (1997): Customer Value Change in Industrial Marketing Relationships: A Call for New Strategies and Research, in: Industrial Marketing Management, Vol.26, pp. 163-175.

[9] Slater; Stanley; (1997): Developing a Customer Value-Based The-ory of the Firm, in: Journal of the Academy of Marketing Science, Vol.25 (Spring), pp.162-67.

[10] Wang, Y.; Lo, H. P.; Chi, R. et al. (2004): An Integrated Framework for Customer Value and Customer-Relationship-Management Performance: A Customer-Based Perspective from China, in: Managing Service Quality Vol.14, pp.169-82.

[11] Kimita, K.; Shimomura, Y.; Arai, T. (2009): A Customer Value Model for Sustainable Service Design, in: CIRP Journal of Manufacturing Science and Technology, Vol. 4, No. 1, pp. 254-261.

[12] Hara,T.; Arai, T.; Shimomura, Y.; et al.(2009): Service CAD system to integrate product and human activity for total value, in: CIRP Journal of Manufacturing Science and Technology, pp. 262-271.

[13] Grocer, P. (1964): How much do customers know about retail prices? (February), pp. 103-106.

[14] Dodds, W. B.; Monroe, K. B.; Grewal, D. (1991) : Effects of Price, brand and store Information on Buyers' Product Evaluations, in: Journal of Marketing Research, Vol. 28pp. 307-319.

[15] Sheth, J. N.; Newman, B. I.; Gross, B. L. (1991): Why we purchase what we buy: A theory of consumption values, in: Journal of Business Re search, Vol.22, pp.159-170.

[16] Nam P. Suh (1998): Axiomatic Design Theory for Systems, in: Research in Engineering Design, Vol.10, pp.189–209.

[17] Griffin, A.; Hauser, J. (1992): Patterns of communications among marketing, engineering and manufacturing – A comparison between two new product development teams, in: Management Science, Vol. 38, No. 3, pp. 360-373.

[18] Yang, M.; Lin, Z. (2004): Optimization model of GOQ considering continuous and discrete technical characteristics, in: Chinese journal of mechanical engineering, Vol. 40, No. 3, pp. 111-114.

[19] Jones, T. O.; Earl Sasser Jr. W.(1995): Why satisfied customers defect, in: Harvard Business Review, pp. 56-69.

Integrating PSS Design Methods with Systems for Customer Value Management and Customer Satisfaction Management

Amir Taabodi[1] and Tomohiko Sakao[1]

[1] Department of Management and Engineering, Linköping University, Sweden

Abstract

Manufacturers today are increasingly focusing on services and concepts such as Product Service Systems (PSS) and Integrated Product Service Engineering (IPSE) have been getting more attention. On the other hand, there are various existing methods/systems in companies. Some aim at managing customer value or/and customer satisfaction. Based on some empirical results obtained from an attempt to design PSS at a manufacturer in Sweden Customer Received Value (CRV) is defined as an index to integrate Customer Value Management (CVM) and Customer Satisfaction Management (CSM). The results reveal the benefits of using PSS as an approach to manage CVM and CSM.

Keywords:

Implementation; Empirical Study; Customer Received Value

1 INTRODUCTION

Manufacturers today are increasingly focusing on services [1], and concepts such as Product/Service Systems (PSS) [2] and Integrated Product Service Engineering (IPSE) [3] have been getting more attention. In line with this trend, several methods for designing PSS effectively and efficiently have been developed and verified for use in manufacturing industries (e.g. [4, 5, 6, 7]). Industry is indeed interested in methodical support by PSS design methods, since they want to improve their performance efficiently. On the other hand, there are various existing methods/systems in companies. Some aim at managing customer value (e.g. value map, performance-importance (P-I) matrix) or/and customer satisfaction measures (e.g. expectancy-disconfirmation model [8]). There are different models that try to describe the link between customer value and customer satisfaction, and even with other performance measures such as customer loyalty and competitiveness. But obviously, measuring customer satisfaction and customer value is something and managing them is something else. Customer value management (CVM) and customer satisfaction management (CSM) are important in order to build and sustain competitive advantages over product lifecycle as well as the loyalty of customers, especially for those products with long life cycle and high technological complexity, e.g. industrial investment machines. In spite of the obvious connection and potential benefits through integrating two types of methods/systems such as prioritizing R&D projects in order to sell new products and/or keeping current customer satisfied as well as managing the internal relation between different departments and interaction with customers through a unified and systematic method, little research has investigated that possibility.

This paper discusses a potential way of integrating these two. To do so, some empirical results obtained from an attempt to design PSS and relate a PSS design method to existing methods for CVM/CSM at a manufacturer in Sweden are presented. Section 2 introduces related works in the fields of PSS design, CSM, and CVM, especially in order to shed light on the connection between them. Section 3 proposes a potential way of integration, while Section 4 reports some empirical results from applying the idea at a Swedish manufacturer. Then, Section 5 presents a discussion and Section 6 concludes the paper.

2 RELATED WORK

2.1 Customer satisfaction management

Measuring customer satisfaction has become increasingly common in the last two decades as a strong predictor of behavioural variables such as repurchase intention, word-of-mouth or loyalty [9]. The expectancy–disconfirmation model defines customer satisfaction as a result of a subjective (or direct) comparison between customer expectation and perception. When the product or service performance is equal to the customer expectations, customer satisfaction will be achieved. The lower or higher performance of the product will result in dissatisfied or very satisfied customers, respectively. Thus, the performance of a product will determine the customer satisfaction [8][10]. Although this model is popular among scholars, the nature of satisfaction is not sufficiently clear. Satisfaction is a state of mind, and it clearly arises through a cognitive process which is basically comparing perceived performance against some comparison standard [9].

CSM groups customers in three categories: very satisfied, satisfied and dissatisfied. CSM usually establishes a baseline to a target year on a year improvement, and then measures the level of satisfaction and dissatisfaction of customers. Measures are useful, as the first two or three of such a survey yield insight and action (and hopefully improve customer satisfaction), but the scores of the customer survey soon flatten out and initiative wanes. At this point, practitioners look for improved measures and examine the relationship between customer satisfaction and loyalty intention of customers by using criteria such as "willingness to repurchase" or "willingness to recommend" [11].

2.2 Customer value management system

In its most basic and simplest form, CVM can be considered as a kind of enhancement of CSM through including price and value.

J. Hesselbach and C. Herrmann (eds.), *Functional Thinking for Value Creation: Proceedings of the 3rd CIRP International Conference on Industrial Product Service Systems, Technische Universität Braunschweig, Braunschweig, Germany, May 5th - 6th, 2011*,
DOI 10.1007/978-3-642-19689-8_19,

CVM not only connects customer satisfaction with every aspect of product and service, but also measures satisfaction related to the price paid by customer. In addition, CVM provides a base for comparative value assessment across the marketplace, your customers and competitors' customers. It is possible to have highly satisfied customers but a diminishing market share. This case could imply that the overall need of the marketplace was not satisfied [11].

Thus, CVM clarifies how customers prefer one supplier to another and focus on making gains in both customer and market share. It is shown that value is a concept varying from one consumer to another [12]. Identification and prioritization of customers' value for each market segment is necessary to increase customer value. This requires the determination of answers to three basic questions:

1. What are the key factors that customers value when they choose between competing offerings in a marketplace?
2. How is the organization's performance rated on each factor relative to each of its main competitors?
3. What is the relative importance of each of these components of customer value?

The answers of these questions provide a ground to construct a weighted index of customer value for the company and its competitors [11].

The reference [13] lists the use of the Performance-Importance (P-I) Matrix as a tool to evaluate a firm's competitive position in the market, to identify improvement opportunities, and to guide strategic planning. The P-I Matrix utilizes customer satisfaction data, where for each product (or service) attribute, the average score of performance is plotted against the average importance score using a 2 * 2 matrix.

An interesting method has been developed [14]. One idea adopted in this method is that customers recognise, in such promising spaces, high importance but low satisfaction. Thus, this method asks customers to give ratings to importance and satisfaction of several potentially desired outcomes.

According to [15], the Kano model [16] contains value components, which are dissatisfiers (must-be), satisfiers (more is better), and delighters (exciters). It means, when a certain value of a customer is not fulfilled properly by product-service attributes, customers evaluate the performance of the product/service below his/her expectations, which in turn causes dissatisfaction. On the other hand, product/service attributes that deliver higher value than customer expectations cause higher satisfaction, e.g. the delighter. Therefore, customer satisfaction seems to be an important indication whether the customers, during the use of a product, regard the attributes of a product as valuable or not.

A purchasing decision and perceived value, e.g. exchange value, are associated to each other. According to Zeithaml [17], perceived value is the result of several factors such as perceived quality and price. The reference [18] categorizes the constituents of perceived value in three groups, e.g. product-related quality, service-related quality, and promotion-related quality.

The perceived value is defined as "the consumer's overall assessment of the utility of a product (or service) based on perceptions of what is received and what is given" [17].

The reference [19] defines customer-perceived value as the ratio between perceived benefits and perceived sacrifice:

$$\text{Customer-perceived value} = \frac{\text{Perceived benefits}}{\text{Perceived sacrifice}} \quad (1)$$

The perceived sacrifice includes all the costs the buyer faces when making a purchase. The perceived benefits are some combination of physical attributes, service attributes and technical support available in relation to the particular use of the product, as well as the purchase price and other indicators of perceived quality.

The reference [9] lists some characteristics of customer-perceived value such as:

- Pre-/Post-Purchase Perspective
- Strategic Orientation
- Present and Potential Customer
- Suppliers and Competitors' Offerings

Customer-perceived value has a strategic orientation which aims at assessing how the value offered by a supplier can best meet customer requirements.

Lapierre [20] has developed a conceptual model to clarify the process of customer value creation over time in a B2B market. Two levels of value have been defined, value exchange and value in use. The exchange value criteria consist of four main parts: technical quality, functional quality, relational variable and image.

In the usage phase, the value has four main criteria: financial, social, operational and strategic [20]. After the usage phase, customers gain a total received value. Customer-received value (CRV) is the ultimate value that the customer gains after using the integrated product and service offered. In addition, customers accomplish goals through acquisition and ownership of product.

3 INTEGRATION

3.1 Motivation and overview

CVM aims at improving competitive advantages of a company by increasing perceived value in the purchasing phase, while customer satisfaction is a construct with a post-purchase perspective and a tactical orientation which provides a guideline of action for current product/service offerings [9]. Thus, CSM aims at filling a gap between perceived performance and customer expectations in the usage phase.

Since CVM has a strategic point of view and CSM has a tactical one, generally, different divisions with various experiences have the responsibility to manage customer value and customer satisfaction.

In order to keep the customer satisfied over product life cycle, we should apply a holistic point of view into the customer-supplier relationship. The CSM cannot be successful if the CRV is not considered. However, the customers may have a high perceived value when they buy the product.

Although CSM tries to recognize the unsatisfied customer and remove the causes of dissatisfaction, it does not consider the ultimate value of using product and service as a main goal. Therefore, the customer may be satisfied for a short time, while the ultimate received value will not be high enough. This is due to the fact that CSM is generally designed based on customer expectations rather than customer value.The notions of customer satisfaction and customer value, however, cannot be disconnected [21, 9]. It becomes either the predecessor of value [22, 15] or the successor of value [23][24, 9].

In other word, it is needed to balance between efforts to increase perceived value for selling new product and received value after usage phase. Sustainable growth of a company not only depends on selling new product but also depends on keeping current customer satisfied by offering higher service in usage phase.

Generally, different divisions have various approaches to customers. Some just try to sell more spare part and new services at higher price to increase monetary benefits for companies, while others focus on increasing competitive advantages in the market. It is needed to develop a framework to not only support competitive advantages of a company but also increase monetary benefits of the company as well as customer satisfaction level.

We define customer-received value (CRV) as the extent that customers can achieve their own goal through the acquisition and usage of product/service. Then, CRV can be used as an index for the integration of CSM and CVM. Instead of separating the product development and service development under CVM and CSM, we need a concept to integrate all managerial and tactical efforts through a concrete methodology.

3.2 Procedure using a PSS design method

PSS is a concept that integrates products and services in one scope for planning, development and delivery, thus for the whole lifecycle with a focus on CRV. Specific business models (see [25]) are used to reach that aim and bound customers to their providers over long periods. In order to integrate CSM and CVM through a unified procedure, we adopt a PSS design method developed by the authors and verified at a German company in the field of industrial investment machines [7]. This procedure was developed with a focus on value and cost of each market segment in order to support a differentiation strategy in the same line with PSS as strategy (e.g. [2]). Figure 2 describes the steps schematically.

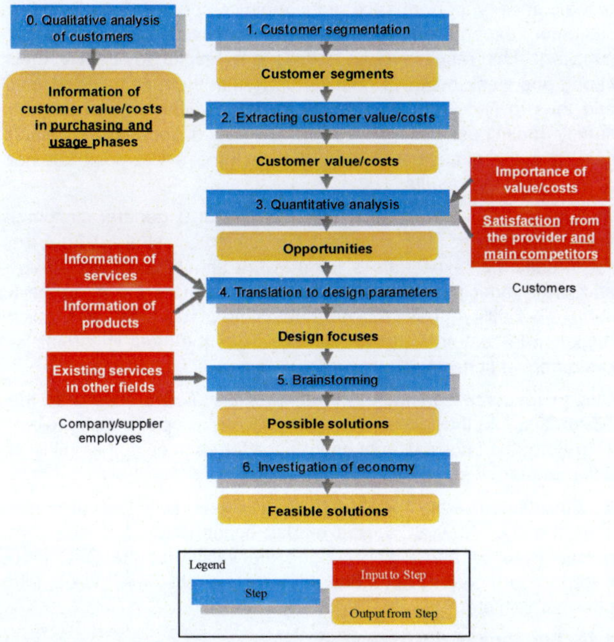

Figure 1: PSS design method integrated with CVM and CSM.

Note: "Output from Step" in the legend is also an input to the next step (except for the final output).

In order to capture the customer value/cost over the product lifecycle, we categorize the customer's information into two groups:

1. Customer value /cost in the acquisition phase.
2. Customer expectations in the usage phase.

Following the working steps, the output of Step 2 is a complete list of customer value over the product lifecycle. This list not only contains the value which affects the first purchasing decision, but also the value behind the customer expectation in usage phase. In order to increase customer satisfaction in the usage phase, we focus on the value behind the expectation instead of the expectations themselves, which has some advantages.

Customer expectations can be a list of solutions and services which have already been developed by other competitors. However, in order to provide new services with higher added value, designers should focus on customer value. In addition, different customer expectations may refer to the one value from different aspects. Furthermore, customer expectations can be too general or vague, while customer value does not have these kinds of negative attributes.

The outcome of Step 3 is the strength and weakness of product and service which can feed product development for future and potential customers as a part of CVM. In addition, it can support CSM to increase customer satisfaction by improving product/service attributes which cause dissatisfaction. In order to increase customer satisfaction, the value behind customer expectations in the usage phase should be addressed. In addition, the satisfaction is considered as a function of importance and performance of each value. This procedure starts by customer segmentation, so the Importance–performance matrix evaluates the value of product/service offered to each market segment. Then, the certain value which weighs more in dissatisfaction of customers (e.g. ones with higher importance/lower performance) can be addressed segment by segment.

In Step 4, the relation of product and service with each customer value is defined within a hierarchy. The outcome of this step is a list of product and service components which affect customer value and subsequently customer satisfaction. Some factors have higher impact, which can be used as starting point for improvement.

The outcomes of Steps 5 and 6 are some feasible solutions for each market segment.

This procedure integrates product/service as a unified offering to the customer and increases customer-received value. The outcomes of this procedure not only increase the satisfaction of current customers, but also can be seen as a strategic guide to improve product/service offerings for future customers.

In other words, the list of product/service components containing correlation between these components and customer value provides a strong foundation to guide CSM and CVM under an integrated framework. As a result, the balance between CSM and CVM can be better managed by clarifying the weight of product/service components in CRV. The R&D efforts can also more effectively support the CSM and CVM

4 CASE STUDY AT A SWEDISH MANUFACTURER

4.1 Business situation and strategy of the company

This company, with 400 employees in its own organization, is one of the major international manufacturers in its field. An instance of episodes indicating such characteristics of the firm is as a follower. This company once ventured into a new market segment, and thus they developed a new machine and tools, which have some competitive advantage over existing ones. Since the company faced high technological uncertainty, the huge investment was put on R&D to realize the technological competitive edge with little concern for services.

Over the past years the company has tried to sell new products with focus on increasing perceived value in purchasing decisions. By selling new products, the company can be sure about the financial source for future R&D as the result of selling more spare parts.

On the other hand, the company has the goal of "achieving reasonably satisfied customers", since creating prerequisites for a delighted customer would be overly expensive for the company.

Thus, there is not a balanced view over the product lifecycle regarding customer value. This is a strategic issue for the company: how much money they should invest on perceived value compared to received value. It seems that the company is trying to move from the position of market follower to market leader by shifting from investment on customer-perceived value to investment on customer-received value.

4.2 Existent concepts at the company

At the company, there are three main concepts which aim to satisfy customers' needs over the product lifecycle.

- Customer Value management and Market Adapted Standard (CVM and MAS)
- Regionalization
- Customer satisfaction management

CVM and MAS: A certain department has responsibility for forecasting the trend of the market. In addition, the competitive positioning of the company in different market segments by using the P-I matrix is a part of this department's responsibility. However, they developed the P-I matrix just for customer value, which affects purchasing decisions and is related to product attributes. Although there are several channels to capture customer voices, the content of this matrix is solely based on the perception of top managers, which does not necessarily reflect the customer's perception regarding performance and importance of product attributes. In addition, ***Market Adopted Standard*** is a concept developed to increase the flexibility of fulfilling customer needs in purchasing decisions. Market Adapted Standard (MAS) was commenced in 1996. The overall purpose with the program is to create standardized modules that can be combined in order to meet a variety of customer requirements without the necessity for development of each order.

MAS is intend to bring benefits in the firm of shorter development and delivery times, improved reliability and reusability of technologies and processes, deliveries at the right moment, and the possibility for an improved profit margin for different products.

Regionalization: This policy was initiated in 2009. The overall purpose with the program is to capture and fulfil customer needs of those customers who have already bought a product, through a close dialog with them through local agents. Regarding the technical requirements of customers, e.g. upgrading different features of a product, the results of the annual meeting with the agents is disseminated to the different divisions of the organization. In addition, the company intends to develop technical competency in each region in order to increase the service quality offered to customers. This business model is in its first phase and is not yet sufficiently mature. However, the general framework is set up and approved.

Customer satisfaction management: As an overall strategic performance management tool, the company internally uses the balanced scorecard (BSC). As a part of BSC, the company runs an annual customer survey which contains an anonymous generic on-line questionnaire. The procedure has two main steps:

In the first, the survey was designed through customer expectation research rather than writing standard questions or gathering information from internal sources like managers, engineers etc. The research team went to about 100 of the company customers before they ran the online questionnaire. The customers were asked about quality and characteristics of an ideal supplier, and their answers made up the questionnaire. Based on the results from the customer survey, an Action Priority Matrix (APM) was made to show the best improvement opportunities. The APM highlights the customer expectations with both the largest ideal and competitive gaps.

The research team recently attempted to utilize a PSS design method that the authors have developed, and obtained positive and useful outcomes as well as finding some obstacles and difficulties in implementing PSS design methods. The biggest obstacle was accessibility to the information of customer value, which in the customer value management system, surprisingly, was not addressed in a manner suitable for the PSS design method. Through this empirical study, there was a discussion of how to implement PSS design methods in an integrated way with customer value/satisfaction systems.

4.3 Process of the application

The authors applied the proposed steps at the company. First of all, we found some existing obstacles in Step 0 in applying the service design method in this company. The biggest obstacle is accessibility to the information of customer value. The annual customer survey contains 21 items of customer expectations. Some of them are potential customer value (e.g. flexibility to meet customer needs); others, however, are how to realize customer value (e.g. having high-quality technical staff). For each of the 21 items, a customer is supposed to compare the levels of the following four values: minimum customer expectation, performance of the company, performance of the main competitor, and maximum customer expectation (i.e. possible expectation on an ideal supplier). The results of the survey are helpful to identify their strong and weak points in relative terms with their main competitor, and thus to find their focal points for improvement. However, the survey doesn't ask customers about the level of importance of each item, and the expectations from the customer's viewpoint are not presented in the results.

Considering this situation, in Step 2 a list of 8 general customer value was created. We extracted this customer value list from the list of customer expectations used in the annual customer survey and some other internal information at the company. Among the 8 items, "Reliability", "available operation time of machine", and "quick response to customer needs" were seen as the most important for customers in general.

Due to resource constraints, there was insufficient access to the customers, and thus Steps 1 and 3 had to be skipped in this project. Therefore, the information of importance for each customer value at different market segments was not available.

In Step 4, experts from different divisions such as business development, R&D, SOS, and market support reflected their own perspectives on and acknowledged the validity of the QFD [26]-based model that is intended to translate customer value into design parameters.

4.4 Findings of the case study

The authors generated a list of parameters connected to different customer value some of which have strategic importance while others have tactical importance in terms of customer-received value. This provides the basis to improve the whole system in both the long term and short term. For instance, in order to increase the reliability of a product in the long term, the factors that should be improved in product and service are defined. These factors cover a wide range, comprising supplier, third party and customer's factors. In addition, the focus is not just on the technological improvement of a product. The role of service and proactive maintenance and corresponding managerial issues are also covered. Thus, in the long term, different aspects of product/service and their corresponding managerial infrastructure can be managed systematically.

On the other hand, regarding certain low certain value which creates the dissatisfaction for customers, improvement action can be initiated based on evaluation of the factors affecting that certain value. For instance, in order to increase the reliability of a product of a certain customer which is below the normal level, by using the models, the related factors should be evaluated and an improvement action plan be initiated. For example, the training hours of the operator of that machine may need to be increased, or the service plan may need to be customized.

As a result, not only can the product/service be managed at a strategic level to increase the competitiveness of the company, but the service can also be customized segment to segment to increase customer satisfaction. Thus, both CVM and CSM are supported systematically through a unified method.

The model provided a strong foundation for generating ideas for new services and confirming the meanings of the new ideas. For instance, two design parameters, "time between analysis of component/function history" and "quality of information about component/function history", were found to have a strong impact on "available operation time of machine" i.e. availability factor (AF), one of the important customer value. A new service idea for implementing an early warning system was created for improving "available operation time of machine" based on the strategy of improving the offerings for the two design parameters.

In addition, the company has already defined some technical and organizational Key Performance Indicators (KPI) such as AF, reliability factor (RF) and time to respond to customer need. The company aims to improve each KPI and reach a pre-defined target level. One concrete result of this method is defining a tree for each customer value while developing different criteria to show different aspects of a product/service offered to customers. For instance, the four different criteria are defined in order to improve the value of "quick response to customer needs" i.e. "time of notification", "time to find the problem", "time to provide solution", "time to implement solution". Then a tree is developed for each criterion. Based on importance of each criterion and its correlation with other factors, the R&D effort as well as organizational and managerial improvement can be targeted properly. In other word, the budget and effort will be assigned to the most important factors which result in save of money and time as well as development of the more effective solution for each customer value. Last but not least, the main finding of this case study was filling a gap which exists between CSM and CVM at the company. CSM is designed based on improving the performance of the company to meet customer expectations which have high priority. However, the customer expectations were a mixture of general and vague statements. We extracted the value behind each customer expectation, and provided a list of customer value over the product's lifecycle.

5 DISCUSSION

5.1 Effectiveness of the method

The method proposed in Section 3 was effective, since it provides a comprehensive picture regarding customer value over the product lifecycle. In addition, the models can be a basis for further discussion and improve the communication between different departments. Furthermore, the top managers can focus on departments which have a critical role and evaluate the relation and status of different departments in their business. In this case, change of process can be handled more easily. The feedback from managers was positive and they found the models useful in handling their processes and confirm the benefits of them.

On the other hand, the words in the model may not have the same meaning for different people. Neither does the correlation between different factors have same weight from different perspectives. Therefore, building a standard language is a must to increase the effectiveness of this procedure.

5.2 Obstacles in applying the service design method at the company

As mentioned in Section 4.3, the researchers had an obstacle in obtaining access to the described information about customer value. Therefore, it was mainly personal knowledge and experiences of the firm's employees that provided the relative importance among different expectations of customers. This means that it is possible that, so far, the company has focused on less important customer value in an inappropriate manner. However, it should be noted that the competence in knowledge and experience of the top managers from different divisions of this company is high. In addition, the working culture of this company is good at sharing information among different employees at different levels. These factors may have compensated for the lack of using a systematic methodology.

In addition, the company doesn't have any systematic framework for CSM. Therefore, it was impossible for us to find such a fundamental foundation which the service design method can be based on. Actually, there is a missing link between R&D and business department in the context of service development. In another words, results of the annual customer survey are not fed back effectively to R&D projects supporting service improvement.

6 CONCLUSION

The paper proposed a method for integrating PSS design method with methods for CVM/CSM. Through a case study at a manufacturer in Sweden, it showed some positive empirical results. At the same time, some obstacles to implement this method at the company were revealed. The authors have begun a project at this company to overcome this obstacle. Future work will include evaluation of the relation between different value. We should define the level of improvement for each value while considering the potential negative effects on another value. In this case, an increase in a customer value may decrease the total value of product/service offering. This means that some product/service attributes may become "too good". Therefore, in the next step we are going to define the correlation between different customer value and evaluate their impact on customer-received value.

7 ACKNOWLEDGMENTS

We extend our sincere thanks to the Swedish company who allowed us to conduct this case study.

8 REFERENCES

[1] Meier, H., Roy, R., Seliger, G. (2010): Industrial Product-Service Systems - IPS², CIRP Annals Manufacturing Technology. Vol. 59, No. 2, pp. 607-627.

[2] Tukker, A., Tischner, U. (2006):. New Business for Old Europe. Greenleaf Publishing, Sheffield.

[3] Lindahl, M., Sandström, G. Ö., Sundin, E., Rönnbäck, A. Ö., Östlin, J. (2008): Learning networks: a method for Integrated Product and Service Engineering – experience from the IPSE project, in: Manufacturing Systems and Technologies for the New Frontier - Proceedings for The 41st CIRP Conference on Manufacturing Systems. M. Mitsuishi, K. Ueda and F. Kimura Eds. pp. 495-500, Springer, Tokyo.

[4] Morelli, N. (2003): Product-service systems, a perspective shift for designers: A case study: the design of a telecentre, Design Studies. Vol. 24, No. 1, pp. 73-99.

[5] Alonso-Rasgado, T., Thompson, G. (2006): A rapid design process for Total Care Product creation, Journal of Engineering Design. Vol. 17, No. 6, pp. 509 - 531.

[6] Sakao, T., Shimomura, Y. (2007): Service Engineering: A Novel Engineering Discipline for Producers to Increase Value Combining Service and Product, Journal of Cleaner Production. Vol. 15, No. 6, pp. 590-604.

[7] Sakao, T., Birkhofer, H., Panshef, V., Dörsam, E. (2009): An Effective and Efficient Method to Design Services: Empirical Study for Services by an Investment-machine Manufacturer, International Journal of Internet Manufacturing and Services. Vol. 2, No. 1, pp. 95-110.

[8] Oliver, R. L. (1981): Measurement and evaluation of satisfaction processes in retail settings, Journal of Retailing. Vol. 57, pp. 25-48.

[9] Eggert, A., Ulaga, W. (2002): Customer-perceived value: a substitute for satisfaction in business markets?, Journal of Business and Industrial Marketing. Vol.17, No.2/3, pp. 107-118.

[10] Parasuraman, A., Zeithaml, V. A., Berry, L. L. (1988): SERVQUAL: A Multiple-Item Scale for Measuring Consumer Perceptions of Service Quality, Journal of Retailing. Vol. 64, No. 1, pp. 12-40.

[11] Daniels, S. (2000): Customer value management, Work Study. Vol.49, pp. 67-70.

[12] Bolton, R. N., Drew, J. H. (1991): A Multistage Model of Customers' Assessments of Service Quality and Value, Journal of Consumer Research. Vol. 17, pp. 375-384.

[13] Garver, M. S. (2003): Best practices in identifying customer-driven improvement opportunities, Industrial Marketing Management. Vol.32, pp. 455-466.

[14] Ulwick, A. W. (2002): Turn Customer Input into Innovation, Harvard Business Review. pp. 91-97.

[15] Khalifa, A. S. (2004): Customer value: a review of recent literature and an integrative configuration, Management Decision. Vol. 42, No. 5, pp. 645-666.

[16] Kano, N., Seraku, N., Takahashi, F., Tsuji, S. (1996): Attractive quality and must-be quality, in: The best on quality. J. D. Hromi Ed. pp. 165-186, ASQ Quality Press, Milwaukee, WI.

[17] Zeithaml, V. A. (1988): Consumer Perceptions of Price, Quality, and Value: A Means-End Model and Synthesis of Evidence, Journal of Marketing. Vol. 52, pp. 2-22.

[18] Ulaga, W., Chacour, S. (2001): Measuring Customer Perceived Value in Business Markets, Industrial Marketing Management. Vol.30, pp. 525–540.

[19] Monroe, K. B. (1991): Pricing – Making Profitable Decisions. McGraw-Hill, New York.

[20] Lapierre, J. (1997): What does value mean in business-to-business professional services?, International Journal of Service Industry Managemnt. Vol. 8, No. 5, pp. 377-397.

[21] Higgins, K. T. (1998): The value of customer value analysis: customer value analysis becomes the starting point in marketing research, Marketing Research. Vol. 10, No. 4, pp. 39-44.

[22] Dumond, E. (2000): Value management: an underlying framework, International Journal of Operations & Production Management. Vol. 20, No. 9, pp. 1062-1077.

[23] Oh, H. (1999): Service quality, customer satisfaction, and customer value: a holistic perspective, Hospitality Management. Vol. 18, pp. 67-82.

[24] Van der Haar, J. W., Kemp, R. G. M., Omta, O. (2001): Creating value that cannot be copied, Industrial Marketing Management. Vol. 30, pp. 627-636.

[25] Tukker, A. (2004): Eight Types of Product-Service System: Eight Ways to Sustainability? Experiences from Suspronet, Business Strategy and the Environment. Vol. 13, pp. 246 – 260.

[26] Akao, Y. (1990): Quality Function Deployment. Productivity Press, Portland, OR.

Problem Definition in Designing Product-Service Systems

Gokula Annamalai[1], Rajkumar Roy[1], Mehmet Cakkol[2]
[1] Manufacturing Department, Cranfield University, UK
[2] School of Management, Cranfield University, UK

Abstract
In these early stages of PSS research, it is necessary to define a holistic problem definition for designing product-service systems (PSS). In this paper, the current industrial practice of the problem definition within PSS is analyzed and presented. This analysis has lent itself to a classification of the steps involved in the problem specification stage into developing the Customer Specification (what the customer wants) and the Customer Demand (how the customer wants it) stage. Also it summarizes the important parameters considered in the problem definition. These results help to identify methods required to solve the industrial problem definition of PSS.

Keywords:
Product-Service System; Problem Definition; Specification

1 INTRODUCTION

In today's competitive and globalising economy, the strategic move from a product- to a service-based enterprise is highly sought especially in business to business environments. Servitizing manufacturers are bundling products and services into an integrated solution in the aim of creating sustainable competitive advantage. One of the examples of this shift in business strategy is Product-Service Systems (PSS). The perceived merits of PSS for manufacturers are to lock the customer into a long-term relationship which will be difficult to imitate by competitors and to generate higher revenues as services provide a steady income over the agreed period of time. Albeit these advantages there are additional risks to the manufacturers in achieving this long term performance. Although PSS domain has received substantial attention from academics for the last two decades, the domain is still argued in its infancy.

Many definitions have been proposed for PSS within the extant literature. Goedkoop et al. [1] define 'a product service-system is a system of products, services, networks of "players" and supporting infrastructure that continuously strives to be competitive, satisfy customer needs and have a lower environmental impact than traditional business models'. Meier et al. [2] define an IPS^2 (Industrial Product Service Systems) is an integrated product and service offering that delivers values in industrial applications. In common, PSS is defined as an integrated product and service offering that delivers value in use. Even though the majority of authors agree on the definition of PSS, the same consensus does not apply to a holistic problem definition for the design of PSS.

In order to leverage the benefits of PSS, various design methodologies are proposed in the literature. The aims and objectives of these proposed design methodologies and problems used to demonstrate the potential of these methodologies differs extensively. At this initial stage of PSS research, it is absolutely essential to define the holistic problem definition considering variations across industrial domains. Explicit and clear definition of PSS problems help customers to articulate their needs and desires in a relatively more structured form. It also aids manufacturers to identify and evaluate superior solutions through which customers' actual needs are satisfied. It is commonly noted in product and service design research that 'needs' and 'desires' are not appropriately transformed into problem statements and requirements. Often problem statements are incomplete, inconsistent, imprecise and ambiguous. Even though problem statements will contain these characteristics during the initial stages of design, a holistic approach which considers the whole lifecycle is usually missing. Therefore, we believe that investigating this problem will not only address the gap in the literature but also will be of benefit to practitioners from various industries.

In this paper, we examine the related literature in this domain as well as the current industrial practice for the problem definition of PSS. In order to do so, initially we conducted an in-depth literature review covering various disciplines such as engineering design, PSS, supply chain with a focus on manufacturing. Secondly, we conducted several face-to-face interviews in a large manufacturer which operates within the transportation industry. Furthermore, we extensively analysed the tender documents in this industry (provided by the customer to the manufacturers to facilitate the bidding process) in order to understand how actually the problem is defined for PSS contracts in real life. For validation of our results, we prepared and circulated a report to industry and according to received feedback we have made the necessary changes. The triangulation of data collected from different resources (e.g. literature, interviews and tender documents) enabled us to thoroughly analyse the literature vs. current industrial practice. In so doing, we managed to illustrate the differences and commonalities between the literature domain and actual practice. We also revealed the differences between the foci of both perspectives. Finally, we show that although industry provides a more comprehensive approach to problem definition than the extant literature does, practitioners are also not able to generate a robust problem definition for the whole life cycle of PSS.

The rest of the paper is structured as follows: Section 2 details an in-depth literature review of various problems used to elaborate various proposed PSS design methodologies. Section 3 illustrates

J. Hesselbach and C. Herrmann (eds.), *Functional Thinking for Value Creation: Proceedings of the 3rd CIRP International Conference on Industrial Product Service Systems, Technische Universität Braunschweig, Braunschweig, Germany, May 5th - 6th, 2011*,
DOI 10.1007/978-3-642-19689-8_20,

research aim and methodology used to identify the problem statement. Section 4 discusses the results obtained from the analyses of tender documents to illustrate current practices adopted in industry to describe PSS problem definition. Section 5 concludes with the current challenges in defining PSS problems and methods required to solve these problems.

2 LITERATURE SURVEY

Within the literature, few methodologies are proposed for designing PSS. In this section the aim(s) of the proposed methodologies and problems used to evaluate these methodologies are discussed. It is important to understand these problems used in the research papers to describe the application of the proposed methodology. This understanding helps to compare current industrial practices with respect to the problems used in the proposed methodologies. It should be noted that the papers discussed with proposed methodologies and evaluated using the design problems are selected and studied in this section. This section describes the problem statement used by the researchers in detail.

Komoto and Tomiyama [3-4] propose 'Service CAD' to generate conceptual design of PSS. PSS design is modelled through goals, quality, activities and environment. They also developed ISCL (Integrating Service CAD with a life cycle simulator) which has functions to support quantitative and probabilistic PSS design using life cycle simulation. They demonstrate Service CAD for the functional upgrading of service in computer-embedded systems to maintain the newness of functionality by replacing some system elements with shorter function lifetime, whilst continuing the utilization of other system elements with a longer functional and physical lifetime. They argue that the value added by functionally upgrading service for products should be performed by incorporating the delivery of ownership, functions, and results of the utilization (product-oriented, use-oriented, and result oriented PSS). To demonstrate ISCL, they focus on services in a life cycle of durable and capital goods such as washing machines, computer embedded systems, and automobiles. They argue that these services can provide additional value to product users and additional profit to the manufacturers of products other than the ownership delivery of products. They emphasize the question how often this is provided differs with the module type. They noted that the design of combinations of these services can decrease life cycle costs by decreasing the need for expensive services by increasing inexpensive and timely service. This study analyses a method to generate alternative combinations of these services, and evaluate the life cycle costs of the generated alternative services. The economic performances of multiple PSS business models are compared from a life cycle perspective.

Komoto and Tomiyama [3] also analysed another problem which described maintenance during a life cycle whilst varying user behaviour in a competitive environment to study the economic feasibility of integration of maintenance activities with the other services in a product life cycle from an auto manufacturer's perspective. They consider that the feasibility can depend on the user preference for specific service content and purchase timing, and on the service content provided and discounted by competitors. The problem is to design a comprehensive maintenance package considering the diversity of user behaviour if sales are to be raised over those for on-the-spot maintenance.

Sakao and Shimomura [5-6] propose a method for designing service activity and product concurrently and collaboratively during the early phase of product design. They develop a service model consisting of four sub-models: 'flow model (who)', 'scope model (what)', 'scenario model (why)', and 'view model (how)'. They used an elevator operation service problem to describe a method developed for unified representation scheme of human process and physical process in service activity and to evaluate these processes with Quality Function Deployment. This business is chosen for working on heightening the value of the product throughout its lifecycle. In addition to regular maintenance of elevators, they also consider developing novel services including building security control. They used another problem considering a service by a refrigerating facility provider for a logistics firm dealing with fresh meat. Maintaining the temperature in the facility is the target in this case. Developing both a refrigerating facility with low malfunction frequency and a system for technicians to visit the site quickly in case of malfunction are possible measures to fulfil the goal. They also applied this to business cases such as selling washing machines, providing pay-per wash service, and cleaning service of washing machines. They argue that finding a solution for machine performance to cleaning performance allows designers to search for a solution within a wider space.

Welp et al. [7-8] present a model based approach to support an IPS² (Industrial Product Service Systems) designer generating heterogeneous IPS² concept models in the early phase of development. They represent a noun (IPS² object) and a verb (IPS² process) that constitute a function, the combination of 'IPS² object' and 'IPS² process' constitutes an IPS² artefact. To demonstrate their model they used the problem of a customer trying to enter niche markets by offering rotationally symmetric µm-parts, e.g. shafts in mechanical watches. The requirements of the customer are defined as: rotationally symmetric µm-parts which have to be machined on customer's EDM-machine; machining has to be performed by the supplier's personnel; the demanded manufacturing volume can vary and the defined manufacturing time per part has to be observed (maximising productivity). Based on these requirements, the IPS² supplier defines the following restrictions: a specialized process is used for manufacturing rotationally symmetric µm-parts; machine operator has to have a certain qualification; the presence of additional (material) IPS² artefacts, necessary to manufacture µm-parts, remain in IPS² supplier's ownership; local and temporal availability is guaranteed and technical availability is guaranteed.

Maussang et al. [9] develop a methodology to support the design of PSSs starting from the design of the architecture down to the details of physical objects (products) specifications. To demonstrate their proposed methodology they used a problem from transporting energy through superconducting cables. To obtain superconducting characteristics, cables need to have a functioning temperature of between 65 and 70 °K, rather than few Kelvin degrees for older superconducting cables. To ensure this range of temperature, the cables are surrounded and refrigerated by liquid nitrogen (LN2). To ensure the regulation of the temperature of the LN2 coolant, the refrigerator has been envisaged to realise it. On the other hand, the customer wants to pay per watt of refrigeration to the company which will ensure the result.

Muller et al. [10] present a guideline to elicit and analyze requirements on IPS^2 properties and quality. They have clustered the checklist of criteria in terms of lifecycle activities, values, contracts, business and operation models, structure, behaviour, technical artefacts, service, information, communication and actors. To illustrate the usefulness of the proposed checklist they have used the following problem: to retrieve and formalize the requirements for a provider who plans to offer a spindle for a milling process in a manufacturing system in the availability-oriented business model. Even though this checklist details the factors to be considered in defining requirements, no formulation method has been provided to define the complete problem for a particular scenario.

Figure 1 summarizes the elements used in the various problem definitions discussed above. The observations from these problems definitions used in describing and evaluating the proposed methodologies are as follows:

- Even though the value addition due to PSS has been mentioned in the problem, it is not detailed comprehensively. Life cycle cost and productivity are widely used as evaluation criteria.
- The possibility of exploring a variety of business models is not mentioned.
- Mostly the impact and integration of products and services are considered. But the impact and integration along the lifecycle stages are not emphasized.
- The problems related to services are well expressed in terms of frequency, costs and user behaviour.
- Mutual influences between business models and products and services are mentioned.
- The factors concerned by the user, the manufacturer and the environment are mentioned, but not elaborated on and detailed.

From these observations, we could state that the problem definitions used in literature are not comprehensive. More elements need to be added in each layer mentioned in Figure 1 to provide a comprehensive problem definition. PSS design should focus on integrating business models, products and services together whilst considering all the lifecycle stages, which will create an innovative value addition to the system. Even though these problems are taken from industries, it is not clear how industries formulate these problems at present and what the key challenges are to solve them, i.e. real-time industrial problems are not addressed. In general, the complete description of the PSS problem is not replicated in any problems used to describe the proposed methodologies. Especially, value addition, risks and resources are not adequately modelled. The following sections discuss the current definition of PSS problems adopted in an industrial sector and compare these with the observations identified in this section.

3 RESEARCH QUESTIONS

Before we discuss about the research questions addressed in this paper, it is essential to discuss more about three important dimensions which are often not appropriately elaborated upon in the literature. These three dimensions are: customer's intelligence, maturity of products and services, and business context (B2B, B2C, and B2G). In our opinion, it is necessary to define these dimensions firstly before formulating the problem definition. The customer's intelligence could be mapped by being aware of their ability to articulate their needs and requirements and specifying the ways in which these requirements should get satisfied. Intelligent customers know exactly what they want and how they want the solution from the manufacturer. We call these two characteristics as customer specification and customer demand respectively. Alonso-Rasgado et al. [11] notably discussed the different combinations of hardware and services available in Total Care products: novel (new) hardware, adapted (from existing) hardware, new service support system, and adapted service support system. They mention that the challenge is involved in answering how to choose the most suitable combination of products and services to provide the best solution for all parties involved. With the exception of this work, the maturity of products and services are not observed in defining the PSS problem definition.

To understand the current industrial problem of PSS, we analysed PSS type tender documents in the transportation industry. Due to confidentially, the names of the documents and companies are kept anonymous in this paper. The industrial context in which we analysed the current practices of PSS problem definition is:

- Intelligent customer
- Mature products and services
- Predominately B2G with B2B involvement

Since the products are mature in this domain, the customers know exactly what they want and how they want it. The research questions investigated in this study are:

- How is the problem definition of PSS specified in the industrial sector?
- What set of questions should the required methods answer to solve the problem definition?
- What are the commonalities and differences between the problem definition mentioned in literature and the industrial sector?
- What are the missing elements that could be added in the industrial problem definition of PSS?

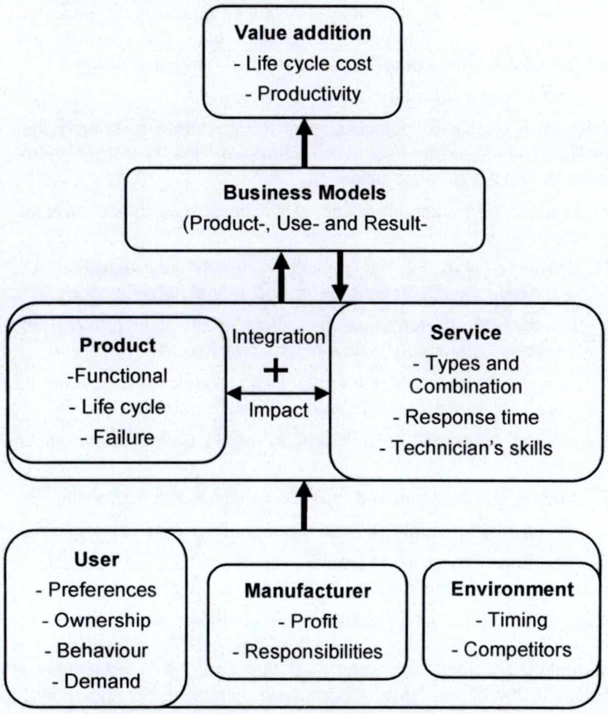

Figure 1: Elements used in the current problem definitions of PSS.

The main aim of this tender is that the customer is solely interested to pay the manufacturer in terms of asset usage hours and the manufacturer is completely responsible for the asset to be functionally available for the stated period (15-30 years). The primary intention of this tender is that the product should provide higher capacity, improved performance and energy efficiency, less weight, higher reliability, improved infrastructure and operational interfaces, an intelligent product (which tells passengers and maintainers what they want to know) and provides a higher quality passenger environment compared with that currently offered. The crucial influencing factors are whole life cycle cost, reduced energy consumption, minimum impact to the infrastructure and minimum maintenance and cleaning costs. The tender has been framed in a way that requires the manufacturer to work closely with the operator and infrastructure of the provider. The customer believes that this collaborative approach will facilitate an optimum overall system design. The main stakeholders involved in this process are the customer (the government), the manufacturer, the operator, the financier, the purchaser, and the infrastructure provider. The manufacturer will also be a service provider. The manufacturer will be responsible for the provision of products to the operator for daily operational service. The next section discusses the results obtained from analysing these tender documents and answers the intended research questions.

4 RESULTS

Following sub-sections answer each of the proposed research questions.

4.1 Industrial problem definition of PSS

This analysis lent itself to a classification of the steps involved in the problem specification stage into developing the Customer Specification (what the customer wants) and the Customer demand (how the customer wants it) stage. Also it summarizes the important parameters considered in the problem definition. The following sub-section explains the important steps mentioned in both stages.

Customer's specification

This stage is sole responsibilities of the customer to specify their needs. The important tasks considered by the customer in the specification are detailed below.

1. Identify the maximum usage of the product for the contractual term.
2. Create a plan for the project - identify key activities and respective execution time for product and service functions.
3. Identify key performance indicators (KPI) for products and services. Quantify KPIs as much as possible.
4. Separate mandatory requirements and expected requirements. Give importance rankings to these requirements.
5. Identify advance technologies to be incorporated to satisfy KPIs.
6. Identify stakeholders and their relationships and responsibilities.
7. Identify higher level risks and share those across stakeholders.
8. Incorporate dynamic characteristics of products and services in this long term contracts (e.g. flexibility, modularity etc.).

Apart from these specifications, the customer performed many exercises to elaborate upon the product specification. The weighting provided for satisfying the product specification is substantially higher (80%) compared to satisfying the service specification (20%). In the product specification, weightings are higher for reliability, general product design and system functions. Reliability is assessed through mean distance between service affecting failures, mission failures, technical delays and design for reliability. Reliability, availability, maintainability and safety are primarily used to assess the product. Also the weighting for technical satisfaction is much higher (70%) compared to the financial deliverability (30%). The customer has shown interest in being aware of the growth of reliability of the product over the contractual period rather than merely maintaining constant reliability. These specifications inform us that the intelligent customer plays a vital role in framing, expanding and communicating their needs without ambiguity to the other involved stakeholders.

Customer's demand

At this stage, the customer demands how their specifications mentioned above need to get satisfied. The important demands considered by the customer are mentioned here.

1. Identify the number of products required to satisfy the maximum usage level and the delivery schedule for these products.
   a. Describe the approach to calculate this number and schedule.
2. Show the achievement of KPIs for products and services and key deliverables throughout the life cycle.
3. Describe an overall approach to defining, monitoring, managing and controlling activities following the commissioning and introduction into service of the new product.
4. Define an approach for delivery of the product's operational performance through the whole life. Represent product availability and reliability (Fault tree analysis and failure mode, cause and effects analysis and redundancy arrangements).
   a. Detail the scope of maintenance activities for the proposed product and the periodicity of each of these activities in an energy-efficient manner and locations in which these activities will be carried out. Describe a strategy to provide an overview of planned and unplanned maintenance.
   b. Demonstrate how quality will be effectively managed, controlled and monitored throughout the contract life.
   c. Describe capacity modelling to demonstrate sufficient capacity for these scheduled activities (e.g. equipment, spares, technicians etc.).
   d. Detail the capacity and capability to recover from disruptions that could create changes to the planned maintenance plan.
   e. Detail the procurement and contracting strategies including supplier selection, financial standby checks and management. Detail supply chain management, reporting and control.
   f. Describe an overall approach to the monitoring, managing and controlling of engineering issues throughout the project life cycle.
   g. Detail training specification: who will need to be trained, how they should be trained, when they should receive their training and how the knowledge retention will be tested.
   h. Details of state of the art technologies to be incorporated within a system which communicates in real time with the infrastructure, enabling information relating to service conditions, connections, interchanges and disruptions to be communicated.
   i. Justify assurance of effective communication between stakeholders.
5. Define the approach and methodology to be used for identifying and managing project risks throughout the life of products - strategy to manage and mitigate such risks.
6. Describe financial deliverability and robustness; and assessment of value; for example fixed rental charge per month (with spares and associated equipment included), maintenance service charge per unit per mileage bands for the stated contractual period.
7. Describe project deliverability - Identify the logical sequence of activities, key milestones, resources, critical activities, information management system and time scales.

Apart from these demands, the customer has also mentioned that the manufacturer should provide additional products planned for zero mileage to provide the operator with operational flexibility. The tender emphasizes that the operator should be intimately involved with the manufacturer in the planning and presentation of units for maintenance and repair, such that fleet availability and performance is optimised to best support the provision of services to passengers. It should be noted that the time scale for the manufacturer to address these demands is six months: this illustrates the pressure of time scales involved in the industrial sector. The manufacturer risks are allotted in the area of design, build, acceptance into service, performance, reliability, quality of the products and maintenance.

The first preference is to meet the customer's specification. Other preferences followed by the requirements specification are assessment of deliverability and the value of the solutions. It should be noted that the customer is not only interested to know about the solutions but also relying on evidence to prove that the particular solutions would be valid for the stated contractual period. There is also a severe penalty regime in place which is subject to a maximum cost deduction of 5% when the performance of the product falls to or below the unacceptable performance level threshold. Also there is a heavy penalty mentioned for each delayed minute to provide the required operational product. These demands highlight the challenges involved in framing solutions for real time PSS type problems. The next section discusses the methods required to solve this problem definition of PSS.

4.2 Methods required to solve the industrial problem definition of PSS

Since real time industrial problem definition of PSS is complex, various supports are required by the customer and the manufacturer to frame and solve the problem. The customer requires a set of methods for framing and explicitly communicating their needs without ambiguity to the other stakeholders involved. A set of questions the required research methods should answer for assisting the customer to detail their specifications based on the discussed industrial problem are stated below:-

- How to identify usage period of the product and its variation patterns for the stated contractual period?
- How to identify and specify KPIs for the products and services?
- How to segregate requirements into mandatory and expected functions and rank those?
- How to identify and predict advances in technology for the length of this period?
- How to identify and assign responsibilities and build relationships between the stakeholders?
- How to identify risks involved in the products and services and share these across the network of stakeholders?
- How to identify dynamic characteristics (e.g. PSS specification needs to incorporate adaptability with changing environment) to be incorporated in the products and services?

The dynamic characteristics have been highlighted and discussed in [12]. Answering these questions should help the customer to frame their specification correctly with ease. Compared to the methods required for identifying the specifications list, substantial efforts and methods are required to solve the specifications through stated demands. A set of questions the required research methods should answer for assisting the manufacturer to satisfy the demands of the customer based on the above industrial problem definition is provided below:

- How to identify the number of products required based on the usage patterns mentioned by the customer?
- How to trade-off between products and services activities to identify the required number of products?
- How to show the meeting of KPIs of the products and services?
- How to represent the approach involved in defining, monitoring, managing and controlling a set of activities involved in the solutions?
- How to identify, manage and mitigate risks involved in the products and services?
- How to find, describe and assess financial deliverability?
- How to represent the delivery of the project?
- How to define an approach for the delivery of the product's operational performance throughout the whole life-cycle?

In this set of questions, primary importance has to be given to the last question. That is defining and representing how the delivery of the product operational performance plays a vital role to structure the remaining parts. Methods helping to answer these questions substantially assist industries in answering PSS problems. The next section compares this problem definition with the problems used in the PSS literature.

4.3 Commonalities and differences between the literature and the industrial sector

This section compares the observations from the PSS literature and the problem definition in the industrial sector. The following points discuss the commonalities and differences between these sections.

- It is clear from these discussions that the scope of problem definition used in the industrial sector is much broader compared to the focused view discussed in the PSS literature.
- The PSS literature should be substantially enlarged to cover the scope of problem definition in industries.
- Conflicting objectives are avoided and clearly represented in the problem definition in industry.
- Sustainability and environment management are stressed in the industrial problem definition whereas it is not strongly emphasized in the PSS literature.
- Compared to the products and services much emphasis has been provided in the system developed which will support and deliver the solutions in the industrial problem definition. Whereas in the PSS literature much emphasis has been given to integration and impact of products and services, the three factors highlighted both in industry and literature are *reliability, availability and maintainability*.
- The issue of time scales for framing problems and solutions is not sufficiently highlighted in the literature compared to the industrial definition.
- The importance of suppliers and supply chain network in the industrial problem definition is stressed as a way to achieve sustainable solutions. The customer is keen to know more of the manufacturer's supplier network. This element is not adequately stressed in literature.
- Depth of coverage in the life cycle stages of the industrial problem definition is substantial compared to literature.

From these points we could argue that the PSS problem definition used in industry is substantially comprehensive compared to the problems used in literature to demonstrate PSS design methodologies. Industry's problem definition is more complex and ahead of current literature. Literature is addressing focused problems which have potential benefits to industry. Nevertheless academic research should widen the scope to encompass the industrial practices in real-time.

4.4 Missing elements that could be added in the industrial problem definition of PSS

Even though the industrial problem definition is comprehensive some of the elements could be added and emphasized more. These points are highlighted below.

- Changing customer requirements need to be mentioned in detail.
- Even though the collaboration process is stressed in the problem definition, it is not detailed and structured to state how this co-creation process (e.g. sharing resources) between stakeholders should evolve.
- It seems uncertain that every customer input involved in this operation is included in the problem definition - because it is not explicitly stated in the problem document. It could be worth mentioning about every customer and bringing them into problem definition and solutions finding stages. The customer could be the capability sponsor (people deciding on capability requirements), the customer responsible for procuring and delivering, and the end user. Integration between them is required because there could be a complete segmentation between them.
- The current industrial problem definition is rigid and solutions are expected for their specifications and demands. Adaptability should be added to these elements. Capabilities of the stakeholders including suppliers and their enhancement during the long contractual period is not mentioned and detailed.

Incorporating these elements into the problem definition could enhance the comprehensiveness of the problem definition. It should be noted that in general the problem definitions and solutions will co-evolve. Even though the current problem definition is detailed, the modification occurring to it during problem solving has to be investigated and incorporated into the problem definition.

5 CONCLUSION AND FUTURE WORK

This paper discusses the current problems used in PSS literature and in an industrial sector. The comparison between the academic and industrial practices leads to an understanding that the scope of PSS problems used in industrial sector is much broader compared to the focused view of academia. Especially in addition to integration and impact of products and services, academics should focus on system development to support the PSS offerings. The analysis emphasizes the need to define the PSS problems appropriately considering the types of customers and industrial sectors involved. Even though the problem definition discussed in this paper is given from the customer to the manufacturer, these problem definitions could be directly transferred from the manufacturer to their suppliers. This study might help to transfer these global specifications to individual suppliers involved in the network. Albeit that the problem definition used in industry is elaborate, there is scope for improvement especially in consideration of the co-creation process and stressing the capabilities of the stakeholders. Also there should be space to exploit innovative business models to identify suitable solutions rather than having a rigid problem definition. The analyses lead to the understanding of various methods required to answer the set of questions framed from the industrial problem definition. The future work involved in this research is to map existing methods, tools and techniques from literature to solve identified problem specifications and demands. This would help to identify the research gaps in literature to solve real-time industrial PSS problems.

6 ACKNOWLEDGEMENTS

The authors would like to acknowledge EPSRC for supporting, via the Cranfield IMRC, the work carried out in the preparation of this paper.

7 REFERENCES

[1] Goedkoop, M., van Haler, C., te Riele, H., and Rommers, P. (1999): Product Service-Systems, ecological and economic basics. Report for Dutch Ministries of Environment (VROM) and Economic Affairs (EZ).

[2] Meier H., Roy R., Seliger G., (2010): Industrial Product-Service Systems - IPS², CIRP Annals - Manufacturing Technology, Vol. 59, pp. 607–627.

[3] Komoto H., (2009): Computer Aided Product Service Systems Design (Service CAD and its integration with Life Cycle Simulation, PhD Thesis, Delft University of Technology, Delft, the Netherlands.

[4] Komoto H., and Tomiyama T., (2008): Integration of a service CAD and a life cycle simulator, CIRP Annals - Manufacturing Technology, Vol. 57, pp 9–12.

[5] Sakao T, and Shimomura Y, (2007): Service Engineering: a Novel Engineering Discipline for Producers to Increase Value Combining Service and Product, Journal of Cleaner Production, 15, pp 590-604, 2007.

[6] Sakao T, Shimomura Y, Sundin E, and Comstock M, (2009): Modeling design objects in CAD system for Service/Product Engineering, Computer-Aided Design, Vol. 41, pp 197-213.

[7] Welp E.G., and Sadek T., (2008): Conceptual design of industrial product service systems (IPS²) based on the extended heterogeneous modelling approach, International Design Conference - Design 2008, Dubrovnik - Croatia.

[8] Welp E.G., Meier H., Sadek T., and Katja Sadek, (2008): Modelling Approach for the Integrated Development of Industrial Product-Service Systems, 41st CIRP Conference on Manufacturing Systems.

[9] Maussang, N., Zwolinski, P., Brissaud, D., (2009): Product-service system design methodology: from the PSS architecture design to the products specifications, Journal of Engineering Design, Vol. 20, Issue 4, pp. 349 - 366.

[10] Müller P., Schulz F., and Stark R., (2010): Guideline to elicit requirements on industrial product-service systems, Proc. of 2nd CIRP International Conference on Industrial Product/Service Systems, Linkoping, Sweden.

[11] Alonso-Rasgado T, Thompson G, and Elfström B, (2004): The design of functional (total care) products, Journal of Engineering Design, Vol. 15, Issue 6, pp. 515 - 540.

[12] Roy R., and Cheruvu K. (2009): A competitive framework for Industrial Product Service Systems, International Journal of Internet Manufacturing and Services, Volume 2, Number 1, pp. 4 – 29.

Small Scale Socio-Technical Experiments as Stepping Stones for Eco-Efficient Product-Service Systems Diffusion: a New Role for Strategic Design for Sustainability

Fabrizio Ceschin[1], Carlo Vezzoli[1], Philip J. Vergragt[2]

[1] Politecnico di Milano, INDACO Department, Unit of Research Design and system Innovation for Sustainability (DIS), Italy

[2] Tellus Institute, Boston MA, USA; George Perkins Marsh Institute, Clark University, Worcester, MA 01610-1477, USA

Abstract

Eco-efficient PSS innovations represent a promising approach to sustainability. However the adoption of such business strategies is still limited because often involves significant corporate, cultural and regulatory barriers. Changes in basic belief, values, and interpretative frames (in other words higher order learning) are required. One way to facilitate this type of learning is through socio-technical experiments. Starting from these considerations, and bringing together insights from innovation sciences, the paper investigates the potential contribution that a strategic design approach can have in designing small scale socio-technical experiments as a way to stimulate and favor the adoption and diffusion of eco-efficient PSS.

Keywords:

Product Service Systems; Socio-Technical Experiments; Design for Sustainability

1 ECO-EFFICIENT PRODUCT-SERVICE SYSTEM (PSS): A PROMISING APPROACH TO SUSTAINABILITY

It is a shared opinion that conditions for sustainability require a system discontinuity, meaning that a radical redefinition of the current structures of production and consumption is needed. In fact, considering the raising consumption levels and the doubling of the world's population, a sustainable society should use 90%-95% less resources than industrialized countries are doing today [1] [2]. Given the dimension of the required change it is therefore clear that innovations on a process and product level, although being fundamental and necessary, are not alone sufficient to obtain the just mentioned radical shift. For these reasons, if we want to effectively tackle sustainability, there is a need to move from a focus on product improvements only, towards a wider systemic approach that takes in consideration new potential ways of satisfying the social demand of wellbeing. In this perspective, as suggested by Stahel [3], we should move from an *industrial economy*, in which the central value is based on the exchange of products to be consumed and in which the growth is strongly linked to resources consumption, to a *functional economy*, in which products are mere means of providing functions (or better still satisfaction). A functional economy is oriented to satisfy consumers through the delivery of functions (e.g. mobility; thermal comfort; having clean clothes) instead of products (e.g. cars; boilers and methane; washing machines and powder). Its economic objective is to create the highest possible use value for the longest possible time while consuming as few material resources and energy as possible; thus it is potentially more dematerialized than the present economy, which is focused on production and related material flows as its principal means to create wealth [3]. It can potentially bring about a reduction in the current levels of resources consumption, without minimizing consumers' level of satisfaction [4] [5].

Within this perspective several authors consider promising to look at the concept of **Product-Service System (PSS) innovation**. A PSS can be described as an integrated system of products, services and socio-economical stakeholders, designed to fulfill a specific client need [6], with the word *system* referring to both the *system of products and services* delivered to the client, and the *system of actors* that produce and offer the combination of products and services. Therefore PSS is not merely selling physical products or services (as in the traditional economic model), but offering a combination of products and services where the focus is the satisfaction of a client demand. PSS is not a new economic concept: several examples of PSS have in fact been implemented in the last decades by various companies [6] [4] [5] [7]. However, the key point to be underlined is the fact that PSS, if properly conceived, can decouple economic value from material and energy consumption. These PSS are defined **eco-efficient** [4] [7], meaning that it is the economic and competitive interest (of the socio-economical stakeholders involved in the PSS offer) that foster an optimization of the environmental resources. As opposed to traditional business models, eco-efficient PSS rewards low resources consumption in use and product longevity (rather than obsolescence and high running costs). In fact, if what is offered is the fulfillment of a satisfaction (e.g. thermal comfort), less material and energy are used by the provider to deliver this satisfaction, minor will be the costs to be sustained and higher the profits.

In the last 15 years the European Union has dedicated special attention to this kind of innovations: a wide number of research projects in the field of PSS and sustainability have been supported by EU funding. These researches brought to clarifying the concept of PSS, understanding its characteristics, potential benefits, drivers and barriers, possible rebound effects, etc., and to developing (and partly testing) different methods and tools to orient and support the design of eco-efficient PSS (see for example the overview made in [8]). Despite all the knowledge accumulated on understanding how to develop eco-efficient PSS, and despite their potential win-win characteristics, it has to be underlined that the application of this concept is still very limited. The reason is that eco-efficient PSS are intrinsically radical innovations, and the adoption of such strategy bring with significant *corporate*, *cultural* and *regulatory* challenges:

- For **companies** the adoption of a PSS strategy is more complex to be managed than the traditional way of delivering products alone. In fact there is the need to implement changes

J. Hesselbach and C. Herrmann (eds.), *Functional Thinking for Value Creation: Proceedings of the 3rd CIRP International Conference on Industrial Product Service Systems, Technische Universität Braunschweig, Braunschweig, Germany, May 5th - 6th, 2011*,
DOI 10.1007/978-3-642-19689-8_21,

in corporate culture and organization in order to support a more systemic innovation and service-oriented business [4]. As consequence new design and management skills are required.

- For **customers**, the main barrier is the cultural shift necessary to value an ownerless way of having a satisfaction fulfilled, as opposed to owning a product [6] [4] [5]. In fact the problem is that solutions based on sharing and access contradict the dominant and well established norm of ownership [9]. Another obstacle is the lack of knowledge about life cycle costs [5], which makes difficult for users to understand the economic advantages of ownerless solutions.
- On the **regulatory** side, actual laws may not favor PSS oriented solutions. Environmental innovation is often not rewarded at the company level due to lack of internalization of environmental impacts [10]. In addition there are difficulties in implementing policies to create corporate drivers to facilitate the promotion and diffusion of this kind of innovations [10] [11].

Schot and Geels [12] consider *radical innovations* always immature when they enter the market because they cope with a dominant socio-technical regime (and its established and stable rules and networks of actors). Eco-efficient PSS innovations are in most of the cases such a radical innovation, and usually encounter the opposition of the existing socio-technical regime (because they challenge existing institutions, customers' habits and lifestyles, companies' organizations and regulative framework). This kind of innovations cannot be implemented in the same way as incremental innovations, which assume the stability of technological, regulatory and market environment. In order to facilitate and speed up the adoption of such radical innovations, what is required is to break down the routine behaviors that are daily reproduced by individuals, groups, business communities etc. A change in basic belief, rules and norms is necessary. In other words, as the literature on changes in socio-technical systems rightly argues, we hypothesize that a deep **learning**, on a scale ranging from individuals to business communities and governmental institutions, to the society at large, is needed.

Starting from these considerations, the paper is structured as follows. Section 2 introduces to the concept of learning, explaining why it is crucial in radical innovations. Section 3 illustrates the importance of small scale scio-technical experiments to induce learning and favour the societal embedding of radical innovations. Section 4 explores the potential role that strategic design approach can have in designing this kind of experiments. Section 5 illustrates an application of this approach in an on going experience: the Cape Town Sustainable Mobility project.

The rationale behind this paper is consistent with that of *design-based research*. The key characteristics of design-based research are that it is focussed on designing interventions in the real context (interventionist characteristic) combined with efforts to understand and improve interventions (process orientation) utilizing state of the art theories whilst the field testing and the evaluation of the consecutive prototypes should contribute to theory building (theory orientation) [13]. In this case the on going Cape Peninsula Mobility project represents a first step of a design intervention through which reflect on theory (I), improve and refine the intervention itself (II), and make explicit the design decisions associated with a design process (III).

2 LEARNING AS A CRUCIAL FACTOR TO FAVOR RADICAL INNOVATIONS

Generally speaking *learning* takes place when individuals assimilate new information and apply it to their subsequent actions [14]. It is therefore crucial in the process of inducing changes towards the adoption and diffusion of eco-efficient PSS innovations (in which new alternative ways of producing and consuming are required). To explain learning processes (in particular in relation to eco-efficient PSS innovations), a distinction can be made between the subject of learning (who learns?), the objects of learning (learns what?), the results of learning (to what effect?) [15], and the process of learning (learning how?) [16].

With regard to the first component **(who learns?)**, several subjects of learning can be distinguished: company/s and related partners (innovation promoters), customers/users, but even governmental institutions, research centers, NGOs, citizens, media, etc.

In relation to the second component **(learns what?)**, there are two forms of learning: *lower* and *higher order learning* [17] [18]. The first one concerns new insights on how to solve a problem, without changes in problem definitions. The second one is related not only with the solutions to a certain problem but also with the problem itself.

Regarding the third component **(to what effect?)**, a learning process can result in a correction or resolution of a certain problem (lower order learning), but it may also lead to a paradigm shift, which refers to a change in the problem definition, basic assumptions, norms, values and interpretive frames which govern the decision-making process and actions of individuals, communities and organizations (higher order learning) [14]. Eco-efficient PSS often entail new and radically different roles for the various involved stakeholders (new ways of providing a solution, new ways of getting a satisfaction, etc.). Thus, both broad learning (implying that a wide range of actors becomes involved in the innovation process) as well as deep learning (implying that both first and second order learning occurs) is necessary [16].

The fourth component **(learn how?)** is related to the processes for stimulating learning. Learning takes place through a *feedback-stimulus* mechanism, when the existing, well accepted and trusted interpretive frames and competences receive feedback on their performance in solving a problem or advancing specific objectives [17]. If, as a result of this feedback, the desired outcomes are not reached, these interpretative frames are reassessed and, if needed, substituted with new ones. In other words actors will reproduce their behaviors over and over as long as nothing happens to change the context of their actions. Giddens [19] speaks about a "structure of rules", shared in a society, that drives the everyday routine activities in that society. A structure that is continuously embedded in the course of the action: structure creates routine actions, that create structure, that creates actions, and so on in a circular process, defined by Giddens "structuration". A feedback-stimulus is necessary to brake down this circular process. In the context of organization, the stimuli can come from threats to survive and success, failures, disaster or other unexpected events [20]. Scenario building, collective visioning, and other group techniques can also generate feedback on the accepted assumptions and behaviors, and induce learning (in particular the higher order one) [21]. Other authors [22] [18] emphasize the role played by interactions among groups with different belief systems and interpretive frames as the means for learning. In general, having something to interact about, such as a specific project or a problem to solve, can represent an effective inducer of the processes leading to learning. And in relation to this it has been argued by several authors that a key role can be played by small **scale socio-technical experiments**, collective experiments driven by a long-term vision, carried out by a coalition of different actors, and aimed at testing and introducing radical new solutions [23] [17]. At this point a crucial question is: how socio-technical experiments should be implemented in order to favor a successful adoption of radical innovations (in particular eco-efficient PSS innovations)?

3 SMALL SCALE SOCIO-TECHNICAL EXPERIMENTS AS AGENTS OF SYSTEMIC CHANGE

It has been argued that small scale socio-technical experiments represent a promising strategy to induce higher order learning processes to favor the societal embedding of radical new alternatives. This is coherent with some recent developments in the innovation science field, in particular the contributions from *Strategic Niche Management (SNM)* and *Transition Management (TM)* approaches. Through historical socio-technical case studies, transition scholars have analyzed how system innovations take place and elaborated a model, called *the multi-level perspective on transitions* [24] through which describe the dynamics that regulate these processes. Three analytical concepts are distinguished:

- the *socio-technical regime*, which can be defined as the dominant way of innovating, producing, distributing, consuming etc. It is made up of different socio-economical stakeholders, practices, shared rules and ways of doing related to a specific field (mobility, energy, etc.). Usually regimes are stable and resistant to change, because the regime rules and institutions guide regime actors in a specific direction discouraging the development of alternatives.
- the *niche*, a protected space "isolated" from the influence of the dominant regime, where radical innovations can be tested, become more mature, and potentially replace regime practices.
- and the *landscape*, that is the relatively stable social, economic and political context in which actors interact and regimes and niches evolve. It represents the background for regimes and niches. It can influence the regime and the niches, but cannot be influenced by them (at least in short term).

Niches are a fundamental part of transitions because they act as "incubation rooms" for radical novelties [24], where experiments and learning processes take place. However, even if niche developments can hold great promise, they do not immediately live up to expectations because they are immature when they enter the market and because they conflict with the way society is organized [25]. In this sense, if immediately exposed to market competition, they have great probability to not survive. For this reasons a protected space should be created, where continuous experiments can bring the innovation to mature [26]. Niches can be used as strategic opportunities for experimenting, learning, and improve the innovation and establish new social networks in order to gain momentum for diffusion or even replace dominant regime practices [27]. In other words these socio-technical experiments can be used:

- as a **Lab**, to test, learn and improve the innovation, and identify implementation and diffusion barriers. In this sense it is important to: verify the technical and usability aspects of the innovations but also to understand how the different social groups (the local community, local administration and institutions, research centers, NGOs, media, etc.) respond to the innovation; improve the innovation in relation to the feedback coming from the various stakeholders.
- as a **Window**, to give visibility and promote the innovation and establish new linkages and synergies with various socio-economical actors;
- and more in general as **Agent of change**, to introduce and diffuse new ideas and knowledge to the community, stimulating various social groups (users, public institutions, companies, etc) to change their perspectives, beliefs, and lifestyles. The experiment represents a stimuli tò induce actors in changing their behaviors and interpretative frames (for example it can bring to attract and stimulate users in testing the innovation, or to push governmental institutions in implementing proper policy measures to favor the innovation itself).

Continuous socio-technical experimentations can bring the innovation to mature, through a transition pathway based on [28]: *deepening* (learning as much as possible from the experiment in a specific context), *broadening* (repeating the experiments in other different contexts), and *scaling-up* (embedding the experiments in the regime, to replace dominant practices).

At this point a proper question is: how experiments in niche can be fostered to favor wider transition processes? Several scholars have investigated more deeply this theme; this research, denominated *Strategic Niche Management (SNM)*, individuated as key important three internal processes [27]:

- firstly, the building-up of a **broad socio-economical network**, including all relevant types of actors, capable to protect and support the innovation incubation and development;
- secondly, the convergence of actors expectations into a **shared long-term vision**, in order to give strategic orientation to the innovation development.
- thirdly, an **effective learning process** between the involved actors, which is recognized crucial because it enables adjustments of the niche innovation and increases chances for a successful diffusion.

It has been argued that small scale socio-technical experiments can act as opportunities for overcoming the regime barriers (cultural, corporate and regulative ones) and gaining momentum for the diffusion of radical novelties [17] [18]. In relation to the specificities of eco-efficient PSS innovations, a case study research (that investigated the innovation journeys made by some companies in introducing their eco-efficient PSS concepts in the market), showed that this implementation strategy is valuable also for managing the introduction and diffusion of this kind of innovations [29]. At this point a crucial question is: what are the implications on the design level? What role could be played by strategic design in facilitating the process of experimentation and diffusion of eco-efficient PSS?

4 DESIGNING SOCIO-TECHNICAL EXPERIMENTS TO FAVOR WIDER DIFFUSION OF ECO-EFFICIENT PSS

We know that several methods and tools have been developed to support designers in ideating eco-efficient PSS concepts (see for example the overview of different methods and tools in [8]). However, if designers want to act as effective agents of change they have to be aware of the mechanisms and dynamics that regulate the implementation and diffusion of this kind of innovations. In other words, we are arguing that *strategic design for sustainability* could have a role not only in generating eco-efficient PSS concepts, but also in supporting and facilitating the introduction and diffusion of the concept itself [29]. In this sense designers could guide and support a company, an institution or a network of actors, in the process of introducing and gradually embed in the society eco-efficient PSS concepts.

As we have seen before, niche experiments have a strategic importance for stimulating transition to eco-efficient PSS innovations. For this reason a potential role for strategic designers is the **design of socio-technical experiments** capable to favor higher order learning and act, as said before, as *labs*, *windows* and *agents of change* (and foster in this way a gradual societal embedding of the solution). As it has been previously argued, for a successful experimentation it is of key importance the presence of a wide and dynamic network of actors, and the presence of a shared long term vision (capable to give a direction to the experimentations and orient stakeholders' actions). These actors could have very different problem definitions, framings of the issues, and even world views, but these differences would not hamper successful collaboration in a small-scale experiment.

For these reasons strategic design could have a role in **identifying and involving the most suitable socio-economical actors** that can support the socio-technical experiment. Not only the actors more tightly linked to the innovation (e.g. producer, partners and suppliers) but also other actors: for instance universities and research centers (that can give scientific support), institutions and public administrations (that can promote the innovation and give political support), and also NGOs and media (that can give visibility to the innovation). In other words it is fundamental to strategically build-up a heterogeneous network characterized by scientific, social, economic, politic and cultural linkages. Moreover strategic design could also act as **facilitator in the building-up of a collective vision** (a vision capable to converge the expectations and orient the actions of the involved stakeholders, and attract new actors).

Socio-technical experiments represent a step of a complex transition pathway that could bring the innovation to be gradually embedded in the society. Thus, strategic designers should take in consideration this "transitional perspective" and be able to strategically define (or better still co-define with all the involved actors) the sequence of socio-technical experiments (and more in general the sequence of actions) that can favor this process. In other words the design of an eco-efficient PSS concept should be accompanied by the **design of the strategic transition path** to gradually incubate, test, introduce, and diffuse the innovation. It has to be underlined that this path is not meant as a fixed roadmap to be strictly carried out as a rigid procedure. It is a flexible path that is continuously adjusted and redefined in relation to the feedback coming during the whole implementation process. As a consequence even the *long-term vision* and the *supporting networks of actors* are not static but evolve in time (they are continuously adjusted and modified along the transition path). Therefore designers should also be capable to dynamically manage the continuous redefinition of the vision, the path towards its achievement and the supporting network.

In synthesis, in addition to designing eco-efficient PSS, the *strategic capabilities* required by a designer in supporting and facilitating the introduction and diffusion of such innovations are:

- to **develop** *eco-efficient PSS concepts* capable to act as **long-term visions** (to orient stakeholders actions);
- to translate this long-term vision in a **transition path** oriented towards the gradually achievement of the vision itself. In this transition paths a key role is played by **small scale socio-technical experiments**; it is therefore needed the capability to design such experiments (experiments capable to act as *Lab*, *Window* and *Agent of change*);
- to **identify** the proper **socio-economical actors** (or transition network) to support the various steps of the transition path;
- to **facilitate** the **strategic conversation** between the involved actors, in order to enable them to participate in the co-definition/ development of the *long term vision*, the transition path (and its socio-technical experiments), and the transition network. Communication/negotiation skills are therefore key important.
- to **dynamically manage** the activities listed above, facilitating the continuous adjustment and refinement of the long term vision, the transition path (and its experiments), and the transition network.

In order to test the hypothesis formulated in section 1 and theoretically founded in the subsequent theoretical sections, this strategic design approach is currently applied in some on-going research projects, coordinated by the unit of research *Design and system Innovation for Sustainability (DIS)* of Politecnico di Milano. One of these, run in collaboration with the South African *Cape Peninsula University of Technology (CPUT)*, is the *Cape Town Sustainable Mobility Project*. The following section briefly describes the design process and approach adopted in the project.

5 CAPE TOWN SUSTAINABLE MOBILITY PROJECT

Background

The background of the project is represented by MULO System, a family of light working vehicles powered by solar, electric and human power, convertible in four variants: freight transport, people transport, green areas maintenance and street seller vehicle[1]. One version of the vehicle was prototyped in August 2006 by the high school *IPSIA "A. Ferrari" Maranello* in collaboration with *DIS* of Politecnico di Milano.

Incubation

In 2008, within the University Chairs on Innovation project (UNIDO umbrella), the Faculty of Design of the Politecnico di Milano activated 3 master degree theses related to the design of sustainable mobility systems (based on the adaptation of the MULO vehicle) conceived for low-income and emerging contexts in Africa; one of this thesis was aimed at designing a system for disabled student transportation in Cape Town (South Africa)[2], in collaboration with the Cape Peninsula University of Technology. During the master degree thesis a first eco-efficient PSS concept was developed. In brief the idea is to give access to mobility to students with disabilities, in particular covering the path between homes and the nearest public transport stops. The concept foresees a business model in which the local producer, rather than selling the vehicles will keep the ownership of them and will offer a mobility service (in this way the producer is economically interested in extending the vehicle lifespan and reducing as much as possible the energy consumption: to reduce costs and maximize profits). The solution is especially conceived for the suburbs of Cape Town, usually characterized by mobility problems (in these areas in fact public transport services are limited and most of the time absent). At this stage some potential local producers were contacted. Two of them, *Shonaquip* (a company producing and commercializing wheelchairs and mobility aids for disabled people), and *Disability Workshop Enterprise Development* (DWDE, an association coordinating several workshops involving persons with disabilities) shown a real interest and started to collaborate providing feedback and suggestions. In September 2009 a series of meetings took place, involving DIS, CPUT, Shonaquip and DWDE, to review and co-design the different elements of the project. The PSS concept and a first hypothesis of transition path were presented by DIS. The discussion that followed the presentation brought to:

- **adjust the PSS concept:** it was decided to extend the mobility service even to elderly people (to move them for example to the hospital or the nearest bus stop). Moreover it was proposed to offer also a transportation service for tourists in the city centre (as a source of revenue). In relation to the vehicle the design specifications were defined.
- **adjust the transition path:** in particular it was agreed that the next step would have been the implementation of a small pilot (to be used to test, learn and give visibility to the project) in a specific suburban area of Cape Town.
- **identify new actors to be involved:** in relation to the pilot to be implemented it was suggested to involve *The Reconstructed Team*, an association (operating in the selected suburban area) aimed at reintegrating into society former drug addicts and

[1] Designed by Fabrizio Ceschin for his master degree thesis, Politecnico di Milano, 2006.

[2] Master degree thesis of Hazal Gumus.

criminals; their role in the project would have been the management of the service. It was also recommended to establish synergies with the municipality (and in particular the local public transport company).

- **identify implementation barriers:** the main problem emerged was the unavailability of economic resources to entirely finance the vehicles production and the pilot implementation.
- **define roles:** agreement on the tasks to assign to each actor.

In October 2009 The Reconstructed Team was contacted by CPUT and decided to take part in the project. DIS presented the project idea to some Italian companies, which decided to provide a sponsorship. The remaining needed economic resources were made available by Politecnico di Milano. In January 2010 started the process to redesign the vehicle in order to meet the local requirements (user needs, technological constraints, local aesthetic preferences, etc.). The activity was coordinated by DIS in collaboration with CPUT and Shonaquip.

Small scale socio-technical experiment

Shonaquip started the construction of the first vehicle prototypes in September 2010. It is foreseen to have the first vehicles ready for December 2010, and implement the pilot in the first months of 2011. The area selected for the socio-technical experiment is the Athlone District. The reason is that The Reconstructed Team (which is located here) is well known in the area because very active in a series of activities to support the local community. As consequence it could be easier to introduce the solution in the community. The socio-technical experiment has been designed in order to:

- **verify the technical and usability aspects of the solution:** the experiment should in fact bring to collect feedback and understand which are the adjustments and improvements to be undertaken in relation to the technical aspects (product level) and usability aspects (service level).
- **identify implementation and diffusion barriers:** identify the different barriers (institutional, regulative, economic, etc.) that can potentially hinder the future implementation and diffusion of the innovation.
- **stimulate higher order learning between participants,** through adjustments of problem definitions and frames of meaning, and re-interpreting the issues to be addressed and solved [17].
- **give visibility to the project:** the objective is also to show and promote the project, in order to raise the interest around the innovation, attract new stakeholders and funds. In relation to this it has been planned to involve local media and NGOs, and organize communication/promotional events.
- **stimulate change:** the experiment will be used not only to collect feedback and define the adjustments to be undertaken, but also to foster change. In particular the pilot will be used to attract potential users (and stimulate them to adopt the solution), the local municipality (to understand which policy measures could be adopted to favor the innovation) and interested companies (to make synergies and increase the availability of economic resources).

More broadly, this experiment would be an initial test for our hypothesis on higher order learning in small-scale socio-technical experiments as potential facilitators for radical innovations and regime shifts.

Next steps?

The next steps are of course subjected to the results of the pilot. Nevertheless the objective, shared by the project partners, is to: (I) develop the pilot in an economically self-standing solution (not dependent from the external sponsorships); (II) extend the service to other areas of Cape Town; (III) differentiate the service (e.g. support to street sellers, delivery of goods in the city centre, etc.).

6 DISCUSSION AND CONCLUSIONS

Eco-efficient PSS are in most of the cases radical innovations, and their introduction and diffusion usually encounters the opposition of existing customers' habits, companies' organizations, and regulative framework. Higher order learning is required is to change the routine behaviors that are reproduced by individuals, groups, business communities etc. Small scale socio-technical experiments could represent an effective inducer of the processes leading this kind of learning, and therefore can be used to favor the societal embedding of radical new alternatives. Starting from these considerations our hypothesis is that there is a new potential role for *strategic design for sustainability* in designing small scale socio-technical experiments as a way to stimulate and favor the adoption and diffusion of eco-efficient PSS. In other terms strategic designers could have a role not only in generating eco-efficient PSS concepts, but also in supporting and facilitating the introduction and

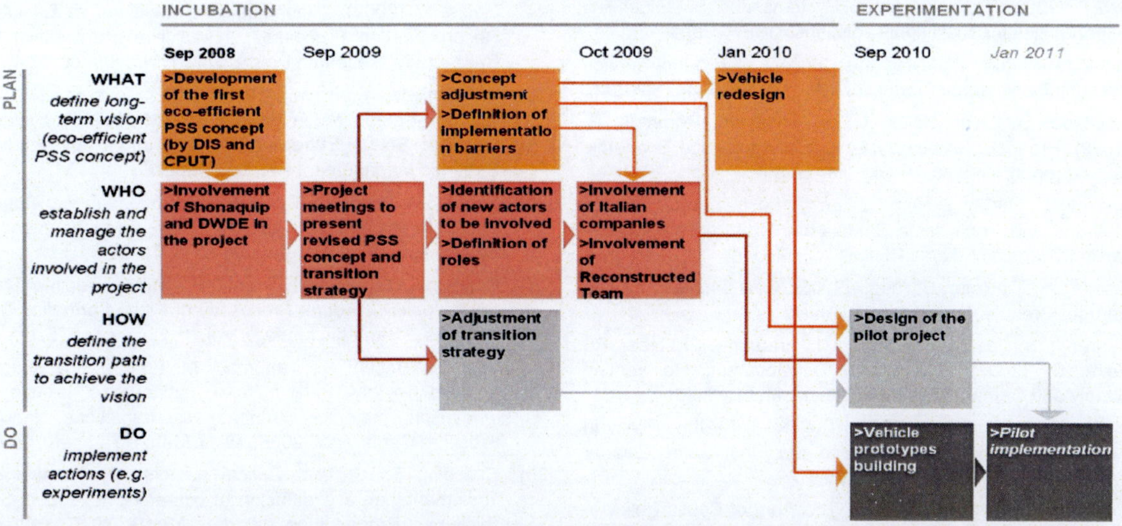

Figure 1: Cape Town Sustainable Mobility project: overview of the different activities undertaken so far.

diffusion of the concept itself (i.e. designing transition paths). It is required to break with the dichotomy between designing and implementing; design, development and implementation are activities carried out simultaneously. This approach is currently adopted and tested in the Cape Town Sustainable Mobility Project. Looking at the design process followed along the project (Figure 1) it clearly emerges:

- **a bifocal design approach:** the design of *long-term visions* is combined with the definition of *short-term actions* (e.g. stakeholders involvement, tests, experiments, etc.).
- **a multi-dimension design approach:** design contributes not only to the ideation and development of the solution (what), but also to the identification of the stakeholders to be involved in the whole process (who), the definition of a strategic transition pathway towards the achievement of the long-term goals (how), and the definition/coordination of the implementation steps (do).

Van de Kerkhof and Wieczorek [16] emphasize the need for a transition manager who facilitates the whole process. Our hypothesis is that strategic designers could be such a facilitator/coordinator. On the basis of the previously hypothesized strategic design approach, the unit of research DIS is currently developing and testing a toolkit to enable strategic designers in defining transition paths for the introduction and diffusion of eco-efficient PSS innovations. It will consists of three parts: a *theoretical* part, presenting the relevant basic knowledge "translated" for strategic designers; a *practical* part, presenting approach, method and tools for designers; and an *exemplificative* part describing examples of applications of the design approach, method and tools.

7 REFERENCES

[1] Schmidt-Bleek, F. (1996): MIPS Book or the Fossil Makers – Factor 10 and More, Berlin, Boston, Basel.

[2] Vergragt, P.J., van Grootveld, G. (1994): Sustainable technology Development in the Netherlands. The first phase of the Dutch STD program, J. of Cleaner Production, Vol. 2, No. 3/4, pp. 133-139.

[3] Stahel, W. R. (1986): Hidden Innovation: R&D in a Sustainable Society, Science and Public Policy, Vol. 13, No. 4, pp. 196-203.

[4] UNEP (2002): Product-Service Systems and Sustainability. Opportunities for sustainable solutions, UNEP, Paris.

[5] Mont, O. (2002): Clarifying the Concept of Product-Service System, J. of Cleaner Production, Vol. 10, No. 3, pp. 237-245.

[6] Goedkoop, M., van Halen, C., te Riele, H., Rommes, P. (1999): Product Services Systems, Ecological and Economic Basics, report 1999/36, VROM, The Hague.

[7] Vezzoli, C. (2007): System Design for Sustainability. Theory, Methods and Tools for a Sustainable "Satisfaction-System" Design, Maggioli Editore, Rimini.

[8] Tukker, A., Tischner, U. (eds.) (2006): New business for Old Europe, Greenleaf publishers, Sheffield.

[9] Behrendt, S., Jasch, C., Kortman, J., Hrauda, G., Pfitzner, R., Velte, D. (2003): Eco-Service Development: Reinventing Supply and Demand in the EU. Greenleaf, Sheffield.

[10] Mont, O., Lindhqvist, T. (2003): The Role of Public Policy in Advancement of Product Service Systems, J. of Cleaner Production, Vol. 11, No. 8, pp. 905-914.

[11] Ceschin, F., Vezzoli, C. (2010): The Role of Public Policy in Stimulating Radical Environmental Impact Reduction in the Automotive Sector: The Need to Focus on Product-Service System Innovation, Int. J. Automotive Technology and Management, Vol. 10, No. 2/3, pp. 321–341.

[12] Schot, J., Geels, F. (2008): Strategic Niche Management and Sustainable Innovation Journeys: Theory, Findings, Research Agenda, and Policy, Technology Analysis & Strategic Management, Vol. 20, No. 5, pp. 537-554.

[13] Plomp, T. (2009): Educational Design Research: an Introduction, in T. Plomp, N. Nieveen: An Introduction to Educational Design Research, SLO, Enschede.

[14] Hall, P. (1993): Policy Paradigms, Social Learning and the State. The Case of Economic Policymaking in Britain, Comp. Polit., Vol. 25, No. 3, pp. 275–96

[15] Bennett, C., Howlett, M. (1992): The Lessons of Learning: Reconciling Theories of Policy Learning and Policy Change, Policy Sci. Vol. 25, pp. 275–294.

[16] Van de Kerkhof, M., Wieczorek, A. (2005): Learning and Stakeholder Participation in Transition Processes Towards Sustainability: Methodological Considerations, Technological Forecasting & Social Change, Vol. 72, pp. 733–747.

[17] Brown, H. S., Vergragt, P., Green, K., Berchicci, L. (2003): Learning for Sustainability Transition through Bounded Socio-Technical Experiments in Personal Mobility, Technology Analysis & Strategic Manag. Vol. 15, No. 3, pp. 291–315.

[18] Brown, H. S., Vergragt, P. (2008) Bounded Socio-Technical Experiments as Agents of Systemic Change: The Case of a Zero-Energy Residential Building, Technological Forecasting and Social Change Vol. 75, pp. 107-130.

[19] Giddens, A. (1984): The Constitution of Society, University of California Press, United States.

[20] Argyris, C., Schön M. (1978): Organizational Learning: A Theory of Action Perspective, Addison-Wesley, Reading, MA.

[21] Senge, P. M. (1990): Building Learning Organizations, Sloan Manage. Rev., Vol. 32, No. 1, pp. 7–23.

[22] Schön, D. A., Rein, M. (1994): Frame Reflection. Towards the Resolution of Intractable Policy Controversies, Basic Books, New York.

[23] Paquet, G. (1999): Governance Through Social Learning, University of Ottawa Press, Ottawa.

[24] Geels, F. (2002): Technological Transitions as Evolutionary Reconfiguration Processes: a Multilevel Perspective and a Case-Study, Research Policy, Vol.31, No. 8/9, pp.1257-1274.

[25] Witkamp, M. J., Raven, R. P. J. M., Royakkers, L. M. M. (2010): Strategic Niche Management of Social Innovation: the Case of Social Entrepreneurship. Technology Analysis & Strategic Management (working paper).

[26] Schot, J.A., Hoogma, R. (1996): De Invoering van Duurzame Technologies: Strategisch Niche Management als Beleidsinstrument, Programma DTO, TU Delft, Delft.

[27] Kemp, R., Schot, J., Hoogma, R. (1998): Regime Shifts to Sustainability Through Processes of Niche Formation, Techn. Analysis & Strategic Manag., Vol.10, No.2, pp. 175-195.

[28] Van den Bosch, S., Taanman, M. (2006): How Innovation Impacts Society. Patterns and mechanisms through which innovation projects contribute to transitions, Innovation Pressure Conference 2006, 15-17 March, Tampere, Finland.

[29] Ceschin, F. (2010): Critical factors for managing the implementation and diffusion of eco-efficient Product-Service Systems, Proceedings of the APMS 2010 conference. Cernobbio-Como, Italy, 11-13 Oct. 2010, Poliscript, Milano.

Towards Sustainability-Driven Innovation through Product-Service Systems

Anthony W. Thompson[1], Tobias C. Larsson[1,2], Göran Broman[1,2]

[1] School of Engineering, Blekinge Institute of Technology, Karlskrona, Sweden

[2] Division of Functional Product Development, Luleå University of Technology

Abstract

Many current sustainability considerations in industry constrain design space by emphasizing reduced material and energy flows across product life cycles. However, there are also opportunities for sustainability awareness to extend design space and drive innovation. Product-service systems (PSS) in particular can be a vehicle through which sustainability-driven innovation occurs. A framework for strategic sustainable development, including a backcasting approach, provides the basis for understanding sustainability in this work and provides insight into how incremental and radical approaches could be aligned within product innovation. This work explores how sustainability considerations can be better integrated into existing product innovation working environments, with an emphasis on opportunities that occur as sustainability knowledge leads to innovation through a product-service system approach. It is demonstrated and ideas are discussed around how sustainability can be used to *drive* innovation processes through product-service systems that companies rely upon, while also supporting global society's movement toward sustainability.

Keywords:

Sustainable Product Innovation; Sustainable Product Development; Strategic Sustainable Development; Ecodesign; Product Service Systems (PSS)

1 INTRODUCTION

Most people that work with product innovation – both product designers and business managers – are in the dominant paradigm that puts short-term profit forward as the primary goal. However, these people are also quickly awakening to the need to more directly include both environmental and social issues in their daily decisions [1]. This is happening for many reasons: customer demand, an expanding regulatory environment, global resource constraints, and perceived opportunities for cost savings to name just a few.

One reason product developers have left sustainability essentially outside of their focus is that there is general confusion in the world around the topic of sustainability [2]. There is general agreement in the scientific community that things need to change [3], and this is often discussed under the term "sustainability." This paper builds upon the foundation that has been developed over the past 20 years into a framework for strategic sustainable development (FSSD) [4]. This FSSD provides an operational definition of sustainability and initial set of strategic guidelines that can be used to provide guidance to decision-makers, e.g. people working with product innovation.

With regard to products, there are two obvious things that can be changed. First, the physical artifacts themselves can be changed, and second, the way that products are managed (including how they are used) over their life cycles can be changed. For the former, more efficiency can be pursued, e.g. material reduction and energy optimization. These are generally good, though alone are not sufficient from a sustainability perspective. They also risk leading to the "rebound effect," which is the idea that improvements on a per-unit basis can lead to greater overall impacts due to increased volume that is enabled by, e.g., reduced cost that stems from the improved efficiency, see, e.g., [5]. While product innovation has traditionally focused on the former with occasional glances toward the latter [6], the movement in industry is now toward the design of artifacts and services together – often referred to as product-service systems (PSS) – and presents an opportunity for these two opportunities to be considered and improved in tandem [6, 7].

1.1 Aim and Scope

This paper endeavors to contribute to answering the following question: *How can sustainability considerations be better integrated into existing product innovation working environments, especially with regards to pursuing a product-service system approach?*

1.2 Method

This paper presents ideas that have been collected through several research projects. As such, it draws from research that included the following methods and techniques:

- A broad survey of literature has provided an opportunity to explore the related topics and specifically focus on the intersection between these key topics, in order to better understand the past and present thinking within the research field.
- Interviews and interaction with people working within the area of product innovation were conducted in order to better understand and describe the state of practice in industry.
- Participation in and facilitation of workshops with development teams with companies involved in these research projects have provided insights in how to aid companies in including sustainability in their thinking around product innovation.

This paper is based on the idea that research into design processes cannot be re-created or tested with a control group. In practical research terms, every design project is unique because of a unique set of needs in an ever-changing global context being addressed by a single design group. Furthermore, there is no "correct" or even "best" solution, as this will change from user to user or context to context. This is the essence of the idea of "wicked problems" introduced by Rittel and Webber relating to planning with regard to social problems, where they see "social processes as the links tying open systems into large and interconnected networks of systems... it has become less apparent where problem centers lie, and less apparent where and how we should intervene even if we do happen to know what aims we seek" [8]. In this regard, case study research

J. Hesselbach and C. Herrmann (eds.), *Functional Thinking for Value Creation: Proceedings of the 3rd CIRP International Conference on Industrial Product Service Systems, Technische Universität Braunschweig, Braunschweig, Germany, May 5th - 6th, 2011*,
DOI 10.1007/978-3-642-19689-8_22,

is valuable because it allows for research topics to be defined broadly while potentially considering multiple variables and relying on multiple sources of evidence [9]. In recognition of the need for a new approach to this type of research, Blessing and Chakrabarti put forth a Design Research Methodology (DRM) [10]. The DRM and its methods provide guidance for planning and implementing design research, thus providing a more rigorous approach to research.

2 RELATED AREAS

Concepts in this paper draw upon three broad topics: (1) sustainability, (2) product innovation, and (3) product-service systems. This section presents briefly each of these, as well as an additional area that is emerging as a combination of them: sustainable product innovation.

2.1 Sustainability

In recent decades, numerous reports, studies, theses, articles and books have been published documenting impacts and opportunities, e.g., species loss [3], resource constraints [11], and the business opportunities for those aware of sustainability issues [12]. The Brundtland definition of sustainable development [13] puts forth an attractive vision, but leaves a significant gap for the business need to be operational. This has lead to many attempts to clarify the concept of sustainability; one of which is a framework for strategic sustainable development (FSSD) based upon a five-level model that can be used to plan in any complex system. When it is used to provide guidance toward a sustainable human society (i.e. "human society within the biosphere"), it is referred to as the framework for strategic sustainable development [4].

Three key aspects of the FSSD make it well-suited for use in both strategic and operational contexts.

- Five level structure, clearly distinguishing between the system, the definition of success, strategic guidelines, actions, and tools
- Unique definition of success in basic principles for sustainability
- Backcasting from a desired future (contrasted with forecasting current trends only)

Combining backcasting with this unique definition of success results in "backcasting from sustainability principles" allowing for strategic decision making that promotes flexibility, movement toward a sustainable future, and appropriate allocation of resources.

The term "sustainability" in this paper, then, refers to global socio-ecological sustainability. It does not, unless specifically stated, refer to the sustainability of some other (sub-) system, e.g., a company.

2.2 Innovation

Innovation, generally, refers to new products, processes or ideas that are put into use in the world. "Innovation" differs from "invention" which is the creation of those new products or processes, in that innovation implies inventions that are put into practice. Schumpeter lists five types of innovation: new products, new methods of production, new sources of supply, exploitation of new markets, and new ways to organize business [14]. In this paper, the term "product" is in line with the ISO definition and refers to "what is sold" and thus not only the physical artifact:

> *A product is an output that results from a process. Products can be tangible or intangible, a thing or an idea, hardware or software, information or knowledge, a process or procedure, a service or function, or a concept or creation.*

Innovation literature frequently comes from the social sciences with roots originating with, e.g., Schumpeter. Innovation references also originate from within the field of engineering, e.g., [15, 16]. One related observation is presented by Kline and Rosenberg:

> *Economists have, by and large, analyzed technological innovation as a "black box" – a system containing unknown components and processes. They have attempted to identify and measure the main inputs that enter that black box, and they have, with much greater difficulty, attempted to identify and measure the output emanating from that box. However, they have devoted very little attention to what actually goes on inside the box; they have largely neglected the highly complex process through which certain inputs are transformed into certain outputs.*
>
> *Technologists, on the other hand, have been largely preoccupied with the technical processes that occur inside that box. They have too often neglected, or even ignored, both the market forces with which the product must operate and the institutional effects required to create the requisite adjustments to innovation. [17]*

It is often challenging to arrive at a shared vocabulary between these different perspectives. This paper attempts to draw on literature from social science and engineering perspectives.

2.3 Product Service Systems

The concept of product-service systems (PSS) emphasizes a shift in focus from selling only a physical artifact or service to selling the result of a combination of the two. Definitions of PSS typically include reference to increased competitiveness of PSS providers and often refer to reduced environmental impacts.

Tukker presents eight types of PSS, divided into three categories: product-oriented, use-oriented, and result-oriented [18]. Tukker, Tischner and Verkuijl have explored the opportunities for environmental improvement with regard to these eight types of PSS, suggesting that all eight types are usually, but not necessarily, associated with improved environmental performance [19]. Of the eight types, some are believed to have the opportunity for more significant environmental improvement than others, with the function-oriented type having the most significant opportunities. This eighth type, functional result-oriented PSS, leads into the idea of "functional product development" described by [6] as having the objective of "developing the solution (i.e. any combination of hardware, software, services, etc.) to customer needs that create value for the customer."

2.4 Sustainable Product Innovation

There is significant research in a variety of areas closely relating to sustainability in product innovation. Recently, work was done at Imperial College focusing on Sustainable Product and Service Development (SPSD) that reviewed many approaches to sustainability in product development and resulted in an approach emphasizing functional and systems thinking [7]. Ecodesign emphasizes bringing ecological issues into the product innovation process; see [20] for an introduction, as well as closely-related concepts like Design for Environment (DfE). The present work differs from those by utilizing the framework for strategic sustainable development mentioned in 2.1, thus providing a different perspective with regard to the sustainability component with potentially different results.

There are various approaches to design (more broadly than product development) that also bring in sustainability-related thinking, e.g., Cradle-to-Cradle or Biomimicry. Here, emphasis is placed on radical innovation through outside inspiration. Cradle-to-Cradle, with the mantra "waste equals food," emphasizes the need for technical systems to operate in cycles, and highlights the concept that "eco-efficiency only works to make the old, destructive systems a bit less so" [21]. Biomimicry suggests that nature has been innovating for millions of years, and that there is a huge amount of inspiration to be explored by human designers [22]. These two examples are mentioned because they strongly relate to the idea of using sustainability thinking to drive innovation.

3 TOWARDS SUSTAINABILITY-DRIVEN INNOVATION THROUGH PRODUCT-SERVICE SYSTEMS

Section 3.1 presents views on how sustainability considerations are currently included in product innovation processes. Section 3.2 develops the case for using, and then presents ideas for how to work towards, sustainability-driven innovation through product-service systems.

3.1 Observations on Sustainability in Swedish Product Innovation

This section first reflects on some motivations for companies wanting to include sustainability in their product innovation processes, followed by some ways that they are including sustainability in those processes and some of the justifications they provide for doing so.

Motivations for Including Sustainability

Companies include sustainability criteria in their product innovation processes primarily for one of these reasons: 1) legislation, 2) cost reduction (e.g. resource efficiency), or 3) employee interest in "doing good."

Certainly the Swedish companies involved in this work include sustainability aspects at least to the extent that they must in order to comply with legislation. Sustainability criteria that overlap with cost savings (e.g. efficiency of resource use or energy) are also very likely to be considered. Other sustainability criteria that do not have direct effects on cost are much less likely to be considered; though aspects that can have indirect impacts on company success, e.g., through the company's image are being considered with greater frequency.

The origination of sustainability from legal requirements or employee interest often leads to sustainability considerations being perceived as an extra expense, i.e. one more requirement that competes for resources.

In both business-to-consumer (B2C) and business-to-business (B2B) situations, customers are increasingly demanding sustainability be considered. In B2C it is often in the form of eco-labels or other identifying factors that provide peace of mind to the consumer, while in B2B situations it frequently relates to procurement demands by the purchasing company to reduce risks, e.g., of not being in compliance with environmental legislation.

Ways of Including Sustainability

In response to the way sustainability aspects are beginning to be required of companies, sustainability aspects are being added into product requirements, e.g., compliance with materials lists that say certain substances are not to be used in a product itself or the manufacturing processes for the product; carbon emissions over the life of the product must be estimated and held at or below a certain level; or the working conditions of suppliers must meet certain requirements. Some of these have been around for decades (e.g. material lists), while others are more recent (e.g. social aspects at suppliers).

Innovation processes must then take these additional requirements into consideration. This further limits (e.g. beyond technical limitations) the design space in which product developers are able to create solutions, and may draw resources away from other types of improvements that could be made. This adds to the cost of the innovation project, putting additional constraints on the already stretched allocation of resources.

Justification for Including Sustainability

These approaches typically lead to attempts to show how sustainability efforts reduce costs or increase revenues, and to argue that when fully considered sustainability aspects do not increase overall costs for the company. Theoretically, this is done at a product level through, e.g., life-cycle costing (LCC), total-cost accounting (TCA), or full-cost accounting (FCA); see, e.g., [23, 24]. Willard has written on the effects of sustainability at the firm level, and suggests that there are significant economic impacts on a company's bottom line from incorporating sustainability aspects that relate to, e.g., staff retention, attraction of the best talent, etc. [12]. Many companies that have been involved in these research projects are aware of these approaches, but do not appear to have them integrated into standard procedures.

Summary of Observations

The chain of thought presented in this section suggests that companies include sustainability considerations either because they are required by legislation, out of some sense of greater good, or in order to attract or retain customers and staff. All of these are fine reasons to include sustainability considerations, and likely contribute to a company's success. However, this chain of thought does not get directly to the main motivation for industry: profitability. Rather, there is an indirect journey that leads back to profitability. As with the FSSD referred to in section 2.1, "sustainability" is at the "success" level. However, for companies, "success" is not "global socio-ecological sustainability," but rather "profitability." Awareness of sustainability issues and the strategic use of them can certainly support a company's efforts to be profitable.

Innovation is a significant factor in profitability: the ability to identify and successfully take to market new products, to find new and better ways to produce physical artifacts and to deliver services, etc. directly support competitiveness and profitability for firms. And, innovation that supports development of society towards a future that 'can be' (i.e. a sustainable society) should have an inherent overall advantage over innovation towards a future that 'cannot be'. Assuming this is true, there is an opportunity for sustainability to drive innovation processes in companies that leads to profitability. What is missing, then, is the competence to use sustainability, and especially a strategic sustainability perspective, to guide and accelerate innovation processes.

3.2 Sustainability as Driver of Innovation

The Case for Sustainability as Driver

Sustainability as described in section 2.1 asks what is necessary in order for human society to not systematically degrade the social and ecological systems that it depends upon, and suggests that society ought not to do things that potentially risk long-term existence. This way of thinking about sustainability can be used to drive innovation by guiding incremental and radical innovations in either products or processes (e.g. reduced material or energy use by a product or increased efficiency in production processes). This thinking can also drive incremental and radical innovations in business models, market conditions and societal institutions which opens up for the meeting of human needs in ways that mean significantly reduced negative environmental or social impacts.

Sustainability-driven innovation is different than "innovation for sustainability," which implies that the innovators are interested in pursuing sustainability as an end goal. This is not typically the way companies do, or even legally can, define success. Rather, knowing about sustainability issues can help companies to be more successful on an increasingly sustainability-driven market.

Using a product-service system approach provides an opportunity for companies to reconsider how their artifacts, services, and combinations of these create value and generate revenue. Pursuing a PSS does not necessarily explicitly demand a sustainability focus or even awareness, and it does not necessarily imply an improved sustainability profile. Rather, a PSS-approach opens up to new ways of thinking which are inherently in less contradiction to a sustainable society than more traditional approaches focused only on generating revenue from the sales of physical artifacts. This is because a PSS-approach opens the possibilities to generating revenue based on the provision of specific functions that meet needs rather than generating revenue based on the sales of those physical artifacts. Revenue based on function is further enhanced through sustainability-related initiatives such as dematerialization, consideration of closed-loop product life cycles, minimization of operating costs that are often indicative of negative environmental or social impacts, etc.

Making Sustainability the Driver

Section 3.1 presents observations of the day-in and day-out of sustainability aspects in product innovation in some Swedish companies: there exists a core product, there is a desire to improve the product both in terms of meeting evolving customer needs and in terms of profitability, and there is an interest or a need to maintain or improve sustainability performance. With that in mind, and also keeping in mind Section 2.4, which briefly introduced other existing innovation-based design approaches (with Cradle-to-Cradle and Biomimicry as specific examples), the following are thoughts on how innovation processes can become more sustainability-driven.

Backcasting when Developing Support

The challenge when developing support for innovation processes is that with regard to sustainability, there is a sense of needing the radical changes that can be inspired by more radical concepts. On the other hand, the challenge of integrating support into existing product innovation working environments is that there are established routines and tight timeframes for innovation projects; asking for a radical re-thinking of how a product should or could function is simply not possible given limited resources. Product developers ask for a simple tool that guides them to the right material choice; e.g. aluminum requires more energy to produce than steel, so steel should be used. This, of course, is a gross over-simplification of the life cycle impacts of the different materials, and is precisely why simple, well-intentioned guidance is problematic: the questions seldom have simple answers. People understand this: aluminum is lighter than steel, so using aluminum instead of steel in some applications will recover the extra energy used in production, eventually having a better overall performance with regard to energy use. However, the best design may depend significantly upon user behavior, thus an apparently simple question becomes a wicked problem as described in section 1.2.

Support concepts must acknowledge the reality of the present product innovation working environment, including resource (e.g. time) constraints as well as product performance obligations. This naturally tends towards an incremental approach to improving the sustainability performance of products. At the same time, there is an urgency to provide support that is capable of meeting the ever-higher demands of the global context. In light of this, there is an opportunity to use a backcasting approach when developing support tools and methods. This would entail developing support that considers both the immediate decisions that product developers are being asked to make, and also using that support to lead the product developer's thinking into new areas.

The short-term steps involve providing support tools and methods that companies need to continue exploring a PSS mindset may not result immediately in function-based innovation (since function-oriented products are only one type of PSS). The long-term is about working toward function-based innovation so that revenue streams can evolve to be based on sales of function – with its associated potential benefits for global socio-ecological systems.

Here the suggestion is that the backcasting approach should be used by researchers to develop support methods and tools. Furthermore, based on the assumption that pursuing function-based products is a very attractive opportunity that combines society's need to pursue sustainability with the business need to be profitable, the suggestion is that the vision that is backcast from should be a product innovation working environment that is focused on functional product innovation.

Expand from Sustainability Constraints to Sustainability-driven Innovation

As described in 3.1, sustainability is often incorporated into product innovation working environments as an "add on," e.g., through product or process requirements that serve as filters to reduce the number of ideas or concepts until only the "more sustainable" (i.e. the options with the fewest known environmental impacts) remain.

To a greater or lesser extent, adding sustainability-based design requirements and incorporating methods and tools to existing product innovation processes are ways of comfortably introducing sustainability into those environments. However, as the easy opportunities for improvement with regard to sustainability are implemented (i.e. the "low-hanging fruit" are "harvested"), continued improvement with regard to sustainability aspects is more difficult. After the easy opportunities are exhausted, then there is a need/opportunity for sustainability to proactively drive innovation.

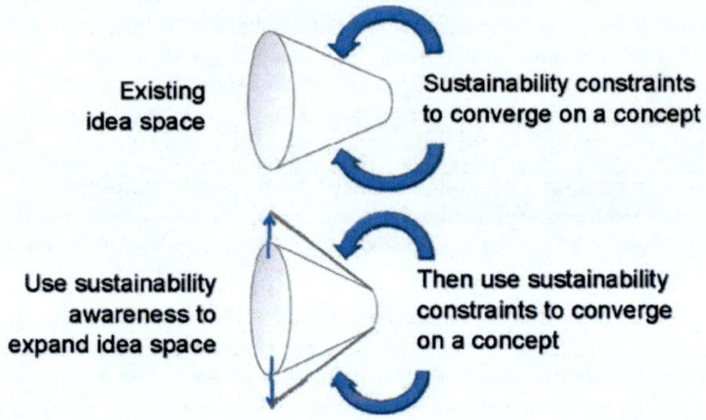

Figure 1: Using sustainability knowledge to expand idea space prior to using sustainability knowledge to constrain selections.

Here it is suggested that sustainability can drive innovation by opening up the idea space during idea generation, i.e. contributing to the "divergence" that occurs in earlier stages, before sustainability aspects are used as a filter to "converge" into a final product. An example of this is provided by Ny et al. through a waterjet cutting machine, where they describe a way in how to do this by modeling a current system and then looking at it in an anticipated future in which the market is increasingly sustainability-driven.

Create value by optimizing at a broader system level

Expanding the peripheral perspective of people working with innovation can lead to opportunities for capturing value that is otherwise outside of their scope. This is because there is frequently an emphasis on the optimization of sub-systems, while higher level systems remain sub-optimized: focus is on tweaking the details of lower-level systems, while opportunities for significant higher-level system improvement are missed. This is in line with what Bey and McAloone [25] suggest when discussing the role of ecodesign and LCA in PSS development: that a PSS approach inherently promotes thinking at a higher system level.

A waterjet cutting optimization project illustrates this: the first efforts in the project related to building detailed technical models of the machines and machines parts, which were used to better optimize the weight of the parts, and thus improving the energy efficiency of the machine, e.g., an opportunity to reduce the weight of moving parts by 30 percent lead to overall system improvement [26].

Additionally, outside of the scope of those early technical improvements, was the opportunity to optimize the broader system with regard to use of sand as an abrasive in the process. The weight optimization of machine parts is at a more focused system level, thus involving a smaller number of actors, and thus easier to modify. The opportunity to optimize the abrasive was out of the scope of the initial focus, and when explored, involved a significantly larger number of actors. There is, however, economic value to be captured and environmental improvement to be made specifically by reducing transportation related to the sand. One can assume the current situation happens as it does today because it optimizes the economics at a certain level. However, as the market becomes increasingly sustainability-driven (e.g. increased transportation costs due to energy price increases, carbon-related taxes, etc.) opportunities to optimize at a higher system level will become more economically rational.

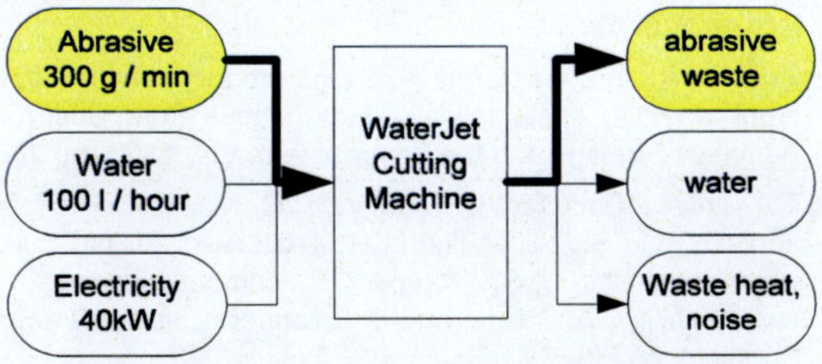

Figure 2. Expanding the view of the Waterjet Cutting Machine to include other required energy and material flows in the cutting process. The abrasive is highlighted due to having the greater environmental impacts.

This example is provided as an illustration of how broadening system boundaries can lead to improvement: first the machine itself was optimized. Then this example broadened system boundaries to consider how to optimize consumables related to the machine, The next opportunity to broaden system boundaries involves a move toward a PSS based offer (selling the function of cutting, instead of selling machines) by further extending system boundaries, e.g., to better consider how users interact with the waterjet cutting machine and the specific contexts of how the machine is used.

Capturing the value created by optimizing at higher system levels is challenging, particularly with business models focused on the sales of physical artifacts. However, Thompson et al. present an example that appears poised for capturing value at this higher level: long-life light tubes that reduce total cost of ownership by eliminating operating costs associated with changing the light tubes at the end of their useful life [27]. This value is not typically considered in the development of physical artifacts, and communicating it to customers is also challenging. The PSS approach, however, opens possibilities for win-win-win situations for the light-tube providers, users and (because of improved sustainability-performance) society.

Innovate the offer, not the artifact

The case presented by Thompson et al. also shows that there is an opportunity to use an existing product to focus on a new approach to providing the function that customers want. In the case of long-life light tubes, the physical artifacts have a specific attribute (a working life several times longer than the average light tube) that (potentially) offers a significant sustainability advantage. This case suggests that in order for the sustainability advantage provided by the attribute of that product to also be made into an economic advantage, the business model around the product needs to shift toward a function-based offer of providing light, rather than remaining focused on selling the physical artifact.

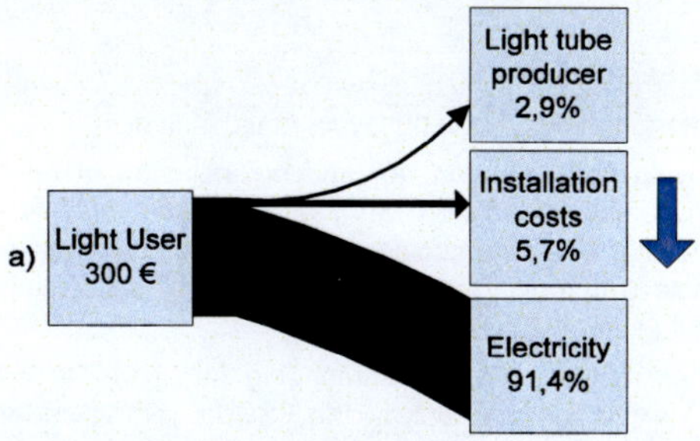

Figure 3a: Total customer costs for light during 12 years with 4 standard-life light tubes sold as the product.

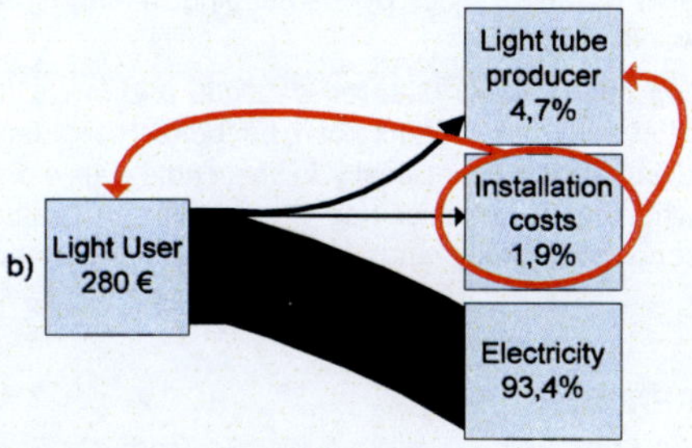

Figure 3b: Total customer costs for light during 12 years with one long-life light tube as the basis for a PSS-offer.

4 RESULTS AND CONCLUSION

This paper aims at supporting the inclusion of sustainability considerations in the product innovation process by articulating how sustainability can be a driver in the innovation process, specifically through a product-service system approach. This work contributes to understanding with regard to theory about how sustainability-driven innovation can occur through product-service systems within the broader research field, as well as how to apply that understanding to the state of practice in industry today in order to realize more sustainable PSSs.

There is an opportunity to improve sustainability performance of a PSS and increase customer value by broadening the scope in product innovation working environments to allow for increased consideration of opportunities in two areas highlighted in this paper. First, technical systems can be optimized at higher system levels, e.g., as in the case of abrasives with the waterjet cutting case. Second, business models can be modified to focus on communicating value through function, especially with regard to products that have sustainability attributes as demonstrated in the case of long-life light tubes. For those working in the area of sustainability, "sustainability" often becomes the primary motivating factor. While a business cannot be sustainable if it is part of an unsustainable society, the perspective is different from within the business world. Thus, companies are typically more interested in innovation than sustainability, for the ability to innovate is what allows the company to sustain itself: changing customers, offering new products/services, expanding into new markets, etc. In recognition of this, this paper has demonstrated and developed ideas around how sustainability can be used to drive those innovation processes through product-service systems that companies rely upon, while also supporting global society's movement toward sustainability.

5 FUTURE WORK

Future efforts building upon this work could include:

- Clarifying the argument for shifting inclusion of sustainability aspects from a "do less bad" approach that only emphasizes quantifying and reducing known negative environmental impacts, and moving toward a methodology where sustainability is driving innovation processes;
- Further reviewing and summarizing sustainability aspects of PSS, with a specific look at the FSSD's role in understanding and analyzing the value PSS can bring to global sustainability work;
- Continuing to support working toward sustainability-driven innovation through PSS by developing methods, tools and frameworks; and
- Exploring how socially oriented aspects, and in particular user interaction with the product, can be better considered during product innovation, specifically in the context of a PSS based offer, where user behavior has very significant implications for the economic viability and sustainability performance of the offer.

6 REFERENCES

[1] Porter, M.E. and van der Linde, C., (1995): Toward a new conception of the environment-competitiveness relationship, The Journal of Economic Perspectives, Vol. 9(4), pp 97.

[2] Johnston, P., Everard, M., Santillo, D., and Robert, K.H., (2007): Reclaiming the definition of sustainability, Enviro. Science and Pollution Research, Vol. 14(1), pp 60-66.

[3] Millenium Ecosystem Assessment, (2005): Living beyond our means. Natural assets and human well-being. Statement from the board, J. Sarukhán and A. Whyte, Editors.

[4] Holmberg, J. and Robèrt, K.-H., (2000): Backcasting from non-overlapping sustainability principles--a framework for strategic planning, International Journal of Sustainable Development and World Ecology, Vol. (7), pp 291-308.

[5] Binswanger, M., (2001): Technological progress and sustainable development: what about the rebound effect?, Ecological Economics, Vol. 36(1), pp 119-132.

[6] Isaksson, O., Larsson, T.C., and Ronnback, A.O., (2009): Development of product-service systems: challenges and opportunities for the manufacturing firm, Journal of Engineering Design, Vol. 20(4), pp 329-348.

[7] Maxwell, D., Sheate, W., and van der Vorst, R., (2006): Functional and systems aspects of the sustainable product and service development approach for industry, Journal of Cleaner Production, Vol. 14(17), pp 1466-1479.

[8] Rittel, H. and Webber, M., (1973): Dilemmas in a General Theory of Planning, Policy Sciences, Vol. 4, pp 155-169.

[9] Yin, R.K., (2003): Applications of case study research, Sage Publications, Thousand Oaks.

[10] Blessing, L. and Chakrabarti, A., (2009): DRM, a Design Research Methodology, Springer.

[11] Gordon, R.B., Bertram, M., and Gradel, T.E., (2006): Metal stocks and sustainability, Proceedings of the National Academy of Sciences of the United States of America, Vol. 103(5), pp 1209.

[12] Willard, B., (2002): The Sustainability Advantage: seven business case benefits of a triple bottom line, New Society Publishers.

[13] Brundtland, G.H., ed. (1987): Our common future: [report of the] World Commission on Environment and Development, Oxford University Press: Oxford, UK.

[14] Fagerberg, J., Mowery, D., and Nelson, R., eds., (2006): The Oxford Handbook of Innovation, Oxford University Press: Oxford.

[15] Roozenburg, N.F.M. and Eekels, J., (1995): Product Design: Fundamentals and Methods, John Wiley & Sons Ltd, Chichester, England.

[16] Ulrich, K.T. and Eppinger, S.D., (2003): Product Design and Development, 3rd ed, McGraw-Hill/Irwin.

[17] Kline, S. and Rosenberg, N., (1986): An Overview of Innovation, in The Positive Sum Strategy: Harnessing Technology for Economic Growth, R. Landau and N. Rosenberg, Editors. National Academy Press: Washington, DC. pp 275-306.

[18] Tukker, A., (2004): Eight types of product-service system: eight ways to sustainability? Experiences from SusProNet, Business Strategy and the Environment, Vol. 13(4), pp 246.

[19] Tukker, A., Tischner, U., and Verkuijl, M., (2006): Product-services and Sustainability, in New business for old Europe: Product-Service development, competitiveness and sustainability., A. Tukker and U. Tischner, Editors. Greenleaf: Sheffield. pp 72-97.

[20] Karlsson, R. and Luttropp, C., (2006): EcoDesign: what's happening? An overview of the subject area of EcoDesign and of the papers in this special issue, Journal of Cleaner Production, Vol. 14(15-16), pp 1291-1298.

[21] McDonough, W. and Braungart, M., (2002): Cradle to Cradle: Remaking the Way We Make Things, North Point Press, New York.

[22] Benyus, J.M., (1997): Biomimicry: Innovation inspired by nature, Harper Perennial, New York.

[23] Norris, G.A., (2001): Integrating life cycle cost analysis and LCA, The International Journal of Life Cycle Assessment, Vol. 6(2), pp 118-120.

[24] Shapiro, K.G., (2001): Incorporating costs in LCA, The International Journal of Life Cycle Assessment, Vol. 6(2), pp 121-123.

[25] Bey, N. and McAloone, T.C., (2006): From LCA to PSS - making leaps towards sustainability by applying product/service-system thinking in product development, Proceedings of LCE 2006, 13th CIRP International Conference on Life Cycle Engineering, Vol., pp 571-576.

[26] Byggeth, S.H., Ny, H., Wall, J., Broman, G., and Robèrt, K.-H., (2007): "Introductory Procedure for Sustainability-Driven Design Optimization", International Conference on Engineering Design (ICED).

[27] Thompson, A.W., Ny, H., Lindahl, P., Broman, G., and Severinsson, M. 2010: Benefits of a Product Service System Approach for Long-life Products: The Case of Light Tubes, in: Proceedings of the 2nd CIRP International Conference on Industrial Product-Service Systems (IPS2), pp. 83-89, Linköping University, Sweden.

A New PSS Conceptual Design Approach Driven by User Task Model

Xiuli Geng, Xuening Chu
School of Mechanical Engineering, Shanghai Jiao Tong University, Shanghai, China

Abstract
Product Service System (PSS) has become a widely-accepted concept in the business world. The user task model traditionally used in user-centered design is employed in analyzing user's goals to drive the PSS conceptual design. The proposed PSS conceptual design approach has three models: user task model, function model and conceptual service blueprint. The conceptual service blueprint is proposed to model the PSS concept. PSS concept is constituted by service activities and product modules. Product behavior is introduced to bridge the gap between function and physical module. A case study is carried out to demonstrate the effectiveness of the approach.

Keywords:
Product Service System (PSS); User Task Model; Service Blueprint

1 INTRODUCTION

Service is playing a more and more important role in the economy society, and accounts for a large proportion in the national economy as a whole. Service can substitute product to satisfy user's need with less environment impact. In the industrial field, manufacturers are shifting from providing physical products to providing service-based solutions. Product Service System (PSS) has emerged as a new competitive strategy in the environmental conscious and customer-centered business society. PSS is supported by physical product together with corresponding services [1, 2], and it can provide higher values to users not merely depending on material and technology investment.

The conceptual design plays an important role in the traditional product development, and so it does in the development process of PSS. In traditional design theory, user's requirement is the beginning of conceptual design. The PSS concept is process-oriented and correlates tightly with user's using process and scenario. However, in the existing PSS conceptual design approaches, there is no systematic method to express user's requirements considering the PSS using process. Requirements are always described as a set of discrete parameters. Great gap exists between the discrete requirements expression with the process-oriented PSS concept expression. In fact, users really require the use or the result of PSS, and PSS design emphasizes realizing users' goals or value. Therefore, user's goals or value is the basis and driving force of PSS conceptual design. Analyzing user's business process and tasks is of significant importance in generating PSS concept being up to user's using scenario. Consequently, we adopt user task model to express the whole structured requirements for realizing user's goals, and to drive the PSS conceptual design.

Task is an important concept in user-centric design, e.g. in the discipline of software design. Tasks define how the user can reach a goal in a specific application domain. Montabert et al. [3] pointed out that adopting task model to capture requirements can bridge the gap between scenarios and parameters benefiting software design. Considering the process nature of goals realization and design objects, the user task model used in driving PSS conceptual design is process-oriented, and constructed in the form of a network involving task interacting relationships and hierarchical decomposing relationships.

The PSS design objects consist of products and services, and the PSS design space is broader than simple product design and service design. Functions are generally used as the medium between requirements and concept. Functions can also act as the bridge between user task model and PSS concept, and they don't distinguish product functions from service functions. Traditional static and hierarchical function model can not reflect the flow information in the user task model and provide process relationship data to describe PSS concept. In this paper, function model is constructed in the form of a network as user task model.

The tough issue in the PSS conceptual design process is how to describe the PSS concept to combine products and services seamlessly. Product concept is always described as a tree structure, and the leaf node is product module or component. Service concept is much more abstract than product concept. Service activity occupies the central role in the concept, because it realizes function directly. In this paper, PSS concept is constituted by service activities and product modules. Product modules are designed aiming at implementing functions in function model or functions required by service activities. Product behavior is incorporated to generate product module indirectly to bridge the gap between function and physical module. Service blueprint technique [4] is a traditional service description approach. Service blueprint can be extended to include product system [5]. We propose the conceptual service blueprint to model PSS concept based on the extended service blueprint [5] in this paper. User's actions, service activities, product behaviors and modules, and the relationships among them are described in the conceptual service blueprint.

The PSS conceptual design approach proposed in this paper has three steps: (1) analyze user's goals and construct user task model; (2) design functions and construct function model; and (3) design service activities and product modules, and construct conceptual service blueprint. The rest of this paper is organized as follows. Section 2 reviews the related product design, service design and

J. Hesselbach and C. Herrmann (eds.), *Functional Thinking for Value Creation: Proceedings of the 3rd CIRP International Conference on Industrial Product Service Systems, Technische Universität Braunschweig, Braunschweig, Germany, May 5th - 6th, 2011*,
DOI 10.1007/978-3-642-19689-8_23, © Springer-Verlag Berlin Heidelberg 2011

PSS design approaches, and gives out an explanation of user task model application. Section 3 presents the PSS conceptual design framework, and depicts user task model, function model, and conceptual service blueprint, respectively. Section 4 uses a case of PSS design for elevator to demonstrate the effectiveness of the developed PSS conceptual design framework. Section 5 gives conclusions achieved in this research.

2 LITERATURE REVIEW

Most of the early studies on PSS development were primarily conducted from the viewpoint of marketing and management [6, 7]. Few systematic engineering approaches for PSS development are proposed. Aurich et al. [8] has proposed that the PSS design should integrate with traditional design methods, which cannot be a solely design problem.

2.1 Product and Service Design Methods

Function-Behavior-Structure (FBS) method [9, 10] and Axiom Design (AD) method [11] are two widely-used methods for product conceptual design. The FBS framework represents designing by a set of processes linking function, behavior and structure together. According to AD, every design object can be depicted in four design domains: the customer, the functional, the physical and the process domains.

Service design emerged as a research field in the 1980s [12]. The characteristic of service design is that the design object is a process or activity. Service blueprint technique has been widely used to define service delivery system. It is generally a useful tool for the service operation managers and the marketing managers. Service Engineering was proposed in the mid 90's in Germany and Israel [13], is a technical discipline concerned with the systematic development of services using suitable models, methods and tools. Service Engineering attempts to use existing engineering methods of product development to develop innovative services.

2.2 Existing PSS Engineering Design Methods

Some PSS researchers adopt approaches borrowed from product development and soft engineering to design PSS. Aurich et al. [8] claimed that life-cycle engineering can be used for PSS design, and introduced a systematic design process to deliver product-related technical services based upon process modularization. Aurich et al. [14] developed a life cycle-oriented configuration model for PSS development. In the model, one sub-model is product which is modeled in the traditional form of tree structure. The other sub-model is service which is modeled including three modules: result modules, process modules and resource modules. Kimita et al. [15] adopted the approach of AD to detect and avoid the conflicts considering the complex design process of PSS, and applied the modeling methods from Service Engineering to represent the PSS structure. Chen [16] proposed TRIZ based eco-innovative design methodology to support designers to develop PSS.

Shimomura and Tomiyama [17] defined a service as "an activity that a provider causes, a receiver to change from an existing state to a new state that the receiver desires, where both service contents and service channel are means to realize the service". In their research, product can be deemed as a part of service. Sakao and Shimomura [18] proposed a new service model which consists of four sub-models: flow model, scope model, view model and scenario model, and developed a service CAD to support the service/product design. Shimomura et al. [5] extended the original service blueprint to include physical product processes, enabled service designers to construct service activity by connecting the blueprint with the view model with the consideration of human process and physical process.

2.3 Application of User Task Model

Task analysis was first proposed in the work of Gilbreth [19] and Taylor [20]. Later on, task analysis was introduced to the design fields due to the worthy aspects of the concept of "task". Many task-based design methods were proposed, e.g. hierarchical task decomposition [21], task-action grammars [22], and task-based design [23]. Redish and Wixon [24] proposed that task analysis included users, tasks and environments, and argued that all of these were necessary inputs to the design of any product. Gould and Lewis [25] summarized the main principle of user-centered design was focusing on users and tasks early. The typical application area of task model is User-Interface (UI) design, whose idea is user-centered design. In the interaction system design, 50% codes or so have relationships with UI, so the UI design has important significance to the success of a system. Model-based UI development approach has been popularly recognized. User task model has been recognized as a starting point for UI design. It is used to understand user, user's work environment, and user's goals and tasks. The abstract presentation model of user tasks are refined to generate concrete presentation model, and the executive UI of application system can be generated finally.

PSS conceptual design should be considered based on user-centered design approach the same as UI design, and it should pay attention to user, environment, and user's goals. User task model is employed to be applied in driving PSS conceptual design. With it, user's goals can be analyzed and expressed completely and systematically as the driving force of PSS design.

3 PSS CONCEPTUAL DESIGN FRAMEWORK

3.1 The Presentation of PSS Conceptual Design Approach

According to the AD theory, the product conceptual design has three domains: customer domain, functional domain, and structure domain. The PSS conceptual design approach is proposed based on it. Since design objects are different, the design frameworks will be different. The essential point of PSS conceptual design differing from product conceptual design is that the design driving force is the realizing user's goals not satisfying user's product requirements. Design space is not limit to product domain, and expanded to include service domain. The analysis of the corresponding three domains and three design steps for PSS conceptual design is as follows:

1. Traditionally, customer domain generally contains static requirements for product. The corresponding customer domain of PSS conceptual design should contain user's tasks for realizing user's goals. Correspondingly, the first PSS conceptual design step is analyzing user's goals to construct user task model. User task model can bridge the gap between user's goals and PSS functions. Considering the process nature of goals realization and PSS concept, the user task model is constructed by process-oriented approach.

2. Tasks are achieved by functions which are derived by function design. The contents in the functional domain of PSS conceptual design are still functions. However, the functions are not limited to be implemented by products, but can also be implemented by services. The second PSS conceptual design step is designing functions to construct function model. Driven by the process-oriented user task model and required to provide design data to the concept with process nature, the function model is required to be constructed in the form of a process structure.

3. Functions are exhibited by products or services. Therefore, two design activities are required to generate the PSS concept: designing service activities, and designing product structures. Service activity may generate new product function requirements to

support activity execution. Therefore, some product structures have to be designed aiming at implementing these functions. Corresponding to the structure domain in product design, which contains product modules, the concept domain of PSS contains service activities and product modules. PSS concept should be modeled in a process structure involving user, service staff, and product system. The service blueprint can be modified for expressing user actions, service activities and product modules. The modified service blueprint expresses PSS from design perspective but not from operating perspective, and it is called conceptual service blueprint in this paper. Correspondingly, the third design step is designing product structures and service activities to construct conceptual service blueprint.

The idea of proposing the PSS conceptual design framework is shown in Figure 1. The PSS conceptual design process and corresponding supporting models are shown in Figure 2.

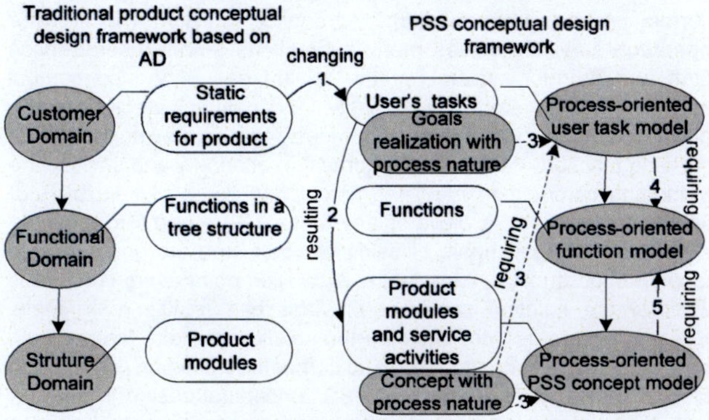

Figure 1: The idea of proposing the PSS conceptual design framework.

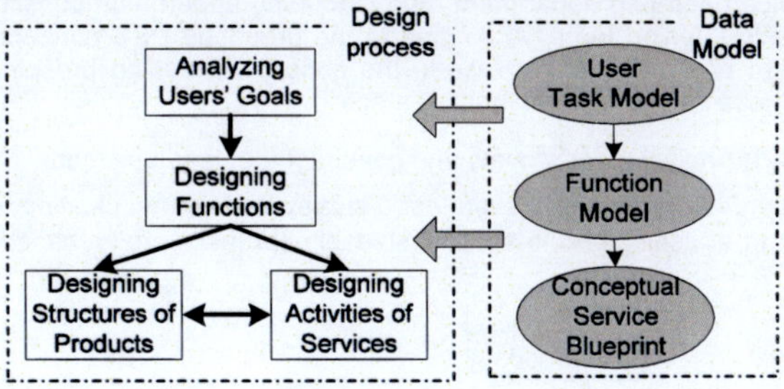

Figure 2: The PSS conceptual design process and corresponding supporting models.

3.2 User Task Model

User task model defines how the user's goals can be realized, describes the user's tasks and their sub-tasks, and gives out the conditions of task implementation. The acquisition methods of user task model have two categories: explicit method and implicit method. The explicit method acquires user task model through user's own definition and modification. This method requires user having in-depth domain knowledge. User is not expertise in the real design case, so this method is randomly used. The implicit method acquires user task model through interacting with user about goals' realization process and/or following user's task process. This method can be commonly used in the real PSS development cases.

A set of user tasks is identified by analyzing user's goals and scenario. The relationships among these tasks may be temporal or logic sequence, information transferring, material supporting, or energy supporting. The tasks are systematically broken down into sub-tasks which are manageable to formulate functions, until the detailed level of granularity is attained. The partial form of user task model excluding the hierarchy task decomposing structure, which reflects the process-oriented modeling method, is shown in Figure 3. Tasks are modeled as boxes, and various relationship flows are models as arrows. The symbols for representing these flows among tasks are given out in Figure 3.

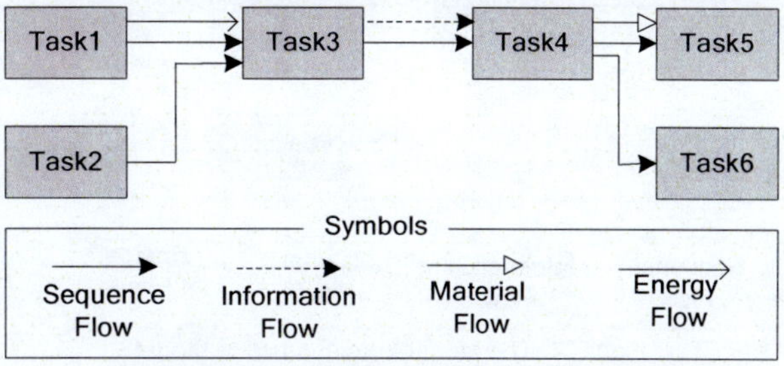

Figure 3: The partial form of user task model.

3.3 Function Model

Two definition methods are generally used to express function. One defines function in the form of "verb+noun" based on the nature language. The other one defines function as a transformation between input flows and output flows. The first kind of definition method, which is popularly adopted, can describe function distinctly and support decomposing and recomposing function. It is convenient for storing and processing information. However, this method can not describe the relationships among functions. Static function model does not conform to the process-oriented user task model, and cannot provide relationship information within the PSS concept. We can combine the advantages of the two function definition methods mentioned above to construct the function model. Function is expressed in nature language as "verb+noun", which is expressed in a box. The input flows and the output flows among the functions are expressed as arrows. The arrows in function model are categorized into four classes: sequence flow, information flow, material flow, and energy flow. The functions are systematically broken down into sub-functions which are manageable to formulate service activities and product modules, until the detailed level of granularity is attained. The partial function model excluding the hierarchy function decomposing structure, is shown in Figure 4, which reflects the process-oriented modeling method of function model. The symbols for representing various flows among functions are given out in Figure 4, which are the same as the symbols for user task model.

3.4 Conceptual Service Blueprint

The nature of service is process. Service is mainly described as activities. Service activities are constrained and supported by service strategies (e.g. product maintenance strategies: preventive maintenance or conditional maintenance) and resources (e.g. manpower, tools, or spare parts). Service activities do not require to be described in detail in the conceptual design. The conceptual service blueprint is proposed to define user's actions, service activities and product structures, and the link among them to model the PSS concept. On one hand, the interaction between user and PSS in the conceptual service blueprint can reflect the user's role in completing tasks. On the other hand, the service blueprint can facilitate the PSS operators and marketers making strategies.

The conceptual service blueprint has three levels: user level, service staff level, and product system, as shown in Figure 5. Line of interaction separates the user level and the service staff level. User's actions are in user level. Service activities are in service staff level. The product system level is divided into two parts: product behavior level and product module level. Product behavior is analyzed to bridge the gap between functions and modules.

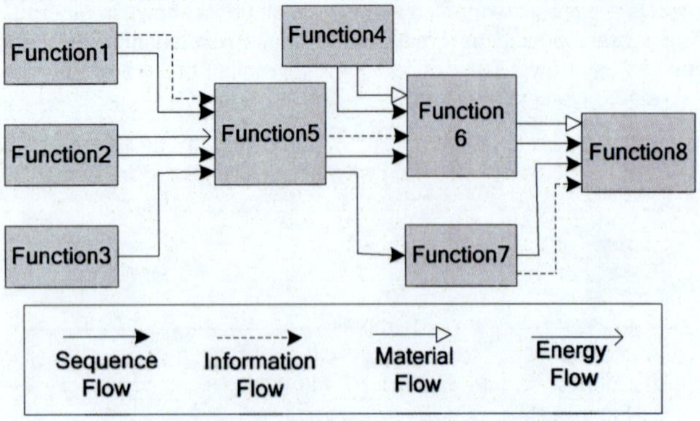

Figure 4: The partial form of function model.

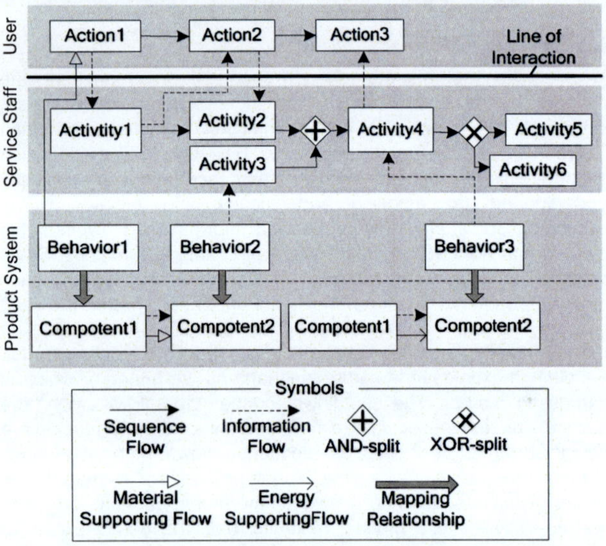

Figure 5: The general form of conceptual service blueprint.

and the function model, so that the PSS conceptual design framework is described as a unified whole. Mapping relationships between product behaviors and product modules are also modeled as arrows. The symbols for expressing relationship of "AND-split" and "XOR-split" among service activities are employed from the Business Process Modeling Notation (BPMN) [26]. The general form of conceptual service blueprint is shown in Figure 5.The conceptual service blueprint can facilitate designers to design detailed services and products, and help designers to analyze service process and make strategies.

4 CASE STUDY

The proposed PSS conceptual design framework is applied to an elevator operation service used in paper of Shimomura et al. [5]. Considering the elevator-PSS, user tasks are easy to identify and correlate with product function and service function. The elevator operation service involves many interactions among user, service staff and product system, so the advantages of the conceptual service blueprint are easy to show. The design objective of the elevator company is not restricted to designing an elevator satisfying carrying function. With the development of economy and society, the marketing competition becomes fiercer and fiercer. Users are not satisfied with a single elevator any more. Users begin to evaluate solutions from their goals of using elevator, and the goals involve facets about emotion, experience, etc. The point users concern is whether the solution company provides can realize their goals. However, users can not describe the solution to theirs fancy. Thus, how to design a user satisfying and competitive solution is the tough task of company. The proposed PSS conceptual design framework including user task model, function model and conceptual service blueprint, are used to analyze the PSS design process for elevator.

Experts from elevator design department, service design department, service operating department, and marketing department constitute the PSS design team. According to the proposed PSS conceptual design process and framework, the conceptual design process in this case can be divided in three steps.

Step 1: Analyze user's goals and construct the user task model.

Through carrying out questionnaire survey and communicating with the target user group, a set of first-level user tasks are identified by

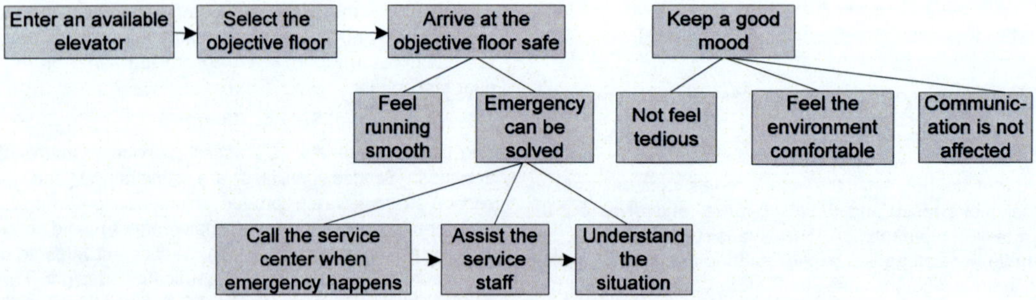

Figure 6: The user task model for elevator PSS conceptual design.

In order to facility designers who are always experts from various departments, to understand and use the conceptual service blueprint without special knowledge, the conceptual service blueprint should be designed to be clear and distinct. User's actions, service staff's activities, product behaviors and product modules are modeled as boxes in the conceptual service blueprint. Sequential relationships and interacting relationships existing among user's actions, service activities and product behaviors, and these relationships are modeled as different arrows among boxes. The meaning of these arrows is the same as that of the user task model

the design team: enter an available elevator, select the objective floor, arrive at the objective floor safe, and keep a good mood. Sequence flows exist among the first three tasks. The last two tasks can be decomposed into sub-tasks, as shown in Figure 6.

Step 2: Design functions and construct the function model.

Through analyzing the user task model, the design team proposes functions to realize user's tasks. For example, aiming at the task "enter an available elevator", two functions are identified by the design team: move vertical to carry people, and maintain the moving function. In this case, the task of "emergency can be solved" is highlighted. The related function model derived from analyzing this task is given out in Figure 7. The function model in Figure 7 is only a part of the whole function model in the elevator PSS conceptual design.

Step 3: Design service activities and product modules and construct the conceptual service blueprint.

Functions in the function model are analyzed by the design team, and service activities and product modules can be both designed to implement these functions. In this step, the function model in Figure 7 is analyzed. The constructed conceptual service blueprint is depicted in Figure 8. The user level contains user's actions, and this level can connect with the user's tasks in user task model to reflect how the user's tasks are completed. The arrows in red in Figure 8 reflect the connections between the user's tasks and the conceptual service blueprint.

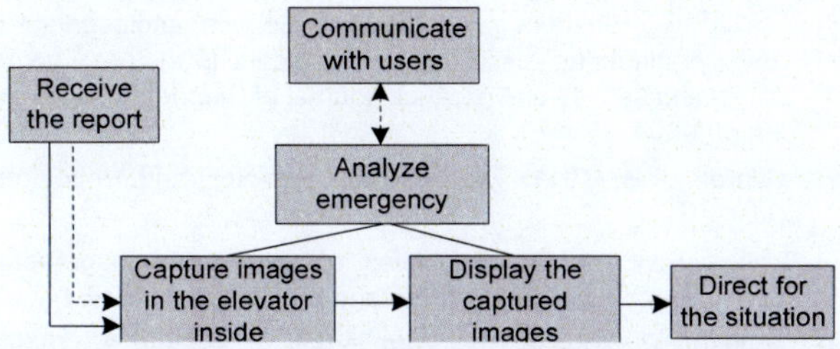

Figure 7: The function model derived by analyzing "emergency can be solved".

With the conceptual service blueprint, the detailed design of elevator emergency service in the whole elevator-PSS can be carried out.

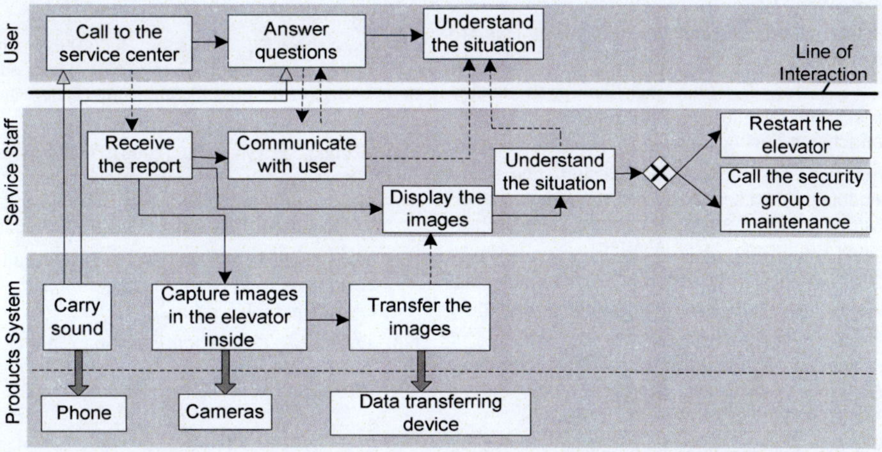

Figure 8: The conceptual service blueprint derived by analyzing the function model in Figure 7.

The emergency service involves the product-service system constituted by phone, cameras, data transferring device, and a series of user's actions and service activities. The conceptual service blueprint will drive the detail design of these physical modules of elevator and emergency service activities operating plan. A process of parameter transferring and decision-making also accompanies the PSS conceptual design process, which constraints the three models in the PSS conceptual design framework. A set of constraint requirements for tasks are identified with the user task model. These requirements are translated into design requirements to constrain the function model, and then the function requirements are translated into module characteristics of product and service to constrain the conceptual service blueprint. Decision-making approaches are incorporated to generate an optimal specific PSS concept finally.

5 CONCLUSIONS

Service attracts much more attention of manufacturers due to its value-added capability. Manufacturers are shifting from selling product to providing service. PSS approach has become a widely-accepted strategy in the fiercely competitive market and environmental conscious society. PSS conceptual design plays a critical role in PSS development, and it is not a completely new problem. A new PSS conceptual design approach is proposed including constructing user task model, function model, and conceptual service blueprint. The PSS concept includes service activities and product modules, and product modules are designed aiming at satisfying user's task directly or supporting service activities.

The major characteristics of this research are summarized as follows.

(1) The necessity of proposing a new modeling method to analyze and express user's goals to drive the PSS conceptual design process is analyzed.

(2) Considering the successful and wide application of user task concept in user-centered design, the user task model is proposed to analyze the user's goals and bridge the gap between user' goals and functions.

(3) User task model is constructed in a process structure considering the process nature of PSS concept and realization process of user's goals. Function model is also in a process structure, because it is driven by the user task model and required to provide process data to the PSS concept.

(4) Conceptual service blueprint is proposed to model PSS concept, which includes three levels: user, service staff, and product system. First, introduction of product behavior can bridge the gap of abstract function and physical module. Second, the description of user actions can reflect the user's role as the PSS provider.

Compared with the existing PSS engineering design methodologies, the proposed PSS conceptual design approach can comprehend the starting point of PSS conceptual design thoroughly to guide the designers to acquire and analyze the proper PSS design requirements. Furthermore, the systematic design structure based on AD can use the Independent Axiom and Information Axiom to assist the decision making of design concept.

The future work will focus on developing a systematic decision-making approach to support parameter translating process accompanying PSS conceptual design framework. The whole theory system of PSS conceptual design will be established to generate specific PSS concept according to different preferences of different user groups.

6 ACKNOWLEDGEMENTS

The project was supported by National Natural Science Foundation, China (No.51075261), Shanghai Science and Technology Innovation Action Plan (No.09dz1124600, No.10dz1121600), Shanghai Jiao Tong University Innovation Fund For Postgraduates. The authors would also like to express their grateful appreciation to the anonymous referees for their helpful comments to improve the quality of this paper.

7 REFERENCE

[1] Mont, O. K. (2002): Clarifying the concept of product-service system. Journal of Cleaner Production, 10(3), 237-245.

[2] Maussang, N, Zwolinski, P, Brissaud, D. (2005): Design of product-service systems. The 10th ERSCP, Antwerp, Belgium.

[3] Montabert, C., Scott McCrickard, D., Winchester, W. W., Perez-Quinones, M. A. (2009): An integrative approach to requirements analysis: How task models support requirements reuse in a user-centric design framework. Interacting with Computers, 21(4), 304-315.

[4] Shostack GL. (1984): Designing services that deliver. Harvard Business Review, 62(1):73-78.

[5] Shimomura, Y., Hara, T., Arai, T. (2009): A unified representation scheme for effective PSS development. CIRP Annals – Manufacturing Technology, 58(1), 379-382.

[6] Manzini, E., Vezzoli, C. (2003): A strategic design approach to develop sustainable product service systems: examples taken from the “environmentally friendly innovation” Italian prize. Journal of Cleaner Production, 11(8), 851-857.

[7] Williams, A. (2007): Product service systems in the automobile industry: contribution to system innovation? Journal of Cleaner Production, 15(11-12), 1093-1103.

[8] Aurich, J. C., Fuchs, C., Wagenknecht, C. (2006): Life cycle oriented design of technical product-service systems. Journal of Cleaner Production, 14(17), 1480-1494.

[9] Umeda, Y., Ishil, M., Yoshioka, M, Shimomura, Y., Tomiyama, T. (1996): Supporting conceptual design based on the function-behavior-state modeler. AI EDAM, 10(4): 275-288.

[10] Gero, J S., Kannengiesser, U. (2004): The situated function-behavior-structure framework. Design studies, 5:373-391.

[11] Suh, N.P. (1990): The principles of design. New York, Oxford University Press.

[12] Shostack, G.L. (1982): How to Design a Service, European Journal of Marketing, 16/1: 49–63.

[13] Bullinger, H., Fahnrich, K. P., Meiren, T. (2003): Service engineering-methodical development of new service products. International Journal of Production Economics, 85, 275–287.

[14] Aurich, J.C., Wolf, N., Siener, M., Schweitzer, E. (2009): Configuration of product-service systems. Journal of Manufacturing Technology Management, 20(5), 591-605.

[15] Kimita, K., Akasaka, F.,Hosono, S., Shimomura, Y. (2010): Design method for concurrent PSS development. Proceedings of the 2nd CIRP IPS^2 Conference, Linkoping, 283- 290.

[16] Chen, J.L., Li, H.-C. (2010): Innovative design method of product service system by using case study and TRIZ method. Proceedings of the 2nd CIRP IPS^2 Conference, Linkoping, 299-305.

[17] Shimomura Y, Tomiyama T. (2002): Service modeling for Service Engineering. In Proceedings of The 5th International Conference on Design of Information Infrastructure Systems for Manufacturing 2002 -DIISM2002-, Osaka University, Japan, 309-316.

[18] Sakao, T., Shimomura, Y. (2007): Service engineering: a novel engineering discipline for producers to increase value combining service and product. Journal of Cleaner Production, 15: 590-604.

[19] Gilbreth, F.B. (1911): Motion Study. Van Nostrand, Princeton, NJ.

[20] Taylor, F.W. (1912): Principles of scientific management, Scientific Management. Harper and Row (reprinted 1967).

[21] Johnson, P., Johnson, H., Waddington, P., Shouls, A. (1988): Task-related knowledge structures: analysis, modeling, and applications. In: Jones, D.M., Winder, R. (Eds.), Cambridge University Press, pp. 35–62.

[22] Payne, S. (1984): Task-action grammars, Proceedings of IFIP Conference on Human–Computer Interaction: INTERACT'84, vol. 1. North-Holland/Elsevier Science, London/Amsterdam, pp. 139–144.

[23] Wilson, S., Johnson, P. (1995): Empowering users in task-based approach to design, Proceedings of ACM DIS '95 Symposium on Designing Interactive Systems, MI, USA, pp. 25–31.

[24] Redish, J., Wixon, D. (2003): Task analysis, in J.A. Jacko & A. Sears (eds) The human-computer interaction handbook: Fundamentals, evolving technologies and emerging applications, Lawrence Erlbaum Associates, Mahwah, NJ.

[25] Gould, J.D., Lewis, C. (1983): Designing for usability: Key principles and what designers think, Proceedings CHI'83, ACM Press, New York.

[26] White SA. (2004): Introduction to BPMN, BPTrends, July.

Product-Service Systems (PSS) Design Process and Design Support Systems

Y. S. Kim[1], S. W. Lee[1], H. Jin[1], J. H. Shin[1], J. A. Park[1], Y. G. Lee[2], C. D. Kim[3], B. H. Seo[3], S. J. Lee[4]

[1]Creative Design Institute, Sungkyunkwan University, Suwon, Korea

[2]Department of e-Business, Korea Polytechnic University, Siheung, Korea

[3]T & C Partner Co., Ltd., Seoul, Korea

[4]D&Tech 7 Consulting Co., Ltd., Seoul, Korea

Abstract

This paper presents a new PSS design process and associated design support systems being developed by the PSS Design Project Consortium sponsored by the Korean Ministry of Knowledge Economy. Six procedural steps are included in the PSS design process. For design support systems, the activity modelling system is developed for a complete description of activity with context elements. The PSS function modelling system can support the definition of functions and service providers/receivers and their decompositions. The modified service blueprint system is also developed to address relations among activities and functions. Finally, the PSS DesignScape is used to capture and visualize the PSS design process. Sample case example is presented to show the effectiveness of the PSS design support systems.

Keywords:

Product-Service Systems (PSS); PSS Design Process; PSS Design Support System; Context-Based Activity Modelling System; PSS Function Modelling System; Modified Service Blueprint System; PSS DesignScape

1 INTRODUCTION

A significant attention has been paid to Product-Service Systems (PSS) to address diverse values of consumers by providing integrated solutions of products and services. Goedkoop et al. firstly introduced PSS to simultaneously reflect ecological and economic issues, and they defined PSS as a marketable set of products and services, jointly capable of fulfilling a user's need [1]. PSS was also defined by Mont as a system of products, services, supporting networks and infrastructure that is designed to satisfy customer needs and have a lower environmental impact than traditional business models [2]. In addition, Manzini defined PSS as an integrated body of products and services and communication strategies that was conceived, developed and promoted by (a network of) actors to generate values for society [3].

The research on the methodological framework for PSS design has been recently studied. Morelli proposed PSS design framework by considering major functions and requirements and then connecting them with the elements of products and services [4]. Aurich et al. studied PSS design process in terms of a life-cycle oriented view by addressing technical services related to products [5]. Matzen and McAloone proposed the structured modeling framework to differentiate and categorize tasks of PSS development towards product/service oriented business with considering the case of maritime equipment [6]. Substantial studies on service design methods have been conducted by Shimomura's group in their service engineering research [7][8][9]. In their service model, flow model, scope model, view model and scenario model were included, and a receiver state parameter (RSP) was also defined to describe value and cost. More recently, they have carried out a number of various researches on analyses and evaluations of services from the viewpoints customer importance and economic cost [10]. In addition, they studied the PSS development method incorporating functions, service activities, product behaviors and attributes of entities with the avoidance of conflicts [11]. The Creative Design Institute (CDI) in the Sungkyunkwan University has been very active to develop the systematic PSS design process including following six procedural phases: (1) requirement identification and value targeting, (2) stakeholder activity design, (3) PSS function modeling, (4) function-activity mapping and PSS concept generation, (5) PSS concept detailing and (6) PSS concept prototyping [12].

Computerized design support systems have also been developed to effectively manage a number of design information during design process. Two representative examples are a design repository system and Service Explorer. A design repository system was proposed and developed to support product design process by Bohm and Stone [13][14]. Their design repository system archived a number of product data with functions and flows in particular. With their design repository system, designers can easily search products realizing specific functions and can come up with various design concepts more effectively. In the service engineering research, Shimomura's group also proposed and developed the computerized system, referred to as Service Explorer, to support service design [15]. In Service Explorer, they could understand and represent service target, describe realization structure and evaluate service.

Meanwhile, the PSS design process proposed by CDI includes many complicated issues including stakeholders, requirements, values, activities, functions, product/service elements, and so on. In addition, the above information should be appropriately generated, recorded, retrieved, retrieved and used. Therefore, proper computerized design support systems should be necessary to manage a great deal of design information during the PSS design process.

In this paper, we address the computerized design support systems for effectively applying our PSS design process including context-based activity modelling system, PSS function modelling system and PSS modified service blueprint system. In addition, the PSS DesignScape is also discussed as the system for capturing and visualizing PSS design process. Sample case example using above design support systems is shown for examining their effectiveness.

J. Hesselbach and C. Herrmann (eds.), *Functional Thinking for Value Creation: Proceedings of the 3rd CIRP International Conference on Industrial Product Service Systems, Technische Universität Braunschweig, Braunschweig, Germany, May 5th - 6th, 2011*,
DOI 10.1007/978-3-642-19689-8_24,

2 PSS DESIGN PROCESS [12]

This section introduces the systematic methodology to design PSS developed by CDI. The PSS design method is composed of 6 phases, as described in the previous chapter. Figure1 schematically shows the overview of the PSS design process.

In the first phase, the life-cycle step analysis is conducted to extract needs and wants of various stakeholders, and they are connected with *E3 values* composed of economical, ecological and experience aspects [16]. The needs and wants of various stakeholders are interpreted as requirements once target values are determined.

In the second phase, various stakeholders' activities are designed. Those activities are defined and their relationships are analyzed and visualized via service blueprint. Then, each activity is detailed by context-based activity modeling,and they are combined and sequentially arranged to generate alternative PSS scenarios.

Overall PSS function and service provider/receiver are defined, and they are decomposed into sub-functions and sub-providers/receivers respectively in the third phase. The overall function and service provider/receiver are represented as a block and two nodes attached upper left and lower right corners of the function block. The overall function is decomposed into a number of critical sub-functions, and they are linked by logical relations of flows of energy, material and information. Likewise, overall service provider and receiver are decomposed into sub-service providers and receivers and they appropriately assigned to each sub-function block.

When generating PSS concepts, functions, stakeholders, activities and product/service elements should be considered as a whole. Thus, in the fourth phase, the modified service blueprint method is proposed by inserting the layer of function between customer activity layer and on-stage service provider activity layer to link various sub-activities via sub-functions and to generate alternative PSS concepts. With the modified service blueprint, alternative PSS concepts can be compared by linking different activities conducted by different stakeholders to realize same functions.

In the phase of PSS concept detailing, product and service elements are linked with various external stakeholders to elaborate PSS concepts. In addition, PSS operation sequence can be defined as flows of operational activities of service providers with product/service elements and constraints. In the final phase of PSS concept prototyping, PSS concepts are realized by prototypes. The PSS prototypes can be physical structure or story/scenario describing crucial functional requirements and their attributes.

PSS includes several product elements and service elements that are closely related to each other. A new PSS can be generated from a single product, as schematically shown in Figure 2.

In Figure 2, a product P is further divided into several product elements P_i. New product elements P_j can be added and new values V_k are identified. Then, new service elements S can be added and combined with product elements in an appropriate fashion to achieve the identified target values. As a result, new PSS can be generated. The generation of a new PSS shown in Figure 2 can be effectively realized by applying the proposed PSS design process.

Figure 3 shows the schematic diagram of the associated design tools and design information for generation of a new PSS from a single product with the proposed PSS design process. In Figure 3, rectangles represent the design tools or methods, and words connected by arrows mean the associated design information.

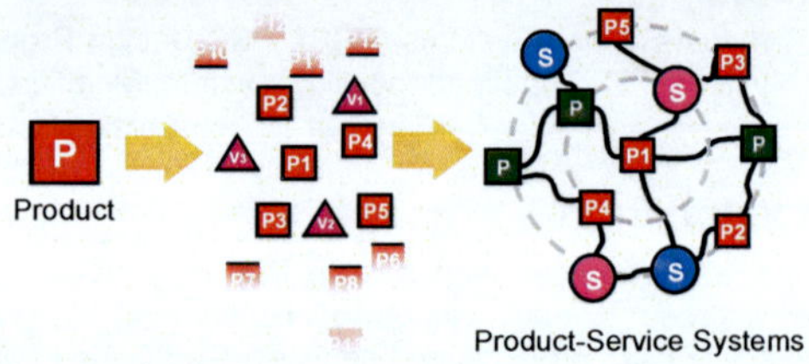

Figure 2: Schematic Diagram of Generation of new PSS [17].

The life-cycle step (LCS) analysis is applied to a single product, and the life cycle steps are defined. In each life cycle step, many associated stakeholders are also defined and then a great number of requirements are extracted. In addition, the E3 value framework is applied to identify the set of target LCS and values. The target LCS and values are used to design activities and functions, and they are used to create the modified service blueprints, which describe alternative new PSS concepts. As depicted in Figure 3, in

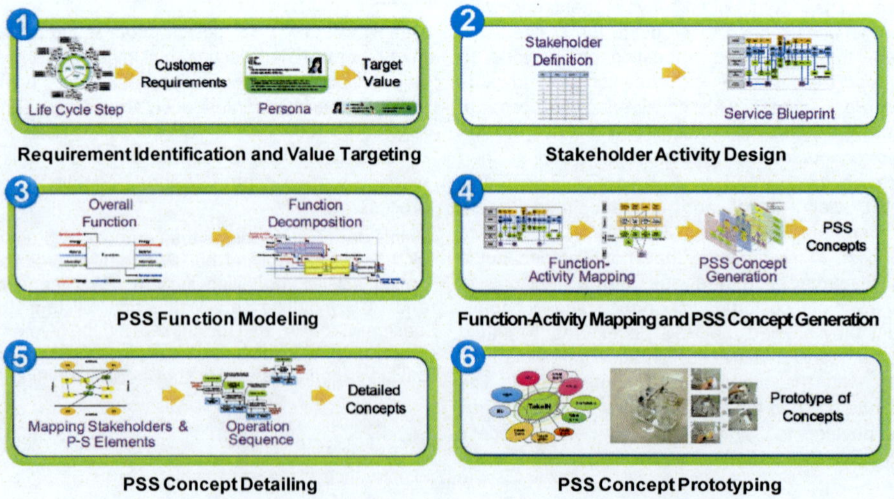

Figure 1: Product-Service Systems (PSS) Design Process [12].

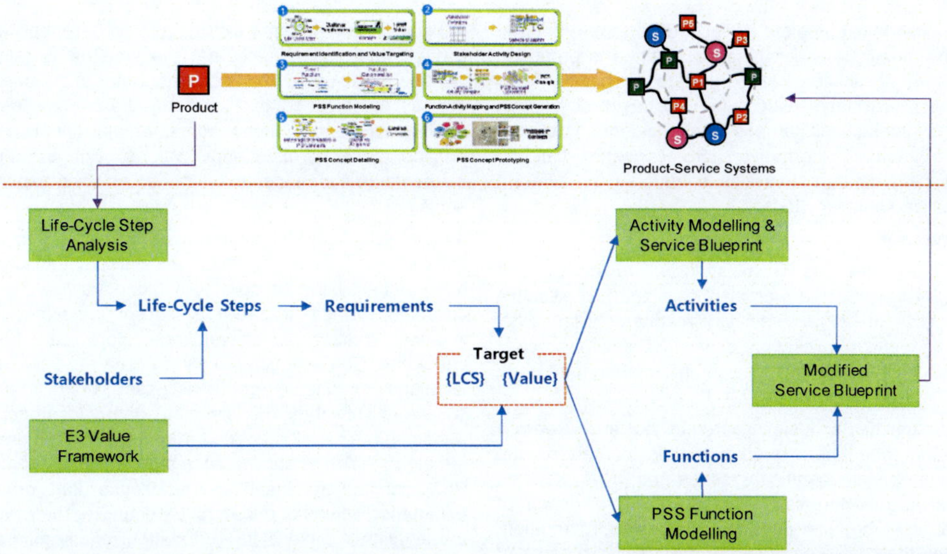

Figure 3: Schematic Diagram of Design Tools/Methods and Associated Design Information during PSS Design Process.

PSS design process, a great deal of various design information is generated. These much information should be properly handled for effective PSS design. Therefore, design support systems for effectively saving, retrieving and transferring these design information are necessary.

3 PSS DESIGN SUPPORT SYSTEMS

To support PSS design process, several computerized design support systems have been developed. These systems can be used for designing activities and functions and combining them to generate alternative PSS concepts. In more detail, when designing activities and functions, the activity modelling system and PSS functional modelling system were developed. In addition, the modified service blueprint system was developed for relating activities and functions. In addition, PSS DesignScape system was developed for the overall management of whole design process.

3.1 Activity Modelling System

Designing activities of various stakeholders reflecting design strategies formed using value proposition is a critical step in PSS design, which is addressed in the second phase of PSS design process. Interactions between service providers and receivers are designed through association of their activities. For this, service blueprint originally proposed by Shostack [18] is used. Each activity is detailed by reflecting its various contexts. To appropriately describe activities in more detail, the context-based activity model has been proposed [12] by borrowing some meaningful elements from [19], and its schematic diagram is given in Figure 4.

As can be seen in Figure 4, seven elements are defined such as Activity, Actor, Object, Tool, Event, Context and Environment. The element of "activity" is a goal-oriented behavior of "actors", and the "actor" is composed of "active actor", "passive actor" and "third-party actor". The "object" is the subject of "activity" and "tool" is a means to accomplish the goal. The element of "event" drives and changes the "activity" in the various "context". The element of "context" consists of goal context, relevant structure, physical context and psychological context. The "activity" occurs in the "environment", which includes inner space, outer space and virtual space.

Figure 3 shows the schematic diagram of the associated design tools and design information for generation of a new PSS from a single product with the proposed PSS design process. In Figure 3, light green rectangles represent the design tools or methods, and blue words mean the associated design information.

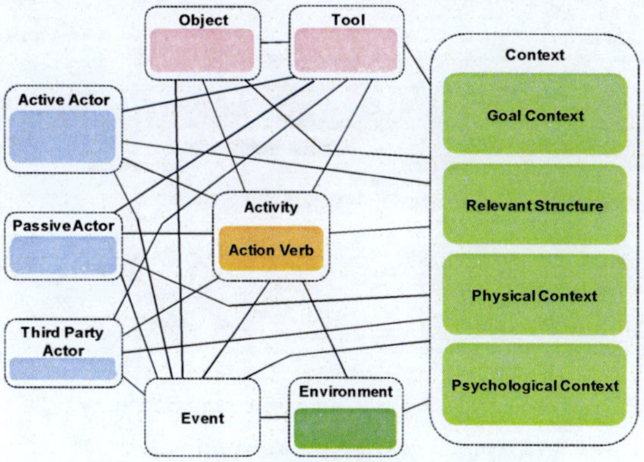

Figure 4: Context-Based Activity Model [13].

3.2 PSS Function Modelling System

In PSS function modelling, the overall PSS function and service provider/receiver are defined, and the subsequent decompositions are conducted. Similar to the case of product design, the expression of the function block and associated flows of energy, material and information are used. To represent service provider and receiver in the PSS function, they are expressed as folded lines in upper left corner and lower right corner of the function block, respectively. The PSS Function Based Design (FBD) system to support PSS function modelling activity was developed.

3.3 PSS Modified Service Blueprint System

When generating PSS concepts, functions, stakeholders, activities and product/service elements should be considered as a whole. The functions, stakeholders and activities can be linked together with product and service elements to form a specific PSS concept. The mapping between sub-functions and sub-activities is completed via associated stakeholders, and product and service elements are assigned to sub-functions to complete PSS concept generation and representation. The modified service blueprint method is proposed by inserting the layer of function between customer activity layer and on-stage service provider activity layer. The layer of function can represent the interactions among stakeholders. With the modified service blueprint, alternative PSS concepts can be compared by mapping different activities conducted by different stakeholders to realize same functions.

3.4 PSS DesignScape

A prototype PSS design process visualization and review system, referred to as PSS DesignScape, was also developed to visualize the design process in more intuitive manner. With PSS DesignScape, designers can quickly grasp the PSS design process patterns and easily access to the design information and their relationships.

The PSS DesignScape has a design capturing tool and a design process visualization tool. The capturing tool is located on the screen when designers write significant issues and draw sketches using graphic software. Whenever designers finish a period of distinct design process, they can capture their design information. They can also code their activities based on the coding scheme of PSS design process. This system can keep track of all the information, and store them into a form of XML file. Once designers terminate the capturing tool, the visualization tool gets ready to display the design process. The design process is represented as a successive chain of color rectangle nodes along the timeline. Figure 5 shows the snapshot photo of the developed PSS DesignScape.

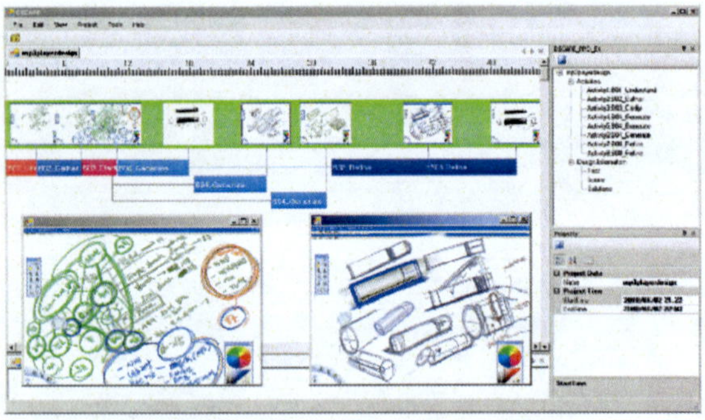

Figure 5: Snapshot of PSS DesignScape.

4 CASE EXAMPLE – CLOTHES TAKEIN PSS

As an illustrative example of the PSS design process and associated design support systems, the design case of 'Clothes TakeIN PSS' developed at the Creative Design Institute is described in this section. TakeIN is the PSS series name for the effort to develop PSS that provide methods for desirable reuse of products, indicating taking them back in use rather than throwing away. In the life-cycle step analysis on clothes, the step of 'reuse' was considered and the current situation of used clothes bin was observed. The users don't feel happy to go out to donate their clothes to the current used clothes bin that is usually located in a secluded corner of back alley streets with trash bins since they feel unsafe and unpleasant. Negative reactive emotional experience values obtained from the bins in such physical contexts would reduce your active emotional experience value for your voluntary donation of clothes. Therefore, these negative value attributes encountered in the current used clothes bins was selected as target values that need to be converted into positive ones.

When context-based activity model is used, the activity for a housewife donator, for example, to approach the clothes bin with the goal to donate used clothes can be described with the physical context information such that the weather is rainy and that the location of the bin is in a secluded corner of the back alley, and with the psychological context such that the donator feeling unsafe and unpleasant. Under these contexts, the donation activity cannot provide positive experience values. But by replacing various elements of those contexts, rather positive experience could be achieved.

If the physical context of the donator activities is considered, we can think of the location with good accessibility for donators as well as good accessibility of collectors using vehicles to pick up those collected clothes. In order to satisfy these requirements of the location context, a convenience store can become a good candidate. Convenience stores in Korea are very well distributed at locations to ensure good accesses both from donators and collectors with very friendly atmosphere to attract customers. By replacing the location context from the secluded areas in back alley streets to inside of convenience stores, we can come up with a new PSS concept for used clothes reuse that drastically improve experience values of donators. By changing the physical contexts of the activity of used clothes donation, totally different experiences and activities/behaviors of donators are to be designed. Donators' experiences and behaviors as if they gave presents to their nephews, for example, by giving the used clothes of their sons would be enabled. They even would like to provide further services like packaging, repairing and cleaning used clothes as presents. Such a scenario of used clothes TakeIN PSS is depicted in Figure 6.

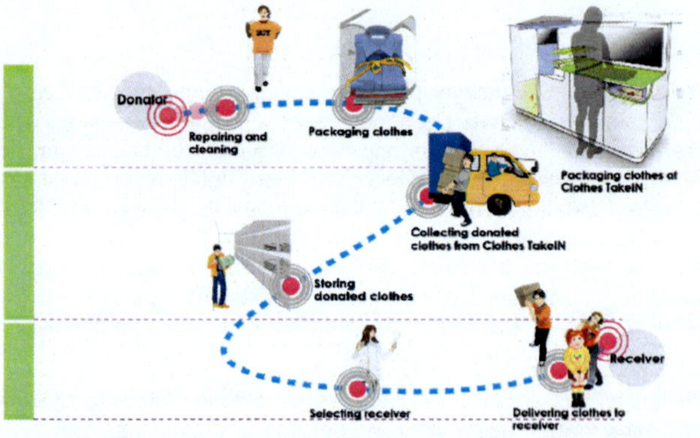

Figure 6: Scenario of the New Clothes TakeIN PSS [12].

In designing such PSS, our proposed modified service blueprint plays important roles in associating PSS functions and activities of stakeholders. A new PSS concept with more functions such as repairing and cleaning services added and carried out by donators. Note that the used clothes TakeIN PSS station needs to be designed to enable desirable packaging experiences of donators.

In PSS design process of clothes TakeIN PSS, the developed design support systems were effectively used to represent context-based activities, to decompose and identify critical functions and to map activities and functions in the modified service blueprint. Figure 7 shows the snapshot of the modified service blueprint to describe the abovementioned concept of the clothes TakeIN PSS. Each activity and function in the modified service blueprint is linked with AMT and PSS FBD system. For example, the donor's activity of 'package clothes' and the clothes TakeIN staff's activity of 'provide packaging space' are linked with the function of 'provide easy package handling', which are shown in the magnified view.

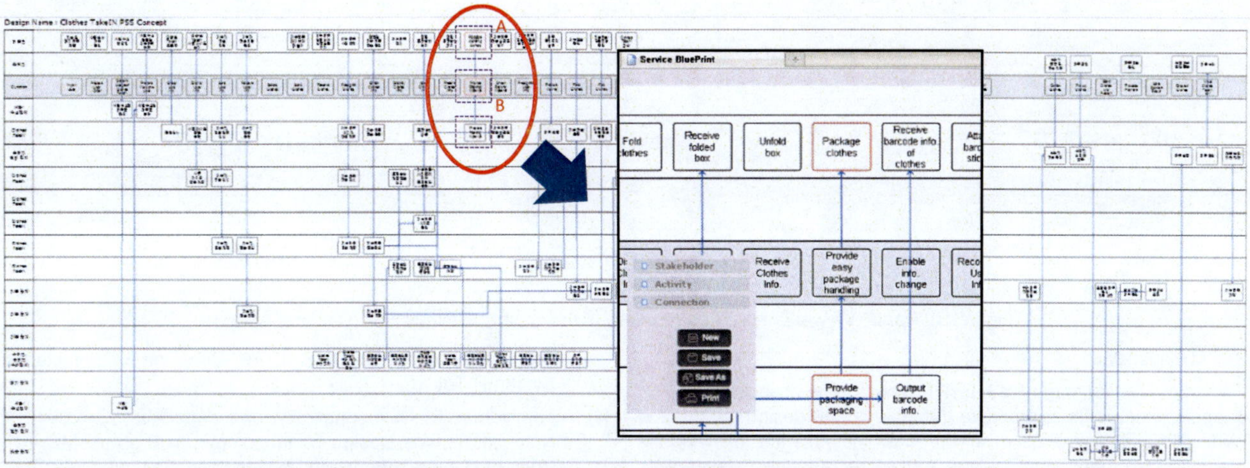

Figure 7: PSS Modified Service Blueprint of the New Clothes TakeIN PSS Concept.

More detailed descriptions of the activity of 'package donating clothes' – part A in Figure 7 – created in AMT and the function of 'provide packaging materials' – part B in Figure 7 – created in PSS FBD system are given in Figures 8 and 9, respectively.

As can be seen in Figure 8, the donor (active actor) packages (activity) the clothes (object) with the packing materials (tool) after preparing donating clothes (event) to donate clothes (goal context) in the convenience store (physical context) with pleasant mode (psychological context).

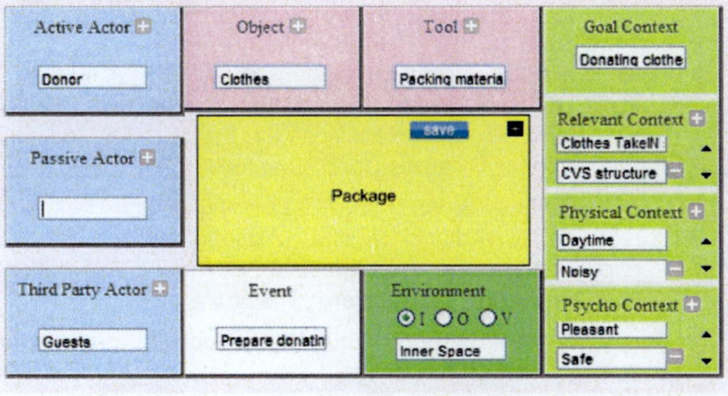

Figure 8: Snapshot of Context-Based Activity of 'Package Donating Clothes'.

Figure 9 shows the function decomposition diagram of the new clothes TakeIN PSS concept, which was generated in the PSS FBD system. Total 27 critical functions have been identified and they are connected together with various flows. The function of 'provide easy handling' has the service provider of CT (clothes TakeIN staff) and the service receiver of US (user), who is a donor, which is shown in the magnified view of Figure 9.

Figure 10 shows the partial view of the design process of the clothes TakeIN PSS design project. The vertical axis represents six PSS design phases shown in Figure 1, and horizontal axis represents time. The snapshot given in Figure 10 denotes the design process pattern for first 15 days. As can be seen in Figure 10, the PSS design process was visualized and its pattern could be intuitively understood by PSS DesignScape users.

Overall, each design support system can be used for supporting desingers' activities during actual PSS design process, and they are connected together. When generating the modified service blueprints that represented the alternative PSS concepts, AMT and PSS FBD systems have been actively used. In particular, the design information coming from various PSS design support systems were inter-connected. The PSS DesignScape system could entirely manage the PSS design process including a number of design information generated from various design support system, and the patterns of design activities of designers could be monitored and captured for the review and evaluation.

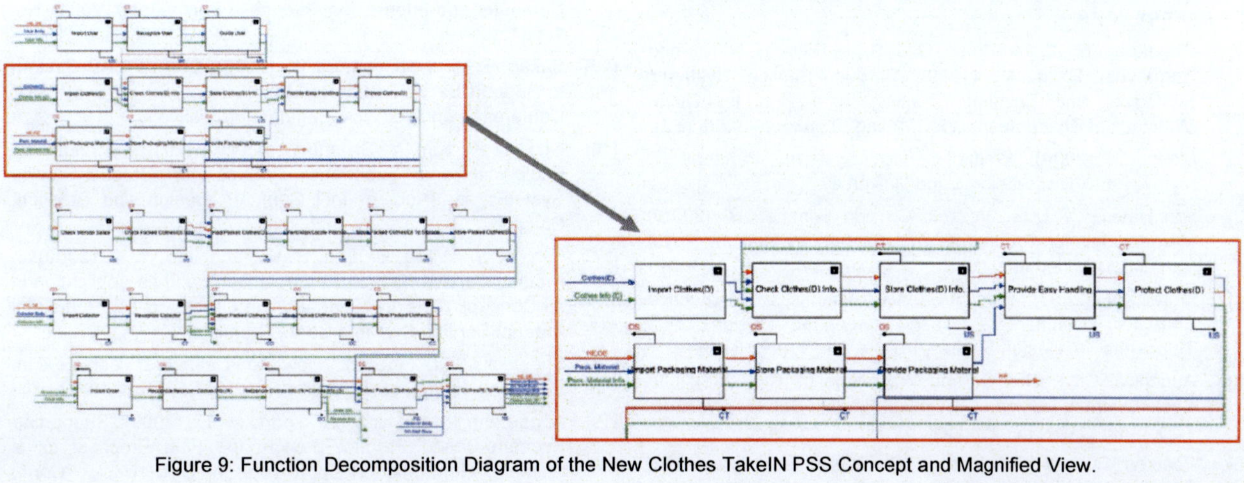

Figure 9: Function Decomposition Diagram of the New Clothes TakeIN PSS Concept and Magnified View.

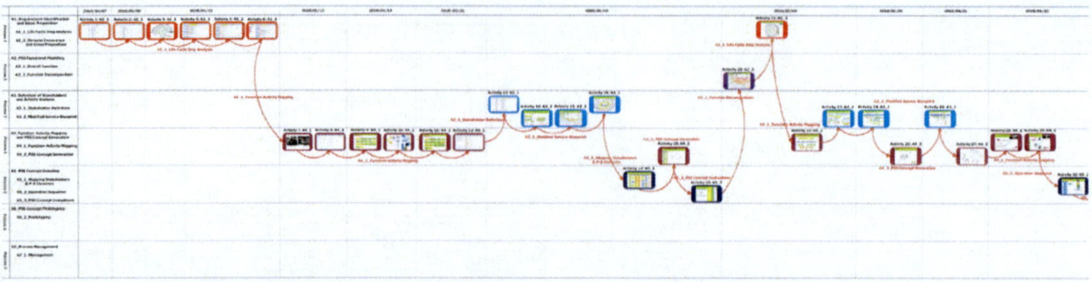

Figure 10: Partial View of the Process Pattern of the Clothes TakeIN PSS Design Case.

5 SUMMARY

This paper presents the PSS design process and associated design support systems for appropriate management of design information during PSS design. PSS design process includes the following six steps: requirement identification and value targeting, stakeholder activity design, PSS function modelling, function-activity mapping and PSS concept generation, PSS concept detailing and PSS concept prototyping. Several design support systems were developed for systematic management of a great deal of associated design information. The context-based activity modelling system, referred to as activity modelling tool (AMT), was developed. It enables a complete description of activity with various context elements. The PSS function modelling system, referred to as PSS function-based design (FBD) system, and was also developed to support the definition of PSS functions and related service providers/receivers and their decompositions. In the third, the modified service blueprint system was developed to establish relations among activities and between activities and functions. It could effectively be used for generation and comparison of various PSS concepts. The AMT and PSS FBD systems were linked in the PSS modified service blueprint system. Finally, the PSS DesignScape was developed for capturing and visualizing the PSS design process information. It could facilitate designers to review and manage the PSS design process much more systematically. Sample case example was considered, and the results showed their effectiveness during the used clothes TakeIN PSS design.

6 ACKNOWLEDGMENTS

This research was supported by the Korean Ministry of Knowledge Economy under the Strategic Technology Development Program.

7 REFERENCES

[1] Goedkoop, M. J., van Halen, C. J. G., te Riele, H. R. M., and Rommens, P. J. M. (1999): Product Service Systems: Ecological and Economic Basics, in: Report for Dutch Ministries of Environment (VROM) and Economic Affairs (EZ).

[2] Mont, O. (2004): Product-Service Systems: Panacea or Myth?, Ph.D. Dissertation, Lund University.

[3] van Halen, C. J. G., Vezzoli, C., and Wimmer, R. (2005): Metododlogy for Product-Service System Innovation, Royal Van Gorcum, Netherlands.

[4] Morelli, N. (2003): Product-Service Systems, a Perspective Shift for Designers: A Case Study: the Design of a Telecentre, in: Design Studies, Vol. 24, No. 1, pp. 73–99.

[5] Aurich, J. C., Fuchs, C., and Wagenknecht, C. (2006): Life Cycle Oriented Design of Technical Product-Service Systems, in: Journal of Cleaner Production, Vol. 14, pp. 1480–1494.

[6] Matzen, D. and McAloone, T. C. (2008): From Product to Service Orientation in the Maritime Equipment Industry – a case study, in: Proc. 41st CIRP Conf. on Manufacturing Systems, Tokyo.

[7] Tomiyama, T., Shimomura, Y., and Watanabe, K. (2004): A Note on Service Design Methodology, in: Proc. ASME Int'l. Conf. of Design Theory and Methodology, Salt Lake City.

[8] Sakao, T., Shimomura, Y., Comstock, and M., Sundin, E. (2005): Service Engineering for Value Customization, in: Proc. 3rd Int'l. World Congress on Mass Customization and Personalization (MCPC), Hong Kong.

[9] Sakao, T., and Shimomura, Y. (2007): Service Engineering: a Novel Engineering Discipline for Producers to Increase Value Combining Service and Product, in: Journal of Cleaner Production, Vol. 15, pp. 590–604.

[10] Kimita, K., Hara, T., Shimomura, Y., and Arai, T. (2009): Cost Evaluation Method for Service Design Based on Activity Based Costing, in: Proc. CIRP Industrial Product-Service Systems Conf., Cranfield.

[11] Hara, T. , Arai, T., and Shimomura, Y. (2009) : A Method to Analyze PSS from the Viewpoints of Function, Service Activity, and Product Behavior, in: Proc. CIRP Industrial Product-Service Systems Conf., Cranfield.

[12] Kim, Y. S., Lee, S. W., Kim, M., Lee, W. J., and Park, J. A. (2010): A Systematic Design Methodology for Product-Service Systems and a Design Case, in: Proc. of International Service Innovation Design Conf., Hakodate.

[13] Bohm, M. and Stone, R. (2004): Product Design Support: Exploring a Design Repository System, in: Proc. of IMECE'04, IMECE2004-61746, Anaheim.

[14] Bohm, M., Vucovich, J., and Stone, R. (2008): Using a Design Repository to Drive Concept Innovation, in: Journal of Computer and Information Science in Engineering, Vol. 8, No. 1, 014502-1-8.

[15] Sakao, T., and Shimomura, Y. (2007): Service CAD System to Support Servicification of Manufactures, in: Proc. of CIRP Conference on Life Cycle Engineering, Tokyo.

[16] Cho, C. K., Kim, Y. S., and Lee, W. J. (2010): Economical, Ecological and Experience Values for Product-Service Systems, in: Proc. of Int'l Conf. on Design and Emotion, Chicago.

[17] Kim, Y. S., Wang, E., Lee, S. W., and Cho, Y. C. (2009): A Product-Service System Representation and its Application in a Concept Design Scenario, in: Proc. CIRP Industrial Product-Service Systems Conf., Cranfield.

[18] Shostack, G.L. (1982): How to Design a Service, in: European Journal of Marketing, Vol. 16, No. 1, pp. 49-63.

[19] Anggreeni, I., and van der Voort, M. C. (2009): Supporting Scenario-Based Product Design: the First Proposal for a Scenario Generation Support Tool, in: CIRP Design Conf., Cranfield.

Analyzing the Practical Usability of the Heterogeneous Modeling Approach for Conceptual Product-Service System Development

Tim Sadek[1], Matthias Köster[1]

[1] Institute Product and Service Engineering, Chair of Engineering Design, Ruhr-University of Bochum, Germany

Abstract

Product-Service Systems are characterized by the integration of technical products and services along their lifecycle. To generate principle solutions of a PSS that meet customer-specific requirements, it is important to consider interdependencies between products and services already at an early phase of development. Thus, a suitable model-based approach for conceptual PSS development has been developed. Due to the complexity of this modelling approach, deficiencies regarding its practical usability could be revealed. This contribution aims at presenting initial requirements and optimization objectives concerning the practical usability of the modelling approach. Therefore, two PSS scenarios with practical relevance have been analyzed.

Keywords:

Product-Service Systems (PSS); Conceptual PSS Development; Heterogeneous PSS Concept Modeling

1 INTRODUCTION

During recent years, the increasing variety of customer demands, technological advances and strong dynamics of global markets prompt enterprises to shift their business focus from merely designing and selling technical products to offering product and service integrated solutions. In scientific publications concerned with the issue of bundling technical products and services, the fact that value creation can be increased through new business models is frequently discussed [1] [2]. The combination of products and services reaches from product after sales services (e.g. maintenance) up to Product-Service Systems (PSS). PSS are characterized by an integrated and mutually determined process of planning, developing, delivering and using of products and services containing their immanent software components [3].

A PSS constitutes a customized solution, in which providing a performance is set above the purchase of a mere technical product. It can comprise any combination of product and service shares. Once such an offer is planned, the PSS concept development phase generates principle solutions of a Product-Service System. Considering interdependencies already in the early phase of PSS development is especially important to ensure the synergetic interaction of products and services during their lifecycle as well as to offer an integral customer-specific solution. Thus, a novel modeling approach has been developed, called heterogeneous concept modeling [12] [13].

2 SCOPE

Currently the heterogeneous PSS concept modeling approach provides a suitable generic basis for the modeling of PSS concepts; however, a possible drawback regarding the practical usability of the modeling approach could be the rising complexity. Furthermore, the approach has not yet been evaluated in practice. The aim of this article is to determine initial requirements and optimization objectives concerning the practical usability of this modeling approach. To fulfill this aim, two exemplarily PSS scenarios with practical relevance have been analyzed. These PSS scenarios consider typical PSS issues and are based on existing concepts. The focus laid on PSS specific concerns, like how to consider shared risks and interactions between a PSS provider and a customer as well as the dynamics during the delivery and use of PSS. To show the feasibility of the novel modeling approach, it has been implemented as a computer-aided tool. Meeting new requirements concerning the practical usability of the modeling approach as well as the achievement of optimization goals are particularly necessary to optimize the computer-aided modeling tool in order to enable a transfer into the industrial practice.

This article is divided into the following sections. Based on the introduction, sections 3-5 present the theoretical fundamentals of Product-Service Systems and the specific modeling approach for conceptual PSS development. In section 6 and 7 the PSS scenarios are described and analyzed. Specific requirements to improve the modeling approach are determined in section 8. This paper concludes with a summary and an outlook.

3 PRODUCT-SERVICE SYSTEMS (PSS)

In academia, the term Product-Service System has been established to describe an integrative customer solution combining products and services [4]. Thus, a PSS is a marketable set of products and services capable of jointly fulfilling a user's need. It can enclose products plus additional services or even a service plus an additional product. In extension to this definition, Roy [5] includes the aspect of the allocation of property rights. Likewise, the growing influence of ecological awareness in the development of PSS is emphasized [6]. The key to sustainable PSS is that they are designed and marketed to provide customers with a particular result or function (clean clothes, mobility, warmth etc.) without having to own or buy physical products, such as a washing machine, a car or a heater. Following the scientific understanding, Product-Service Systems (PSS) relate to consumer markets. To specify a problem solution solving business-to-business market issues, the term industrial Product-Service System (IPS^2) is used [12] [13].

A PSS features a particular model of business. Regarding the

J. Hesselbach and C. Herrmann (eds.), *Functional Thinking for Value Creation: Proceedings of the 3rd CIRP International Conference on Industrial Product Service Systems, Technische Universität Braunschweig, Braunschweig, Germany, May 5th - 6th, 2011*,
DOI 10.1007/978-3-642-19689-8_25,

business model most classifications make a distinction between three main categories of PSS, the product-oriented PSS, the use-oriented PSS and the result-oriented PSS [4] [7]. The application of PSS is based on contracts. According to Burianek and Reichenwald [8], relational contract situations (incomplete contracts) are characteristic to provide flexibility to a customer over a long period of time. The distribution of generated value is determined by the selection and design of the related revenue model [9]. In the relevant literature on PSS, various approaches to classify revenue models have been found, such as i) functional, ii) usage-based and iii) performance-based revenue models [8] [10].

To summarize and extend the aforementioned characteristics, a Product-Service System meets individual customer requirements by integrating multidisciplinary technical subsystems and human artifacts [12]. Besides, subsystems with superordinate properties, for example controlling or logistics do also exist [13]. Thus, from a provider's point of view a PSS is a long-term, socio-technical, economic commodity that targets generating added value. A PSS features innovative business models as well as long-term business contracts. Partial substitution of constituents of a PSS allows its dynamic adaptation to changing restrictions or influencing factors during its long-term delivery and use phase. This specific feature of a PSS is considered as "changeability" [20] [23].

4 CONCEPTUAL PHASE OF PSS DEVELOPMENT

The conceptual design of a PSS follows the phase of strategic planning [11]. The business contract and the PSS requirements are the basis for the development of PSS concepts. The conceptual phase of PSS development focuses on the determination and selection of principle solutions. As a result, the PSS concept includes all selected principal solutions and describes the structural interaction between them as well as their logical functionality on a medium level of abstraction. Consequently, the interaction of service and product components during the value creation and usage is predetermined in the early phase of the PSS development. A PSS concept serves as a qualitative or quantitative basis to evaluate the system regarding technical and economic reference criteria. In addition, a PSS concept is the basis for the subsequent design of PSS modules and components.

5 HETEROUGENOUS PSS CONCEPT MODELING

A novel modeling principle has been developed to support model-based conceptual PSS and IPS² development. This principle, called heterogeneous concept modeling is proposed in the following paragraph. It is particularly suitable for integrating product and service modeling, but is also transferable to other multidisciplinary development issues. The integration ranges from the combination of PSS artifacts to linking model elements on arbitrary levels of detailing, abstraction and formalization to form a PSS concept. A systematic and detailed deduction of this approach is described in [12].

According to Sadek [12], a PSS is basically composed of a combination of PSS artifacts that fulfill required functions. The introduction of the PSS artifact aims at both dissolving fuzzy distinctions between product and service as well as consolidating multiple perspectives in PSS concept development. To characterize PSS artifacts, five constitutive characteristics (specificity, dominant transformation, scale of integration, capability for partial substitution and connectivity) have been developed, whose detailed description is presented in [12] [13].

The existential origin of a PSS artifact is a function. According to Pahl and Beitz [14] such a function is defined by a combination of "noun" and "verb" (cf. Figure 1). In this context, a noun represents the operand and a verb the operator of a solution element. According to mathematics an operand represents the structural basis of a function which can be transformed by an operator. Hence, an operand is determined by its possible state, whereas the operator is characterized by its ability to change an operator's state. The distinction between operand and operator is generally applicable to all types of functions. To use this as a generic principle for integrating different types of elements on a medium and lower level of abstraction rather than just as a functional description, the terms "noun" and "verb" are extended to the terms "PSS object" and "PSS process". Whereas "noun" and "verb" constitute a function, the combination of "PSS object" and "PSS process" constitute a PSS artifact.

A PSS object equals the material or immaterial operand of a PSS artefact that possesses definable states. PSS processes complement PSS objects. On one hand, they can be regarded as operators that effect PSS objects and their respective states. On the other hand, PSS processes are regarded as operators that are able to coordinate and to control the interaction of PSS objects. However, only the combination of PSS objects and PSS processes can generate a functional behavior.

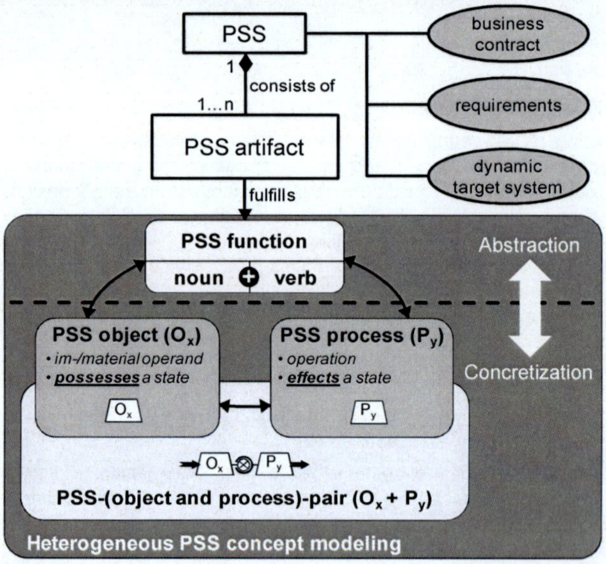

Figure 1: Heterogeneous PSS concept modeling.

Based on the definition of modeling elements, the PSS function, the PSS object and the PSS process, a functional, an object-related and a process-related development perspective can be distinguished in order to fully grasp the complexity of heterogeneous PSS concept modeling. This leads to the definition of three system-coherent modeling layers. A detailed characterization of the modeling layers as well as the description of corresponding modeling elements and relations between these elements is described in [12].

The transfer from planning a PSS to its systematic conceptual development is carried out on the PSS function layer. A PSS problem solution can be modeled abstractly and solution-neutral by defining and linking PSS functions. Apart from modeling PSS functions this layer is used to specify and model the distribution of risks that need to be shared between customer and provider regarding a business contract [3] [8]. Therefore, the PSS function layer is divided into the PSS provider and the customer modeling region. By placing a PSS function in one of these regions, its related risk can be clearly assigned to provider or customer. Both modeling regions are logically connected via a provider-customer relationship. Content of the underlying business contract with relevance for PSS concept development is defined in this meta-relation. This includes the distribution of property rights, the definition of the underlying revenue model as well as a certain portfolio of options, for instance.

The deliberately omitted concretization of a PSS problem solution on that layer is transferred to adjoining modeling layers. The PSS object layer is intended for modeling a material or immaterial operand of a PSS artifact. Three different types of PSS objects have been defined. While technical PSS objects are artificially generated by the combination of mechanical, electronic and information technology elements, human PSS objects represent human PSS individuals. In contrast, superordinate PSS objects are defined to control and to affect technical and human PSS objects. Contrary to human and technical PSS objects these system elements possess properties of an "organizing collective" (e.g. logistics, controlling, planning). In contrast, the PSS process layer has been defined to model operations that effect PSS objects. The PSS process layer is also used to model interactions between provider and external factor (customer). Therefore, two modeling regions are defined and separated by a "line of interaction". In this case, the line of interaction is equivalent to the correspondent line in service blueprints, according to [15].

Furthermore, the connection of the modeling elements to related relevant aspects of the development context, such as requirements and technological aspects can be modeled by contexts elements [12]. Thus, context elements are not part of the system itself and they do not show any functional and physical behavior.

The definition of the aforementioned modeling principle enables a PSS designer to develop PSS concepts in a successive way. Step-by-step PSS functions, PSS objects and complementing PSS processes can be developed. In doing so, interdependencies between PSS artifacts can be determined in an early phase of development. The combination of all three modeling views constitutes a heterogeneous PSS concept model.

6 SCENARIO 1: SUSTAINABLE INNER-CITY MOBILITY

6.1 Scenario description

The first PSS scenario that has been analyzed deals with a novel mobility concept. Instead of selling means of transportation, like a car or a bicycle, inner-city mobility could be provided, for example as a solution for commuters in a metropolis like London. An example of a seamless, flexible and sustainable mobility solution for Greater London is the "TFL bikeshare network" concept designed by Gabriel Wartofsky (http://gabrielwartofskydesign.com/). As depicted in Figure 2, this concept comprises an e-bike that can be rode from home to storage modules, which are distributed all-over the city.

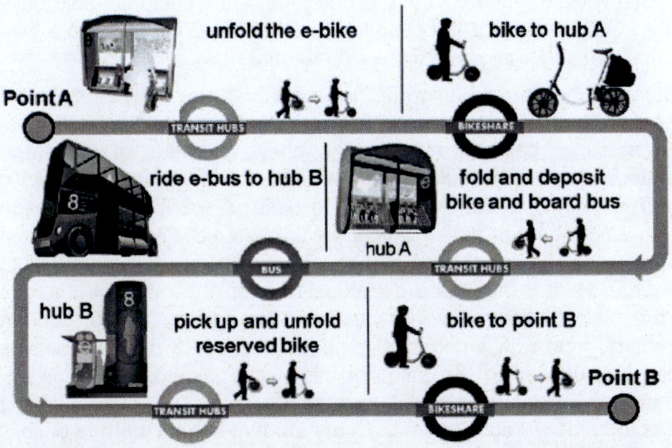

Figure 2: "TFL Bikeshare network": System and components.

Commuters access the bikeshare using an identification card, and lease the service on a per trip or monthly basis. The e-bike has been designed with airless tires, robust tubular construction and shaft drive, to ensure reliability, while the laminated front bar and dazzling graphics increase the visibility and safety of the user. The e-bike is also designed for the user to quickly and intuitively fold and deposit it into the storage modules upon arrival to a transit hub, so that the user can get on an e-bus bike-free. The user's identification card gets charged via RFID technology according to the amount of time the e-bike was rented. When docked, the e-bikes serve as hub seating and recharge for the next user. Upon arrival at the destination, another hub will be there to rent another bike. The transit hubs are designed to optimize the commuter's transitional experience between bikeshare and bus. Each hub provides commuters with access to bikeshare, real-time transit scheduling, cultural information about the surrounding neighborhood et cetera. Powered primarily by an on-board battery and ultra-capacitors, the e-busses emit no exhaust. During the time passengers get on a bus, its batteries get charged until the next stop via a small arm extended from the hub. The buses also collect and disperse e-bikes according to demand, effectively diminishing the need of third party bikeshare distributors. The electrical modular buses aim to optimize the needs of next generation commuters by providing ultimate convenience for public transfer. Specialized interior zones address the varied needs of passengers. Short stop and standing passengers are located downstairs, while long-distance travelers and tourist can enjoy the city from the panoramic second floor.

6.2 PSS concept modeling

The PSS "TFL bikeshare network" is modeled via the heterogeneous modeling approach.

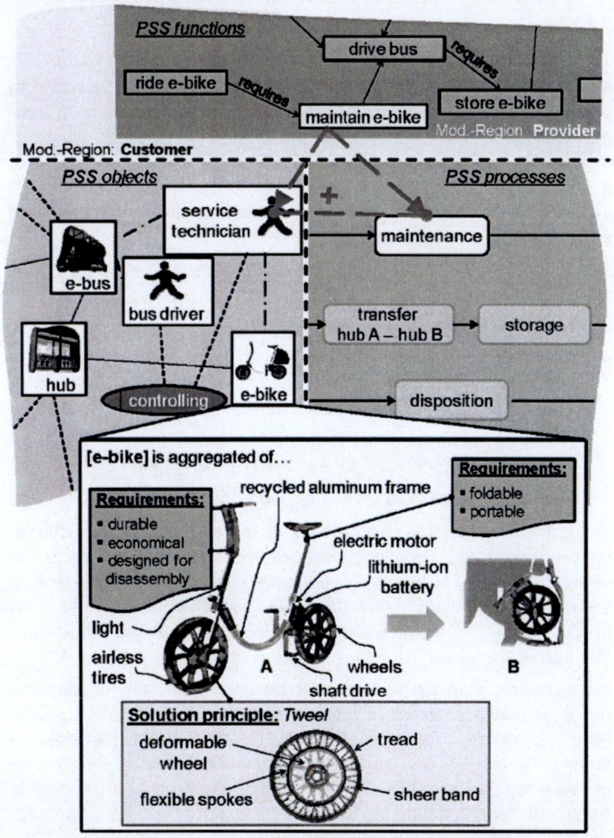

Figure 3: PSS concept model of the "TFL bikeshare network".

An extract of the concept model is depicted in Figure 3 in a simplified representation. The figure represents the three modeling views, the PSS functions, the PSS objects and the PSS processes. The combination of all three modeling views constitutes a heterogeneous PSS concept model of the "TFL bikeshare network". The concretization of a PSS function into a PSS object and a corresponding PSS process and their combination to a PSS-(object and process)-pair is illuminated by the PSS function *maintain e-bike*, the PSS object *service technician* and the PSS process *maintenance*. In contrast to the layer-specific relations that are exclusively defined to link modeling elements on a certain modeling layer, associations are used to link system elements across modeling layers [12] [13].

To solve the required PSS functions, various PSS objects and PSS processes are needed and have to be modeled on the corresponding modeling layers. As represented in Figure 3, the basic PSS objects on the lowest level of detail are the *e-bus*, the *e-bike* and the *transit hub*. Furthermore, the human PSS objects *bus driver* and *service technician* are required to drive the *e-bus* and to maintain the technical PSS objects. In addition, the human resources need to be coordinated by the superordinate PSS object *controlling*. The detailing of a PSS object is shown by the example of the e-bike in the lower part of Figure 3.

Furthermore, various context elements are used to provide relevant information for other developers or third party model-users in order to be able to understand the currently modeled system. As an example, the Tweel-Technology should be applied to this PSS concept to increase the reliability of the e-bike.

7 SCENARIO 2: EFFICIENT USE OF WASTED ENERGY

7.1 Scenario description

To operate industrial processes like the provision of hot water and steam, the melting and heating of materials, drying and de-watering processes as well as distillation and evaporation procedures, thermal heat is usually required. In all these processes, a waste of thermal energy occurs, which is referred to as waste heat. In order to reduce the primary energy consumption and environmental impacts due to CO_2 emissions, the wasted heat has to be recovered efficiently.

Academical and industrial studies reveal different ways to reuse waste heat for other applications [16]:

- Feeding into a district heating network
- Converting the wasted thermal energy into electrical energy, for example by using the thermoelectrical effect [17] (Seebeck effect) or by applying the organic Rankine cycle technology [18]
- Adding additional energy to recover a usable temperature level for industrial processes, for example by using heat pumps
- Storing of waste heat for a later use

Companies which are concerned with the application of technologies for waste heat recovery, are however confronted with risks and uncertainties (cf. Figure 4) [16]. These are for example a lack of know-how in this field of technology, risks involved with investments in these technologies as well as other technological and administrative barriers.

Furthermore, if waste heat cannot be used internally, it could be made available to external customers. Waste heat can be used to heat and cool office or public buildings for instance. In this case, the supplier takes the risk to supply customers with heat, which requires adequate logistics. In addition, the possibility to store heat is needed if there is a time lag between heat supply and heat demand, as in the case of seasonal demand fluctuations.

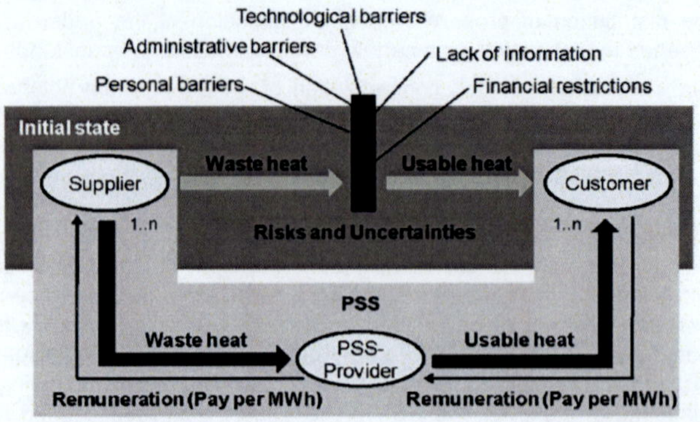

	Supplier	PSS-Provider	Customer
Initial state	Benefit of the wasted heat: $P_{WH} = 0$		Conventional energy supply: **Costs =** $C_{Heat,C}$
PSS	Benefit of the wasted heat: $P_{WH} > 0$	Benefit from the PSS: P_P **= Revenue - (**C_{Heat}**+**C_{PSS}**)**	$C_{Heat,PSS} < C_{Heat,C}$

Figure 4: An approach to the efficient use of wasted energy.

One way to efficiently recover waste heat is discussed in the following PSS scenario (cf. Figure 5). The PSS concept is based upon the existing concept "TransHeat" (http://www.transheat.de/) [16]. As depicted in Figure 4, the scenario consists of a PSS provider, who takes the risk for the storage and the transfer of the waste heat to the customer.

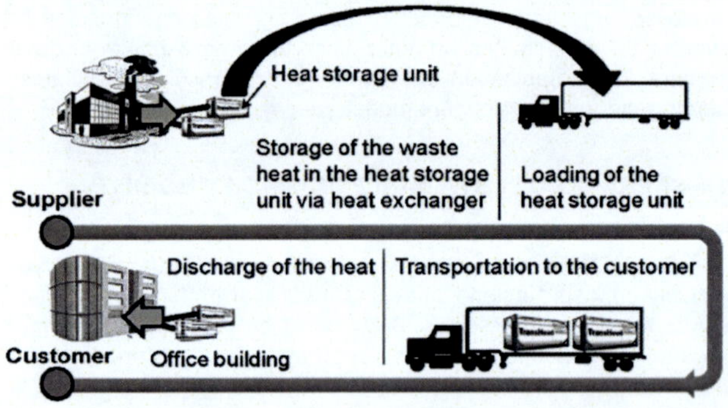

Figure 5: "TransHeat": System and components.

The customer uses the heat storage unit to supply an office building with heat. On one hand, the supplier gets paid per stored waste heat (MWh) by the PSS provider, which is equivalent to a value-based revenue model. One the other hand, the customer also pays the PSS provider for the supplied heat (MWh). This leads to a win-win situation for all stakeholders (cf. Figure 4).

Storage and transportation of the waste heat are realized by latent heat storage units, which can be loaded onto a truck. A latent heat storage can be achieved through phase changes of a storage material (e.g. solid-liquid, solid-gas) [19]. For this purpose, specific phase change materials (PCM) are used. A PCM is a substance with a high heat of fusion which, melting and solidifying at a certain temperature is capable of storing and releasing large amounts of energy. Heat is absorbed or released when the material changes from solid to liquid and vice versa. Through the use of sodium acetate, heat with a constant temperature of 58.8 degrees Celsius can be supplied to the office building until the stored heat is exhausted. To ensure a continuous heat transfer, the office building has to be delivered with a necessary amount of heat storage units.

7.2 PSS concept modeling

As represented in Figure 6, the basic physical PSS objects are the *heat storage unit* and an appropriate vehicle to transport the unit from the supplier to the office building and vice versa. Furthermore, the human PSS object *fitter* is needed to drive the transport vehicle (*truck*) as well as to perform the installation of the heat storage units at the location of the supplier and the customer. The heat storage unit consist of a *tank*, which is filled with the phase change medium sodium acetate. To store the waste heat, heat transfer oil is heated with a *heat exchanger* and pumped through the sodium acetate. Thereby, the energy carried by the oil is transferred to the PCM. A *pump* provides the required heat flow. The storage of the heat via the sodium acetate is based on the solution principle of latent heat storage, which is also represented in the concept model. The disposition of heat storage units needs to be coordinated mainly by the superordinate PSS objects *logistics* and *controlling*. Requirements and restrictions, such as the maximum storage temperature, the average heat loss and the minimum required number of heat storage units are modeled via context elements.

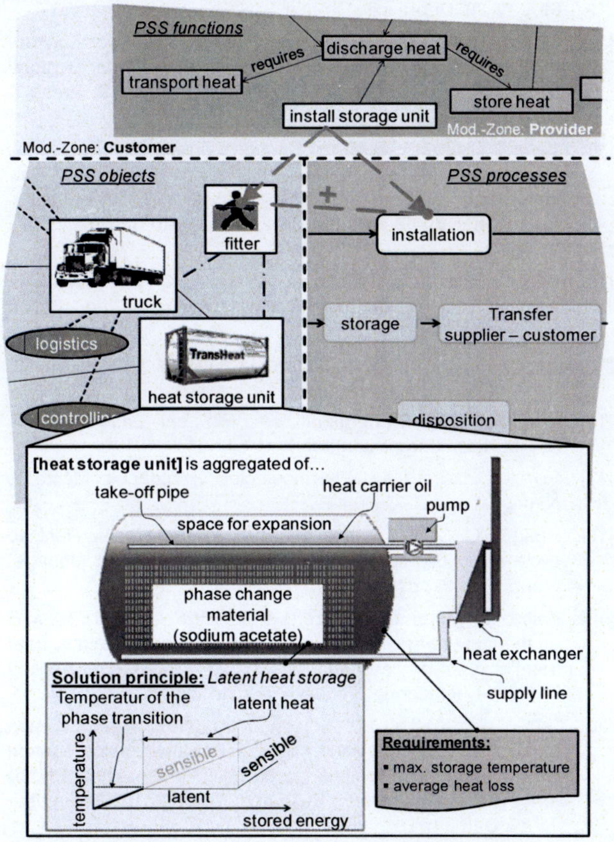

Figure 6: PSS concept model of "TransHeat".

8 ANALYSIS

8.1 Results

The applied modeling approach is adequate to represent and connect the main solution elements of the two considered PSS concepts by combining PSS objects and PSS processes to fulfill a desired PSS function. It can be emphasized, that the distinction between technical, human and superordinate PSS objects on the PSS object layer is advantageous, because all system elements of the considered PSS can be assigned to one of these classes of PSS objects. However, the analysis of the different PSS scenarios has also revealed deficiencies regarding the practical usability of the modeling approach.

In earlier analysis on the theoretical validation of the modeling approach, the system complexity has always been reduced to a 1:1 relationship between PSS provider and customer. However, the first PSS scenario, the "TFL bikeshare network" consists of a 1:n relationship between provider and customer. In contrast to a typical customer in a 1:1 relationship, here there are rather various users who become temporarily part of the PSS. This leads to different questions regarding the practical usability of the modeling approach. So far, it is not yet possible to represent processes carried out by the user on the PSS process layer. In accordance to the rules of the modeling approach, a PSS object "user" also has to be modeled to complete a PSS-(object and process)-pair in order to fulfill a desired PSS function. Furthermore, in the second scenario, the PSS concept "TransHeat" can also be extended to a n:n relationship between PSS provider and customer, if a network of different suppliers is used by the PSS provider to deliver heat to different customers. It can be seen clearly, that the adequate modeling of 1:n or n:n-relationships is essential and has to be implemented in the modeling approach.

Another deficit of the modeling approach is the consideration of the PSS dynamics. To provide flexibility to the customer in an uncertain environment, the PSS has to allow its dynamic adaptation to changing restrictions and influencing factors (e.g. market and technology) during its lifecycle. Therefore, an adaptable architecture of the PSS is necessary and has to be conceptualized. To support this, the currently incorporated changeability of the PSS concept has to be represented in the model [20]. Furthermore, the model has to serve as a basis to anticipate uncertainty and to define specific future scenarios.

The difficulty of the comprehensibility and thus the practical usability of a heterogeneous PSS concept model can be particularly traced back to the complexity and diversity of layer-specific model elements and relations due to the dissection of the model on three modeling layers. In further research, it has to be analyzed if the practical usability of the model can be enhanced by the combination of the three modeling views into one heterogeneous modeling layer, on which PSS functions as well as PSS objects and PSS processes can be modeled equally. Therefore, the current spectrum of various layer-specific modeling elements has to be reduced to a reasonable degree that is usable in practice.

8.2 Initial requirements and optimization objectives

Based on the previously discussed problems regarding the usability of the modeling approach, this section summarizes the main optimization objectives. In a subsequent study, these objectives should first be implemented and then evaluated.

- Development of a possibility to describe the PSS user and the PSS dynamics in PSS concept models
- Design of an application-specific library of modeling elements for heterogeneous PSS concept modeling
- Design of an application-oriented representation to describe the interaction between the PSS provider or rather a network of PSS providers and customer networks in PSS concepts
- Design of a model-based method for the systematic development of the partial substitution in PSS concepts
- Combination of the three modeling layers into one heterogeneous modeling layer and development of an application-oriented representation of the modeling elements and relations on that modeling layer

9 SUMMARY AND OUTLOOK

In this contribution, the practical usability of the heterogeneous modeling approach for conceptual development of Product-Service Systems is discussed. To fulfill this aim, two exemplary PSS scenarios about inner-city mobility and the efficient use of wasted energy have been modeled and analyzed. Based on the developed PSS concepts, different deficiencies regarding the practical usability of the modeling approach could be extracted. Due to these problems, optimization objectives have been determined and need to be solved in further research.

The long-term objective is to transfer the theoretical approach into industrial practice. To do so, two major research activities have to be completed. On one hand, a case-study in cooperation with a company will be performed. This involves the conceptual development of an industrial Product-Service System for surface coating of components with complex geometries through a specific type of galvanization.

On the other hand, a descriptive study will be conducted [21]. Thereby, the practical usability of the modeling approach should be analyzed by two appropriate types of experiments. The first experiments are carried out by a set of test participants, for example students of engineering design (novice designers) without any previous knowledge in the field of conceptual Product-Service System development. Afterwards, expert designers of the PSS research community and the manufacturing industry with an extended knowledge of product and service engineering are used as probands. A successful conduction of this experimental study needs an intensive cooperation of engineering designers with psychologists and sociologists, which is already actively pursued. A similar experiment with different test participants has already been performed. This study included the heterogeneous concept modeling but restricted itself to the classical development of multidisciplinary technical products. [22].

10 ACKNOWLEDGMENT

Financial support from the German Science Foundation (DFG) through SFB/TR 29 is gratefully acknowledged.

11 REFERENCES

[1] Vargo, S.L.; Lusch, R.F. (2004): Evolving to new dominant logic for marketing, in: Journal of Marketing, Vol. 68.

[2] Oliva, R.; Kallenberg, R. (2003): Managing the transition from products to services, in: International Journal of Service Industry Management.

[3] Meier, H., Roy, R., Seliger, G. (2010): Industrial Product-Service Systems - IPS², In: CIRP Annals – Manufacturing Technology 59, pp. 607-627.

[4] Baines, T. S.; Lightfoot, H. W.; Evans, S.; Neely, A.; Greenough, R.; Peppard, J.; Roy, R.; Shehab, E.; Braganza, A.; Tiwari, A.; Alcock, J. R.; Angus, J. P.; Bastl, M.; Cousens, A.; Irving, P.; Johnson, M.; Kingston, J.; Lockett, H.; Martinez, V.; Michele, P.; Tranfield, D.; Walton, I. M.; Wilson, H. (2007): State of-the-art in product-service systems, in: Proceedings of the I MECH E Part B, Journal of Engineering Manufacture, 221(10). pp. 1543-1551.

[5] Roy, R. (2000): Sustainable product-service systems, in: Futures 32. Elsevier Science Ltd, pp. 289-299.

[6] Mont, O. (2002): Clarifying the concept of Product-Service System, in: Journal of cleaner Production 10 (3), Elsevier Science Ltd, pp. 237-245.

[7] Tukker, A. (2004): Eight types of Product-Service System: Eight ways to sustainability? Experiences from Suspronet, in: Business Strategy and the Environment 13, pp. 246-260.

[8] Burianek, F.; Reichwald, R. (2009): Providing Industrial Solutions - Discussed from the Perspective of Contracting, in: Proceedings of the International Conference on Economics and Management of Networks (EMNet), September 3-5, Sarajevo.

[9] Sadek, T; Steven, M. (2010): Design of PSS Revenue Models, in: Proceedings of the CIRP IPS² Conference 2010, Linköping, Sweden.

[10] Tukker, A.; Tischner, U. (2005): New Business for old Europe – Product-Service Development Competiveness and Sustainability, Greenleaf Publication, Sheffield, UK.

[11] Müller, P; Stark, R. (2010): A generic PSS development process model based on theory and an empirical study, in: Proceedings of the international Design Conference 2010, May 17-20, Dubrovnik, Croatia.

[12] Sadek, T. (2009): Ein modellorientierter Ansatz zur Konzeptentwicklung industrieller Produkt-Service Systeme, Doctoral thesis, Ruhr-University of Bochum, Institute of Product-Service Engineering.

[13] Sadek, T. (2009): A Model Based Approach for Conceptual Development of Industrial Product-Service Systems, in: Proceedings of the International Conference on Engineering Design (ICED 2009), August 24-27, Stanford University, CA, USA.

[14] Pahl, G.; Beitz, W.; Feldhusen, J; Grote, K. H. (2007): Engineering Design – A systematic Approach, 3rd ed., Berlin: Springer.

[15] Shostack, G. L. (1984): Designing services that deliver, in: Harvard Business Review, Vol. 62.

[16] Pehnt, M. (2010): Energieeffizienz: Ein Lehr- und Handbuch, Berlin, Heidelberg: Springer.

[17] Rowe, D. M. (2006): Thermoelectrics Handbook: Macro to Nano, 1st ed., CRC Press.

[18] Hung, T.C. (2001): Waste heat recovery of organic Rankine cycle using dry fluids, in: Energy Conversion and Management, Vol. 42, pp. 539-553.

[19] Zalba, B.; Marin J.; Cabeza, L.; Mehling, H. (2003): Review on thermal energy storage with phase change: materials, heat transfer analysis and applications, in: Journal of Applied Thermal Engineering, Vol. 23, pp. 251-283.

[20] Sadek, T.; Köster, M.; Herzog, M. (2010): Initial Decision Making (IDM) in the early stage of Product-Service System Development, in: Proceedings of the 8th International Conference on Manufacturing Research ICMR 2010, September 14-16, Durham, UK.

[21] Blessing, L. T. M.; Chakrabarti, A. (2009): DRM, a Design Research Methodology, London: Springer.

[22] Sadek, T.; Meuris, D.; Wendland, M. (2010): An experimental study on comprehending heterogeneous modeling, in: Proceedings of NordDesign 2010, Gothenburg, Sweden.

[23] Richter, A.; Sadek, T.; Steven, M. (2010): Flexibility in industrial product-service sytems and use-oriented business models, in: CIRP Journal of Manufacturing Science and Technology, Oxford, Volume 3/2, pp. 128-134.

Criteria for assessing the Value of Product Service System Design Alternatives: an Aerospace Investigation

Marco Bertoni[1], Hakki Eres[2], Ola Isaksson[1,3]

[1] Division of Functional Product Development, Luleå University of Technology, Luleå, Sweden

[2] Computational Engineering Design Group, School of Engineering Sciences, University of Southampton, UK

[3]Volvo Aero Corporation, Trollhättan, Sweden

Abstract

Growing from data collected within a major EU research project in the aerospace domain, the paper proposes a candidate approach to assess the value of Product Service System design alternatives in a preliminary design phase. A framework composed of six main families of value criteria is proposed to support the product/service development team in evaluating the responsiveness and trade-off between alternative designs that target lifecycle commitment, with respect to the impact on perceived values and stated needs and expectations. A visualization approach is also proposed to quickly communicate the value contribution of the design alternatives at the decision gate.

Keywords:

Industrial Product Service Systems; Value Driven Design; Value Criteria

1 INTRODUCTION

In the new millennium, engineers are no longer solving only the traditional problems they used to solve. Especially in the aerospace industry, the design of a new aircraft engine cannot merely be reduced to a pure technical activity, such as the calculation of stresses on the blades or on the intermediate case. Thus, engineers are not dealing with "tame" problems only; rather they have to pay increasing attention to "wicked" problems [1] as well, such as developing a "passenger-friendly" airplane [2].

Additionally, it can be observed a move towards extending traditional product-based offers to incorporate more intangible assets, i.e. software and services, taking on lifecycle responsibilities to secure the aftermarket and to satisfy increasingly sophisticated customer needs. Initiatives such as Functional Products [3], Product Service Systems [4], and Integrated Product Service Engineering [5] highlight the opportunity to add value to users and stakeholders by providing a "function" instead of merely selling hardware. In such a perspective, the manufacturer maintains the ownership over the product and becomes responsible for the availability of the function through the entire life cycle, while requesting the customers to pay only for the provision of agreed results.

An illustration of the new "functional product" idea is apparent in the Total-Care® Package [6] offered by Rolls-Royce plc, which aims to deliver "power by the hour" in order to "*...improve product availability and reduce the cost of ownership by tying a supplier's compensation to the output value of the product generated by the customer (buyer)*" [7]. In a nutshell, rather than transferring the ownership of an engine to the airliners, the provider requests the customer to pay only for the use of the product (e.g. number of flight hours) [8].

Empirical evidence suggests that the lifecycle value delivered by these systems is primarily determined at the beginning of development programs. Early on in the engine development process, it is necessary to reason upon how to improve hardware, software and service to add value to the solution space. It is a great challenge, however, to effectively understand the impact of changing design variables at the micro-level on the overall system "value". The concept of "value" refers to different stakeholders and users, belonging to different organizations and groups, sitting in different locations, and who may have totally different perceptions on what "value" entails. This stresses the criticality of good decision-making strategies during early conceptual design. There is a need for better guidance for design choices, a guidance that translates the desires of customers and business developers into terms that are immediately meaningful to the Product Service Systems engineers [9].

2 MOTIVATION AND OBJECTIVES

The main purpose of the research is to contribute in achieving the Advisory Council for Aeronautics Research in Europe (ACARE) [10] targets by radically changing the way product development is initiated in the aerospace industry through the development of innovative mechanisms to 1) capture, model and understand customers' and stakeholders' needs and expectations, 2) to incorporate the value dimension into preliminary design and 3) to identify criteria affecting customer perceived value to be used in preliminary design studies evaluation.

In this paper the authors focus on the latter, investigating and proposing a set of criteria to assess design alternatives' value contribution from a lifecycle perspective. These criteria aim to support the product/service development team in evaluating the responsiveness and trade-off between alternative designs that target lifecycle commitment, with respect to the impact on perceived values and stated needs and expectations.

3 METHODOLOGY

The current study has been performed in the frame of an European Commission's research project within the FP7 programme. The research project has provided access to several aerospace companies (i.e. major aircraft, engine and sub-systems manufacturers and other companies with experience in aerospace

J. Hesselbach and C. Herrmann (eds.), *Functional Thinking for Value Creation: Proceedings of the 3rd CIRP International Conference on Industrial Product Service Systems, Technische Universität Braunschweig, Braunschweig, Germany, May 5th - 6th, 2011*,
DOI 10.1007/978-3-642-19689-8_26,

development projects), which have contributed to the problem domain definition with their empirical data and expertise.

Action research, commonly described as a set of iterative activities performed by practitioners and researchers jointly [11] that follows a particular a diagnosis, invention and reflective learning cycle [12], is in alignment with how the authors have performed the research in the project. Empirical and qualitative data have been generated by the authors' active participation in physical and virtual work-meetings with the industrial partners. The discussions with the aerospace companies in the diagnosis stage have contributed to the definition of the problem area chosen to investigate. Reflective learning has been aided by the continuous participation in debriefing activities by the research team in relation to the work-meetings.

4 DESIGN DECISIONS FOR COMPLEX PRODUCTS

Design teams would like to deliver the best possible design. Hence they follow structured and explicit guidelines, or they base their decisions on engineering judgement or intuition. In aerospace sector, Systems Engineering (SE) [13] is a common approach to ensure that the customer and stakeholder's needs are satisfied in a high quality, trustworthy, cost efficient and schedule compliant manner throughout a system's entire life cycle. SE is a holistic and interdisciplinary methodology for engineering complex system supported by standards such as the EIA 632 [14].

However, a major problem dealing with complex systems decisions in an unpredictable environment is that it is not straightforward to talk about what "best" means. Given two alternative designs, how can reviewers say which is better in an objective, repeatable, and transparent manner? Which criteria should be used for the evaluation?

As a decision aid for decision making in complex environments, Multi-Attribute Utility Theory (MAUT) [15] and Multi-Attribute Value Theory (MAVT) [16] support concept selection dealing with multiple and conflicting attributes. MAUT provides a construct for assessing values and subjective probabilities of individuals in the presence of uncertainty due to risk [17]. MAUT is based upon expected utility theory [18] and can take uncertainty into account, representing it directly into its decision support model. However it requires strong assumptions to ensure additivity; thus is very difficult to apply and few real applications are known. MAUT is also called the strong form of decision making whereas MAVT the weak form. MAVT [19] [20] is closely related to MAUT and can be used to address problems that involve a finite and discrete set of alternative policies that have to be evaluated on the basis of conflicting objectives. For any given objective, one or more attributes or criteria are used to measure the performance.

Considering that every system exists to deliver value to stakeholders, a key challenge for any evaluation technique is using stakeholder needs to drive engineering design decisions [34]. In other words, in order to decide upon the "best" of the proposed alternatives, the analysts need to establish a link between the technical parameters and the customer value.

In spite of the centrality of the value concept, there is a wide diversity of opinions and many speculative assertions on its real meaning and how it is determined by the stakeholders. Hence, the challenge in the early design phase is to understand how to develop solutions that, in a profitable way for the manufacturer, satisfy customers delivering value in a situation where customer expectations evolve and challenge the product/service definition.

First, it is particularly important for the decision maker to highlight how the system design impacts the product profitability. Value Driven Design (VDD) [21] aims to improve the existing systems engineering approaches by introducing economics in the decision making process and enabling optimum solution strategies to be instantiated during the conceptual and preliminary design stages of a product.

In an idealised situation, value driven design methods can be summarised as an overall scoring system that provides an unbiased metric on the 'goodness' of the final product [22]. A value model is essentially a function that accepts a vector of attributes as its argument and assigns a scalar score to system value. This allows designers to run formal optimisation algorithms for a vast number of input parameters and, if needed, to perform what-if and sensitivity analyses easily. A Value Model, therefore, aims to produce a scalar number to rank a design. For a product that generates revenue, such as an aircraft fleet, this value can be a surrogate model for profit (Surplus value), which usually takes the form of Net Present Value (NPV) for most cases.

Value Driven Design assesses design alternatives on the base of Extensive Attributes [21], which are intended as attributes of the system or product being designed, or attributes of its components. Examples of extensive attributes in aerospace are: weight, all performance attributes, reliability, maintainability, safety, and similar supportability attributes, plus all aspects of cost, schedule and technical risk.

Secondly, for a system characterized by high cost, long lifecycles, high complexity, interdependencies with other systems, and dynamic operational contexts, value is also determined by the capability to maintain or improve its functions in the presence of change. Tradespace exploration [23] considers customer value embedded in the customer process context and utilizes "ilities" to evaluate the system robustness under changing process conditions, i.e. using criteria such as survivability, adaptability, flexibility, scalability, versatility, modifiability and robustness. The epoch framework proposed by Ross et al. [24] allows the systematic creation of trade-space model(s) to quantify these "ilities". Other valuation methodologies do exist (e.g., real options for flexibility [25] [26]) but just for a few of these criteria.

Furthermore, value is something perceived by customers rather than objectively defined by a provider [27]. For some part, customer value is not related either to the physical product or to a "changeability" perspective, rather it is depending on the customer as an individual, i.e. customers may perceive both product and service attributes as relatively intangible [28].

The concept of "intangible" value has long generated debate concerning its definition and impact on consumer decision making. Vargo and Lusch [29] have discussed the tangible vs. intangible dichotomy, claiming that there are no pure services or goods, instead goods and services can be arrayed on a continuum of relative tangibility, with goods being more tangible and services more intangible [30]. Intangibles are often associated with knowledge, emotion, and experience; dimensions that cannot be experienced by the customer before using the product.

Steiner [31] has proposed an extended model of customer value based on three main layers: products, services and intangibles. The intangibles can interact with the product layer and the service layer to form a "value platform" from which total customer value is some combination of each layer of the total product. Intangible criteria (such as epistemic, emotional, or image value) should therefore be used for assessing the overall system value in the beginning of a product development project.

Moving towards providing Product Service Systems (PSS), it is increasingly important to move from left to right, i.e. from hardware-centred parameters, to considering criteria closer to the final user perception of the value that the "functional" product provides.

5 PRELIMINARY DESIGN EVALUATION: CURRENT GAPS

Experts in the field of decision making agree that the conceptual design stage of a new system does not typically involve a broad, systematic exploration of design alternatives [32]. In practice, the intuition of experienced designers is usually consulted in the pursuit of feasible solutions (i.e. designs within the acceptability range of stakeholder cost and benefit) [33]. Typically, design teams establish a favourite baseline from which small perturbations are examined in an analysis-of-alternatives. This approach of rapidly converging and optimising on a conceptual stage is susceptible to identifying locally optimal designs that provide only a vague picture of the complexity of the broad space of possible designs [32].

Ross and Hastings [34] illustrates four levels of trade that can be conducted during conceptual design: (1) local point solution trades; (2) examination of a small number of point designs in an analysis-of-alternatives; (3) multi-objective optimisation to identify designs along the Pareto front; (4) full trade-space exploration.

Level 1 requires the least effort, but implies incomplete understanding of the broader trade-space and may eventually lead to selection of sub-optimal solutions. Level 2 may highlight solutions that are non-dominated, while Level 3 lacks the sensitivity of the "optimal" designs to changes in the objective function [35].

The authors' experience in aerospace development projects [36] has shown that Level 1 and 2 approaches are common, especially in the development of aircraft sub-systems. Preliminary design decisions are mainly based on technical considerations (e.g. about weight, specific fuel consumption, noise, emissions, etc.) and on cost-related criteria (manufacturing costs, maintenance costs throughout the lifecycle, etc.). "Ilities" are also considered to some extent, while a general lack of focus on other aspects was observed. In synthesis, aircraft components are largely valued on weight, performance and cost only. Different dimensions of value, such as monetary value, perceived value, confidence on the product, and requirement fulfilment levels, are not considered in a more holistic value model especially during the early stages of the design process.

A key gap is the ability to judge preliminary designs from a business perspective, taking into consideration all the relevant lifecycle-related issues. It is not only a matter of including manufacturing costs, but also to have an overview on e.g. maintenance costs, recycling opportunities, infrastructural changes and so on. Moreover, conceptual designs are poorly evaluated with respect to their potential contribution to the customer process. Appropriate criteria are needed to display the value associated to a particular technology/component in a customer process "static" scenario, focusing on the activities the product should support, for instance, from an airline perspective. Although criteria such as availability, maintainability and reliability are often considered in preliminary design analysis, the contribution of a new technology to dimensions such as ground operations, upgradeability and timeliness, is far less evident to the designers in an early stage, although they eventually drive the airliner's choice to buy or not to buy the PSS. Nowadays it is also difficult to highlight the value contribution of a product/service in front of radical changes in the customer value scale due to changes in the surrounding context. Such a long-term is considered mostly implicitly and should be made more evident.

Furthermore, the latest aircraft development projects' delays [37][38] have outlined that the value of a solution is also strictly related to the capability to address the contractual constraints. Timely deliveries are crucial in a highly competitive market. As expectations are transferred into actual specifications on the sub-systems, there is no time to develop the best solution from scratch. Technology need to be mature and already in place to fit into the parent system. The gap is evident in two ways; lead time for transmitting requirements on overall system to sub-system, and the loss of intent in the same process causing non-appropriate design conditions. In short, pre-development design iterative capability between overall system and sub-system introduces a risk of delay and/or increased cost on sub-system solutions.

Eventually, the emotional aspect of value is typically underestimated and poorly translated into engineering requirements, in spite of the fact that aspects like feeling or identity may eventually determine the end user's decision to buy or not to buy the product/service.

6 CRITERIA FOR VALUE ASSESSMENT – A CANDIDATE APPROACH

Studies have shown an inverted U-shaped relationship between the efficiency of the decision making task and the amount of information provided [39]. Hence, dealing with 5 or 6 main parameters may result to be as effective as using 200 criteria, but considerably less effective than using 25-30 parameters. In order not to overwhelm decisions makers with unnecessary details at the gate, the number of criteria to be considered for a preliminary value evaluation has to be kept to a minimum, while providing enough information to regarding strengths and weaknesses of proposed alternatives.

Emerging from these considerations, from the three main "complementary" approaches for value evaluation highlighted in literature, and from the discussion with the industrial partners, a framework composed by six main value criteria has been proposed to guide lifecycle-oriented decisions in a preliminary design phase. The main purpose of the approach is to outline the.

The framework is based on 6 high-level criteria that aim to provide a complete picture of the value associated to a set of alternative designs at the gate, both from external and internal stakeholders perspective, namely: *Performance attributes*, *Risk*, *Profitability*, *Operational Performances*, *Ilities*, and *Intangibles*.

Performance attributes, *Risk* and *Profitability* look into the manufacturer value scale, while *Operational performances*, *Ilities* and *Intangibles* have the customer process in focus and extend this view to encompass a customer of customer perspective.

In some design situations it might be enough to have a synthesis assessment, in others some aspects have to be detailed with more rigor and each attribute should be cascaded down to a lower level to allow experts to recognize "patterns" of behaviour in the trade-off study. Each of these high-level criteria can be specified in terms of *value dimensions*, specific for a given hardware.

Figure 1 exemplifies value criteria and dimensions emerging from an example taken from an aircraft engine sub-system manufacturer. Concept A and concept B represent two intermediate compressor case (IMC) alternatives to mounted on an engine offered as part of a Total-Care® package. A and B are benchmarked against the 6 main value criteria to provide a high-level assessment of their value contribution.

Each criterion, such as *Operational Performances*, is further specified in terms of value dimensions such as *Availability*, *Logistic* or *Ground personnel management* to communicate, in detail, the overall contribution of a design from an airliner point of view.

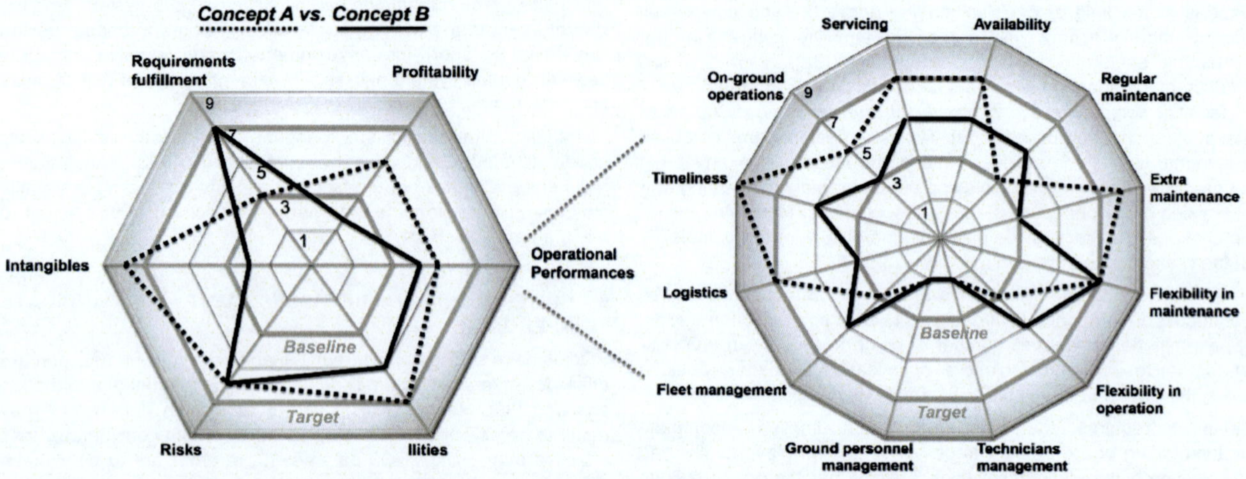

Figure 1: Spider diagrams displaying the value contribution of an IMC concept A (dashed line) and B (continuous line).

Requirements fulfilment plays a strong role in the decision of the right design alternative, because product requirements are specified in strict contractual agreements and are linked to penalties, hence the value associated to the achievement of the contractual thresholds has to be highlighted at the time of taking a decision. Early on, the decision makers should also be aware of the expected gains in terms of product realization, marketing, use, maintenance and disposal, i.e. of the expected *Profitability* associated to a hardware/technology, which might depend from system-level economies of scale.

The *Operational performances* dimension spotlights, instead, the value contribution of a given solution in a Business-to-Business (B2B) situation. In an aerospace context, product availability, maintainability and reliability are very important factors that have a direct impact on the value perceived by the airlines. Some of these process-related parameters are already considered today in an early phase, while others, such as, for instance, ground operations, fleet management and schedule, tend to remain hidden and need to be further leveraged when dealing with "wicked" problems.

Ilities are used to evaluate the value robustness of a solution in a more "dynamic" scenario, i.e., when the customer value scale changes due to changing environmental and market conditions. Furthermore, the framework highlights the *Risk* for the present and future projects, in terms of feasibility, knowledge gains, novelty, and maturity of the solution. Eventually, Intangibles gather those value dimensions related to the value perceived by the end customers (e.g. passengers), as well as dimensions such as brand acknowledgement, charm factor, social welfare or environmental impact.

Due to the difficulties in obtaining quantitative figures for all these criteria, and in the spirit of stressing the value contribution of radical designs, the approach aims to highlight the "delta" compared to a given "baseline" and "target". Once a value study is performed, the outcome of the analysis is compared against a baseline (i.e., the minimum expected value) and the target (i.e., the desirable outcome, from historical data) to obtain a single scalar, from 1 to 9, that represent the degree to which the design satisfies these two benchmarks. Setting the baseline equal to 3 and the target equal to 7, the approach aims to primarily show is a design is: 1) below the baseline (No-Go area, modify), 2) between baseline and target (Go-area, adjustments needed) and 3) above the target (Go area, proceed).

Many stakeholders have expressed a preference towards a single numerical metric for value, to make an easier comparison between dimensions very different in nature. Scalars are used as common denominators to merge the outcomes of several value studies (through a weighted average), and to generate a high-level assessment of the value contribution of a component. The weights reflect the customers' expectations and needs, as well as the company's long-term strategy, its culture (e.g. aversion to risk), the market and financial situation, and the specific contractual agreements (e.g. in presence of very strict agreements characterized by high penalties).

Spider diagrams have been proposed (Figure 1) to communicate the value contribution of a design alternative at the decision gate, highlighting strengths and weaknesses from a lifecycle perspective and spotlighting integration opportunities. Spider diagrams intend to support a more intuitive comparison between two or more concepts, facilitating the recognitions of "patterns of behaviour" to an expert eye.

To make an example, the study has shown that, nowadays, high/low cycle fatigue, limit/ultimate load capability, hale ingestion, strength and stiffness (E-modulus), corrosion, oxidation and creeps are the main criteria used for the evaluation of IMC concepts at the gate [40]. Looking at the diagram, both the designs are above the minimum acceptance criteria for the project in terms of *Requirements fulfilment*. What seems to be a very straightforward choice for the decision makers (concept B, in fact, shows very desirable characteristics, while A is just above the threshold) might be very controversial considering the other criteria defined in the framework. Choosing A, in fact, the manufacturer could benefit from a more profitable cash-flow distribution and lower risk. Airliners could also receive significant gains in terms of availability and logistics, while passengers could enjoy lower noise levels and more comfort in the cabin. Depending on the complexity of the task, the decision makers can further cascade down each dimension to a lower level, evaluating value contribution with more detail. In Figure 1, the *Operational Performances* criteria has been cascaded down to 12 value dimensions to provide a more detailed picture of the value contribution of A and B (for simplicity reasons, the same weight has been assigned to all the value dimensions in the picture), to show the areas where the value contribution is the lowest/greatest.

7 DISCUSSION

The criteria proposed in the paper aim to enable the assessment of design concepts from a system of systems perspective and, in this way, to highlight the value contribution of radical concepts that would be rejected if merely evaluated from a technical viewpoint. Moreover, the framework serves as a means, in the early design process phase, to improve the communication and dialogue between stakeholders, system architects, designers, and analysts.

The establishment of proper weights in the matrix represents the main issue for this approach. How to compare gains in terms of Survivability against losses that refer to Flexibility in Manufacturing? The final use of the product basically dictates the scope of the weights to be employed. For example, for mobile phones it is more important to consider functional capabilities of the product along with a measure on how the product is perceived by the end users, but, perhaps, the total life cycle cost or maintenance cost of the product is of less importance. On the contrary, for products like a satellite system, *Ilities* should be weighed more compared to other intangible factors [23].

Moreover, it might not be straightforward in many situations to boil everything down to a single number representing the overall value associated to a given choice. The definition of the value metric can be ambiguous when revenue is not the main measure of the product. The reliability of the figures contained in the table might be questionable and it might be also difficult to set the baseline, because in the beginning of a radically new project the team has nothing in existence to compare the solution concepts with. The emphasis should be, therefore, on covering all possible aspects with reasonable accuracy by using high-level attributes of the product, but not on trying to build perfect models that require a considerable amount of data (which may not exist during the early design stage) and long running computations.

8 CONCLUSIONS

This paper has proposed an approach to support PSS development teams in evaluating, in an early stage, the responsiveness and trade-off between alternative designs that target lifecycle commitment, with respect to the impact on perceived values and stated needs and expectations. The empirical data gathered from the aerospace partners have shown that it is a common practice to assess the design "goodness" of a product/system from a "requirements fulfilment" point of view, not taking the bigger picture in consideration. A framework composed of six high-level value criteria has been proposed to support more explorative design studies and to stimulate more radical - lifecycle-oriented - innovations.

An enhanced degree of openness, trust and cooperation throughout the supply chain is a key enabler to leverage the companies' ability to reason from a system-of systems perspective. This encompasses the ability of not only sharing technical data, but also to: 1) find the right people to talk with to make the "right" guesses, 2) codify value-related information in a way that can be shared with low threshold across the enterprise, functions and departments, 3) visualize value contribution in a way that makes sense to the engineers and that relates to the engineering work, impacting on their everyday decisions.

A main concern, dealing with value studies, is about raising the designers' awareness on those higher-level issues that tend to be neglected if focusing on at system-level only. Radical designs that normally would be rejected when merely considered at technical performances may eventually be preferred, if the design problem is undertaken at a super-system level. What is important is to be able to communicate the lifecycle value contribution of each design solution early in the development program. In this spirit, an approach to visualize the value contribution of a given part or assembly directly in a 3D CAD model (on the basis of the 6 high-level criteria described in this paper) is under development [41].

The measurement of the customer satisfaction is crucial to refine the need scale and to base decisions upon fresher and more validated data. This directs the attention towards the development of seamless link between stakeholders and engineers, monitoring the customer satisfaction level over time and requires a new means to capture and share downstream knowledge (i.e. knowledge form the later lifecycle phases) and to feed it back in the early design process steps. Lightweight knowledge sharing technologies - Web 2.0 style - are currently explored to enable seamless capturing of value-related knowledge across function and companies, lowering the threshold for stakeholders to formalize information that could trigger more value-oriented decisions at the design gate.

9 ACKNOWLEDGMENTS

The research leading to these results has received funding from the European Union Seventh Framework Programme (FP7/2007-2013) under grant agreement n° 234344 (**www.crescendo-fp7.eu/**).

10 REFERENCES

[1] Rittel, H., Webber, P. (1973): Dilemmas in a General Theory of Planning, Policy Sciences, Vol. 4, pp.155-169.

[2] Boeing, (2006): Technology Redefines Joy of Flight, **http://www.boeingcapital.com/p2p/archive/06.2006/techredefinesjoy.htm**.

[3] Isaksson, O., Larsson, T.C., Rönnbäck, A.Ö. (2009): Development of product-Service Systems: Challenges and Opportunities for the manufacturing firm, Journal of Engineering Design, Vol. 20, No.4, pp. 329-348, 2009.

[4] Baines, T.S.; Lightfoot, H.W.; Evans, S.; Neely, A.; Greenough, R.; Peppard, J.; Roy, R.; Shehab, E.; Braganza, A.; Tiwari, A.; Alcock, J.R.; Angus, J.P.; Bastl, M.; Cousens, A.; Irving, P.; Johnson, M.; Kingston, J.; Lockett, H.; Martinez, V.; Michele, P.; Tranfield, D.; Walton, I.M.; Wilson, H. (2007): State-of-the-art in product-service systems, in: Proceedings of the Institution of Mechanical Engineers, Part B: Journal of Engineering Manufacture. Vol. 221, No. 10.

[5] Lindahl, M., Sundin, E., Rönnbäck, A.Ö., Ölundh, G. and Östlin, J. (2006): Integrated Product and Service Engineering – the IPSE project, in Proceedings of Workshop of the Sustainable Consumption Research Exchange (SCORE!), April 20th - 21st, Copenhagen, DK.

[6] Shehab, E.M., Roy, R. (2006): Product service-systems: issues and challenges, in Proceedings of 4th International Conference on Manufacturing Research (ICMR 2006) September 5th - 7th, Liverpool, UK.

[7] Kim, S-H., Cohen, M.A., Netessine, S. (2007): Performance Contracting in After-Sales Service Supply Chains, Management Science. Vol. 53, No. 12, pp. 1843—1858.

[8] Harrison, A. (2006): Design for service - Harmonising product design with a services strategy, in Proceedings of the ASME Turbo Expo 2006, Vol. 2, May 06th-11th Barcelona, E.

[9] Cooper, R. G., (2008): Perspective: The Stage-Gate® Idea-to-Launch Process-Update, What's New, and NexGen Systems*. Journal of Product Innovation Management, Vol. 25, pp. 213-232.

[10] ACARE Group (2002): Strategic Research Agenda Volume 1, www.acare4europe.org/docs/es-volume1-2/volume1.pdf.

[11] Stringer, E.T. (1999): Action Research, 2nd Edition. Thousand Oaks, CA: Sage Publications Inc.

[12] Avison, D., Lau, F., Myers, M., Nielsen, P.A. (1999): Action Research, Communications of the ACM, Vol. 42, No.1 pp. 94–97.

[13] Bahill, A.T., Gissing B. (1998): Re-evaluating systems engineering concepts using systems thinking, IEEE Transaction on Systems, Man and Cybernetics, Part C: Applications and Reviews, Vol. 28, No. 4, pp. 516-527.

[14] EIA632 (1998): Processes for Engineering a System, ANSI/EIA-632.

[15] Dyer, J., Fishburn, P., Steuer, R., Wallenius, J., Zionts, S. (1992): Multiple criteria decision making, multiattribute utility theory: The next ten years, Management Science Vol. 38, pp. 645-654.

[16] Von Winterfeldt, D. and Edwards, W. (1986): Decision Analysis and Behavioural Research, Cambridge University Press, Cambridge.

[17] Keeney R. and Raiffa H. (1993) Decisions with multiple objectives: preferences and value trade-offs. Cambridge University Press.

[18] French, S. (1988): Reading in Decision Analysis, Chapman and Hall, London.

[19] Fishburn, P.C. (1967): Methods of estimating additive utilities, Management Science, Vol.13, No.7.

[20] Beinat, E., Nijkamp, P. (1998): Multicriteria Analysis for Land-Use Management, Kluwer Academic Publishers, Dordrecht.

[21] Collopy, P., Horton, R. (2002): Value Modeling for Technology Evaluation, AIAA-2002-3622, American Institute of Aeronautics and Astronautics.

[22] Collopy, P., Hollingsworth, P. (2009): Value Driven Design, in Proceedings of the 9th AIAA Aviation Technology, Integration, and Operations Conference, September 21st-23rd, Hilton Head, South Carolina.

[23] Ross, A., Hastings, D., Warmkessel, J., Diller, N. (2004): Multi-Attribute Tradespace Exploration as Front End for Effective Space System Design, Journal of Spacecraft and Rockets, Vol.41, No.1, pp. 20-28.

[24] Ross, A., Rhodes, D., Hastings, D. (2008): Defining changeability: Reconciling flexibility, adaptability, scalability, modifiability, and robustness for maintaining system lifecycle value, Journal of Systems Engineering, Vol.11, No.3. pp. 246-262.

[25] Saleh, J.H., Lamassoure, E.S., Hastings, D.E. (2003): Flexibility and the Value of On-Orbit Servicing: New Customer-Centric Perspective, Journal of Spacecraft and Rockets, Vol. 40, No. 2, pp. 279-291.

[26] Wang, T., de Neufville, R. (2006): Using a Coupled-Design Structure Matrix Framework to Screen for Real Options "In" an Engineering System, in Proceedings of the 16th INCOSE Symposium, July 9th-13th, Orlando, FL.

[27] Kauppinen, M., Savolainen, J., Lethola, L., Komssi, M., Töhönnen, H., Davies, A. (2009): From Features Development to Customer Value Creation, in Proceedings of the 17th International Requirements Engineering Conference, August 31st–September 4th, Atlanta, GE, pp. 275-280.

[28] Iacobucci, D. (1992) :An Empirical Examination of Some Basic Tenets of Services, in Advances in Services Marketing and Management, T.A. Swartz, D. E. Bowen, and S. W. Brown, Eds. Greenwich, CT: JAI, 1992, pp. 23-52.

[29] Vargo, S.L., Lusch R.F. (2004): The Four Service Marketing Myths: Remnants of a Goods-Based, Manufacturing Model, Journal of Service Research, Vol. 6, pp. 324-335.

[30] Swartz, T.A., Bowen, D.E., Brown S.W. (1992): Fifteen years after breaking free: services then, now, and beyond, Advances in Services Marketing and Management, Vol. 1, pp. 1-21.

[31] Steiner, F. and Harmor, R. (2009): The Impact of Intangible Value on the Design and Marketing of New Products and Services: An Exploratory Approach, in Proceedings of PICMET 2009, Portland, Oregon USA.

[32] McManus, H.M., Richards, M.G, Ross, A.M., Hastings D.E. (2007): A Framework for Incorporating "ilities" in Tradespace Studies, AIAA Space 2007, Long Beach, CA.

[33] Maier, M. and Rechtin, E. (2002): The Art of Systems Architecting, CRC Press, Boca Raton.

[34] Ross, A.M. and Hastings, D.E. (2006): Assessing Changeability in Aerospace Systems Architecting and Design Using Dynamic Multi-Attribute Tradespace Exploration, AIAA Paper 2006-7255, American Institute of Aeronautics and Astronautics, Reston, Virginia.

[35] Richards, M.G. (2009): Multi-attribute tradespace exploration for survivability, PhD Dissertation, Massachusetts Institute of Technology.

[36] CRESCENDO (2010): D2.2.2 Benchmark of candidate approaches for value driven design, EU FP7 CRESCENDO project public deliverable, 2010, available at www.crescendo-fp7.eu.

[37] Greising, D., and Johnsson, J., (2007): "Behind Boeing's 787 Delays," Chicago Tribune, December 8th.

[38] AIRBUS (2006): Airbus confirms further A380 delay and launches company restructuring plan, AIRBUS press release, http://www.airbus.com/en/presscentre.

[39] Zanakis, S.H., Solomon, A., Wishart, N, Dublish, S. (1998): Multi-attribute decision making: a simulation comparison of select methods, European Journal of Operational Research Vol. 107, pp. 507-529.

[40] Gustavsson, R. (2006): Design Study of Advanced Metallic Structural Integrated Fan Outlet Guide Vanes in Civil Jet Engines, Master Thesis, Luleå University of Technology, Luleå, Sweden.

[41] Bertoni, A., Bertoni, M., Isaksson, O. (2011): Communicating the value of PSS design alternatives using color-coded CAD models, accepted to the 3rd CIRP IPS² conference, Braunschweig, D.

Design of a Product-Service Systems Business Model: Strategic Analysis and Option Generation

Joris Van Ostaeyen[1], Bart Neels[2], Joost R. Duflou[1]

[1] Centre for Industrial Management, Katholieke Universiteit Leuven, Leuven, Belgium

[2] Sirris, Leuven, Belgium

Abstract

Many manufacturers are considering the adoption of a business model in which the use or the function of a product is sold instead of the product itself, namely a Product Service System (PSS). This paper presents the first two phases of a generic method to design a PSS business model for a manufacturer of investment goods. The first phase consists of the analysis of the company's current business model, its competitive environment and the function provided by the investment good under consideration. The second phase focuses on the generation of PSS business model options, based on the formulation of value creation options over the life cycle of the investment good in some specifically chosen customer segments. The method presented is illustrated with a case study conducted for an elevator manufacturer. It is demonstrated that, in the two first phases of this PSS design method, it is essential to have a correct representation of the customer segments and to create within each segment an understanding of customer specific requirements related to the function of the product.

Keywords:

Business Model; Design; PSS Design

1 INTRODUCTION

Many manufacturers of investment goods see a potential to create a competitive advantage through business model innovation. Specifically, business models in which the use or the function of a product is being sold, instead of the product itself, seem an interesting option. Such business models are known in academic literature as Product-Service Systems (PSS) and are gradually gaining importance in industry. In a PSS, a manufacturer assumes more responsibility for products throughout their use phase, increasing the value or reducing the costs for the user while being able to generate potentially profitable downstream activities. In the context of eroding profit margins for new product sales due to increasing competition and subsequent commoditization, additional opportunities to create value in the installed base can have a substantial earning potential [1].

Although many manufacturers perceive a PSS as an attractive option, evidently a fundamental change to a company's business model is often a risky undertaking. Adopting a PSS business model can have a profound impact on a company's strategic position, profitability and fundamental modus operandi within its value network of customers, suppliers and other stakeholders. Therefore, there is a need for systematic analysis techniques that both assist companies in identifying and specifying PSS options and aid in assessing their feasibility and impact on profitability. This is where PSS design methods can apply.

This paper presents a PSS design method developed within the BOSS (Business Opportunities in Service Systems) research project at the Centre for Industrial Management of K.U.Leuven, in cooperation with Sirris, the Collective Centre of the Belgian Technology Sector. The goal of this project is to develop methods to investigate the economic potential of a PSS from the point of view of a manufacturer of investment goods. BOSS targets Flemish manufacturers of investment goods, with a focus on small and medium enterprises.

The outline of this paper is as follows. In Section 2 a concise overview of some relevant literature on Product-Service Systems, business models and design methods is presented. In Section 3 the general outline of the proposed PSS design method is provided. The first two steps, strategic analysis and option generation, are elaborated in Sections 4 and 5. These detailed descriptions are illustrated with an industrial example of an elevator manufacturer. A summary is provided in Section 6.

2 LITERATURE BACKGROUND

2.1 Introduction

Over the last decades, the terms 'business model' and 'Product-Service Systems' have generated an extensive interest, as well from academic researchers as from industrial practitioners. Both concepts are important for this paper, but since there is no consensus among researchers on their definitions and the existing proposals vary significantly, a clarification on how the authors understand these terms is obligatory.

2.2 Business Model

A 'business model' is a company's core logic for creating, delivering and capturing value [2, 3]. It forms an intermediary level between the strategy and business process level within a company's management structure [4]. Many authors view a business model as a multidimensional concept, which can be decomposed into several 'atomic' elements or 'building blocks'. A wide variety of such decompositions into building blocks is available throughout the academic literature: for an overview see [5] and [6]. In Section 3.1 a new decomposition into building blocks is presented specifically for business models related to investment goods manufacturers. Reference to the relevant literature that has inspired the development of this framework, will be made there.

At this stage two building blocks in particular are already discussed because they are often confused with the business model as a whole. One is the revenue mechanism, which determines how the revenues are generated within the business model. For an investment good, examples of different revenue mechanisms are a product-based revenue mechanism (the customer pays per product

J. Hesselbach and C. Herrmann (eds.), *Functional Thinking for Value Creation: Proceedings of the 3rd CIRP International Conference on Industrial Product Service Systems, Technische Universität Braunschweig, Braunschweig, Germany, May 5th - 6th, 2011*,

DOI 10.1007/978-3-642-19689-8_27, © Springer-Verlag Berlin Heidelberg 2011

bought) and a use-based revenue mechanism (the customers pays per time unit that the product is being used). Notwithstanding the importance of a revenue mechanism in differentiating between business models, most authors consider the revenue mechanism to be only one element of a business model.

The other building block that is often confused with the business model as a whole is the value proposition, which is the combination of all benefits that are delivered through products and/or services by the vendor to the customer in return for the customer's payment. The value proposition is essentially what the business offers to the market and thus obviously a very important element of a business model, but it does not form a complete description of a company's business model.

2.3 Product-Service Systems

In this paper, Product-Service Systems are seen as a specific type of business model in which the value proposition is a combination of products and services that delivers value in use. Throughout the PSS-literature there exists some confusion on whether a PSS must be understood as a specific type of business model or as a specific type of value proposition. In this paper the first approach is chosen, which is consistent with the definition of a PSS in references [7] and [8], and in contrast to the definition provided in reference [9] in which a PSS is viewed as a type of value proposition.

2.4 PSS Design Methods

A variety of PSS design approaches exists in academic literature. Most authors focus on the design of the value proposition within a PSS, the combination of products and services that are offered to the customer. Aurich et al., for example, propose a service design method based on Life Cycle Engineering [10]. Other methods focus more on the development of products for the full life cycle spanned by the PSS. A comprehensive overview of all recent PSS design methods can be found in reference [11].

3 PSS DESIGN METHOD

3.1 Design domains and PSS building blocks

In Section 3.2 a generic method in five steps is presented that facilitates the design of a PSS business model for an industrial investment good manufacturer. In the following sections the first steps of this method will be elaborated. But first, in this section, the concept of 'design domains' and 'PSS building blocks' will be introduced, because they form the conceptual framework in which the PSS design method is rooted.

The design of a PSS business model involves design activities within four domains:

1. In the **customer domain** the value proposition that is targeted towards specific market segments is designed, which can be reached through selected distribution channels.

2. In the **technological domain** the physical architecture is drawn of the products and systems that are instrumental in the value creation within the business model, as well as how these systems are interfaced to their physical environment and what their function is.

3. In the **organizational domain** the value network is designed of all organizations and individuals that are involved in the value creation process, how tangibles and intangibles are interchanged, as well as the competencies and resources of the actors involved.

4. In the **financial domain** the formal structure of revenue flows and costs is described, as well as the description of which actor is responsible for which investments and assumes which risks within the business model.

Thus in order to design a business model, each of these four domains must be addressed. In each domain, a blueprint is made of the business model design. The combination of the blueprints in the four domains forms a comprehensive description of the business model. Within each domain, several PSS building blocks can be discerned. All the building blocks together constitute a complete description of the business model. An overview of all building blocks can be found in Figure 1. The building blocks were identified based on the existing ontologies of business models such as those of references [4] and [6]. Within each domain one building block can be designated as a predominantly internal, one as an external and one as an interface-related characteristic. For example, in the customer domain the value proposition is developed internally and interfaced through the distribution channel to the external customer segments. In the organizational domain, the internal competencies are translated through processes in the external value network.

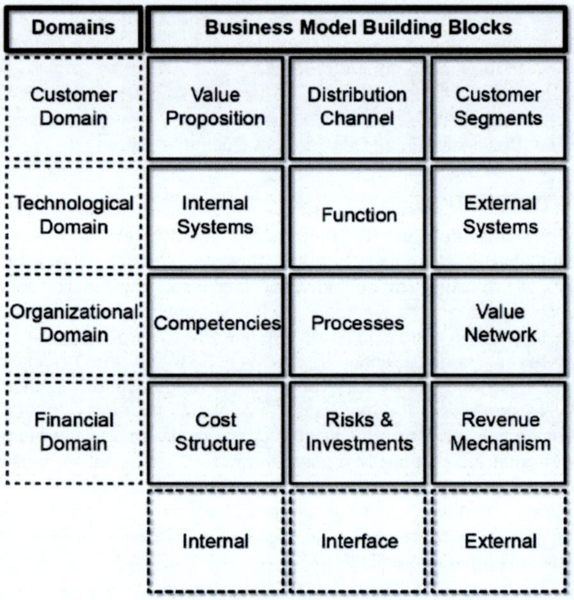

Figure 1: Business Model Domains and Building Blocks.

3.2 General structure of the PSS design method

The proposed PSS design method follows a sequence that is analogue to a strategy formulation process and aims to answer the following central questions:

1. What is the optimal PSS business model, in terms of profitability and feasibility, for an investment good manufacturer?

2. What is the complete description of this PSS business model design such that it can be implemented in practice?

3. What are the key success factors that define the profitability of the optimal PSS business model?

An answer to these questions is sought by running through the following steps (cfr. Figure 2):

1. **Strategic analysis.** In the first step an analysis is made of the current business model under which the company is operating. In addition to that, an environmental scan is performed such that all factual and subjective information is gathered concerning the business environment in which the company is operating. The current business model is then further analyzed in the technological and in the organizational domain.

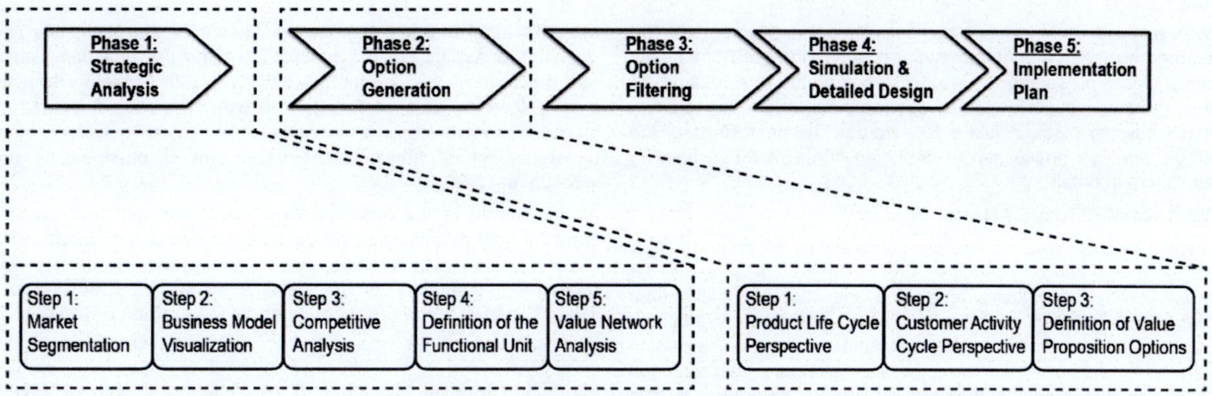

Figure 2: Generic PSS Design Method.

2. **Generating options.** In the second step all PSS business model options are listed in a structured way. During this phase, by systematically running through the different life cycle phases of the investment good, ideas for new value creation options are detected and translated into product and/or service ideas.

3. **Filtering options.** In the third step, out of the variety of value proposition options identified in the previous step, the realistic and attractive ones are filtered, based on the strategic preferences and capabilities of the company on the one hand and the attractiveness of each option on the other hand. After this step, a few value proposition options are retained that can be put under profound scrutiny in order to identify the optimal business model.

4. **Simulation and detailed design**. In the fourth step, the few retained value proposition options are thoroughly investigated, first by specifying the design of all business model building blocks for these options. Subsequently a financial simulation model is constructed that allows an a priori evaluation of the profitability of each option. A final choice is made between the remaining business model options.

5. **Implementation plan.** After simulation and detailed design, an implementation plan is elaborated in order to prepare the roll out of the new business model.

The five step design method introduced here involves design activitities in the four different domains of Section 3.1. The designs within each domain are developed in parallel rather than consequently. Based on the practical experience of applying this method in a number of case studies, a further decomposition and discussion of the first two steps (strategic analysis and generating options) is presented in the next Sections 3.3 and 3.4.

3.3 Strategic analysis

The purpose of the strategic analysis is to gain an understanding of the company's current business model and to collect all relevant information related to the company's competitiveness within its business environment. The strategic analysis as conducted within the BOSS project runs through the following five steps (cfr. Figure 2):

1. Market segmentation
2. Visualization of the current business model
3. Analysis of the competitiveness of the current business model
4. Specification of the functional unit of the investment good
5. Visualization and analysis of the value network

STEP 1: MARKET SEGMENTATION

In the first step the current business model is analyzed in the customer domain. The actual and potential customers the company hopes to serve are segmented into distinct groups with similar needs. Market segmentation is a central method of marketing research and has received considerable interest since this concept was introduced by Smith in 1956 [12]. It is acknowledged that industrial market segmentation is often more complex than consumer market segmentation due to, among other reasons, the fact that customers are more heterogeneous and that offerings are more complex and often involve co-creation [13]. Another specific challenge arises here because in the context of Product-Service Systems, it is necessary to derive one consistent method of segmenting the market for both the new product business and the aftermarket business within a company, while in practice it occurs that separate segmentations are defined.

In order to divide a market into segments, a segmentation basis is needed, which is the criterion that allows assigning potential customers to homogenous groups. Segmentation bases are classified as observable (if they can be measured directly) or unobservable (assumed) and on the other hand general (not related to specific products or services) or product-specific [14]. An unresolved academic discussion is ongoing about which segmentation bases are the most applicable in which situations (see [13] en [15] for an overview) and thus a pragmatic approach is advisable.

From the experience of the case studies within the BOSS project, two types of segmentation bases were found to be most helpful; on the one hand a segmentation based on the sector of the customer, in which it is often possible to combine sectors that have similar requirements, and on the other hand a segmentation based on the type of distribution channel (e.g. public tender or direct sales).

STEP 2: VISUALIZATION OF THE CURRENT BUSINESS MODEL

In the second step, a general overview on the business model is developed within the four domains. The visualization of the company's current business model by mapping it on a 'business model canvas', similar to the method proposed by Osterwalder [16], facilitates a common understanding of the business model by all parties that are involved in the business model design process. Osterwalder's model is a generic framework applicable for a wide variety of companies. In the context of investment goods, it lacks certain aspects that are considered crucial in understanding the value creation logic for a manufacturer of investment goods. Therefore, an altered business model framework has been constructed that is inspired on Osterwalder's business model canvas and the other existing business model ontologies [5]. The presented business model framework contains the business model building blocks presented in Figure 1. Mapping the business model in this framework is performed by concisely describing the essence of each building block in just a few words. The next steps of the strategic analysis phase aim at exploring some of these building blocks more in detail.

STEP 3: COMPETITIVE ANALYSIS

A qualitative analysis of the competitiveness of a company's business model can be conducted on three hierarchical levels: the internal level, the industry level and the level of the macro environment. The main purpose of this step is to identify all factors on these three levels that are instrumental in choosing a certain business model over another. In the strategic management literature, many methods and tools exist that can be applied for this purpose, a few of which will be highlighted here.

The internal level of the competitive analysis focuses on the 'core competencies' of a company. The notion of core competencies was introduced by Prahalad and Hamel in their seminal article [17]. A core competence is a corporation's fundamental strength and is widespread in a corporation, crossing the boundaries of strategic business units. A practical approach to help formulate core competencies is provided in reference [18]. Three tests can be applied to confirm the identification of a core competence:

1. It provides potential access to a wide variety of markets.

2. It contributes significantly to the perceived customer benefits of the value proposition.

3. It is difficult for competitors to duplicate.

At this point the company's current business model, as visualized in the previous step, can be evaluated in relation to the competences of the company.

On the industry level the analysis is aimed at determining the objective industry attractiveness, i.e. the opportunity to generate sustainable and stable profits with a certain product-technology combination. This can be achieved by investigating on the one hand the growth rate of the industry and on the other hand the intensity of competition within that industry. A common practice to determine the intensity of competition is the application of the framework of Porter [19] by assessing the five forces that drive competition within an industry: rivalry among existing competitors, the bargaining power of suppliers and customers and the threats of new entrants and substitute products or services.

Competitive analysis on the level of the macro environment encompasses identifying political, economic, social and technological (PEST) trends that could influence the profitability or feasibility of the business model from the outside. A useful and popular framework in this respect is offered by PEST or STEP analysis [20].

It is clear from the previous paragraphs that this step of the strategic analysis, the analysis of competitive factors, can become a quite comprehensive study in itself. But as a guideline it must be kept in mind that it is the main purpose of this step to ensure that all relevant factors are identified and listed in a structured way, without delving too deep in each issue separately.

STEP 4: FUNCTIONAL UNIT

In the fourth step, the current business model is analyzed in the technological domain. During Step 2, all internal and external systems were already listed. In this step the function of the internal systems is expressed. The function of a product is an abstraction of its intended behavior and can be expressed in terms of the delivered 'functional unit.' This key concept originates from the Life Cycle Assessment (LCA) literature. Within the context of LCA a system's functional unit "names and quantifies the qualitative and quantitative aspects of the function(s) along the questions what, how much, how well, and for how long." [21, 22]. As a PSS involves a shift towards sale of function or use of products instead of products themselves, it is essential at this point to have a clear understanding of the actual function that is provided by the investment good.

As mentioned in the definition of a functional unit, this concept contains both quantitative and qualitative aspects. The quantitative aspects can be broadly categorized into two basic types: aspects related to the duration of use (in time units) of the function and aspects related to the quality or extent of the function provided [23]. It can be helpful to keep the four questions mentioned in the definition in mind while defining the quantitative aspects (what, how much, how well, for how long). In the context of an elevator, some examples of quantitative aspects of the functional unit are:

- handling the transportation of maximum X passengers between the specified floors of a building
- at a nominal elevator speed of 1 m/s
- with an uptime of 97%
- respecting a maximum noise level of 35 dBA

The combination of qualitative aspects constitutes the non-quantifiable description of how the function is provided and of other qualities of the product. These other qualities are often perception aspects, related to the user's perception of the value of the product [24]. For example for an elevator important perception aspects are:

- the aesthetics of the elevator car and doors (e.g. the materials used, the degree of finishing, the general design)
- the ease of use of the elevator controls

Apart from the division in quantitative and qualitative aspects another division in obligatory, positioning and market irrelevant properties of the solution is possible [21, 23]. Obligatory properties are the properties that must be provided in order to be considered as a relevant alternative. For example, for an elevator satisfying all legal requirements related to elevator safety is an obligatory property. Positioning properties are considered as nice to have by the customer and as such can differentiate the solution from the competition. For an elevator, an example of a positioning property is the response time of a service intervention. Positioning properties can be scored subjectively to indicate their relative importance to customers. Market irrelevant properties form a third category. As is obvious from their denomination, they are properties that are of no importance to the customer within that specific market segment.

From the experience of the case studies performed within the BOSS project, the description of the functional unit is very much market segment dependent. The quantitative and qualitative description of the functional unit and the allocation of properties to the abovementioned categories can be fundamentally different from segment to segment. Therefore it is imperative that the description of the functional unit as discussed here is performed for each market segment for which a PSS business model is considered. The result of this step is a worksheet with a textual description of the functional requirements for each relevant customer segment.

STEP 5: VALUE NETWORK

The last step of the strategic analysis phase of the proposed PSS design method comprises the analysis of the value network in which the company is operating. A value network is a model of exchanges of tangibles and intangibles, such as products, services, information and benefits, between organizations and individuals. In the proposed business model representation model of Section 3.1, it is the main building block of the organizational domain of a business model, thus clarifying the logic of value creation and capture in terms of the interaction between different organizations.

The term value network originated as an extension of the value chain framework, which was introduced by Porter [24] and has become the main framework for describing firm-level value creation. Stabell and Fjeldstad highlighted the shortcomings of the value chain concept, especially in the context of services [25]. Increasingly, non-linear concepts are proposed in the strategic management literature, such as value networks, value grids or value webs.

During this step of the strategic analysis, a method of value network analysis is conducted in order to facilitate understanding of the inter-organizational dynamics present within the current business model. The method applied during this step is inspired by Allees Value Network Analysis method [26] and is a topic for further research.

3.4 Option generation

During the 'option generation' phase of the PSS design method, the goal is to generate new value proposition options. During this phase a list of value creation possibilities is generated over the life cycle of the investment good and over the customer activity cycle. The value creation possibilities are translated into possible value propositions and these options will be filtered during the next phase, through the assessment of their feasibility and profitability.

Thus the option generation phase consists of these steps:

1. **Product Life Cycle Perspective**: analysis in the technological domain of the business model to find value creation options over the product life cycle of the investment good
2. **Customer Activity Cycle Perspective:** analysis in the customer domain of the business model in order to find value creation options over the customer activity cycle
3. **Defining Value Propositions:** translation of the options generated during step 1 and 2 into specific value propositions

STEP 1: PRODUCT LIFE CYCLE PERSPECTIVE

A first dimension along which value creation options are sought is the Product Life Cycle (PLC) of the investment good. If necessary, this analysis can be performed for each of its constituting subsystems. The PLC consists in general of four phases: design, production, use and end-of-life treatment. It is further decomposed into an activity hierarchy, containing all activities that are performed by different actors of the value network over the PLC of the investment good.

Subsequently, for each activity in this structure, ideas for value creation are written down. These ideas can either be related to a reduction in the consumption of resources during that activity (thus a cost reduction) or to an increase of the benefits that are delivered through the system for one or more actors of the value network.

These possibilities for cost reduction or value increase are not only sought internally within the system, but also in all external systems that are interfaced to the investment good. For example, integration of the elevator with a ventilation system in the elevator shaft can result in a significant reduction of heat losses through the shaft and thus in this case the cost reduction is realized in an external system.

The value creation options are broadly categorized in two classes: options that can be realized through adaptation of the product design and options that can be realized through the offering of a service. In some cases though, this distinction is not evident, as some options require both product and service innovations.

STEP 2: CUSTOMER ACTIVITY CYCLE PERSPECTIVE

Another dimension along which value creation options can be listed is the customer activity cycle (CAC), a concept introduced by Vandermerwe [30]. While the product life cycle follows the product through its different life stages, the customer activity cycle takes the customer as a reference point and follows him through the different activities in which the acquisition and usage of the investment good are embedded. At each point in time, analogue to the previous step, value creation options through product or service innovations are sought. Applying the Customer Activity Cycle as a useful concept in the context of PSS development was proposed by Matzen and McAloone in reference [28].

Vandermerwe discerns three phases in the CAC:

1. During the pre-transaction phase, "the customer is still deciding what to do to get the desired result" [30]. In the case of an elevator, in this phase the customer is deciding how many elevators he should acquire, where he should locate them in the building and – possibly – what kind of technology he would prefer (hydraulic, traction with or without machine room).

2. The transaction phase starts when the product is acquired by the user and he starts using the product in order to get the desired result.

3. In the post-transaction phase, the focus is on maintaining the result: reviewing, renewing, upgrading, updating and extending [30].

As can be seen from these descriptions, the customer activity cycle will coincide for a considerable part with the product life cycle, mainly during the transaction and the post-transaction phase. As this step is included to offer a complimentary perspective, the focus is mostly on the activities that fall outside the product life cycle. For building installations specifically, the activities during the pre-transaction phase, the design of the building and the preparatory studies by engineering firms and architects, are often determinants of the concept of all installations in the building, and offering additional value during that phase can be substantial.

The customer activity cycle was originally proposed for B2C markets and thus a specific complexity of B2B investment goods must be taken into account. Often suppliers of such equipment do not face one single individual within the client company who takes every decision concerning the investment good. For example the decisions concerning the elevators in a hospital environment are partially taken by the technical maintenance supervisor, partially by external engineering firms, partially by the construction project development company, partially by administrative staff and partially by the hygiene department. Thus it is important to keep all different stakeholders within the client company in mind and look for possibilities to increase value within the activity cycle of each stakeholder separately.

STEP 3: DEFINING VALUE PROPOSITIONS

During this step, the value addition possibilities that were identified in the product life cycle and the customer activity cycle are translated into specific value propositions. A value proposition is an offering that can be put on the market, thus a bundle of products and/or services that can be offered to a customer, either in return for a payment or for free. In the example of company A, offering 3D models of standard elevator designs to architects is an illustration of a value proposition that can be put on the market for free, in the hope that the architects will favorably influence the decision process of new elevator purchase to the advantage of company A.

The ideas of the previous steps are thus combined or treated separately as a potential value proposition. An example of a combined value proposition is an administrative service in which the responsibility for all administration related to the elevator is being carried out by company A, including keeping all certification papers and maintenance reports up to date. Assessment of the feasibility and profitability of these options is only carried out in the third phase, option filtering.

4 SUMMARY

In this paper the first two phases of a PSS design method developed within the BOSS project of K.U.Leuven and Sirris are presented. Starting from the assumption that a PSS is essentially a type of business model, a framework for describing business models in four domains with twelve building blocks is introduced. This is tailored to manufacturers of investment goods, who are the target group of this research.

The PSS design method presented in this paper follows a similar approach as a strategy formulation process. In the first phase, namely the strategic analysis, the goal is both to develop a common and comprehensive understanding of the current business model of the company and to analyze more deeply three building blocks: market segments, function and value network. The authors demonstrate that it is imperative to pay attention to a correct market segmentation, as functional requirements can vary significantly from segment to segment. In order to define the function of an investment good, the method incorporates the functional unit description used within Life Cycle Assessment theory.

During the option generation phase, both the product life cycle and the customer activity cycle form the two fundamental concepts, as they offer a perspective to systematically identify value creation potential. At the end of this phase, a list of value proposition options is retained that can be filtered in the next phase. The following phases of this PSS design method form the subject of future research.

5 ACKNOWLEDGMENTS

The authors express their sincere thanks to the companies that are participating in the case studies of this research project and to the Institute for the Promotion of Innovation by Science and Technology in Flanders (IWT) for the financial support to carry out this research through sponsored Project No. IWT 095063.

6 REFERENCES

[1] Wise, R; Baumgartner, P (1999) Go Downstream: The New Profit Imperative in Manufacturing, Harvard Business Review, Vol. 77, No. 5., 133-141.

[2] Linder, J, Cantrell, S (2000) So What Is a Business Model Anyway: Accenture Institute for High Performance Business.

[3] Magretta, J (2002) Why Business Models Matter. Harvard Business Review, 80(5), 86-92.

[4] Osterwalder A (2004) The Business Model Ontology: A Proposition in a Design Science Approach. Ph.D. thesis, HEC Lausanne.

[5] Pateli, A, Giaglis G (2004) A research framework for analysing eBusiness Models, European Journal of Information Systems, 13, 302-314.

[6] Ballon, P, (2009) Control and Value in Mobile Communications: A Political Economy of the Reconfiguration of Business Models in the European Mobile Industry. Ph.D. thesis, Free University of Brussels.

[7] Tukker A, Tischner U, Editors (2006) New business for old Europe. Product services, sustainability and competitiveness, Greenleaf Publishing Ltd., Sheffield.

[8] Mont, O (2002) Clarifying the concept of product service-systems. Journal of Cleaner Production, 10(3), 237-245.

[9] Baines TS, Lightfoot HW, Evans S and Neely A (2007) State-of-the-art in Product Service Systems. Proceedings of the Institution of Mechanical Engineers Part B. Journal of Engineering Manufacture 221(10):1543–1552.

[10] Aurich J, Fuchs C, DeVries MF (2004) An Approach to Life Cycle Oriented Technical Service Design. CIRP Annals - Manufacturing Technology 53(1):151–154.

[11] Meier H, Roy R, Seliger G (2010) Industrial Product-Service Systems—IPS², CIRP Annals - Manufacturing Technology, Volume 59, Issue 2: 607-627

[12] Smith, W (1956) Product differentiation and market segmentation as alternative marketing strategies. Journal of Marketing, Vol. 21, No. 1, 3-8.

[13] Boejgaard, J, Ellegaard, C (2010), Unfolding implementation in industrial market segmentation, Industrial Marketing Management, In Press, Corrected Proof.

[14] Wedel, M, Kamakura, W (2000), Market segmentation: Conceptual and methodological foundations, Kluwer Academic Publishers, Boston.

[15] Foedermayr, E (2008), Market Segmentation in Practice: Review of Empirical Studies, Methodological Assessment, and Agenda for Future Research, Journal of Strategic Marketing, Vol. 16 (3), 223-265.

[16] Osterwalder, A, Pigneur, Y (2009), Business model generation. A handbook for visionaries, game changers, and challengers. Amsterdam: Modderman Drukwerk.

[17] Prahalad, CK, Hamel G (1990). The core competence of the corporation, Harvard Business Review, 68(3), 79–91

[18] Mansour, J (1998) Core competence: What does it mean in practice?, Long Range Planning, 31(1), 60-71

[19] Porter, M (1980) Competitive Strategy, Free Press, New York

[20] Fahey, L and Narayanan, VK (1986) Macroenvironmental Analysis for Strategic Management, West Publishing, St Paul, MN.

[21] European Commission (2010) ILCD Handbook (International Reference Life Cycle Data System), Specific guide for generic Life Cycle Inventory data sets, European Commission.

[22] ISO 14040 (2006) Environmental management - Life cycle assessment – Principles and framework

[23] Weidema B et al (2004) The product, functional unit and reference flows in LCA. Danish Environmental Protection Agency. Environmental News No. 70

[24] Porter, M (1985) Competitive Advantage, Free Press, New York

[25] Stabell, CB, Fjeldstad, Ø (1998) Configuring value for competitive advantage: On chains, shops, and networks Strategic Management Journal Vol. 19, 413–437

[26] Allee, V (2000) Reconfiguring the value network. The Journal of Business Strategy 21(4): 36.

[27] Vandermerwe, S (1993) Jumping into the customer's activity cycle: a new role for customer services in the 1990s, Columbia Journal of World Business, 28 (2), 46-65.

[28] Matzen, D and McAloone, TC (2006), A tool for conceptualizing in PSS development, in Meerkamm, H (Ed), 17 Symposium Design for X, Lehrstuhl für Konstruktionstechnik,Friedrich-Alexander-Universität-Erlangen-Nürnberg, Neukirchen, p. 131.

Integrated Evaluation of a PSS Business Case and a PSS Design Method – Application of the PSS Layer Method to an industrial Drilling Solution

Tomohiko Sakao[1], Svante Paulsson[1], Patrick Müller[2]
[1] Department of Management and Engineering, Linköping University, Sweden
[2] Division of Virtual Product Creation, Fraunhofer IPK Berlin, Germany

Abstract
A multitude of PSS design methods has been published but industrial application is sparse. The paper introduces a new case of integrated product-service offerings in the construction business. Then, we apply the PSS Layer Method developed earlier to evaluate this method further on and to publish new experiences. We show that the method is a helpful tool to describe integrated offerings, thus product-service systems, holistically. Generalized findings on what industry is not addressing enough in PSS development will be given.

Keywords:
PSS Design Dimensions; External Communication

1 INTRODUCTION

1.1 Opportunities for PSS in Construction Sector

Construction machines look "clumsy" and old-fashioned to many people, even to many engineers. However, construction solutions based on machines (see Figure 1) and related services have high potential of being designed and offered as Product/Service Systems (PSS). Construction machines typically are complex systems including mechatronic subsystems, hydraulics, tough designs, and precisely working equipment. Such systems often are mobile and supporting fast reconfigurations. Customer-oriented configurations and also pooling and sharing are not unusual and driven by the use cases. Moreover, such systems underlie complex use conditions, due to operation aside infrastructures and stable power grids. The use conditions (loads, weather, dust/dirt, etc.) are less precisely known, than in many other applications. As construction is typically carried out under tough time constraints, machine availability is highly demanded. Service like spare parts provision, system maintenance, repair, and overhaul (MRO), user training, renting, or even process data analyses are vital to secure efficient machine use. Although such systems fulfil many characteristics that are promising to implement PSS business, cp. [1], PSS design and business models (as promoted by research) are not state of the art in this sector.

Common sense in research is that Product/Service Systems (PSS) is a concept to integrate products and services in one scope for planning, development, delivery, use, and EOL (end of life) treatment, thus for the whole lifecycle. Predominantly used in academia so far, the concept now increasingly penetrates industry, even if explicitly integrated planning and development of products and services is rare. Some providers consider themselves as solution providers who essentially offer solutions including products and services. An extension of business models to incorporate PSS is an ongoing movement, like in the construction machines sector. PSS often means that a provider takes over more responsibility in product operation (the products may even remain in the ownership of the provider) and what a PSS customer actually buys or pays for is the functionality or performance of the products in a form of service. The same approach fits our investigated case in order to offer added value to the customer, cp. references [2, 3, 4, 5, 6, 7, 8, 9].

Figure 1: Examples of construction machines/applications.

1.2 PSS Design Methods' Evaluation

PSS Design and Development Processes

PSS design and PSS development process are not standardized so far. There are generic development process models in literature (e.g. [10, 11, 12, 13, 14]), but industrial practice is far away from

J. Hesselbach and C. Herrmann (eds.), *Functional Thinking for Value Creation: Proceedings of the 3rd CIRP International Conference on Industrial Product Service Systems, Technische Universität Braunschweig, Braunschweig, Germany, May 5th - 6th, 2011*,
DOI 10.1007/978-3-642-19689-8_28, © Springer-Verlag Berlin Heidelberg 2011

explicitly integrated development processes of products and services. Industry implicitly performs product and service packaging in operation and sometimes by implicitly related products and services already in development. Missing is a holistic system model including the business model, products, service processes and resources, environmental infrastructure, contracts, customer needs and finally values provides / implemented.

PSS Design Methods

In PSS research we are facing a continuously and fast growing variety of PSS design methods. This fact creates a need of consolidation in order to make communication and dissemination on PSS research and practice more convenient and easier [15]. This is especially crucial for practitioners in industry. For instance, a consolidated model would facilitate them to learn from other cases with more ease. However, at the same time, it would be effective also for researchers to share different concepts on PSS design among different research schools.

In an earlier attempt by the authors [15], a PSS system model (meta-model) was defined based on the PSS Layer Method [7] so that it is suitable for different levels of abstraction and is applicable to design and evaluate PSS. The PSS Layer Method has been applied successfully to a PSS case in the energy sector [7] and some other cases. For a profound validation there is a need to validate the PSS Layer Method further in more cases and on more contexts.

Target

PSS is modelled as a marketable offering. This is a crucial way to view PSS. Thus, this paper investigates the applicability of the PSS Layer Method to model a PSS offering communicated between a provider and its customer. An example is taken from the real market and provided by a Sweden-based manufacturer. The first aim is to evaluate its usability to communicate the case concept internally and externally. The second aim is to briefly illustrate a new real PSS case in the research domain.

1.3 Research Method

Approach

A PSS in construction has been chosen due to the substantial difference to the previous sectors of the PSS Layer Methods' application. The case is a PSS provided by a Swedish manufacturing company offering construction machinery and the related services. This case will be used as a reference example to discuss advantages and shortcomings of the PSS Layer Method. With these conditions, the followings are set as particular research questions of this paper.

RQ1. Are the nine dimensions of the PSS Layer Method sufficient and necessary to cover important concepts of a PSS?

RQ2. Is the granularity of the dimensions of the PSS Layer Method appropriate?

RQ3. Is the PSS Layer Method effective in representing a PSS?

Materials

The materials utilized in this case study are a 47-page brochure and other detailed materials of the services associated with a rock drilling equipment and their performance available internally from the provider.

2 PSS LAYER METHOD - BASICS

This section gives an overview of the PSS Layer Method [7]. Nine obligatory design dimensions (dim.) are used structure the following design rationale: Customer needs (dim. 1) are satisfied by customer value (dim. 2), which a customer perceives. Such value has to be generated by deliverables (dim. 3), which creates value for a customer. The deliverables are results of delivery processes, i.e. lifecycle activities (dim. 4). To implement a chain of lifecycle activities resources are needed. Actors (dim. 5), core products (dim. 6) and periphery (like IT infrastructure or public transport systems) (dim. 7) are such resources. Contracts (dim. 8) frame the entire value creation process, including billing (dim. 9), offerings, and finally the entire business model.

The following subsections describe all nine dimensions in detail.

Customer needs (customer view)

This dimension summarizes customer needs. The idea is to capture non-solution-oriented needs, for instance the need for access to broadcast information. Nevertheless some needs will be solution or context related. The need to operate a TV set and radio with a solar home system is an example [7]. This dimension does not contain requirements and specifications, which are descriptions towards how a system function has to be designed, cp. [7].

Customer value (customer view)

"Value is what I get for what I give" according to one finding which Zeithaml retrieved based on an empirical study [16]. We assume that from an economic viewpoint, the value can be expressed as monetary benefit in the end. However, a differentiation of value types is helpful to show differences in how PSS ideas and concepts work to meet customer needs. In our view, the customer value is equal to the benefit a customer gains by a deliverable. The major four types of benefits are economic, environmental, social, and technical benefits (e.g. saved time). Less precisely defined benefits are information and knowledge advantages, health preservation, protection or enhancement, prestige, or advanced process robustness, agility, flexibility etc. Briefly, the protection and enhancement of a customer's market position belong to this dimension. To link such values to customer satisfaction, scales and target ranges should be defined for each value. (Using the example above, the bandwidth and the time range to access broadcast information can be defined, measured, and compared to the customers' actual state, in order to capture the customer value and the satisfaction of his need.)

Deliverables

Deliverables is what the PSS provider delivers to its customer. Deliverables can be material or immaterial. Technical artefacts, software, information, or knowledge are the main deliverables. It is important, that a deliverable is a result of an activity or an activity chain, which can be interpreted as part of a service or business process. Not every deliverable has value for a customer. For instance, the delivery of out-dated information might be contra-productive for a customer. Thus, it is important to differentiate between deliverables and customer values.

Lifecycle activities

This dimension contains activities performed by the PSS provider and/or the customer. As afore mentioned, activity chains result in deliverables or change or optimise already existing deliverables.

Actors

Single actors (players), stakeholders, enterprises and enterprise units or divisions as well as even software agents are classified as actors. Actors participate in activities and have an aim and a perception of delivered values. Software agents also have aims and interact although they are not physical. Allowing software agents as actors might be important in case of replacing a manually executed service by a technical artefact communicating with a providers' software agent platform.

Core Products

Products, which have to be designed or at least offered in a package by the PSS provider, are captured by this dimension. The most important aspect of core products is that they have high relevance for the final PSS value generation. Products, which have to be designed, adopted or configured viz. those where conventional engineering tasks have to be performed by the PSS provider (network), are captured within this dimension.

Periphery

Support equipment, technical periphery, tools, infrastructure, or PSS execution systems, which are type of a platform, outer condition, support, or constraint for the PSS delivery, should be captured in this dimension.

Contract

As the contract is one basic element of the PSS, remarks on the contract design should be made early, to detail out the business model. Examples are implementations of obligations, options, exception handling, duration, fines, regulations of payment, take-back conditions, warranty, transfer of ownership, responsibilities etc. An offering has the same meaning.

"Finance" (Billing, Revenues, Monetary dim.)

This dimension shows when a customer is paying for deliverable(s) and how much. For instance, flat rates, pre-paid, scheduled or incremental (down-)payment and payment on tickets may be possible. (The name of this dimension is finally not fixed and thus set in ticks. In a software prototype, this design dimension was earlier called "Finance".)

3 CASE STUDY

The main aim is to investigate whether the nine dimensions of the before introduced PSS design method are necessary and sufficient suitable to describe and document a PSS for and while its planning and development. The PSS investigated is already launched on the market. Thus, the existing PSS was kind of reverse engineered and the outcome use to validate the PSS concept and to verify the documents, assets and processes prepared to perform the PSS concept in the market.

Case Description – The Company

The company investigated develops, manufactures and delivers drilling equipments for e.g. construction business. Training, spare parts delivery and maintenance, repair and overhaul are part of the company's service portfolio. The PSS investigated was among their major services provided. Instead of selling physical products alone, spares and additional services, the company designed and offered a service offering in addition to a physical product, a drilling machine. The authors took details from a marketing brochure and some additional internal information on that service offering and modelled it with the nine design dimensions defined by the so-called PSS Layer Method (see Müller et al. in [7] and [15]).

Case Description – The PSS Offering

The motivation of this company to provide PSS is to create higher value for customers/users. In providing PSS, the company recognizes possibilities of improvement and attempted to apply the Layer Method to their offering. Table 1 shows a description of the offering using the nine dimensions. It can be traced from this table that this company provides additional services on top of a drilling machine as a core product with a customer who wants to increase the performance. Figure 2 is an overview of a portion of this offering represented in a schematic figure, while its small part is enlarged in Figure 3 for better readability.

Important information on the marketing brochure could be described on any of the dimensions. For instance, productivity of the equipment user, an element of the value, as well as warranty of quality, as a part of deliverables, are crucial to recognize in this PSS. In addition, "pay per uptime" as a way to charge on the service part is also important to grip the service concept. On the other hand, every dimension was used to represent a part of the PSS. This means, in responding to RQ1., that the nine dimensions are sufficient and necessary to represent the information of the PSS communicated externally with customers. Thus, the result of this case study was found to serve as another evidence of the nine dimensions being useful to represent important elements of a PSS.

To answer the question if the nine dimensions are sufficient, necessary and suitable in more detail we are conducting interviews with several experts in the company. As for RQ2., the granularity of the method was also found good.

In addition, the authors found that describing some more specific information of the relations between different dimensions is not facilitated. For instance, the information about the task allocation of lifecycle activities between different actors would support better clarification. This was uncovered on the brochure about the concerned service, especially because some commitment was demanded for a customer to receive some of the deliverables in this service. E.g. daily and weekly service procedures of the drilling machine are supposed to be achieved by an end user. I.e. those activities by a drill user are included in a contract with the provider, and the user must carry out such a daily check of drill conditions in order for the user to receive the service. With the current PSS Layer Method, these activities are described in a mixture of those by the provider. By implementing the capability of describing information of the relation between the actors and the lifecycle activities, the method would be improved. In the first software tool prototype, some features to represent such relations are included. The next version of the software is required to support such relations and a breakdown more sophisticated.

In responding to RQ3., the form of a table is hard to get the connection among different information represented on any of the nine dimensions. For this purpose, the graphical way of representation is beneficial. This will solve the problem mentioned in the previous paragraph at the same time by showing the links between an actor and its lifecycle activity. Nevertheless the tabular representation is helpful to capture PSS ideas before going into detail with the graphical representation. This is helpful to address different levels of design abstraction in early development phases.

Furthermore, it enables a more detailed discussion of lifecycle phases, episodes, and activities. Related resources, contract sections, deliverables, and customer value become more easily traceable. For elements modelled in the graphical representation, design and quality requirements can be retrieved in a methodical way.

Customer Needs	• High drilling performance • Focus on its core business • Easy projection of its own cost
Customer Value	• Productivity and availability • Saved MRO (maintenance, repair and overhaul) time • Reduction of drilling interruptions, • Reduction of spares stock (just "fall back spares" to stay agile in case of delivery problems, highly reliable availability level) • Product quality • Easy commissioning • Reduction of cost • Higher delivery precision • Information: Recommendations on operation of drilling equipment to reach high performance • Precise matching to needs
Deliverables	• Drilling equipment • Warranty of quality • Cleaning equipment • Original spare parts in time • Early information on next MRO activities • Grease and oil in adequate quality • Service Binder
Lifecycle Activities	• Focus on use phase and there on drilling and MRO; early fault detection (daily and weekly service procedures of the drilling machine), • MRO prognostics and MRO execution including scheduling • Transport of spares to the field and take back of rotatables and broken parts
Actors	• Drill owner • Drill provider • Service technician • Drill operator
Core Products	• Drilling system • Monitoring solution • Original manufacturer provided spare parts to fulfil contracted MRO activities • Service Binder
Periphery	• Grease and oil • Logistical / Transport systems to put spare parts in place • Monitoring and IT infrastructure to send machine data to remote monitoring systems
Contract	• Performance contracting (availability as measurement criterion)
Finance	• Pay per uptime / drilling hours for the service part (i.e. excluding the drilling machine)

Table 1: Elements of the concerned service described with the nine dimensions of the PSS Layer Method.

4 DISCUSSION

Based on the case study, we showed the PSS Layer Method is suitable in representing a PSS. The improvement by describing the relations between different dimensions pointed out above could be supported by the important characteristic of PSS: the mixture of actors in lifecycle activities is quite common in PSS, since co-creation by a provider and a user/customer is an important characteristic of PSS. For the moment, the lifecycle activities are modelled in a linear sequence in the graphical representation of the method with simplification. One can go into detail, modelling a layer plot for each episode or activity chain in a lifecycle. This helps to partition the lifecycle and to investigate critical phases, episodes, or activities. A software tool is under development to support this. What was found from this case study is regarding the information of a final solution, which is a result of a design process. The final solution required the nine dimensions to be used. Considering that the information of the solution was created somewhere in the whole design process, the nine dimensions are addressed somewhere. Finally, it is of interest, which dimensions are addressed with more importance at different stages of the whole PSS development.

5 CONCLUSION AND FUTURE WORKS

Through applying the PSS Layer Method to one PSS case provided by a manufacturer, this paper showed the method is effective in representing important information of a PSS communicated with a customer/user. Concretely, the nine dimensions of the method were found to be sufficient and necessary to cover important elements of the PSS. It was also shown that a graphical representation of the information, especially about the links of different elements, facilitates a representation that is more specific. One benefit of the method confirmed by this application is the ability to grasp the whole set of important information of a given PSS. In other words, this is powerful in organizing relevant information of a PSS. Immediate future work is getting insights about the information addressed during the development process from practitioners of the provider in our case. Another future work is asking the same questions as RQ1. by investigating other PSS cases described with other schemes in literature.

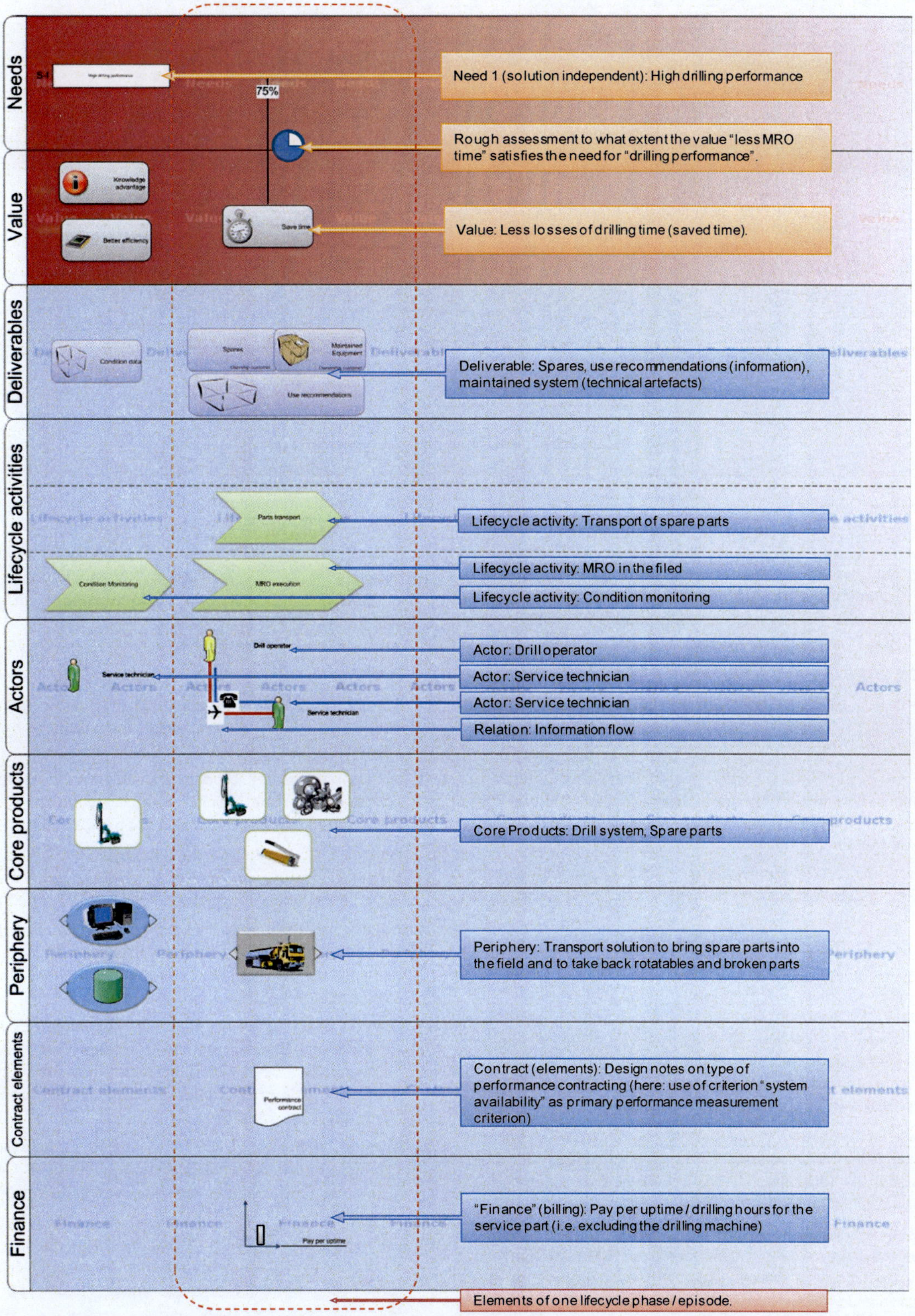

Figure 2: Extract from a software supported PSS Layer Model for the concerned service: Lifecycle oriented representation.

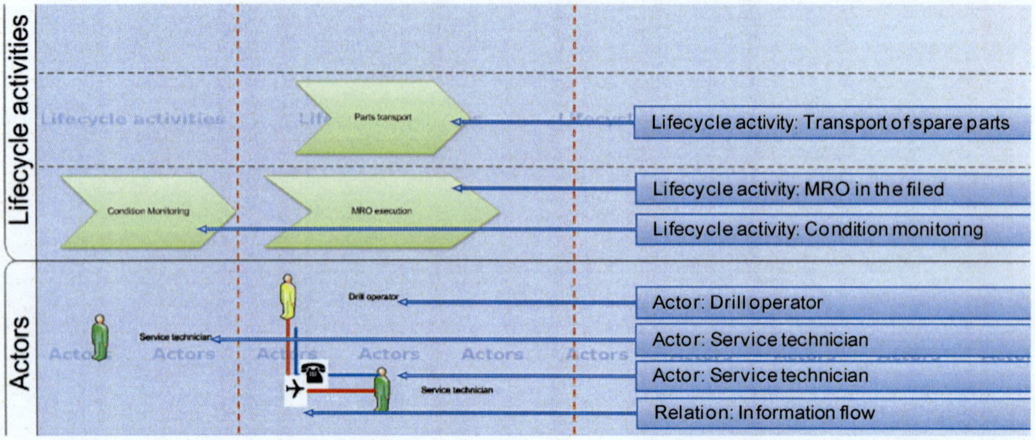

Figure 3: Model extract illustrating multiple actors and activities.

6 ACKNOWLEDGMENTS

We express our sincere thanks to all who contributed to the findings captured in this paper, especially to the PSS-providing company. We thank our colleagues from the International PSS Design Research Community [17], a.o. from Linköping University for valuable discussions. Finally, we thank all readers.

7 REFERENCES

[1] Müller, P. (2011): Opportunities for enhanced Project Management in PSS Development, in: 3rd CIRP IPS² Conference, Braunschweig, in print.

[2] Tomiyama, T. (2001): Service Engineering to Intensify Service Contents in Product Life Cycles. Second International Symposium on Environmentally Conscious Design and Inverse Manufacturing, Tokyo, IEEE Computer Society.

[3] Mont, O. K. (2002): Clarifying the concept of product–service system, Journal of Cleaner Production. Vol.10, No.3, pp. 237-245.

[4] Baines, T. S., Lightfoot, H. W., Evans, S., Neely, A., Greenough, R., Peppard, J., Roy, R., Shehab, E., Braganza, A., Tiwari, A., Alcock, J. R., Angus, J. P., Bastl, M., Cousens, A., Irving, P., Johnson, M., Kingston, J., Lockett, H., Martinez, V., Michele, P., Tranfield, D., Walton, I. M. Wilson, H. (2007): State-of-the-art in product-service systems, Proceedings of the Institution of Mechanical Engineers - B. Vol.221, pp. 1543-1552.

[5] Matzen, D., Sakao, T. Sandström, G. Ö. (2007): Design Research on Manufacturing Firms Moving from Products to Services, in: 16th International Conference on Engineering Design, CD ROM, Paris.

[6] Ericson, Å., Müller, P., Larsson, T. Stark, R. (2009): Product-Service Systems – From Customer Needs to Requirements in Early Development Phases. CIRP IPS² Conference, pp. 62-67, Cranfield.

[7] Müller, P., Kebir, N., Stark, R. Blessing, L. (2009): PSS Layer Method - Application to Microenergy Systems, in: Introduction to Product/Service-System Design. T. Sakao and M. Lindahl Eds. pp. 3-30, Springer, London.

[8] Sakao, T. Lindahl, M., Eds. (2009): Introduction to Product/Service-System Design. Springer's global publishing programme in engineering and management. Springer, London.

[9] Salvendy, G. Karwowski, W., Eds. (2009): Introduction to Service Engineering. Wiley, New Jersey.

[10] Morelli, N. (2003): Product-service systems, a perspective shift for designers: A case study: the design of a telecentre, Design Studies. Vol.24, No.1, pp. 73-99.

[11] Alonso-Rasgado, T. Thompson, G. (2006): A rapid design process for Total Care Product creation, Journal of Engineering Design. Vol.17, No.6, pp. 509 - 531.

[12] Aurich, J. C., Fuchs, C. Wagenknecht, C. (2006): Life cycle oriented design of technical Product-Service Systems, Journal of Cleaner Production. Vol.14, No.17, pp. 1480-1494.

[13] Sakao, T. Shimomura, Y. (2007): Service Engineering: A Novel Engineering Discipline for Producers to Increase Value Combining Service and Product, Journal of Cleaner Production. Vol.15, No.6, pp. 590-604.

[14] Sakao, T., Birkhofer, H., Panshef, V. Dörsam, E. (2009): An Effective and Efficient Method to Design Services: Empirical Study for Services by an Investment-machine Manufacturer, International Journal of Internet Manufacturing and Services. Vol.2, No.1, pp. 95-110.

[15] Müller, P. Sakao, T. (2010): Towards Consolidation on Product-Service Systems Design, in: CIRP IPS² Conference, T. Sakao, T. C. Larsson and M. Lindahl Eds. pp. 219-225, Linköping.

[16] Zeithaml, V. A. (1988): Consumer Perceptions of Price, Quality, and Value: A Means-End Model and Synthesis of Evidence, Journal of Marketing. Vol.52, No.July, pp. 2-22.

[17] The International PSS Design Research Community. Retrieved Nov. 15, 2010, from **www.pssdesignresearch.org**.

Simulation of Life Cycle Costs of a Product Service System

Gisela Lanza[1], Benjamin Behmann[1], Patrick Werner[1], Stephan Vöhringer[1]

[1] Institute of Production Science (wbk), Karlsruhe Institute of Technology (KIT), Karlsruhe, Germany

Abstract

Concerning Industrial Product Service Systems (IPS^2) in machine engineering, the challenge for the supplier exists in the determination of the expenses for the provision of the included services over the life cycle of the IPS^2 before the start of contract. Thus, in this paper an approach is presented which explains how a supplier of an IPS^2, which consists of a machine together with a warranty of availability, can determine his expenses for the warranty. In doing so, the underlying actual operating- and load-dependent failure behavior is taken into account to estimate the supplier´s serviceability to provide the warranty.

Keywords:

Serviceability; Life Cycle Management for IPS^2; Reliability of Echnical Systems

1 MOTIVATION

In German plant and machine engineering, the technical service provided by the manufacturers of investment goods plays an important role as it contributes a great share to the corporate revenue. As can be seen from Figure 1, the services spare parts supply, repair, failure elimination and maintenance are the most profitable.

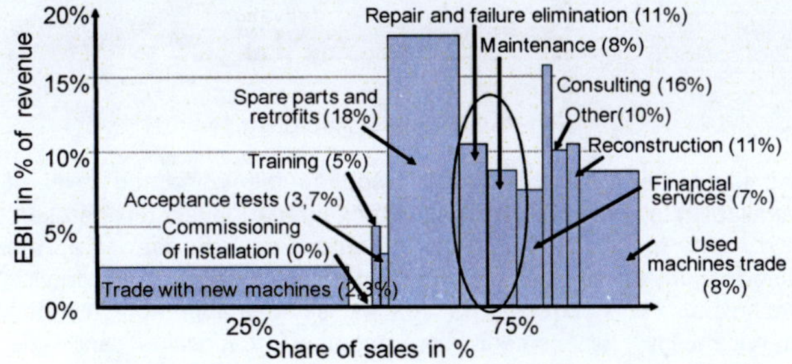

EBIT Earnings Before Interests and Taxes

Figure 1: EBIT of technical services [1].

The gross earning of the company per service product or sale of new machines can be read off the surface of the grey fields in the figure. Thus, it is to state for the record that the trade with new machines constitutes a sectoral average of over 30% of the turnover. However, this portion of the turnover produces earnings before interest and taxes of only approximately 2,5%. Compared to the services provision of spare parts and maintenance, which comprise significantly lower turnover proportions, the operating result, however, is considerably improved due to both services. From an economical point of view a combined offer of both product types (machine and service) is indicated if additional synergetic effects are considered between the single product types. Thus, experiences obtained in service business can be useful to optimize hardware and accordingly new technologies or machines are again useful to develop new service products.

If there are technical services offered in combination with a capital good as inseparable unit one refers to it as a hybrid product or Industrial Product Service System (IPS^2). By the fusion of goods and services to inseparable units in mechanical engineering suppliers of hybrid products are confronted with new challenges. One challenge exists in the fact to provide the technical service included in the hybrid product in the promised quality also over the period of performance, e.g. during the whole life cycle of the machine or plant, and to forecast the expenses for immaterial product components. This life cycle of the hybrid product is defined as product service life cycle in the following.

An economically significant hybrid product example in mechanical and plant engineering is the sales of machines combined with a promise of performance, e.g. technical availabilities of 96% and higher. For this purpose, the machine manufacturer has not only to be able to design and produce machines in high class quality but also to forecast the reactions of components under possible operating and load conditions over the life cycle. As to the fabrication of wind power plants promises concerning the availability of the plants are common. In addition to that, long-term service contracts with duration of 2 years and more are sold together with the wind power plant [2].

Regarding the cost types, which occur during the operating period of the wind power plant, the costs for maintenance and repair (servicing) cause 26% of the total life cycle costs of wind power plants (Figure 2).

J. Hesselbach and C. Herrmann (eds.), *Functional Thinking for Value Creation: Proceedings of the 3rd CIRP International Conference on Industrial Product Service Systems, Technische Universität Braunschweig, Braunschweig, Germany, May 5th - 6th, 2011*,
DOI 10.1007/978-3-642-19689-8_29,

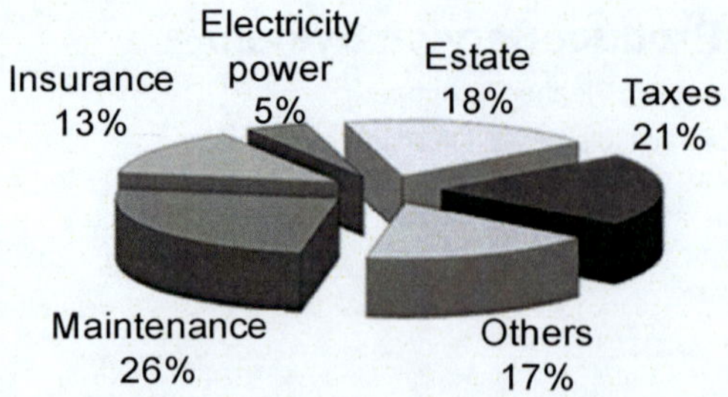

Figure 2: Life cycle costs of wind turbines [3].

The cost types "Others" are split into personnel costs for the company and contributions to trade associations and lobbies. In this context it is important to note that the operating costs do not incur in a linear manner over the life cycle of the plant but, especially the costs for maintenance because of deterioration, are subject to the annual growth rates, which are included in Figure 2. From this it follows that the operator of a wind farm or of a machine tool tends to be interested in a manufacturer's warranty about the benefit of the capital good in the form of a promise of performance over the life cycle of the technical system. One possibility exists in the transaction of a hybrid product which consists of the capital good and a promise of performance. Within this framework the supplier of the hybrid product has to provide for the technical service under consideration of the operating and load influences at the capital good in order to keep the promise of performance within the scope of the product service system.

This brings about the challenge that the promised performance and the service delivery to achieve the performance is time shifted. Therefore, it is impossible for manufacturers to determine potential service expenses in future accurately. Furthermore, the random, load-dependent failure behavior of the mechanical components is generally not factored into the calculation, which additionally increases the uncertainty with regard to potential service expenses. Given the yet inadequate failure prognosis in services, there is still a lot of room to optimize the efficiency of resources employed in order to provide the services. These findings were confirmed by a survey of economic potential looking into technical services of international operators and suppliers of machines and plants [4]. In conclusion, the outcome of the study states that customers require an improvement of the quality sustainable services and of the profitability of services. The resulting challenge consists in the development of a suitable approach which ensures that the promised performance will be delivered taking into consideration the random load-dependent failure behavior of mechanical components and ensuring the cost-optimized employment of resources for the provision of the maintenance service [5].

2 STATE OF THE ART

Previous approaches in the field of Industrial Product Service Systems give their attention to the development and optimization of service engineering (e.g. [6]) or business models (e.g. [7]) from a strategy point of view. Additionally Nesges researched the relationship between different service products and their influence on the technical availability of machine tools [8]. Duffuaa developed a simulation for determining the maintenance staffing in an industrial environment. Hereby Duffuaa focused on maintenance teams for the shop floor in one production line without integrating the relationship between load-dependent reliability and service resources [9].

Two further research approaches in the field of resource planning of IPS^2 are developed by Krug [10] and Vasen [11]. Krug describes in his work, a method to calculate the needed resources (e.g. service technicians) to provide different service products. He developed a method which is adaptable to a variety of products and services. His method uses a division of an overall system (e.g. machine) into sub-systems to identify the influence factors and to derive the resource requirements [10].

The approach of Vases with the focus on resource planning for the technical customer service results in a service action plan which has to be configured with physical and mathematical relationships to determine the resources for a specific reference [11].

3 OBJECTIVES AND GENERAL APPROACH

This paper presents a method which describes how an IPS^2, can be simulated over the life cycle of a machine. The provider of an availability oriented promise has to be enabled to forecast the costs which occur over the term of a contract in order to be able to keep the promise of performance. Furthermore, the method has to give information about the serviceability of the organization of the supplier and to allow conclusions to which extend changes in the organization (e.g. recruitment of a new service employee) have effects on the overall costs. Figure 3 gives an overview of the general approach which has been developed together with an industrial partner to assess his serviceability.

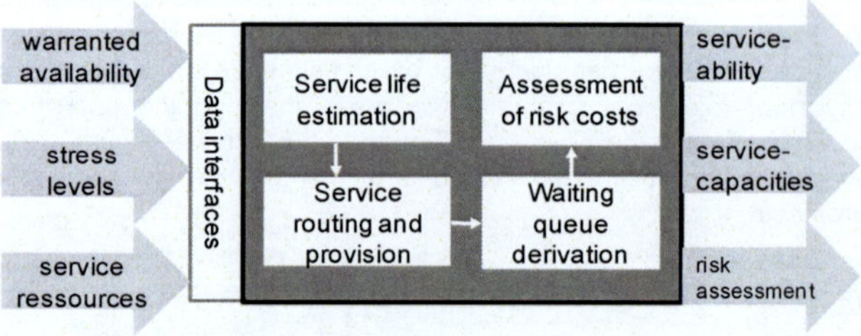

Figure 3: General Approach.

As input the general approach requires the warranted level of availability and the continuously changing load levels of operation, e.g. with the use of intelligent components. Furthermore, the information about the present service resources, e.g. service technician, is needed. The results of the approach are the serviceability assessment by smoothed capacities and the probability for breaching the warranty. The methods to determine the results are explained in detail in the following.

The first used method of the approach represents the service life estimation how under consideration of the former real load of a component and the forecasted load the remaining service life and service life time of the component can be estimated. This method connects the statistical examination of failure behavior by means of a Weibull distribution with the results of the diagnostics of the machine condition (how the concept works in detail compare chapter 4.1). As a benefit of these combined methods, diagnostic together with stochastic reliability, the pre-warning time of a failure can be extended to pool and route the service orders in order to reduce the service costs. For that reason a route optimization was implemented which regards the local distances and the failure risk at the same time.

If a service order cannot be dispatched due to the lack of a service technician, the service order has to wait in a waiting queue for service provision. Because of different failure behaviors of the components behind the service orders a waiting queue discipline considers the criticality of an imminent failure. Parallel, during the waiting time, the risk for a failure increases and is summed up in the risk cost model. The risk cost model contains besides the maintenance costs (e.g. personnel costs, spare part costs, and travel costs) also an assessment of the costs which could incur possibly because of a production stop. As result the supplier receives an approach for the assessment of the expenses for an IPS^2 by means of a cost-optimized maintenance service taking into consideration the operating and load conditions of the critical components.

In addition, the approach was implemented by simulation to determine the serviceability and to illustrate the results. Due to a regular extension of the original data by current operation conditions of the considered components the precision/accuracy of the results can be improved. To test the simulation results for their robustness in reality, unpredictable failures were additionally generated within the simulation and dispatched to maintenance. The time and cost parameters which arise until a steady state is reached again give information about the capacity of service organization of the IPS^2 supplier.

4 ASSESSMENT OF THE SERVICEABILITY

The approach for the determination and later simulation of serviceability starts with an application of those instruments which **estimate the components´ service life** (Figure 4). The results are an identification of the components which lead in the case of failure to a breakdown of the entire technical system. The background is that only such critical components justify preventive servicing. All other components are provided by planned servicing within the model.

The parameters and results taken from service life time estimation are continuously adapted to the actual conditions in the field (**Simulation of the serviceability with present conditions**) and that not only the results of an examination concerning reliability but also the optimal provision of service building on that in order to keep the promise of performance.

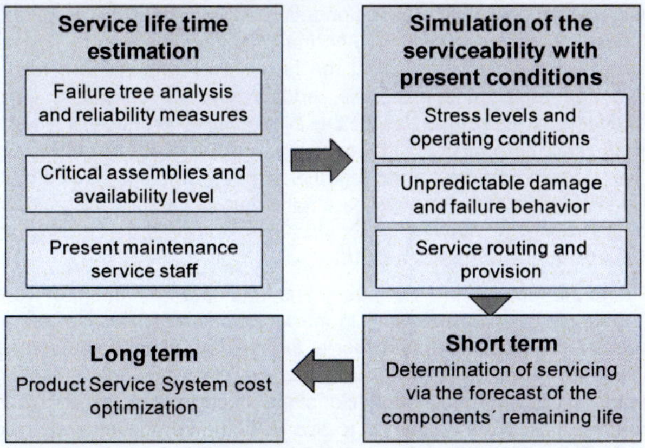

Figure 4: Course of Action.

4.1 Service life time estimation

In order to determine the servicing dates, which are necessary in order to keep a promise of performance, the failure behavior of the most important components has to be described as accurately as possible in order to estimate their remaining service life. At first these components with a high failure potential have to be identified, i.e. components whose failure leads to a breakdown of the whole machine. The identification is carried out by means of a fault tree analysis [12] which allows to divide the entire system in single subsystems and to define for each subsystem a conclusion concerning reliability and criticality of failure.

Then, the effects which determine the service life of the single components are determined. As to mechanical components these effects can be mechanical (e.g. forces, torques), chemical (corrosion) and thermal. The effects, which affect the life time of components significantly, can either occur apart or in a combined manner. In addition, they are not constant over the service life.

For the critical components the failure parameters are estimated by means of a statistical reliability analysis. In this approach, the Weibull distribution is used as statistical distribution, which is the most applicable instrument to describe the failure behavior of machines´ components [13]. The Weibull distribution is extended by the general log-linear model in order to be able to consider several parallel loads and by the cumulative damage model in order to be able to consider temporally unsteady loads [14, 15].

Using these models the remaining service life distribution can be calculated and adapted to the actual load conditions by load monitoring in the components. This remaining service life distribution indicates the conditional probability for the fact that a component survives a certain time in the future under a projected load on condition that it has withstood during a known period of time under (at least partial) known loads. Even if only indirect loads are known, e.g. wind speed at a wind energy plant, this model can possibly help to improve prediction accuracy for the remaining life cycle.

Due to the fact that the system is able to calculate the remaining service life distribution at any time during the life cycle, it allows a long-term planning of economically reasonable servicing dates. Since the model does not consider the real state of the component directly but only the external factors which have an effect on the state it is completed by the use of condition monitoring measures for short-term servicing. The servicing dates caused by these systems are automatically prioritized compared to the dates determined by the reliability model.

Besides, for the determination of the servicing dates it is important to know which effects an unplanned failure has, especially in comparison to a planned servicing which would prevent this failure. For this purpose, all costs for the unplanned failure must be included. Besides labor and spare part costs, which are mostly the same in case of both types of failure, production failure costs and the costs incurred by consequential damages at components or concerning the environment have to be considered. The higher the additional costs of an unplanned failure are compared to a planned servicing the lower is the probability of failure which the supplier of the product service system is willing to tolerate before the preventive measure is carried out.

Along with the information regarding current service capacities the serviceability of the supplier can be determined afterwards taking into account the service processes in order to forecast the expenses of the IPS^2 until the end of the operating time and to optimize them continuously by means of an improved data set.

4.2 Simulation of the serviceability with present service operation conditions

By the simulation of serviceability the supplier receives a basis for decision making within a short time. This basis helps to find out how the technical service, bearing in mind the remaining service life of the components, should be provided and how the service capacities should be distributed to the cases of servicing. Proceeding in this way, the supplier is able to schedule and balance his service capacities in time as well as he can track his expenses continually within the Industrial Product Service System.

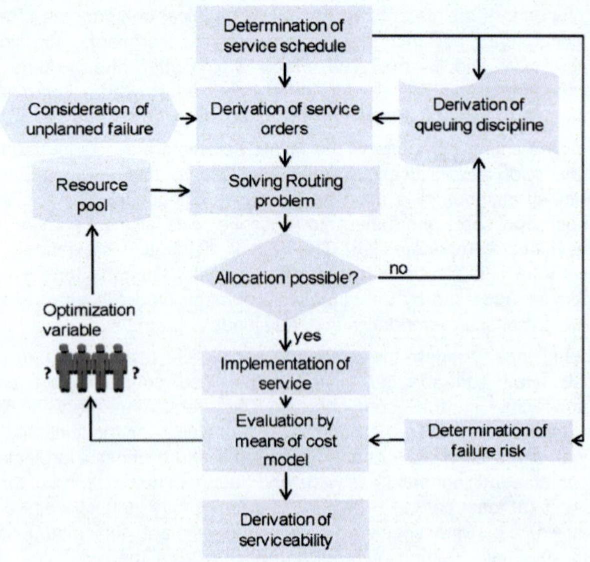

Figure 5: Simulation Flow Chart [5].

The simulation flow starts with the derivation of the service schedule. This concept uses the results of the service life estimation and considers the occurrence of unpredictable failures due to human misuse. After that the service schedules are extended by information about location of the customer, needed tools for service provision, drawings etc. to derive service orders.

If there are service orders which cannot be allocated because of a lack of resources, e.g. spare parts or service technicians (Figure 5), they have to wait. The consequences can be machine failure during the waiting period (case 1), or an increase of the failure risk of the components for which the service order was created as a result of the delay (case 2). Case 1 will be handled and entered into the system as an unplanned failure, whereas case 2 will be dealt with through the integration of a queuing discipline that defines the queuing sequence for the service order clusters that are waiting.

The parameter used to determine the waiting queuing discipline is the assessed failure rate λ(t), which represents the measure for the risk with which a component is going to fail which has survived until the time under consideration. The failure rate itself is defined as the relationship of the failure density function f(t) and the survival probability R(t). Both functions are calculated analytically on the basis of the dynamic lifetime distribution for the determination of service times (4.1). The failure rate for each component, while the service order cannot be allocated, assessed with the imminent production loss because of a failure represents the waiting queuing discipline. By comparing assessed failure rates not at the time of the planned service operation but one time unit after that, a queuing sequence is established that gives higher handling priorities to higher assessed failure rates. As sometimes several orders in several clusters must be compared by means of the failure rate and imminent production loss, the most urgent order of every cluster is used for the comparison.

If there are waiting times because of resource bottlenecks, the impending risk costs, the probability with which they will be incurred and the direct service costs must be identified. Given that the direct service costs can generally be easily identified ex post, the focus is placed on the impending risk costs. Risk costs μ are determined as an expectancy value that results from the failure probability under consideration of operating and load influences. They are then multiplied with the production downtime costs, which are identified by means of the contribution margin reduction C method (1-2):

$$\mu = \int_{t_{ist}}^{t_{soll}} f(t \mid t_{soll})(t - t_{soll}) \cdot C \, dt \tag{1}$$

In formula (1) μ describes the expectation of the production failure costs during the downtime of the machine caused by a not operated service order. In the process, the variable C considers the failed production turnover multiplied by the contribution margin subject to the remaining capacity. The remaining capacity defines itself from:

$$remaining\ capacity = 1 - \sum_{maschine} \frac{t_{actual} - t}{net\ time\ of\ usage} \cdot utilization \tag{2}$$

If, for example, on half of the time of the production shift a failure occurs which lasts for the rest of the working time, the remaining capacity amounts to (according to formula 2) 0,5.

Finally, the risk costs and the direct costs are totaled in order to optimize personnel costs and in order to be able to assess the serviceability.

5 SIMULATED RESULTS WITH INDUSTRY

After the development of the approach to simulate a specific product service system over the life cycle the simulation was tested based on the operator data of a machine tool manufacturer. Actually, it is about long-established production processes concerning machine tools, which are established in the market as well as to the customer and which are not expected to fail because of start-up difficulties.

In order to calculate the service lifes the load and service data of two machine tools identical in construction and operation conditions were analyzed. This included the determination of the failure behavior of the critical components and the description of the service provision with a Log-normal distribution. The field data contains the description about the failure time and reason, repair time and replaced spare parts. Additionally, the field data were extended by information about the cycle time to produce one work piece, shift time model, units per shift and the costs of production loss. With this field data set together with the present service staff the simulation software was parameterized, especially the method to determine the service schedules, the service order routing as well as the present service staff.

Due to the fact that the field stress and load data from the past were incomplete, it was necessary to introduce different load classes to forecast the breakdown behavior of the critical components over the simulated product life cycle. Because the machine tools are used under similar operation conditions the documented failure times from the past were combined to ascertain the components specific Weibull distribution to derive the service schedule in the future.

By means of presumed load classes over the product service life cycle of one year at a warranted technical level of availability of 98% the serviceability was simulated by a Monte Carlo simulation based on the failure critical assemblies. The exemplary results are illustrated in Figure 6.

Serviceability measured in service costs per assembly in the field describes the relation between the extent of service organization and the technical systems to be supervised in the field. In addition to the failure behavior of the system the provision of service was taken into consideration and that bearing in mind the model concerning the waiting queue described in chapter 3.2 as well as taking into account the cost model. In doing so, the cost model comprises the direct service expenses and the pending risk costs if the technical service is not provided in time. By means of serviceability an allocation of both determinants population and resources can be found at optimal costs for the special hybrid product consisting of machine and promise of availability.

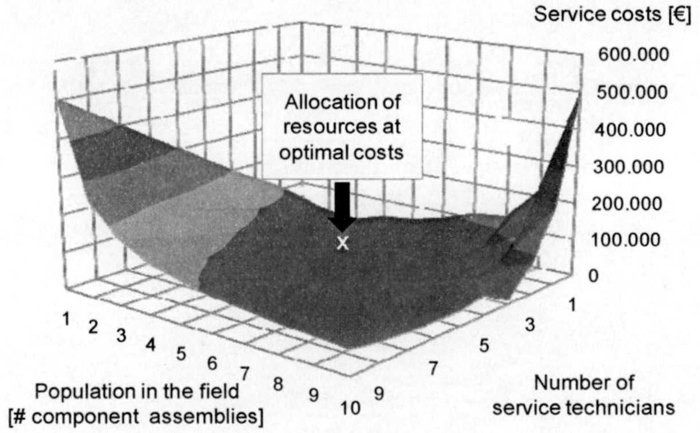

Figure 6: Serviceability measured in service costs per assembly for a period of the IPS^2 life cycle.

6 SUMMARY AND OUTLOOK

Industrial product service systems are economic products in German mechanical and plant engineering and thus especially attractive for suppliers if expenses of the product service systems can be calculated over their life cycle by the suppliers. In addition, the manufacturers combine and sell with their investment goods different technical services to integrated product service systems, e.g. the combination of a machine tool hand in hand with ambitious performance oriented service promises. The suppliers of these product service systems are faced by the challenge to assess the expenses for integrated technical service over the product service life time at the beginning of contract. On the one side this cost assessment is determined by the operation- and load- dependent failure behavior of the components, on the other side also by the availability of resources for the provision of services.

In order to set a price, the service organization and the resources required to provide the quality sustainable services must be planned as detailed as possible. Hence, the paper presents a method to estimate the manufacturer´s serviceability. The serviceability includes the cost optimal allocation of service technician and population of technical units for achieving the assemblies´ technical availability level with regards to the stochastic failure behavior of the assemblies. In order to improve the accuracy of the stochastic failure prediction, the method continually adds load-dependent data of load monitoring systems on statistical reliability distribution to schedule maintenance service adapted to stress impacts.

The results based on the analysis of real field data in cooperation with a manufacturer for machine tools has shown that such a relation exists and an allocation of both determinants – population and service resources – can be determined at optimal costs. Thereby, this allocation describes the service effectiveness of the supplier of the product service system. However, in order to verify the overall conclusion it is necessary to apply the method over a longer period of time in practice. Thus, simulation can be tested with additional real data. This requires that the flow of information and data between manufacturer and customer is installed via standardized interfaces into simulation to increase the level of automation significantly and to reduce the manual effort at the same time. One possibility exists via wireless transmission of current machine conditions within shop floor tours or via remote technology to a central service control station.

7 ACKNOWLEDGMENTS

The survey of economic potential looking into technical services and the development of the above described approach were carried out as part of the research project "Knowledge transfer and the preparation of basics for technical services in machine and plant engineering" (WiTal) which is funded by the Baden-Württemberg Stiftung und supported by the German Aerospace Center. The Institute of Production Science highly appreciates and acknowledges their commitment.

8 REFERENCES

[1] Mercer Management Consulting (2003): Umfrage unter 200 Unternehmen des Maschinen-Anlagenbaus.

[2] Sched, W., Pfeiffer, K. (2004): Wissensbasierte Instandhaltung bei Windparks, in: e&i Elektronik und Informationstechnik, Vol. 121, Issue 4, pp. a15-a18.

[3] Deutsches Windenergie Institut GmbH (2002): Studie zur aktuellen Kostensituation der Windenergienutzung in Deutschland, Wilhelmshaven, 2002.

[4] Lanza, G., Behmann, B., Peters, S. (2010): Verfügbarkeitswirksame technische Dienstleistungen - Potenzialanalyse bei KMU im Maschinen- und Anlagenbau, in: ZWF - Zeitschrift für wirtschaftlichen Fabrikbetrieb, Vol. 5, pp. 421-425.

[5] Lanza, G., Behmann, B., Werner, P. (2010): Enhancement of the serviceability in the machine and plant industry, in: Proceedings of the APMS International Conference on Advances in Production Management Systems (APMS 2010), digital, Cernobbio, Como, Italy.

[6] Bullinger, H., Scheer, A. (2005): Service Engineering: Entwicklung und Gestaltung innovativer Dienstleistungen, Springer Verlag, Heidelberg.

[7] Meier, H. (2004): Dienstleistungsorientierte Geschäftsmodelle im Maschinen- und Anlagenbau, Springer Verlag, Heidelberg.

[8] Nesges, D. (2005): Prognose operationeller Verfügbarkeiten von Werkzeugmaschinen unter Berücksichtigung von Serviceleistungen, Forschungsberichte aus dem wbk Institut für Produktionstechnik, Band 131, Shaker Verlag, Aachen.

[9] Duffuaa, S.O., Raouf, A. (1992): A simulation model for determining maintenance staffing in an industrial environment, Simulation 59, Issue 8, pp. 93-99.

[10] Krug, C. (2010): Framework zur strategischen Kapazitätsplanung Hybrider Leitungsbündel, Dissertation Ruhr Universität Bochum, Shaker Verlag, Aachen.

[11] Vasen, J. (2003): Einsatzplanung für den technischen Kundendienst im Maschinenbau mit Bildung von Auftragsreihenfolgen durch ein kombiniertes Prioritätsregelverfahren, Dissertation, RWTH Aachen, Shaker Verlag, Aachen.

[12] Haasl, D.F., Roberts, N.H., Vesely, W.E., Goldberg, F.F. (1981): Fault Tree Handbook. Nuclear Regulatory Commission, Washington, DC, USA.

[13] Abernethy, R. (2000): The New Weibull Handbook – Reliability & Statistical Analysis for Predicting Life, Safety, Survivability, Risk, Cost and Warranty Claims. 4th edition, North Palm Beach, USA.

[14] Nelson, W. (1990): Accelerated Testing: Statistical Models, Tests, Plans and Data Analysis. Wiley, New York.

[15] Lanza, G., Werner, P., Behmann, B. (2010) : Prognosis of machine tool warranty costs considering arising risks through weak data sets and unknown load profiles, in : Proceedings of the European Safety and Reliability Conference (ESREL 2010), Rhodes, Greece, p.1014-1018.

Flexible Resource Planning in the Context of Dynamic IPS² Business Models

Horst Meier[1], Birgit Funke[1], Mario Boßlau[1]

[1] Institute Product and Service Engineering, Chair of Production Systems, Ruhr-University Bochum, Bochum, Germany

Abstract

Industrial Product-Service Systems (IPS²), combined with a flexible planning and scheduling process, represent a long-term competitive advantage. The value provision via IPS² comprises several degrees of freedom, such as the partial substitution of product and service shares, as well as the integration of customers' resources. Thus, the IPS² scheduling represents an optimization problem, flexibly considering the relevant aspects of the current business model.

The article first describes the scheduling approach for IPS² which considers the IPS² potentials for optimization. Furthermore, the correlation between the individual specifications of the business model features and the planning procedure is described in detail.

Keywords:

Advanced Planning and Scheduling; IPS² Business Models; Value Provision

1 INTRODUCTION

Due to reinforcing market dynamics and globalization as well as shorter product life cycles, the difficulty to attain lasting competitive advantages is currently increasing. In contrast to nowadays product offerings, Industrial Product-Service Systems (IPS²) imply the integrated development of product and service shares; IPS² are mainly specified by delivering value in use to the customer [1].

As approved by a survey within the project Ogemo.net [2] the main focus for the customer and at the same time the widest scope of offered industrial services exist during the delivery phase. New technologies and new customer demands, deregulation and internationalization of the markets as well as further causes influence the business models of industrial service providers lastingly. Existing business models cannot fulfill the requirements of the dynamic market and will get obsolete. Hence, these business models must be critically analyzed by the enterprises and, if necessary, new innovative business models considering these dynamic demands have to be developed.

Moreover, it is apparent that the existing approaches focusing business models are insufficient for the special demands of IPS² analyzed by Meier [3]. Hence, the future challenge is to examine industrially and intersectorally existing business models theoretically as well as empirically with regard to the suitability for IPS². In addition, the integrated consideration of product and service also adds complexity to the IPS² delivery phase due to the long-term relationship to the customer. To reduce this complexity IPS² specific partial models have to be derived empirically as well as analytically with regard to their dynamic and flexible behavior.

The integrated consideration of products and services also leads to new chances and challenges during the delivery phase of this product bundle [4]. To offer the combination of products and services advantageously to all partners, according to Meier a flexible resource planning is necessary to realize the individual specifications of the adapted business models [5]. Thus, the aspects dynamics and flexibility have to be taken into account to fulfill all IPS² specific requirements.

2 DYNAMIC IPS² BUSINESS MODELS

2.1 Current Approaches and Definitions

A good business model is essential to every firm, whether it is a new venture or an established player [6] because it positions the firm within its value network, shows how it transacts with customers and suppliers, and highlights the products and services that are exchanged. Moreover, a business model reveals the underlying economic logic that defines how the firm creates value [7]. The business model concept is used by many authors as a typology and systematization of innovative ideas of business [8], for the modeling and analysis of business strategies [9] as well as for the qualification of business success [10]. Nevertheless, most of the concepts indicate that a firm's business model should explain how a firm creates value.

Timmers defines a business model as an architecture for the product, service and information flow, including a description of various business actors and their roles. Furthermore, it is a description of the potential benefits for the various business actors as well as a description of the sources of revenues. [11]

Osterwalder describes a business model as a conceptual tool that contains a set of elements and their relationships and allows expressing a company's logic of earning money. It is a description of the value a company offers to one or several segments of customers and the architecture of the firm as well as its network of partners for creating, marketing and delivering this value and relationship capital in order to generate profitable and sustainable revenue streams. [12]

Amit and Zott outline a business model as an architectural configuration of the components of transactions designed to exploit business opportunities. A transaction component refers to the specific information, service, or product that is exchanged and the parties that engage in the exchange, respectively. The architectural configuration depicts and characterizes the linkages among the components of transactions and describes their sequencing [13].

Magretta states a business model as a set of assumptions about how an organization will perform by creating value for all the

J. Hesselbach and C. Herrmann (eds.), *Functional Thinking for Value Creation: Proceedings of the 3rd CIRP International Conference on Industrial Product Service Systems, Technische Universität Braunschweig, Braunschweig, Germany, May 5th - 6th, 2011*,

DOI 10.1007/978-3-642-19689-8_30, © Springer-Verlag Berlin Heidelberg 2011

stakeholders on whom it depends, not just its customers. In essence, a business model is a theory that is continually being tested in the marketplace. [6]

In the concept of Müller-Stewens and Lechner a business model defines how a firm's particular configuration of its value chain is made concrete by adopting a capitalization perspective which answers the question of how to make money in this business. The business model closes the gap between strategic and operative management by answering the questions: What services shall be offered to what customers? How and within which structure shall these services be offered? How do I win, foster and keep appropriate customers? How shall the revenue model be defined concretely? [14]

The contemporary discussion of the business model concept in the engineering and economics literature is insufficiently structured. Some authors analyze certain business models, e.g. operator models [15], performance contracting [16], revenue sharing [17] etc. Both theoretical and empirical work has been published here. These results refer to a certain business model and generally only to a line of business. Consequently, conditions under which the examined business model can be profitable for the supplier or user are identified in empirical studies for the respective line of business. Furthermore, there are a variety of case studies which analyze individual case specific, practice-oriented business models to identify success or also failure factors. The definitions examined in this paragraph show that the term 'business model' is used with different meaning and intent.

2.2 Decomposition and reduction in complexity of business models

As a further structuring approach the decomposition of business models is introduced by many authors with the aim to reduce complexity. Different morphological concepts are presented which provide partial models, such as value proposition, value architecture, revenue model, marketing model etc. A partial model enables the analysis of central dimensions of a business system and supports a differentiated examination in regard to the underlying research question [8].

Analyzing the technical literature it becomes apparent that up to now there has been no agreement in regard to the design and the elements of business models. Nevertheless, in most cases the value proposition, the value architecture as well as the revenue model are represented as partial models. On the one hand these partial models can be represented in an integrated business model; on the other hand they offer the possibility to describe central dimensions isolatedly in more detail.

2.3 Dynamic Aspects of business models in the context of Industrial Product-Service Systems

New forms of value creation can be seen in close relation to the differentiation of new business models. On the one hand, these business models are the result of organizational transformation processes; on the other hand, new creation of value forms requires the transformation of organizational structures and processes. Hence, the central driver of the differentiation of business models is the exposure with economic uncertainty [18].

The change in time of quantities indicates the dynamics of a system. A dynamic system exists if the time variables are connected to each other by difference or differential equations. Furthermore, economic dynamics can be understood as „... the study of economic phenomena in relation to preceding and succeeding events." [19] The possibility of the adaptation to changing customer needs is one of the central characteristics of IPS². It is necessary that business models can change over time and adapt to the respective requirements flexibly. Flexibility describes the ability of a system to adapt rapidly and with low financial overhead to changed influence factors. Moreover, the concept flexibility covers the control of scenarios which are already taken into account at the planning phase.

Particularly the motives for the change of business models as a reaction to changed market conditions have to be considered in the context of IPS². The way a business model will behave over time is difficult to predict because of the complex feedback dynamics inherent in business [20]. The challenge in context of a flexible planning process is to provide decision-supporting tools which reveal central changes of management ratios and the underlying business model features from a strategic perspective.

3 FLEXIBLE RESOURCE PLANNING FOR IPS²

3.1 Resource planning including IPS² specifics

During the IPS² delivery phase, unpredictable events may cause planning problems. This applies, if the due date of the planned processes is too near in future, i.e. adjustments of capacity are no longer possible. For example, if a machine's break down requires an additional technician's assignment in short time and due to that other processes are finished too late. Further planning problems can occur, if a required resource fails or an employee is affected by a disease.

Compared to classical factory or service planning, where simple heuristics or dispatching rules apply to create a solution (e.g. [21]), an optimization strategy is needed to adapt the IPS² resource planning for emergent processes. Furthermore, in distinction to autonomous or negotiations based approaches (e.g. [22]), this optimization approach has to be central and real-time-capable to exploit the concept of central planning and decentral control within the IPS² network. [23]

Additionally, Meier has evaluated, that the long-term relationship based on IPS² business models enables comprehensive statistics on machine load, operating hours, etc. [4]. These statistics can be used to reduce uncertainty during the delivery phase by making predictions as to the service processes in future. This leads to a more certain, and thus less complex, planning situation, compared to classical production planning.

Planning horizon

The long term capacity planning has a wide planning horizon; hence there are many degrees of freedom. But this variety of possibilities also implies the optimization regarding several aims. On the other hand, the operational resource planning especially has to deal with unscheduled jobs, which leads to changing priorities. In this case the optimization objectives are changed to enable the execution of all necessary processes; i.e. costs become second most important on this short term issue. [24]

When unscheduled jobs cause a planning problem at first a quick solution is needed; as a planning problem in most cases means the dissatisfaction of a customer it should be solved as soon as possible. In this case the operational resource planning uses the IPS² potentials for optimization (see below) to solve the planning problem. Whereas, planning alternatives can be considered that were unfeasible during the strategic planning in relation to a capacity adjustment.

A process in this context means every action to be done by the provider, e.g. the execution of a service process as well as the delivery of a spare part or sending a software update. A resource means all required machines, material, personnel, tools et cetera. A solution in this context means a certain combination of resources and processes including the whole scheduling (i.e. the complete planning) for all IPS² to be planned.

Potentials for optimization

IPS² are especially characterized by delivering value and fulfilling customer demands based on a variable combination of products and services [3]. Thus IPS² imply some specific potentials for optimization [25], what further differentiates the IPS² resource planning from normal production or service planning. The main potentials for optimizations are:

- variance in time;
- variance of resources;
- variance of processes;
- variance of allocation time;
- service distribution;
- partial substitution of product and service shares;
- integration of customers' resources;

If an additional job has to be planned, or another planning problem is to be solved, the resource planning can make use of these IPS² specific optimization potentials. According to [26] they are applied consecutively to the relevant resources and processes to generate interim solutions for evaluation. Whereas, the potentials for optimization each induce several opportunities to solve a planning problem. Due to this wide scope of variables the IPS² resource planning is a multi-dimensional optimization problem and becomes more complex by interrelations and dependencies between the variables. [26]

Starting the planning all process-resource-combinations are chosen regarding cost minimization etc., so reducing transition times is the main criteria of optimization. If there are delayed processes or additional processes being executed soon, the resource planning has to use the specific optimization potentials to solve these problems in real time.

3.2 The heuristic resource planning approach

According to the evaluations in [27] the IPS² resource planning is an optimization problem that cannot be assessed by direct mathematic methods: The relationship between two planning solutions can be described by applying the IPS² potentials for optimization, but the relation between planning changes and solution improvements (if → then) cannot be modelled mathematically. Thus, a heuristic approach has been chosen by MEIER. [27]

The analysis in [26] shows that the resource planning is a multi-objective problem. This occurs, because solutions have to be compared regarding the current objectives of the planning. There are at least three objectives (punctuality, costs, and work load) and, depending on the planning horizon, the priorities are changing.

To deal with the multidimensional and multiobjective planning of service processes within IPS², a real-time capable optimization strategy is needed. Thus, the IPS² resource planning approach presented in [28] uses an individual hybrid metaheuristic to optimize the resource planning. Therefore, a combination of Evolutionary Algorithms, Simulated Annealing and Brute Force Search is used in connection with the structured application of the IPS²-specific optimization potentials.

Within this approach the potentials for optimization are structured referring to their effectiveness and efficiency. Therefore, the potentials are ranked according to the caused costs, preferring low cost potentials. Additionally the resources and processes are ranked according to their effectiveness; those directly linked to a problem are preferred.

The potentials for optimization are mainly categorized by the direct costs they cause. The application of the **variance in time** (i.e. postponement of processes) does not directly cause any costs. The use of **alternative resources** does not cost any direct costs, either only unused, existing resources are considered; however, it can indirectly cause some costs, because e.g. an alternative worker is not familiar with the job and thus needs longer. The use of **alternative processes** might cause costs, as during the development already the most cost-effective process has been chosen for each task. To **request necessary resources to the network** will cause a significant amount of costs, as the external partners have to be included, coordinated and supported as well; also it is a very direct but slow way to replace a necessary resource. At least the **adjustment of the allocation speed** is considered; it directly causes costs when using faster but more expensive means of transportation.

3.3 Evaluation of planning alternatives

A basic requirement for every heuristic optimization is a calculable fitness function to compare solutions and move towards the optimum. Therefore, the quality of a planning solution has to be measured by calculating key performance indicators (KPI). According to the three main objectives (punctuality, costs, and work load) three KPIs are used. The single terms normally tend to the optimum "1", while lower values represent worse KPIs.

Punctuality

The total punctuality *z* considers delays or earliness of all processes to be planned. It puts strength on large delays and includes weighting by process' length and time variance.

Cost efficiency

As a KPI on relative costs the cost efficiency *ce* is calculated. It is the ratio of planned total costs and the total costs of a new planning solution. Due to unpredicted, additional events, the new costs will exceed the planned cost, thus *ce* will be < 1.

Work load

The activity of all resources *a* combines the work load of all resources, and the smoothness of the work loads distribution.

3.4 Adjustment of the planning to the preferences

To handle the three optimization objectives according to the above mentioned KPIs, a weighted sum approach was presented by Meier to avoid comprehensive pareto-optimization [28]. Therefore, after calculating the single KPIs, they are combined to a fitness function for the optimization result.

$$f_i = w_1(t_i) \cdot z_i + w_2(t_i) \cdot ce_i + w_3(t_i) \cdot a_i \quad (1)$$

The summation of the weights is 1. The total fitness *f* is always > 0 and normally < 1; the greatest *f* represents the best solution. Only exceptionally a fitness > 1 can occur. This situational combination of the three main KPIs allows reducing the optimization complexity as there is now only one objective: the optimization of the total fitness.

What affects these weighting factors?

The weights are adjusted dynamically according to the position within the planning horizon t_i. Thus the planner can vary the focus of the planning between punctuality, costs and work load. During the operational planning, which has to react on short-term tasks e.g. due to a machine breakdown, the punctuality for all jobs including the emergent process is the main objective.

Additionally, main preferences of the IPS² provider can be included in these weighting factors by setting planning preferences. For example, the mainly sold business model will influence the decision, whether costs or punctuality should be preferred. This will be further discussed in chapter 4.

4 CORRELATION BETWEEN DYNAMIC IPS² BUSINESS MODELS AND THE FLEXIBLE RESOURCE PLANNING

Considering a single IPS², its business model defines which preferences it implies during the delivery phase. If an IPS² is sold in an availability-based business model, much emphasis has to be placed on its punctuality. In a function-oriented business model the pricing for additional service processes might be negotiated regularly (short-term process execution is priced higher) so its preferences can be adapted depending on the current situation.

Considering many IPS² and their corresponding business model, the preferences during the delivery phase depend on much more than the main partial models (value proposition, value architecture, revenue model) of individual business models. A huge number of availability-oriented business models again imply much effort in punctuality. But, on the other hand, the preferences during planning have to consider further details of the business model and its contract details, respectively. If short reaction times and high availability are contractually guaranteed, and all risks according to that are undertaken by the IPS² provider, of course the preference during the delivery phase will be on the accurate, punctual execution of all processes (Figure 1 a).

Market conditions that demand for low investments can induce varying business models: On the one hand, a function-oriented business model with low flexibility; on the other hand, a result-oriented business model, where the provider undertakes most of the risks, with a revenue model of low invest but higher operating costs on customer side; both lead to the preference of low-cost planning solutions during the delivery phase (Figure 1 b).

To model those relationships between planning preferences, business model features, and contract details, it is necessary to extend the previous business model concept by further dimensions (Figure 2).

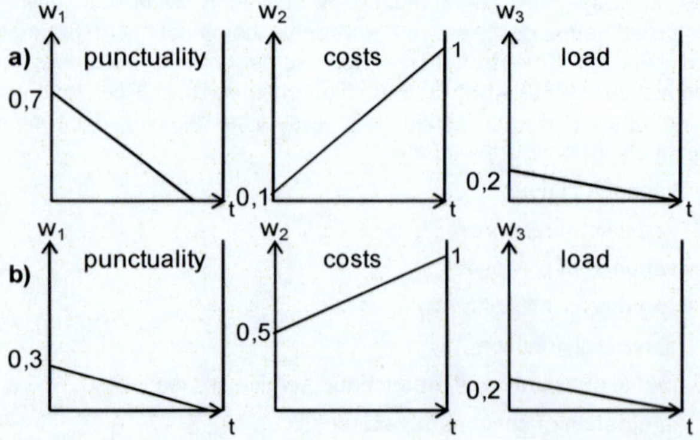

Figure 1: Weighting factors depending on the preferences.

Based on a comprehensive theoretical and empirical analysis, a morphology of the identified attributes and characteristics of business models will be developed. To reduce complexity and get a differentiated view of the identified business models, partial models for analyzing the central dimensions will be generated. The partial models investigated till now (Value Proposition, Value Architecture, Revenue Model) have to be extended by assigning the previously identified IPS² specific attributes. With the help of a partial model library dynamic IPS² business models can be configured and simulated with respect to the three main objectives punctuality, costs and work load from a strategic perspective. Thus, decisions on dynamic customizations of the IPS² business model during the delivery phase can be derived. These decisions are used for the flexible planning and scheduling process as a proactive customization.

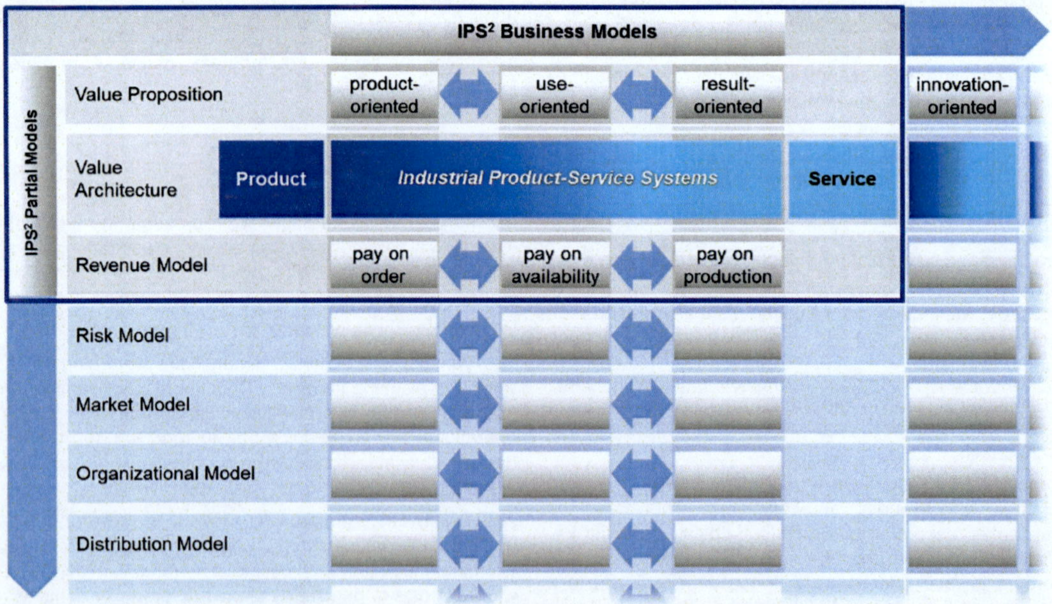

Figure 2: IPS² business models – extension by further dimensions.

5 SUMMARY AND FUTURE PROSPECTS

A number of deficits can be derived from the current business model discussion which was described in this paper. The heterogeneous concept understanding, the various partial models as well as different taxonomies cause a hardly structured discussion of the business model concepts in the technical literature. By means of a comprehensive analysis of existing business model approaches a uniform reference framework for a better concept understanding has to be created. Therefore, the future research focus lies on the development of a comprehensive business model concept, in which neither a specific industry, nor a specific kind of business model will be focused. A synopsis of the existing business models in theory and practice should be generated. Previous research results have a rather static character. Approaches including dynamic aspects, e.g. the innovation degree or the position in the value-added chain, only occasionally exist. Furthermore, the factors dynamics and flexibility are often not taken into account explicitly. Moreover, the term business model is mentioned regularly and used as a catchword both in the scientific theory and in the operational practice. The existing heterogeneous notation forms for the modeling and design of business models show the necessity to create a uniform systemic notation approach considering the aspects dynamics and flexibility.

6 ACKNOWLEDGMENTS

We express our sincere thanks to the Deutsche Forschungsgemeinschaft (DFG) for financing this study within the Collaborative Research Project SFB / TR 29 on Industrial Product-Service Systems – dynamic interdependencies between products and services in the production area.

7 REFERENCES

[1] Meier, H., Roy, R., Seliger, G. (2010): Industrial Product-Service Systems - IPS², In: CIRP Annals - Manufacturing Technology 59, pp. 607-627.

[2] Meier, H., Lanza, G. (Ed.) (2009): Kooperative Geschäftsmodelle zur Integration von Sach- und Dienstleistungen, VDMA Verlag.

[3] Meier, H., Kortmann, D. (2007): Leadership – From Technology to Use, 14th CIRP Conference on Life Cycle Engineering, 159-163.

[4] Meier, H., Völker, O.; Funke, B. (2010): Industrial Product-Service Systems (IPS²) - Paradigm Shift by Mutually Determined Products and Services. In: The International Journal of Advanced Manufacturing Technology, Special Issue Product-Service Systems, Springer Verlag.

[5] Meier, H., Krug, C. M. (2008): System for Planning of Resources in IPS²-Delivery, 41st CIRP Conference on Manufacturing Systems, Tokyo, Japan.

[6] Magretta, J. (2002). Why business models matter, in: Harvard Business Review, Vol. 80, pp. 86-92.

[7] Grasl, O. (2010): Professional Service Firms, Business Model Analysis – Method and Case Studies, University of St. Gallen, St. Gallen.

[8] Wirtz, B. W. (2001): Electronic Business, 2. Auflage, Wiesbaden, Gabler.

[9] Bieger, T.; Rüegg-Stürm, J.; Rohr, T. v. (2002): Strukturen und Ansätze einer Gestaltung von Beziehungskonfigurationen, in: Bieger, T.; Bickhoff, N.; Caspers, R.; Knyphausen-Aufseß, D. z.; Reding, K.: Zukünftige Geschäftsmodelle: Konzept und Anwendung in der Netzökonomie, Berlin, Heidelberg, New York, Springer.

[10] Krüger, Q.; Bach, N. (2001): Geschäftsmodelle und Wettbewerb im e-Business, in: Buchholz, W., Werner, H. (Hrsg.), Supply Chain Solutions, Schäffer-Poeschel.

[11] Timmers, P. (1998): Business Models for Electronic Markets, Electronic Markets, in: International Journal of Electronic Commerce & Business Media, Vol. 8, Nr. 2, pp. 3-8.

[12] Osterwalder, A. (2004): The Business Model Ontology, A Proposition in a Design Science Approach, Ph. D. thesis, University of Lausanne.

[13] Amit, R., Zott, C. (2000): Value Drivers of e-Commerce Business Models, in: INSEAD Working Paper, INSEAD, Fontainebleau.

[14] Müller-Stewens, G., Lechner C. (2005): Strategisches Management - Wie strategische Initiativen zum Wandel führen, Schäffer-Pöschel.

[15] Meier, H. (2004): Dienstleistungsorientierte Geschäftsmodelle im Maschinen und Anlagenbau - Vom Basisangebot bis zum Betreibermodell, Springer Verlag.

[16] Meinefeld, M. (2004): Strategische Erfolgsfaktoren für Contracting-Angebote von Energieversorgungsunternehmen, University of Paderborn, Deposit.ddb.de 974405094.

[17] Koulamas, C. (2006): A Newsvendor Problem with Revenue Sharing and Channel Coordination, in: Decision Science, Vol. 37 (1), pp. 91-100.

[18] Spath, D. (2007): Kooperative Geschäftsmodelle, Eine Herausforderung für produzierende Unternehmen und deren Dienstleister, in: Bullinger, H.-J.: Outsourcing in Deutschland: Rahmenbedingungen, Konzepte und Best Practices, Schäffer-Poeschel.

[19] Schwarz, R. (2002): Controlling-Systeme, Eine Einführung in Grundlagen, Komponenten und Methoden des Controlling, Gabler.

[20] Sterman, J. (2000): Business Dynamics, Systems Thinking and Modeling for a Complex World, McGraw-Hill International Edition, MIT, Boston.

[21] Kung, L.-C., Chern, C.-C. (2009): Heuristic factory planning algorithm for advanced planning and scheduling, Computers and Operations Research, 36/9.

[22] Scholz-Reiter, B., Höhns, H., Hamann, T. (2004): Adaptive Control of Supply Chains: Building blocks and tools of an agent-based simulation framework, CIRP Annals - Manufacturing Technology 53/1: 353-356.

[23] Meier, H., Uhlmann, E., Völker, O., Geisert, C., Stelzer, C. (2010): Reference Architecture for dynamical organization of IPS² service supply chains in the delivery phase, 2nd CIRP Conference on Industrial Product-Service Systems, Linköping, Sweden, April 14-15.

[24] Dyckhoff, H., Spengler, T. (2007): Produktionswirtschaft: Eine Einführung für Wirtschaftsingenieure, Springer Verlag.

[25] Meier, H., Krug, C. M. (2006): Strategische Kapazitätsplanung hybrider Leistungsbündel, PPS Management 11, 48-51.

[26] Meier, H., Funke, B., Krings, R., Krug, C. (2008): Case Study Based Service Process Library Supporting the Resource Planning for Industrial Product-Service-Systems (IPS²), 6th CIRP International Conference on Intelligent Computation in Manufacturing Engineering (ICME), Naples, Italy, 93-97.

[27] Meier, H., Funke, B. (2010): Resource Planning of Industrial Product-Service Systems (IPS²) by a Heuristic Resource Planning Algorithm, 2nd CIRP Conference on Industrial Product-Service Systems, Linköping, Sweden, April 14-15.

[28] Meier, H., Funke, B. (2010): Heuristic Resource Planning Approach for Industrial Product-Service Systems (IPS²), In: Proceedings of the 7th CIRP International Conference on Intelligent Computation in Manufacturing Engineering (ICME), Naples, in press.

Learning About Product Service Systems – Conveying Information to Improve the Benefit Evaluation and Stabilize Preferences

Mario Rese[1], Wolf-Christian Strotmann[1], Markus Karger[1], Judith Gesing[1]

[1] Marketing Department, Ruhr-University Bochum, Germany

Abstract

Especially for Industrial Product Service Systems (IPS²) which are very new and can therefore be classified as Really New Products (RNPs) the stability of customer preferences is a crucial success factor. Against this background it is essential to identify forms of knowledge transfer which enable customers to learn about IPS² and help them stabilize their preferences. In this study, we set up a large scale practitioner experiment to test the effect of combinations of information and learning mechanisms on preference stability. We use an innovative device for remote service provision. Our results indicate the importance of conveying information combinations for learning about IPS². We therefore conclude, that only by regarding both, information and learning mechanisms, and determining appropriate information strategies can effects on customer preferences really be controlled for in the best possible manner.

Keywords:

Really New Products; Industrial Product Service Systems; Preference Stabilization

1 INTRODUCTION

The development and marketing of Industrial Product Service Systems (IPS^2) is a strategic priority for many companies, as they offer differentiation from competitors and long term customer relationships. However most IPS² are very new to customer can be classified as Really New Products (RNPs). RNPs are innovations that revolutionize product categories or define new categories. Most RNPs are manufactured using a new technology or offer new technologies to the customer and induce significant behaviour changes [1]. Thus, innovative solutions, such as innovative IPS², fail at a very high rate at the market [2] [3]. One reason for this is their high degree of newness, which places a large learning burden on customers [4] [5] [6], resulting in high uncertainty when customers build their preferences [2] [7] [8]. This uncertainty leads to instability in customers' revealed priorities among design features when market research is used to evaluate the product concepts of RNPs. This lack of preference stability poses a major threat for companies and is a significant barrier to RNP success [7]. Knowing which product idea has potential and which product configuration customers prefer is a crucial discriminator for success or failure of a new product development [9]. If configured based on instable preferences, RNPs and therewith also very innovative IPS² might not meet true customer needs and find no market acceptability.

Hence, identifying means of making customers understand RNPs, especially its concepts and design features, is a key challenge for companies [10] [11]. As the only source of information about a RNP is the supplier [7], customer understanding solely depends on the RNP description provided by him. Hence, providing adequate information about RNPs is the most crucial factor for preference stabilization, since inappropriate selectivity and lack of information are two of the major faults impeding well-constructed preferences [12]. Despite its importance, this topic has found only limited consideration in literature.

To provide a first step in closing this research gap, this study aims at determining which type of information can be most effectively combined with different learning strategies to generate stable customer preferences and therewith reliable customer feedback for IPS² development. In a large-scale experimental study with practitioners we use combinations of how-to-knowledge and principles-knowledge together with analogies and mental simulations to test how these information combinations differ in generating stable preferences for such solutions.

2 PREFERENCE STABILITY

Customers have stable preferences, if they have a stable frame of reference to evaluate a product across contexts and time [13] [14] [15] [16]. Such a frame of reference enables customers to identify the relation between product attributes and their overall satisfaction with a product [17] and to identify the attributes and their weights that are important for their decision [15] [18].

When building preferences customers usually rely on prior experience. The more experience customers have in a product category, the more elaborated and stable their preferences are [17]. A new mobile phone e. g. is evaluated based on known mobile phones. The customer knows which importance a calendar function or a camera has for him and can thus evaluate a new phone easily. However, RNPs are totally new and cannot be evaluated based on existing product categories [4]. Since customers do not have any prior experience with these innovations, no product category understanding has been established. Therefore, no frame of reference yet exists for RNPs [19]. As these innovations also induce a lot of changes [20], understanding RNPs requires extensive cognitive thinking [4] and induces significant learning costs [3] [5] [6].

Hence, educating customers is especially relevant in the case of RNPs [21] and it is a central aspect of building stable preferences for RNPs. As customers decide between descriptions of products instead of the products themselves [22], the types of information provided by RNP suppliers is one aspect of educating customers which warrants thorough examination.

J. Hesselbach and C. Herrmann (eds.), *Functional Thinking for Value Creation: Proceedings of the 3rd CIRP International Conference on Industrial Product Service Systems, Technische Universität Braunschweig, Braunschweig, Germany, May 5th - 6th, 2011*,
DOI 10.1007/978-3-642-19689-8_31, © Springer-Verlag Berlin Heidelberg 2011

3 THE RELEVANCE OF DIFFERENT TYPES OF INFORMATION FOR PREFERENCE STABILIZATION

So far no research at all has investigated the effects of different types of information on preference stability for RNPs. However, two different types of information are essential to educate customers about innovations in general, how-to-information and principles-information [23].

How-to-information explains how an innovation works and is necessary to use an innovation properly. It provides a stepwise explanation of all phases of usage and therefore draws a clear picture of benefits and drawbacks of the product. It is argued that the more complex an innovation is from the customer perspective, the greater the amount of how-to-knowledge needed is going to be. If the amount of how-to-information provided to customers is too low, the innovation is likely to be rejected [23] [24]. Principles-information explains why an innovation works, i.e. the functioning principles underlying an innovation. In most cases principles-information is going to contain a greater amount of technical details as these are necessary to explain why an innovation works [23] [24]. It is argued that while the provision of how-to-information is essential for innovation adoption, it might be possible to adopt an innovation without providing principles-information. However, principles-information increases individuals' competence of judging the effectiveness of an innovation and reduces the danger of misusing it [23].

At first sight, both types of information might be necessary to form stable preferences for RNPs. On the one hand, RNPs are characterized by a high degree of technological newness [20] [25]. They are relatively complex and often combine several functionalities [21]. Consequently, one could argue that it is important to provide product attribute information and explain technical functionalities and why a RNP works (principles-information). As their high degree of (technological) newness results in customers' difficulty of understanding RNP benefits, explaining these benefits by showing how a RNP works (how-to-information) might also be sensible.

At second sight, however, principles-information is likely to contain a high degree of quantitative, technical attribute information, since it describes the functioning principles underlying a RNP. For customers it is a strenuous, difficult task to process and understand such information [26] [27] [28]. Both the information processing as well as trying to understand the RNP therefore requires a substantial amount of cognitive resources. Customers tend to be "cognitive misers" [29] [30] reluctant to engage in extensive cognitive thinking [7]. If cognitive resources are scarce, customers use heuristics. When deciding for RNP design features the use of heuristics negatively affects preference stability [15]. This is because heuristics are context-specific decision rules [31]. Thus in a new context a new heuristic would be used and preferences build in this way are not portable to new decision contexts. Hence, reducing cognitive resources by providing principles-information is likely to negatively affect customers' preference stability. Furthermore, explaining product characteristics and functioning principles forces customers to recognize the benefits of the RNP themselves and assess their risks. They are therefore left with a lot of room for deliberation. If customers are granted too much room for deliberation, however, they might focus their cognitive resources away from the most important information, with negative consequences for preference stability [32]. Consequently, providing principles-information might not serve to stabilize preferences.

In contrast to principles-information, how-to-information contains less technical details. How-to-information serves to explain to customers how to use a RNP [23]. It focuses on RNP benefits for customers, leaving less room for deliberation. Additionally, how-to-information is mostly qualitative information which is comparatively easy to process [26] [28]. Hence, the cognitive effort to process the information is lower [33].

Providing how-to-information to customers therefore should leave them with sufficient cognitive resources to form stable preferences and at the same time should not leave too much room for deliberation and speculation. Providing customers with principles-information leaves much room for deliberation and elaboration on RNP attributes, possibly resulting in a high cognitive strain. This cognitive strain might impede positive effects on preference stability, thus:

H1: Providing how-to-information has a greater positive impact on preference stability than providing principle-information.

4 THE RELEVANCE OF DIFFERENT LEARNING STRATEGIES FOR PREFERENCE STABILIZATION

In addition to different types of information, learning strategies might help stabilize customer preferences for RNPs. Such learning strategies are used naturally when customers decide about adopting a product [7] [34]. Hence, any type of information provided is likely to be combined with some form of learning strategy and develop a joint effect on preference stability. Therefore, it is the effect of such *information combinations* on preference stability which warrants examination.

Two prominent learning strategies, which could aid customers in constructing their knowledge structures for RNPs, have been given considerable attention in literature, mental simulations [7] [10] [13] [35] [36] [37] and analogies [7] [10] [11] [25] [38].

A mental simulation is the imitative mental representation of an event or a series of events [36]. The use of mental simulations is well established as a cognitive tool for making product evaluations and product adoption decisions and help customers learn about RNPs [5] [39] [40]. Such processing mechanisms enable customers to imagine a product purchase, simulate a product experience and to understand better the consequences of product usage [5] [40].

Mental simulations can act as a surrogate experience for customers [7] [10]. They envision themselves using the product and are therefore able to understand the product and its consequences [34] [39] and to recognize how the RNP fits into existing usage patterns and whether it requires behavioral changes [7]. Thereby customers can predict if the consumption of a product will lead to satisfaction [34]. Hence, mental simulations help to educate customers and enable them to make more informed decisions [39]. Hitherto, research has shown that mental simulations can stabilize customer preferences [13] and can do so also for RNPs [7].

Since no research exists which gives reason to question the usefulness of mental simulations in combination with different types of information in a preference stabilization context, we hypothesize:

H2a: Combining how-to-information with mental simulations enhances preference stability.

H2b: Combining principles-information with mental simulations enhances preference stability.

In an analogy, information from a familiar domain (the base) is used to understand a novel domain (the target) [11]. Analogies have been shown to help customers learn about and understand RNPs [11] [25] [41]. Since RNPs cannot be classified into existing product categories [1] [20], customers are often unable to understand their benefits. Analogies help customers to link the attributes of a RNP with the benefits it delivers by drawing inferences from a familiar product or product category [38]. Furthermore, analogies reduce the perceived complexity of the RNP by enabling the customer to link it with something known [10] and are a means of integrating the RNP into existing knowledge structures [11]. Hence, analogies reduce the cognitive load of customers and therefore lead to a reduced use of heuristics. Additionally, as they help customers to understand the RNP and derive its benefits there is less room for deliberation. Nevertheless, analogies cannot be combined with each type of information. El Houssi et al. (2009) show that combining attribute information (technical information) with an analogy can have negative impacts on product evaluation. Evaluating technical characteristics of a RNP is a very resource demanding task and less attention can be dedicated towards the analogy. Consequently, the beneficial effect of the analogy is diminished or even nullified [21]. Analogies are more persuasive when substantial resources are devoted to their processing than when resources are limited [41]. Thus providing analogies in combination with principles-information might not have a positive impact on preference stability, as principles-information consists of mainly technical details.

In line with the positive effects of analogies on customer learning for RNP identified by the literature so far, we hypothesize:

H3a: Combining how-to-information with analogies enhances preference stability.

Drawing on the findings by El Houssi et al. (2009) which show that combining analogies with attribute information does not deliver optimal results, we hypothesize for the combination with principles-information:

H3b: Combining principles-information with analogies has no positive effect on preference stability.

5 EMPIRICAL STUDY

5.1 RNP used in the sudy

Imagine sending a service technician to a location anywhere in the world and have her carry out a service previously unknown to her correctly via guidance from the service headquarter. The mavus®-system of the German company Heitec makes this possible. It is a device to enable the remote service of machines in Business-to-Business markets. Hence, this system is an example of a IPS² and will be used in this study to discuss the topic at hand, preference stabilization for innovative IPS². The mavus®-system is a vest containing a technical complex head-mounted display and -camera system (see Figure 1).

Figure 1: The mavus®-system.

With the mavus®-system, maintenance and service processes can be conducted by technicians with a lower degree of expertise, as they are guided and instructed by experts in the service headquarters. The mavus®-system offers a new way of doing maintenance. It can neither be categorized as a traditional maintenance device nor as a standard remote service device, as it by far exceeds every possibility provided to customers so far. Hence, it can be classified as a RNP. In the case of the mavus®-system the new technology which is typically for RNPs consists of a new kind of camera, able to deliver high resolution images to the service headquarters under all circumstances without time lags.

5.2 Experimental sample and setting

Participants of this experiment were 219 practitioners which were recruited at various colleges for service technicians. From these 219 participants a total of 143 took part in the whole experiment and could be used in the final analysis. The average age of the participants was 25 years; their work experience was an average of 6 years. The sample consisted of 4% females and 96% males.

Before the experiment the participants were divided into six experimental groups, resulting in a 2 (principles-information vs. how-to-information) x 3 (no learning strategy vs. mental simulation vs. analogy) mixed design. In stage one participants were presented with different combinations of information about the mavus®-system according to their experimental group. They then had to fill out a questionnaire and do a full-profile conjoint ranking of twelve product profiles consisting of four product attributes (comfort, resolution of the head display, video quality, weight) with three levels each. This ranking allows us to estimate the relative importance of each attribute for the customer when evaluating a product and therewith the preferences.

In stage two the participants did a one-day-practical workshop where they worked with the mavus®-system in a simulated tele-maintenance situation. The participants experienced the use of the mavus®-system as a technician who gets instructions via headphone and head-display as well as a expert who gives the instructions and monitors the maintenance process from his computer. The workshop was conducted in cooperation with technicians from Heitec who provided the technical background and answered questions of the participants. After the workshop the subjects again had to fill out a questionnaire and repeat the conjoint ranking task. By comparing the relative importance given to the attributes of the mavus®-system before and after trying it we can measure the stability of participants' preferences (see Figure 2). If these are relatively stable the information combination given in the beginning was appropriate to inform the participant about the mavus®-system. If preferences change after trying the system, the participant got a false impression of the mavus®-system beforehand and the information combination was inappropriate.

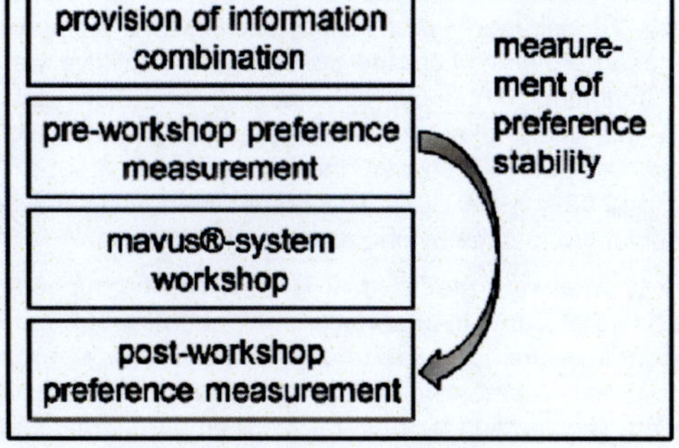

Figure 2: Experimental setting.

5.3 Information Stimuli

Because of the complexity of the product we designed the information stimuli used in the experiment carefully based on literature, open interviews with practitioners and critical dialogue with Heitec. To make sure that the information provided really was how-to-information and principles-information we confronted 70 students with an engineering background with both types of information stimuli. The respondents had to identify which stimuli represented how-to-information and which represented principles-information. As 83 % of the participants correctly identified the stimuli as intended, they were used in the experiment.

For the mavus®-system, how-to-information lets the participants know how to put on the vest, secure it, establish a connection with the service headquarters and confer with an expert via the head mounted display, speakers and the camera. Principles-information explains the underlying principals of remote maintenance and augmented reality. The participants get to know how the connection between the service technician and the headquarters is established on a technical basis, why the camera is able to always deliver sharp images without time lag and which technological background enables the functioning of the head mounted display.

In the mental simulation condition, participants were encouraged to mentally process 7 different aspects. The following example shows the mental simulation for the risk factor "learning costs":

Imagine you use the mavus®-system for the first time. Think about how difficult it is to learn the usage of the mavus®-system and potential problems that may occur.

In the analogy condition we used the following analogy:

The mavus®-system works like a personal assistant. You can always ask a personal assistance for advice, in order to be informed about something unknown to you. If you do not need the assistant you just do not require his service. Therewith you can always get advice and support when it is needed and you do not possess sufficient knowledge. Most work processes can therefore be done quickly and effectively with the help of an assistant. With the mavus®-system it is the same way. Just like with a personal assistant you can always ask for information and support from the expert in the service headquarters and you can get it in the required form. Through the help of this individually adjusted support you can carry out maintenance and service processes without possessing all the knowledge which is required.

5.4 Results

To test for preference stability we used a standard ANOVA (analysis of variance) with preference stability as dependent variable and the six different experimental groups as independent variables. To measure preference stability we compared the differences in pre-workshop and post-workshop relative weights of product attributes. Thus we calculated if and how differences between the information groups emerged. The ANOVA shows significant differences between the experimental groups (F(5) = 2,582, p = 0.029, η^2= 0.087). The results comparing the different experimental groups one by one are shown in table 1.

Contrary to what was predicted in H1, we find no support for the hypothesis that how-to-information is superior to principles-information in terms of preference stability (t(46) = 1.134, p = 0.263). When combining learning strategies with how-to information, respondents who were informed via mental simulation had more stable preferences than respondents using no learning strategy, if only weakly significant (t(48) = 1.775, p = 0.082), which supports H2a. Albeit the difference between the combination of how-to-information with analogies and purely providing how-to-information does not reach significance (t(43) = -1.511, p = 0.138), therefore H3a must be neglected.

		no learning strategy how-to-information	mental simulation how-to-information	analogy how-to-information	no learning strategy principles-information	mental simulation principles-information
no learning strategy	how-to-information	-	-	-	-	-
mental simulation	how-to-information	**1.775** *	-	-	-	-
analogy	how-to-information	1.511	**3.592** ***	-	-	-
no learning strategy	principles-information	1.134	**3.023** ***	0.266	-	-
mental simulation	principles-information	0.568	**2.288** **	0.806	0.512	-
analogy	principles-information	0.975	**2.488** **	0.123	0.084	0.484

*** p <0,10; ** p < 0,05; * p < 0,1

Table 1: Results of pair wise ANOVA.

Concerning principles-information, our results show that combinations with mental simulations do not improve preference stability (t(45) = 0.512, p = 0.609), thus H2b must be neglected. Combining principles-information with analogies does either not improve stability as hypothesized (t(45) = -0.084, p = 0.933), hence H3b prevails.

Considering all possibilities, one information strategy seems to be best suited to improve preference stability. The combination of how-to-information with mental simulations reaches a positive statistical difference in terms of preference stability compared to all other five information strategies (vs. how-to-information no learning strategy, t(48) = 1.775, p = 0.082; vs. principles-information no learning strategy, t(56) = 3.023, p < 0.01; vs. how-to-information analogy, t(53) = 3.592, p < 0.01; vs. principles-information analogy, t(47) = 2.488, p = 0.016; vs. principles-information mental simulation, t(47) = 2.288, p = 0.027).

6 DISCUSSION

6.1 Summary

Developing new products or services or product service combinations in form of IPS^2 is a challenging task for companies. To do this successfully companies must gather feedback from their customers concerning the configuration of the IPS^2. This feedback however is not reliable if customers do not understand the IPS^2 presented. The goal of this paper therefore was to contribute to a better understanding of customer learning processes in the context of IPS^2, with a specific focus on generating stable customer preferences. The aim is to improve IPS^2 development by granting reliable feedback from the market. Additionally our results can help in marketing IPS^2 and reaching acceptance for them, as we show how to increase customers` understanding of IPS^2. Our study thereby contributes not only to IPS^2 literature but also to the innovation research, as IPS^2 can be characterized as RNPs.

The results of our study draw a clear picture: it is of the utmost importance, to identify suitable combinations of types of information and learning strategies to help customers stabilize their preferences. This is because some combinations are better suited to stabilize preferences than others or than simply providing certain types of information without learning strategies. None of the types of information provided solely led to a significantly higher stability of participants' preferences than any of the information combinations did. On the contrary, combinations of how-to-information and mental simulations generated significantly more stable preferences than the respective use of the sole type of information.

Our results also clearly indicate which type of information should be used in combination with which learning strategy. As our study shows, analogies neither help to stabilize preferences in combination with how-to-information, nor are they beneficial when used in combination with principles-information. These results are in line with previous findings from Hoeffler (2003). This might be due to the problems analogies hold despite their beneficial effect for customer learning. Analogies lead to a selective processing of new product information. Attributes that base and target domain share might be overrated, whereas characteristics that are only present with the RNP might be overlooked [10] [11]. Thus, analogies can be beneficial for consumer learning about RNPs in some cases, but are problematic with regard to preference stabilization, as preference stability depends on an evaluation of product attributes.

Contrary to analogies, mental simulations are well suited to complement certain types of information with the goal of preference stabilization for RNPs. When combined with how-to-information mental simulations were able to generate significantly more stable preferences for both the technically complex and less complex RNP. Hence, this information combination seems suitable to inform about RNPs and generate a reliable feedback.

6.2 Theoretical and Managerial Implications

This paper contributes to product innovation and marketing theory in two major ways. Foremost in this study is the finding that the focus adopted in literature so far – either focusing on the type of information to convey or focusing on learning strategies to help customers understand RNP - does not capture the complexity of developing and marketing RNP. Our results show that alternative combinations of different types of information and learning strategies used in marketing communication for RNPs can have significantly different effects on RNP learning and customer preference stability. To our best knowledge this study is the first to examine the effects of such combinations on preference stability for RNPs.

Second, this study adds evidence to and enhances the findings of the growing body of literature which examines the power of learning strategies such as analogies and mental simulations in preparing customers for new product acceptance. Such learning strategies are necessary to enable customers to learn about RNPs, as these are so new that customers do not possess any useful knowledge which could be applied. This study pioneers in investigating the usefulness of the analogies and mental simulations as complements to information stating how and why RNPs work, in the context of preference stabilization. It could be shown, that it strongly depends on the type of information with which they are combined, whether learning strategies help stabilize preferences.

This research provides valuable insights into customer preference formation for RNPs and especially innovative IPS², which product managers may want to consider in the development and marketing. The type of information provided and the learning strategies used to enable learning about such solutions strongly influence the way people assess their design features. Hence, companies should always carefully consider the outcome of their providing different types of information and ways of presenting them. Thus, the key findings this paper provides support the findings of Feiereisen et al. (2008), that from a managerial standpoint marketing communication strategies for RNPs need to be carefully considered. The fact that for RNPs comprehension levels may be very low if the strategy used to help customers understand RNPs is not appropriate is supported by the significantly differing preference stabilities evoked by altering information strategies.

Our results suggest which kind of information strategy to use to generate stable customer preferences and gain reliable customer feedback. Marketers should explain how the IPS² works, doing so by using mental simulations.

7 REFERENCES

[1] Aggarwal, P., Cha, T., Wilemon, D. (1998): Barriers to the Adoption of Really New Products and the Role of Surrogate Buyers, in: Journal of Consumer Marketing, 15. Jg. (4), pp. 358-371.

[2] Hoeffler, S. (2002): Conceptualizing Newness and Positioning Really New Products, in: Advances in Consumer Research, 29. Jg. (1), pp. 267.

[3] Gourville, J. (2006): Eager Sellers and Stony Buyers: Understanding the Psychology of New Product Adoption, in: Harvard Business Review, 84. Jg. (6), pp. 98-106.

[4] Zhao, M., Hoeffler, S.,; Dahl, D. W. (2009): The Role of Imagination-Focused Visualization on New Product Evaluation, in: Journal of Marketing Research, 46. Jg. (1), pp. 46-55.

[5] Dahl, D. W.; Hoeffler, S. (2004): Visualizing the Self: Exploring the Potential Benefits and Drawbacks for New Product Evaluation, in: Journal of Product Innovation Management, 21. Jg. (4), pp. 259-267.

[6] Moreau, C. P.; Markman, A. B.; Lehmann, D. R. (2001): "What is it?" Categorization Flexibility and Consumers' Responses to Really New Products, in: Journal of Consumer Research, 27. Jg. (4), pp. 489-498.

[7] Hoeffler, S. (2003): Measuring Preferences for Really New Products, in: Journal of Marketing Research, 40. Jg. (4), pp. 406-420.

[8] Herzenstein, M.; Posavac, S. S.; Brakus, J. (2007): Adoption of New and Really New Products: The Effects of Self-Regulation Systems and Risk Salience, in: Journal of Marketing Research, 44. Jg. (2), pp. 251-260.

[9] Cooper, R. G. (1975): Why New Industrial Products Fail, in: Industrial Marketing Management, 4. Jg. (6), pp. 315-326.

[10] Feiereisen, S.; Wong, V.; Broderick, A. J. (2008): Analogies and Mental Simulations in Learning for Really New Products: The Role of Visual Attention, in: Journal of Product Innovation Management, 25. Jg. (6), pp. 593-607.

[11] Gregan-Paxton, J.; Hibbard, J. D. ; Brunel, F. A.; Azar, P. (2002): "So That's What That Is": Examining the Impact of Analogy on Consumers' Knowledge Development for Really New Products, in: Psychology & Marketing, 19. Jg. (6), pp. 533-550.

[12] Payne, J. W.; Bettman, J. R.; Schkade, D. A. (1999): Measuring Constructed Preferences: Towards a Building Code, in: Journal of Risk and Uncertainty, 19. Jg. (1-3), pp. 243-270.

[13] Zhao, M.; Hoeffler, S.; Zauberman, G. (2007): Mental Simulation and Preference Consistency Over Time: The Role of Process- Versus Outcome-Focused Thoughts, in: Journal of Marketing Research, 44. Jg. (3), pp. 379-388.

[14] Kardes, F. R.; Cronley, M. L.; Kim, J. (2006): Construal-Level Effects on Preference Stability, Preference-Behavior Correspondence, and the Suppression of Competing Brands, in: Journal of Consumer Psychology, 16. Jg. (2), pp. 135-144.

[15] Amir, O.; Levav, J. (2008): Choice Construction Versus Preference Construction: The Instability of Preferences in Learned Context, in: Journal of Marketing Research, 45. Jg. (2), pp. 145-158.

[16] Hsee, Ch. K.; Yang, Y; Gu, Y.; Chen, J. (2009): Specification Seeking How Product Specifications Influence Consumer Preference, in: Journal of Consumer Research, 35. Jg. (6), pp. 952-966.

[17] Hoeffler, S.; Ariely, D. (1999): Constructing Stable Preferences: A Look into Dimensions of Experience and their Impact on Preference Stability, in: Journal of Consumer Psychology, 8. Jg. (2), pp. 113-139.

[18] West, P. M.; Brown, Ch. L.; Hoch, Stephen J. (1996): Consumption Vocabulary and Preference Formation, in: Journal of Consumer Research, 23. Jg. (2), pp. 120-135.

[19] Veryzer, R. W. (1998): Key Factors Affecting Customer Evaluation of Discontinuous New Products, in: Journal of Product Innovation Management, 15. Jg. (2), pp. 136-150.

[20] Urban, G. L.; Weinberg, B.; Hauser, J. R. (1996): Premarket Forecasting of Really New Products, in: Journal of Marketing, 60. Jg. (1), pp. 47-60.

[21] El H., Amina A.; Morel, K. P. N.; Hultink, E. J. (2009): Less is More When Learning by Analogy: The Disruptive Impact of Attribute Information on Consumers' Benefit Comprehension of Really New Products in: Advances in Consumer Research, 36. Jg., pp. 467-472.

[22] Tversky, A. (1996): Contrasting Rational and Psychological Principles of Choice, in: Zeckhauser, R. J.; Keeney, R. L.; Sebenius, J. K. (Hrsg.): Wise Choices: Decisions, Games, and Negotiations, Boston: Harvard Business School Press, 1996, pp. 5-21.

[23] Rogers, E. M. (2003): Diffusion of Innovations, 5te Auflage, New York, London, Toronto, Sydney: The Free Press, 2003.

[24] Binsack, M. (2003): Akzeptanz neuer Produkte. Vorwissen als Determinante des Innovationserfolgs. Dissertation - Technische Universität Berlin, 2002. 1. Aufl. Wiesbaden: Dt. Univ.-Verl.

[25] El H., Amina A.; Morel, K.P. N.; Hultink, E. J. (2005): Effectively Communicating New Product Benefits to Consumers: The Use of Analogy versus Literal Similarity, in: Advances in Consumer Research, 32. Jg. (1), pp. 554-559.

[26] Yalch, R. F.; Elmore-Yalch, R. (1984): The Effect of Numbers on the Route to Persuasion, in: Journal of Consumer Research, 11. Jg. (1), pp. 522–527.

[27] Anderson, R. E.; Jolson, Marvin A. (1980): Technical Wording in Advertising: Implications for Market Segmentation, in: Journal of Marketing, 44. Jg. (1), pp. 57-66.

[28] Burkell, J. (2004): What are the Chances? Evaluating Risk and Benefit Information in Consumer Health Materials, in: Journal of the Medical Library Associaton, 92. Jg. (2), pp. 200–208.

[29] Bettman, J. R.; Luce, M. F.; Payne, J. W. (1998): Constructive Consumer Choice Processes in: Journal of Consumer Research, 25. Jg. (3), pp. 187-217.

[30] Fiske, S. T. and Taylor, S. E. (1991): *Social Cognition*, New York: McGraw-Hill.

[31] Tversky, A.; Kahnemann, D. (1974): Judgment under uncertainty: Heuristics and Biases, in: Science, 185. Jg., pp.1124-1131.

[32] Nordgren, L. F.; Dijksterhuis, A. (2009): The Devil Is in the Deliberation: Thinking Too Much Reduces Preference Consistency, in: Journal of Consumer Research, 36. Jg. (6), pp. 39-46.

[33] Stone, D. N.; Schkade, D. A. (1991): Numeric and linguistic information representationin multiattribute choice: *Organizational Behavior and Human Decision Processes*, 49. Jg..

[34] Phillips, D. M.; Olson, J. C.; Baumgartner, Hans (1995): Consumption Visions in Consumer Decision Making, in: Advances in Consumer Research, 22. Jg. (1), pp. 280-284.

[35] Taylor, S. E.; Pham, L. B.; Rivkin, I.D.; Armor, D. A. (1998): Harnessing the Imagination – Mental Simulation, Self-Regulation, and Coping, in: American Psychologist 53. Jg. (4), pp. 429-439.

[36] Taylor, S. E.; Schneider, S. K. (1989): Coping and the Simulation of Events, in: Social Cognition 7. Jg. (2), pp. 174-194.

[37] Pham, L. B.; Taylor, S. E. (1999): From Thought to Action: Effects of Process- Versus Outcome-Based Mental Simulations on Performance, in: Personality and Social Psychology Bulletin, 25. Jg. (2), pp. 250-260.

[38] Gregan-Paxton, J.; Roedder J., D. (1997): Consumer Learning by Analogy: A Model of Internal Knowledge Transfer, in: Journal of Consumer Research, 24. Jg. (3), pp. 266-284.

[39] Phillips, D. M. (1996): Anticipating the Future: The Role of Consumption Visions in Consumption Behavior, in: Advances in Consumer Research, 23. Jg. (1), pp. 70-75.

[40] Shiv, B.; Huber, J. (2000): The Impact of Anticipating Satisfaction on Consumer Choice, in: Journal of Consumer Research 27. Jg. (2), pp. 202-216.

[41] Roehm, M. L.; Sternthal, B. (2001): The Moderating Effect of Knowledge and Resources on the Persuasive Impact of Analogies, in: Journal of Consumer Research, 28. Jg. (2), pp. 257-272.

A Simplified Decision Making Model for Employing Product Service System in Industry at a Preliminary Planning Stage

Rachata Khumboon[1], Sami Kara[1], Suphunnika Ibbotson[1]

[1] Life Cycle Engineering & Management Research Group, University of New South Wales, Sydney, Australia

Abstract

Many successful companies employ Product Service Systems (PSS) into their business model mainly due to an economic incentive. However, there is no simple and structured approach to ensure an improvement of PSS's environmental performance at an early decision making stage in lieu selling product. This paper presents a preliminary simplified decision making model for manufacturer's view to design service parameters of a potential PSS strategy at an early planning stage. The model based on Life Cycle Assessment, Life Cycle Costing, hierarchical clustering and goal programming optimisation methods. Two forklift case studies demonstrated the model applications for leasing and renting strategies.

Keywords:

Product Service Systems Classification; Life Cycle Assessment; Life Cycle Cost

1 INTRODUCTION

Over the past years, the global trend of shifting towards selling service is widely recognised among manufacturing firms due to the economic incentive [1]. Many businesses have started to provide both products and services to their customers. This new business model is defined as Product Service System (PSS) [1][2].

In general, there are three PSS types namely product-oriented PSS, use-oriented PSS and result-oriented PSS [3]. The product ownership is shifted increasingly from customers towards manufacturers across the three PSS types. The product-oriented PSS concerns a product that a customer owns a product, while the service is provided by a manufacturer. Therefore, the ownership is with the customer such as a product under a warranty period. The use-oriented PSS is when a customer leases or rents, therefore the ownership remains with the manufacturer. The result-oriented PSS is for a case of outsourcing. A customer uses the service without owning or operating the product such as a limousine service, where a car manufacturer providing limousines with chauffeurs.

Clearly, it is not only the shifting of ownership but also the manufacturer responsibility from the sourcing material, fabricating activities to cover the entire product life cycle. The additional activities of the product include operating, maintaining and remanufacturing, and disposing activities. In this context, PSS allows manufacturers to have a full control of the entire life cycle activities. This creates an opportunity for manufacturers to provide sustainable PSS for a customer. Manufacturers gain additional revenues over the fixed production and service providing costs by maximising the utilisation of products. The maximisation is achieved by the value added from extending product's service lifetime such as employing end-of-life strategies like reuse, repair, refurbishing and reconditioning [1].

PSS is also recognised as a sustainable strategy because it minimises numbers of new products being manufactured to satisfy the same number of customers. The environmental impact of the product life cycle is reduced because less material and energy are consumed. Nevertheless, PSS in practice may not straightforwardly minimise environmental impacts [5]. Additional environmental impact may incur due to the changing circumstances of the extended service activities and conditions of product use. For instance, the use-oriented PSS such as a product rental service offer a product and delivery service to a customer. In this case, one product is shared by many customers. This reduces dramatically the number of purchased products but fuel consumption is increased from the delivery service. Therefore, PSS ecological improvement can be surpassed by higher environmental load from extra transportation of product delivery to customers during servicing periods [6].

Evidently, PSS strategies affect environmental impact and costs associated with the entire product life cycle. A number of studies have attempted to assess either Life Cycle Assessment (LCA) or Life Cycle Costing (LCC) on a case-by-case basis [1][2]. One research also assessed both LCA and LCC but only for one specific case study [7]. These studies indicate the complexity of implementing sustainable PSS because it involves with many product and service factors. Uncertainties of these factors can be found across the product life cycle activities. On this remark, it can be quite a challenging estimation for manufacturers to identify sustainable PSS potentials for their products. This is particularly true when manufacturers are unfamiliar with PSS concept as well as lack of an associated input data.

Thus, this paper aims to deliver a preliminary simplified model for manufactures to make a decision in implementing sustainable PSS at their early planning stage. The model development is presented in the methodology section which is followed by results and discussion section to present the model applications. The conclusion section summarises the content and a future work of this paper.

2 METHODOLOGY

2.1 Methodology overview

On the whole, the model was aimed to provide a simplified decision making model. Manufacturers can use the model in assessing the

J. Hesselbach and C. Herrmann (eds.), *Functional Thinking for Value Creation: Proceedings of the 3rd CIRP International Conference on Industrial Product Service Systems, Technische Universität Braunschweig, Braunschweig, Germany, May 5th - 6th, 2011*,
DOI 10.1007/978-3-642-19689-8_32,

PSS potentials and their service parameters during the early planning where experience and input data are limited. Fundamentally, the proposed model was developed through four main steps. These steps are: 1) identify PSS characteristics; 2) collect PSS case studies; 3) cluster PSS case studies into groups based on their identified PSS characteristics; and 4) specify service parameters that are associated with the PSS characteristics.

Firstly the product (internal) and business (external) characteristics that influence both LCA and LCC of PSS were reviewed from associated literature. Secondly, the characteristics of vast successful 36 PSS cases were collected majorly from various successful PSS companies and extensive literature [8][9][10][11][12].

Thirdly, the hierarchical clustering technique was subsequently used to study the commonalities of the collected data of the characteristics. This was because numerous qualitative and quantitative data of the characteristics were retrieved but no meaningful conclusion could be directly obtained. As a result, eight clustered groups with the similar characteristics were produced. Hence, the model was a case-specific model that developed from literature and vast PSS case studies. These eight simplified groups are quite suitable for manufacturers to use at the preliminary stage, where accuracy and input data are limited.

Finally, the goal programming optimisation was employed to optimise the PSS characteristics, that predominantly influence the LCA and LCC of PSS. A sensitivity analysis as well as two LCA and LCC parametric models were used in this step to identify the dominant PSS characteristics.

On the occasion, the LCA parametric model was aimed to estimating the environmental impacts of the entire product life cycle using the simplified LCA approach [19]. The LCC parametric model also used the same approach and estimated the cost from the manufacturer's point of view. These means only costs that are likely to be associated with the manufacturers were considered in the model. This was due to two main reasons. The first reason is the ease of cost data collection. The included cost elements are simple for manufacturers to collect at the preliminary planning stage as they are often readily available at the manufacturing plants.

The second reason is that it can be used to estimate fixed costs for future product design thresholds and pricing strategy. Moreover, it can also be included with a social cost component such as willingness-to-pay, to estimate the total economic value as described in environmental economics. For instance, a manufacturer can use the estimated LCC to redesign the product by reducing its net weight and to set a suitable PSS price to maximising its profit margin from the potentially high PSS operation costs. As a result of the fourth step, four highly sensitive service parameters were discovered namely 1) the sharing factor; 2) total delivery distance of a product; 3) total distance for maintenance service and 4) total distance for remanufacturing.

For the ease of the paper presentation, a preliminary version of the model is demonstrated in the next section as a framework of the proposed model. The first two service parameters for the use-oriented PSS are included in the model presented in this paper.

2.2 PSS characteristics and clustering

Identification of PSS characteristics

The business and product characteristics of PSS were identified from literature using the four following criteria, PSS characteristics:

- must represent product and service attributes,
- must link to the service activities that influence environmental impacts,
- must link to the service activities that are the product cost drivers,
- must be sufficient to differentiate between different PSS concepts.

As a result, 16 final PSS characteristics as listed in Table 1 were statistically identified from a number of PSS characteristics found in literatures. The PSS characteristics in Table 1 were classified into external (business) and internal (product) characteristics of PSS. The external characteristic mainly includes use and technology characteristics. The internal characteristic largely covers product, maintenance and remanufacturing characteristics.

External characteristic	Example of collected data
1. Usage behaviour of customer	Either >3 times/week, 1-3 times/week or < 3 times/year)
2. Location of usage	Either customer or service provider site
3. Single or multiple customers using a product	Either single or multiple
4. Technology cycle	1 to 10 years
5. Redesign cycle	1 to 5 years
6. Reason of redesign	5 possible reasons eg. original design, feature and function
Internal characteristic	**Example of collected data**
7. Service life	Varies from 2 to 12 years
8. Weight	Varies from 0.03 to 1292 kg
9. Dimension	Varies from 3E-04 to 8.92 m^2
10. Energy consumption for operation	Either consume or not consume energy
11. Percentage of total maintenance cost to product's price	Varies from 5% to 40%
12. Percentage of reused part after reconditioning process	Varies from 0% to 40%
13. Number of parts Items	Varies from 1 to 14,000 parts
14. Number of modules Items	Varies from 1 to 20 modules
15. Number of fasteners Items	Varies from 0 to 500
16. Percentage of parts reused after remanufacturing	Varies from 0% to 70%

Table 1: PSS characteristics and examples of collected data.

PSS case studies collection and clustering analysis

The data of the identified internal and external characteristics were collected from 36 PSS of leasing and renting from literatures and companies [13][14][15][16][17][18]. Company data was collected by posting a questionnaire to PSS companies and interviewing company owners and experts. The questionnaire for the entire product life cycle used in selling service comprised of characteristics, product data, service activities and associated cost. Examples of collected data from the case studies are also given in Table 1. Such data is from the renting of a DVD movie and a hand tool as well as the leasing of an air-conditioning and a car. Noticeably, the data has categories, discrete and continuous data types that varies significantly.

Thereafter, the PSS characteristics of the collected PSS case studies were clustered using the hierarchical clustering method from SPSS version 12.0. The primary goal was to form simplified and meaningful PSS groups that have similar PSS characteristics.

The method was chosen as it is suitable for a small sample size of data that is moderate less than 1,000.

As a result, eight clustered PSS groups were obtained as listed in Table 2 with their descriptions. The product characteristics of each PSS group are in the variation of:

- size: small (< 3kg and < 0.3m^3), medium (3-100kg and 0.3-5m^3)and large (< 100kg and > 5m^3) product;
- energy used during the usage stage: active or energy-based and passive or material-based product;
- maintenance requirement (percentage of total cost during life span to the product price): low (<10%), medium (10-30%) and high (>30%) product;
- remanufacturing eligibility: low (<20%) and high (>20%) product.

Examples of case studies for each group are 1) car leasing, 2) car renting, 3) DVD movie renting, 4) table renting, 5) baby cot leasing, 6) television leasing, 7) coined washing machine and 8) hand tool renting. As it can be observed, the eight clustered groups present only the common PSS case studies found. Other possible use and product characteristics at this stage are unavailable due to the limitation of case studies. However, a manufacturer may use the LCA and LCC parametric models in the next section to assess the optimisation of the service parameters for a specific product such as a small active product with high maintenance cost.

PSS Groups	1	2	3	4	5	6	7	8
Use characteristic: Frequent (F) >3 times/week; Normal (N) <3 times/week; Special (S) <3times/year								
Use frequency: F/N/S	F	S	S	S	F	F	N	S
Product characteristic: size, energy used, maintenance level and remanufacturing eligibility								
Size: Small (sl); Medium (md); Large (lg)	Lg	Lg	Sl	Md	Md	Md	Md	Md
Energy used: Active (A); Passive (P)	A	A	P	P	P	A	A	A
Maintenance: Low (L);Medium (M); High (H)	H	H	L	L	M	M	M	M
Remanufacturing eligibility: Low (L); High (H)	H	H	L	L	H	H	H	H

Table 2: Clustered PSS groups of the use-oriented PSS type.

2.3 Specification of service parameter's values for the PSS Clusters

This step aimed to specify the values of service parameters for each PSS clustered groups using the LCA and LCC parametric models. This is to provide a quantitative estimation for manufacturers to use as a guideline in implementing the sustainable PSS. The specification was started by revisiting the service parameters for all PSS characteristics found in literatures and case studies. The characteristics include 25 cost-related parameters and 37 environmental impact-related parameters, which were subsequently used to develop two parametric models for evaluating LCA and LCC of PSS. The parametric model of LCA for the entire life cycle of PSS is given in Equation (1). It was based on the SLCA approach [19]. In this case, the quantitative data of the 37 environmental impact-related parameters were used with the SLCA drivers.

$$I_{SS} = I_M + I_P + I_U + I_{EOL} \qquad (1)$$

Where

I_{SS} is environmental impact indicator of selling service

I_M is environmental impact indicator of material acquisition stage

I_P is environmental impact indicator of production stage

I_U is environmental impact indicator of product use stage

I_{EOL} is environmental impact indicator of end of life stage

The LCC parametric model from the manufacturer's point of view in Equation (2) was developed in a similar manner as the LCA parametric model. Thus, 25 cost parameters were used in Equation (2) to estimate the life cycle costs of PSS.

$$C_{SS} = C_A + C_O + C_D \qquad (2)$$

Where

C_{SS} is cost of selling service (US dollars)

C_A is acquisition cost (US dollars)

C_O is operation and support cost (US dollars)

C_D is disposal cost (US dollars)

The equation was calculated by the acquisition, operation and support, and disposal costs from the manufacturer's view in a present value. The acquisition cost is the cost involved in manufacturing the product including material, labour and process overhead costs. The operation and support cost is the cost incur while using the product. This cost considered delivery, energy, maintenance and remanufacturing costs. The disposal cost is the salvage value at the end of product life when the product is disposed. Thereafter, a sensitivity analysis was performed using the two developed parametric models to identify the service parameters, that have significant impact towards the produced LCA and LCC results. As a result, four service parameters were identified namely: 1) the sharing factor; 2) total delivery distance of a product; 3) total distance for maintenance service and 4) total distance for remanufacturing.

Sharing factor and the total delivery distance are the two service parameters presented as a demonstration for the development of the decision making tool. The definitions of these parameters are given as follows. The sharing factor, 'S' as shown in Equation (3) was defined as a multiplication of the usage ratio and the product life span ratio. The usage ratio in Equation (4) entails the number of customers that share the same product. It is a ratio of the 'product capacity for use per year' [2] and customer's usage intensity per year. For example, more customers share the product when either the 'product capacity for use' is high or when customers have low usage intensity due to they use the product occasionally.

The product life span ratio in Equation (5) represents an additional number of customers that can share the PSS product during its possible product life span. It is the ratio between the total possible life span of the PSS product and the life span of the standard product.

$$\text{S} = \text{Usage ratio} \times \text{Product life span ratio} \qquad (3)$$

$$\text{Usage ratio} = \frac{\text{Product capacity for use per year}}{\text{Customer's usage intensity per year}} \qquad (4)$$

$$\text{Product life span ratio} = \frac{\text{The total possible life span of PSS product}}{\text{Life span of a standard product}} \qquad (5)$$

The total delivery distance of a product, 'd' was defined as the total distance during one year service, which manufacturers travel to delivering the product to their customers. The longer the delivery distance is, the higher environmental impact and costs are. The total delivery distance is a service parameter that depends largely on other PSS characteristics. For instance, total delivery distance is usually increased for a short rental period that has a high delivery frequency. Finally, a goal programming which is a multi-objective optimisation method was employed to assess the optimised values of the highly sensitive service parameters. The optimisation was conducted using the two LCA and LCC parametric models under the constraints of the two optimisation objectives. The following subsections provide a brief explanation for the optimisation method.

2.4 Multi-objective optimisation using goal programming

Goal programming method was used to optimise the values of the two specified service parameters namely the sharing factor and total delivery distance. This multi-objective decision making approach simultaneously optimised two objectives (goals) using the two parametric models. The objectives were related to the environmental and economical aspects which are described as follows.

- Objective 1: Cost of selling service is less than or equal to that of the selling product (s)

This objective was set to guarantee that the cost of selling service is less than or equal to that of the selling products. This term can be specified mathematically by Equation (6). The sharing factor (s) presents the reduction in the number of new products. The cost of selling service (C_{ss}) is generated by the maximum total delivery distance which is one of the service parameters.

$$C_{SS} \leq \; s \times C_{SP} \tag{6}$$

Where

C_{SS} is cost of selling service

C_{SP} is cost of selling product

s is sharing factor

- Objective 2: Environmental impact of selling service is less than or equal to that of the selling product (s)

The second objective is to guarantee that the environmental impact of selling service is less than or equal to that of the selling products. This term can be specified mathematically by Equation (7). The sharing factor (s) presents the reduction in the number of new products. The maximum total delivery distance is a service parameter in the environmental impact of selling service (I_{ss}).

$$I_{SS} \leq \; s \times I_{SP} \tag{7}$$

Where

I_{SS} is environmental impact indicator of selling service

I_{SP} is environmental impact indicator of selling product

s is sharing factor

These two objectives were utilised by the goal programming to produce the optimal sharing factor and the total delivery distance for each clustered group. Therefore, the optimised service parameters of PSS can be used to ensure that its LCA of the entire life cycle and the LCC from the manufacturing's point of view are less than or equal to that of the selling product business mode.

2.5 PSS decision model

Henceforth, the optimised values of the sharing factor and the total delivery distance were interpreted in simple set of equations as listed in Table 3. At the preliminary planning stage, these equations in Table 3 can be used by manufactures as a guideline for two applications. The first application is when a manufacturer whom would like to introduce PSS into their current selling business model. The second application is for a PSS manufacturer whom would like to apply a new PSS strategy into their current PSS business model. Manufacturers may apply this proposed model during their preliminary decision making by following two steps. Firstly, the manufacturer matches their product and the use characteristics to the PSS groups in Table 2. Secondly, the manufacturer calculate the optimised values of either the sharing factor or the total delivery distance by using the corresponded equation of the matched PSS group in Table 3. The selection of the equation depends on which value is known or can be estimated.

PSS use-oriented group	Total delivery distance, d (km per year)	Sharing factor, s (units)
G1	d = 332s-930	s = 0.003d+2.8012
G2	d = 432.4s-840	s = 0.0023d+1.9426
G3	d = 86.3s-843	s = 0.0116d+9.7683
G4	d = 219.8s-1756	s = 0.0045d+7.9891
G5	d = 318.2s-738	s = 0.0031d+2.3193
G6	d = 217.33s-347.67	s = 0.0046d+1.5997
G7	d = 149.4s-204	s = 0.0067d+1.3655
G8	d = 747.08s-10347	s = 0.0013d+13.849

Table 3: PSS optimised equations for eight PSS groups.

3 APPLICATIONS

Thus far, the methodology of the model was presented which produced eight PSS groups and their PSS optimised equations for the service parameters. This section shows the two prescribed model applications using two forklift cases studies in Table 4. The cases represent the selling and renting business areas for a relatively complex product (large, active and, high maintenance and remanufacturing ability). It would require extensive input data if traditional LCA and LCC are to be assessed for the PSS potentials.

The first case study shows the model application where a manufacturer whom would like to introduce PSS into their current selling forklift business model. It has a product life span of 8 years. This means the current business model only involves with making and delivering a product to a customer. The second case study shows the model application for a manufacturer whom would like to improve the sustainability of their current PSS business model. In this case, the company already involves not only with making and delivering a product to a customer but also with operating, maintenance, remanufacturing and disposing the product. The product life span of Case studies 1 and 2 extends to 24 years due to the remanufacturing process that is conducted every 8 years.

Life cycle activities	Case study 1	Case study 2
Materials	Core materials such as steel, aluminium, and rubber	
Production	Electricity consumption	
Delivery	Transportation	
Operation	Not applicable due to as a manufacturer point of view, these activities are conducted by a customer.	High electricity consumption
Maintenance		Spare parts and transportation
Usage: Remanufacturing		Spare parts, Electricity consumption, and transportation
Disposal		Landfill and recycling

Table 4: A manufacturer's point of view on current life cycle activities of two case studies.

3.1 Case study 1 results and discussion

Case study 1 presents the application of a manufacturer who currently sells forklift would like to introduce the leasing forklift business model to its customers. The manufacturer selects G1 group from Table 2, because G1 group's product characteristics are matched with the forklifts. The use characteristic of G1 group is also suitable for a customer who regularly uses as expected by the company's customer requirement records.

Thus, the company is able to estimate the sharing factor of the forklift as 3. This number is derived from the 3 high usage intensity customers (use regularly for five times a week) are expected to lease the forklift during its extended life span of 24 years. The extended years are due to 2 remanufacturing processes that triple the original life span of selling forklift. Subsequently, the sharing factor of 3 was used to calculate the total delivery distance of 66 km per year by using G1 equation (d = 332s-930) in Table 3.

The LCA for the entire life cycle and the LCC from the manufacturing's point of view for Case study 1 (selling product) and the PSS potential business that has total delivery distance of 66 km per year (Applied G1) are compared in Figure 1. The functional unit of these two scenarios is to use the forklift 5 days per week during 24 years. In this case, Case study 1 requires 3 forklifts but the Applied G1 required 1 forklift with 2 remanufacturing processes.

The environmental impacts of the applied PSS G1 (leasing product) is 1,524 points which is considerably smaller than 2607 points of Case study 1 (selling product) by 41.5%. The LCC result of this forklift in Figure 1 is also significantly reduced by 34% when employing the leasing strategy in lieu of selling the 3 forklifts.

The life cycle environmental impacts and costs of each life cycle stages were compared further in Figure 2 to highlight the reduction sources of the leasing scenario. According to Figure 2, both environmental impacts and costs of leasing product are significantly decreased even though certain environmental impacts and costs are gained from the additional maintenance and remanufacturing activities. This is largely due to the significant saving from the material and production stages of the leasing scenario.

Therefore, at this preliminary planning stage the manufacturers can improve their cost and environmental impacts by employing PSS business model with sharing factor of 3 and the total distance of 66 km.

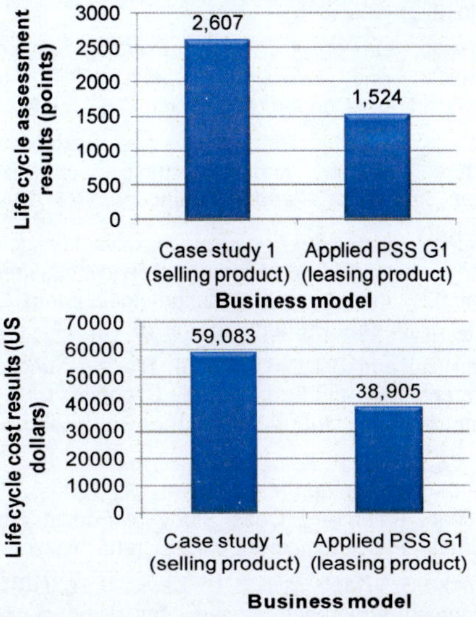

Figure 1: LCA and LCC results of Case study 1.

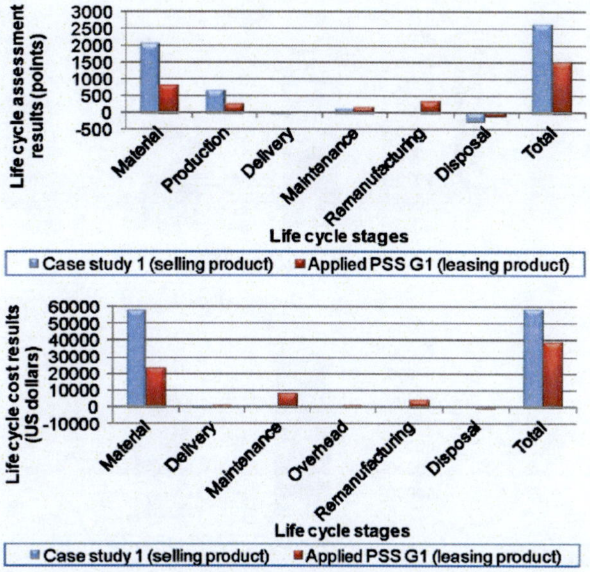

Figure 2: Product life cycle stage results for LCA and LCC of Case study 1.

However, if the total distance is known and is likely to be more than 66 km, a manufacturer has to either increasing the sharing factor or redesigning the product and increasing the leasing price. Otherwise in this non-complied situation, selling product may be a better option than establishing the PSS system.

3.2 Case study 2 results and discussion

Case study 2 assumed that another manufacturer has already implemented a renting service for a forklift. The company would like to improve the environmental impacts and the operation costs of the current renting service system. Therefore, this company reviews a renting record of their existing customer during the 24 years of PSS life span. The record reveals that the current renting forklift has been rented by five customers who are a low usage intensity customer (use 10 days per year during two week period). The total delivery distance is 2,000 km per year.

According to the information, the manufacturer can improve the existing service system by selecting G2 group from the PSS groups in Table 2. G2 group is selected as its product characteristics are matched with the rental forklift and customers use the forklift sporadically or for special occasion such as using for loading the delivered goods. The sharing factor of 5 is used to calculate the total delivery distance using G2 equation (d = 432.4s-840) from Table 3. This number is the 5 sharing customers from the renting record. The optimised total delivery distance of new renting service is calculated as 1,322 km per year from the proposed model.

The LCA and LCC of the PSS's new distance (Applied PSS G2) are compared with Case study 2 in Figure 3. The functional unit of these cases is to use a forklift for 50 days per year over 24 years. The results show that the current rental forklift, Case study 2 can reduce significantly by 29.5% of LCA result and 23% of LCC result when Applied PSS G2 that applies the optimised total delivery distance. Likewise, if in occasion where 1,322 km is estimated to reach before the end of the year, then at this preliminary planning stage, the manufacturer have to seek for a solution to either increasing the sharing factor or redesigning the product and increasing the leasing price. Otherwise, the manufacturers can be aware of that although the current PSS system is profitable but its environmental impact is not as good as selling 5 forklifts.

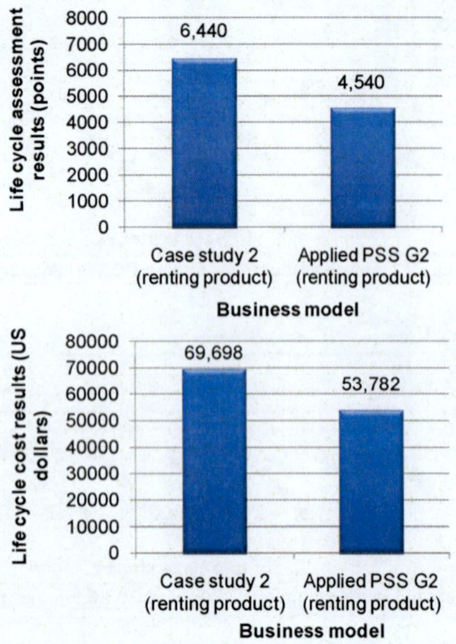

Figure 3: LCA and LCC results of Case study 2.

4 CONCLUSION

This paper proposed a simplified model for manufactures to make a decision in implementing sustainable PSS at their preliminary planning stage. This partial model was developed based on actual PSS case studies to identify and cluster PSS characteristics using the hierarchical clustering method. Two parametric models of LCA for product life cycle and LCC from the manufacture's point of view were developed for the sensitivity analysis and goal programming optimisation method. The model outcome was the PSS optimised equations for eight PSS groups. The model applications were shown using leasing and renting forklift cases. The results shown that the model can offer the optimised total delivery distance and the sharing factor values, which reduce the estimated LCA and LCC. The model can be used as a guide for manufacturers at a preliminary stage to sustainably move from selling product to PSS and improve their current PSS. However, it does not always leading to the better solution, if manufacturers cannot comply with the suggested values. In this unfortunate case, the manufacturers may have to either adjusting other service parameters or redesigning the product and resetting the service price. Thus, the proposed model is not a substitute model for a full LCA and LCC at the detailed planning stage. Future work can be continued in several areas. Another two parameters namely total distance for maintenance service and total distance for remanufacturing should be included in the optimisation process. The model can be extended to cover the result-oriented PSS type. Other cost elements can be added such as the social cost. Lastly, more case studies can be further collected to enhance the model's product and use characteristics.

5 ACKNOWLEDGMENTS

The authors would like to express our sincere gratitude to companies who provided the valuable data for this study. The warmest thanks also go to our colleagues at the Life Cycle Engineering & Management Research Group at UNSW, Australia for their helpful suggestions. Lastly, special recognition must be given to the UNSW and the Dhurakij Pundit University, Thailand for their financial support.

6 REFERENCES

[1] Meier, H., Roy, R., Seliger, G., (2010): Industrial Product Service Systems–IPS², Manufacturing Technology, in: Annals of the CIRP, Vol. 59, No. 2, pp. 607-627.

[2] Mont, O., (2004): Product Service-Systems: Panacea or Myth? PhD Thesis, Lund University, Sweden.

[3] Tukker, A., and Tischner, U., (2006): New Business for Old Europe, Sheffield, Greenleaf Publishers, UK.

[4] Wong, M., (2004): Implementation of Innovative Product Service Systems in the Consumer Goods Industry, PhD Thesis, Cambridge University, UK.

[5] Mont, O., (2000): Product Service-Systems, Final Report, IIIEE, Lund University, Sweden.

[6] Mont, O., (2002): Clarifying the Concept of Product Service Systems, Journal of Cleaner Production, Vol. 10, No. 3, pp. 237-245.

[7] Haapala, K., Jordan, K., Sutherland, J., (2008): A Life Cycle Environmental and Economic Comparison of Product Service Systems, Transactions of NAMRI/SME, Vol. 36, pp. 333-34.

[8] Rose, M., (2000): Design for Environmental: A Method for Formulating Product and End-of-Life Strategies, PhD Thesis, Stanford University, the United States of America.

[9] Zwolinski, P., (2008): Remanufacturing strategies to Support Product Design and Redesign, Journal of Engineering Design; Vol. 19, No. 4, pp. 321-335.

[10] White, A., Stoughton, M., Feng, L., (1999): Servicizing: The Quiet Transition to Extended Product Responsibility, U.S. Environmental Protection Agency.

[11] Fishbein, B., McGarry, L., Dillon, P., (2000): Leasing: A Step towards Producer Responsibility, INFORM, New York.

[12] Sundin, E., Ostlin, J., Ronnback, A., Lindahl, M., Sandstrm, G. (2008): Remanufacturing of Products Used in Product Service Systems Offerings, The 41st CIRP Conference on Manufacturing Systems, pp. 537-542.

[13] Mont, O., Dalhammar, C., Jacobsson, N., (2002): A New Business Model for Baby Prams Based on Leasing and Product Remanufacturing, Journal of Cleaner Production, Vol. 10, No. 3, pp. 237-245.

[14] Goedkoop, M., et al., (1999): Product Service systems, Ecological and Economic Basics, PRè consultants/ PricewaterhouseCoopers N.V/ Storm C.S., the Netherlands.

[15] Besch, K., (2005): Product Service Systems for Office Furniture: Barriers and Opportunities on the European Market, Journal of Cleaner Production, Vol. 10, No. 3, pp. 1083-1094.

[16] Williams, A., (2007): Product Service Systems in the Automobile Industry: Contribution to System Innovation? Journal of Cleaner Production, Vol. 15, No. 11, pp.1093-1103.

[17] Sundin, E., and Bras, B., (2005): Making Functional Sales, Environmentally and Economically Beneficial through Product Remanufacturing, Journal of Cleaner Production, Vol. 10, No. 3, pp. 913-925.

[18] Sundin, E., Larsson, M., Nielsen, A., (2005): "Design for Functional Sales – a Case Study of Forklift Trucks at BT Industries", Proceedings of EcoDesign-05, Tokyo.

[19] Manmek, S., Kaebernick, H., Kara, S. (2010): Simplified Environmental Impact Drivers for Product Life Cycle, International Journal of Sustainable Manufacturing, Vol. 2, No. 1, pp. 30-65.

Indoor Location Analytics for Designing a Location-Based Product-Service System

Sun K. Kim[1], Jin-Wook Ro[1]

[1] Graduate School of Systems Design and Management, Keio University, Yokohama, Japan

Abstract

As location-related technologies and devices have become increasingly available, service offerings based on location-context information are gaining much attraction. This paper describes a new location-based service model, Indoor Location Analytics, a network-based service that helps store owners or building managers of a structure or a shopping mall, to analyze the traffic information of customers. Just as existing web analytics services benefit website managers, Indoor Location Analytics benefit store owners, building managers, and shoppers by providing a common data platform service.

Keywords:

Location-Based Services; Web Analytics; Service Platform

1 INTRODUCTION

With maturity of location technologies such as GPS, Assisted GPS, mass adoption of Smart phones, mobile networks and supportive policies such as E911 [1], which is a seamless caller's position data transmission service through 911 calls, outdoor location-based services are proliferating. In the future, emerging technologies such as Indoor Messaging System [2], Wi-Fi, RFID, Infrared Ray, and Visible Light [3] will pave the road for indoor location-based services. Based on this trend, this paper describes a service concept—Indoor Location Analytics. Indoor Location Analytics is a service that provides store owners and building owners with data on visitors of an indoor structure.

In the process of developing a viable business model for the novel Indoor Messaging System (IMES) technology, our previous research [4] shows business-to-business model is financially more viable compared to business-to-customer model in the early stages of technology adoption. In other words, due to lack of infrastructure and possibility of slow customer adoption rate in the beginning phase, the financial sustainability of the service relies better when the technology is acquired and used between businesses. One target of the service model using indoor GPS or IMES technology was Indoor Location Analytics. Applying a systems design framework, Scenario-based Amorphous Design [5], the research team sought to find value for the stakeholders who would invest upfront in the infrastructure for location-based services. The key stakeholders were building managers and store owners because they would make the decisions whether to set up the infrastructure or not. One way to provide value to these key stakeholders was by helping them to increase the profitability of their current businesses. Seeing how web analytics can help website owners to improve their business and visibility, the research concluded that similar principles would apply to the physical domain. Web analytics benefit owners and managers of websites, but for owners of physical stores or buildings, such services do not exist yet. Whereas Internet services have analytic tools to visualize the traffic flow or their marketing performance, physical stores cannot take advantage of such a service.

2 THE ANALOGY OF A BUILDING TO A WEBSITE

In terms of structure, an indoor structure or a building can be compared to a website. As the visual comparison shown on Figure 1, a typical building structure consists of multiple floors. In a floor, there are many rooms. Similarly, a website consists of multiple web pages. Each web page has numerous clickable icons in most cases.

If a person chooses to enter a building, this behavior is analogous to a person entering a website either by typing in the URL address or clicking on search results or a link. A webpage corresponds to a floor of the building. Just as a website consists of many sub pages, a building also consists of multiple floors. If a person views a page, then he or she is observing information on the page and is in the process of deciding the next step for a click. Finally a room corresponds to a clickable icon. Simplified table view of this concept is shown in Table 1.

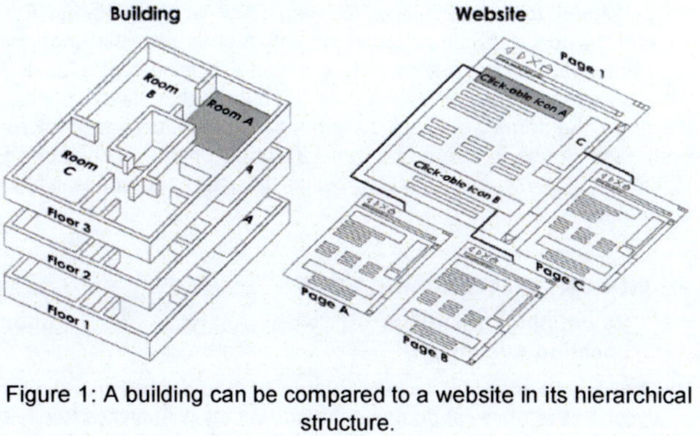

Figure 1: A building can be compared to a website in its hierarchical structure.

J. Hesselbach and C. Herrmann (eds.), *Functional Thinking for Value Creation: Proceedings of the 3rd CIRP International Conference on Industrial Product Service Systems, Technische Universität Braunschweig, Braunschweig, Germany, May 5th - 6th, 2011*,
DOI 10.1007/978-3-642-19689-8_33,

Indoor Location Analytics	Web Analytics
Building	Website
Floor	Page
Room	Click-able icon

Table 1: How definitions of metrics, dimensions and features translate from Web Analytics to Indoor Location Analytics.

3 REVIEW OF WEB ANALYTICS

Web analytics is a common tool used for collecting web user's data on Internet websites. The primary method of gathering data is by tracking and recording the users' actions. This data is later used for optimizing the performance of a website [6]. Recently, the trend of usage of web analytics has spread to generating profit through managing advertisements [7]. Major information technology companies such as Google and Yahoo are servicing web analytics to web page owners. Owners are able to select various types of analytics depending on the kind of service they are providing.

There are mainly two methods of collecting data – logfile analysis and page tagging. Logfile analysis method records visitor's status and movement between pages on the web server. The web server will count the number of requests of particular page and log them into a file. In the early days of Internet, page views and visits were introduced as measurements in the logfile [8]. As the Internet services grew, technologies such as web proxies, dynamic IP addresses, cache, etc. became available and made the Logfile obsolete due to technical complexity in tracking unique visitors.

Page tagging method is a more advanced technology compared to the logfile analysis. Instead of web server carrying out the logging, each page has a distinct tag that counts the number of time the page has been opened then sent it to the server through JavaScript [9]. The process is more accurate since the counter is page-oriented rather than server-oriented. Every movement between pages conducted by the visitor is temporarily stored in the visitor's computer. As a result, the number of page views and movement path of unique visitors are significantly more accurate than would be by using logfile analysis.

Another highlight of page tagging is the usage of cookies. Cookies are assigned to unique visitor to keep track of their status. For example, when a visitor performs an online shopping, the shopper puts potential purchase items into a virtual shopping cart. Cookies will record the items so the next time when the same visitor re-visits the page, the items would still be in the cart. Since, cookie provides both ease of use to users and more accurate unique user-tracking to the website owners, many websites have already adopted this technology.

4 INDOOR LOCATION ANALYTICS

4.1 Re-defining Elements of Web Analytics for Indoor Location Analytics

By applying analogous principles, elements, and dimensions of web analytics to the physical domains, Indoor Location Analytics service can benefit building owners, store owners and shoppers by providing location information. Table 2 shows how some metrics and dimensions of web analytics [10] would translate into those of Indoor Location Analytics. Naturally, not all metrics and dimensions can be translated into the physical location domain but this section describes a few of the common ones.

The first column lists the metrics, dimensions and features that Indoor Location Analytics can borrow from web analytics. The second column and the third column show the difference between the definitions of the corresponding metrics..

Elements	Definition in Web Analytics	Definition in Indoor Location Analytics
Bounces	Counts single-page visit to the website.	Indicates a customer who enters an area, but leaves in a short period of time.
Entrance (visits)	Counts entrances (visits) to the website. An entrance consists of a single-user session.	Indicates number of enters that are made to the area. This data can be used for calculating customer flow rate, capacity fill rate of the area, distribution of customers etc.
Exit	Counts exits from the website.	Indicates number of customers who leave the area. This data can be used for calculating customer flow rate, capacity fill rate of the area, distribution of customers etc.
New Visit	Counts first-time visits to the website.	Indicates the customer who came in the area for the first time. This data can be used as direct measurement of the growth of the business and effectiveness of the advertisement.
Time on Page	Records how long a visitor spent on a page.	Time on Area indicates the customer's total time spent inside a selected area. This can be used to control customer density in the area.
Visitors	Counts the total number of visitors to the website.	Indicate the total number of customers in the area. A single customer is eligible to make multiple entrances or exits. This data can be used for calculating area's optimal usage, and profit per customer.
Advertisement Click	Counts number of clicks made to the advertisement linked to the website.	Advertisement Visit indicates the customer who saw the advertisement prior to the entrance to the area. This data can be used to find the optimal location, type, and number of the advertisement.
Cost per Click	Cost to advertiser per click.	Cost per Visit indicates the cost needed to pay to the advertiser depending on the effectiveness. This is to give incentives to the advertiser, so they would want to advertise the area.

Elements	Definition in Web Analytics	Definition in Indoor Location Analytics
Search Depth	The average number of subsequent page view after using internal search feature.	Indicates average number of areas that the customer visited before he or she arrives at the destination. Requires a search feature. This data can be used to improve the layout of area.
Search Exits	Count number of exits on the site that occurred following a search result from the internal search feature.	Indicates the number of customers who could not find their search entities. This data is capable of showing the needs of the customers.

Table 2 : Corresponding elements between Indoor Location Analytics and web analytics.

In web analytics, bounce is a counter for single-page visits to the website. In Indoor Location Analytics, bounce indicates a customer who enters an area, but leaves in a short period of time. A bounced visitor may have been curious initially but soon lost interest and was not led to goals such as transactions. Another possibility is that the visitor came to the place unintentionally. Reducing bounce can make efficient traffic flow among customers that will lead to transactions and can indicate the efficacy of advertisement.

An entrance, in web analytics, is a counter for the number of visits to the website. Similarly, in Indoor Location Analytics, entrance indicates the number of visits to a certain area. This data can be used for calculating customer flow rate, capacity fill rate of the area, distribution of customers, etc. This metric is useful when combined with other dimensions to determine which area served as an entrance to the building or a store.

Exit in web analytics is the number of exits from a website. In Indoor Location Analytics, exit indicates the number of customers who leave an area. This data can also be used for calculating customer flow rate, capacity fill rate of the area, distribution of customers etc. in conjunction with the metric, entrance.

New visit, in web analytics, is a counter for the number of first-time visits to the website. In Indoor Location Analytics, new visit also refers to the number of customers who came into the area for the first time. This data can be used as direct measurement of the growth of the business and effectiveness of the advertisement, if location information is linked with transaction information.

Time on page, in web analytics, records how long a visitor spends on a web page. The corresponding metric in Indoor Location Analytics is Time on Site, which indicates the customer's total time spent inside the selected dimensions of an area. This data can be used to control customer density in the area.

Visitors, in web analytics, is the total number of visitors to the website. Similarly in Indoor Location Analytics, visitors is the total number of customers to the area. Since a single customer may make multiple entrances or exits, there are various algorithms to calculate this metric. This metric is used to calculate the optimal usage of an area and profit per customer.

Advertisement click, in web analytics, is the number of clicks of an advertisement linked to the website. The corresponding metric in Indoor Location Analytics is advertisement visit, indicating the number of visits by customers who saw the advertisement prior to the entrance to the area. This data can be used to find the optimal location, type, and number of advertisement campaigns.

Cost per click is the cost a website owner must pay to the advertisement hosting company for a customer's click of their advertisement. The corresponding metric in Indoor Location Analytics is cost per visit, meaning the cost required to pay to the advertiser depending on the effectiveness. This is to give incentives to the advertiser, so they would want to advertise the area. For example, a small amount of money is given to the customer who informs a friend about the place.

Search depth, in web analytics, is the average number of subsequent page views after using internal search feature. In Indoor Location Analytics, search depth means average number of areas that the customer visited within a building. Collecting this metric requires a search feature for the shoppers. Search depth can be used to improve the layout of the area.

Search exits, in web analytics, is the number of exits following a search result from an internal search feature. In Indoor Location Analytics, search exits shows the number of customers who could not find what they are interested in. This data is capable of showing the needs of the customers.

Click path, in web analytics, is the sequence of hyperlinks website visitors proceeds through on a given site. A visitor's click path may begin from the website or from another website. These series of clicks through pages creates a click path. Visit path in Indoor Location Analytics refers to the sequence of visits that customers follow, from one store to the other.

4.2 Visualizing Indoor Location Analytics

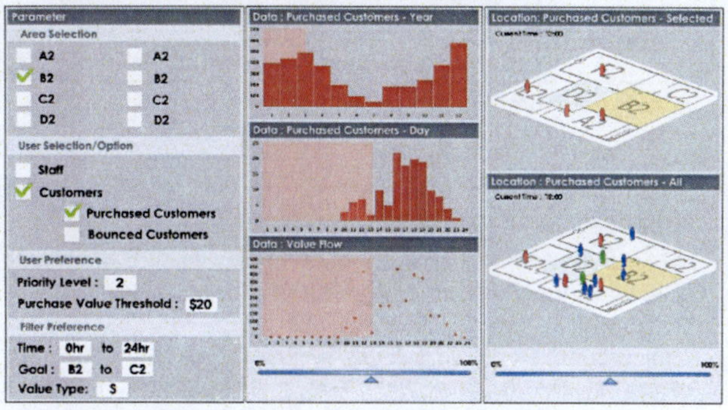

Figure 2: Concept Image of Dashboard of Indoor Location Analytics.

The first process for data mining is to set up the proper parameters. Setting the proper parameters will generate more accurate data, which leads to better results. When selecting parameters, there are two main fields to consider for analyzing location-based services – Area and User.

Area is an arbitrarily defined geographical boundary inside an indoor map. The status and interaction of user is recorded based on the selected area. A user outside of the area is neglected although they satisfy the selected user criteria. When the decision factor of selected user requires a form of interaction, such as purchasing an item, only the user who makes that interaction within the area will be counted as selection. Multiple areas may be selected for applying same user criteria.

User is an arbitrarily defined unique group of users with strict criteria. The distinct criteria of user selection will define their entities. The user groups are allowed to have sub-user groups. Although the entities are defined arbitrarily, there are some rules to follow when making criteria or rules. Failing to do so will confuse the results by having different number of total number of unique users.

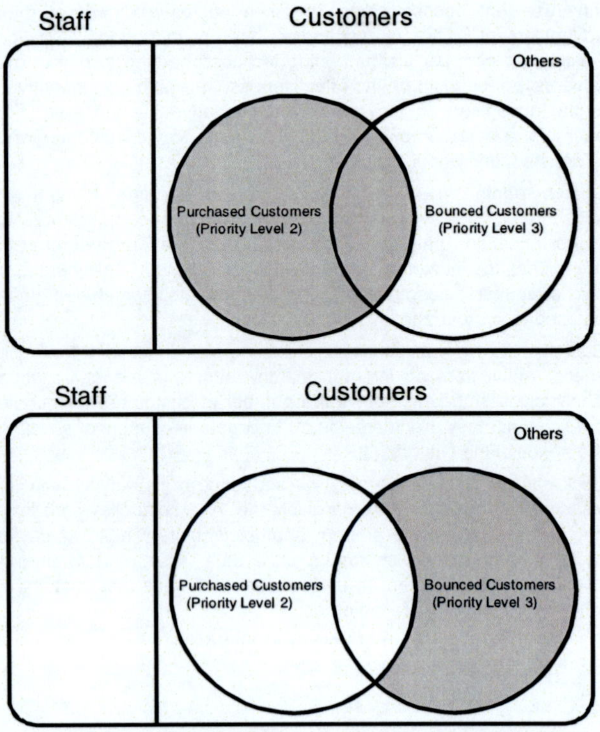

Figure 3: Set representation of Users to prevent calculation error.

The rules are (see Figure 3):

- The highest level of user type must be mutually exclusive to each other. (i.e. staff and customers in Figure 3)
- Set of sub-user must have a complementary set of sum of other sub-user of same level. For instance, complementary set of purchased customers is bounced customers and Others.
- Union selection of two sub-users must satisfy both criteria for selection. For instance, when both purchased customers and bounced customers are selected with Union option, the union set of purchased customers and bounced customers is selected.
- When a sub-user has multiple entities, priority level is set, higher priority will overwrite the lower priority entity. For instance, when a bounced user from area B2 comes back to purchase a product, the user will become a purchased customer (priority level 2) instead of bounced customer (priority level 3). Resulting selection of higher priority user vs. lower priority users are shown in Figure 3.
- A sub-user may have multiple criteria (rules) to define the entity, but the rules must not be a subset rule of one another (i.e. currency-value purchase threshold cannot be applied to bounced customers).
- Criteria or rules are set by value factor. (i.e. purchased customers are defined by customers who purchased over 20 US dollars)

Some other parameters that should be considered are Time, and Goal. Time is arbitrary chosen length of duration of data collection. Time can be divided into discrete units. Goal is arbitrary chosen criteria, which is created by multiple specific users' action or movement. A goal also requires multiple area – the areas are not required to be within selected boundary of an area. For example, selecting a group of users who started from area B2 and arrived in C2 is considered a goal.

Visualization of data is significant because it provides information that is easy to understand. Metrics of data can be selected depending on the user type, goals, and, value factor (within the criteria of user selection). However, all metrics must be associated with time to show the progress and trend of the data. Therefore a graph with metrics vs. time is the most adequate method of visualization. Another useful method is visualizing the floor map and indicating the location of selected users. This enables sensing individual's movement within the group of users.

4.3 Tracking of Visitors in Indoor Location Analytics

Some technologies, such as Wi-Fi, Bluetooth, and IMES are capable of tracking visitors in an area. These technologies are yet to be mature, however, the resolutions these technologies provide are enough to perform Indoor Location Analytics.

To acquire data of visitors, there are few prerequisites that need to be fulfilled. The building should be equipped with an indoor positioning technology along with a main server that collects the data. The visitors must carry corresponding technology's handset unit with an unique ID that can transfer data to the main server. For example, IMES technology can send position data to smartphone's GPS sensor and trigger the smartphone to send corresponding data to the server through Wi-Fi. The MAC address, which is a unique ID for smartphones, may be utilized to distinguish between user entities such as new visitors.

Nonetheless, the technology is just a tool to perform Indoor Location Analytics; despite of type of technology that is being used the analogy is still valid and applicable.

There are advantages of Indoor Location Analytics compared to web analytics when it comes to having correct total number of unique visitors. The most significant advantage comes from the fact that actual individual is performing on the restricted territory. Unlike web pages where users can land on any type of pages, Indoor Location Analytics has particular entrances and exits where the users are created and disappeared. Therefore, when one user enters the system the total number of unique user increases and does not decrease unless they proceed to the selected exit area.

Another advantage comes from each user having their personalized handsets. Web analytics also uses cookies to make unique identifiers to visitors, but the risk of the cookies being erased and disabled by JavaScript makes the result imprecise. Although there is a risk of users exchanging handsets, occurrence of this is low and if their financial information (i.e. credit card) is stored in the handset, the effect would be minor. These advantages make tracking unique visitors significantly simpler. Physical constraints and clearer criteria in defining the entity of users solve the dilemma of inconsistency in calculating the number of users.

5 CONCLUSION AND FUTURE WORK

Service offerings based on contextual location information are gaining much attraction as location-related technologies and devices have become increasingly available. We proposed Indoor Location Analytics, a location-based service model that helps store owners or building managers of a structure or a shopping mall, to analyze the marketing data of customers. Just as existing web analytics services have benefited website managers, Indoor Location Analytics will benefit the users by providing a service data platform that enables creation of hybrid commerce models.

Further research is in progress to collect empirical data, to refine the technology and design new business models in order to offer new values to location-based services.

6 ACKNOWLEDGMENTS

Part of this research was carried out with the support of the MEXT Global COE Program "Center for Education and Research of Symbiotic, Safe and Secure System Design." We thank members of Keio Aerospace and Intelligent Systems Laboratory for their contributions to this project.

7 REFERENCES

[1] Wireless 911 Services, **http://www.fcc.gov/cgb/consumer facts/wireless911srvc.html**.

[2] Kogure, S., Maeda, H., Ishii, M., Manandhar, D., and Okano, K.(2008): The Concept of the Indoor Messaging System, The European Navigation Conference.

[3] Tanaka, T., Haruyama, S. (2009): New Position Detection Method using Image Sensor and Visible Light LEDs, Second International Conference on Machine Vision, pp. 150-153.

[4] Kim, S. K., Minato, N., Busser, D., Kohtake, N. (2010): Transforming Seamless Positioning Technology into a Business using a Systems Design Approach—Scenario-based Amorphous Design, IEEE International Systems Conference, INCOSE Track.

[5] Kim, S. K., Ishii, K., Beiter, K. (2008): Scenario-based Design for Amorphous Systems, Proceedings of ASME International Mechanical Engineering Congress and Exposition, Vol. 4, pp191-201.

[6] The Official WAA Definition of Web Analytics, www.webanalyticsassociation.org/?page=aboutus.

[7] Google Ad Manager, **https://www.google.com/dfp/login/en_US/ features.html**.

[8] Sen, A., Dacin, P. A., Pattichis, C. (2006): Current trends in web data analytics, Communications of the ACM, Vol. 49 No. 11, pp 85-86.

[9] Neuman, J. (2007): The complete Internet Marketer, **http://www.thecompleteinternetmarketer.com/articles/article_howwebanalyticsworks.html**.

[10] Google Analytics Dimensions and Metrics Reference, **http://code.google.com/apis/analytics/docs/gdata/gdataReferenceDimensionsMetrics.html**.

Opportunities for enhanced Project Management in PSS Development

Patrick Müller[1], Rainer Stark[1, 2]

[1] Division of Virtual Product Creation, Fraunhofer IPK Berlin, Germany
[2] Chair of Industrial Information Technology, TU Berlin, Germany

Abstract

Development processes in product and service engineering are foremost carried out as projects. One characteristic of projects is their uniqueness. As Product-Service Systems (PSS) are intended to be customized solutions, unique development projects are natural for PSS. Nevertheless, there is no specific approach to support an integrated management of product and service development projects. In our paper, we focus on opportunities, a solution approach, and software support for PSS development project management. We provide a set of opportunities how to adapt existing approaches or how to develop new methods for the purpose off fully integrated project-based PSS development.

Keywords:
Process-Based Project Management in PSS Development; PSS Design; Software Support

1 INTRODUCTION

Development processes in product and service engineering are foremost carried out in project form. One characteristic of projects is their uniqueness. As Product-Service Systems (PSS) are catered to be customized solutions, unique development projects are natural for implicit and explicit PSS (a definition is given in section 2.2). Nevertheless, there is no particular method or software to support an integrated and sophisticated management of product and service development projects, which especially addresses PSS development projects.

1.1 Motivation and Problem Statement

An interview study with thirteen German companies offering products, services and integrated solutions has shown that project-based development is a means to offer customer-oriented solutions and actually their daily business (all offering business-to-business solutions) [1]. In an inquiry 10 of 17 German mechanical engineering companies answered that they have carried out integrated development projects. In another question of the same inquiry 12 of 19 agreed that there is clear potential to enhance project-management methods for an integrated PSS development. This is an indication for the need of research on PSS specific project management methods. Furthermore, such methods are needed to enable a breakdown of generic PSS development process models, e.g. as introduced by our own work [1] or by Aurich [2], Lindahl [3], or Shimomura et al. [4], to an executable project form, viz. to an operational level. A process and project management that embeds PSS specific design methods by intention and successfully is not available, nor mentioned in important literature like [5] or [6]. A clarification of opportunities how to adapt existing project management approaches and software, in order to enhance PSS development projects, has not been published. A specification how to share and synchronize generic PSS development processes is outstanding.

As far as we can oversee it, there is no project (except our own), focusing on these issues on a detailed level. Moreover, questions on how PSS design methods can become operable in industrial practice remain unanswered. This has been addressed in an industrial presentation held by a representative of the organization "BAE Systems" on the 1st CIRP IPS^2 Conference 2009 in Cranfield. Highlighted was the great challenge to link systems engineering, project management and the service-dominant logic [7]. During work in the standardization group of DIN PAS 1094 [8] a corresponding research need became visible. The most obvious signal was received in interviews on product, service, and systems development processes in German industry. Project-based development is an expression of customer-oriented development and common for complex systems' development. Both conditions are natural for PSS development, cp. sections 2.3 and 3.2. Complex business and operation models (e.g. in case of performance contracting), which include customer-specific commitments and system configurations, request a successful PSS project management during development and system implementation and ramp-up for productive use. Generic development process models alone cannot satisfy such needs.

In systems engineering literature [9] [10] we find standards addressing development project management and integration of service providers, but detailed information on how to address service integration are lacking. The standard V-Model XT [11] [w1] offers interesting ideas, which are borrowed and adopted in our work, see section 2.4.

1.2 Target

The main target is to make our previous research results on generic PSS development processes (see section 2.4) and PSS design methods operable. For this purpose PSS development project management (see section 3) is a prioritized means.

1.3 Research Approach

This paper is type of a straightforward approach: We directly include earlier work results. We transfer experiences form basic and applied research and industry projects. We illustrate our activities of software customization.

J. Hesselbach and C. Herrmann (eds.), *Functional Thinking for Value Creation: Proceedings of the 3rd CIRP International Conference on Industrial Product Service Systems, Technische Universität Braunschweig, Braunschweig, Germany, May 5th - 6th, 2011*,
DOI 10.1007/978-3-642-19689-8_34, © Springer-Verlag Berlin Heidelberg 2011

An inquiry and questionnaire study in German industry gave insight into the processes and thereby project execution of the development of products, services and combined offers. This served as an input for our research on PSS development project execution. A review of process and project management literature in the domains of product development, systems engineering, service engineering and PSS engineering was used to find gaps of / in PSS development. An investigation of (engineering) software including functions for project management, methodical trial outs of (best-in-class) software solutions and customizing trail outs were carried out to clarify the state-of-the-art capabilities and opportunities for enhancement.

Research Questions

- Which are the basic (industrial) needs to enhance project management in PSS development?
- What are essential opportunities to enhance project management in PSS development?
- How can such opportunities been taken by customization of a best-in-class software solution for engineering project management?

Result: Opportunities of PSS Development Projects

The result is a set of opportunities how to adapt existing approaches or how to develop new methods for the purpose off fully integrated project-based PSS development. Finally, we provide some insight in our current software development activities in the area of project-management support for PSS development.

2 PSS BASICS

2.1 Product-Service Systems (PSS)

Product-Service Systems (PSS) is a concept to integrate products and services in one scope for planning, development, delivery, use, and EOL (end of life) treatment, thus for the whole lifecycle. Predominantly used in academia the concept PSS is now increasingly penetrating industry. Solutions integrating products and services are attracting attention, even if not explicitly planned and developed in integrated processes. Some providers consider themselves as solution providers who essentially offer solutions including products and services. An extension of business models to incorporate product-service systems is an ongoing movement, for instance, in manufacturing industry. Furthermore, PSS often means that a provider takes over more responsibility in product operation (the products may even remain in the ownership of the provider) and what a PSS customer actually buys or pays for is the functionality or performance of the products in a form of service. Therefore, this business scenario is thought of as a special case of *servitization* or *servicification*, where an integrated product and service offer brings *added value* to the customer. Finally, many researchers consider PSS as a means to implement sustainability due to its lifecycle orientation and because PSS does not leave out control after the development, cp. references [12] to [19].

2.2 A PSS Definition

To sum up we apply our following definition of PSS published in 2010 in [20]:

[Necessary criterion] Product-Service Systems (PSS) are customer, lifecycle, and foremost sustainability oriented systems, solutions, or offers, integrating products and services.

[Sufficient criterion] Business models framed by contracts align incentives of the customer and the provider, aim at assuring functionality throughout system lifetime, and aim at implementing added value to satisfy customer needs.

[Phenotypes] (i) Explicit PSS are planned, developed, delivered, and utilized in integrated processes. (ii) Implicit PSS are not explicitly planned, developed, delivered, and utilized in integrated processes, but already existing in today's markets.

2.3 Typical Characteristics of IPS^2

Despite promising features, PSS are not successful in every kind of market or application area and PSS are not the smartest solution in every case. One should ask:

- Which characteristics do PSS and especially IPS^2 have?
- Which effect do these characteristics have on development projects?

Based on project experiences a finding is that (implicit) IPS^2 had at least one of the following characteristics (expressed as statements in the following list):

- *IPS^2 are foremost complex systems.* Example: Navigation devices and services including customer best-practice routes via a specific platform (cp. Falk Navigation Systems).
- *IPS^2 underlie complex use conditions.* Examples: Offshore energy conversion solutions, agriculture and construction machinery.
- *IPS^2 predominantly underlie frequent changing requirements during operation and needs for adaption.* Examples: IT client applications and solutions.
- *IPS^2 often have characteristics of »prototypes and pilots«.* Examples: Offshore wind turbines, customized security concepts for buildings.
- *IPS^2 are foremost systems of individual character.* Examples: Plants and material flow solutions.
- *IPS^2 often base on knowledge intensive systems.* Examples: Aircrafts, plants, nautical systems.
- *Distributed use (pooling, sharing) are basic ideas to transfer products to PSS.* Examples: Car sharing, agricultural machinery in collaborative use.
- *IPS^2 are often systems with extreme requirements on availability.* Examples: Energy solutions, aircrafts, manufacturing solutions.

2.4 PSS Development Processes

Industrial PSS development is not standardized and foremost implicit [1] [20], so far. There are generic development process models in literature, but industrial practice is far away from explicitly integrated development processes of products and services. Industry implicitly performs product and service packaging in operation and sometimes by implicitly related products and services already in development. Missing is a holistic system model (PSS meta-model) including the business model, products, service processes and resources, environmental infrastructure, contracts, customer needs and finally values provided / implemented. In our own research we have developed a generic PSS development process model [1] and nine major design dimensions [17] [20] for PSS design, both opposing the customer view and the provider / engineering view.

Generic Process Model Description

The generic development process model provides three main functions:

1. Setting a framework to include the PSS related findings retrieved in our study (descriptive)
2. Setting a framework for new PSS planning and development methods (prescriptive)
3. Defining a communication basis for future PSS development processes and projects (prescriptive)

The generic development process model in Figure 1 bases on findings made in an interview study and literature review [1] and on calibration in a research project [w4]. From the V-Model we "borrowed" the shape, the validation and verification iterations and the breakdown form a system level to domain specific design. From the V-Model XT ® [11] [w1], we adapted the idea to extend the models shape horizontally on the left and right branch, to capture activities which are not pure engineering activities. Characteristic for the new model is the strong emphasis on the planning and project definition at the left branch and the fade out from development into delivery on the right branch. Both phases are iterative and have high PSS relevance [1]. The upper levels "customer / market / environment" and "value" address customer needs and customer values, which are essential PSS and IPS^2 design dimensions, cp. [8] [17] [20]. Additionally, the model highlights the simultaneous activities (i) brief conceptualization, architecture and business model design, (ii) specification on system level, and (iii) development project definition. Together all three activities lead to the offering of a provider towards a customer. The phase "integrated concept and embodiment design" is addressing new PSS design approaches for an integrated design of products and services on a sub-systems level. The breakdown of system-level requirements to product and service level requirements is a straightforward orientation at product and service quality issues and supporting the application of domain specific design methods in the later phase. The pilot run integrating external factors (customer, environment) highlights the influence of services and is type of prototyping. The configuration on the value level underlines the customer orientation even in late development phases. The requirements management is modeled as an accompanying process in order to satisfy today's approaches of requirements management cp. [21] and [22]. The possibility to feed back knowledge from the delivery and use phase is illustrated by arrows in the right part of the model. The separator line between the value and the system level annotated with "contracts" (development project contract; service level agreements (SLA)) considers the influences of customer specific PSS development projects, the appointment on a business model or operation model and e.g. maintenance contracts.

To apply and deploy our generic development process model (Figure 1) in future PSS development projects, the approach of "process-based project management" will be used. The next sections provide deeper insight on this approach.

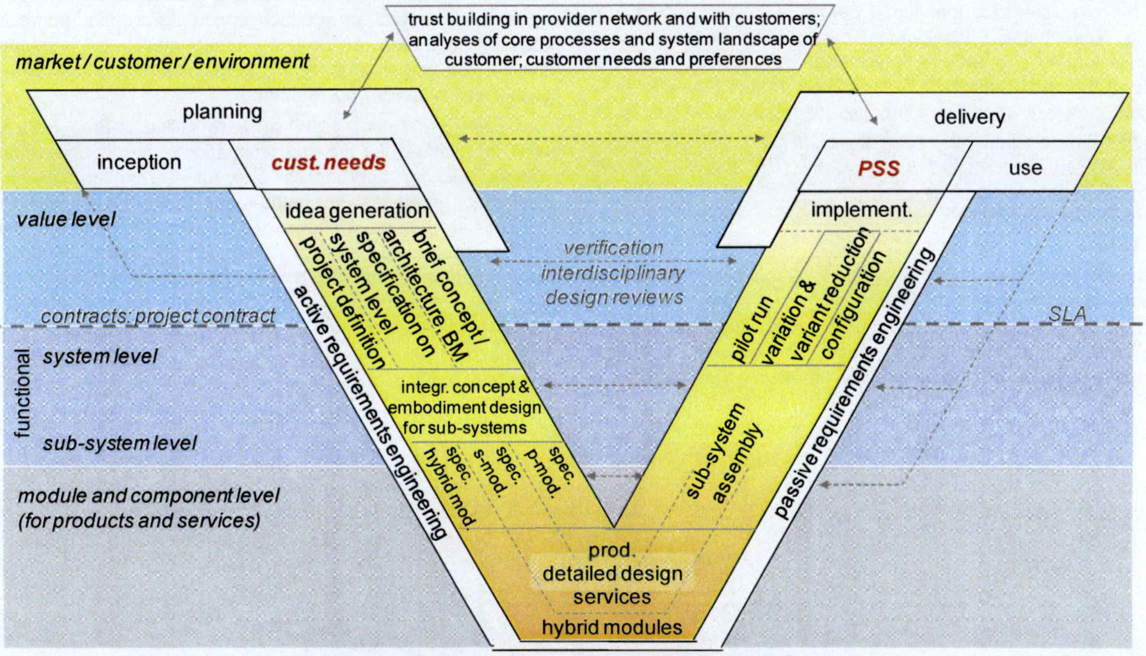

Figure 1: Generic PSS Development Process Model (own model after [1]).

3 PSS Development Project Management

3.1 Project Management in General

Typical issues in project management are time and cost planning, the definition of milestones, milestone deliverables, and responsibilities (e.g. of project managers or engineer assigned to perform project tasks, etc.). Features that are needed and supported by state-of-the-art and best-in-class software are the following: Process-based project management, dynamic tailoring, user / client management, job assignment, reporting, and sometimes integrated requirements management or model-driven development.

One challenge is to integrate core, support, and management processes and to remain being flexible. The core process in our case is the PSS engineering process (stochastic and dynamic). Management processes are, for instance, change, requirements, risk, or quality management processes (highly formalized). Supporting processes are, for instance, knowledge management processes or administrative processes (e.g. staffing, IT administration).

3.2 Project-based Development – Why?

Referring to the system characteristics that fit to PSS (see section 2.3) each development of such systems comes along with unique customer needs, requirements, design targets, and boundary conditions. The provider network is often heterogeneous and integrating various competencies of different companies. Thus, it is natural that system development is carried out in project form the more the characteristics mentioned before fit to a specific future solution. Nevertheless, development project efficiency remains a driving factor. An approach to stay efficient, process-oriented and flexible at the same time is "process-based project management". The core idea is to replicate a process for repeating project types (e.g. internal development service or system implementation project, system customization project etc.). This helps to deploy predefined project types faster, to scale up projects, and to realize savings by "repetition effects".

3.3 Target

The major target is to make our generic process (see Figure 1) operable among partners in a PSS development project. To do so, we adapt the principle of process-based project management. As prerequisite we have the process model (see section 2.4), which now has to be tailored for deployment. This means that appropriate project types, roles, and task blocks have to be defined to build up a tailoring rationale. The steps are:

- Collection and definition of PSS design task blocks as "bricks" to fill in generic development process phases.
- Usage of such task blocks to tailor the generic development process model.
- Customization of approved project management methods and tailoring approaches and integration of those in a state-of-the-art project management software solution.

In the next subsections, opportunities for the tailoring approach are discussed in more detail.

3.4 Opportunities to enhance Project Management for PSS Development

To organize multi-project management for complex, multi-side development projects, effective software support in inevitable. Thus, we focus on software-supported project management. In our scope were project management features of Document Management Systems (DMS), Product Data Management (PDM) Systems and project management software. One finding is, that project management support can be described in form of five major levels, see Figure 2. The lowest level is a simple data and information exchange platform. The next level enables formalization of information and data exchange to make change traceable. Level three becomes specific for the system to be developed. Specific templates and information structures can help to enable a level three support. Level four is addressing processes being specific for a company, project or product/service. Level five addresses a process integration of project management and model-driven development (today performed in IT development and coming up in product and systems engineering).

I our case, we selected the software called in-Step [w2], which builds the platform for the implementation of our PSS development process tailoring approaches and for the integration of PSS development methods (to enable model-driven PSS development). This software has been chosen, because it already offers methods for process-based project management and model-driven development originating from the IT development domain. Specific PSS design support is not implemented, but the interfaces to customize this software are supportive for our targets.

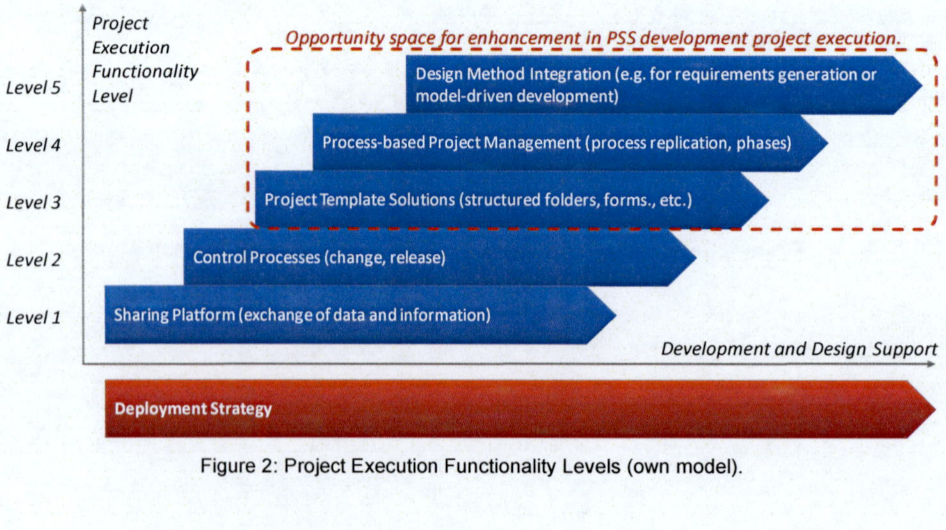

Figure 2: Project Execution Functionality Levels (own model).

3.5 Opportunities on Level 1, 2 and 3

Much project management software is on the market [23]. Most tools provide functions that satisfy level one, two and three to a sufficient extent. At least just on level three the definition of templates that fit domain, company or product specific requirements is necessary. An opportunity on level three would be to define PSS idea templates, e.g. after [20], which support typical PSS elements like contractual conditions, service assets etc.

3.6 Opportunities on Level 4

Customized Project Types and Contracts

In conventional product development, specifications define a system to be developed. In a PSS context, this is too limited (see section 2.3). Instead we propose three types of development project types and contracts:

1. *Needs-based development contract:* Such kind of contract is reasonable if a customer makes a contract with a PSS provider. The customer is interested in a solution and not necessarily in how this is realized.
2. *Requirements-based development contract:* To integrate a new PSS into an existing technical environment, such as a manufacturing facility, functional and non-functional requirements have to be defined more precisely as in the previous case. The customer needs a specific solution and thus the solution space is smaller than in the first case.
3. *Specifications-based development contract:* In some cases, system integrators define hard specifications of how or by which working principles the solution has to be implemented. Examples for such specifications in IT development projects are preselected interface types, technologies or architectures.

Figure 3 illustrates such project/contract types by extension of a model after [16].

New Roles and Project Partners

For PSS development we need new committed roles. A "PSS Architect" could be such a role performing the following tasks: Moderation of PSS idea generation; documentation of PSS ideas and concepts in an adequate form; management of component manufacturer, service and MRO (Maintenance, Repair, and Overhaul) supplier integration; quality and PSS validation management. Such a role is missing in industry. This hinders the PSS design method deployment, a holistic discussion of PSS ideas and concepts, and an adequate requirements and quality management. Such a role is necessary to implement a skilled project owner to drive a PSS development project.

New Task Blocks and Routes

In a review of project templates and literature we collected more than 200 project activities, which we structured in 21 categories, so far. These categories will be used as part of a rational to configure (tailor) design projects. The reviews of the project templates had shown that foremost economic and business planning tasks are captured. Engineering tasks generally are weakly represented. Specific PSS design activities are almost entirely unrepresented. To assign tasks in PSS development projects such tasks should be selectable in the project management solution. A project owner needs to assign design tasks to (PSS) engineers and those should be able to handle and breakdown such tasks in their view in the project management solution. Towards the upper management level a reporting mechanism is needed to trace the progression of projects carried out in parallel.

In a next step we will extend the list by PSS specific development activities. A strong focus will be on use-case definition and requirements generation. The selection of project activities to be performed by project managers and engineers will depend on the project type chosen in the beginning.

3.7 Opportunities on Level 5

To enable level five support we planned to integrate a PSS requirements checklist [24] and a graphical PSS design method [17], which we defined in earlier work, in the project management solution. The aim is to enable the specification of a PSS with an embedded use-case design. In IT development projects this approach is common [w2], but in PSS development such solutions do not exist, so far.

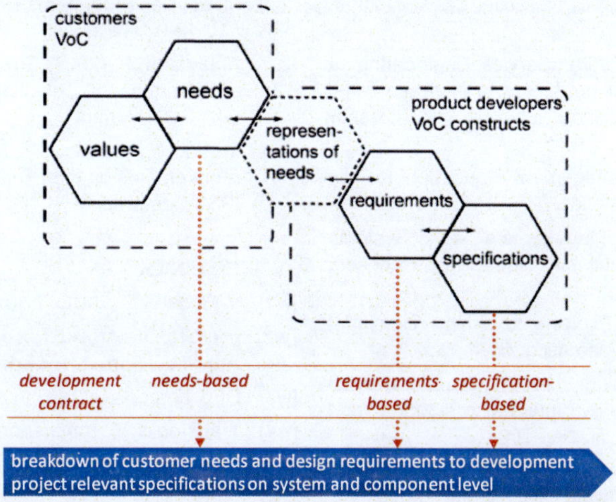

Figure 3: PSS Design Requirements Base Lining according to applied Project Forms and Development Contracts (after [16], extended).

4 DISCUSSION – SUMMARY AND OUTLOOK

Best-in-class project management software seems being capable to support PSS development projects, but for efficient application customization is necessary. The opportunities to do so have been described. The next steps will be a detailed definition of the rationale for the aforementioned task blocks. Outstanding are methods for a dynamic tailoring and milestone planning in running projects. This will be part of the industrial evaluation of our approach. At the same time, we plan to evaluate the generic PSS development process model.

Evaluation

By the beginning of 2011 we start evaluating our customized software solution and our methods in the basic research project Transregio 29 "Engineering of industrial Product-Service Systems" [w4]. The evaluation case is a collaborative, multi-disciplinary development of a manufacturing solution for IPS^2 business (including machining tools and business plans). By mid 2011 we aim to start transfer projects with industrial partners to evaluate our solutions in real life development projects.

5 ACKNOWLEDGMENTS

We express our sincere thanks to the German Research Foundation (Deutsche Forschungsgemeinschaft, DFG [w1]) funding our research. We thank our colleagues form TR29 [w4] and members of microTOOL [w2] for many valuable discussions. Finally, we thank all readers and hope that we provided interesting and useful material.

6 REFERENCES

[1] Müller, P., Stark, R. (2010): A generic PSS development process model based on theory and an empirical study. International Design Conference, DESIGN 2010, Croatia.

[2] Aurich, J.C., Fuchs, C., Wagenknecht, C. (2006): Life cycle oriented design of technical Product-Service Systems. Journal of Cleaner Production, Elsevier.

[3] Lindahl, M., Sundin, E., Shimomura, Y., Sakao, T. (2006): An interactive design model for service engineering of functional sales offers. International Design Conference, DESIGN 2006, Croatia.

[4] Sakao, T., Shomomura, Y. (2006): Service Engineering: a novel engineering discipline for producers to increase value combining service and product. Journal of Cleaner Production, p. 5.

[5] Litke, H.-D. (2005): Projektmanagement Handbuch für die Praxis. München, Carl Hanser Verlag.

[6] Kerzner, H. (2006): Project Management – A Systems Approach to Planning,Scheduling and Controlling. Hoboken, New Jersey, John Wiley & Sons, Inc.

[7] Vargo, S., Lusch, R. (2004): Evolving to a New Dominant Logic for Marketing. Journal of Marketing, 68, 1-17.

[8] DIN PAS 1094 working group. (2009): Product-Service Systems – Value Creation by Integrating Goods and Services. Public Available Specification 1094

[9] ISO/IEC 12207. (2008): IEEE Std 12207-2008, Systems and software engineering – Software life cycle processes. International Standard, Second Edition.

[10] IEEE Computer Society. (2004): IEEE Std 15288™-2004, Adoption of ISO/IEC 15288:2002 Systems Engineering – System Life Cycle Processes.

[11] Authors of the V-Model XT®. (2009): BUNDESREPUBLIK DEUTSCHLAND: V-Model XT, (cp. [w1]).

[12] Salvendy, G., Karwowski, W., eds. (2009): Introduction to Service Engineering. Wiley. 720.

[13] Sakao, T., Lindahl, M., eds. (2009): Introduction to Product/Service-System Design. Springer's global publishing programme in engineering and management. Springer.

[14] Baines TS, Lightfoot HW, Evans S, Neely A, Greenough R, Peppard J, Roy R, Shehab E, Braganza A, Tiwari A, Alcock JR, Angus JP, Bastl M, Cousens A, Irving P, Johnson M, Kingston J, Lockett H, Martinez V, Michele P, Tranfield D, Walton IM, Wilson H. (2007): State-of-the-art in product-service systems. In Proceedings of ImechE, Vol 221, Part B: J. Engineering Manufacture:1543-1552.

[15] Mont, O. (2002): Clarifying the concept of product-service system, Journal of Cleaner Production, 10: 237-245.

[16] Ericson, Å., Müller, P., Larsson, T., Stark, R. (2009): Product-Service Systems – From Customer Needs to Requirements in Early Development Phases, 1st CIRP IPS2 Conference, Cranfield.

[17] Müller, P., Kebir, N., Stark, R., Blessing, L. (2009): PSS Layer Method - Application to Microenergy Systems, In: Sakao, T., Lindahl, M., eds. Introduction to Product/Service-System Design. Springer.

[18] Tomiyama, T. (2001): Service Engineering to Intensify Service Contents in Product Life Cycles, Research in Artifacts, Center for Engineering, The University of Tokyo.

[19] Matzen, D., Sakao, T., Sandström, G. Ö. (2007): Design Research on Manufacturing Firms Moving from Products to Services. In 16th International Conference on Engineering Design, Paris.

[20] Müller, P., Sakao, T. (2010): Towards Consolidation on Product-Service Systems Design. 2nd CIRP IPS2 Conference.

[21] Kruchten, P. (2006): The rational unified process: an introduction. 3rd ed., Addison-Wesley.

[22] Hoffmann, H.-P. (2006): HARMONY-SE / SysML Deskbook, Model-Based Systems Engineering with Rhapsody®. Rev 2.0 / 10-12-06, Telelogic.

[23] Ahlemann, F. (2007): Project Management Software Systems – Requirements, Selection Process and Products. 5th Edition, OXYGON Verlag.

[24] Müller, P., Schulz, F., Stark, R. (2010): Guideline to elicit requirements on industrial product-service systems. 2nd CIRP IPS2 Conference.

7 WEB LINKS

[w1] V-Model XT®, IABG: **http://www.v-modell.iabg.de/.**

[w2] microTOOL (provider of software tool in-Step) homepage: **http://www.microtool.de/instep/de/index.asp**

[w3] DFG homepage: **http://www.dfg.de/en/index.jsp**

[w4] Transregio 29, Engineering of Industrial Product-Service Systems, project homepage: **http://www.tr29.de**

[w5] Fraunhofer Innovation Cluster on MRO, project homepage: **http://www.innovationscluster-mro.de/**

Last request of all websites on January 18th, 2011.

Toward Establishing Design Methods for Cloud-Based Business Platforms

S. Hosono[1], K. Kimita[2], F. Akasaka[2], T. Hara[3], Y. Shimomura[2], T. Arai[3]

[1]Service Platforms Research Laboratories, NEC Corporation, Kanagawa, Japan

[2]Department of System Design, Tokyo Metropolitan University, Tokyo, Japan

[3]Department of Precision Engineering, The University of Tokyo, Tokyo, Japan

Abstract

'Cloud computing' enables system integrators to develop business platforms from existing system by combining software and hardware functions provided by virtualized servers and resources. This functional substitution is the essence of cloud computing and it is the embodiment of the concept of Product-Service Systems in the IT industry. Bearing this in mind, this paper proposes design methods for cloud-based business platforms. By applying Axiomatic Design, Design Structure Matrix and Design-for-X, these design methods can be streamlined, thereby identifying better alignment/separation of functionalities and allocating them to physical servers and resources. Accordingly, the methods will facilitate the shift to service-based business in the IT industry.

Keywords:

Axiomatic Design; Design Structure Matrix; Design for X

1 INTRODUCTION

'Cloud computing' [1] brings about a paradigm shift in the IT industry, as software and hardware products are replaced with equivalent functions from cloud services: Software-as-a-Service (SaaS), Platform-as-a-Service (PaaS) and Infrastructure-as-a-Service (IaaS). In this context, the IT solutions, which system integrators use for building their clients' business platforms, can be regarded as Product-Service Systems (PSS). The computing model simplifies the conventional way of systems integration for business platforms by building the business applications, not from scratch source codes but from choosing and combining various functions from the cloud services.

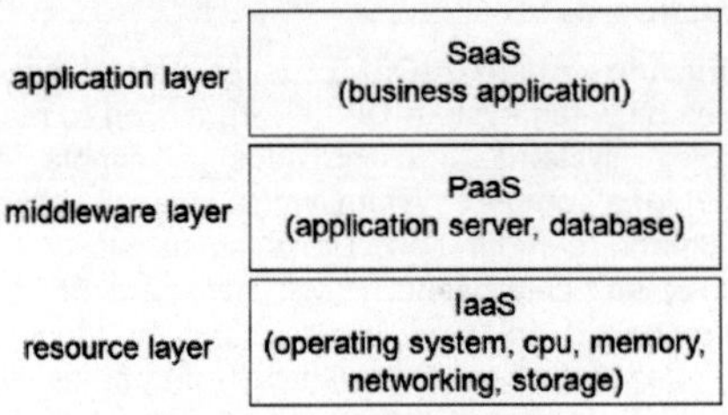

Figure 1: Cloud Services.

These services have been established by server/storage virtualization technologies and middleware, which is distributed on a large scale. However, these technologies have a cognizance difficulty with regards to non-functional requirements (NFR); estimating performance, robustness of infrastructure, and assurance of security levels. This is because hardware functions are separated from the hardware resources and virtualized on them. Therefore, several issues have to be considered - how to identify functionalities and modularity of business application and how to deploy them on physical servers.

The next section discusses how a PSS concept can be identified in cloud computing and technical challenges are determined. Section 3 provides overviews of related work. Then, methods to design architecture for cloud-based business platforms are proposed in section 4. Section 5 shows a case study, and the effectiveness is discussed in section 6.

2 PSS CONCEPT IN CLOUD COMPUTING

The emerging cloud computing is made possible by maturity of information technologies; multi-core CPUs, trends for lowering prices of hard drives/network equipments, and virtual technologies for these hardware. These abundant and high-performance hardware resources are available to meet IT systems' demands. For example, the number of virtualized CPU cores or volume of storage can be allotted logically to an OS for the IT system. This means that hardware products are disintegrated into the number of CPU core functions, the volume of memory or storage function, and IT systems can utilize these spare resources for a certain period.

With these technological backgrounds, IT vendors can now provide cloud services. Hardware and OS are provided as IaaS. PaaS services are provided by adding middleware, e.g. database and application server, on these IaaS services. On top of that, SaaS services are provided by adding business applications. In other words, these cloud services are regarded as re-composition of the hardware/software products. Iaas is the re-composition of hardware and OS functions, PaaS is that of middleware, and SaaS

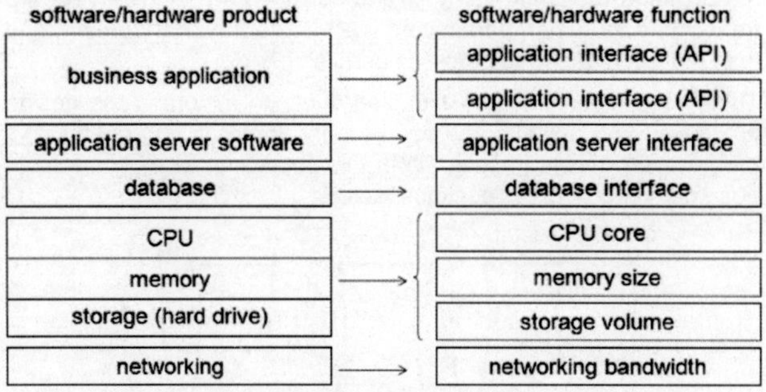

Figure 2: Functional Disassembly of Software/Hardware.

J. Hesselbach and C. Herrmann (eds.), *Functional Thinking for Value Creation: Proceedings of the 3rd CIRP International Conference on Industrial Product Service Systems, Technische Universität Braunschweig, Braunschweig, Germany, May 5th - 6th, 2011*,
DOI 10.1007/978-3-642-19689-8_35, © Springer-Verlag Berlin Heidelberg 2011

is the publication of each business application function. In short, the essence of cloud computing is a dis-assembly of software/hardware into functions (Figure 2) and the reassembly of them. Traditionally, users have purchased hardware, OS, database and middleware products and developed their own business application. However, the emergence of cloud computing have replaced a part of their own business platforms with combinations of functions provided from cloud service vendors. This will lead a paradigm shift from software/hardware product selling to its function selling business. Hence, this is an embodiment of PSS concept in the IT industry.

Emerging Difficulty in Building Business Platforms

Under this cloud computing environment, the number of CPU cores, the size of memory or the volume of hard drives is represented as logical functions. Their functionalities will be explicitly separated from physical resources, and become new design interfaces. Therefore, the number of design interfaces will increase when designing business platforms. This will become an obstacle for system integrators to identify the best combination of these new functionalities as they have to achieve the NFRs, e.g. performance or maintainability.

3 SIMPLIFYING ARCHITECTURE DESIGN

In order to reduce complexity in architectural design, it is essential to simplify the design process and minimize dependency. Axiomatic Design is responsible for the foundation of simplifying the design process and Design Structure Matrix gives a strategy to reduce dependency.

3.1 Minimizing Dependencies

Axiomatic Design

As we have seen Suh has already introduced the Axiomatic Design (AD) approach to design process [2]. This is an approach to simplify design mechanism through abstraction of decision making and thinking process. Contrary to algorithmic approaches e.g. design for assembly (DFA) and design for manufacturability (DFM), it is suitable to embrace several requirements for product design. The AD approach realizes them by specifying the four domains in design process. The domains consist of 1) customer domain, 2) functional domain, 3) physical domain and 4) process domain. Figure 3 describes the relationships between them. The left domain indicates to the right domain what kind of targets should be accomplished, and the right domain shows design solutions, which are how the targets can be realized. In the customer domain, customers' requests or customer attributes (CA) are identified. In the functional domain, the customers' requirements are transcribed to functional requirements (FR) and constraints (C). The FRs are transcribed to design parameters (DP) in the physical domain, and the DPs are mapped to process variables (PV).

This design approach has extracted common factors in the design processes and actions, and has brought about an improvement in progress for design phases. With this approach, two axioms have been obtained from the common factors.

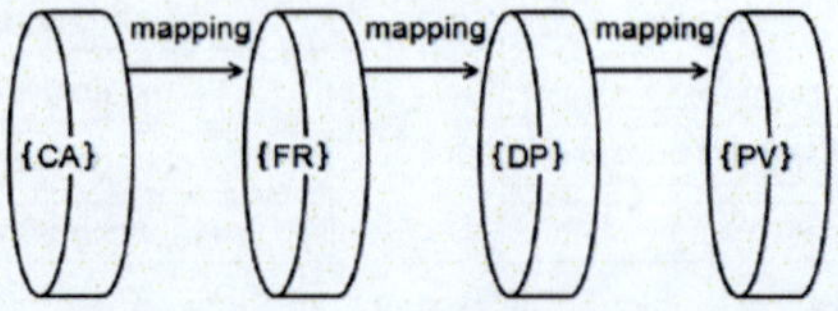

Figure 3: Four Domains in Design Process.

The first axiom is Independence Axiom. This is defined by the functional requisites, which are minimum sets of independent requisites describing design objectives. The second axiom is Information Axiom. It is described as the best solution for design has minimum design information where the independency must be retained under this condition.

Based on the axioms as the basis of abstracting the design mechanism, the relationships of design domains can be described in mathematical forms. A set of requirements can constitute FR vectors in the functional domain. In the same way, a set of design solution in the physical domain can constitute DP vectors. Then, the relationship is represented as the equation:

$$\{FR\} = [DM]\{DP\} \quad (1)$$

where [DM] is a design matrix. For instance, when both FR and DP are three items, the DM is given as follows:

$$DM = \begin{bmatrix} dm_{11} & dm_{12} & dm_{13} \\ dm_{21} & dm_{22} & dm_{33} \\ dm_{31} & dm_{32} & dm_{33} \end{bmatrix}$$

In the same way, the relationships between the physical and process domains are represented as:

$$\{DP\} = [DM]\{PV\} \quad (2)$$

In this equation, the values in the design matrix must be laid out in a diagonal line or in a bottom-left triangle, if they satisfy Independence Axiom. When the values are in a diagonal line, they are 'uncoupled design'. When the values are in a bottom-left triangle, they are 'decoupled design', which guarantees the independence of FR when DP is determined in proper order, and other design choices are 'coupled design'. Therefore, designers only focus on embracing a strategy to align the matrix, whose values are in a diagonal line or in the bottom-left rectangle.

	DP_1	DP_2	DP_3	DP_4
FR_1	X	0	0	0
FR_2	X	X	0	0
FR_3	0	0	X	0
FR_4	0	0	X	X

Figure 4: Design Matrix in AD.

Design Structure Matrix

Design Structure Matrix (DSM) is a tool that manages dependencies between system DPs [3, 4], aiming to manage large scale complex systems. It provides a simple and visual representation of a complex system, which supports better solutions by partitioning and re-integration. Browning classified DSM into two types [5]; static and time-based DSMs, and four DSM applications: 1) component-based DSM, which is useful for modeling system component relationships and facilitating appropriate architectural decomposition strategies; 2) team-based DSM, which is beneficial for designing integrated organization structures, which account for team interactions; 3) activity-based DSM, which is advantageous for modeling the information flow among process activities; and 4) parameter-based DSM, which is effective for integrating low-level design processes based on physical parameter relationships.

In particular, this paper has focused on the component-based DSM as it is more appropriate when discussing the architecture level. Parameters are represented in a square matrix form. An 'X' in the matrix indicates a dependency between two parameters; the row-parameter depends on the column-parameter. A better modular design can be achieved through clustering inter-dependent parameter to form larger blocks.

3.2 Considering Design Perspectives Concurrently

The AD and DSM approaches have contributed significantly in simplifying dependencies in the design process. However, these dependencies are not direct relationships of calling or referencing functions, but also indirect relationship, such as dependencies of security, reliability, maintainability, or performance, i.e. NFRs.

These design perspectives can be considered concurrently with the following approach.

Design for X

Design for X (DfX) is one of the most effective approaches to implementing Concurrent Engineering, focusing on a number of vital elements at a time [6, 7]. DfX is a systematic activity at the early phase of product development. It analyzes features of a project and gives guides for the following development stages. This design technique allows available resources to be put into best use.

3.3 Related Work

Earlier study has adopted the AD, DSM, and DfX approaches in product/service design. Kimita et al. have brought the concept of AD into designing services [8, 9]. They identified the four domains of AD in generic service development. The work discussed the high level of service system management such as organizations of human resources. The work has to be enhanced to identify functionalities in infrastructure layers for IT-enabled services.

Meanwhile, Albers and others focused on the determination of the product architecture in the early phase of product development [10]. They have proposed a model to describe the relationship between a function and embodiment, the Contact and Channel Model (C and CM), which defines product architecture and its corresponding implementation. The model consists of two basic elements – Working Surface Paris (WS) and Channel and Support Structures (CSS). The WSP are pair-wise interfaces and the CSS are a physical component, which links two WSPs. They introduced DfX perspectives into DSM, where the sum total of all the important factors should be equal to 1.0. Referring to Table 1, the functional interaction is weighted at 0.4, the design for A is weighted at 0.3, and that of B is 0.3, and they are visualized in a DSM form of Figure 5. This DfX approach provides a better insight toward product design. However, there are still ambiguities in applying method for service infrastructure, as DfX aspects in cloud infrastructure have not been yet systematized.

	Important factor	CSS in module1	CSS in module2	CSS in module3
Fuctional Interaction	0.4			
Design for A	0.3	3, 5	4, 6, 7, 8	1, 2, 9, 10
Design for B	0.3	4, 5, 6, 7, 8	1, 2,3, 9, 10	-

Interactions: $I_{39} = W_{39} + S_{39A} + S_{39B} = 0.4 + 0 + 0.3 = 0.7$

Table 1: Important Factors [10].

	1	2	3	4	5	6	7	8	9	10
1		1.0	0.3						0.5	0.5
2	1.0		0.3						1.0	1.0
3	0.3	0.3			0.7				0.7	0.7
4					0.7	0.6	1.0	0.6		
5			0.7	0.7		0.7	0.3	0.3		
6				0.6	0.7		1.0	0.6		
7				1.0	0.3	1.0		1.0		
8				0.6	0.3	0.6	1.0			
9	0.6	1.0	0.7							0.6
10	0.5	1.0	0.7						0.6	

Figure 5: Evaluated C and CM Matrix [10].

The study in [8, 9] focuses on the designing process and [10] discusses dependency of functionalities in a design phase, without reference to design process On the other hand, Lim has a comprehensive vision of AD and DSM, which can be applied to design process [11]. He states that both AD and DSM are complementary relationships and they can be tools for identifying dependencies.

By extracting the facts from the above work, in the next section, the authors propose to design methods for business platforms.

4 DESIGN METHODS FOR CLOUD-BASED BUSINESS PLATFORMS

4.1 Architecture Design Method

As a premise for cloud-based business platform, system integrators have to consider not only the business logic in the application but also the lower layers of system functions; virtualized servers and storages, comprehensively. In order to simplify this design process, the authors introduce a model, which integrates AD and DfX (Figure 6). FR and NFR are regarded as FR in AD. NFRs are deemed as DfX aspects, which are specified as availability, performance, maintainability, assimilation, security and environment. Software functions correspond to design parameters (DP1), while virtual hardware functions correspond to design parameters (DP2). These functions are allocated to hardware resources, which are regarded as process variables (PV).

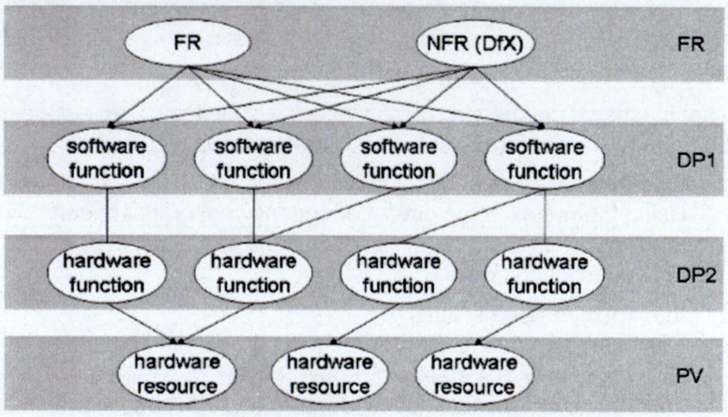

Figure 6: AD and DfX Integration Model.

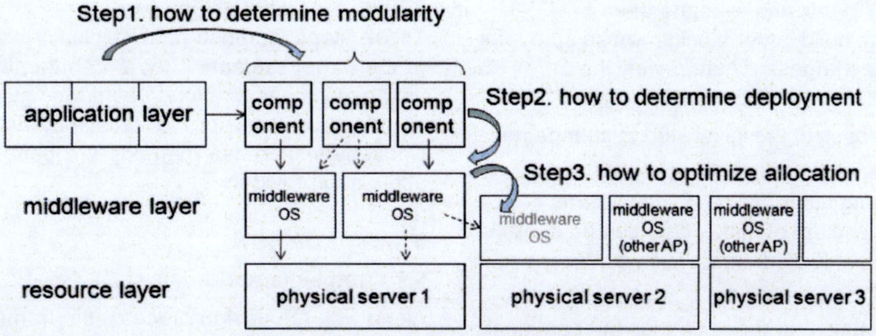

Figure 7: Architecture Design Process.

With this model, the design process can be streamlined into three major steps (Figure 7); (1) design and implement software components, (2) plan to put these components to one or more virtual servers, (3) load memory images of virtual server on one or more physical servers and deploy the components on them. Each step utilizes a DSM, and the best solution from each DSM will be shown as an input for the next step.

The procedure of architecture design is summarized below:

Step1. Establishment of Functional Modularity

Step2. Transference to Virtual Platforms

Step3. Resource Management

4.2 Architecture Design Framework

The steps identified in the previous section are implemented with the architecture design framework in Figure 8. The framework consists of (1) direct dependency editor, (2) design perspectives editor, (3) design data store, (4) 1st DSM: functional modularity matrix (FMM), (5) 2nd DSM: module deployment matrix (MDM), (6) 3rd DSM: resource allocation matrix (RAM), and (7) matrix realignment calculator.

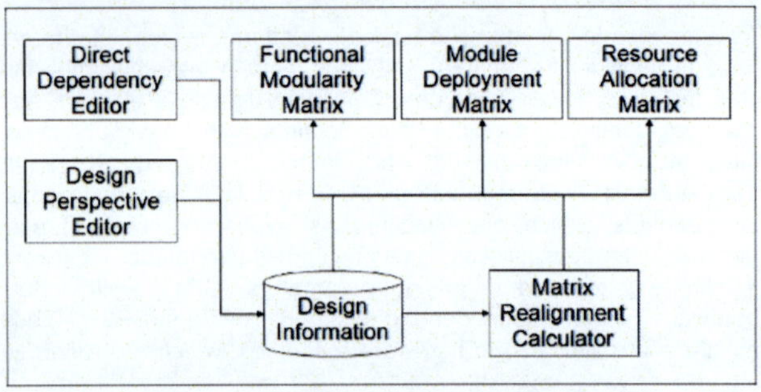

Figure 8: Architecture Design Framework.

Step 1: Establishment of Functional Modularity

The functional modularity of software, i.e. the logics of business platform, will be determined by the following steps (Figure 9).

1. Input relationships of direct function calls.
   Using the matrix in the direct dependency editor, users can mark an 'X' on the crossing cell when the row function calls the column functions or vice versa.
2. Input design perspectives.
   The users can define DfX parameters by using the design aspects data with a standardized NFR list [12]. Then, the importance of functional dependencies and NFR aspects are weighted (Table2).
3. Store design data of step 1 and 2.
4. Build larger blocks in the matrix and evaluate them.
   Using the FMM, the users build larger blocks, whose sum total in each block becomes the largest. Then, review the modularity.
5. Decide to proceed or go back to the initial step to change the weight of parameters.
6. Continue realignment using the matrix and find patterns of larger blocks. This replacement of matrix data can be done automatically by using genetic algorithms, such as described in [13].
7. Confirm the realigned matrix, and then use it as the best design choice.

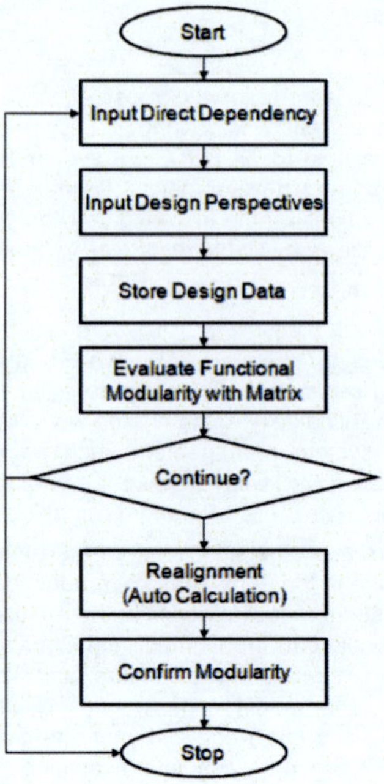

Figure 9: Modularity Fitting Procedure.

	Importance	Module 1
Functional Dependency	0.5	-
Performance	0.3	Func1, Func2, Func3
Security	0.2	Func1, Func2, Func3

Table 2: Dependency of DfX.

Step 2: Transference to Virtual Platforms

This step is to set up the modules to identify the number of needed virtual machines, which consist of an operating system with middleware, defined in step 1. The module fitting procedure is equivalent follows to step 1, though it uses the MDM instead of the FMM. The matrix in this step is used for visualizing the appropriateness of deployment and how many business components should be deployed.

Step 3: Resource Management

The virtual machines determined in step 2 can be run on physical servers. The fitting procedure is also similar to step 1, though it uses the RAM instead of the FMM. The development of hardware of servers has been progressing rapidly and they may have multi-core processors. Therefore, hardware resource can be divided into several virtual machines.

These steps increase cost efficiency as well as the maintainability of the target system. Step 2 can identify minimum set of required virtual machines for cloud users, and step 3 can reduce the number of physical servers to accommodate the virtual machines. Consequently, these methods are beneficial for both cloud service users and providers.

5 EVALUATION

5.1 Implementation

Along with the design process, the framework will be utilized by the service practitioners - application designers, application implementers, system architects and datacenter operators. The

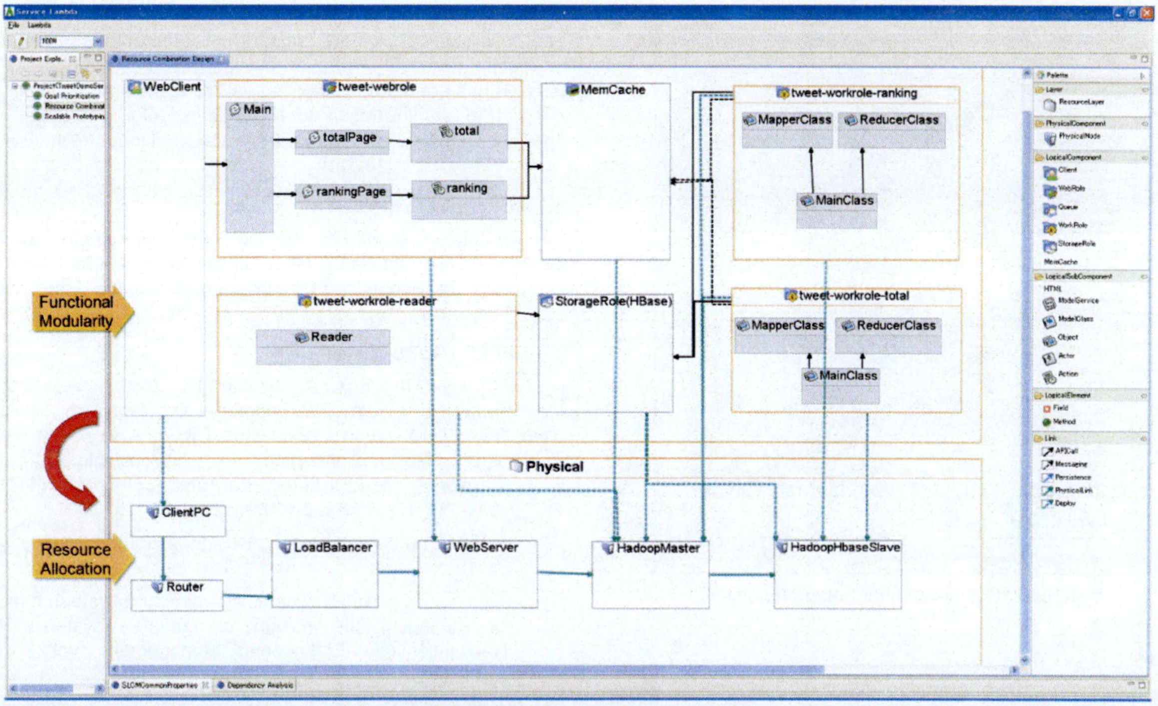

Figure 10: Architecture Design Tool.

application designers and implementers will be in charge of for step 1, system architect will handle step 2, and datacenter operator will manage step 3. In the design support framework (Figure 8), the step 1 and 2 are implemented as an integrated development tool.

Figure 10 is a screenshot of the tool. It can design and evaluate the modularity as well as their allocation. It can edit functional modules of business application software on the upper part of the editor, and functional modules of resources on the bottom part.

Function calls are depicted as sideway arrows between components and relationships between the logical and physical components are depicted as lengthways arrows. DfX aspects between the components on the editor are not depicted as arrows but they are given as their data attributes.

5.2 Case Study

The framework has been partly applied to development of a simple scalable application. The application visualizes trends on Twitter - a social networking and micro-blogging service. Figure 10 depicts the structure of the application. The crawling module collects tweets from a Twitter server and stores them in a data store. When web clients send a count-up request, the front-end module calls backend modules, which analyzes the number of times a word appears in the text to identify the current topics. Finally, the count data are summed up and the results are returned to the web client. The bottom of Figure 10 shows the resource layer. The WebServer, HadoopMaster, HadoopHBaseSlave are virtualized resource, i.e. virtual machines, which consist of CPU, memory, storage, OS and data processing middleware (web server and 'Apache Hadoop' data processing middleware). The memory images of virtual machines will be loaded on the memory of the physical server when it comes to the production stage (Figure 11).

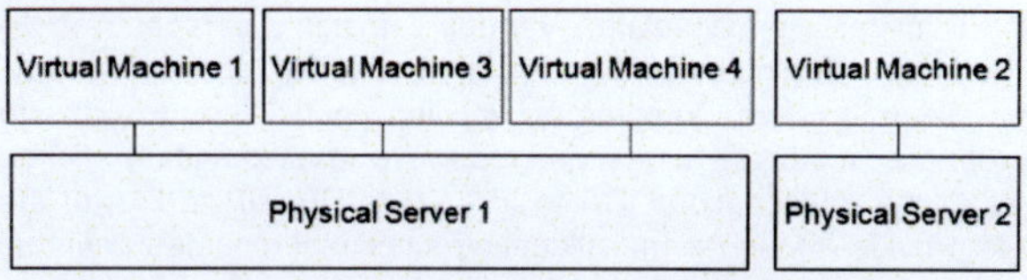

Figure 11: Structure of Virtual Machines and Physical Servers.

Design perspectives are chosen from the guiding template (Table 3) and added as attributes for the matrix. Figure 12 shows a screenshot of the original and evaluated matrices. In this case, the allotment of direct functional relationship is 0.4, that of performance factor is 0.3, and that of maintenance factor is 0.3. The darker colored blocks in the matrix suggest that the set of these functional modules is a better solution, considering performance and maintainability.

Design Aspect	Item
Performance	Data Transaction
Performance	File I/O
Performance	...
Maintainability	Concurrent Server Shutdown
Maintainability	Duplicate Owners
Maintainability	...
Security	Data Isolation
Security	Virtual Machine Isolation
Security	...
Availability	Extent of Failure Impact
Availability	Fault Tolerance
Availability	...
Transitivity	Middleware Update
Transitivity	Module Hot Update
Transitivity	...

Table 3: NFR Template.

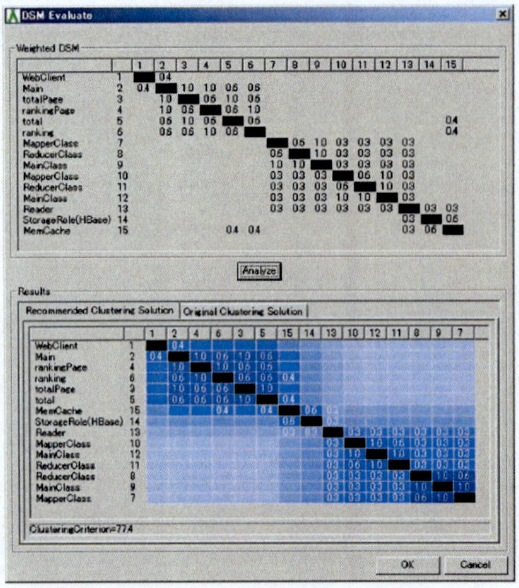

Figure 12: Optimization for Better Modularity.

6 DISCUSSION

The case study exemplifies how to improve application architecture for virtual machines in cloud environment.

The matrices encapsulate various design perspectives - performance, maintainability, security, reliability and transitivity. This gives system integrators guiding principles to design appropriate modularity. The steps are systematically linked together, as transcriptions, CA to FR, FR to DP, DP to PV, of the AD to provide a foundation. Allocating functional modules to virtual machines is a concrete form of transcription from FR to DP. In the same way, the physical servers are determined for creating virtual machines in a transcriptional form from DP to PV. The output from each evaluation method becomes the input for the next evaluation method. Combining these steps into a single unit, the methods will cover the whole architecture design process for cloud-based business platforms. Hence, such methods can solve the vagueness in the emerging cloud environment, and they can be a standard for designing cloud application architecture.

Therefore, the methods will modernize the design techniques based on 'service-oriented architecture' (SOA), which has been used for a component-based de facto design approach for designing traditional business platforms.

7 CONCLUSION AND FUTURE WORK

This paper discussed cloud computing as an emerging paradigm, which is underpinned by a PSS concept. Though the concept brings certain obstacles in the business platform development, the proposed methods systematize the design process with three major steps and show concrete procedures of architectural design. The methods utilize the matrices of the design principles and clarify design constraints and dependencies between functionalities. Such matrices with DfX perspectives succeed in formalizing non-functional dependencies between functions, and quantifying them along with functional dependencies. This will help in better identification of alignment or separation of functionalities and how to allocate them to physical servers or storages. Therefore, the methods can assist system integrators and practitioners in formulating cloud-based business platforms for their clients, and it will lead to a pervasion of the PSS concept in the IT industry.

The case study shows the feasibility of the methods to streamline the design process. However, the last resource optimization step (step 3) has not been exemplified yet, as it should be implemented as part of administration tools for datacenter operators. The applicability of it for providers will be evaluated in future studies.

8 ACKNOWLEDGMENTS

We would like to thank Min Min, Hu Huang, Dehua Lin, Lin Li, Xin Zhao, Tao He, Jiafu He, Xuemei Liu of NEC China for discussions to realize our concepts and for implementation of the prototypes.

9 REFERENCES

[1] Armbrust, M., Fox, A., Griffith, R., Joseph, A., Katz, R., Konwinski, A., Lee, G., Patterson, D., Rabkin, A., Stoica, I., Zaharia, M. (2009): Above the Cloud: A Berkeley View of Cloud Computing, Available at: **http://www.eecs.berkeley.edu/Pubs/TechRpts/2009/EECS-2009-28.pdf** (accessed 10th Nov. 2010).

[2] Suh, N.P. (2001): Axiomatic Design, Oxford University Press, 2001.

[3] Steward, D.V. (1981): The design structure system: a method for managing the design of complex systems, IEEE Transactions in Engineering Management, vol.28, no.3, pp.71-84.

[4] Eppinger, S.D. (1997): A planning method for integration of large-scale engineering systems, In Proceedings of the 11th Int'l Conf. on Engineering Design (ICED1997), pp.199-204.

[5] Browning, T.R. (2001): Applying the Design Structure Matrix to System Decomposition and Integration Problems: A Review and New Directions, IEEE Transactions on Engineering Management, Vol. 48, No. 3, 2001, pp.292-306.

[6] Miller, G.A. (1956): The Magical Number Seven, Plus or Minus Two: Some Limits on Our Capacity for Processing Information, Psychological Review, 63, pp.81-97.

[7] Paul, G., Beitz, W. (1996): Engineering Design – A systematic Approach, 2nd Edition, London: Springer, 1996.

[8] Kimita, K., Akasaka, F., Hosono, S., Shimomura, Y. (2010): Design Method for Concurrent PSS Development, In Proceedings of CIRP, IPS^2 Conf. 2010, pp.283-290.

[9] Kimita, K., Akasaka, F., Hosono, S., Shimomura, Y. (2010): Design Method for Life Cycle Oriented Product-Service Systems Development Design, In Proceedings of the 43rd CIRP Conf. on Manufacturing Systems, pp.281-288.

[10] Albers, A., Sedchaicharn, K., Sauter, C., Burger, W. (2009): A Method to Define a Product Architecture Early in Product Development Using Contact and Channel Model, In Proceedings of the Int'l Conf. on Engineering Design 2009, pp.241-252.

[11] Lim, Ys. (2007): Differing Roles of Axiomatic Design and Design Structure Matrix in Reducing System Complexity, In Proceedings of the IEEE Industrial Engineering and Engineering Management (IEEM 2007), pp.994-998.

[12] Information-Technology Promotion Agency (2010): Standardizing Grades of Non-Functional Requirements, Available at **http://sec.ipa.go.jp/reports/20100416.html** (accessed 10th Nov. 2010, In Japanese).

[13] Whitfield, R.I, Smith, J.S, Duffy, A.H.B (2002): Identifying component modules, In Proceedings of the 7th Int'l Conf. on Artificial Intelligence in Design (AID'02), pp.15-17.

Sharing Expertise: Easier Said Than Done

Johan Holmqvist[1], Johan Wenngren[1], Åsa Ericson[1], Christian Johansson[1], Peter Thor[1]

[1] Division of Functional Product Development, Luleå University of Technology, Sweden

Abstract

An extension towards service provision takes place in manufacturing industry. Inclusion of softer service aspects indicates that the common view on knowledge management to control and monitor a technical process have limitations. Sharing expertise is an additional way of managing knowledge particularly with the intentions to make experience based knowledge organizational available. By studying product developers' daily work, especially how they perceive that they apply and share knowledge, we problematize knowledge activities in product-service development to discuss the established knowledge management activities. The paper suggests some considerations to support the development of a knowledge base for product-service design.

Keywords:

Knowledge Management; Knowledge Based Engineering; Web 2.0

1 INTRODUCTION

Today, most companies in a business-to-business environment act on a global market. Besides distance and different time zones, such work is also challenged by organizational and cultural differences. In this turmoil, an industrial move towards Product-Service Systems (PSS), i.e. an integration of products and services aspects in early development where the outcome is provided as a function in a service way, is going on. The integration of products (tangible goods) and services (intangible activities) elements is not straightforward due to being based on completely different logics [1]. These challenges make knowledge work and knowledge management an interesting topic from a research perspective, while from an industrial perspective capabilities to handle the situation could be a matter of survival.

There is also a variety of computer based tools that aid technical development. Due to the explicit nature of knowledge the ways of managing technical product related knowledge are well established development activities and processes. So, on one hand, technical product related knowledge could be categorized as explicit since it can be captured and formalized. On the other hand, services are based on experience based knowledge, i.e. tacit knowledge, for example customer perceived value [2], thus such knowledge can as well reside outside the company. Thereby, some vital aspects of experience based knowledge are difficult to pinpoint and to articulate. Commonly, services are described as activities that are co-produced with customers at their time and choosing [3].

Experience based, tacit knowledge is since long ago recognized as an important part of product development [4], though its ambiguousness is evident. For instance, gut feeling or intuition is not a valid decision base, yet they are recognized as affecting decisions. Consequently, this type of knowledge is not made explicit and is not documented to feed input into new projects. The fact that tacit knowledge is not simply identified, captured, formalized or reused contributes to this situation. It can be argued that there are reasons for either ignoring to incorporate such knowledge in the organizational mind, or for failing in doing so. For instance, tacit knowledge cannot be collected in a similar manner as explicit knowledge; hence it needs other methods than those established in traditional product development. Due to its stickiness [5], i.e. it is context dependent [6], it is difficult to transfer into written text. And, it has to be interpreted in a holistic manner. Written text is a common way to capture, for example lessons learned in product development. When tacit knowledge is collected, it usually results in a vast material; consequently it is a tedious work to identify candidate information. Further, due to its contextual dependency, attempts to capture and reuse it risk to detach it from its background. Thus, the good intentions can result in building knowledge silos [7].

Basically, tacit knowledge is embedded in peoples' minds and senses, thus it is not only hard to capture and formalize, but it also follows the employees when they leave the company. When this happens, it takes time for the company to replace what was lost; hence it can cause expensive delays in projects etc.

A way to make experience based knowledge an organizational asset is via socialisation [8], e.g. when people talk about their abstract know-how and share their experiences with colleagues. Stories about how you, partners, colleagues and so forth have dealt with a situation, successfully or detrimentally, support the creation of an organizational culture, but are also a way to share tacit knowledge. However, in technical product development, sharing of experience-based knowledge might not be supported. We have studied technical product developers' work to understand how knowledge is shared. Based on this, the purpose of the paper is to discuss established knowledge management activities to suggest some considerations to support the development of a knowledge base for product-service design.

2 METHODOLOGY

The result presented in this paper is built on an empirical study from manufacturing industry. This is an exploratory study from which it is difficult to draw general conclusions. The data is part of a more extensive study, which aims to prescribe methods and tools for expertise sharing in technical product development, whereas this

J. Hesselbach and C. Herrmann (eds.), *Functional Thinking for Value Creation: Proceedings of the 3rd CIRP International Conference on Industrial Product Service Systems, Technische Universität Braunschweig, Braunschweig, Germany, May 5th - 6th, 2011*,
DOI 10.1007/978-3-642-19689-8_36,

paper describes some of the industrial practice. Hence, the paper present a glimpse of an ongoing study performed in two joint industrial/academia research projects. And, the generated data have been used as a part of a pre-study for two research projects.

The two companies, share the same parent company, are active at two different markets, but have a common product development process. Both of the research projects focus knowledge sharing in technical product development, but differ in terms of particular interest in team-based innovation and tools for knowledge sharing. Background material has been generated in company visits and workshops at the companies.

The companies have assisted in finding interesting respondents based on a description of interest areas. A questionnaire has been answered by 7 respondents and 3 interviews have been performed at company A. The interviews were performed with engineers from the methods development domain. At company B, 8 respondents have been interviewed. The interviews followed an interview guide with predefined questions. All questions were focusing on the topics knowledge, engineering knowledge and approaches for knowledge sharing. Intentionally, we have not defined knowledge in our questions, rather the respondents were free to base their answers on what they perceive as knowledge. And, all interviews were semi-structured, i.e. the respondents could formulate answers freely and the researcher can pose follow up questions on interesting topics.

3 KNOWLEDGE MANAGEMENT

Commonly, product development is described as knowledge intensive work. Though, depending on the prevailing perspective, what *kind* of knowledge it involves differs. Particularly for PSS situations it is concluded that there are differences between an industry and technologist view versus a service research and service perspective point of view [9]. For manufacturing companies, moving towards PSS, there is a challenge in integrating the service aspects in the development process and so far many companies that offer services do not fit these with the existing Product-Service System [10].

A comparison of competing paradigms of product development in academic communities have recognized marketing, organization, engineering design and operations management as main conflicting perspectives [11]. Particularly the view on products differs between the perspectives (p.3). From a marketing perspective the product is *"...a bundle of attributes"*, making customer preferences of those attributes important knowledge. From an organizational perspective the product is considered to be *"...an artefact resulting from an organizational process"*, making knowledge of creating internal workflow vital. Engineering design views the product, first as something discrete and manufactured [12], second as a *"...a complex assembly of interacting components"* [11], making knowledge about the technical device and its interfaces a priority. Finally, the operations management point of view defines a product as *"...a sequence of development and/or production steps"*, indicating a focus on process knowledge. And, all these divergent views on knowledge insist on different performance metrics, decision variables and success factors [11]. Hence, the knowledge management insists being performed differently. Within adjacent areas, earlier studies have been done but from another perspective [13].

Knowledge is commonly described as contemporary organizations' most important resources. Though, in product development literature the view on knowledge is typically described from an engineering design perspective. Thereby, formalizing and storing of knowledge for the purpose to reuse it is in focus [14]. The engineering design expertise can be formulated as "rules of thumb" and applied in knowledge based systems applications [14].

Knowledge management can be divided into two main approaches [15][16]. The first focuses on gathering, collecting, storing and reusing knowledge, mainly as a method to control and monitor, for example, a development process [15]. The second focuses on knowledge creation, integration and sharing, mainly applied to bolster human components of knowledge [17]. The second approach is commonly called knowledge sharing or expertise sharing [15], rather than knowledge management. In expertise sharing, the problems of cross-boundary knowledge work and social aspects are recognized [17]. The aim for knowledge sharing is to support communication, learning and organizational knowledge [17], consequently it gives knowledge management the task to transform tacit knowledge into explicit knowledge, as well as transform individual knowledge into organizational knowledge. Commonly, such a transformation is explained to depend on a change in mindset, similar to the change into "Toyotism" or "Just in Time" required a change of mentality rather than a change in "machinery" [18].

4 PRODUCT-SERVICE SYSTEMS

In industry, several different terms are used to describe the phenomena of PSS; soft products, total offers, through life solutions, and service 2.0 offerings are some examples.

Tukker et al [19] provide a list of 9 definitions where it can be noticed that the definitions deal with explaining that product-service systems are a mix or a combination of products and services, but give limited explanation of the system part. The system view of PSS seems to address other aspects than an integrated product view does, but this is not further explained.

In a similar way as Krishnan and Ulrich [11] explains a product from different perspectives, PSS could be assumed to be seen from different points of view, but all definitions presented in Tukker et al [19] seem to describe a strategic business point of view where the emphasis is on what is sold to customers, cf. a marketing point of view of what a product is. One example of a definition is that PSS is the result of an innovation strategy, shifting business focus from designing and selling physical products only, to selling a system of products and services which are jointly capable of fulfilling specific client demands [20]. Another example is that PSS are a specific type of value proposition that a business (network) offers to (or coproduces with) its clients [19], this is also based on a marketing/organization business perspective.

Two cornerstones of PSS are outlined to provide guidance for development of solutions [21]:

1. The point of departure for development of PSS should be the functionality or the objective that the user/customer is striving to achieve, rather than to start with a product that fulfills certain functionality.
2. A "green" strategic thinking for the whole business and the company should continuously be modified, instead of taking existing structures and the company's position for granted.

From this guidance, it can, firstly, be interpreted that a development process for PSS always have to start at the customer's place and that no idea of a product to solve that situation should be in mind. Consequently, PSS development has similarities with wicked [22] or ill-defined problems [23], i.e. *"...before designers can solve a design problem they need to understand some basics – such as what they are designing, what it should do and who should use it and in what circumstances."* [24] (p.2). If so, the product developers' technical knowledge base is not encompassing enough. And, to achieve PSS, the company's capabilities of radical, rather than incremental innovation, need to be refined. Second, an interpretation of the guidelines can also be that companies are challenged in a much more extensive way than just figuring out how to mix or combine product and service solutions.

A base for innovations is that some aspects of new knowledge are added to the existing knowledge base [25], such knowledge can be about markets, processes, ideas, products, structures etcetera.

5 KNOWLEDGE MANAGEMENT IN THE INDUSTRIAL CASES

The companies in the study are moving towards PSS, but do not have an integrated process for services. Offered services are typically offered as add-on to products. One of the companies has tried PSS on contract level but not on a development level.

For the companies included in this study, knowledge based tools play an important role to validate the included technologies, the manufacturing processes and the use of the product. These tools make storing, retrieving and dissemination of technical information manageable. Also, the tools provide traceability of performed activities, as well as provide the possibility to transfer production and usage information into earlier stages. These tools have to be built on verified knowledge, for example defined governing processes, design practices, method descriptions and standards. In early development, these tools are valuable for simulating and visualizing potential solutions before developing physical prototypes, thus, decreasing both risks and cost.

One of the companies is performing in house development of some knowledge based tools, which is referred to as method development. For a method developer, applications such as the formalization of knowledge into information or expert systems are at the core. But, also, they have to understand how to implement, how to educate users and scalability aspects of the application.

Method development is mainly pursued on a need basis at the company, where for example a project leader identifies an activity that needs improvement or support. The method developer investigates the situation, analyzing the purpose of the task, the steps included, the input information etcetera. Thereby, part of the method developer's work is to identify, justify, capture and formalize knowledge. This work is guided by a knowledge based engineering lifecycle (cf. [14]). The method developer has an extended responsibility to package and activate the application in collaboration with the IT-department. The method developers emphasize that the success of the application depends on how well it fits into the established development process, its usability and if it is accepted by its users. Besides collaboration with IT-department, the method developers work closely with people that are considered as domain experts, such as process owners. These people possess deep and focused knowledge about certain activities and provide valuable input to the method development.

Sometimes the method development follows a technology push approach, where the method developer locates an activity that is not supported (automated) yet, but have the potential of being improved. In such a case, the demands on the developer are different. Usually, the method developer identifies the governing processes and the documented instructions for the activity. Since all activities are not documented and described at the company, the method developers have to solve problems as they emerge. Thus, feeling comfortable with a "learning by doing" approach is part of the task. There is a team of method developers that can seek support among each other. For example, reoccurring meetings, as well as documented guidelines and instructions for method development found in collaborative repositories, such as collaborative workspaces and networked disk drives, are seen as knowledge sharing activities.

The respondents find that knowledge sharing in product development is closely linked to the project and group activities. In this context, verbal communication is a common way to share experiences. Respondents stressed that team members or colleagues commonly ask for guidance or help. Email and chat conversations are mentioned as a way to contact people within the organization, commonly this is done to set up a face to face meeting. Project portals, team portals and chat applications are used for written text, pictures, presentations, CAD models and so forth. These support systems make it possible to form dispersed teams or a dispersed project organization.

Company servers are used to disseminate formal and organizational documents, for instance written instructions, lessons learned, standards, best practice and design practice. The staff can search the servers to find relevant information, though it is expressed that finding information is perceived as time consuming. The engineers prefer to ask a colleague or find a former team member to ask, this is an easier way to retrieve information.

5.1 How often do you share knowledge?

The very open-ended question "How often do you share knowledge (within a certain context)?", gave, as expected, different degrees of occurrence. We used follow up questions, like, "with whom?", "why?" and "how?" to gain more detail in the answers.

How often the respondents shared knowledge within the organization was said to depend on their position in the company. For example, those who worked as project leaders perceived that they did share knowledge within the organization; such activities were explained to occur many times a week and concerned mainly project issues, 13% in Figure 1. The 12% of knowledge that was shared on daily basis was explained as reporting progress in projects to other involved or personnel higher in the hierarchy.

69 % of the respondents perceived that they seldom shared knowledge within the organization, i.e. outside their department and outside their projects. None of the respondents agreed that knowledge sharing occurred constantly (see Figure 1).

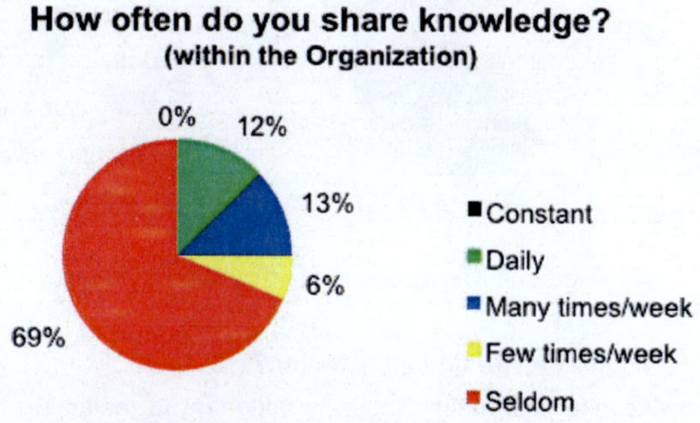

Figure 1: Knowledge sharing within the organization.

Knowledge sharing with colleagues at a department level visualizes other responses, see Figure 2.

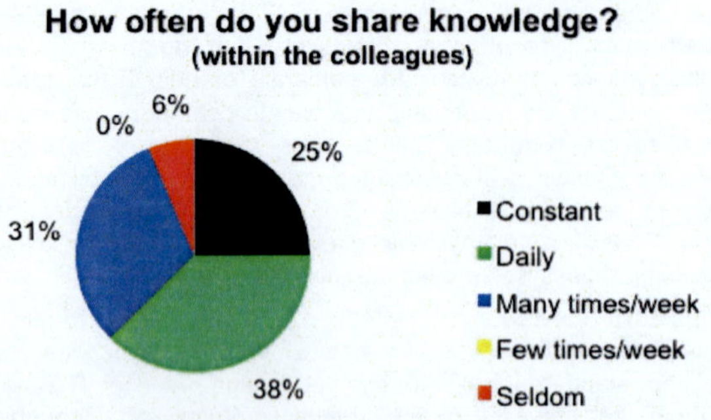

Figure 2: Knowledge sharing with colleagues.

In Figure 2, constant (25 %), daily (38 %) and many time/week (31 %) indicates that knowledge sharing is perceived as occurring more frequently when viewed in the context of colleagues.

The projects in this study are branded by the company as technical projects. There was a mix of expertise in the team, but that mix was overall from within the engineering field.

In Figure 3, sharing knowledge at the project level is shown. 13 % of the team members perceived that they shared knowledge within the project constantly, 19 % expressed that they did share on a daily basis, and 31 % shared knowledge many times/week. Also, 31 % perceived that they shared knowledge only a few times/week, and 6 % expressed that they seldom shared knowledge within the project.

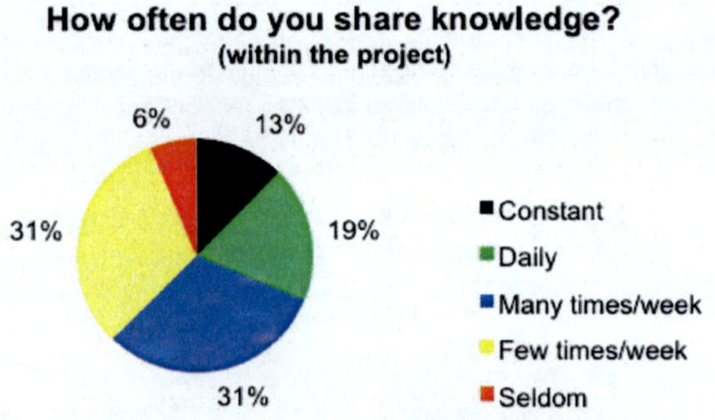

Figure 3: Knowledge sharing within the project.

5.2 Where do you find information?

Knowledge or information reuse is important to maintain effective project work, to support this; the members have access to information systems as for example lessons learned systems, project archives, product assurance plan. Lessons learned systems are used to capture experience based knowledge from the projects, documents related to the project work are stored in the project archive, and the product assurance plan defines the delivery criteria for opening gates in projects according to the product development process.

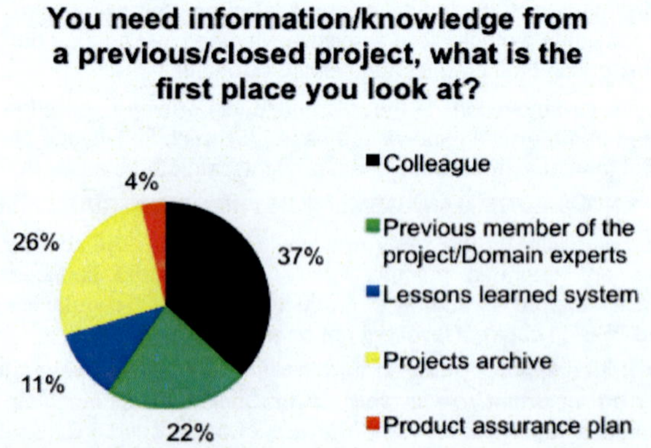

Figure 4: Where do you search for information or knowledge.

In Figure 4, 37 % of the respondents answered that they firstly asked a colleague, which is a person from the same department when they are seeking for information from a previous project. 22 % identified a previous project member or domain expert to find information. Only 11 % looked first into lessons learned systems, and 26 % in the project archive.

One reason for information systems applications as not being the first and most natural choice was explained by the respondents as related to the difficulties to search for the information. To be able to search you have to have some idea (words, project name, project members etcetera.) of how the information is tagged and stored, this is found in the case studies and supported by the literature [26].

6 IMPLICATIONS FROM THE STUDY

There are a number of implications from the study that needs to be discussed, though a contribution from it is that it provides an overview from which further research activities can embark. For industrial practices, the discerned indications can provide a basis for reflection.

The generation of data did not focus on finding out individual's perceptions of *'what'* was shared, i.e. their definitions of knowledge or information. Instead the study is built on the respondents' views of *'if'* they share or not, meaning that the respondents were free to interpret what is knowledge and what is information. We have in previous studies found that defining knowledge is not straightforward and those attempts did not provide guidance for how to support the sharing activities. In those previous studies the respondents have been able to, theoretically, define knowledge as something else than information, but when they start to describe their work they apply a more practical definition, i.e. what they need to know to do their work is expressed as knowledge. Within the engineering practice area, the perception is that some of the companies' applications are built upon information (facts, figures, calculations, test results and so forth). And, it is the human and contextual interpretation of that displayed information that transforms it into knowledge. Still, this is found as a theoretical discussion rather than a practical one.

However, in the study presented here there seems to be a difference between that practical view of knowledge and another view of knowledge, which can be categorized as 'experiences'. Thus, making the perception of shared knowledge to align with explicit knowledge and information, and making the distinction that what is not so frequently shared is experiences. If so, the continuation of these research projects could benefit from studying experiences, what it is, how it affects project work and how it can be made organizationally available.

The lessons learned systems are tools to capture and disseminate experiences, but as indicated in this study the respondents did not have trust in finding the information needed, i.e. those systems were not their first choice of knowledge source. The study presented here does not provide insights into the reasons, but indicates that the information retrieval from lessons learned systems is perceived as bringing limited benefits in project work. If so, it might be possible to suspect that the information input could be sparse also. Experiences often come in 'eureka moments' [27] from which lessons learned has to be reflected on. Is so, the team have to have the possibility to reflect in practice in the 'eureka moment', rather than after project closure to avoid hindsight and construction of experiences. Yet, how can such activities be supported and performed in practice? And, the support for capturing these reflections in action, how would that look like? Reflections in teams usually do not come in written text, rather in dialogues.

Before the study, it was assumed that knowledge sharing activities should be most frequent within the project and that knowledge sharing should be less frequent on an organizational level. The empirical data indicates that the assumption has some relevancy, but provides no insights to why a situation like this could occur. One explanation could be that the purpose for overall organizational activities is more or less to pass on information from high level management, for example to communicate what is happening in the company as a whole or to transfer information about the business environment. If so, there is a need of an organizational strategy to support other forms of meeting that support knowledge sharing. It seems useful to; for example, figure out how to provide structured and focused feedback from a larger audience and how to evaluate the impact of an informative meeting in order to improve the larger meeting format for knowledge sharing. Though, it is important to note that all larger meetings should not turn into knowledge sharing activities, information transfer is also needed. Basically, there is a need to consider how to find dynamic and useful meeting formats for any specific agenda.

7 DISCUSSING KNOWLEDGE MANAGEMENT

In general, companies emphasize that their knowledge base are their most valuable asset. A key in this perspective is if the asset is considered to be protected (not shared) or if it is considered to be accessible to evoke, for example, innovations and collaboration. A basic PSS cornerstone is that it should be a strategic and holistic effort in collaboration with partners [21].

Typically, earlier research efforts within the knowledge management field focus on the first approach, i.e. how to store and reuse factual and explicit knowledge [15]. But, as the areas of computer technologies mature and are more capable to deal with automation of knowledge work the interest is turning towards perceived and experience based knowledge. Still, in companies, there is an awareness of the challenges to transform the stored information into actions and the staff into knowledgeable workers.

For manufacturing companies, the extension towards PSS provision seem to change the core of the knowledge base since service experience based knowledge have to be considered in early development. The engineers experiences are seldom shared in formal reports but are a necessity in the development of services, i.e. the experiences is shared verbally and is therefore often missing in activates related to knowledge management work. Donald Rumsfeld, US Secretary of Defense during the first years of 2000s, has received much biting wit from media, but pinpointed important knowledge aspects when he coined the concepts of *'known knowns'* (the facts that we know something about), *'known unknowns'* (those things we know that we miss information about) and, *'unknown unknowns'* (for us unknown information that we do not even realize that we miss). Modica and Rustichini [28] mentioned the concept of 'conscious uncertainty', a state of knowing that we do not know something, as opposed to 'unawareness' that denotes that we do not know something and we do not know that we do not know it.

The analogy of PSS is that manufacturing companies generally can perceive a shift as ill-defined and wicked [22] [23], hence a stepwise extension of business models might be recommended as an approach. Contemporary knowledge based systems, such as knowledge based engineering tools and other repository based knowledge tools are useful for developing physical artifacts, but they do not in their as-is state support PSS as a whole. The ambiguous and uncertain nature of development activities that often goes hand in hand with PSS and service provision means that the knowledge base have to be built upon a 'learning by doing' approach. If this is the case, striving for providing, for example leasing, would build up service knowledge, subsequently how to develop suitable physical products for a leasing contract is one step towards PSS. And, from these experiences how to change the product development process can be understood.

A consideration for PSS is its appropriateness for any kind of products and any type of companies has to be done. PSS, it seems, has become a wider concept than intended, branding all the variants of traditional sales of things (development as usual) to the view that all products remain in the ownership of the manufacturer and that company sell the service of operations in the customer processes [9]. Today, engineers and product developers have to possess a wide range of knowledge, though still that vast knowledge base is about the physical product. PSS adds more relational complexity into that situation, since for development issues the engineers and developers have to expand their knowledge domains into the area of customers.

Knowledge sharing, i.e. second approach of knowledge management [15], need to be built on strategic intentions and allocation of appropriate means. To be made into an organizational knowledge asset, the transformation of information into actionable expertise needs an approach of interpretation, digestion, 'eureka moment' and reflection. These kinds of activities are hard to support in project work. Further, such knowledge sharing occurs more verbally, in sketches, simple pictures and even by using gestures. So, sharing expertise is easier said, than done.

8 CONCLUDING REMARKS

This paper outlines results from a study of technical product developers' work performed to describe and understand how knowledge is shared. Besides interaction with company representatives in the research projects, Interviews that composed of predefined questions have been performed. This data provide a background for the purpose of this paper to discuss the established knowledge management activities to suggest some considerations to support the development of a knowledge base for product-service design.

Intentionally, the paper takes a stance to problematize contemporary management of knowledge for the purpose of PSS introduction in the company. Thereby, an interpretation can be that the paper describes a hopeless situation where little can be done. On the contrary, by seeing something as problems, solutions can be found.

The considerations for PSS that we propose need further attention are:

- The differences between factual knowledge about artifacts and experiences (tacit knowledge) from the team (for example about the thing, its users or its use). How to assure that more aspects than mere technical end up in a knowledge base?
- The multifaceted approach of knowledge sharing, e.g. in dialogues (planned/formal, not planned/informal and so on). How to support different types of meetings?
- The wide range of which knowledge is expressed (text, pictures, drawings, sketches, gestures and so forth). How to provide tools and methods that build on all these aspects?
- The wickedness of the term knowledge and how it is differently applied at distinct levels of the organization. How to create a knowledge sharing culture?

9 ACKNOWLEDGMENTS

We would like to acknowledge the ProViking research programme from SSF (Swedish Foundation for Strategic Research), and FFI - Strategic Vehicle Research and Innovation (a partnership between the Swedish government and automotive industry) supported by VINNOVA. We would also like to thank the case companies and the respondents of the interviews that this paper is based on.

10 REFERENCES

[1] Vargo, S.L., Lusch, R.F. (2004): Evolving to a New Dominant Logic for Marketing. Journal of Marketing. 68: 1-17.

[2] Bovik, C., (2004): Customer-Perceived Value in Business Relationships, Doctoral thesis 2004:44, Division for Business and Economics, Department of Business Administration, Karlstad University, Sweden.

[3] Grönroos, C. (2000): Service Management and Marketing: A Customer Relationship Management Approach, Wiley, Chichester.

[4] Hubka, V., Eder. E. (1996): Design Science: introduction to the needs, scope and organization of engineering design knowledge. Springer Verlag, London.

[5] Ipe, M. (2003): Knowledge Sharing in Organizations: A conceptual framework. Human Resource Development Review, Vol. 2, No. 4, pp. 337-359

[6] Nonaka, I. (1991): The knowledge-creating company, Harvard Business Review, pp. 97-104.

[7] Tomkins, C. (2001): Interdependencies, trust and information in relationships, alliances and networks, Accounting, Organizations and Society, Vol. 26 No. 2, pp. 161-191.

[8] Nonaka, I., Toyama, R., Konno, N. (2000): SECI, Ba and Leadership: a Unified Model of Dynamic Knowledge Creation, Long Range Planning, Vol. 33, No. 1, pp. 5-34.

[9] Fransson, M. (2004): Verkstadsindustrins tjänstefiering – en explorativ resa i det nya affärslandskapet (In Swedish). Licentiatuppsats 2004:24, Institutionen för ekonomi, Centrum för tjänsteforskning, Karlstads Universitet, Sweden.

[10] Fischer, T., Gebauer, H., Gustafsson, A., Witell, L. (2009): Managerial Recommendations for Service Innovations in Different Product-Service Systems. In Sakao, T., Lindahl, M., Editors. Introduction to Product/Service-System Design, Springer Verlag, London.

[11] Krishnan, V., Ulrich, K.T. (2001): Product Development Decisions: A Review of the Literature, Management Science, Vol. 47, No. 1, pp. 1-21.

[12] Ulrich, K.T., Eppinger, S.D. (2008): Product Design and Development, McGraw-Hill, Inc, Boston.

[13] Herrman, C., Kuntzky, K., Mennenga, M., Royer-Torney, M., Bergmann, L. (2009): Joint Framework for Product Service Systems and Life Cycle Management, in: Proceedings of the 2th CIRP International Conference on Industrial Product/Service Systems, pp. 453-459, Linköping, Sweden.

[14] MOKA Consortium (2001): Managing Engineering Knowledge. MOKA: Methodology for Knowledge Based Engineering Applications. Edited by Melody Stokes. Professional Engineering Publishing Limited, London.

[15] Ackerman, M., Pipek, V., Wulf, V., Editors. (2003): Sharing Expertise, Beyond Knowledge Management, Massachusetts Institute of Technology, USA.

[16] Swan, J., Newell, S., Scarborough, H., Hislop, D. (1999): Knowledge management and innovation: networks and networking, Journal of Knowledge Management, Vol. 3, No. 4, pp. 262-275.

[17] Erickson, T., Kellogg, W.A. (2003): Knowledge Communities: Online Environments for Supporting Knowledge Management and Its Social Context. In Ackerman, M., Pipek, V., Wulf, V., Editors. Sharing Expertise, Beyond Knowledge Management, Massachusetts Institute of Technology, USA.

[18] Castells, M. (1996): The Information Age, Economy, society and culture, The rise of the network society, Blackwell Publishers, USA, Vol. 1, pp. 153-172.

[19] Tukker, A., Van den Berg, C., Tischner, U., (2006): Product-services: a specific value proposition. In Tukker, A., Tischner, U., Editors. New Business for Old Europe: Product-Service Development, Competitiveness and sustainability, Greenleaf Publishing, UK, pp. 22-34.

[20] Manzini, E., Vezzoli, C. (2001): Product-Service Systems and Sustainability: Opportunities for sustainable solutions, Report UNEP.

[21] Tukker, A., Tischner, U. (2006): Note from the field: Product-services as a research field: past, present and future. Reflections from a decade of research, Journal of Cleaner Production Vol. 14, No. 17, pp. 1552-1556.

[22] Rittel, H., Webber, M. (1973): Dilemmas in General Theory of Planning, Policy Sciences, Vol. 4, pp. 155-169.

[23] Hyman, B., (1998): Fundamentals of Engineering Design, Prentice Hall, Upper Saddle River, NJ.

[24] Randall, D., Harper, R., Rouncefield, M., (2007): Fieldwork for design – theory and practice, Springer-Verlag, London.

[25] Popadiuk, S., Choo, C.W. (2006): Innovation and Knowledge Creation: How are these concepts related?, International Journal of Information Management, Vol. 26 pp. 302-312.

[26] Nergård, H. (2009): Knowledge engineering models as experience carriers, Doctoral thesis, Division for Functional product development, Lulea University of Technology, Sweden.

[27] Mintzberg, H., Westley, F. (2001): Decision Making: It's not what you think, MIT Sloan Management Review, Spring, pp. 89-93.

[28] Modica S. and Rustichini A. (1994): Awareness and partitional information structures, Theory and Decision, Vol. 37, No. 1, pp. 107-124.

Measuring Innovation Capability – Assessing Collaborative Performance in Product-Service System Innovation

Johanna Wallin[12], Andreas Larsson[3], Ola Isaksson[12], Tobias Larsson[1]

[1] Luleå University of Technology, Luleå, Sweden

[2] Volvo Aero Corporation, Trollhättan, Sweden

[3] Lund University, Lund, Sweden

Abstract

This paper reports on a qualitative study, carried out at a Swedish aero engine manufacturer. The study was initiated to explore key indicators related to innovation capability in a Product-Service System (PSS) context. Developing PSS changes the dynamics of collaboration, since the offering of such systems usually involves a network of partners sharing the responsibility for a delivered function over a full lifecycle. In particular, this paper focuses on describing aspects related to external and internal collaboration, and it further discusses how to measure the company's collaborative performance, taking into account both activity and effect measures.

Keywords:

Innovation Capability; Performance Measurement; Collaboration; Product Service Systems

1 INTRODUCTION

Innovation capability, broadly defined as the ability to routinely achieve innovative outcomes, is crucial for industrial companies today in order to be competitive on the market over time. They need to trust their capability to innovate again and again, and by deploying appropriate metrics of innovation capability, they could (1) establish the current baseline when it comes to innovation capability, (2) establish useful indicators related to potential future outcomes, and (3) establish intervention mechanisms to increase the innovation capability when and where needed.

Manufacturing industries are undergoing a transition towards integrating more services into the traditional product concepts. On the one hand, this creates new opportunities for the companies. On the other hand, vanishing boarders between services and hardware changes the way companies plan, develop and produce such Product-Service Systems (PSS) [1]. Innovations where services are a key ingredient of the solution typically require skills and competencies that may not reside fully within a manufacturing-focused organization.

To achieve innovation in the comprehensive PSS domain, companies need to reassess how they collaborate both internally and externally, considering the roles, responsibilities, competencies and skills of an increasingly diverse and distributed set of stakeholders. Key questions when assessing the innovation capability within the context of collaboration are, first, how effective are current collaborative practices? Second, how can these collaborative practices be improved? Answering these two questions introduces a third one – how can the collaboration capability be measured? These were the guiding questions behind this research project being conducted at a Swedish engine component manufacturer in the aerospace industry, and part of a large group in the transport business. The company has a history of technology development, product development, manufacturing and in-service support, but has recently positioned itself as a provider of solutions, including products, technologies and services. Such a move implies that the companies need to improve their capability to develop and offer PSS solutions [1].

On a product development level, innovation capabilities are relatively well known for hardware issues, but service integration implies development of new business models, new competences and collaborative partnerships – both internally and externally. Careful strategic leadership is needed since this is a transition challenging tradition and mindset. To reach the desired future state of a high innovation capability and successful development of product-service systems, several steps need to be taken. One of these steps relates to measuring the innovation capability, and establishing metrics with regard to external and internal collaboration is the topic of this paper. Developing and offering PSS solutions require close business collaboration with other companies and organizations throughout the lifecycle and along the value-chain of the solution. The changing conditions related to offering solutions rather than products require a better understanding of the company innovation capability.

Previous work [2] at the company has identified six areas of importance when measuring innovation capability at the company: *Project selection*, *Customer involvement*, *Interaction between functions*, *Innovation methodology*, *Team climate* and *Innovation rewards*. Furthermore, it was found that dividing the metrics into '*Activity*' and '*Effect*' enabled the organization to deepen the understanding of both real-time performance indicators, which can give a more or less 'instant' feedback to ongoing activities, and the after-the-fact performance measures, allowing a more careful analysis of how particular activities relate to particular outcomes.

Two of the identified areas were found to be of particular importance when developing PSS as opposed to developing 'only' products or technologies. These areas are *Customer involvement* and *Interaction between functions*, where both describe collaboration and networking, the first focused on external collaboration, and the second focused on internal collaboration. A PSS commitment requires an exceptionally close contact with the customer throughout the whole lifecycle, and it also increases the demands on close collaboration between product development functions and service development function.

J. Hesselbach and C. Herrmann (eds.), *Functional Thinking for Value Creation: Proceedings of the 3rd CIRP International Conference on Industrial Product Service Systems, Technische Universität Braunschweig, Braunschweig, Germany, May 5th - 6th, 2011*,
DOI 10.1007/978-3-642-19689-8_37,

2 MEASURING INNOVATION

According to Olsson et al [3], how you define innovation affects the ways to measure it. An innovation is often described in literature as something more than a great idea, since ideas essentially need to be realized and add value for both customers and firms to be an innovation [3][4]. Schramm [4] signifies the importance of measuring both input and output to innovation. Andrew et al [5] include, apart from input and output, processes (which act on and transform the inputs). Chiesa et al [6] note that focusing on input and output indicates rather than explains performance and in order to understand the innovation performance one must include investigate innovation capability and the processes involved in developing and exploiting innovations. Cordero [7] summarizes measures of innovation performance in firms and highlights that performance evaluation in many organizations focuses mainly on resources and outputs, such as R&D expenditure, speed to market, market share, and the number of new products, thus tending to ignore the processes in-between.

Thus, there is a fundamental problem with focusing on, for example, financial metrics since past performance is often a poor indicator of future success. A successful business does not necessarily imply that a company has a high innovation capability. Poor innovation practices can lead to good outcomes and good innovation practices can lead to poor outcomes. Rosenzweig [8] notes, for instance, that just because *"...a given choice didn't turn out well doesn't mean it had been a mistake. It is therefore necessary to examine the decision process itself and not just the outcome."* Muller et al [9] also use the same three categories: input, output and processes, and further suggests that metrics are tailored in three views – Capability, Resources and Leadership. Olsson et al [3] highlight the importance and need of every organization to figure out what is important to measure in view of their specific circumstances. Werner et al [10] discuss the use of objective/subjective as well as quantitative/qualitative metrics and claim that a combination of multiple methods would reduce biases, take advantage of multiple dimensions of excellence and provide built-in checks and balances to capture the full range of the R&D process. It is frequently stated that metrics can lead organizations to successfully capitalize on innovations [4][11] and it can also be seen as a way for companies to better understand if it is worth pursuing potential high-risk projects. Measuring innovative capability is important from a change management perspective and is crucial to the successful identification and selection of high impact ideas for product development.

3 DATA COLLECTION

This paper reports on an ongoing study of the company's current state-of-practice in terms of performance measurements for innovation capability in general and for product-service systems in particular. Five project managers, responsible for advanced product and technology development projects at the early stages of the innovation process were interviewed together with five other individuals involved in service, business development and/or technology development. The rationale for selecting these individuals is that they are organizationally distributed across the company so that they could serve as, and provide, representative examples of the established culture and work methods, and thus provide insight into what factors they believe affect the innovation capability in product-service system innovation at the company, and take part in a discussion around how this capability could be more effectively measured and improved.

The methodology used to capture empirical data focuses on semi-formal interviews, one with each of the selected individuals which also included open discussion on the topics, and close observations of meetings regarding the development of product-service system solutions. The findings from interviews and observations are discussed in reference to the findings from the literature review.

4 PRODUCT-SERVICE SYSTEM DEVELOPMENT IN AEROSPACE

It is a pre-requisite to develop a deep understanding of the contextual situation in order to establish relevant metrics [6]. The aerospace industry is characterized by high technical complexity, high development cost, long product lifecycles, and increasing service integration in offers such as TotalCare® by Rolls-Royce, [12] where airlines pay for service in a package with the product. Since the product development is mainly contract driven, binding contracts are normally signed before development work starts and risks are shared.

Many industries are focusing on traditional lifecycle services (e.g. supply of spare parts, repairs and maintenance) to generate higher profits [13], and extending the PSS components in such business models can offer significant growth potential. Service offers provide an opportunity to differentiate against competitors, build customer loyalty, distribute the revenue profile during the lifecycle and increase the demand for the core product. Integrating services and physical products into a single 'package' comes with a more complex character of the offer. Managing such complexity may cause an extra burden on the information systems required to support the sales and maintenance of such 'packages'. Since addressing these challenges might not be the core expertise of the company [14], important questions relate to whether the company should look towards external partners to realize the idea on the market, or start growing the knowledge internally.

The company's history of technology development and manufacturing has given it a strong product focus even though service and maintenance also have been important parts of the business for a long time. The customer for commercial product development projects is the Original Equipment Manufacturer (OEM) of the engine (i.e. General Electric, Rolls-Royce, Pratt & Whitney), whereas the company is the OEM of the engine on the military market, with the Swedish Air Force as the customer. Recently the company has positioned itself as a provider of solutions and the increase in solution offers requires ownership of the product/technology for the full lifecycle of the solution. Such solutions require a close collaboration with customers, and could be described as a risk and reward sharing partnership rather than as a customer-supplier relationship. The company offers an extensive range of services, including sales of spare parts for aircraft engines and aircraft, as well as maintenance, repair and overhaul of aircraft engines and gas turbines. The military business side has come far in integrating services into the product offers and it is a sector where there already exist complete product-service system offers. Most of the service development is also on the military side, the cradle of the core business, which makes service development in this area easier. However, on the commercial market where aircraft engines components are developed in partnership with the OEM, the services provided have been separate from the total product offers.

Product-service system development also requires a close interaction between the different functions and departments within the company, since product development and service development involve different parts of the organization. Further, it can also involve collaboration outside company borders, such as cooperation within the group or other collaboration partners.

The company uses the term *Soft Products* to describe products and services that enhance the customer experience and satisfaction beyond the core product, and the biggest Soft Product is for some considered to be their product development organization, since the customer, the OEM of the engine, not only gets a hardware product, but a partner who develops it, puts it in production and then supports it. Also, the technology development is for some considered to be a form of service, since it adds value to the customer and on the space market the company sometimes works as a consultant. Consultancy could also be performed in other business areas, although the organization is not built for these kinds of services, given that the company is used to large, multi-year projects rather than a certain amount of consultancy hours. The Soft Product strategy designed by the company mandates that the 'soft' products or services that are developed should be connected to the core business of the company. Things that are not included in the core business should be outsourced to other companies within the group or to external companies. However, it is rare that the company collaborates with another group company and the main reasons for this, according to informants, is that the aerospace industry is unlike other transport businesses within the group, and that other group companies are more truck-oriented, and not used to the airworthiness requirements of the aerospace industry. However, when it comes to product-service systems, other group companies have great experience, knowledge and resources regarding diagnostics, handling large amounts of computer data, making service interfaces and customizations, which are important for service development. On this issue the company has taken a collaborative leap, and for the first time several companies within the group are working together to develop a product-service system for military aircrafts.

5 CUSTOMER INVOLVEMENT – EXTERNAL COLLABORATION

Literature often points to the importance of customer involvement in the product development process in order to fully understand the needs and expectations of the customer [3][15]. Cao et al [16, p.364] identified five dimensions that make up supply chain collaborative advantages: process efficiency, offering flexibility, business synergy, quality and innovation. Cao notes: "*The definition and measures of collaborative advantages can help managers to define specific actions to be taken collaboratively to improve shared supply chain processes that benefit all members*". Eppinger et al [17, p.22] states: "*Best practice in product development (PD) is now rapidly migrating from local, cross-functional collaboration to a mode of global collaboration*".

On the commercial market, the company has specialized in certain components of the aircraft engine and their customers/partners have specialized in others. Dealing with the interfaces between the different components put increasing demands on collaboration, but the roles of *customers* or *suppliers* are blurry. The customer can be described as "*part of the team*" or, as one of the informants put it: "*If we are wrong, they are wrong, if we are right, they are right. We are not always right, sometimes we fail together*". Several departments communicate directly with the customer and this communication is not only on a management level, team members communicate directly with the customer and this is important for the collaboration. The partnership also means that the customer has higher requirements on the way of working than the average customer of a consumer product would have. Since the way of working directly affects the trustworthiness within the partnership and this is especially important the more radical the project is. Certain critical areas, such as engine mounts and acoustics, also involve collaboration with the aircraft manufacturer.

Product development always has an apparent external customer, but this is not always the case for technology development. The external customer creates a stronger driving force for expected results, but it can also create boundaries for the innovation capability. Technology development teams that only have an internal customer can develop in the direction that is most profitable for the company alone. The external customers are considered to be more pushing and active than the internal customers. This points out the importance of keeping both internal and external customers close. The customer involvement is as important, if not more important, when developing product-service systems. However, when providing services the company does not only have to look to their usual customers, but also to the customers of the customer, e.g. the airlines. Here, it is also important to note that this essentially means that they may become competitors with their customers/partners, which provide similar services. The key is to provide services that are close to their own core business but in the periphery of their customers' business. If services and maintenance are to be provided on an engine, a contract is needed with the OEM of the engine.

One informant described the importance of infusing confidence to the customer by coming up with ideas, being proactive, showing the customer their way of working and developing their methods. The company needs good ideas in order to 'push' the customer. This points to the importance of a methodology and a process, which makes the most out of the relationship between the team and the customer, making it as profitable as possible for both company and customer. There is much to gain by going outside the company borders for collaboration. This is how one project manager described it: "*As long as we are just small teams, we cannot afford to be more than six, seven people in the small projects; we are so smoked, if we cannot use the outside world. Usually other people have great ideas, better knowledge, we know our application, but we need to go out and find those other pieces that are needed. /.../ As long as we just sit in our small teams internally, sure we can come up with the world's best solution but the probability is quite low. If we constantly work with the world around us the likelihood is much greater*". This points to the importance of not only collaborating externally with customers but also with suppliers and university researchers.

6 INTERACTION BETWEEN FUNCTIONS – INTERNAL COLLABORATION

The importance of collaboration in product development and innovation is often stated in literature [18][19][20]. For instance, Björklund [19] identifies collaboration as one of the key factors when improving project performance. Sosa et al [21] point out the importance of functions involved talking to each other to anticipate the unexpected problems to save time and cost. The company organization consists of different business functions, such as marketing, purchasing, production and product development, and each business function consists of several departments. The development teams consist of people from different departments and different parts of the organizations and this collaboration is essential for all projects.

There are differences in opinion whether or not the collaboration and communication at the company is satisfactory. The informants who considered the communication to be satisfactory also described the company as 'small'. There are advantages with this, according to one informant: "*I think it works. We are such a small company. The advantage is that things go pretty fast and it's quite informal. The ceiling is quite high when it comes to bouncing ideas.*"

The collaboration within one business function is always described as easier than the collaboration between them. The informants who considered the communication to be unsatisfactory usually had a hard time describing why this is the case. Several of them described the importance of having a common interest when collaborating with other departments even though there often were disagreements on what was important. The collaboration between departments and functions involves many compromises and if many people are involved it can slow down the pace because of the large amount of input, but this collaboration is still described as very important since ideas appear from many sources. Although the projects consist of people from different functions, the focus is on developing products (even in technology development projects). Services are not taken into consideration, except possibly reparations of the product. The reason for this, according to one informant, is considered to be the educational background of the employees: "*We are mechanical and electrical engineers all of us. It is that simple.*"

One important collaboration activity for the development projects are the gate reviews; this is where many important decisions are made. The product and technology development processes are stage-gate processes with decision gates to pass through based on thorough reviews of the project status. Each gate has its checklist and control questions, but these are normally very product focused, and no checklist items in any gates involve services, or poses questions such as: is it possible to refine this product to include a service? Opportunities for service development can be lost when there is no demand in exploring them. If a project is not fulfilling the gate criteria, it is not allowed to pass through the gate. However, when the project has a deadline towards a customer and risk receiving late fees if not delivered, it is sometimes not feasible to make the decision to stop the project at the gate. These gate reviews in early stages are arranged by the engineering organisation and the focus is on technical performance rather than, for instance, the ease of manufacturing. One project manager thought that the collaboration and the handover of projects would perhaps be easier if the responsibilities at these gates would alternate between persons from different functions. Further, if the person responsible for manufacturing aspects would have a stronger vote, the project might not pass through the gate so easily.

Lack of time within the teams can hinder innovation and collaboration. Sharing a good idea can sometimes backfire on the person who created the idea, in the form of an expectation to also follow-through with the implementation of the idea. This can hinder a person from sharing an idea unless this person has time to develop it him/herself. This points to the importance of collaborating within the team when ideas erupt, implying that ideas should not necessarily be owned by the individual but by the whole team, which is also collectively responsible for its development.

There are various tools for external communication and collaboration, such as Microsoft Office Live Meeting and conference calls, which are widely used within the company. Also, video conference software and equipment is available, but is not so commonly used. Internally within the group there are also commonly used online tools, such as chat, intranet and blogs. Internal blogs are often used by the development projects to let the rest of the company know the status of the projects. It is one way for the project teams to communicate to the rest of the company. However, these blogs can be hard to get started: "*There is a resistance to writing; you don't see the point in it. On the other hand you see the point in it when others are writing so I can read it*". One project manager expressed the need to improve this way of communicating and suggested an internal version of Facebook as a way for individual employees to report status updates: "*It is the little things that doesn't mean anything by themselves but together gives the whole picture. /.../If we had Facebook here and I would know what everyone in [PROJECT X] were doing, it would lead to discussions and in the end I would understand what's going on in the project and perhaps I would know when I could help or we could help or they could help us*".

This highlights the importance of not only having tools for *direct* communication between people, but also deploying tools that allow people to communicate *indirectly*, for instance by overhearing communication between other individuals and groups, and observing traces of information left by other people. In these days of online social networking, many people are used to this way of communicating and screening information.

Job rotation is another important aspect related to the collaborative performance of the internal network, and in particular the job-rotation between business functions. It gives the organization a greater knowledge about internal functions and creates connections between departments. Even though many of the informants have had jobs in other business functions, they still considered it to be a rare thing to change jobs between business functions at the company; a more typical career path is within a business function.

7 MEASURING INNOVATION CAPABILITY

As noted earlier, the aerospace industry is characterized by long lead times and product lifecycles in, and considering that it takes years, even decades, for an idea to fully realize its potential and become an innovation on the market, it is very difficult for a company in such an industry context to know already in the early stages that they are on the path to innovation. Knowing how innovative they are long before the product has reached the market implies a need to measure the innovation capability related to 'work in progress'.

The majority of informants were positive to the idea of measuring the innovation capability at the company. One person was a bit hesitant regarding the use of the outcome of the measure and considered the involvement in the engine programs and the technology development for example to be 'proof' enough that they are innovative; otherwise the OEM would not choose them as partners. There were also expressed needs for 'simple' metrics, that changes could be observed in a fairly short period of time and that one would see the immediate effect of one's own work. There was also an expressed need to highlight the good examples of company innovations, those products that were developed twenty years ago and are still on the market.

The metrics that are used today in order to measure innovation capability at the company are (1) customer satisfaction, (2) brand platform, (3) return on investment, (4) number of patents, and (5) number of patent applications. The area of patents is not discussed in this paper since it is not directly linked to collaboration and network innovation. The collaboration with the customer will of course affect both the customer satisfaction and the brand platform, but these metrics do not give the complete picture of the collaborative innovation capability.

All different types of collaboration need to be measured to give a justified result of the company. In previous sections of this paper, seven types of collaboration have been identified from the point of view of the cross-functional project team (See Figure 1, below):

1. *Collaboration within the department:* individuals within a team collaborate with colleagues who are found within their department, but who are not part of the team.
2. *Collaboration within the business function:* individuals within a team collaborate with colleagues who are found within their business function, but who are not part of the team.
3. *Collaboration within the company:* individuals within a team collaborate with colleagues from different business functions.
4. *Collaboration within the Group:* individuals within the company collaborate with individuals in other companies within the Group.

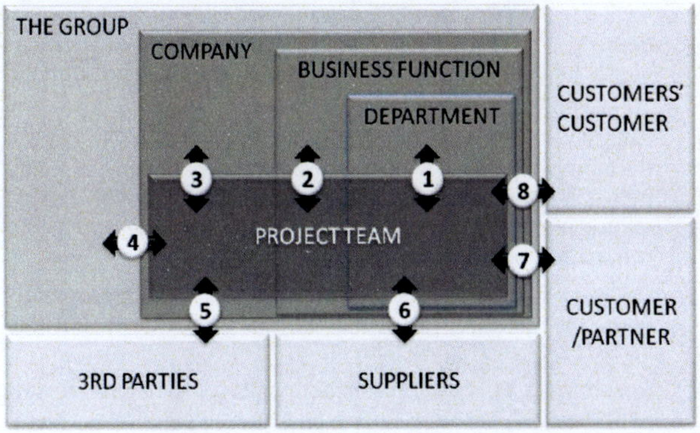

Figure 1: Different types of internal and external collaboration.

5. *Collaboration with 3rd parties such as universities:* the team collaborates externally with e.g. universities.
6. *Collaboration with suppliers:* the team collaborates externally with suppliers and potential future suppliers.
7. *Collaboration with customers/partners:* the team collaborates with customers/partners or potential future customer/partners.
8. *Collaboration with customers of the customer:* the team collaborates with the customers' customer, e.g. aircraft manufacturers and airlines.

The suggested metrics in this paper are divided into '*Activity*' and '*Effect*' so that the focus is not only the outcome, but also on the ways of working in the organization and the activities that are performed (see Table 1, below).

Both quantitative and qualitative metrics are used in order to capture the full range of the innovative process. Each activity metric needs to be measured for each of the eight types of collaboration that were identified at the company.

8 CONCLUSIONS

In this paper eight different types of collaborations (both internal and external) are identified and a set of metrics, divided into 'Activity' and 'Effect', are proposed. The set of metrics was based on the findings through interviews and literature, collectively supporting that innovation for product service systems is linked to collaborative performance.

Since the company has a strong division through in service business, and another in hardware focused business on the commercial engines side, collaboration within the company as well as with new types of external partners needs development to become product service providers. Also, within already established collaborations, the role and type of collaboration need to be improved. It is evident from the military side that the unique and tight connection to the customer is one success factor to that service integration in product offers is more mature. It is recognized that the aerospace business has long product lifecycles, which is likely to impact the rate of change in new innovation schemes.

Regarding activity metrics, the pure number of collaborative activities aims to measure the degree of collaboration, where more activities would indicate a more active collaborative work. It is important to recognize that not only established partner collaboration is measured, but also collaboration with partners contributing to future PSS solutions, such as universities, new type or suppliers and customers. A qualitative measure to explore the degree of ideation in the collaboration activities is proposed in combination. Activity measures to evolve the internal collaborative performance include measuring the quality of responsibility handover through the lifecycle, the use of available tools for collaboration and the perceived satisfaction during internal collaboration.

It is expected that effect measures will show change over longer time. Most important is to include the PSS aspect into the customer satisfaction measures already conducted. Quantitative effect measures include pure counting of ideas, business proposals and concluded new businesses.

ACTIVITY	Description / Examples	EFFECT	Description / Examples
Quantity of collaboration activities/connections	External: number of customers' customer meetings, number of customer/partner meetings, number of supplier meetings, number of 3rd party meetings Internal: number of cross-boundary (team, department, function, company) meetings	Customer satisfaction	A Customer Satisfaction Index establishes whether or not customers are satisfied with the PSS solution.
Quality of collaboration activities/connections	Did the meeting involve collaborative ideation or only exchange of information?	Return on Investment (ROI) on innovation projects	Is the company earning money on their PSS solutions?
Subjective assessment of project handovers	What word would summarize the handover? (e.g. stressful, smooth,)	Project/products delivering on time	Number of delayed delivery days
Subjective assessment of the communication activities	Availability of tools for communication activities, for both direct and indirect communication	Product quality	Requirement fulfillment
Satisfaction of methodology for collaboration activities	Subjective assessment of the methodology used for collaboration	Large amount of ideas	Number of ideas from different sources, number of ideas in total, number of ideas lead to concluded business

Table 1: Metric divided into *Activity* and *Effect*.

Some conditions found that need to be considered in developing new type of innovative capabilities are:

- Customer relation with manufacturing OEM's need to evolve, since services are consumed by airlines
- Aerospace regulation and long life-cycle of products and business contracts may act as conservation mechanism when introducing new innovation models.
- Long term customer relation is significant for building trust and reducing risk in new innovations with service content.
- The company in the study has opportunities to use capabilities within the Group, where complementary skills exist
- True exploitation of internal and external capabilities to combine service expertise and hardware expertise require senior level ambitions and directives since the established way of working not naturally has this focus.
- Common business focus within different functions supports internal collaboration
- Innovations including services for product development should be included in advance of contractual development.

Changing to a service-focused mindset in the organization could mean a need for shorter decision processes. It would also include increased internal collaboration. However, a systematic decision process is already in place at the company, serving to evaluate the decisions whether or not to join the large engine programs.

The internal collaboration is easier when the individuals have the same business perspective, i.e. belong to the same business function within the organization. Job rotation can provide a new perspective; it gives the organization a greater knowledge about internal functions and creates connections between departments.

Measuring innovation may be one of the useful useful tools in this process as it would provide useful insights and facts on how innovative the company is long before the product or service has reached the market.

9 ACKNOWLEDGEMENTS

Financial support from Vinnova through PIEp, and Volvo is greatly acknowledged. The authors are also grateful for the time and participation from interviewees and colleagues.

10 REFERENCES

[1] Baines, T. S.; Lightfoot, H. W.; Evans, S.; Neely, A.; Greenough, R.; Jeppard, J.; Roy, R., Shehab, E.; Braganza, A.; Tiwari, A.; Alcock, J. R.; Angus, J. P.; Bastl, M.; Cousens, A.; Irving, P.; Johnson, M.; Kingston, J.; Lockett, H.; Martinez, V.; Michele, P.; Transfield, D.; Walton, I. M.; Wilson, H. (2007): State-of-the-art in productservice systems, J. Engineering Manufacture, Vol. 221 Part B.

[2] Wallin, J.; Isaksson, O.; Larsson, A.; Larsson, T. (2010): Measuring Innovation Capability in Technology-Focused Development, 17th International Product Development Management Conference, Murcia, Spain, June 13-15.

[3] Olsson, A. (Ed.) (2008): PIEp, Product Innovation Engineering Program, Innovationsförmåga.

[4] Schramm, C. (Ed.) (2008): Innovation Measurement – Tracking the State of Innovation in the American Economy, A report to the Secretary of Commerce by The Advisory Committee on Measuring Innovation in the 21st Century Economy.

[5] Andrew, J.; Sirkin, H.; Haanaes, K.; Michael, D. (2007): Measuring Innovation 2007, A BCG Senior Management Survey, The Boston Consulting Group.

[6] Chiesa, V.; Frattini, F.; Lazzarotti, V.; Manzini, R. (2009): Performance measurement in R&D: Exploring the Interplay between Measurement Objectives, Dimensions of Performance and Contextual Factors. R&D Management, Vol. 39, Issue 5, pp. 487-519.

[7] Cordero, R. (1990): The measurement of innovation performance in the firm: an overview, Research Policy, Vol. 19, pp.185-192.

[8] Rosenzweig, P. (2007): Misunderstanding the nature of company performance: the halo effect and other business delusions, California Management Review, Vol. 49, No.4, pp. 6-20.

[9] Muller, A.; Valikangas, L.; Merlyn, P. (2005): Metrics for innovation: guidelines for developing a customized suite of innovation metrics, Strategy & Leadership. 33(1) pp. 37 – 45.

[10] Werner, B.; Souder, W. (1997): Measuring R&D Performance – State of the Art, Research Technology Management: ProQuest Science Journal: Mar/Apr, Vol. 40, No. 2; pp. 34-42.

[11] Chan, V.; Musso, C.; Shankar, V. (2008): McKinsey Global Survey Results: Assessing innovation metrics, The McKinsey Quarterly.

[12] Harrison, A. (2006): Design for service – Harmonising product design with a services strategy, Proceedings of the ASME Turbo Expo 2006, Vol 2, Barcelona, Spain, May 06-11.

[13] Beauvilliard, A.; Jentzsch, A.; Kurmann, D.; Obring, K.; Ulrich, P. (2009): Achieving Excellence in After-Sales Services – A Primer for Industrial Companies, The Boston Consulting Group. September.

[14] Schaper, M. (2010): Tackling IT complexity in product design, McKinsey Quarterly, McKinsey & Company.

[15] Sivathanu Pillai, A.; Joshi, A.; Srinivasa Rao, K. (2002): Performance measurement of R&D projects in a multi-project, concurrent engineering environment, International Journal of Project Management 20, pp. 165-177.

[16] Cao, M.; Zhang, Q. (2010): Supply chain collaborative advantage: A firm's perspective, Int. J. Production Economics 128, pp. 358-367.

[17] Eppinger, S.; Chitkara, A. (2006): The New Practice of Global Product Development, MIT Sloan Management Review. Summer 2006, Vol. 47 No.4, pp. 22-30.

[18] Brown, T. (2008): Design Thinking, Harvard Business Review. June 2008, pp. 84-92.

[19] Björklund, T. (2010): Enhancing creative knowledge-work: challenges and point of leverage, International Journal of Managing Projects in Business, Vol.3 No.3, pp. 517-525.

[20] Amabile, T. (1998): How to Kill Creativity, Harvard Business Review. September-October 1998, pp. 77-87.

[21] Sosa, M.; Eppinger, S.; Rowles, C. (2007): Are Your Engineers Talking to One Another When They Should?, Harvard Business Review. November 2007, pp. 133-142.

Descriptive Service Product Architecture for Communication Service Provider

Oliver Budde[1], Julius Golovatchev[2]

[1] Institute for Industrial Management (FIR), RWTH Aachen, Aachen, Germany

[2] Detecon International GmbH, Bonn, Germany

Abstract

The product structure plays a major role in the PLM as an integral design element for coping with product complexity. As a prerequisite for enabling product structuring, a descriptive model of the product is needed, a product model. Product models for the service industry have been subject for scientific examination nowadays, but so far a general and practical definition, as for tangible goods, is still missing. This paper proposes a general product architecture for services in the telecommunication industry by reflecting on existing work from the manufacturing industry. These theoretical findings have been evaluated by an international study to provide indications for its adequacy.

Keywords:

Product Architecture; Telecommunication Industry; Product Lifecycle Management

1 MOTIVATION

The manufacturing industry has anticipated the benefits of product models for a long time now for coping with complexity problems [1]. Although there are still some problems unsolved, like information consistency of product models across the whole product lifecycle, IT-systems support and corresponding processes have been established in the industry for a while now. Due to the high popularity of product models in the manufacturing industry it stands for a reason to transfer the idea of product modularity to services.

In the following paragraphs we will propose the cornerstones of a product model for product and service systems for ccommunication service provider (CSP). The need for product models here is similar to the manufacturing industry: coping with product complexity. In order to derive such a model from pre-existing work e.g. from the manufacturing and service industry, we need to define the special characteristics of this industry first. Thereby we define the subject of matter, the product in network industry, on a descriptive basis. Building on that the specific layers of a product model can be identified and distinct design elements can be extracted. As a final step in this paper these design elements have been examined in an international study regarding their implementation degree. Based on these empirical findings we derive the usefulness of our model.

2 DEFINITION OF THE SERVICE PRODUCT

Before proposing a product model for the network industry, the main characteristics that determine the product model needs to be described. For our purpose we suggest the following four characteristics that apply for CSP products.

Network effect

Network industries, like the Communication Service Industry, are characterized by the existence of direct or indirect network effects that have specific requirements concerning the adoption and diffusion of innovations in the market. Besides the original product benefit, telecommunication companies have to consider the derivative product benefit, which can be controlled by them only to a limited extent [2].

Value Net

Network industries are based on a special value configuration: the value net [3]. Typical for this value configuration is a reciprocal production technology and a system technology (the physical net), which imposes special requirements on the coordination of value adding partners involved in the product lifecycle.

Service characteristics

The simultaneity of production and consumption as well as the integration of the external factor are characteristics of telecommunication products. This involves the need to describe CSP products in 3 dimensions: potential, process and resource [4].

Immateriality

A further characteristic of the telecommunication industry is the immateriality of the product. This leads to specific product characteristics: simple mutability, reproducibility and first-copy-effects. In contrast to physical goods and services, high fixed and low variable costs of production and sales are combined. New business models with zero marginal costs are possible only with digital goods [5].

Based on these general characteristics we have derived a detailed scheme for a further concretization of the CSP product. This definition of the product provides essential perspectives on the product that lead to consequences in the defining a corresponding product model.

The following scheme compiles the classification schemes from Corsten [4] for a general product classification, from Gerpott [6] for network industries, from Stabell [3] for value configuration, from Meffert [7] for services characteristics and from Mowshowitz [8] for immateriality. According to this compiled scheme we concretize the CSP Product for the purpose of this paper. The universality of this scheme can not be scientifically proven within the given scope of this paper. This work has to be undertaken in the future.

J. Hesselbach and C. Herrmann (eds.), *Functional Thinking for Value Creation: Proceedings of the 3rd CIRP International Conference on Industrial Product Service Systems, Technische Universität Braunschweig, Braunschweig, Germany, May 5th - 6th, 2011*,
DOI 10.1007/978-3-642-19689-8_38, © Springer-Verlag Berlin Heidelberg 2011

	Attribute	Specification				
General product characteristics	Typology of goods	Physical goods	Service	Information	Legally protected good	
	Scope of the product	Single component	Range of components			
	Focus of the product	Consumer focused (B2C)	Business focused (B2B)			
Network Effects	Network effects	direct	indirect	none		
	Benefit of the product	derivative	derivative and original	original		
	Temporal stability of the product benefits	shifting	partly constant	constant		
	Network effect multiplier	Intensity of Usage	Number of sold items	none		
	Importance of Installed Base	very high	high	low		
Value Network	Value Configuration Model	Value Chain	Value Shop	Value Network		
	Type of Network	Virtual Network (Supply Chain)	Road network	Communication network	Energy network	
	Heterogeneity of components	low	average	high		
Service Characteristic	Degree of technical integration	low	average	high		
	Customer lasting relationship	spontaneous	Average (< 1 Month)	Contract (Subscription, >1 Month)		
	Degree of individualisation	individual product (craft work)	Building block based	Standard product		
immateriality	Character of the information good	Core	Storage	Processing	Distribution	Presentation
	Digitalization degree	low	middle	high		
	Importance of the information core	low	average	high		

Figure 1: Classification scheme for a CSP product.

Based on this scheme the CSP product can be characterised as network-based product service systems that focuses on the retail market. The service fulfilment depends on the existence of network intermediary providing a communication infrastructure for exchanging information between value partners and customers. The components of a network based service product will be described in paragraph 5.

In the next section the challenge of complexity management will be described and linked to the subject of matter the product architecture.

3 COMPLEXITY MANAGEMENT: AN RISING ISSUE FOR TELCOS

Having brought some light into the CSP product definition, the question arises why complexity is an important issue for CSP that needs to be managed adequately. Based on the consulting experiences from several PLM implementation projects in this area, the authors have identified three fields that need to be tackled in the telecommunication industry.

Product Complexity

The telecommunication product is constituted by a big variety of interdependent product components (e.g. Network access, end user device, information goods and IT services). Moreover the individual life cycle of each product component needs to be synchronised with each other. If it happens otherwise, problems may arise like in the case of the introduction of MMS, where a limited market diffusion of appropriate end user devices impedes a fast success of corresponding services. Therefore a high product complexity is a general phenomenon for CSP.

Process Complexity

The process complexity, as the second source of complexity, results from the fact, that in the service industry the complexity of the service processes, which can be categorized in fulfilment, assurance and billing process, increase with the number of subscribers and the variety of the product bundles. A more complex product has other requirements on the service processes than an old fashioned telephony product (e.g. the billing process is more complex, if more partners are involved, like content providers in case of a triple play product). Additionally the need for the execution of mass service processes sets constraints on the organisation, e.g. distributed call centres, which has an impact on the process complexity in terms of establishing the same quality of service standards.

IT-Complexity

The business application landscape of a typical CSP is dominated by individual software, since standard software like Enterprise Resource Planning (ERP) for CSP has generally not been available. Beside this heritage of heterogeneity of the IT-Systems, a second reason for IT complexity arises through the existence of many different data sources for managing the product related data and a missing alignment between these data silos. This diversity on the IT-system level as well as on a data level is primarily responsible for the IT-Complexity, which has led to high cost for the maintenance and for the functional extension of these systems.

4 PRODUCT ARCHITECTURE MANAGEMENT IN THE PLM

For coping with these complexity issues PLM has been proven for being a valid instrument in general, and the product architecture in specific [1]. Budde et al. have motivated the integral role of the product architecture within a holistic PLM-model [9]. They propose three layers that have to be taken into consideration, when designing the product architecture (cp. Figure 3). These three general layers will be described briefly before specifying those in the context of the characteristics of CSP.

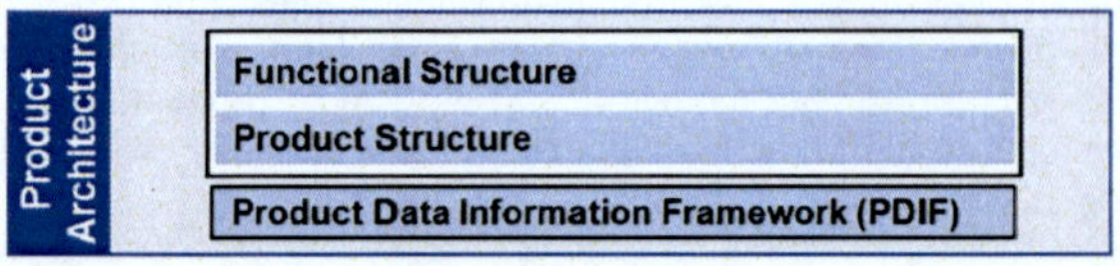

Figure 2: General design areas of a Product architecture.

Product Specification and Function Structure

- Description of the product from a functional perspective taking into account the specific requirements from the branch of industry, e.g. in the service industry the functional description of the product focuses on the specification from a market perspective [10].
- Description of the interrelation between the function structure with the construction and working structure [11]. For the service industry this includes the specification of a process model [10].

Especially for the manufacturing industry there exists a distinct differentiation between the product structure and the function structure [12]. For the goal to have a generalized model being adoptable for specific industries, we integrate the product specification with the function structure in order to emphasize the differentiation between an abstract product concept and its physical representation.

Product Structure from a technical perspective

- Definition of the construction and working structure [11]. The construction structure includes the definition of parts structure
- The specification of the interfaces among interacting physical components [13].

Product Data Information Framework (PDIF)

- Information model scheme that integrates all relevant information of a product for all stakeholder in the PLM [14].
- Facilitating the generation of different views on the product like design structure, purchasing structure, manufacturing structure, order management structure, spare parts structure, service structure [12].

Based on this general structure and the main characteristics of the CSP product a descriptive Service-Product Model will be proposed in the next section.

5 DESCRIPTIVE SERVICE PRODUCT MODEL

Having mentioned the importance and the popularity of Product models in the manufacturing industry, a trend in the service industry towards service modularization becomes visible as well. Thereby most of the authors try to transfer the concepts from the physical goods industry to the service industry (cf. [10]; [15]). But resulting from the extent of heterogeneity of services in general, a widely accepted descriptive model has not been evolved yet.

Our classification scheme from chapter 2 points out the service characteristics of the CSP product. Against this background it becomes evident to examine the general dimensions of a service: structure, process and outcome under consideration of the CSP-characteristics. According to [15] these dimensions have to be taken into account whenever services are to be developed. Consequently the authors suggest using distinct models for each dimension in order to describe the service product in a holistic way. In specific the authors refer to a resource, process and product model. In the next paragraphs these three models will be specialised taking the characteristics of CSP industry into account.

5.1 Product, Process and Resource-Model

Product-Model

According to [15] a product model describes the outcome dimension, the "*what*" of the service. It typically comprises a definition of the service content and a structural plan of the service products. The relevance for such a product model for CSPs results from the dynamic by which the product portfolio changes compared to other service industries (according to [16] the CSP industry belongs to the most dynamic ones). In order to anticipate the need to respond to the high market dynamic a structuring of the range of offerings helps to align the product portfolio to specific customer segments, capturing the complexity of portfolio changes, by a flexible re-arrangement of product modules to new offerings. Thereby the comprehensiveness of the product description on an aggregated level as well as on a module level from a customer perspective is crucial and needs to be taken into account by product manager. Beside this functional description of the product, reflecting the original understanding of the "*what*"-dimension, for CSP a complementary model has to be added to cover the outcome dimension: the price model. This is in line with [11], who also concludes that the outcome dimension is not completely covered by the product model, like proposed by [15]. Within the price model the end customer prices have to be defined and specific rules for calculating the price under consideration of rebates or socio-economic aspects like age / location need to be defined and applied on the product model. Nowadays the customer of a CSP is typically confronted with a huge number of tariffs. These tariffs need to be managed separately from the product description in order to conduct changes on the product model without interfering with the tariffs and vice versa. From an operational point of view the tariff is often perceived as the "CSP Product", which is why the number of those kinds of "products" is continuously increasing, but actually this is only a modelling shortcoming, which results into a growing catalogue of products.

Process Model

According to [15] the process model describes how the outcome of a service is achieved towards a customer. Therefore the process model defines explicitly activities and their execution sequence as well as the interfaces between activities [11]. Corsten distinguishes between two processes in this regard [17]. First, the service agreement phase, in which a customer orders is initiated and the configuration of the service product is agreed upon under consideration of the customer specific settings (e.g. line capacity at the habitation). Second, the service fulfilment phase, in which the service is provided towards the customer and the service outcome, fulfils the expectation.

The service agreement phase corresponds to the fulfilment process in the CSP industry according to the reference process model enhanced Telecom Operations Map (eTOM). This process needs to be designed during the product development phase and consequently implemented in the IT-systems in order to ensure a high automatisation degree for ensuring the delivery of mass processes through different sales channels (e.g. online, on-site, reseller). Simply because of the existence of different sales channel, process variants for the fulfilment process are thinkable and probably necessary. This has to be taken into account in the conception phase and later on in the PLM.

The service fulfilment phase differs in the Telco industry from the general or better saying abstract service industry, due to the fact that the service delivery is automated. In comparison to e.g. a hairdresser that has to execute a manual and personal task in order to deliver the service that results into an outcome for the customer, the service delivery of a CSP is in the regular case completely automated, since it is focused on the establishment of a connection between two or more parties or by providing an IT-service to one customer. In consequence a process execution, which involves resources other than IT-systems, is not needed. Therefore we assume in this paper, that the conceptualisation and implementation of the service delivery process can be characterized as an IT development project that fits better into the Resource Model, which is defined in the next paragraph.

Since the CSP product is in the most business models, at least in West-Europe, delivered on a contractual basis, a long term relationship between the customer and the service company is a prerequisite for the service delivery. This is similar to the insurance business, but different to the hair dresser example. Therefore the CSP needs to provide a service assurance process during the whole customer lifecycle in order to ensure a constant service quality and, in case of disturbance; processes are needed to resolve the customer problem (e.g. technical fault clearance). Consequently the service fulfilment process, proposed by Corsten, has to be extended by a service assurance process.

Another consequence of the long term relationship can be identified in regards to the billing of a CSP product. Since the direct customer interaction is normally limited to the service agreement phase (only accidently in the case of the service assurance process), the billing process plays a major role for CSP in order to transfer the relationship into a prosperous one for both sides. CSP use the billing process, which is normally carried out every month, to interact with the customer e.g. for cross-selling purposes. Another aspect why the billing process, at least nowadays, needs to be designed, is the high error rate, which often involves a manual interaction by a customer agent, either via a call centre or an onsite shop. These processes that occur in the course of the billing need also to be considered within this process model layer of CSP product architecture.

In summary three processes need to be considered in a process model for product architecture: service fulfilment process, service assurance process, and service billing process. The eTom-process-model provides valuable information about the process design.

Resource-Model

According to [15] the resource model describes the outcome of the product development that is necessary for the provisioning of a service. In the resource model the resources have to be described that are consumed in the "production"-process for the provisioning of the services.

In the context of the CSP product multiple resources are needed for the provisioning of the service. Obviously the foundation of the service provisioning is the communication network, as a material resource. In order to establish and use communication services user equipment is needed like mobile devices. According to [4] these resources can be classified as merchandise goods, since their production is mostly not in the scope of Telco Companies, but specialised companies like customer-premises-equipment-providers e.g. Nokia. These devices consume ICT-services on the network and have to be provided either by the network operator or by third parties. In the context of the convergence trend other resources need to be considered. Most importantly this refers to multimedia content, that can also be generated and provided by the network operator or, more often, by third parties like for movies or ring tones. In order to ensure the service assurance, service processes are required like on-site installation services e.g. for WiFi installations. Weber defines another resource in the context of complex product bundles, the legally protected interest good [18]. In case of CSPs this refers to accommodations and warranties. Since the service quality is a critical success factor, CSP have to guarantee response times for service inquiries and are willing to provide accommodations like anytime minutes for compensating faults on the network or errors in the billing process.

In the following table the resources and their classification according to [4] have been listed with a corresponding symbol being used in figure 4.

Resource	Category	Symbole
Network	Operating supplies	—
User equipment	Merchandise good	
ICT-Services	Immaterielle Ressource	
Multi-Media Content	Immaterielle Ressource	
Service	Human Labor	
Legally protected interest	Immaterielle Ressource	

Figure 3: Classification of resources of the CSP-Product.

Figure 4 illustrates the three layers of the product architecture and their interrelations.

5.2 Product Data Information Framework (PDIF)

The product data information framework manages all product related data consistently across the product lifecycle and offers different perspectives on the product depending on the user's role.

Product data information frameworks are widely known in the manufacturing industry and have been the subject of special IT-solutions: Product Data Management Systems (PDMS). For the service industry general frameworks are not that common. Instead special product data models have been elaborated like e.g. for the insurance sector [19]. An overview for product data model for services is given in [10].

For CSP two distinct requirements need to be considered.

The first requirement is about the differentiation between the product concept and its instantiation for a specific customer. A product, described in the aforementioned dimensions, is instantiated by the customer in the moment of the first use, not by the purchase of the product in contrast to the manufacturing industry [2]. This instantiation is characterised by the configuration of the external factor (the customer) e.g. by the installation of the WiFi and the integration of different devices like PC or TV. That is why the specific instantiation

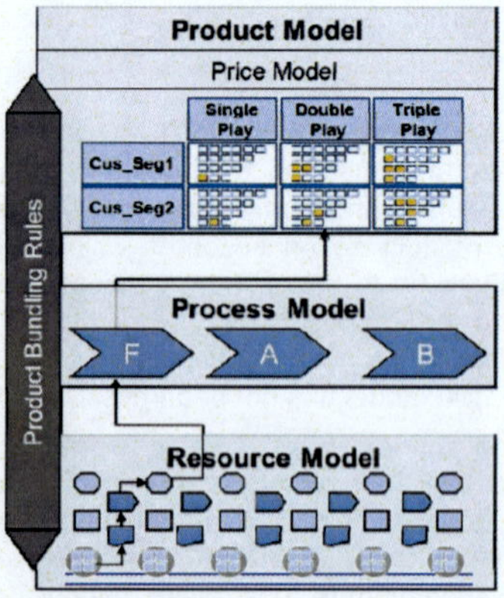

Figure 4: Product Architecture for CSP without the PDIF.

settings as well as the information about the purchased product itself need to be managed side by side in a product data management system. Therefore we distinguish between the product architecture, as a conceptual description of the product features, and the product inventory, that links the product with a specific customer and its individual settings.

The second requirement deals with the definition of the stakeholders for a product model, which differs e.g. from the manufacturing industry, which focuses on production, assembly and sales [20]. For CSP different perspectives of the product including the product inventory are needed. We distinguish between the following four perspectives:

1. Technical Product Development

From a product development perspective the elements of the product and the structure including the relationships between the elements in the three dimensions need to be available. The product development of a CSP focuses from a technical point of view on the network resources and specific ICT-services. In some cases the network resources have been outsourced to a third party like with Bharti Telecom in India. But in general the development and at least the maintenance of the network stays with the network operator. For modelling the different resources the industry association TM Forum has published a reference information framework (SID), which provides valuable guidelines and models for the specification of the resources.

2. Sales

From a sales perspective the range of the product offerings needs to be defined and the sales process for specific products needs to be determined. Regarding the range of the offerings, the sales needs to define the components of a product bundle as well as the price. In terms of the sales process, this includes the selection of the appropriate sales channels as well as the determination of customer segments, which are addressed by a specific product bundle. These information need to be managed, so that a rather automatic service fulfilment can be enabled (e.g. by releasing a new product from the sales perspective, the product should be consistently available in all sales channels offline or online). Similar to the perspective of the technical product development, the SID standard provides guidelines and models, which are subsumed under the customer facing specification

3. Billing

The billing is crucial for CSP and currently one the biggest fields of concern in this industry due to problems with the increasing number and variety of billing events. This is one reason why the topic of revenue assurance is at present very popular in this industry (see Google Trends results on revenue assurance). From our understanding, this relates to an insufficient modelling of the product from a billing perspective. Within this perspective the pricing terms and conditions that have been determined from a sales perspective needs to controlled and monitored throughout the customer lifecycle. Resulting from quite complex billing schemes (e.g. a price is dependent on the time, habitation and the number of optional products) the monitoring and controlling tasks put major requirements on subsequent IT-systems. Since these systems are predominately legacy systems, which have been laid out for much simpler billing schemes, dating back from a monopoly market, these requirements can not be fulfilled, which results into hot fixes and workarounds, that cause the problems in the revenue assurance.

4. Customer Care

From a customer care perspective a complete view on the product and the customer is needed in order to resolve a problem efficiently and effectively. The granularity of the information required differs from the service concept. In case that a CSP has installed a first and second level support, the second level support needs to have detailed information on the technical resources like network capacity, number of open ports on a DSLAM (Digital Subscriber Line Access Multiplexer). The first level support needs access to the product configuration and customer data like account details.

These four perspectives need to be integrated and operate on the same product data, so that a consistent view on the product and the customer is provided. Currently major CSPs are struggling to fulfil this rather simplistic sounding requirement. In the following figure the four perspectives on the product architecture have been illustrated. In this figure the product inventory as a complementary element for managing the customer configuration and the link with the product has been included.

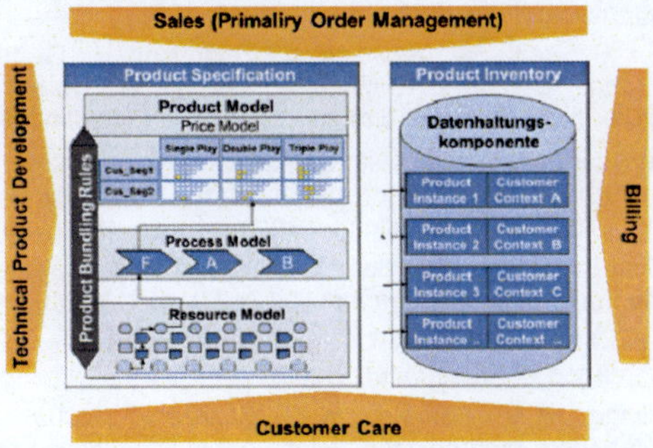

Figure 5: Complete Product Architecture including PDIF.

In the next paragraph the results from an international PLM study regarding the product architecture of CSP will be presented.

6 VALIDATION

6.1 Empirical study

Between May 2009 and October 2009 FIR and Detecon conducted an international study about the Status Quo of PLM in the Communication Service Provider Industry. The goal of this study was to explore PLM design elements in the first place and secondly to identify interdependencies between the implementation degree of the PLM design elements and a PLM target achievement. As a moderating variable the external complexity level was chosen, which has been measured according to [21]. The classification label "Global Elephants" describes a high complexity level, whereas "Emerging Zebras" represents less complex companies.

The questionnaire of this study included a dedicated section for the design elements of product architecture and only this section has been evaluated for the purpose of this paper. In this study 47 CSP worldwide have been interviewed either on-site or through web-conference sessions. The average duration of the interview did last approximately 2hrs, whereby the relevant product architecture part amounts to ca. 25% of the whole interview. For details regarding the research methods please refer to the study document, which can be downloaded from Detecon´s corporate website. Through this empirical investigation we wanted to explore the implementation degree of the proposed product architecture concept. Based on our theoretical findings we derived indicators that have been evaluated through a factor analysis.

Since the study addressed mostly product managers, we did not explore in detail the implementation of the product inventory, since its configuration and implementation could have been answered presumably only by the IT-department.

In the next section four factors will be presented. The implementation degree for each factor and its associated indicators will be presented for three complexity groups that have been identified through a cluster analysis.

6.2 Empirical findings

The empirical findings are graphically represented in a radar chart. Each spoke within the chart represents a design element that has been statistically validated. The data length of a spoke ranges from 0 to 100%, whereby 0% represents the lowest value and 100% represents the highest factor value found in the sample. A data point refers to the average implementation degree of one group for a given design element.

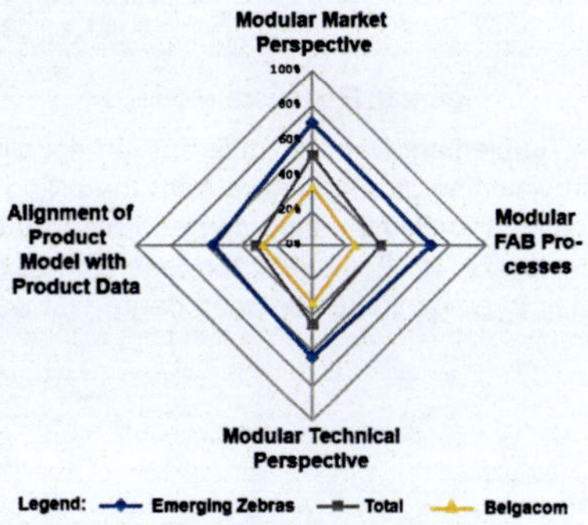

Figure 6: Radar chart.

From the radar chart it can be concluded that even companies facing a high external complexity (termed as Global Elephants) have an average implementation degree just above 60% for the three dimensions of the product architecture. Regarding the ability to provide different perspectives on the product, it can be stated, that the implementation is considerably low, even with the highest complexity group.

Considering the implementation degree of the specific indicators for the factor Product Model, it is interesting to notice, that especially the ability to provide cross strategic business unit (SBU) bundles has a comparatively low implementation degree.

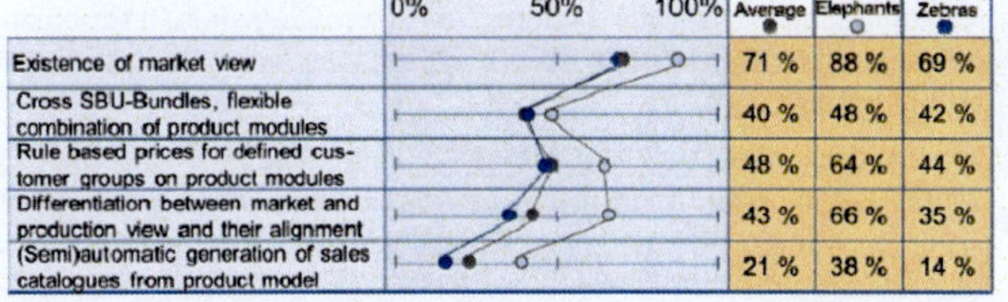

	0% 50% 100%	Overall Average	Global Elephants	Emerging Zebras
Existence of market view		71 %	88 %	69 %
Cross SBU-Bundles, flexible combination of product modules		40 %	48 %	42 %
Rule based prices for defined customer groups on product modules		48 %	64 %	44 %
Differentiation between market and production view and their alignment		43 %	66 %	35 %
(Semi)automatic generation of sales catalogues from product model		21 %	38 %	14 %

Figure 7: Product Model.

In terms of the fulfilment, assurance and billing (FAB) processes the indicators reflect the consideration of process modularity in the service fulfilment, service assurance and service billing processes.

The empirical findings show that there is a considerable gap between the different complexity groups, but still the average implementation of the highest complexity group reaches only almost 60%.

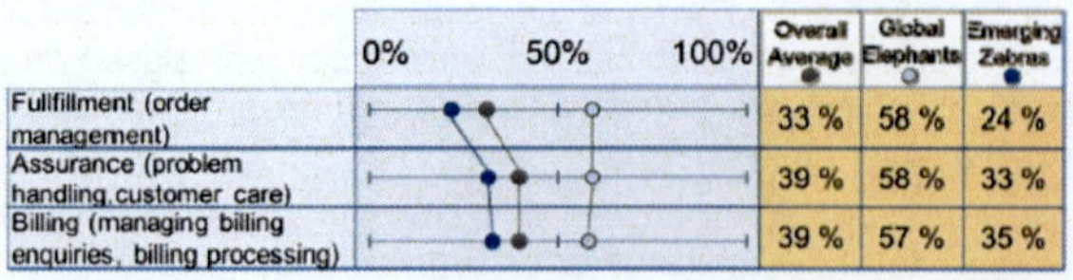

Figure 8: Process Model.

In regards to the Resource model three indicators could be confirmed with the factor analysis. From a modelling perspective the existence of a production view has an average implementation degree in most complex group of 70%. The biggest shortcomings have been identified in considerably low specification communality in the product design. This indicates that for new products the specifications will be reworked again, although for some aspects like the network, the specification can possibly be reused, since the lifecycle of these product elements are rather long.

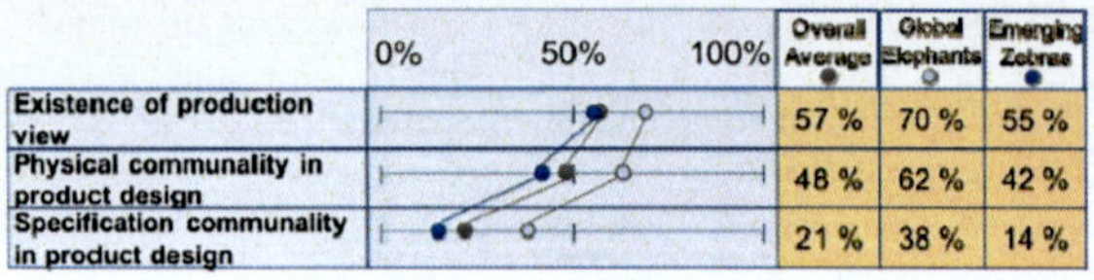

Figure 9: Resource Model.

Regarding the implementation of different perspectives on the product, the provisioning of one invoice from the billing perspective achieves the highest implementation degree among the three indicators shown in figure 10. Especially the less complex companies indicate that their average implementation degree for each indicator is below 50%.

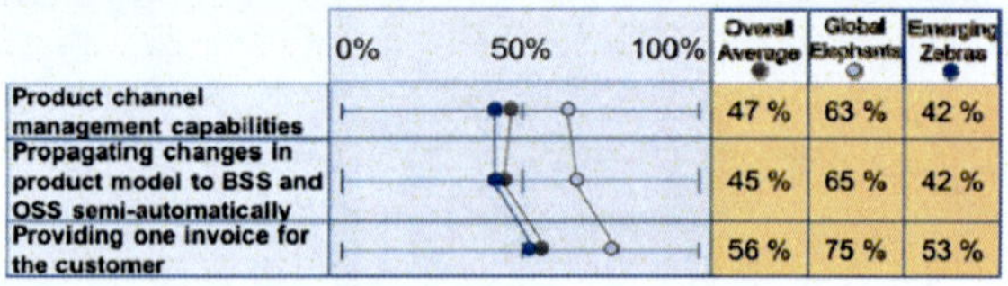

Figure 10: Different perspectives on the product.

7 DISCUSSION AND OUTLOOK

As the results from the empirical study indicate, the implementation degree of specific aspects of product architecture for CSP remains relatively low. But from the interviews with decision makers of CSPs in the context of our study, it was clearly indicated that the benefits of modular product architecture in the understanding of this paper are feasible. The biggest obstacles towards such modular product architecture seem to be the high IT-complexity with diverse data sources and diverse data schemes. Coping with the IT-complexity by eliminating old legacy systems and executing data harmonization projects, seems to be from our perspective a good start to a modular future.

Due to similarities between the telecommunication and the utility sector, our findings might be applicable for upcoming products in the area of smart metering and e-mobility. Future research activities have to explore the adaptability in this direction.

8 REFERENCES

[1] Schuh, G.; Assmus, D.; Zancul, E. (2006): Product structuring - the core discipline of product lifecycle management. In: Proceeding 13th CIRP International Conference.

[2] Weiber, R. (1995): Systemgüter und klassische Diffusionstheorie - Elemente einer Diffusionstheorie für kritische Masse-Systeme. In: Stoetzer (Editor.): Die Diffusion von Innovationen in der TK. Berlin: Springer, p. 39–66.

[3] Stabell, C. B.; Fjeldstad, O. D. (1998): Configuring value for competitive advantage: on chains, shops, and networks. In: Strategic Management Journal 19 (5), p. 413–437.

[4] Corsten, H.; Gössinger, R. (2007): Dienstleistungsmanagement. 5., vollst. überarb. U. ext. Edition. München: Oldenbourg.

[5] Anderson, C. (2008): The long tail. Why the future of business is selling less of more. Rev. and updated. New York:

[6] Gerpott, T. (2005): Unternehmenskooperationen in der Telekommunikationswirtschaft. In: Zentes,J. (Editor.): Kooperationen, Allianzen und Netzwerke. Grundlagen - Ansätze - Perspektiven. 2., Edition.: Gabler, p. 1205–1230.

[7] Meffert, H. (1994): Marktorientierte Führung von Dienstleistungsunternehmen–neuere Entwicklungen in Theorie und Praxis, Die Betriebswirtschaft, 54. In: Die Betriebswirtschaft 54 (4), p. 519–541.

[8] Mowshowitz, A. (1992): On the market value of information commodities. I. The nature of information and information commodities. In: Journal of the American Society for Information Science 43 (3), p. 225–232.

[9] Budde, O.; Schuh, G.; Uam, J-Y (2010): Holistic PLM Proceedings of International Conference on Product Lifecycle Management, Bremen Germany.

[10] Mörschel, I. C. (2005): Produktmodelle für Dienstleistungen. Mögichkeiten zur Strukturierung und Beschreibung von Dienstleistungen. In: DIN (Editor.): Wege zu erfolgreichen Dienstleistungen. 1. Edition. Berlin: Beuth, p. 46–115.

[11] Pahl, G.; Beitz, W.; Blessing, L.; Feldhusen, J.; Grote, K.-H.; Wallace, K. (2007): Engineering Design. A Systematic Approach. Third Edition. London: Springer London Limited

[12] Svensson, D.; Malmqvist, J. (2002): Strategies for product structure management at manufacturing firms. In: Journal of Computing and Information Science in Engineering 2, p. 50.

[13] Ulrich, K. (1995): The role of product architecture in the manufacturing firm. In: Research policy 24 (3), p. 419–440.

[14] Grabowski, H.; Anderl, R.; Polly, A.; Warnecke, H. J. (1993): Integriertes Produktmodell: Beuth.

[15] Bullinger, H.-J.; Fähnrich, K.-P.; Meiren, T. (2003): Service engineering--methodical development of new service products. Structuring and Planning Operations. In: Inter. Journal of Production Economics 85 (3), p. 275–287.

[16] Schläffer, C.; Arnold, H. (2007): Media and network innovation. technological paths, customer needs and business logic. In: e & i Elektrotechnik und Informations-technik 124 (10), p. 317–322.

[17] Corsten, H.; Gössinger, R. (2003): Gestaltungsdimensionen von Dienstleistungen: Lehrstuhl für Produktionswirtschaft.

[18] Weber, R. (2005): Leistungsbündel als Angebotsform für konsumentengerichtete Technologiegüter-Konzept.

[19] Schönsleben, P.; Leuzinger, R. (1996): Innovative Gestaltung von Versicherungsprodukten. Flexible Industriekonzepte in der Assekuranz. Wiesbaden: Gabler.

[20] Rapp, T. (1999): Produktstrukturierung. Komplexitätsmanagement durch modulare Produktstrukturen und -plattformen. Wiesbaden: Gabler [u.a.]

[21] Budde, O.; Golovatchev, J.; Taner, S.: Evaluating Product Complexity in Network Industries. Lugano (ICE 2010 The 16th International Conference on Concurrent Enterprising).

Knowledge Feedback to the IPS² Development

H. Bochnig[1], E. Uhlmann[1], R. Gegusch[2], G. Seliger[2]
[1] Chair Machine Tools and Manufacturing Technology
[2] Chair Assembly Technology and Factory Management
Institute for Machine Tools and Factory Management, Technische Universität Berlin, Berlin, Germany

Abstract
In order to cope with the increased complexity of the integrated IPS² development, an IT based assistance system aids the developer during the entire process. It facilitates defining configurations of product and services fulfilling customer needs and business promises. To ensure an efficient development and to provide decision support, existing knowledge or experience from previous developments and IPS² provision phases should be used. This knowledge can be automatically generated by the Virtual Life Cycle Unit from the spatially and temporarily distributed data along the IPS² lifecycle. This paper introduces feedback mechanisms and concepts of the knowledge generation to support the development processes and its influences.

Keywords:
IPS² Assistance System; Virtual Life Cycle Unit (VLCU); Knowledge Feedback

1 INTRODUCTION

In business-to-business markets there is a tendency towards combined offers of products and services, which are sold in one package to fulfil customer needs. These combinations of products and services are called Product-Service Systems (PSS) or Industrial Product-Service Systems (IPS²) in case of industrial application [1]. The business models of Industrial Product-Service Systems are based on the idea of selling functionality, availability or results instead of selling machines.

The IPS² development process is split in various phases: acquisition of customer needs, determination of IPS² requirements, generation of functional and conceptual models, development of Product-Service Modules (PSM) and configuration of the final IPS². In order to cope with the increased complexity of the integrated development of product and service shares, an IT based assistance system aids the IPS² developer during the entire process. It facilitates defining configurations of product and services fulfilling customer needs and business promises, e.g. availability or result oriented aspects.

To support a competitive design of IPS², knowledge needs to be generated about correlations between product and service shares in order to fulfil the required needs and demands. This knowledge is crucial for a successful combination of product and service shares, whose influences are very complex and not always known. The IPS² life cycle data, especially from the operation phase, is the source to discover this knowledge. Using information technology, the IPS² product- and process data can be made locally available, data which are spatially and temporally distributed along the life cycle of IPS² in different media and formats, and processed to knowledge in an appropriate form - as required by the respective application.

Commonly, implementations of knowledge feedback are realized as expert systems. Expert systems support development processes by the use of knowledge that is manually entered by humans. This knowledge has been generated from their personal subjective experience. Therefore, the use of information technology facilitates to objectify the collected data and the reasoning based on it. IT tools also enable to enlarge the space of observation, since a manual approach always confines its application due to complexity or necessary effort.

In this paper the feedback mechanisms and concepts of the knowledge generation to support the IPS² development processes and influences on the process will be introduced. The knowledge generation is linked with the IT based assistant system to aid the development process and to provide decision support for the IPS² developer. Also, requirements and benefits for a competitive IPS² development are given.

2 DEVELOPMENT PHASES

For the development of physical products as well as IT systems, a number of methodologies exist which define the development process from abstract to detailed levels. The most widespread methodologies are the VDI guideline 2221, the methodology of Pahl/Beitz and for software the V-Model [2,3,4]. The service development and its delivery are not well defined. In [5] a very rough and abstract phase model for service development is introduced. Service classes in the engineering field are defined in [6] and service delivery described in [5]. A concept for service design with a life cycle context is outlined in [7]. Detailed process models for service development are missing at present.

In Industrial Product-Service Systems the product and service shares depend on and influence each other and therefore need to be developed together as an integrated system. The integrated development is characterized by high demands in regard to technical knowledge from different domains, to the process complexity and organisational complexity. The design of a physical artefact, such as a machine part, determines the necessary pertinent services like maintenance, logistics, training, etc. and as such the resulting low or high lifecycle costs. Vice versa, the offering of certain services requires certain physical components or a special design. Therefore, it is necessary to take the interdependencies into account with concurrent and iterative

J. Hesselbach and C. Herrmann (eds.), *Functional Thinking for Value Creation: Proceedings of the 3rd CIRP International Conference on Industrial Product Service Systems, Technische Universität Braunschweig, Braunschweig, Germany, May 5th - 6th, 2011*,
DOI 10.1007/978-3-642-19689-8_39,

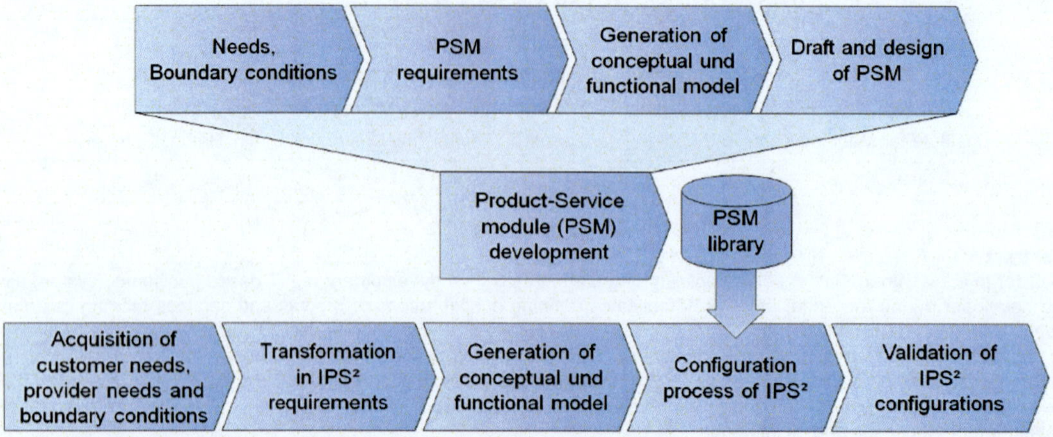

Figure 1: Phases of the IPS² development process as implemented in the IPS² assistance system.

development (sub-)processes until contradictions are eliminated and the best solution is reached. Another key point in IPS² development is the customer integration in order to provide customized solutions with a business model that suits the customer needs. The types of use models range from function oriented, availability oriented to result oriented with a rising service share of the IPS² [8].

For the implementation of an IT based assistance system that supports the developer during the entire IPS² development, a general process has been adapted and employed [9]. It loosely follows the VDI guideline 2221 and the V-Model and adds the configuration of IPS² enabling several ways to go through the development process. The IPS² development process consists of the following phases: acquisition of needs, definition of requirements, function allocation, parallel development processes of Product-Service Modules (PSM), the IPS² configuration process and the validation of the result set. Figure 1 shows a linear view on the phases of the development process.

The IPS² development process starts at the customer by acquiring his needs. The customer integration in the development process is a key point for providing a customized solution. The main interface between the IPS² provider and the customer is therefore the interactive acquisition of his needs. By giving the customer feedback on possible solutions, a knowledge transfer takes place which enables the customer to articulate his needs more precisely. Apart from customer specific factors also external factors like legislation, infrastructure, market, competitors, etc. have to be considered at this stage. These aspects are often not explicitly mentioned, or the customer may even not be aware of them, but nevertheless have to be considered for IPS² development.

The IPS² requirements are derived from the obtained needs in the next step [10]. While the explicitly or implicitly expressed needs formulate a problem description that is preferably solution-neutral, a concrete solution is described by its requirements. The result is the IPS² requirement specification analogous to a product requirement specification at the conventional product development side. Also, a qualified use model that benefits both the customer and the IPS² provider is determined. The function oriented use model may for example include a maintenance contract in order to ensure the functional capability of the machine tool over an agreed time period. With an availability oriented use model, the IPS² provider additionally guarantees a level of availability. By this he adopts business processes of the customer on his own authority and carries a part of the manufacturing risk. He is responsible for all processes that ensure availability like repair or preventive maintenance. The result oriented use model assigns the responsibility for the manufacturing result to the IPS² provider because the customer only pays for faultless parts. On the basis of the customer needs and the IPS² requirement specification, a recommendation for one of these use models is made.

From the inputs IPS² requirement specification and use model, a functional and a conceptual model are generated that provide the basis for configuration. Especially the functional structure is important, because the functions are used to access the Product-Service Modules for the IPS² configuration. PSMs consist of variable shares of tangible products and non-tangible services and fulfil one or more functions. The general concept of IPS² assumes an interchangeability of Product-Service Modules with varying product and service shares for the same function to the extent that a PSM with only service processes may be replaced by a physical artefact (e.g. automated solution) and vice versa.

In the parallel development process, Product-Service Modules are developed if they are not included yet in the PSM library. The applied methodology resembles the development of the IPS² on a higher level. For each subprocess the single phases acquisition of needs, requirements definition and function allocation are mapped, translated and instantiated from the IPS² level to the level of Product-Service Modules. Finally, the product and service shares of the PSMs are drafted and designed.

The configuration of the Industrial Product-Service System generates and evaluates variants of possible solutions. The configuration process bases on the available PSM library. It accesses predeveloped Product-Service Modules and combines them to a complete IPS² solution.

A software based IPS² assistance system is developed at the Institute of Machine Tools and Factory Management (IWF) of the Technische Universität Berlin to support the integrated planning and design process of Industrial Product-Service Systems [9]. This prototypical software system should guide and support the developer along the concurrent development processes of products and pertinent services and constantly consider their mutual interrelations starting at the earliest development phase. The chosen approach of the modular system architecture is shown in Figure 2. It consists of currently 16 modules, of which two are cross-sectional. The project database module on the left hand side stores and manages data of the partial models from other modules such as data about the customer, the market, competitors, the IPS² provider, subcontractors, etc.. The module "system maintenance"

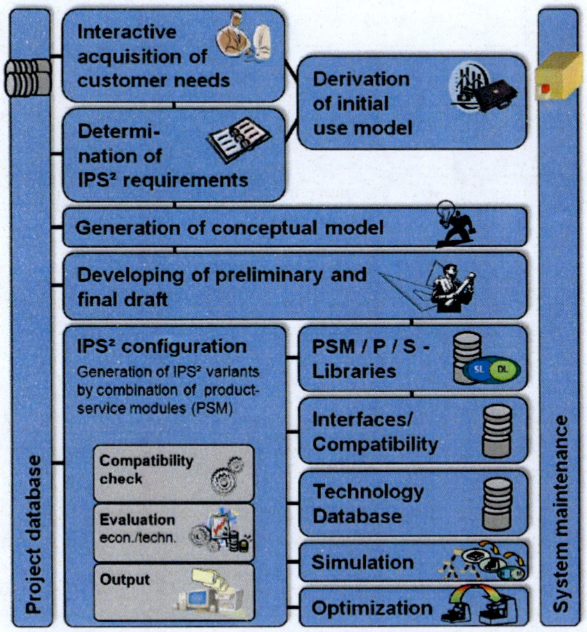

Figure 2: Modular system architecture of the IPS² assistance system.

on the right hand side comprises of tasks like software and data administration and maintenance, the inclusion of feedback and knowledge from the IPS² use phase and the extension and adaptation to different applications. It also permits the inclusion of Product-Service Modules from new manufacturers and providers together with the necessary update of the interface and compatibility module.

The PSM libraries and the databases with the interactions and compatibilities between the Product-Service Modules store the data basis for the configuration module. A technology database provides knowledge about manufacturing processes, their applicability and parameters. Interfaces to simulation and optimization tools are prospects for future work.

3 KNOWLEDGE FOR COMPETITIVE DEVELOPMENT

To define decisions on the single development processes as described before, knowledge on interdependencies between the product and service shares towards the usefulness of the IPS² is required. There are many definitions for the term knowledge, but the definition of [11] has become more and more popular within the last years. NORTH defines knowledge as the appropriate linking of information. Knowledge results from the processing of data and information, acquired in certain situations. So data and information are the raw material for knowledge. The value of knowledge will be shown, when it is referenced to an application and the ability for acting is facilitated. The acting itself needs an intention and if acting correctly and original or inventive, the competitiveness is given, see Figure 3. As shown in the definition, knowledge is crucial for competitive decisions, which means it enables a competitive development of an IPS².

In the development process of an IPS² the initial solution space of all possible IPS² configurations is determined by all combinations of the existing PSMs, see Figure 4. This results in a degree of complexity that exceeds the manageability of the relevant data and knowledge, especially if for each combination it has to be generated

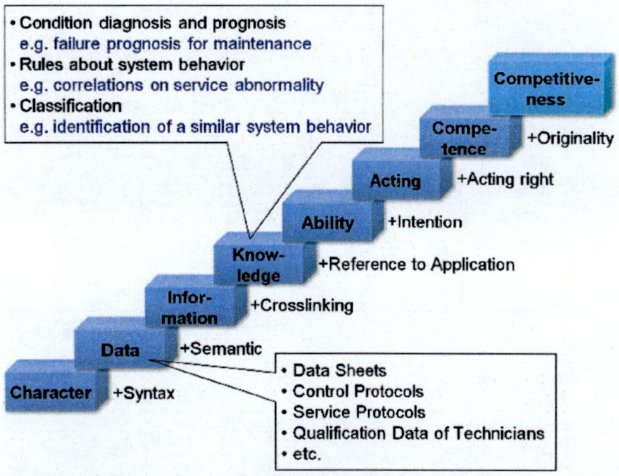

Figure 3: Definition of knowledge by NORTH [7].

and provided to the IPS² developer in order to support his design decisions. During the development steps, the PSM combinations are reduced to a manageable number.

In the configuration process concrete PSM combinations will be selected, conforming to the functional and conceptual model. Then, the solution space is again limited to a smaller number of IPS² configurations by restricting the considered PSMs to the formulated requirements. To decide which configuration fulfils the customer and provider needs best, it has to be verified with experience knowledge from IPS² instances in operation.

The verification at this time of the development makes sense, as the solution space is limited to a manageable number of PSMs with their product and service shares (see Figure 4, innermost ellipse). Concrete interdependencies can be verified now towards their influences on the use of the IPS². This experience knowledge on interdependencies in the selected configuration enables to eliminate unseen interdependencies, which may endanger the designed functionality, availability or result of the IPS². In the earlier steps in the development phase, knowledge, that has been acquired or generated beforehand, supports the decision processes. This so called "old knowledge" is documented, e.g. in an IPS² product-life cycle management system, while the knowledge in the verification step is new. This means it has to be generated from the latest available, current data. The feedbacks and knowledge inputs are depicted in Figure 5.

4 KNOWLEDGE GENERATION

Due to the modern IT based engineering, business communication and documentation, most data along the life cycle of an IPS² are already digitally available and via internet worldwide accessible,

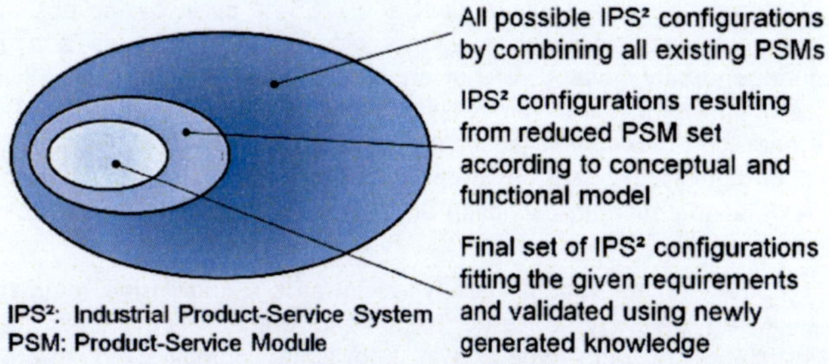

Figure 4: Solution space for IPS² configuration.

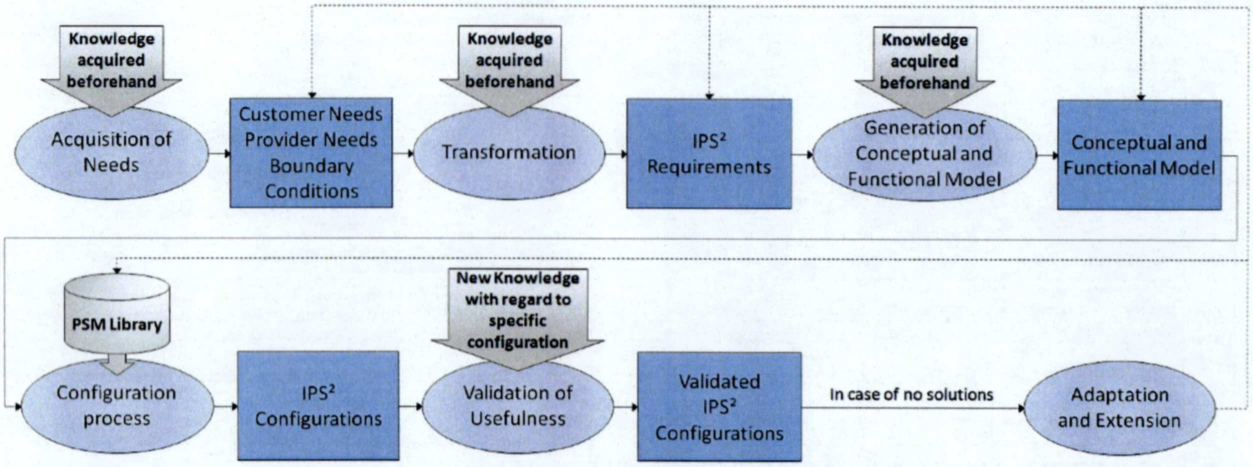

Figure 5: Knowledge feedback loops.

e.g. machine datasheets, service protocols or personal database. The challenge lies in the uniform access, acquiring and communication to process it to knowledge and support the decision-making processes along the IPS² development. The data sources are in different formats and media along the life cycle of IPS² in the databases of the IPS² partners available.

To generate knowledge from the IPS² lifecycle data, an information technology concept has been developed as a support tool for decision processes in the IPS² life cycle. The so called Virtual Life Cycle Unit (VLCU) [12] is an information technology network, which appears to humans or machines as a single unit to support decision processes with automated generated knowledge in the whole IPS² life cycle.

The knowledge processing of the VLCU consists of the following steps:

- Receive Knowledge generation task,
- Data acquisition,
- Data preprocessing,
- Data collection,
- Knowledge generation,
- Sent knowledge.

The knowledge generating task with its specifications on knowledge type and scenario details, like machine types, come from the application or the IPS² actor, e.g. developer. In our case of the development process the IPS² configurations need to be analysed. Therefore, interdependencies between their parameters in order to fulfil e.g. the required availability rate are needed. The VLCU accesses the data sources along the IPS² life cycle by the use of software agents. Software agents are software modules acting independently, based on given rules. The agents build the communication and processing network for the knowledge generating information technology concept of the VLCU. Basically, each agent is responsible for the data sources of one IPS² partner. A detailed concept description on the data acquisition step can be found in [13].

To structure the various data types towards a knowledge request task, an ontology is being used. The ontology describes the structures of data types to a knowledge domain, e.g. micro production. Thus, based on a preselection by this pre-knowledge, the selection of a limited number of data sources relevant for the knowledge generation is facilitated, see Figure 6. The collected data are filtered and cleaned by a preprocessor and stored in the VLCU data warehouse for the processing phase. Knowledge discovery data mining algorithms are being used to generate decision trees with association rules about correlations of parameters towards the demanded usage, e.g. availability of the IPS². An example for a decision tree algorithm is the C4.5 by Quinlan [11].

The generated knowledge can be communicated in different knowledge model types. For the human a graphical representation, e.g. a decision tree, is useful for a quick understanding, while machines require an adequate textual representation. Depending on the inquirer the knowledge will be transformed in the appropriate model type and transmitted to the IPS² actor or software to support the decision process.

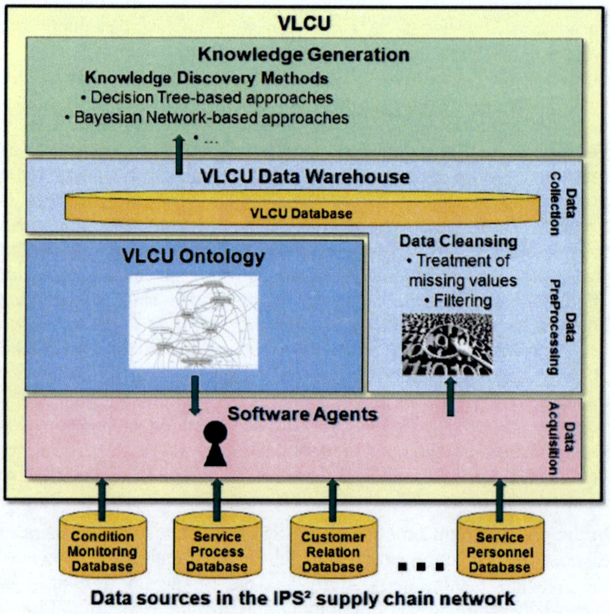

Figure 6: Concept for the acquisition, generation and provision of knowledge from IPS² use [13].

The generated knowledge is a conclusion on the actual available acquired data in existing and running IPS². This allows an evaluation of the theoretical defined configurations of product and service shares towards the customer need. The feasibility to fulfil the required demands on the IPS² of concrete PSM configurations can be verified.

Furthermore, it supports to identify concrete chances to a better fulfilment of the IPS² by supporting the redesign with inferences on combinations of product and service shares influencing the use in a bad way. This means that automated generated knowledge set the foundation for a competitive development of IPS².

5 EXEMPLARY APPLICATION

In order to illustrate the shown theoretical approaches, a practical example shall be given. We assume that a watch manufacturer would like to provide added value to his customers by offering individual, customized products. The end consumer will be able to define an engraved writing or pattern on the watch. Due to the value chain of the manufacturer, the engraving shall be completed within a timeframe of one day. The required use model is result oriented. The IPS² could be realized as an automated, semi-automated or manual process. Due to the higher priced status of the watches, the end customer expects a manual or semi-automated process. By this, the needs of the IPS² customer and the boundary conditions for the IPS² are outlined.

Possible concepts are the integration of the engraving process within the production line of the IPS² customer (case 1) or an engraving at an external site (case 2) which requires additional logistics. By pre-acquired knowledge (existing knowledge from statistical analysis of data obtained with other IPS² instances) it is known, that in the case of a manual process at the customer factory (case 1), local workers are preferred, because the illness rates of commuters are higher. Therefore, the IPS² provider needs to hire new operating personnel instead of sending some of his permanent staff to the customer's production facility.

In the case of an off-site facility, statistical analyses resulted in pre-acquired knowledge which says that the transport distance should not exceed 250 km in order to meet the timeline. This knowledge limits the suitable production locations. An additional factor to consider is that the engraving in the country of the customer is more expensive than in the bordering countries X and Y. With this knowledge conceptual models can be created and the necessary functions allocated that build the base for the configuration process of the IPS².

The detailed, configured IPS² alternatives finally need to be evaluated and validated with new knowledge from IPS² instances in use (current, newly generated knowledge, which has not been available before). Now the number of options to be considered is usually low, which allows a comprehensive validation. For case 1 it may be revealed that local workers need special training despite officially having the required qualifications. For instance because of cultural differences or different regional education standards, the local engravers are not familiar with a certain style of pattern design. This can lead to necessary adaptations in resource and time planning in order to meet the IPS² requirements.

For case 2 knowledge has been generated, that watches are often stolen during transport in country X, which means that manufacturing in country X should not be chosen without high security measures. In country Y statistics show that there is a higher transport time, than theoretically evaluated, which influences the time schedules.

6 SUMMARY

In this paper the IPS² development process has been characterized by its various phases. A prototypical IPS² assistance system bases on the introduced development process. The development of IPS² is a knowledge intensive process. Knowledge supports in different feedback loops the decisions for the effective and competitive design of an IPS². The presented concept of the VLCU is an enabler to generate the required knowledge in an automated way from the IPS² life cycle data to verify developed IPS² configurations. By this an effective and competitive IPS² design can be achieved.

7 ACKNOWLEDGMENTS

We express our sincere thanks to the Deutsche Forschungsgemeinschaft (DFG) for funding this research within the Collaborative Research Project SFB/TR 29 on Industrial Product-Service Systems – Dynamic Interdependencies of Products and Services in the Production Area.

8 REFERENCES

[1] Meier, H.; Uhlmann, E.; Kortmann, D. (2005): Hybride Leistungsbündel - Nutzenorientiertes Produktverständnis durch interferierende Sach- und Dienstleistungen, wt Werkstatttechnik online, 95. Jahrgang, 7/2005, Springer-VDI-Verlag, p. 528-532.

[2] N.N.: VDI guideline 2221, Systematic approach to the development and design of technical systems and products. Verein Deutscher Ingenieure. Berlin : Beuth Verlag, 1987.

[3] Pahl, G.; Beitz, W. (2006): Pahl/Beitz Konstruktionslehre. Grundlagen erfolgreicher Produktentwicklung. 7. Aufl., Berlin : Springer.

[4] N.N. (2009): Das V-Modell. Internet: **http://v-modell.iabg.de/**, June 2009.

[5] N.N. (1998): DIN-Fachbericht 75. Service Engineering, Entwicklungsbegleitende Normung (EBN) für Dienstleistungen. 1. Aufl., DIN Deutsches Institut für Normung e.V., Berlin : Beuth Verlag.

[6] N.N. (2006): VDI guideline 4510, Engineering services and requirements to engineering service providers. Verein Deutscher Ingenieure. Berlin : Beuth Verlag.

[7] Aurich, J.C.; Fuchs, C.; DeVries, M.F. (2004): An Approach to Life Cycle Oriented Technical Service Design, CIRP Annals - Manufacturing Technology, Volume 53, Issue 1, p. 151-154.

[8] Tukker, A., Tischner, U. (2005): New Business for Old Europe – Product-Service Development, Competiveness and Sustainability, Greanleaf Publishing, Sheffield, UK.

[9] Uhlmann, E.; Bochnig, H. (2009): Computer Aided IPS² Development, Proceedings of the 2nd International Seminar on IPS², 23-24 March 2009, Berlin, Germany : Shaker Verlag, p. 31-36.

[10] Uhlmann, E.; Bochnig, H.; Stelzer, C. (2010): Characterization of Customer Requirements in IPS² creation, Proceedings of the 2nd CIRP IPS² Conference 2010, Linköping, 14-15 April 2010, Linköping, Sweden, p. 399-405.

[11] North, K. (2002): Wissensorientierte Unternehmensführung. Wertschöpfung durch Wissen. 3. Aufl., Gabler, Wiesbaden.

[12] Seliger, G.; Gegusch, R.; Bilgen, E. (2007): Wissensgenerierung in hybriden Leistungsbündeln - Wissensgenerierung zur Steigerung der Wirtschaftlichkeit in hybriden Leistungsbündeln. In: wt Werkstatttechnik online, Jahrgang 1997, Nummer 7/8, p. 522-525.

[13] Gegusch, R.; Gestrich, K.; Seliger, G. (2010): Software Agents for Automated Knowledge Generation in IPS², 2nd IPS² conference on Industrial Product service Systems, Linköping, Schweden.

Enhancement in Industrial PSS Design based on TRIZ: a Case Study

Giuditta Pezzotta[1], Daniele Regazzoni[2], Sergio Cavalieri[1], Caterina Rizzi[2]

[1]CELS, Research Center on Logistics and After Sales Services - Department of Industrial Engineering, University of Bergamo, Viale Marconi,5 – Dalmine (BG) Italy

[2]Department of Industrial Engineering, University of Bergamo, Viale Marconi,5 – Dalmine (BG) Italy

(giuditta.pezzotta@unibg.it, daniele.regazzoni@unibg.it, sergio.cavalieri@unibg.it)

Abstract

Despite the emergence in literature of novel contributions in the Product-Service Systems (PSS) Engineering, there is still a lack of systematic knowledge in this area. Most of the proposed methodologies are mere adaptation of traditional engineering approaches to the PSS domain. In this context, one of the most analysed methods is the theory of inventive problem solving (TRIZ). Purpose of this paper is to propose a way to use the Laws of Technical Systems Evolution (LTSE) and the Evolution Trends to predict what are the most likely improvements that can be made to a given PSS. A case study makes evidence of the potential behind the adoption of the LTSE.

Keywords:

Product-Service System; TRIZ; Laws of Technical Systems Evolution

1 INTRODUCTION

Nowadays, the distinction which links products to something tangible and services to something intangible is vanishing. Many service outputs rely also on substantial material components and, conversely, many artefacts have embedded intangible attributes. Customers' purchase habits have progressively changed [1]: they do not buy more goods or services in the traditional sense, perceiving them as two separated entities. They rather look for a solution, whose value (may) encompasses many intertwined components.

In this context, the Product-Service System (PSS) concept finds its root. The basic idea behind the PSS concept is that it ensues from an innovation strategy, shifting the business focus from the design and sales of physical products to the design and sales of a more complex system. PSS is generally defined as a 'product(s) and service(s) combined in a system to deliver required user functionality in a way that reduces the impact on the environment' [2].

A Product-Service System may be decomposed into several components. These components are of different natures and involve many actors along the whole life-cycle of a PSS (e.g. producer, service providers, consumer, maintenance facilities, take back and recycling facilities, etc.). Mont [3] [4] decomposes a PSS into four dimensions that the consumer can apprehend: Product (in terms of physical products or technologies), Services, Infrastructure and Supporting Networks. Aim of a PSS is to generate profits, not only at the point of sale but also throughout the whole life-cycle. Designing and developing a PSS seems a very complex task due to the systematic movement along the product life-cycle and the number of interactions that exist between the different actors involved and the components constituting it.

Therefore, the profit generation and success of a PSS critically depend on its conceptualisation and design, even if this notion has been largely neglected for a long time. It is only recently that a fundamental change in this situation has become evident, coinciding with the increasing practical relevance of the topic, as numerous recent publications have confirmed [5]. In particular, the service activities are trying to be incorporated into the design space, which has been traditionally dominated by the physical products in manufacturing industry [6].

Service Engineering (SE) [7] conceived for answering to these needs, can be termed as "*a technical discipline concerned with the systematic development and design of services using suitable models, methods and tools*". It is characterized through an interdisciplinary integration of expertise in the use of constructive methods from engineering, business, computers science, etc. Although SE also embraces aspects of service operations management, its main focus is on service design and development. [8]. Even though some PSS Engineering models, methodologies and tools have been recently developed, there is a lack of systematic knowledge in this research area. In most contributions the SE approach derives from the adaptation of traditional engineering approaches to the Service System (SS) or Product-Service System (PSS) sector [8]. Moving towards them, many papers describe how to implement these methods and tools in the PSS field [9]: Quality Function Deployment (QFD), Property-Driven Development/Design (PDD), the theory of inventive problem solving (TRIZ), Service Blueprinting, the Failure Mode and Effect Analysis, Design for Serviceability, Taguchi's Methods, Service CAD, etc.

In the last years, one of the most analysed methods is the *theory of inventive problem solving* (TRIZ). Initially proposed for supporting technical innovation, it has been recently also applied to a range of non-technical fields such as social relations, business and management [10] [11] [12] and, in particular, to services [13] [14] [15] [16] [17] [28]. Currently, the application of TRIZ in the SS o PSS context is mainly related to the oldest and most common essentials, as the 40 inventive principles. Only Mann and Jones [18] proposed a paper on the development of sustainable services and systems through the adoption of systematic innovation methods.

Aim of this paper is to propose a way to use the Laws of Technical Systems Evolution (LTSE) and the Evolution Trends to predict what are the most likely improvements that can be made to a given PSS

J. Hesselbach and C. Herrmann (eds.), *Functional Thinking for Value Creation: Proceedings of the 3rd CIRP International Conference on Industrial Product Service Systems, Technische Universität Braunschweig, Braunschweig, Germany, May 5th - 6th, 2011*,
DOI 10.1007/978-3-642-19689-8_40, © Springer-Verlag Berlin Heidelberg 2011

and to apply them in a real case.

The paper is organized as follows. In section 2 TRIZ, LTSE and Evolution Trends are described. In section 3 an analysis on the TRIZ application in the Service and PSS area is reported. section 4 is devoted to the proposal and discussion of the most peculiar PSS Evolution Trends explained both by theory and by the analysis of a real case. In section 5 conclusions and the further development of the work are reported.

2 TRIZ AND LTSE CHARACTERISTICS

TRIZ has evolved for decades and a huge human effort, 1500 person years [19], has been provided to create the most comprehensive systematic innovation and creativity methodology available nowadays. Studying the paths of evolution of technical systems (i.e. a product, a process or a technology), has been a primary research method of TRIZ since its foundation. By the way, until the 1970s the patterns of evolution were not consolidated into a separate section of TRIZ and were scattered among other sections. In that period Altshuller described and formalized the recurrent patterns into the *"Laws of Technical Systems Evolution"* [27]. Unfortunately, although this topic is a fundamental of the TRIZ theory, not all the original Russian documents are available or have been translated in English. Usually, most of the publications about TRIZ in Asian and Western communities have been focused to explain and disseminate problem solving topics rather than forecasting methods, so ignoring a deeper investigation of the Laws. Moreover, TRIZ Laws of Evolution are not perceived from all the TRIZ community as a real tool but rather as an elegant theoretical complementary to other tools, as the Evolution Trends.

2.1 TRIZ Laws of Evolution

In his pioneering work of 1975, Altshuller subdivided the laws into three categories [27]:

Static (Laws 1-3) describing criteria of viability of newly created technical systems.
Kinematic (Laws 4-6) laws defining how technical systems evolve regardless of external conditions (technical or physical factors).
Dynamic (Laws 7, 8) laws defining how technical systems evolve according to technical or physical factors.

In the following each law will be shortly described.

L1. The law of the completeness of the parts of the system
Every technical system should consist of four parts or components: engine, transmission, control unit and working unit. A technical system requires all these components and their minimal capability to perform system functions. If any component is missing, the technical system does not exist. If any component fails, the system does not survive.
Corollary: to make a technical system controllable, at least one of its components should be controllable. Such a system is capable of changing properties in the way required by the operator.

L2. The law of energy conductivity
Prerequisite to viability of a system is the free flow of energy through all system parts. As every technical system transforms energy, this energy should circulate freely and efficiently through its four main parts. Energy can flow by means of substance (i.e. mechanical means), field (i.e. electro-magnetic), or substance-field (i.e. charged particles).
Corollary: part of the control unit is required to provide power conductivity between a given part and the control unit.

L.3.The law of harmonizing the rhythms of parts of the system
Prerequisite to the viability of a technical system is coordination (or purposeful de-coordination) of vibration frequencies (periodicity of operation) of all parts in a technical system.

L.4. The Law of increasing the degree of Ideality of the system.
All systems evolve towards the increase of degree of Ideality. An ideal technical system is a system that does not exist but whose function is delivered to fulfil the design intent. *Ideality* is one of the main pillars in the TRIZ philosophy and it relies on the capability of increasing benefits and reducing harms to the extreme so that the system performs its function(s) in a perfect way without any kind of drawbacks. The practical aspects of Ideality are the Ideal Final Result (IFR) and the Level of Ideality. IFR is the final state in which the system should be and correspond to perfect ideality, i.e. the system disappear not to generate harms but its functions are still performed. IFR provides a formulation of the right goal to fulfil and of the direction for achieving it. The Level of Ideality can be estimated as the ratio of useful functions performed by the system and related harms. The result obtained is qualitative but suitable for comparing different solutions to a given problem or different releases of the same system.

L.5. The law of uneven development of parts of a system
The development of parts of a system proceeds unevenly; the more complicated the system, the more uneven the developments of its parts. Uneven development of the parts of a system is a reason for the occurrence of technical and physical contradictions and hence inventive problems. Corollary: elimination of contradictions in less developed parts allows the system to continue to improve.

L.6. The law of transition to a supersystem
During evolution, technical systems merge and form bi- and poly-systems being either homogeneous or heterogeneous. When a system exhausts the possibilities of further significant improvement, it is included in a super-system as one of its parts. As a result, new development of the system becomes possible.

L.7. The law of Transition from macro to micro level
The development of working organs proceeds, at first, on a macro and then on a micro level. What is changed is not the size of components but the scale of interaction between the system and the object it acts on.

L.8. The Law of Increasing the S-Field involvement
Substance-Fields (S-Fields) models are a synthetic representation of a technical system made up of two Substances interacting by means of a Field. In case the Field or a substance is not present, the system is not complete, or it is a non S-Field system. According to the 8th, Law non S-Field systems evolve to S-Field systems. Within the class of S-Field systems, the fields evolve from mechanical fields to electro-magnetic fields according to a precise path. The dispersion of substances in the S-Fields increases. The number of links in the F-fields increases, and the responsiveness of the whole system tends to increase.

The Laws of Evolution provide a high-level set of indication on the way each system evolves, or, on the other hand, the way we should design new systems to gather successful results. By the way, to help in everyday design activity some empirical simple trends can be drawn as well. Actually, a number of *Trends of Evolution* exists to help finding the right way to design and develop a system. Trends provide evolutionary indications at a lower level, i.e. less abstract, and closer to the system, but their relevance is not universal. Thus, anyone is allowed to add a trend of evolution. What is requested is to recognise a repetitive pattern in the way a number of systems change over time and define clearly some intermediate steps. On the other hand, it may happen that an evolution line loses its relevance and the corresponding trend may be discarded.

Lists of trend of evolution are included in most important TRIZ based software such as Goldfire Innovator [A] and Creax Innovation Suite [B]. In Figure 1 some trends are shown.

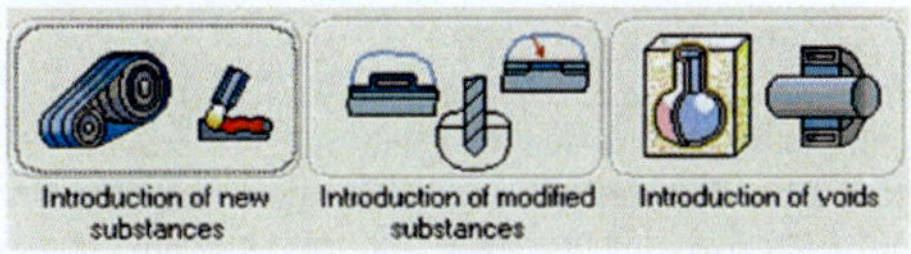

Figure 1: Fragment of a list of Trends of Evolution (Image courtesy from Invention machine).

Noteworthy an example of Trend of Evolution is [20] *Dynamization*. This trend deals with the way things move relatively to other things. In other words, the system becomes more and more adaptable to the environment and working conditions by shifting from a solid static system to a solid made of jointed parts to a completely flexible system. Moreover, to gather a better dynamicity solid is substituted with a fluid and at last stage of evolution no more substance is required but a field is capable of performing the system's functions (see Figure 2).

Figure 2: Main evolution step of Trend Dynamization.

2.2 Trend and Evolution Potential

The evolution potential is a concept developed to evaluate the overall degree of evolution of a system by means of trends. Operatively, the technical system must be compared against each trend of evolution. The result is that some trends are not relevant, while others are useful in analyzing any system. The non-relevant trends (which will shift from one application to another) are eliminated from the analysis. Each of the remaining trends is then compared to the system under evaluation to assess how far the system has gone along the trend. In particular, the concept of evolutionary potential can be defined as "the difference between the limits of each of the evolution trends and the relative maturity of the current system with respect to these trends" [21]. The resulting data are normally plotted in a radar diagram in which the inner area is a measure of the system's progression. [19].

3 PSS ENGINEERING AND TRIZ

TRIZ is nowadays adopted in the SS and PSS for idea generation. Zhang et al, [13] in their paper refer to "a recent debate on the issue of new service design centers on whether successful new services come about as a result of intuition, personal fancy, or inspiration, or are more likely the outcomes of a formal development process. To address this issue, more research findings must be obtained to validate the arguments from either side of the discussion".

Creativity in engineering is usually defined as the generation of ideas, whereas innovation is about putting these into action. Creativity is about divergent thinking, whereas innovation requires convergent thinking [15]. McAdam and McClelland [22] illustrate the vital importance of innovation in engineering, especially, service systems by indicating that *"idea generation is a key component of creativity"*. As [11] and [10] suggest, TRIZ "40 inventive principles" have been successfully applied to service system design with the aim to generate ideas. For these reasons, the main applications of TRIZ in the PSS area are in the 40 Inventive Principles (as evident in Table 1). By the way, Inventive Principles are more suited for solving technical problems at a high level of detail rather than to assist the servitization process of a system. Inventive Principles constitute a powerful means for problem solving and should be adopted each time a technical problem arises, but they can hardly help designers finding the right direction of evolution of the overall system.

Aim of this work is to overcome the mere contextualization of the TRIZ tools to inventive problem arising from Service Engineering contradictions. What is proposed is to add a method and practical tools to analyse and assess a system evolution. LTSE and Trends of Evolution can be borrowed and adapted to forecast and ease the evolution from a traditional product to an innovative Product PSS.

An application in the Copy Printer Machine industry is proposed supporting the Trends of Evolution identified through the PSS and TRIZ theory.

Low et al. 2001 [25] The work proposes an extension of the TRIZ principles in the PSS area to include sustainable service design factors. It focuses on the systematic building and innovative clustering of products as a means of delivering an eco-centred service. A framework for potential TRIZ applications in service was developed.
Mann and Jones, 2002 [18] They report an application of the TRIZ-based systematic innovation process to a PSS case study. The concept of *Ideality* to explain the different kinds of PSS is described.
Chai, et al., 2005 [14] Based on the principles of TRIZ, the authors present an approach to systematic service design that encompasses three main stages. The input to the method is an identified service problem. In particular 40 inventive principles and 4 separation principles are adopted to identify and eliminate the contradictions. The final output is a list of possible innovative conceptual solutions to service design.
Karni and Kaner, 2009. [17] In this work TRIZ is used as one of the tools adopted to optimize the Service delivery processes. In particular the authors develop a methodology to support the step of methods for modifying processes. The framework proposed by the authors comprises and integrates 21 goal determinants to be achieved through process redesign; 32 best practices describing possible process modifications; 40 TRIZ inventive principles for generating new improvement ideas; and case-based reasoning (CBR) for retaining and reusing past improvements.
Abdalla et al., 2004 [28] The paper aims to benchmark the performance of TRIZ in delivering solutions, guidance and consultancy for the design and development of a new PSS. In particular different TRIZ tools are implemented to find solutions ideas for the development of the PSS. The "40 principles " are the most implemented but also inventive principle and patterns of evolution are taken into consideration in the analysis.

Table 1: Main literature contributions on TRIZ application in PSS.

4 EVOLVING THE COPY PRINTER MACHINE USING LTSE AND TRENDS OF EVOLUTION IN PSS

The proposed case study, namely in the Copy Printer Machine industry, has the aim to support the identification and explanation of the Trend of Evolution of a PSS and of a PSS Engineering process.

In order to identify the most important evolutionary trends of the PSS, an analysis of the specific issues, which are peculiar to the PSS engineering approaches, has been carried out. In particular, from the analysis it is evident that:

(i) An increasing gap between product function and customer value emerges, due to the multifaceted nature of the perceived value. In order to fill this gap, a more upstream conceptual design in the early stages of design and a major involvement of the customer is required. [23]

(ii) In the design of a PSS, its contents are delivered as a customer benefits a package that combines physical goods and non-physical goods [23] while its channel requires a combination of infrastructures and networks. In this sense the engineering of a PSS should be based on a poly-system approach.

(iii) In the transformation process of a PSS, a structured feedback iteration process to collect and use all the feedback information coming from the customer in each instant of the life cycle, is fundamental to be able to answer properly to customers' needs since the Design phase. In fact, customer's needs in the PSS context continually change.

(iv) analogously to a generic system, also a PSS has to evolve in the direction of "*ideality*". Given the main nature of a PSS an increased level of benefits in the area of eco-efficiency and eco-friendliness is required.

Starting from these considerations, it is evident that the main evaluation trends in the area of PSS are related to the customer involvement in the Design phase, the integration of Products and Services, the feedback/iteration approaches and the potential sustainable benefits.

In the followings, each of the PSS specific trends will be shortly described. For each trend, the LTSE are referred and the most important ones for each case are put into evidence, in order to provide a high-level set of indication on the way a PSS evolves. Moreover, an evaluation of the stage in which the analysed case is positioned, is also reported. Finally, to summarize and support the identification of the evolutionary potential, a radar plot is reported.

4.1 The Trends of Evolution

The case study under investigation represents the PSS design processes of a B2B printer manufacturer. The company works in a global context, and has offices both at continental and national levels. Its customers are public administrations, companies, which use the printer/copier as an office tool and companies which use the printer/copier as a working tool. In particular, the new business of the companies is related to the sale of the service of the copied papers by the number of photocopies instead of the number of copy machines. The engineering process and the solution provided are hereafter examined by the trends of evolution specifically developed for the analysis of PSS Engineering process.

User Role – In the classical engineering literature, the customer is only considered as a mere source of information. If Service is, as reported in the Service Dominant Logic [24], co-created between a provider and a receiver, how can a service be designed without input and representation from Service users acting as co-creators? This evolution trend deals with the way customers are integrated in the Design phase. The system becomes more and more servitized if the customer shifts from being a mere source of information to be a co-designer of the PSS and, as already happens in the Software engineering area, to let him act as the designer of the system itself. In this way, as suggested by the LTSE, the system (PSS and Customer) should be able to develop itself without the producer's support.

Figure 3: Main Evolution steps in the PSS User Role.

In the case analyzed, the users are not really involved in the process. They are a *source of information* for the requirements' identification. *Dialog activities* are carried out only with internal stakeholders. The designers are the owners of the process, and they are only influenced by the internal stakeholders and in particular by the personnel of the company operating in the local markets.

Integration of Products and Services – What is nowadays clear is that to optimize service value to customers, functions of product and service activities should be integrated transparently. Both products and service activities are meant for value creation, and different combinations of them can fulfil the same needs. The evolution started when After Sales Services were perceived as a necessary evil, to be guaranteed with the offered product. in This scenario, the product remains the main element of the transaction. Then, in the Extended product, Service, as an added value, is provided with the Product. However, the Product is still the main element of the transaction and of the Design process. In the PSS concept, the main element of the transaction is the entire Solution, where the tangible and intangible elements are not distinguishable and the design is done simultaneously. This evolution is in line with the law of transition to a super-system because of the shifting of functions from the product to a complex system of a higher level.

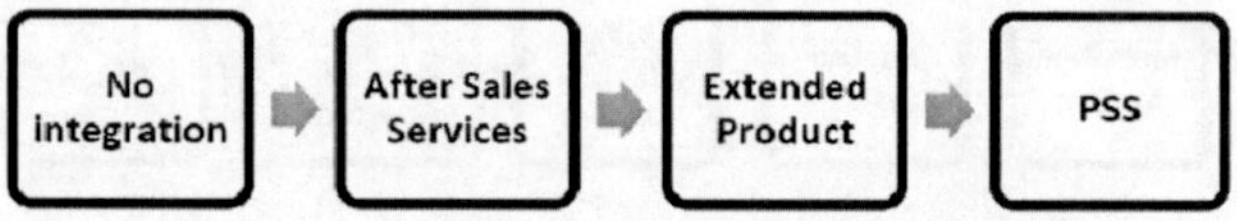

Figure 4: Main Evolution steps in the PSS Integration.

In the case under analysis, it is important to underline that the PSS offer is well integrated. In particular, people involved in the engineering of the PSS do not distinguish the "product" from the "service" elements. They refer to their offer as a "solution".

Ownership and use of the PSS – According to the trend, when a company starts its Servitization process, a traditional concept of Own and Use is adopted: customers own and use the solution they buy. Then, in the most servitized context the producer is the owner and the user of the system, while the customer only receives the functions requested. This concept is based on the law of Increasing *Ideality* stating that a system evolves towards pure functionality, i.e. a system that does not exist but its function is delivered. The way this may be carried out can be explained by the law of transition to supersystem.

Figure 5: Main Evolution steps in the PSS ownership and use.

As mentioned before, the company analyzed does not sell copy machines anymore, but it provides all the tangible and intangible elements necessary to sell the single copy. In this context, the users/customers do not own the product, but they use the solution provided. This kind of PSS offer is referred in the PSS literature as Product Utility Service.

Feedback Iteration - The iteration process in the Design of a PSS is fundamental in order to change a solution considering all the feedback coming from the customer. This evolution trend concerns with the change of the models adopted to collect the information and to iterate it in the Design process. The evolution starts with Linear Models, where the development of a designed entity proceeds step by step and the iteration process is not modelled. The second step of the evolution is represented by Linear Models with Iteration for Validation. Feedback information is collected only in the last steps of the Design model, during the Verification and Validation phase, and is merely considered at the beginning of the design process. Spiral Models, the third step of the evolution, allow for deliberated controlled iteration. Even if each phase is assumed to be broadly composed of the same activities (e.g. Requirements, Design, Implement, Evaluate) there is an explicit specification and revision of attributes depending on the feedback obtained. The last step of the evolution is given by the Multi Entry Models: there are multiple entry points into Design activities, which are dependent on the starting point for design and evaluation activities. The feedback information can enter in the different multi-entry points of the process, in different stages of the design and development of PSS.

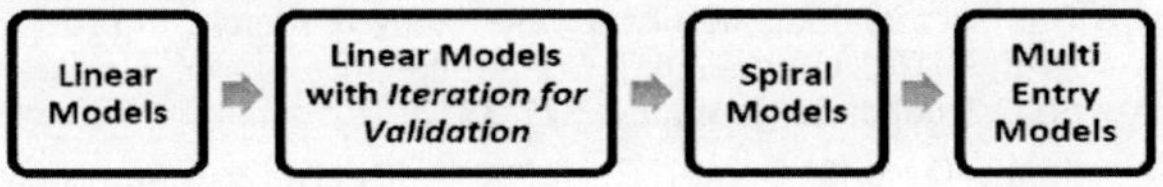

Figure 6: Main Evolution steps in the PSS Feedback iteration.

In the Engineering process of the case analysed, a multitude of feedbacks is collected during the Sell, the Use and the Assist phases. These feedbacks are collected in a systematic way, and then are also well considered in the requirements' identification, in the idea development and in all the phases of the design process. The structure of the feedback process can then be assimilated to a Spiral Model.

Mono-Bi-Poly: the traditional engineering of products generally is related to the design and development of the tangible elements. In order to engineer properly a PSS the design and development of all the four elements characterising it (Product, Service, Network and Infrastructure) is required. Once defined the main useful function that the system must perform, this trend show how the system evolve from a mere product to a complex system that includes new actors and elements also from the supersystem. All the elements should interact and cooperate to achieve the final result.

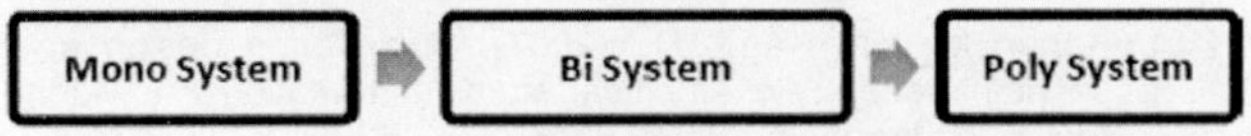

Figure 7: Main Evolution steps in the PSS system development.

In the case under analysis a strong integration of the tangible and intangible elements of the provided solution is already accomplished. A lot of efforts have been also provided in terms of infrastructure and network integration, even if, due to the fragmented service chain (more actors involved), not all the process phases can be explicitly defined and, thus, accurately monitored.

Design Emphasis: along with the evolution of the market offer from Product-Oriented, to Product-Utility, to Product-Results Services, the design emphasis, with a sustainability perspective, has to shift from first-cost costs, to life-cycle cost to a maximum benefit perspective [18]. This shift is supported by the *Ideality* concept.

In particular, in a Product-Oriented Service, where the product is owned and the service is added to the product, designers are commonly oriented to opt for a first-cost solution [18].

Level of Ideality = (acceptable benefits) /
(lowest first cost + compliant harm)

Where: "Acceptable benefits" are the benefits that the customers expect from the system, while "compliant harm" is related to the environmental legislation.

In a Product Utility Service scenario the provider no longer sells a product but rather its use (as in the case study under analysis). In this context the design paradigm shifts from a lowest first cost to a lowest life-cycle cost. Since the provider retains the ownership of the solution, his main purpose is to get the minimum cost during the whole life cycle [18].

Level of Ideality = (acceptable benefits) /
(lowest life cycle cost + compliant harm)

A provider of Product Result Service guarantees the customers with certain results and performances regardless how they are provided. The provider has the best interest to achieve the optimum benefits from the solution life cycle with the lowest life cycle cost [18].

Level of Ideality = (maximum benefit) /
(lowest life cycle cost + compliant harm)

To move towards *the IFR* in the design of a PSS, there is also a fundamental shift of emphasis in the design of the whole super-system.

Figure 8: Main Evolution steps in the Design Emphasis.

4.2 Trend and Evolution Potential

The Evolution Potential, represented with a radar plot, gives a sharp image of the level of development reached so far compared to the maximum achievable. Each branch of the plot corresponds to a trend that is described through a number of intermediate steps of evolution. The last step for each branch corresponds to the best situation we can imagine and is formulated by means of the concept of *Ideality* and *Ideal Final Result*. By comparing the features of a real PSS with those described in the trends we can carry out an assessment of the the system according to each branch and finally draw the area representing the actual value of the PSS. In Figure 10, the case study analysis previously reported for each branch is summarized by the radar plot.

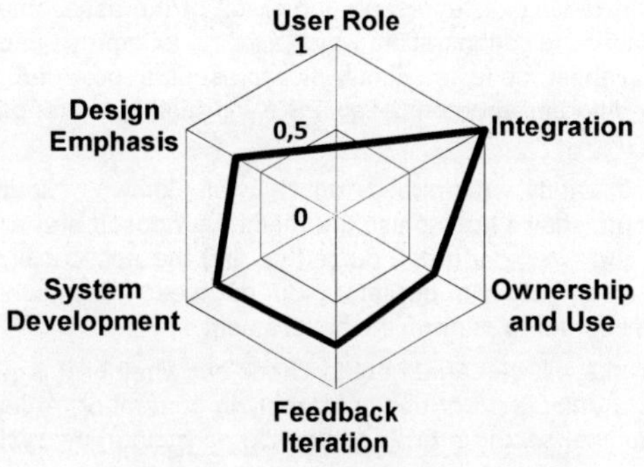

Figure 9: Radar plot of Evolution Potential.

The area of the polygon is a qualitative measure of the overall level of evolution of the system. The evaluation can be functional to two main purposes:

- To improve system value: the plot highlights which are the weakest directions to work on in order to balance the evolution level of each branch (according to the Law of uneven development of parts of a system) and then make it grow further. In the case analysed, the User Role is the less evolved branch: as a result, designers and the management should work in order to understand how to improve it.
- To compare two or more different systems and find out which are the main reasons determining a PSS better than others and, as a consequence, better finalise a winning development strategy.

5 CONCLUSION AND OUTLOOK

Service Engineering (SE) represents a young discipline, whose significance in academic and practical work grows continuously. One of the main challenges of the SE approach resides on the possibility to extend and adapt some of the traditional engineering approaches to the service sector.

TRIZ was born in former USSR in late '40s but it started spreading in western countries only in '90s. Only recently it has been fully recognized as a powerful means to innovate and predict future generations of products. In 2004, Fey [24] gave some reasons why

"TRIZ flies but do not soar" and a relevant one was "insufficient integration with other productivity enhancement tools". In the last years, several attempts have been carried out in this direction. By the way SE was marginally considered mainly because of the number of lifecycle steps that separate it from conceptual design where TRIZ in normally applied. Moreover, the efforts to merge SE with TRIZ have involved only classical tools for solving problems, neglecting the high potential of forecasting tools.

In this paper some pillars of TRIZ are briefly introduced: Laws of Technical System Evolution, Trends of Evolution and Ideality. The introduction of such concepts in a PSS Engineering context consist in providing a means to assess and guide the evolution of systems in their progression of servitization. Actually, the capability of evaluating the level of development of a system towards some specific trends of evolution allow to make a breakthrough in system design. TRIZ Trends of evolution are traditionally formalized according to the need of enhancing products and processes, but not services. In order to achieve a valuable result for PSS as well, some new Trends are proposed and the results of their application to the case study are provided.

The concept of Evolution Potential allows to comparatively assess how systems progress towards a predetermined set of evolution steps. The radar plot represents the result of the assessment and also provides indications about which aspects to improve in order to get the highest benefits. Finally it represent a powerful tool to compare different and competing PSS and highlight strengths and weaknesses.

The case study, performed on a well known example of servitization, shows how to use the method proposed and allows to evaluate the easiness of the procedure and the applicability in the PSS context. Actually, designers do not need to master TRIZ concepts in order to perform the assessment.

Future works will be carried out to check out whether it is possible to define a unique set of trends to apply in general or, at least, for each industrial sector in order to provide an even more systematic and easy way to assess the evolution potential.

6 REFERENCES

[A] http://www.invention-machine.com/

[B] http://www.creaxinnovationsuite.com/

[1] Lovelock, C. and Gummesson, E. (2004): Whither Services Marketing? In Search of a New Paradigm and Fresh Perspectives. Journal of Service Research Vol.7 No1 pp20.

[2] Baines, T.S. et al. (2007): State-of-the-art in product-service systems. Journal Engineering Manufacure, Vol 221(B), 1543.

[3] Mont, O., (2002): Clarifying the concept of product–service system. Journal of Cleaner Production, Vol10, pp.237-245.

[4] Mont, O. and Plepys, A., (2004): What is behind meagre attempts to sustainable consumption? Institutional and product-service system perspective. In Proc. of International Workshop on Driving Forces and Barriers to Sustainable Consumption. University of Leeds, Leeds, UK, pp. 108-126.

[5] Pezzotta G., Cavalieri S., and Gaiardelli P. (2009): Product-Service Engineering: State of the Art and Future Directions, In: Proc. of 13th IFAC Symposium on INformation COntrol problems in Manufacturing (INCOM), Moscow (RU), June 3-5.

[6] Sakao, T, Napolitano, N, Tronci, M, Sundin E. and Lindahl M., (2008): How are product-service combined offers provided in Germany and Italy? Analysis with company sizes and countries, J. of systems science and systems Eng., Vol 17.

[7] Bullinger, H., Fahnrich, K. and Meiren, T., (2003): Service engineering—methodical development of new service products. Int. J. of Production Economics, Vol 85 No 3.

[8] Aurich, J.C., Mannweiler, C. and Schweitzer, E., (2010): How to design and offer services successfully. CIRP Journal of Manufacturing Science and Technology. In printing.

[9] Bandinelli, R. and Gamberi, V., (2009): Servitization in Oil and Gas sector: outcomes of a case study research. In proc. Of 11th MITIP International Conference, Bergamo: Università degli Studi di Bergamo, pp. 327-334.

[10] Mann, D. and Domb, E. (1999): 40 inventive (business) principles with examples. The TRIZ Journal, September (electronic journal, unpaginated).

[11] Retseptor, G., (2003): 40 inventive principles in quality management. The TRIZ Journal, (March).

[12] Mann, D., (2007). Hands-on systematic innovation for business and management. 2nd Edi., Clevedon UK: IFR Press.

[13] Zhang, J., Tan, K. and Chai, K., (2003): Systematic innovation in service design through triz. In Proceedings of the EurOMA-POMS 2003 Annual Conference. Cernobbio, Lake Como, Italy, pp. 1013-1022.

[14] Chai, K., Zhang, J. and Tan, K., (2005): A TRIZ-Based method for new service design. Journal of Service Research, Vol 8 No1, pp.48-66.

[15] Karni, R. and Kaner, M., (2007): An engineering tool for the conceptual design of service systems. In D. Spath and K. Fähnrich Advances in Services Innovations. Springer Berlin Heidelberg, pp. 65-83.

[16] Kaner, M. and Karni, R., (2007): Engineering design of a service system: An empirical study. Information Knowledge Systems Management, Vol 6, pp. 235-263.

[17] Kaner, M. and Karni, R., (2009): An Ideation Framework for Service Process Improvement. Int. J. of Service Science, Management, Engineering, and Technology, 1-33.

[18] Mann D., and Jones E, (2002): Sustainable Services and Systems (3s) through systematic innovation methods, The journal of sustainable product design, Vol2, N 3-4,

[19] Mann, D. (2002): Manufacturing technology evolution trends, Integrated Manufacturing Systems, Vol. 13 Iss: 2, pp.86 - 90

[20] Mann, D. (2004): Fan Technology: Evolutionary Potential and Evolutionary Limits. In IMechE International Conference on Fans, London, 9-10 November.

[21] Mann, D., Dewulf, S., (2002): Evolutionary-Potential in Technical and Business Systems, The TRIZ Journal.

[22] McAdam, R. and McClelland, J., (2002): Individual and team-based idea generation within innovation management: organizational and research agendas. European Journal of Innovation Management, Vol. 5 No. 2, pp.86–97.

[23] Hara, T. et al. (2009): Service CAD system to integrate product and human activity for total value. CIRP Journal of Manufacturing Science and Technology, (In Press).

[24] Lusch, R.F., Vargo, S.L. and Wessels, G., (2008): Toward a conceptual foundation for service science: Contributions from service-dominant logic. IBM Systems Journal, Vol 47 No.1.

[25] Low, M.K. et al. (2001): Manufacturing a Green Service: Engaging the TRIZ Model of Innovation. IEEE Transactions on Electronics Packaging Manufacturing, Vol 24 No 1 .

[26] Fey V., (2004): Why does TRIZ fly but do not soar? In Proc. of the 4th TRIZ Future Conference, Florence 3-5 Nov..

[27] Altshuller, G., (1984): Creativity as an exact science, Gordon and Breach Publishers.

[28] Abdalla, A., Bitzer B., Morton, D. (2004): Benchmarking TRIZ in the field of Product Service systems "PSS, In: Proc. of the 4th ETRIA TRIZ future Conference, Florence, Italy, 3-5 Nov.

An Ontology for Product-Service Systems

Gokula Annamalai[1], Romana Hussain[2], Mehmet Cakkol[3], Rajkumar Roy[1], Stephen Evans[1], Ashutosh Tiwari[1]

[1] Manufacturing Department, Cranfield University, UK

[2] School of Engineering, Cranfield University, UK

[3] School of Management, Cranfield University, UK

Abstract

Industries are transforming their business strategy from a product-centric to a more service-centric nature by bundling products and services into integrated solutions to enhance the relationship between their customers. Since Product-Service Systems design research is currently at a rudimentary stage, the development of a robust ontology for this area would be helpful. The advantages of a standardized ontology are that it could help researchers and practitioners to communicate their views without ambiguity and thus encourage the conception and implementation of useful methods and tools. In this paper, an initial structure of a PSS ontology from the design perspective is proposed and evaluated.

Keywords:

Ontology; Product-Service Systems; Methodology

1 INTRODUCTION

In this competitive globalizing economy, industries are changing their strategies from Product-centric to Service-centric approaches. An example of these approaches is Product-Service Systems (PSS) which is widely defined as an integrated product and service offering that delivers value in use. There is an increasing interest towards PSS in order to develop theories, methodologies, tools and techniques to understand the concept as well as to support industries and designers in developing these offerings [1]. Nevertheless, this research is in its infancy which necessitates the development of ontology for the PSS domain. Ontology is commonly defined as an explicit formal specification of the terms in the domain and the relations amongst them [2]. Although ontologies may be used for various purposes and applications, they are most commonly used for sharing, navigating, searching, indexing and retrieving domain knowledge. Furthermore, these purposes are used widely to validate the proposed ontology. Advantages to develop an ontology for a domain in its initial stage are:

- Effort required to develop an ontology is modest,
- There is an opportunity for progressive updating as the understanding evolves and
- It also facilitates the adaptability of the domain structure

The importance and applications of ontologies have been widely discussed in literature and are emphasized in many domains. The immediate advantages of developing PSS ontology are:

- To provide a platform for stakeholders to communicate and share their concepts with each other effectively and without ambiguity,
- It helps to understand the uniqueness of research outcomes
- And it aids to validate the research outcomes.

Schlenoff et al. [3] also stress that a domain ontology is helpful for unambiguous communication, standards-making and semantic-alignment efforts as well as future industrial information infrastructures. In this paper, we intend to explain the initial structure developed for PSS ontology from the design perspective as well as the methodology followed in this development process. The following section details comprehensive literature on the methodology used in developing an ontology, a product ontology, a service ontology, and an initial structure framed for PSS ontology and finally the validation process.

2 METHODOLOGIES TO DEVELOP ONTOLOGY

Many ontology development methodologies are proposed in extant literature. This section reviews and summarizes these methodologies and elaborates the steps followed to develop a PSS ontology. Ahmed et al. [4] propose that a methodology for creating ontologies for a particular purpose consists of six stages. They combine a number of methods from social science and computer science to develop this methodology. The stages are: identifying the root concept of the taxonomies that form an engineering design ontology; identifying existing taxonomies for each of the root concepts from the previous stage; creating taxonomies if no existing taxonomy was found; testing the taxonomies for the particular application; building a thesaurus for the integrated taxonomy and; refinement of the integrated taxonomy. It is important to note that they stress methodological development through empirical analyses rather than from documentation. They also stress that it is often necessary to acquire domain knowledge from experts when building the ontology. Interviews, literature reviews, document analysis and map instances to taxonomy are proposed research methods for these steps.

Jayaram and Jayaram [5] suggest following steps for the process of developing the ontology: decide which domain the ontology will cover and refer to the related resources to determine what terms to use from the domain, define the classes hierarchically and define the associated properties for these classes, define constraints for the properties, create the instances of the classes and create the axiom definitions in ontology for reasoning purposes. They suggest a layered structure to build engineering ontologies for product design and analysis. The 3-tier ontology structure consists of: General Domain Ontology (GDO), Domain Specific Ontology (DSO)

J. Hesselbach and C. Herrmann (eds.), *Functional Thinking for Value Creation: Proceedings of the 3rd CIRP International Conference on Industrial Product Service Systems, Technische Universität Braunschweig, Braunschweig, Germany, May 5th - 6th, 2011*,
DOI 10.1007/978-3-642-19689-8_41,

and Application Specific Ontology (ASO). Pinto and Martines [6] summarize five stages used for ontology building: specification, conceptualisation, formalisation, implementation, and maintenance. In the specification stage, the purpose and scope of the ontology are determined and subsequent conceptual descriptions of the ontology are made.

Eris et al., [7] propose a methodology for constructing the ontology which consists of: discussing the purpose and appropriateness of applying an ontological approach to product development projects in small teams, conducting a literature review of bases, purposes, and methods of identification and classification in other sciences, formulating tentative ontological frameworks, conducting internal validations, and making the frameworks accessible to researchers, discussing the frameworks with colleagues from related fields and developing criteria and evaluation systems for testing the validity, utility, and reliability of the proposed frameworks. Uschold and King [8] outline a methodology for developing and evaluating ontologies considering informal techniques concerning issues such as scoping, handling ambiguity, reaching agreement and producing definitions as well as a formal approach. They propose a comprehensive methodology for developing ontologies to include the following: identification of the purpose and scope, building the ontology (this encompasses ontology capture, ontology coding and the integration of existing ontologies), evaluation, documentation and the guidelines for each phase. They identify the following criteria to design ontologies: clarity, coherence, extensibility, minimal ontological commitment and minimal encoding bias.

Gruninger and Fox [9] propose an approach to develop engineering ontologies based on experiences in the development of TOVE (Toronto Virtual Enterprise). The approach consists of motivating scenarios, informal competency questions, terminology specification, formal competency questions, axiom specification and completeness theorems. Noy and McGuinness [10] propose a methodology for developing an ontology which includes: determination of the domain and the scope of the ontology, the consideration of reusing existing ontologies, the enumeration of important terms in the ontology, definition of the classes and the class hierarchy, definition of the properties of classes—slots, definition of the facets of the slots and the creation of instances. They argue that there is no one correct way to model a domain and that ontology development is necessarily an iterative process.

Jones et al. [11] analyse various ontology development methods and conclude with following suggestions: many of the methodologies take a task which is found useful as a starting point; if the purpose and requirements are clear at the outset, a stage-based approach would seem more appropriate whereas if no clear purpose had been identified, an evolving prototype model may be more applicable. By analysing these methodologies, the steps followed to develop PSS ontology are illustrated in Figure 1. The purpose of developing this ontology is to better understand the PSS domain from the design perspective. The subsequent sections elaborate these steps.

3 IDENTIFYING EXISTING TAXONOMIES

This section analyses various ontologies proposed in the product and service literature and also elements used in product-service systems. These analyses aim to identify research gaps to develop ontology for PSS. Each sub-section below addresses these topics individually.

3.1 Product Ontology

A number of representations have been proposed to characterize the artefact being designed. Chandrasekaran et al. [12] presented Functional Representation (FR). Structure, Behaviour and Function (SBF) [13]. Hubka and Eder [14] describe a technical system and the transformation process it creates in terms of process, function, organ and component structures.

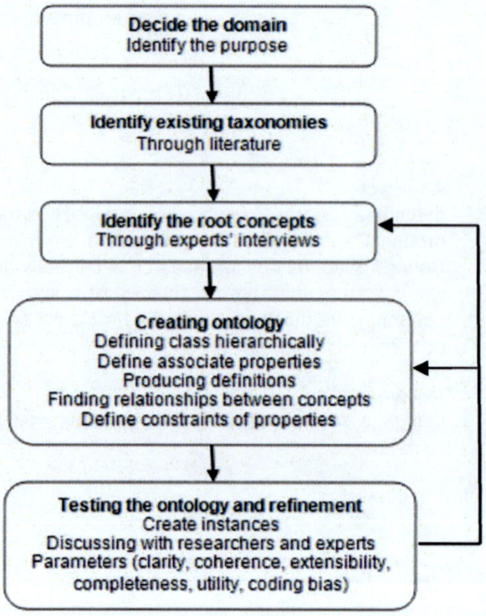

Figure 1: Steps in developing PSS ontology.

Andreasen [15] argues that design specifications and structures are linked by causal relations: the process determines the functions, which are created by the organs, which are materialized by the components. Chakrabarti et al. [16] proposed the SAPPhIRE model of causality with the following constructs: state, action, part, phenomenon, input, organ, effect and their relationships.

Szykman et al. [17] represent product knowledge as requirements, specifications, artefacts (sub-artefacts, functions, forms and behaviours), design rationale, constraints and relationships. The taxonomy referred to as the functional basis with a set of functions (verbs) and flows (nouns) was developed by Hirtz et al. [18] and this integrates the efforts of Sykman et al. [17] with those of Stone and Wood [19]. Ahmed et al. [4] proposed an Engineering Design Integrated Taxonomy (EDIT) which consists of several taxonomies and their relations. They argue that it may not be possible to find an ontology for engineering design that suits all requirements. They postulated that a difficulty in identifying an engineering ontology that is generic to the product or system being designed is usually because it is specific to a particular company or project.

STEP (Standard for the Exchange of Product Model Data) [20] is an inter-lingua for defining and specifying products. The primary motivation for STEP is to achieve inter-operability and to enable product data to be exchanged amongst different computer systems and environments associated with the complete product lifecycle. Lin at al. [21] present an ontology for representing requirements that supports a generic requirements management process in the engineering design domain. Objects included in the ontology are parts, features, requirements, and constraints.

Schlenoff et al. [3] analyse various ontologies (CYC, Enterprise Ontology, TOVE (Toronto Virtual Enterprise)) through typical manufacturing scenarios. They conclude that all three packages were approximately equal in their ability to represent manufacturing information based on the information which already existed and their ability to specialize this information to make it appropriate to the manufacturing field. But they noted that the inferencing capabilities in CYC proved to be a bit more mature than the other two packages.

Eris et al. [7] identified four categories under the product development project entry: project input and character, product development process, project output and character and project phases. They classified the product development process category into five overall dimensions: actors, activities, information, physical artefact, and environment. Li et al. [22] develop an engineering ontology to represent the established design and manufacturing knowledge for engineering information retrieval. Process Specification Language (PSL) developed by NIST [23] treats more general (discrete) "processes" such as manufacturing process. It includes core-concepts such as activity, time point and objects as well as relations. Yoshioka et al. [24] explore ontological models of theories of engineering tools and their integration for KIEF (Knowledge Intensive Engineering Framework). KIEF consists of basic concepts which are categorized into entity, relation, attribute and physical phenomena and physical law.

The development of a standard engineering language has been a widespread and major objective for decades throughout the engineering design community. Even though various ontologies are proposed in literature for representing artefacts, a standardized language has not emerged. It should be noted that it may not be possible to find an ontology for engineering design that suits all requirements. However, this leads to an interoperability issue which is a huge concern across all industries. In this context, developing an ontology for the new domain of PSS has added more complexity. The next section describes the ontologies proposed in the service domain.

3.2 Service Ontology

The number of service ontologies is less than the number of product ontologies proposed in literature. This section reviews various published service ontologies. Silvestro et al. [25] argue that although many service classification schemes have been proposed before, no categorization has been either as pervasive or as useful as the process type classification provided in the production management literature. They have summarized six service dimensions from literature which are used in service operations management literature. These are: equipment/people focus, length of customer contact time, extent of customization, the extent to which customer contact personnel exercise judgment in meeting individual needs, the source of value added, front office or back office and product/process focus. Bullinger et al. [26] argue that a typical service can be characterised by three different dimensions: a structure dimension (the structure determines the ability and willingness to deliver the service in question), a process dimension (the service is performed on or with the external factors integrated in the processes) and an outcome dimension (the outcome of the service has certain material and immaterial consequences for the external factors). They show the integration and interaction of a product model, a process model and a resource concept within a basic service model.

Boyt and Harvey [27] categorize services through characteristics: replacement rate, essentiality, complexity, personal delivery and credence properties. Cho and Park [28] argue that factors frequently adopted for the classification of services from literature are tangibility or intangibility, interaction or customer contact, customization, availability of service outlets at single or multiple sites, and product / process. Bakrir [29] presents a typology of services founded upon a classification based upon the following criteria: the consumption of the service (internal or external), the association of the service to an object (to a product, to a service or none and the entity), subject of the service (persons or companies). He used manufacturing production systems typologies to analyze the common characteristics between products and services from the view of a production system: the degree of uncertainty and the service relation to customers, the level of standardization of characterizing products, and the level of diversification. Wemmerlov [30] argue that one of the major problems in the study of service production is of a taxonomical nature and deals with the identification and classification of service systems in a meaningful way. He summarizes the following decision variables tied to the design and operation of service systems: strategic role, service facility, process design, goods, workforce, planning and control systems, marketing programme and relationship to other service processes. He argues that a taxonomy should be related to these decision variables in a meaningful way. The descriptive variables selected for the taxonomy are the nature of the customer/service system interaction, the degree of routinisation of the service process, and the objects towards which the service activities are directed.

Even though few ontologies are proposed in the service sector, standardization is again a bigger challenge. Many applications need to be generated around these ontologies to validate and prove their usefulness. The next section summarizes the terminologies commonly used in the PSS domain.

3.3 Elements used in Product-Service Systems

The terminologies proposed for the PSS domain are reviewed in this section. Business models are widely discussed in this sector. Tucker and Tischner [31] propose three models: function-, availability-, and result-oriented use models based on the customer-supplier relation. Meier and Massberg [32] differentiate business models by: production responsibilities, supply of operating personnel, service initiative, ownership, supply of maintenance personnel and service turn model. Roy and Cheruvu [33] identified different IPS² (Industrial Product-Service Systems) contract types from the literature and various industries. Datta and Roy [34] noted that main parameters considered in the contracts are responsibility, cost of performance and incentives. Risks and uncertainties are other important parameters which are very much referred to [35]: Understanding and estimating the true whole life cost of an IPS² contract is required [36]. Datta and Roy [34] classify the key cost elements into recurring cost, non-recurring cost, overheads and hidden costs.

Apart from business issues, elements used in the design of PSS are reviewed. Factors distinguishing products and services such as intangibility, inseparability, heterogeneity, perishability, simultaneous production and consumption a well as ownership are discussed [37]. They have also noted that the development of PSS is influenced by several factors such as partners, organization, benefits for the IPS² provider, benefits for the IPS² customer, the environment, social considerations as well as the intention to use IPS², interactions of system/users and system life-cycle phases. It has been commonly noted that integrating the business model, technical artifacts, service activities, the IPS² lifecycle, aspects of the system's context and resources to deliver added value is a major challenge. Doultsinou et al. [38] argue that product design processes are well-structured, whereas the service design process is highly intuitive. They identified important variables which generate the difference in service issues were size, location, budget, number of projects in parallel, and personnel availability. Mahnel [39] stresses the quality of the service as an essential factor for customer retention.

Brunner and Wagner [40] have identified quality criteria for services: presentation and ambiance, reliability, accuracy, correctness, competence, politeness, friendliness, cooperativeness, understanding, authenticity, security, accessibility and availability and ability to communicate and sociableness. Baxter et al. [41]

depict an upper level PSS structure that enables the description of a combined product- and business- system (Figure 2). The central class 'life cycle system' is comprised of three classes: product, process and resource. Kim et al. [42] propose graph and ontological representations of PSS, consisting of values, product and service elements, and their relations. Shen and Wang [43] define product service ontology as the conceptualization of the product service and proposed basic ontological representation of product service.

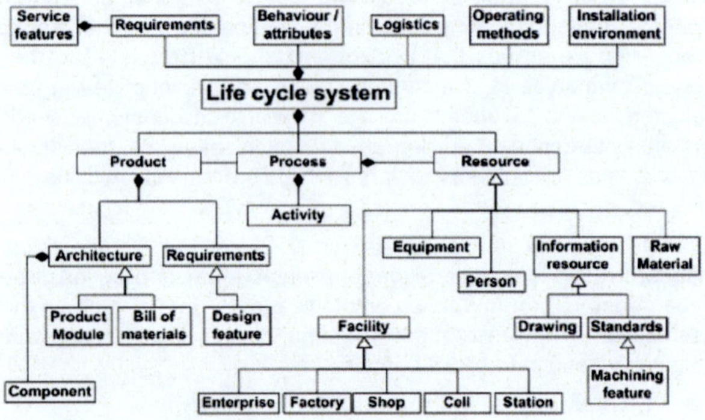

Figure 2: Integrated knowledge framework structure [41].

The ontology development for PSS is in its infancy. The various ontologies proposed are to be evaluated in-depth in terms of inclusiveness and exhaustiveness. Due to the involvement of many stakeholders within PSS, the semantics for each term should be defined accordingly. The challenge is not in building various information technologies but to develop a common representation across domains. This will subsequently expedite knowledge retrieval, discovery, editing, sorting and also automate reasoning with minimal implementation and maintenance effort.

4 IDENTIFYING ROOT CONCEPTS AND CREATING PSS ONTOLOGY

In order to identify the root concepts of PSS, thirteen explorative interviews which were conducted with various experts to elicit the definitive processes and challenges within PSS were analysed (please refer to the acknowledgement regarding the research team which conducted these interviews). This research team provided us transcribed interviews which were conducted with three different industries in UK who are heavily involved with developing PSS concepts. These interviews were shared equally to each of three researchers to extract the concepts uttered by the interviewees which were then consolidated, discussed, filtered and refined to develop the root concepts. These identified root concepts were compared with concepts proposed within literature and then further refined (Figure 3). This lent itself to an industry as well as academic perspective for PSS ontology.

Therefore mentioned root concepts aim to describe PSS comprehensively. These are: customer needs, stakeholders, PSS-Design, product life cycle, use phase, infrastructure, business elements, business models, supply network and benefits.

The various PSS related processes are very diverse and often used interchangeably within literature. To avoid this ambiguity, three processes are considered: to describe the design of PSS (PSS-Design), stages of products (product life cycle) and customer activities revolving around the products (use phase). Although business models are greatly discussed within literature, business elements are not sufficiently addressed.

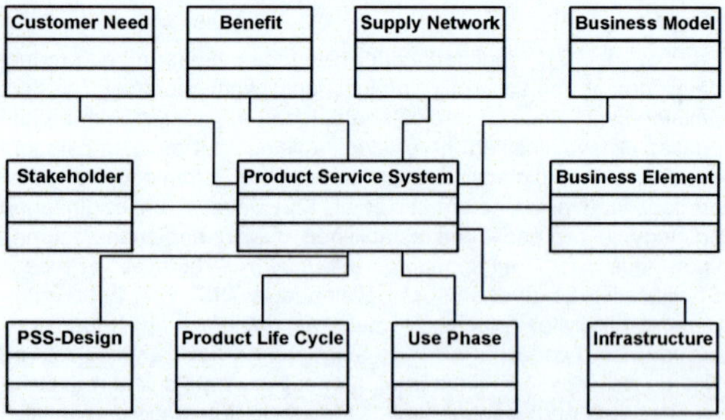

Figure 3: Root concepts to describe ontology of PSS.

Business elements describe the processes and issues involved in businesses which influence PSS. Since a viable PSS is not possible without a proper infrastructure and supply network, these concepts were also selected. While the importance of stakeholders is adequately stressed in literature, an extensive list of members involved in PSS has not been identified. To emphasize this, the concept of stakeholder was added to the root concepts. Since variety of business models in PSS depend upon the inclusion of products and services in the offerings, the product and service ontologies discussed in the literature have been incorporated in the concept business model.

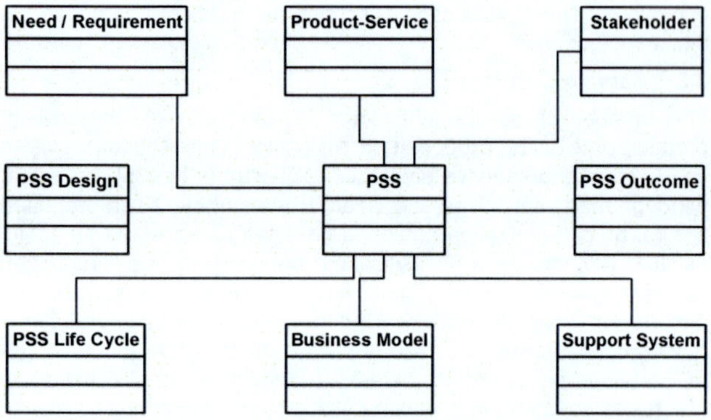

Figure 4: Revised root concepts to describe ontology of PSS.

Whilst generating these root concepts, sub-concepts associated with each concept were also simultaneously identified and then hierarchically categorized along with their respective properties. In total, 214 sub-concepts and properties have been identified. We used Protégé™ software to develop these concepts and sub-concepts. As the definitions for many concepts were not fully articulated in the interviews, these concepts were then framed by referring to the literature and discussion amongst the researchers. The next section describes the important observations from the validation of the proposed ontology. The modifications for the top concepts from the validation are represented in Figure 4.

5 VALIDATION

Figure 5 illustrates the steps followed to validate the proposed ontology. The first step for validation was to find a common agreement between the three researchers who extensively worked in developing this ontology. It was an arduous process to arrive at a common agreement because each researcher had different perspectives from different domains (product design, information usage and supply chain). Nevertheless, this encouraged the development of a comprehensive ontology. For the second step, exhaustive, individual discussions were conducted with four experts (academic researchers) who are engaged in PSS research. From this, we received valuable comments and suggestions to improve the developed ontology. It is interesting to note that all these researchers have provided positive feedbacks about the structure of the ontology. Their comments and suggestions were noted and incorporated into the ontology. The chief comments received from them are as follows:

- Need clarification in classifying processes,
- Not to use properties and characteristics alternatively,
- To describe process characteristics in terms of quality, cost, time and risk,
- The sub-concept risk missed in the business model and
- To include risk reduction in the 'benefit' concept.

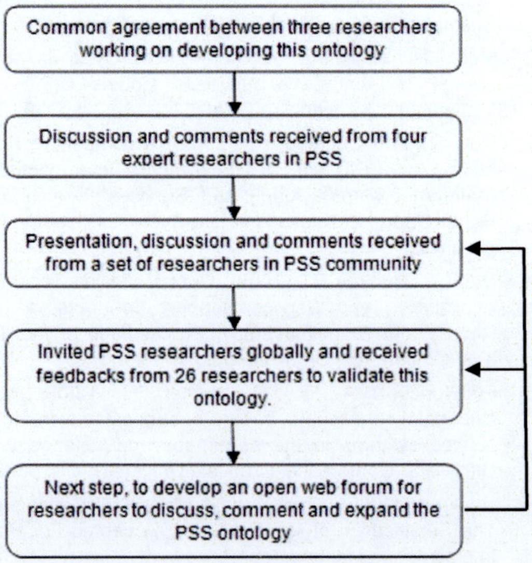

Figure 5: Steps involved in validating PSS ontology.

Such positive responses begged for a demonstration presentation to which a group of PSS academic researchers were invited. This provided more positive feedback to this ontology and also generated an opportunity for more researchers to participate in this ontology development. An invitation was then sent to PSS researchers globally to validate this ontology from which we have received feedback and comments from 26 researchers from various universities in different countries (UK, Japan, Belgium, Sweden, France, Germany, Italy, Netherlands and US).

Apart from renaming some of the concepts, major modifications are: the addition of PSS life cycle (both product life cycle and use phase are included within this); infrastructure and supply network are now grouped to form a single concept 'support system'; and business element is incorporated into 'business model'. To emphasis value proposition 'product-service' and 'business model' are kept separate. The next step will be to develop a web forum through web protégé to allow international researchers to participate and further develop the globally elaborated, comprehensive ontology. The current step in ontology enhancement is to find relationships between concepts, define constraints of properties and to test this ontology by creating instances for different applications.

6 DISCUSSION AND FUTURE WORK

In this paper, the initial ontology for the PSS domain has been developed from interviews with experts and from literature. An initial demonstration of this ontology was positively responded by researchers. The identified root concepts were almost found to be complete. The ontology mainly needs to be evaluated for its completeness, consistency and intuitive appeal to users. This could be carried out by involving many researchers for expansion, discussion and developing many applications from this ontology. The collaborative approach adopted should provide diverse view points to strengthen this ontology. Protégé^{TM} software was used to develop this ontology as this provides a well established platform for collaboration; Web Protégé is in development which will support the users to create and discuss the ontologies collaboratively over the Internet. Importantly, longitudinal studies are required to develop an exhaustive ontology because retrospective interviews may not provide a complete picture. The ultimate goal of this work is to develop a single PSS ontology which aims to understand the commonalities and differences between research groups and between industries and helps industries to develop viable PSS by providing good communication between the stakeholders.

7 ACKNOWLEDGEMENTS

The authors would like to acknowledge EPSRC for supporting, via the Cranfield IMRC, the work carried out in the preparation of this paper. It should be noted that interviews analysed in this work were conducted by a research team headed by Prof. Steve Evans at Cranfield University. We are grateful to this team for providing valuable interview data.

8 REFERENCES

[1] Meier H., Roy R., Seliger G., (2010): Industrial Product-Service Systems - IPS². CIRP Annals - Manufacturing Technology, Vol. 59, No. 2, 2010, pp. 607-627.

[2] Gruber, T.R. (1993): A Translation Approach to Portable Ontology Specification. Knowledge Acquisition, Vol 5, pp. 199-220.

[3] Schlenoff, C., Ivester, R., Libes, D., Denno, P., Szykman, S., and Libes, D. (1999): An Analysis of Existing Ontological Systems for Applications in Manufacturing, Proc. ASME Design Engineering Technical Conference/Computers in Engineering.

[4] Ahmed, S., Kim, S., and Wallace, K. M. (2007): A Methodology for Creating Ontologies in Engineering Design, Transactions of the ASME of Journal of Computing in Information Science in Engineering, Vol. 7(June), pp.132-140.

[5] Jayaram and Jayaram, (2010): http://vrcim.wsu.edu/pages/ontology/knowledge.html.

[6] Pinto, H. S., and Martins, J. P. (2004): Ontologies: How can They be Built?, Knowledge and Information Systems, Vol. 6, No. 4, pp. 441-464.

[7] Eris, O., Hansen, P. H. K., Mabogunje, A., and Leifer, L. (1999): Toward a Pragmatic Ontology for Product Development Projects in Small Teams. Proc. International Conference on Engineering Design, Munich, pp.1645-1950.

[8] Uschold, M., and King, M. (1995): Towards a Methodology for Building Ontologies, Proc. Workshop on Basic Ontological Issues in Knowledge Sharing, IJCAI, Montreal.

[9] Gruninger, M. and Fox, M.S. (1995): Methodology for the Design and Evaluation of Ontologies, IJCAI-95 Workshop on Basic Ontological Issues in Knowledge Sharing, Montreal, August 19-20th.

[10] Noy N.F., and McGuinness D.L., (2000): Ontology Development 101: A Guide to Creating Your First Ontology, Stanford University.

[11] D. M. Jones, T.J.M. Bench-Capon, and P.R.S. Visser. (1998): Methodologies for ontology development. In Proc. IT&KNOWS Conference of the 15th IFIP World Computer Congress, Budapest, Chapman-Hall.

[12] Chandrasekaran B, Goel A, and Iwasaki Y, (1993): Functional Representation as Design Rationale, IEEE Computer, Vol. 26, pp. 48–56.

[13] Goel A, (1991): Model Revision: A Theory of Incremental Model Learning, Proc. of 8th Intl Conf on Machine Learning, Chicago, IL. Morgan Kaufman, pp. 605–609.

[14] Hubka V, and Eder W, (1984): Theory of Technical Systems: A Total Concept Theory for Engineering Design, Springer-Verlag, London, UK.

[15] Andreasen M.M., (1980): Synthesis Methods as Basis for System: Contribution to a Design Theory, Ph.D. Thesis, Lund Institute of Technology.

[16] Chakrabarti A, Sarkar P, Leelavathamma B, and Nataraju B.S., (2005): A Functional Representation for Aiding Biomimetic and Artificial Inspiration of New Ideas, Artificial Intelligence For Engineering Design Analysis and Manufacturing, Vol. 19, No. 2, pp. 113-132.

[17] Szykman S, Racz J.W, Sriram R.D, (1999): The representation of function in computer-based design. In: Proceedings of the 1999 ASME design engineering technical conferences, Las Vegas.

[18] Hirtz J, Stone R.B, McAdams D.A, Szykman S, Wood K.L, (2002): A functional basis for engineering design: reconciling and evolving previous efforts. Res Eng Des, 13, pp. 65–82.

[19] Stone R.B, Wood K.L, (2000): Development of a functional basis for design. J Mech Des 122, pp. 359–370.

[20] Initial release of international standard (ISO) 10303.1994. Technical Report IS10303, International Standards Organization.

[21] Lin, J., Fox, M. S., and Bilgic, T. (1996): A Requirement Ontology for Engineering Design. Concurrent Engineering: Research and Applications, 4(3), pp.279--291.

[22] Li z., Raskin V., and Ramani K., (2007): Developing Ontologies For Engineering Information Retrieval, Proceedings of the ASME 2007 International Design Engineering Technical Conferences & Computers and Information in Engineering Conference, IDETC/CIE, Las Vegas, Nevada, USA.

[23] ISO TC184/SC4/JWG8 2003 Process Specification Language, **http://www.tc184-sc4.org/SC4_Open/SC4_Work_Products_Documents/PSL_(18629)/**.

[24] Yoshioka, M., Umeda, Y., Takeda, H., Shimomura, Y., Nomaguchi, Y., Tomiyama, T., (2004): Physical Concept Ontology for the Knowledge Intensive Engineering Framework, Advanced Engineering Informatics, 18(2), pp. 95-113.

[25] Silvestro R., Fitzgerald L., and Johnston R., (1992): Towards a Classification of Service Processes, International Journal of Service Industry Management, Vol. 3 No. 3. pp. 62-75.

[26] Bullinger H.-J.; Fahnrich K.-P.; Meiren T. (2003): Service engineering-methodical development of new service products, International Journal of Production Economics, 85 (3-11), pp. 275-287.

[27] Boyt T., and Harvey M., (1997): Classification of Industrial Services - A Model with Strategic Implications, Industrial Marketing Management, 26, pp. 291-300.

[28] Cho S., and Park K., (2002): Empirical taxonomy of services and service products in electronic commerce Electronic Commerce Research and Applications, pp. 339–350.

[29] Bakrir, R. (2003): Towards a New Service Design Approach Assisted by Computer Tools:A typology of services and a post sale service case study in the automotive industry. In: IEEE International Conference Systems, Man and Cybernetics, 2003, vol. 4, pp. 3899–3904.

[30] Wemmerlov U., (1990): A Taxonomy for Service Processes and its Implications for System Design, Journal of Service Management, 1 (3), pp 20-40.

[31] Tucker, A., Tischner, U., (2005): New Business for Old Europe - Product-Service Development, Competiveness and Sustainability, Greanleaf Publishing, Sheffield, UK.

[32] Meier, H., Massberg, W., (2004): Life Cycle-Based Service Design for Innovative Business Models CIRP Annals - Manufacturing Technology, Vol. 53, Issue 1, pp. 393-396.

[33] Roy, R., and Cheruvu, K., (2009): A competitive framework for Industrial Product Service Systems, International Journal of Internet Manufacturing and Services, Special Issue on Product Service Solutions in Life-Cycle Activities, Vol. 2, No.1 pp. 4 – 29.

[34] Datta P. P., Roy R., (2009): Cost Modelling Techniques for Availability Type Service Support Contracts: a Literature Review and Empirical Study, Proceedings of the 1st CIRP IPS² Conference, Cranfield, pp. 216-223.

[35] Erkoyuncu, J. A., Roy, R., Shehab, E., Wardle, P., (2009): Uncertainty challenges in service cost estimation for product-service systems in the aerospace and defence industries, CIRP IPS² Conference, Cranfield University, UK, pp. 200-207.

[36] Neely, A, (2009):. Exploring The Financial Consequences of The Servitization of Manufacturing, Accepted For Publication In Operations Management Review.

[37] Maussang, N., Zwolinski, P., Brissaud, D., (2009): Product-service system design methodology: from the PSS architecture design to the products specifications, Journal of Engineering Design, Vol. 20, Issue 4, pp. 349 - 366.

[38] Doultsinou, A., Roy, R., Baxter, D. I. and Gao, J. X., (2009): Developing a Service Knowledge Reuse Framework for Engineering Design, Journal of Engineering Design, Volume 20, Issue 4, pp. 389 – 411.

[39] Mahnel, M., (2007): Fit for best Service, Stellhebel für den Service von morgen - lernen von den Servicechampions, Impuls Management Consulting.

[40] Brunner, F.J., Wagner, K. W., (2008): Taschenbuch Qualitätsmanagement, Carl Hanser Verlag München, Wien.

[41] Baxter, D., Roy, R., Doultsinou, N. and Gao, J. X. (2009): "A knowledge management framework to support Product-Service Systems design", International Journal of Computer Integrated Manufacturing. 22 (12), pp 1073 – 1088.

[42] Kim Y. S., Wang E.,. Lee S. W., and Cho Y. C, (2009): A Product-Service System Representation and Its Application in a Concept Design Scenario, Proceedings of the 1st CIRP IPS² Conference, pp. 32-39.

[43] Shen J., and Wang L., (2007): A New Perspective and Representation of Service, International Conference on Wireless Communications, Networking and Mobile Computing, WiCom 2007.

Data Stream Mining for Increased Functional Product Availability Awareness

Ahmad Alzghoul[1], Magnus Löfstrand[1], Lennart Karlsson[1], Magnus Karlberg[1]

[1] Luleå University of Technology, Division of Computer Aided Design, Sweden

Abstract

Functional Products (FP) and Product Service Systems (PSS) may be seen as integrated systems comprising hardware and support services. For such offerings, availability is key. Little research has been done on integrating Data Stream Management Systems (DSMS) for monitoring (parts of) a FP to improve system availability. This paper introduces an approach for how data stream mining may be applied to monitor hardware being part of a Functional Product. The result shows that DSMS have the potential to significantly support continuous availability awareness of industrial systems, especially important when the supplier is to supply a function with certain availability.

Keywords:

Functional Products; Data Stream Mining; Availability

1 INTRODUCTION

A functional product is a system consisting of hardware which, together with a support system, delivers the agreed-upon function. The support system consists of software and various services. It keeps the whole system, including hardware, operational and able to deliver not merely hardware functionality, but available productivity for the customer. There are several ways to view the development, sale and procurement of total offers, product service systems or functional products, whatever we choose to call the system that delivers the agreed-upon function. One may view the system as consisting of Hardware (HW), Software (SW) and Services (Ss) as do Brännström, Elfrström and Thompson [1]. In this paper we adopt the view of the system that delivers the function proposed by Alonso-Rasgado, Thompson and Elfström [2] i.e., a system consisting of hardware and a support system. Product Service Systems (PSS) is a research area related to functional products. Műller and Sakao [3] state that "consolidation of PSS approaches among research projects and industrial branches is rudimentary". According to Baines et al. [4], it is clear that the PSS area is growing, indicating an increased industrial need for such offerings. Furthermore, Meier et al. [5] identify the following recent developments in industrial product service systems: business models, sustainability contribution, risk management, knowledge management, design, development, delivery and use. Hopefully, these results and others will lead to further consolidation in the future. For offerings such as FP or PSS, availability is a key concept. One challenge in prediction of FP availability is that the hardware and the support system must be accounted for coupled and concurrently. The reliability of hardware at both component and system level has been studied for years. For a non-redundant (series) system, Henley and Kumamoto [6] showed that the reliability of the individual components must be much greater than the system reliability requirement. Safety-critical industries such as the nuclear and aerospace industries, where failures can result in fatality, have always been a prominent driving force for the development of reliable systems. The study of component reliability has resulted in distribution of failure times such as the Weibull distribution [7]. Several techniques have then been used to relate the component-level performance to that of the system. One of the simplest of these is FME(C)A, (Failure Mode Effects (and Criticality) Analysis) [8]. For complex systems, structure methods such as Fault Tree Analysis [9] and Markov analysis are commonly applied [10].

While hardware availability has been studied for decades, the availability of support systems has not. Some notable exceptions include concepts for service reliability that have been reviewed by Li and Thompson [11] who, in addition, studied the character of the service support system for functional products. Li and Thompson [12] further adopted simulation techniques to assess the service reliability in the context of functional products. Also in the context of functional products, Reed et al. [13] presented simulation-based techniques for service support system reliability.

To couple the hardware availability and support system availability Löfstrand et al. [14] describe on a conceptual level how the availability of functional products (consisting of hardware (HW) and a support system) may be predicted and optimized. In order to further improve and support continuous availability awareness for the functional product supplier, the field of data stream monitoring systems is here introduced in that context.

In the field of Data Stream Mining it is clear that data stream mining algorithms have been applied in various fields, such as financial transactions, telecommunication networks, manufacturing systems and scientific data monitoring systems [15]. Data stream mining can be defined as the extraction of patterns from continuous and rapidly arriving streams of data [16]. Data stream mining algorithms can be applied either on the whole or on a part of the stream. The existing data stream mining algorithms can be classified into four categories: Clustering, Classification, Frequency counting and Time series analysis [17]. Understanding and improving the maintenance phase of the lifecycle of a product is crucial for achieving high reliability, maintainability and availability. Many applications of data stream mining exist in the field of operations and maintenance. For example, a framework for the use of multivariate statistical process monitoring technique and data stream mining techniques to detect

J. Hesselbach and C. Herrmann (eds.), *Functional Thinking for Value Creation: Proceedings of the 3rd CIRP International Conference on Industrial Product Service Systems, Technische Universität Braunschweig, Braunschweig, Germany, May 5th - 6th, 2011*,
DOI 10.1007/978-3-642-19689-8_42, © Springer-Verlag Berlin Heidelberg 2011

the change in the sensor data stream was proposed by Karacal [18], [19]. The framework presented by Karacal was useful for machine monitoring and reliability analysis. Kargupta [20] used a mobile and distributed data stream mining system for real-time vehicle-health monitoring. Kargupta developed the VEhicle DAta Stream mining (VEDAS) system, which was used to monitor the vehicle data streams and driver characterization. VEDAS used an on-board PDA-based distributed data stream mining system connected with remote modules through a wireless network to monitor vehicle fleets. A number of data-driven anomaly detection methods were applied on continuous data stream from a solid rocket motor in [21]. The results showed good performance for the suggested methods in detecting the faults. Angelov et al. [22] proposed a model-based approach to detect the faults in the industrial application. The model was tested and verified on both simulated and real data sets from car engine test benches. Karacal et al. [23] developed a sensor stream mining system for tool condition monitoring. They used chemical data from the gases released during the cutting process to monitor the cutting tool condition. A design of multivariate statistical analysis technique for online failure prediction was developed and tested by Youree et al. [24]. The system was designed for monitoring the health of mission equipment, industrial equipment and facilities. Wegerich [25] used a multivariate similarity-based modeling (SBM) technique for machine health monitoring. They used vibration data to detect incipient faults in an H-60 intermediate gearbox (IGB). The results showed that SBM was able to detect the faults early and to monitor the machine health of the gearbox.

With better monitoring of critical hardware systems and system parameters, failures can be predicted, detected, and avoided before damage occurs. This is of course beneficial for the customer in a traditional sale of hardware products. However, systems for monitoring are of even more importance for the supplier of a function, since the supplier (as well as the customer) will benefit from delivering agreed-upon availability. By applying data stream monitoring and filtering techniques on streams of simulation, experimental, and testing data as well as on streams from sensors on industrial equipment in use, operation of functional products is supported. Such an approach allows prediction of potential future errors before they occur, reducing the risk of unplanned stops and thus increasing hardware and function availability [16]. However, little research has been done on integrating technical systems for monitoring (parts of) a FP and thus further improving system availability.

Therefore, in this paper a DSMS based data stream mining approach is introduced in order to monitor hardware being part of a Functional Product. The DSMS approach is used to support continuous availability awareness of industrial systems. Thus, the objective for this paper is to *derive and present requirements for a DSMS which is to be used for supporting continuous functional product availability awareness.*

2 RESEARCH METHODOLOGY

Previous research included a review of the data stream mining algorithms and their applications in the field of operation and maintenance, further presented in [16]. Based on the research in [16], this paper introduces requirements for, and some issues concerned with, data stream mining for increased functional product availability awareness. The actual research methodology consisted of three steps: (1) developing the five-step approach further presented in Chapter 5, (2) identifying system parameters and responses and (3) identifying requirements on DSMS technologies for functional product availability support.

3 ABOUT DSMS TECHNOLOGIES

As identified in the popular magazine The Economist [26], the quantity of generated data and the available storage are continuously increasing. However, the rate of growth of the generated data is faster than that of the available storage capacity. This overload has compelled researchers to find systems and tools that can handle data on the fly i.e., data stream, without the need to store it. A data stream management system has been found to be an effective means of handling continuously generated data.

A data stream management system can be defined as an extension of a database management system which has the ability to deal with a data stream. Figure 1 shows an abstract architecture for a data stream management system [27].

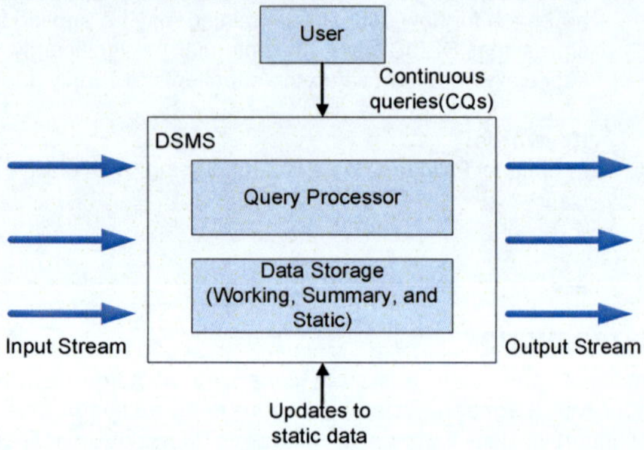

Figure 1: An abstract architecture for a DSMS.

The query processor is responsible for executing the continuous user queries over the input data stream, and then stream the output to the user or to a temporarily buffer. The working, summary, and static data storages are used for temporary working storage, stream synopses and meta-data, respectively [27]. Sensor networks, network traffic analysis, financial tickers and transaction log analysis are examples of DSMS applications [27]. A DSMS has the same structure as a DBMS (i.e., table in a relational database), but a data stream has no associated disk storage; it arrives continuously, its arrival rate may vary from time to time, and the missed data may be lost [28].

4 USE CASE EXAMPLE

In this use case the focus is on hardware monitoring and the usefulness of DSMS techniques for increased functional product availability awareness. The use case is the Swedish company Hägglunds Drives AB (HDAB). Hägglunds Drives AB is a manufacturer of low-speed, high-torque hydraulic drive systems [29] . Their drive systems are used in many industries such as: mining, recycling, pulp and paper, rubber and plastics, offshore, fishing, building and construction. HDAB are interested in improving monitoring to increase availability and reliability of their drive systems.

In [16] the authors have shown that data stream mining together with data stream management systems can be used to increase the availability of HDAB hydraulic drive systems. The tested data stream mining algorithms were able to achieve high classification accuracy. As a result, the algorithms can be used to detect failures at an early stage.

Once a failure is detected the next step is to identify how the system should respond to this failure. Reducing motor speed and shutting down the system are examples of responses which can be applied to protect the HDAB hydraulic motors from extreme damages. In addition, detecting failures at an early stage can reduce or eliminate the downtime of the monitored systems which is beneficial for HDAB and also for their customers.

On the other hand, the output from the monitoring model can be used to update the probability of equipment failure. The updated probability can be used to increase the prediction accuracy of the availability of functional products (consisting of hardware (HW) and a support system).

4.1 Supporting availability of functional products

The selection of the data parameters and the generating speed of the data are important factors for monitoring industrial systems. Therefore, after identifying the essential failures which could affect the system availability, the parameters to be monitored must be selected based on for example information from interviews with experts or through statistical analysis of existing data. Temperature, speed, and pressure are examples of the data parameters which were used to monitor the HDAB systems, further discussed in [16]. On the other hand, collecting data at higher frequency makes the result of the monitoring more accurate. Therefore, the SuperComputer Stream Query processor (SCSQ) [30] data stream management system was used by [16] to control and manage the data stream.

Alzghoul and Löfstrand [16] used the data, which was collected when a HDAB hydraulic motor was running normally i.e. without failures, to train the data stream mining algorithms e.g. Grid-based algorithm. Then the trained algorithm was tested by the authors using artificial failure data. The result showed that the algorithms were able to detect most of the failure data.

The case for using DSMS techniques and approaches in the context of functional product procurement and use may be summarized as stated above: having better monitoring of critical hardware systems and system parameters, failures can be predicted, detected, and avoided before damage occurs. One of the critical issues is to identify sets of parameters to be used for monitoring a data stream (i.e. rules or filters), both normal and abnormal system behaviors need to be described based on parameter sets given by the available sensors. Then required responses need to be decided. These issues will be further discussed in Chapter 5 below. First however, some of the potential effects of using DSMS technologies in the context of functional product availability will be discussed. The supplier decreases his business risk and worry about whether or not the availability he promised to deliver in fact is being delivered. If potential faults (which may eventually lead to hardware failure) are detected in time the supplier may address them without affecting the agreed upon FP availability and customer productivity.

Using DSMS techniques the customer may get continuous assurances that the functional product is working. (While parts of the support system around the hardware core of the functional product may be monitored (spare parts, tools, staff training levels, and number of available staff to name a few), the issues concerned with monitoring the support system are not expressly covered in this paper. Further, the customer may thus be assured of having a certain available production plant, thus significantly supporting his productivity while not having to invest in potentially capital intensive own hardware.

5 DERIVATION OF DSMS REQUIREMENTS

In this chapter, three issues related to the DSMS requirements are presented for the use in the context of functional products: Identifying system parameters and responses, Proposed DSMS structure and Requirements on DSMS technology. Having verified the ability of data stream mining in detecting failures, the next step is to determine how data stream mining can be used to support suppliers of functional products.

5.1 Identifying system parameters and responses

When deriving requirements on the DSMS technology, a five-step approach is proposed. In the first step (1) a usage scenario based on customer needs and the available system is agreed upon, and thereafter (2) the system and the system functionality are described. Step three (3) includes an industrial scenario.

The case study should be aimed at capturing interviewees' design- and operations-related knowledge with the purpose of identifying DSMS filters. Such filters may be defined by distributions of system parameters over time. Having identified system parameters, step four (4) includes modeling and analysis of the system. The filters should be defined on the basis of what kind of information is needed for the industrial company to increase system availability. Step five (5) is aimed at identifying the industrially required system responses; that is, how should the system respond when a certain distribution of system parameters is identified through the DSMS?

5.2 Proposed DSMS structure

Data stream mining can increase the availability of a product by detecting, predicting and avoiding failures. Figure 2 below shows how the services can be obtained using data stream mining. The lifecycle data of a product can be analyzed and interpreted to get the status of the product and also to answer queries from engineers. As presented in figure 2 below, he DSMS is responsible for controlling and managing the generated data stream, whereas the data stream mining algorithms are responsible for information extraction from continuous and rapidly arriving streams of data. The extracted information is then interpreted to obtain relevant knowledge. Finally, that knowledge may be made into a service by being presented in a post-processing interface according to the needs of the functional product supplier or customer.

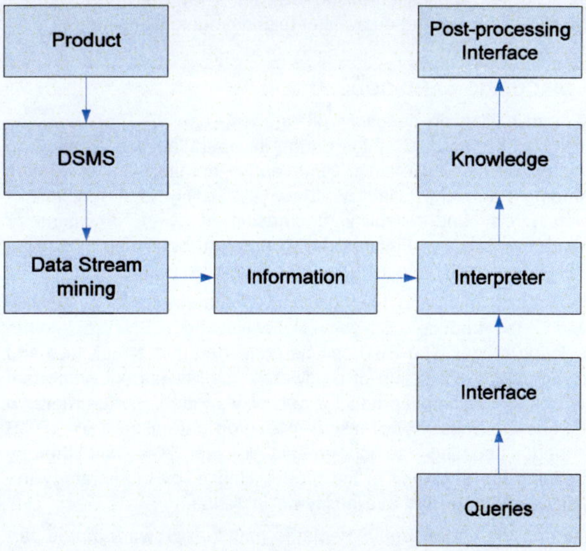

Figure 2: Flow chart showing how services can be obtained using DSMS and data stream mining.

To increase usability and to disseminate the DSMS approach among non-experts in industry, developing a flexible Graphical User Interface (GUI) seems pertinent. In such an interface, the streaming queries ("questions") can be visualized as well as the corresponding results ("answers"), thus presenting information to the user in a simpler, user-friendly way. For instance, a set of queries regarding functional product availability can be presented to an industrial engineer for combination as the engineer sees fit. Also, the interface could also be used to present post-processing data, for example, graphical variances between actual and agreed upon system availability.

5.3 Requirements and examples on DSMS technology

For these characteristics of data stream and to be able to apply the continuous queries on the input data streams the DSMS should fulfill the following requirements [27]:

- The order- and time-based operations should be allowed by the data model and query semantics.
- The streaming query plan must allow the results to be produced before the use of blocking operators.
- The algorithm should use a single pass over the data.
- Real-time monitoring applications should have real-time reactions to unusual data values.
- Long-running queries may encounter different system conditions.
- To achieve scalability, the shared execution of different continuous queries is necessary.
- Uses data summary (cannot store complete stream), which leads to inexact answers.

In order to handle the stream data a special query language, which is compatible with the structure of the DSMS, has to be defined. These queries are applied continuously and permanently on the data stream, producing a new stream or updating a permanent table [28]. SCSQ [30], STREAM [31], TelegraphCQ [32], and Aurora- Medusa-Borealis [33] are examples of general-purpose DSMSs that can be used for more than one purpose or application. On the other hand, Gigascope [34] and Hancock [35] are examples of specialized, application-based DSMSs. They were designed for network monitoring and analysis of telecommunications [28].

6 DISCUSSION AND CONCLUSION

The availability of a functional product can be improved by monitoring the product status during its operating period, allowing for preventive maintenance and early identification of issues regarding hardware reliability. Since availability is a function of maintainability and reliability, decreasing hardware downtime or Mean Time Between Failures (MTB) through monitoring and using a DSMS approach, also increases the hardware availability.

One way of improving the accuracy of the predicted availability would be to introduce a Bayesian Belief Network (BBN) [36] into the simulation model. The BBN can be converted into a fault tree and thereby enable prediction of hardware reliability (as one important part of the functional product availability model). Furthermore, a BBN has the possibility of taking measured data filtered by DSMS as input, updating the parameters in the BBN and thereby increasing the accuracy of the model. Hence, by this strategy the validity of the predicted availability will increase.

Based on the above requirements, to monitor hardware that is part of a functional product a number of sensors need to be installed to collect different data streams from hardware in operation. The data coming from these sensors should be ordered by using, for example, a time-stamp. To search the data stream the applied algorithm must be able to use only one pass over the data stream; otherwise, some data will be lost, which may influence the monitoring operation. Once abnormal data values are detected, real-time (or as fast as possible) reactions are necessary. It should also be noted that the generating rates of the data may affect the monitoring results; having high generating rates could give more accurate results.

By providing the required information on time, data stream mining can mitigate the probability of equipment failures and support continuous availability awareness of industrial systems. Such availability awareness is especially important when the supplier is to supply a function with certain availability such as a functional product.

7 ACKNOWLEDGMENTS

This work was conducted within the projects iStreams (Searching and analyzing on-line high volume industrial streams) and SSPI (Scalable search of product life cycle information). iStreams is funded by VINNOVA (Swedish Governmental Agency for Innovation Systems), Sweden. SSPI is funded by the Swedish foundation for Strategic Research. Further, this work has been partly carried out within the Faste Laboratory, funded by VINNOVA and based at Luleå University of Technology in Sweden. Finally, the authors would like to thank our industrial partners at Hägglunds Drives AB.

8 REFERENCES

[1] Brännström, O., Elfström, B-O., Thompson, G. (2001): Functional products create new demands on product development organisations. ICED 01, Glasgow.

[2] Rasgado, T., Thompson, G., Elfström, B-O. (2004): The design of functional (total care) products. Journal of Engineering Design, Vol. 15, No. 6, pp. 515–540.

[3] Müller, P., Sakao, T. (2010): Towards Consolidation on Product-Service Systems Design. CIRP Annals - Manufacturing Technology, pp. 219-225

[4] Baines, T.S. (2007): State-of-the-art in product-service systems, Journal of Engineering Manufacture Vol. 221, No. 10.

[5] Meier, H., Roy, R., Seliger, G. (2010): Industrial Product-Service systems-IPS² CIRP Annals - Manufacturing Technology, Vol. 59, No. 2, pp. 607-627.

[6] Henley, E.J., Kumamoto H. (1981): Reliability Engineering and Risk Assessment, Prentice-Hall, New York.

[7] Weibull W., (1961): Fatigue Testing and the Analysis of Results, Pergamon Press, New York.

[8] MIL-STD-1629a, (1980), Procedures for Performing a Failure Modes, Effects and Criticality, Analysis, US Department of Defence.

[9] Vesely, W.E. (1970): A Time-Dependent Methodology for Fault Tree Evaluation, Nuclear, Eng Design, Vol. 13, pp. 337.

[10] Billinton, R., Allan, R.N. (1992): Reliability Evaluation of Engineering Systems, Second Edition, Plenum Press.

[11] Li, J., Thompson, G. (2007): Service Reliability: Concepts and Models, Proceedings of ICRMS2007, Beijing, China.

[12] Li, J., Thompson, G. (2009): Simulation-based Reliability Assessment of Services in the Context of Functional Products, J. Safety and Reliab. Soc., Vol. 29, pp. 47.

[13] Reed S., Andrews J., Dunnett S., Karlberg M., Karlsson L. and Löfstrand M. (2010): Modelling Service Support System Reliability, 1st IFAC A-MEST, Lisbon, Portugal.

[14] Löfstrand, M., Andrews, J., Karlberg, M., Karlsson, L. (2011) Functional product system availability: simulation-driven design and operation through coupled multi-objective

optimization, Accepted for Publication, International Journal of Product Development.

[15] Kobayashi, M. (2004): *Data Stream Mining: selected tools & algorithms (Numerical Analysis and New Information Technology).* RIMS Kokyuroku. Vol.1362, pp. 124-132.

[16] Alzghoul, A., Löfstrand, M. (2010). Increasing availability of industrial systems through data stream mining. Computers & Industrial Engineering.

[17] Gaber, M., Zaslavsky, A., Krishnaswamy, S. (2005): Mining data streams: a review. ACM Sigmod Record, Vol. 34, No. 2, pp. 26.

[18] Karacal, S.C. (2006): *Data stream mining for machine reliability* in *IIE Annual Conference and Exhibition*.

[19] Karacal, S.C. (2007): *Mining machine data streams using statistical process monitoring techniques*. in *IIE Annual Conference and Expo*.

[20] Kargupta, H. (2004): *VEDAS : A mobile and distributed data stream mining system for real-time vehicle monitoring.* SIAM Proceedings Series, pp. 300-311.

[21] Matthews, B., Srivastava, A.N. (2008): *Comparative Analysis of Data-Driven Anomaly Detection Methods*. in *JANNAF Conference on Propulsion Systems*. Proceedings of the Joint Army Navy NASA Air Force Conference on Propulsion, Orlando, FL.

[22] Angelov, P., Giglio, V., Guardiola, C., Lughofer, E., Lujan, J.-M. (2006): *An approach to model-based fault detection in industrial measurement systems with application to engine test benches,* Measurement Science and Technology, Vol. 17, No. 7, pp. 1809-1818.

[23] Karacal, C., Cho, S., Yu, W. (2009): *Sensor stream mining for tool condition monitoring.* Computers and Industrial Engineering, pp. 1429-1433.

[24] Youree, R., Yalowitz, J., Corder, A., Ooi, T. (2008): *A multivariate statistical analysis technique for on-line fault prediction*, International conference on prognostics and health management.

[25] Wegerich, S.W.(2004): *Similarity based modeling of time synchronous averaged vibration signals for machinery health monitoring.* IEEE Aerospace Conference Proceedings, 6: pp. 3654-3662.

[26] Data, data everywhere (a special report on managing information), in The Economist. February 27th, 2010.

[27] Golab, L., Özsu, T. (2003): Issues in Stream Data Management, SIGMOD Records, Vol. 32, No. 2,

[28] Fogelman-Soulié, F. (2008): Data stream management and mining. Mining Massive Data Sets for Security: Advances in Data Mining, Search, Social Networks and Text Mining, and Their Applications to Security.

[29] [cited 2010 December, 28]; Available from: http://www.hagglunds.com/.

[30] Zeitler, E. and Risch, T. (2006): Processing high-volume stream queries on a supercomputer. in Proceedings of the 22nd International Conference on Data Engineering Workshops. IEEE Computer Society.

[31] Arasu, A., Babcock, B., Babu, S., Cieslewicz, J., Datar M., Ito, K., Motwani, R., Srivastava, U., Widom, J. (2004): Stream: The stanford data stream management system. a book on data stream management edited by Garofalakis, Gehrke, and Rastogi.

[32] Chandrasekaran, S., Cooper, O., Deshpande, A., Franklin, M., Hellerstein, J., Homg, W., Krishnamurthy, S., Madden, S., Reiss, F., Shah, M. (2003): TelegraphCQ: continuous dataflow processing, ACM.

[33] Abadi, D., Carney, D., Cetintemel, U., Cherniack, M., Convey, C., Lee, S., Stonebraker, M., Tatbul, N., Zdonik, S. (2003): Aurora: a new model and architecture for data stream management. The VLDB Journal, Vol. 12, No. 2, pp. 120-139.

[34] Cranor, C., Johnson, T., Spataschek, O., Shkapenyuk, V. (2003): Gigascope: A stream database for network applications, ACM.

[35] Cortes, C., Fisher, K., Pregibon, D., Rogers, A., Smith, F. (2000): Hancock: a language for extracting signatures from data streams. ACM.

[36] Ben-Gal, I.(2007): Bayesian Networks. In Ruggeri, F., Kenett, R., Faltin, F. (editors): Encyclopedia of Statistics in Quality and Reliability, John Wiley & Sons.

Industry Practices and Challenges in Using Product in Use Data to Inform PSS Conceptual Design

Romana Hussain[1], Helen Lockett[1], Gokula Annamalai Vasantha[2]

[1]School of Engineering, Cranfield University, Cranfield, UK.,

[2]Department of Manufacturing, School of Applied Sciences, Cranfield University, UK

Abstract

Both Product-Service System (PSS) literature and industry express a need to close the design loop by feeding product-in-use data to conceptual design. Nevertheless, how to actually accomplish this for PSS Conceptual Design is largely absent. This paper uses the literature and findings from interviews with industry as the basis for a system which could utilize product-in-use data from the sensors of large, capital-intensive, technical assets along with in-service records as well as receiver and provider needs to aid the creation of PSS Business Models at the PSS Conceptual Design stage.

Keywords:

CIRP International Conference; Life Cycle Engineering; Paper Instructions

1 INTRODUCTION

A requirement for military contracts and, increasingly, wider industry is the fulfilment of availability type contracts where the provider is paid by the customer for various service units such as for each kilometre or day of use of the asset. Such an offering can be delivered with a use-oriented PSS. A definition of PSS is offered by Baines et al [1] '*A PSS is an integrated product and service offering that delivers value in use.*' Use-oriented PSS Business models mean that the provider retains ownership of the asset [2] which can mean that whenever the system is unavailable that they can lose revenue and, for availability contracts, this could occur over a large timescale. Hence, it is necessary for the provider to determine if they have the capability to meet the demands of such contracts profitably. The aim of this research is to ascertain the current understanding, practices and challenges of using product-in-use data, particularly from sensors, for PSS Conceptual Design. From this, the issues are identified so that future steps can be taken to address them. Maintenance is focused upon because, as well as being the remedy, this can also be a cause of unavailability, Although optimising and reducing maintenance would decrease the provider's overall lifecycle costs and hence make offering availability contracts more viable for the provider, for PSS Conceptual Design it is particularly important to determine, given the existing provider capability and new customer needs, *what* needs to be optimised or redesigned in order to make the PSS solution viable in serving the customer's needs. This contrast with determining actually *how* this should be done as this will vary from asset to asset which is the function design. Designing for timely maintenance that can be provided, ideally, outside of the customer's business tasks (or, at least, accommodated during those times) will help to provide the value in use that the customer requires.

Assets containing sensors represent a new way in which the actual second by second performance and use of assets (as well as, perhaps, their contextual pressures) can be continually, accurately and immediately observed and recorded. Nevertheless, how to make use of this data to inform the conception of business models, are demanding challenges for industry and academia alike.

Sensor technology, artificial intelligence techniques, advanced telecommunications and new storage techniques which facilitate data mining as well as statistical analysis all represent innovative ways of capturing, transmitting, housing, immediately analysing, categorising and interpreting data from Infomated assets. Expert systems which utilise such data can also now afford a responsiveness to providers and customers as to how the asset is performing, how it is being used and the contexts within which it is installed. Nevertheless, there is a general absence in the literature regarding how to actually use such product-in-use data for PSS Conceptual design. Similarly, interviews with companies producing large, high value technical assets also delivered a consensus that, whilst this data can now be collected, how to use it for conceptual design has not yet been suitably devised or even fully understood.

This research is based on a series of face-to-face, semi-structured interviews with major stakeholders within companies producing Infomated, capital intensive, technical assets bundled with services. The companies who participated in this study are: aerospace, naval ships, land vehicle systems, an Infomated train and an Infomated truck company. The primary challenges regarding how to use product-in-use data to influence the conceptual design of availability of an asset design have been identified from these interviews across these industries are discussed in this paper.

The literature review presents an overview of the understanding of using sensor data for PSS conceptual design and the gaps therein. The methodology section details how the study of industry was conducted. The findings are presented in the results section. The conclusions from the interviews and literature are drawn under the following section and the discussion section outlines how sensor data could be used, system considerations and next steps to be taken.

2 LITERATURE SURVEY

Neely [3] refers to Infomated products as being those products which have diagnostics and prognostics technologies integrated within them. The term here is expanded slightly to cover all monito-

J. Hesselbach and C. Herrmann (eds.), *Functional Thinking for Value Creation: Proceedings of the 3rd CIRP International Conference on Industrial Product Service Systems, Technische Universität Braunschweig, Braunschweig, Germany, May 5th - 6th, 2011*,
DOI 10.1007/978-3-642-19689-8_43, © Springer-Verlag Berlin Heidelberg 2011

ring devices that can be embedded within a product. Maussang et al [4] refer to the preliminary design phase in PSS as PSS conceptual design and they state that here different solutions should be considered and compared with each other. Isaksson et al. [5] extends this by stating that customer interactions (indicative of needs) should come together with the provider's competences to develop conceptual solutions.

▪ *Closing the Design Loop with Lifecycle Data: The* emergence of new sensor technologies is a relatively recent event which affords the possibility of closing the design loop very closely with a constant stream of extremely detailed data [6]. Information is sparse at the start of design [7] and, although there is much design information at the product Beginning of Life (BOL), it tapers off after this stage. However, Infomated products allow data to be collected and fed back to inform design with regards to modes of use and operational data [6]. Such information can be collected at the middle of life (MOL) and the end of life (EOL) of products and so encourages a life-cycle approach to design as the product (from beginning to end of life) can be continually evaluated [7]. This way, how the availability of the product changes over time and how modes of use, types of use context and maintenance, repair and replacement (MRR) all affect that availability could be ascertained with the help of such data. The expected availability of the asset under these different circumstances could then be known and appropriate MRR offerings conceptualised for each customer and context.

▪ *Using Product- in-Use Data for Design:* Nevertheless, there is an absence within the literature as to how to actually use this data for PSS conceptual design. Although in the wider engineering literature there is a substantial body of research relating to collecting and analysing product usage data, this refers to health monitoring, prognostics and other information collection methods that can be used to acquire feedback about a product in use [8] [9] [10] [11]: here, the emphasis is mainly on predicting when maintenance should be administered and the avoidance of service interruption rather than closing the loop back to design. Furthermore, Sakao et al. [12] state that although user involvement methods can facilitate feedback of information to successfully operationally adapt offerings, this is not currently reflected in existing PSS-development methodologies.

▪ *Sensors Monitoring Context:* Although sensors tend to be incorporated to monitor when maintenance is required, design for incorporating sensors to ensure that feature use and context is also monitored to inform design would be necessary. Contextual data [13] is important as the efficacy of the product is only truly touched upon after the product has been integrated, attuned and familiarized by users. Moreover, the context (such as heat, dust, integration with other assets and so on) can impact upon the availability of the asset and lead to more specific types of MRR to be carried out. Additionally, if, as Ericson et al. [7] state, that people are often not aware of their needs and that their needs are usually embedded in their routines and actions, then how they actually interact with the world around them (including products) could be more indicative of what their underlying needs may be than anything they might say. This is useful for this research which focussed on existing, large, capital intensive, infomated assets; how the receiver actually uses these existing products could help to indicate their deeper needs.

▪ *Upgrades and Updates:* How and when maintenance is delivered could be drastically changed by modular design. There are several theories on how to prepare a common "platform" and exchangeable modules in a product [12]. Such a modular design could allow for the upgrading or updating of a product by switching modules; this way, the selected replacement module could deliver the necessary functionality and performance required by that particular customer or diminish the increasing frequency and time for maintenance across the life of the asset; removed modules could then be sent for MRR and the replacement modules could be ones which are more suitable for the context and use by that particular customer.

▪ *Responsiveness*: It could be considered that an Infomated product affords a channel of communication from the user, through the product to the provider. Therefore, it becomes possible that any shortfall in the product meeting the needs of the user could be detected and adjusted for by the provider. In this way, the relationship that providers and users have with products would become more service-like. Ajaujo and Spring [13] point out that although service relationships should be managed carefully with customers, that this should also be the case for new and complex products. The arrival of sensor technology means that a level of responsiveness can now be afforded through a product which was hitherto only ever seen with services. Grönroos [14] has stated that such responsiveness by the provider allows for more value in use to be created. Vargo and Lusch [15] have stated that it is service that is bought even when a tangible product is offered; this now becomes even more evident with the advent of such Infomated products and availability contracts. For maintenance this means that both the provider and customer could be informed immediately as to when any possible threats to the availability of the asset are occuring and about any corrective action that could be taken to circumvent or minimise this. This could increase availability and reduce the need for maintenance.

▪ *Designing for Value in Use:* Vargo and Lusch define this as '...a customer's outcome that is served directly through the product/service consumption.' [15] Therefore, it is important for the asset to deliver the required level of performance *at* the specific times the customer requires it for that outcome to be served. If the performance becomes so reduced that the asset becomes ineffectual for a specific customer, then, as far as that customer is concerned, the asset has become unavailable. Similarly, if the asset does not have the requisite performance at the exact times that the customer requires it to fulfil business operations, the asset (for that particular customer) could also be regarded as unavailable. In addition to this, if an improved offering can reduce the amount of sacrifices the receiver has to make, then this will free up their resources to be used in achieving their outcomes more easily; hence, such an offering would also facilitate more value in use.

▪ *System Development:* Yang et al. [16] have devised an intelligent product model which is an extension of IVHM-type systems for PSS systems to use lifecycle data to enable product-related services which could, with some modification, be used for Industrial Product Services Systems. The challenge now is to determine the rules to inform PSS Conceptual Design and the requisite statistical techniques to process product in use data to support these rules.

▪ *Requirements for PSS Provision and how Product in Use data could indicate these:* There is a need for the provider to be able to predict the behaviour of the user to enable a good lifecycle cost calculation [17] as well as predict, wherever possible, how to accommodate the changing of customer needs with time. This could be achieved by determining lead users from sensor data by using data mining [18] and predictive analytics (the use of confirmed relationships between variables to predict future events or behaviours) could be used which are techniques used in marketing research for product development [19]. Clustering could also help to group the data [20] into patterns of use and conditions. This could aid the conceptual design of interchangeable modules which are suited to various types of use and conditions. For very long availability contracts, these could prove to be valuable techniques.

The provider may also have to take steps to accommodate and limit more careless behaviour. [17]. Frequent abuse (such as using excessive force in operating the asset) could be calculated from sensor data and such misuse could then lead to the operator being retrained or parts of the asset being conceptually redesigned so that less misuse can occur. For availability contracts, as the price of the asset is recovered in instalments [17], the provider has to pre-finance capital costs. The impetus would be to ascertain if it is feasible for the provider to offer such a contract to a particular customer for their particular use context and what the pricing should be. To accomplish this, the total lifecycle costs need to be aggregated and then divided by the total amount of expected payments. Any savings that could be made on lifecycle costs could be shared by the provider and the customer.

For a pay-per use PSS, the ability to, with the receiver (customer, user and operator), agree upon a set of performance criteria and (to some degree) predict or influence their behaviour [17] is necessary. Influencing the behaviour of the user could be accomplished by providing an automated service to help the user fulfil tasks or help the user to use the machine without damaging it. With regards to determining the performance criteria and then ascertaining if the provider could deliver this, a method needs to be employed which can capture customer needs and then the performance range and availability patterns required for the customer's circumstances; this needs to be agreed upon between the provider and the customer along with other resources or limiting factors that stakeholders such as the receiver would need to offer or stipulate to create an overall solution to be charted. This way, the solution would be co-created by such stakeholders. The provider could ascertain the performances they had previously provided and these could be similarly charted. Statistical trends as to the performance range and the availability of existing assets under various conditions could be calculated and similarly charted which could then be compared to the performance range and availability patterns that the customer requires given their circumstances. The degree of difference between these could be expressed as a percentage to indicate how far away the provider is from supplying the performance range and the availability times the customer requires to fulfil their business operations. Although the Integrated Gap-Model is a valuable tool to describe service quality within an availability-oriented business model [21], another method to chart capability gaps could be more pertinent for PSS Conceptual Design: this should help the provider to ascertain if they could have the capability to offer availability models.

To conclude, closing the design loop can now be achieved very closely and immediately with sensor data but, from the literature, what this data could represent for PSS conceptual design purposes and what data should be collected is largely unknown. The monitoring of the use context is also important for PSS Conceptual Design to fully ascertain how an asset is actually being used and the contextual pressures upon the asset although IVHM (Integrated Vehicle Health Management) tends not to monitor this. Furthermore, modularity could allow for the provider to increase the availability, update and customise the asset to the needs of individual customers and their contexts but how such data could inform conceptual modular designs has not yet been fully identified. The potential responsiveness that could now be afforded to providers and customers as to how the use of the asset and contextual pressures on the asset could be adjusted to increase asset availability and performance is referred to but not , as yet, formulated. An expert system which could be used to process sensor, in-service records as well as receiver and provider needs for PSS Conceptual Design will require rule development and statistical techniques to support those rules. The use of a suitable methodology to chart the capability gaps between what the receiver requires and what the provider can offer is also required.

3 METHODOLOGY

The reviewed literature shows that the conceptual design of PSS is an understudied and nascent topic. The Methodological Fit Model [16] suggests that in studies where theory is nascent or immature, as researchers will not know what issues may emerge from the data, they should avoid hypothesizing specific relationships between variables. Since little is known, rich, detailed, and evocative data are required to shed light on the phenomenon. Therefore, we used semi structured interviews for this research in the aim of obtaining rich and detailed data from the interviews with no former themes.

Semi-structured interviews were held in six companies from a wide range of industries in order to gain a better understanding of the process in a broad industrial context. Another reason for selecting a cross-sectoral case study was to overcome the barriers (such as confidentiality issues) of conducting an in-depth study in one organisation. We approached many servitizing manufacturers of large, technical, capital-intensive, infomated assets and those who were interested were used in the study. All of the companies selected provide PSS to their customers, and all manufacture and support high-value products in business-to-business relationships. Additionally, we ensured that all the offerings of the interviewed manufacturers consisted of highly complex technological systems and subsystems. We focussed on the conceptual design of product-service systems and created cross-industry comparisons.

Table 1 shows the manufacturers and their respective industries which were investigated in this study. For reasons of confidentiality, the anonymity of the manufacturers and interviewees is maintained.

Provider ID	Industry	Interviewee	No. Of Interviewees	Time (hrs)
AeroCo	Defence Aerospace	Senior Management	2	3.0
RailCo	Rail Transportation	Senior Management	3	6.0
TruckCo	Commercial Trucks	Senior Management	5	7.0
LandCo	Defence Land Systems	Senior Management	1	3.0
PlaneCo	Aerospace	Senior Management	4	4.5
NavalCo	Defence Surface Ships	Senior Management	2	4.5

Table 1: Industry Interviews.

An interview protocol was developed and used as the basis for the interviews; however, due to the open-ended nature of the interviews, the questions were not necessarily asked in order. Prior to the data collection, a pilot interview was carried out with three senior academics with practical expertise in engineering design, to ensure that the questions to be asked were clear and comprehensible. As a result, some questions were rephrased and some questions were removed in order to avoid repetition. A total of over 28 hours of semi-structured interviews were conducted with senior managers from each company. All three researchers were present in every interview. The sessions were audio recorded and transcribed. In addition to primary data collection, secondary sources such as archival data, internal documents and brochures were also collected in order to achieve theoretical triangulation [22]. Then, the transcripts and all other documents were read carefully and coded by each researcher using a grounded approach with no a priori definitional codes [23]. This enabled the data to be 'systematically analysed so as to tease out themes, patterns and categories that will be declared in the findings' [24] Then all related codes from each document (including primary and secondary data sources) were listed by all researchers. At the end of this process, the findings were categorized under emerging themes. The results of the analysis were then reflected back to the companies for validation in the form of reports. The reports were then updated to reflect the received comments and re-circulated to the companies.

The results presented in this paper are those which relate specifically to using product in use data from sensors for PSS Conceptual

Design and, most specifically, to maintenance, repair and replacement (MRR) and availability.

4 RESULTS

Below, the main issues from the interviews where Product in Use data could be used to inform the inception and facilitate the on going delivery of PSS Models. There was some feedback (from one or two respondents for each issue) as how this data could generally improve MRR which is shown in Table 4. Nevertheless, very little insight or understanding as to how such data could inform conceptual design was forthcoming (across all respondents) as shown in Table 3. Issues that are common to generally reducing or improving MRR and that also should be addressed for PSS Models to be offered are shown in Table 2.

Issue Identified	Industry Feedback
Incorrect use of the asset can damage it.	For infomated trucks, how the trucks are driven is monitored. For example, drivers who consistently brake harshly are sent to driver training. Driving feedback is given to them periodically rather than immediately.
Fault finding across different assets: analysing data across data Sets	Analysing data across data sets: some systems on some aircraft are quite reliable whereas others are not and it is difficult to find a reason for this. The Environmental Control System is a particular problem in some aircraft but not in others.
Finding the source of faults and correcting them rather than the symptoms	The depiction of emerging trends is desired. Knowledge as to whether a particular item has failed due to aging effects, fatigue, a change in maintenance provider (or member of staff) or spares supplier has changed and so forth. Understanding the source of the fault is imperative. At the moment, the symptoms are often addressed rather than the underlying problem – this sort of 'quick fix' happens quite often as it would take time to diagnose and correct the real fault. If the underlying problem starts to be addressed suddenly the spares required will change. Therefore, reliability as a whole should be considered not just quick-fixes which only temporarily dispel the problem, often at high cost and possible long term damage.

Table 2: Issues common to reducing or improving MRR and for PSS Conceptual Design.

Issue Identified	Industry Feedback
Use context monitoring	For military land vehicles, there is a drive to monitor, along with the vehicle, elements of the use context and other systems which have little bearing on the main asset. Furthermore, as military land vehicles often operate in fleets, there is also a desire to analyse data sets across vehicles. Such monitoring can help determine the conditions under which failures can occur.
Past performance data to inform the inception of PSS business models	For the provider to offer such a business model, past performance data on how available the asset was under various conditions, applied to different uses and operated in various ways is extremely important. If the provider is paid for each kilometre or hour that the asset is used, the cost of providing each unit has to be calculated and it would benefit the provider to reduce this cost.
Using product in use data for conceptual design	The consensus across the industries is that this is largely unknown although it is something that they wish to encourage.

Table 3: How Product in Use data is or could be used for PSS Conceptual Design.

Issues Identified	Industry Feedback
Concern over the customer performing maintenance	Some maintenance takes place at the customer's site. The provider often has concerns as to whether the customers are conducting the maintenance correctly. If not, it could mean that the asset loses availability and it could mean more maintenance work for the provider.
Desire to create fault trees which can be used for interactive diagnosis of faults and linked to maintenance operatives and spares	There are pilots who operate a symptom capture tool. This is a touch screen tablet which allows for Intelligent fault diagnosis - . The tool asks which system/function did not work. The pilot then touches an option and is asked what happened. The pilot is then asked whether there was a warning from that system – the pilot can then choose which warnings were given. In effect, the system builds up a fault code by a series of steps that follows a logic tree. This can help a correct diagnosis to be made. The system can link to the spares ordering system to make a spare available or to order one. When the aircraft lands, the maintenance operative can arrive with the correct spares, maintenance instructions and tools. More development on this system is desired.
When to upgrade or update	When maintenance is performed, it is desirable for upgrades to also be performed at the same time rather than at different times.
General optimisation of maintenance	There is a need to concentrate on activities in maintenance which are adding value and remove waste. Some industries are putting a lot of effort into lean initiatives.
Administering correct maintenance	Maintenance manuals need to be kept up to date; they are not always kept up to date even when upgrades occur. Sometimes maintenance staff do not read the new manuals.
Issues with maintenance training because of high staff turn over	High turnover of maintenance staff can be a problem. This then leads to a need to train new staff along with the loss suffered from experience staff leaving.
Incorrect maintenance can damage the asset	For some assets, it is possible that use of the wrong tools during maintenance could lead to the asset becoming damaged.

Table 4: Issues relating to reducing or improving MRR.

5 CONCLUSIONS FROM INDUSTRY INTERVIEWS

The industry interviews have identified the main challenges for the creation of a method and a system to utilise product-in use data for PSS Conceptual Design. Such a system would require:

- An understanding of what the data could represent and how this could inform PSS conceptual design.
- A method to build-up precise knowledge of the performance of the asset under various circumstances, how it became reduced and the part of the asset which had the fault. This is extremely important for the creation of availability contracts.
- Contextual monitoring (such as temperature and humidity) which currently is often not performed.
- Analysis across data sets of different assets could help to identify possible causes of failure under different contexts.

- To devise effective ways to issue feedback regarding use of the asset which could damage it.

Such a system could also help to make MRR more cost-effective:

- To monitor and link asset availability directly after maintenance with the maintenance operatives who administered the maintenance and why it was administered in that way. This way, the course of action taken and its effectiveness can be determined so that steps could be taken for improvement.
- To, if possible, suggest the tools that should be used for the maintenance job and to identify those that were used.
- Ensure that upgrades are performed, when possible, during maintenance delivery. Condition based maintenance should allow this window of opportunity to be predicted.
- To find a way to ensure that changes in manuals are read by maintenance operatives.
- To ensure that the correct tools are used for maintenance.
- To depict how a fault is built-up as precisely as possible and to link this to the spares ordering system as well as available maintenance operatives.
- To link faults to possible causes and suggest the correct course of action to correct the cause rather than the symptoms.
- To foster effective ways to train maintenance staff.

The responses tended to highlight issues with maintenance which could help in its reduction and improvement and so generally reduce the provider's lifecycle costs. Whilst this is important for PSS, the interview revealed that, in industry, little is truly known about how product in use data could be utilised for PSS Conceptual Design. Knowledge of the customer's business operation times, the required performances, the customern resources that are required to make the offering work for them all need to be gleaned along with precise information as to how the provider's existing assets perform under various contexts. All of this is necessary to facilitate value in use for the customer and thus provide strong PSS availability models. This is particularly important here because, as the customer will not own the asset, they will not be able to modify or heavily integrate the asset to tailor it to their exact needs. Compared with the interviews, the literature indicated much more on how product in use knowledge could be used to support PSS Capability based Conceptual Design. Furthermore, although the literature deems contextual monitoring to be important, this tends not be done in practice.

6 DISCUSSION AND PROPOSALS

6.1 How Product in Use Data could be Used

Expert systems remotely accessing sensor data could be used to enhance existing symptom capture systems. This could automate the whole process and remove operator intervention in locating the source of a fault. This situation would be desirable as this would eliminate operator error as well as not impinge upon the operator's time. Alternatively, a modification of existing IVHM could be used.

Possible ways to improve existing maintenance :

- If electronic manuals were employed, it could be monitored when maintenance manuals were read and by whom. Such manuals could be updated constantly. It would also be possible to link training manuals and suggestions for the correct course of action, recommend tools to be used and also present animations and videos depicting MRR procedures.
- For any defect, a maintenance request could be automatically raised and an available maintenance operative assigned to the task. Ideally, the subsystem behaviour (under the specific conditions) could be linked to appropriate sections in electronic maintenance manuals which could suggest a correct course of action to the maintenance operative. The system could allow the maintenance operative to then log in and if the suggested course of action was not taken this could be flagged and the operative could enter the reason for this. If alternative courses of action were taken consistently for the same fault then this could point to the need for that section of the manual to be revised or point to the presence of another variable that is affecting the asset which the system should then try to model. If the operative is consistently not following the manual, this could point to the need for that operative to be retrained. RFID tags on subsystems or components could allow their replacement to be monitored. Similarly, RFID tags on tools could also allow the monitoring of the tools which were used for each job.
- If asset performance after maintenance as well as the time taken for maintenance could be linked to individual maintenance operatives, then the effect and cost–benefits of experienced staff could be calculated. Also, operatives who had underperformed could be retrained. Online manuals and training could help here as well as, if possible, virtual training and simulation.

Possible ways to improve data capture to inform PSS conceptual design using product-in-use data, particularly from sensors

- If all of the variables that could affect availability could be ascertained then monitored and compared across different assets, it then becomes more possible to discover the circumstances that leads to failure within that subsystem.
- Any asset designed to use product-in-use data should allow a fault tree to be constructed which would show how the fault is built up and where the source of the fault is. Trends of variables that could have affected the asset and thus caused the fault could also be depicted. For example, if it was known (from past performance data) that a particular component tends to develop a defect under high temperatures, then if the sensor data showed that the component was being subjected to such temperatures, then perhaps the provider or customer could try to adjust the temperature before any damage occurs. Of course, any other unusual readings and trends occurring at the same time could also be noted to see if there are consistent correlations with defects. This could help inform the customer and provider as to the conditions which affect the asset. It could also inform conceptual design so that the subsystem affected and corresponding maintenance could be redesigned for specific uses and contexts.

6.2 Challenges for Utilising Product in Use Data:

As the utilisation of product in use data to influence PSS Conceptual Design is a novel idea, there are several challenges to designing systems to collect and interpret such data usefully. The current issues are described below:

Understanding the need for the data: At present, as there appears to be no true understanding as to how product in use data could be used to influence design, it s very difficult to define what data would be useful to collect. To sum up, the need for collecting and the methods to utilise product in use data have not been fully demonstrated as yet. Demonstrating how such data can usefully influence design and provide customised support to customers could win organisational commitment to take further steps.

Understanding of what the data could represent: This is expected to be an ongoing challenge where methods and concepts used for interpreting the data will become increasingly refined with time: at the moment, it is not always clear what the data could be indicating.

Therefore, the validity of any concepts or methods which treat the data may have to be corroborated with other methods.

Platform Considerations: As it becomes more understood as to how to treat the data, finer requirements for the necessary hardware and software are expected. This is seen as a convergence of the understanding of the various stakeholder's needs, the development of theory and techniques to treat product in use data for conceptual design and the advancement of relevant technology (such as sensors and artificial intelligence techniques). It is noted that data mining requires huge amounts of storage and memory capacity so this also has a bearing on platform design.

6.3 Conclusions and Future Work

- Further studies to determine the issues of customers, operators, users and designers are planned. Additionally, the conclusions from the literature with regards to what is required to inform the inception of PSS Conceptual Designs and how statistical analysis of product in use data could support this will be validated with industry.
- Analysis of the literature to evaluate the methods used for manual in-service information for design purposes and to determine as to how this could be adapted for product in use data to inform PSS Conceptual Design.
- Further analysis of the literature to evaluate how conceptual modular design could be informed by product in use data.
- The identification or development of a method to chart capability gaps for PSS Conceptual Design.

7 ACKNOWLEDGEMENTS

This research was undertaken as part of the "Capability based PSS conceptual design (PSS-Design)" project funded by the Cranfield Innovative Manufacturing Research Centre (EPSRC Grant EP/E001874/1).

8 REFERENCES

[1] Baines, T. S., Lightfoot, H. W., S Evans, Neely, A., Greenough, R., Peppard, J., Roy, R., Shehab, E., Braganza, A., Tiwari, A ., Alcock, J. R., Angus, J. P., Bastl, M., Cousens, A., Irving, P., Johnson, M., Kingston, J., Lockett, H., Martinez, V., Michele, P., Tranfield, D., Walton, I. M., Wilson, H. : (2007) : State-Of-The-Art In Product-Service Systems, Proc. IMechE Vol. 221 Part B: J. Engineering Manufacture, pp. 1543-1552.

[2] Tukker, A. and Tischner, U. (2006): New Business for Old Europe, Greenleaf Publishing (UK), pp. 26.

[3] Neely, A. (2007): The Servitization of Manufacturing: an Analysis of Global Trends, 14th European Operations Management Association Conference, Ankara, Turkey.

[4] Maussang, N., Zwolinski, P., Brissaud, D. (2009): Product-Service System Design Methodology: from the PSS Architecture Design to the Products Specifications, Journal of Engineering Design, Volume 20, Issue 4, August 2009, pp. 349-366.

[5] Isaksson, O., Larsson, T. C., Rönnbäck, A., Ö. (2009): Development of Product-Service Systems: Challenges and Opportunities for the Manufacturing Firm, Journal of Engineering Design, Volume 20, Issue 4, August 2009, pp. 329-348

[6] Jun., H., Kiritsis, D., Xirouchakis, P. (2007): Research Issues on Closed-Loop PLM, Computers in Industry, Vol. 58, No. 8-9. (December 2007), pp. 855-868.

[7] Ericson, Å., Müller, P., Larsson, T., Stark, R. (2009): Customer Needs to Requirements in Early Development Phases, Proceedings of the 1st CIRP Industrial Product-Service Systems (IPS^2) Conference, Cranfield University, 1-2 April 2009, pp. 62

[8] Boller, C. (2000): Next Generation Structural Health Monitoring and its Integration into Aircraft Design, International Journal of Systems Science, Vol. 31, No. 11, pp. 1333 – 1349.

[9] Chang, S. E, Changchien, S. W, Huang, R-H. (2006): Assessing Users' Product-Specific Knowledge for Personalization in Electronic Commerce, Expert Systems with Applications, Vol. 30, No. 4, pp. 682-693.

[10] Hoske, M. T. (2006): Healthy Machines [Computerised Monitoring - Preventive Maintenance], Control Engineering, Vol. 53, No. 3, pp. 40-46.

[11] Wang. W, Tse, P.W, Lee, J. (2007): Remote Machine Maintenance System Through Internet and Mobile Communication, International Journal of Advanced Manufacturing Technology, Vol. 31, No.7-8, pp. 783 – 789.

[12] Sakao, T., Fargnoli, M. (2010): Customization in Ecodesign, Journal of Industrial Ecology, Volume 14, Issue 4, pp. 529–532.

[13] Araujo, L., Spring, M. (2006): Services, Products, and the Institutional Structure of Production, Industrial Marketing Management , Vol. 35, pp. 797 – 805.

[14] Grönroos, C. (2008): Service Logic Revisited: Who Creates Value? And Who Co-Creates?, European Business Review Vol. 20, No. 4, pp. 298-314.

[15] Vargo, S. L., Lusch, Robert F. (2004): Evolving to a New Dominant Logic for Marketing, Journal of Marketing, Vol. 68 (January), pp. 1-17

[16] Yang., X, Moore, P. Chong, S. K., (2009): "Intelligent products: From lifecycle data acquisition to enabling product-related services.", Computers in Industry 60184–194

[17] Tukker, A. and Tischner, U. (2006): New Business for Old Europe, Greenleaf Publishing (UK), pp. 50-53.

[18] von Hippel, E. (2005): Democratizing Innovation. MIT Press: Cambridge, MA (Free download by Creative Commons), http://web.mit.edu/evhippel/www/democ1.htm, accessed 2008-06-12, pp.34-44.

[19] Hair, J. (2007): Knowledge creation in marketing: the role of predictive analytics, European Business Review, Vol. 19 No. 4, pp. 303-315

[20] Cook, D. J. (2007): Making Sense of Sensor Data, Pervasive Comp., Vol. 6, No. 2, pp.105–8.

[21] Meier, H., Roy, R., Seliger, G. (2010): Industrial Product-Service Systems - IPS^2, CIRP Annals – Manufacturing Technology, Vol. 59, No. 2, pp. 607-627.

[22] Edmondson, A. C., McManus, S. E. (2007): Methodological Fit in Management Field Research, The Academy of Management Review 32, no. 4 (October 2007), pp. 1155-1179.

[23] Yin, R. K. (2003): Case Study Research: Design and Methods (3rd ed.). Thousand Oaks, CA: Sage, pp. 30-31.

[24] Easterby-Smith, M., Thorpe, R., Jackson, P.R. (2008), Management Research – Third Edition, SAGE Publications, London, pp. 40-45

Pathways of Servitisation in Mobility Systems

Christoph Herrmann[1], Katrin Kuntzky[1], Julian Stehr[1]

[1] Institute of Machine Tools and Production Technology, Product- and Life-Cycle Management Research Group, Technische Universität Braunschweig, Braunschweig, Germany

Abstract

The automotive industry is confronted with an increasing servitisation and rising mobility needs. At the same time the rapid development in information technology sets up new possibilities to create telematic solutions and to strike new paths in the development of mobility concepts.

This paper points out the developments in individual transport of passengers and the progress in information technology development and telematics. These two lines of development are merged into the development of mobility as a system of products and services. Using the method of cladistics, the relationship between various mobility enablers, and telematic solutions is explored. The results of the mutual impacts and co-evolution are analysed.

Keywords:

Automotive Service Systems; Cladistics; Telematics

1 INTRODUCTION

The mobility need (i.e. the need for the shifting of distance [1]) for private transportation is altering rapidly with the technical development and living standards in industrial countries as well as in emerging markets [2]. Therefore, the automotive industrial sector is more intensively exposed to changes and trends than most other sectors. It is particularly confronted with an increasing servitisation, the adding of value through services to the core product [3], due to a rising amount of customers who are no longer interested in owning cars but in the fulfilment of their needs for mobility [4]. Simultaneously, the mobility in terms of the covered distance of each individual is rising significantly [5, 1]. Thus, automotive companies are forced to develop and launch innovative mobility concepts that combine customer required services with their vehicle concepts.

At the same time the rapid development in information technology sets up new possibilities to create telematic solutions in automotive industry. The term telematics integrates telecommunication and informatics, whereas the science telematics means the integration of telecommunication, automatics and information technology [6]. It describes the use of modern technology to increase the efficiency in transportation processes, to enhance security and travel comfort and to reduce the environmental impact. A central task of telematics is the traffic management through information, communication, regulation and monitoring [1].

These overlapping and complementing progresses allow automotive manufacturers to strike new paths in the development of mobility concepts as a combination of the vehicle, telematic systems and service offers.

2 MOBILITY SYSTEMS AND TELEMATICS

2.1 Mobility systems development

Mobility is a basic need for human beings and a basic requirement for economic activities [7]. The goal to move faster, longer distances, with heavy goods and without physical efforts made humans use the available technical options at every time in history. Starting with ships moved by manpower and wind or wagons pulled by animals, the railway caused the 1st revolution in transportation and human mobility. With the development of the car at the end of the 19th century the second transport revolution leads to more individual and flexible mobility [1].

Today, 80% of passenger transportation (in kilometres) in industry nations like e.g. Germany is realised with a car [2]. The individualization of transport has lead to a reduction of service integration in personal traffic. Figure 1 displays the development of passenger traffic in Germany since 1820.

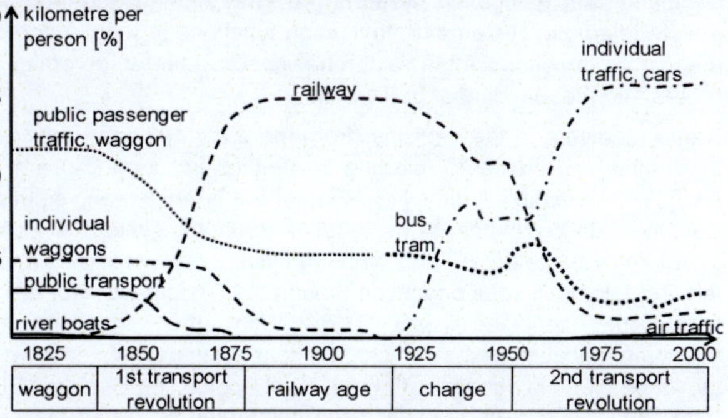

Figure 1: Passenger traffic in Germany since 1820 [1, 8].

But during the last years changes in the choice of transportation means can be identified. In Germany, especially for the younger generation a car is no longer a status symbol or an expression of the life style [4]. Rising ecological awareness vary the choice for transportation means significantly [4]. More and more customers are just interested in the satisfaction of their "mobility needs" instead of possessing the physical product. The trend leads to a demand-driven request of the needed mobility function [9]. The car sharing market in Germany is for example expanding (Figure 2) [4]. High growth rates of 79% in 2007 can also be observed in other important markets like the USA [10].

New mobility concepts, that represent an allocation of the function mobility to the customer, occur on the market. This function is realized with the integrative combination of a mobility-enabling product (e.g. car) and a mobility-providing service (e.g. a sharing service) (Figure 3). It can be understood as a Product-Service-System (PSS) [9].

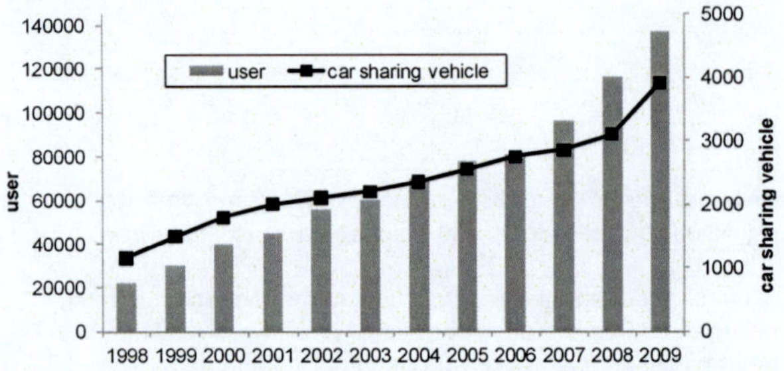

Figure 2 : Development in car sharing market in Germany [4].

The offer of mobility concepts are only rendered possible due to the technical development in information technology and telematics. The availability of control units to monitor vehicle conditions and positions has e.g. fostered the car sharing concept. It assures economic reasonable and user-optimised systems [11].

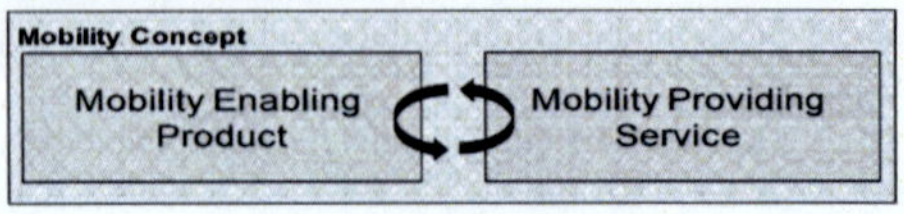

Figure 3: Components of a mobility concept [9].

2.2 Telematic systems development

For telematic systems in general the market was rising with the factor 30 within the last 15 years. The market for applications and services grew very slowly in the first years, but it is increasing enormously since 2000 [12]. 37% of new cars today are already equipped with navigation systems [4]. They extend, together with mobile telematic systems or navigation functions in mobile phones, the car by functions for a faster navigation, shorter or optimized routes and the avoiding of traffic jams.

Radio receivers in the vehicles (from the early 1930ies) and traffic information on the radio (starting in 1960ies in Germany) are the basic for telematic systems [13]. One of the first telematic solutions was the “trip computer” from 1980, a computer system to collect travel and vehicle data. This computer system is identified as one of the first telematic solutions, even though the system was not able to send the collected data [12]. With the integration of telecommunication in the vehicle, first total telematic systems where available [12]. Navigation with the Global Positioning System (GPS) (first available in 1989) [13] and the integration of Traffic Message Channel (TMC) in the early 1990ies [12] further expand the options for the application of telematic systems. With the online capability of modern telematic devices by Wireless Application Protocol (WAP) or General Packet Radio Service (GPRS), the sending and receiving of almost any form of data has become possible [12].

Modern traffic management systems like the Vehicle Information and Communication System (VICS) in Japan measure the current traffic situation and generate real time traffic information for the traffic participants [13]. Systems for commercial vehicles, such as the Brabender R2©, enable the combination of tracking, monitoring and reporting to workshops, navigation, information, communication and order processing as well as time keeping [14].

3 CLADISTICS

Cladistics has its origin in biology and is an approach of systematics. In general, it is a useful approach to organize any comparative data in discrete groups and sub-groups and visualize the results in dendroid graphs, so-called cladograms. It is especially used to reconstruct genealogies of organisms and to construct classifications [15]. The main purpose of cladistic analysis is to hypothesize the phylogenetic kinship of the units of analysis (taxa). The resulting branching diagrams (cladograms; see Figure 4) display the evolutionary relationships between the taxa. Therefore, it is based on Charles Darwin’s concept of ancestors and descendants [16].

Originally, a cladogram does not explain the evolution of the analysed systems and has no implied time axis. It is a graph that summarizes a pattern of character distribution and therewith a statement of relationship but does not include any implication of ancestry and descent. For being a statement of evolution additional assumptions like “evolution is parsimonious” have to be made [16]. Additionally, the characters that determine the groups are not eternal and inevitable. Therefore, a conventional cladistic analysis does not use units of analysis from different times to put it in the classification. Thus, cladistics establishes classifications of units existing at present that are based on evolutionary principles, respectively the process of evolution [17, 18].

Three axioms were postulated to justify the cladistic approach: (1) evolution occurs, (2) one phylogeny (the evolutionary course of a genus) of all organisms exists that is the result of inheritance, (3) properties can be transferred to the next generation via genealogical inheritance (modified or unmodified) [18]. These conditions have to be fulfilled for systems that are analysed using cladistics.

One procedure for the conduction of a cladistic analysis is structured as follows [18, 19]:

1. Selection of investigational group (clade): The selected objects for analysis (taxa) have to represent a clade. This is a group with the common ancestor and all his decedents. Therewith the evolutionary relationships between the taxa can be investigated. The taxa do not have to be biological entities. They just have to show “evolution” and a pattern of common ancestry.
2. Determination and coding of characters: The features of the taxa are selected that can be used to distinguish them among each other and describe the relationships between the taxa. Furthermore, the possible states of each character are established. These states are the values of a character perceived in at least one taxon. The characters can be binary (two states) or multistate (more than two states). Additionally, a polarization (determination of ancestral and derived states for each character) is conducted and the character order and direction (allowed character state transitions) is determined. The state that occurs in the ancestral morphotype is called plesiomorphic and the derived state is called apomorphic (away from the ancestral morphology).
3. Construction of cladogram: The construction of the cladogram can be conducted by using miscellaneous methods. Exact methods as well as heuristics (for large datasets) are applied. In most cases, the target of cladogram construction is to find the cladogram with minimal length because this cladogram has the best representative hypothesis of the taxa relationship as well as parsimony. The length of the cladogram is calculated by the number of steps (character state changes) needed for the generation of a cladogram for the given input data. Sometimes more than one cladogram with the minimum length exists. In

these cases, e.g. a consensus tree can be constructed to get one cladogram with the structure that fits best to the different solutions [16].

4. Interpretation of cladogram: The cladistic approach and the resulting cladogram are used for analyzing the characters and the relationships of the taxa. Based on this, the original objective of cladistics is to draw conclusions on the phylogenetic kinship on the taxa. Divergent from this application in biology, [20] additionally draws conclusions regarding the historical development and evolution of the observed systems. [19] also uses the cladogram to test the different hypotheses about the evolution process of the studied group. Thus, behind this background it could be used to verify whether the conception of the future development is consistent to the pathways of previous development. Additionally, the analysis of the explored area provides a structured overview that could allow a pro-active course of action for the creation of future developments in the area.

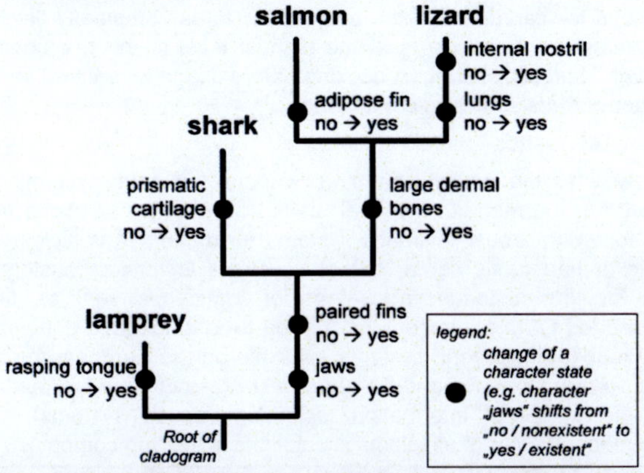

Figure 4 : Example of cladogram (based on [16]).

Despite the fact that the cladistic approach has its origin and most common application in the field of biological systematic [16], it also can be used for analyzing technical systems. An explicit application of cladistics for a technical system (cylinder block) is presented in [21]. The objective of the approach consists in supporting the product design via identification of evolution trends etc. [22] takes up the idea of cladistic classification to group evolving products and product families and faces it with a cladogram for the manufacturing system considering the manufacturing capabilities. The symbiotic interdependency of the two systems is demonstrated and formulated as a co-evolution model. The model shall support an efficient co-development of products and manufacturing capabilities, relates the past generations to the present and fosters the identification of promising and viable future evolution directions.

A further application of cladistics deals with aerospace supply chains. The approach is used to develop an original and innovative benchmarking classification scheme for this sector [23]. [24] applies cladistics for visualizing and explaining the evolution of the hand tool industry. Furthermore, the key strategic decisions for the different types of hand tool manufacturers are identified. [25], [26] and [27] use cladistics for the analysis of manufacturing systems.

There is no clear opinion about the advantageousness respectively fields of application of cladistic analyses for non-biological systems. But it is evident that the systematic and structured classification gained using cladistics enables the organization of data and thoughts regarding reality and therewith is a good tool for mental clarification and communication. Furthermore, the classification allows maximum information retrieval and could have a predictive value [26].

4 CLADISTIC ANALYSIS OF MOBILITY AND TELEMATICS

The investigational groups of street mobility and telematics are classified and analyzed using the cladistic approach. Therewith, technological developments within both areas are displayed. Based on that, interdependencies between the areas are highlighted and opportunities for future mobility concepts offered by telematic technologies are identified in section 5.

4.1 Street Mobility

Various systems that enable mobility can be identified. To meet the cladistic requirements the clade is defined as follows: traffic systems with traffic carrier "street"; no objects with focus on fun or sports (e.g. skateboard); transportation of people from one location to another. The selected taxa are displayed in the rows of Table 2. The common ancestor of all mobility solutions is "walker". A special focus is on mobility systems that can be summarized under the keyword "mobility concepts".

The characters of the systems and their occurring states are shown in Table 1. Besides conventional characters like the velocity and number of passengers, the ownership structure and the estimation of the environmental impact as well as the complexity of service represent newer demands that result for example in innovative mobility concepts. Table 2 presents the mapping of systems and character states that is the starting point for the cladogram construction. The environmental impact and variable costs are rough estimations based on comparisons of the several mobility systems.

		0	1	2	3	4
1	type of drive	muscularity	muscularity (animals)	combustion drive	electric drive	
2	velocity	<10 km/h	<30km/h	<60 km/h	<100km/h	>100km/h
3	ownership structure	ownership	rental of a seat	long-term rental	short-term rental	
4	distance of travel	short distance	mid distance	long distance		
5	passenger	1	<10	>10		
6	point of departure	completely flexible (entry door)	flexible station	fixed station		
7	point of arrival	completely flexible (entry door)	fixed station			
8	time-flexibility	flexible	schedule			
9	driver	self-driven	passenger			
10	initial investment for user	no	yes			
11	variable	none	low	medium	high	very high
12	environmental impact	very good	good	satisfactor y	fair	poor
13	booking	not required	required			
14	complexity of service	none	low	medium	high	
15	route	flexible	fix			

Table 1: Characters and corresponding states of mobility enablers.

Figure 5 shows one solution for the dataset. The cladogram was generated using the software Winclada [28]. It consists of 63 steps (character state changes) and represents one hypothesis of the (ancestral) relationship between the mobility systems.

The "walker" represents the starting point of the development. Then, two big groups can be differentiated: mobility enabling systems where the user is the owner and therefore an initial investment is necessary, and systems with a higher level of servitisation, where the consumer uses the mobility enabler via long-time or short-time rental. Within the group of servitisation highly-developed systems are rental car, car-sharing, and car2go© at which there is a close relationship between rental car and car sharing.

	type of drive	velocity	ownership structure	distance of travel	passenger	point of departure	point of arrival	time-flexibility	driver	initial investment	variable costs	environmental impact	booking	service complexity	route
walker	0	0	0	0	0	0	0	0	0	0	0	0	0	0	0
bicycle	0	1	0	1	0	0	0	0	0	1	1	0	0	0	0
rental bike	0	1	3	1	0	2	1	0	0	0	2	0	0	2	0
rickshaw	0	0	1	0	1	1	0	0	1	0	2	0	0	1	0
horse	1	1	0	1	0	0	0	0	0	1	1	0	0	0	0
motor scooter	2	3	0	1	0	0	0	0	0	1	2	3	0	0	0
motor bike	2	4	0	2	1	0	0	0	0	1	2	3	0	0	0
(public) bus	2	3	1	2	2	2	1	1	1	0	2	2	0	2	1
E-Car	3	4	0	2	1	0	0	0	0	1	3	3	0	0	0
conv. car	2	4	0	2	1	0	0	0	0	1	3	4	0	0	0
taxi	2	4	1	1	1	1	0	0	1	0	4	3	0	2	0
rental car	2	4	2	2	1	2	1	0	0	0	4	3	1	2	0
car-sharing car	2	4	3	2	1	2	1	0	0	0	3	3	1	3	0
segway®	3	1	0	0	0	0	0	0	0	1	1	1	0	0	0
car2go© concept	2	4	3	2	1	1	0	0	0	0	3	3	0	3	0
motor rickshaw	2	2	1	1	1	1	0	0	1	0	2	2	0	1	0
carriage	1	0	0	1	1	0	0	0	0	1	1	0	0	0	0

Table 2: Mapping of systems and characters for mobility enablers.

Both sectors start with bicycle-based mobility enablers, which have low environmental impact, and are driven by muscularity. In the sector of ownership the next development phases are horses and carriages, where the muscularity of animals is used as drive. Carriages come with higher velocity and passengers than horses. The type of drive is changing in the development of this branch to motor based drives; therefore, the environmental impact is rising as well. The branch of segway® has electric drives and is made for short distances, whereas motor scooter, motor bikes and convenient cars use combustion engines. The type of drive is changing to electric engine with the branch of E-car again. Going from left to right, the environmental impact is rising from the segway® to the convenient car as well as the distances of travel, the velocity, the variable costs and the number of passengers.

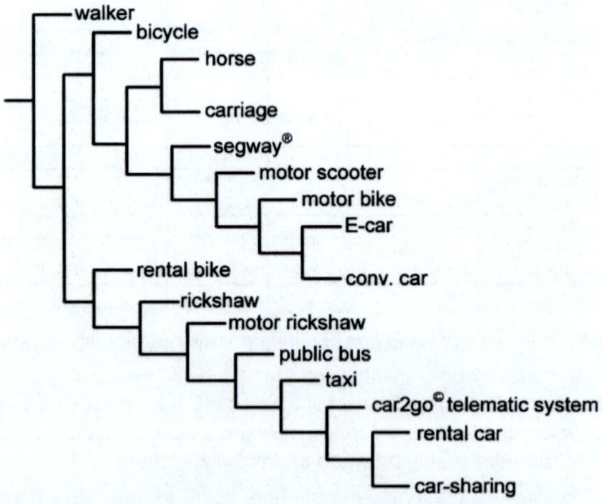

Figure 5: Cladogram of mobility enablers.

In the lower sector, beside the complexity of services and the ownership structure, the point of departure is changing from completely flexible to flexible stations and the variable costs are rising to a medium level. Rental bikes do have fixed stations for departure and arrival, whereas the branches rickshaw and motor rickshaw depart from variable stations, which means the departure point can be wherever a free (motor) rickshaw is found and arrive at completely flexible points. The environmental impact is rising in that sector from left to right, as well as the velocity and the travelled distance. From rickshaw to taxi, the driver status is "passenger" and the owner structure is "rental of a seat", where as the status is turning back to "self-driven" and "short-term-rental" with car2go© and car-sharing and to "self-driven" and "long-term-rental" with rental car. The amount of passengers is rising with rickshaws and again with public busses. With the branch of motor rickshaws, the type of drive is changing to combustion engines. In the branch of busses, the amount of passengers is rising, time flexibility is changing to "scheduled", the characters "route", "point of departure" and "point of arrival" are changing to "fixed" and the distance of travel changes to "long". In the branch taxi and rental car, the variable costs are rising to "very high", whereas car2go© and car-sharing have "high" costs.

Whereas the point of departure and the point of arrival is at fixed stations for car sharing and rental car, the departure is at flexible stations for car2go© and the point of arrival is completely flexible. The service complexity is at the highest level in the branches of car2go© and car-sharing. A booking before usage is required for the sector of rental cars and car sharing.

4.2 Telematics

Likewise to the mobility enablers, various telematic systems are offered in the market. The analysis is based on 25 solutions from the following areas: telematic systems for motorised vehicles with main or part traffic carrier "street", systems for private customers and for fleet customers as well as for commercial vehicles, fixed embedded OEM equipment, back fitted fixed embedded equipment and mobile equipment, systems available on the European market and on the Japanese and Korean market (as cutting-edge markets for the usage of information technology based systems). The common ancestor of all telematic solutions is the "trip computer".

main task of system	communication with ...	embedded or mobile	automatic sending of data	automatic receiving of data
sending inf. about tank level	providing vehicle data (CAN)	congestion alert available	sending/ receiving of changes in road situation	driver identification
Bluetooth connection	voice control	navigation	open vehicle by control centre	open vehicle by PIN
vehicle positioning	route observation	information provision	communication with control centre	job control
safety or danger areas warnings	TV reception	emergency call	IQ Routes (statistics about reachable velocity on route)	Eco-Drive function
velocity	kilometres traveled	open vehicle by access card	booking state send to vehicle	live stream street observation
usage time	CD/ DVD/ USB adapter	hands-free-set	internet in vehicle	status monitoring

Table 3: Characters of telematic systems.

Table 3 summarizes the 35 characters selected for the cladistic analysis of the telematic systems. The software generated 184 cladograms with the length of 104 steps. Therefore, a consensus tree was constructed that consists of the nodes that occur in the majority of the 184 trees (see Figure 6).

The cladogram of telematic systems clearly shows the changes in techniques, starting from the trip computer. The first sector forms mobile devices in the European market. Primarily, their characters concentrate on navigation and automatical reception of required data (e.g. Blue & Me©). The branches on the right hand of Blue & Me© have TMC for traffic jam avoidance available. Partially, the systems are equipped with hands-free equipment, IQ Routes, voice control and vehicle tracking. Vehicle data is not used in this sector. Other ports or communication with a control centre are not available. Data is not sent automatically.

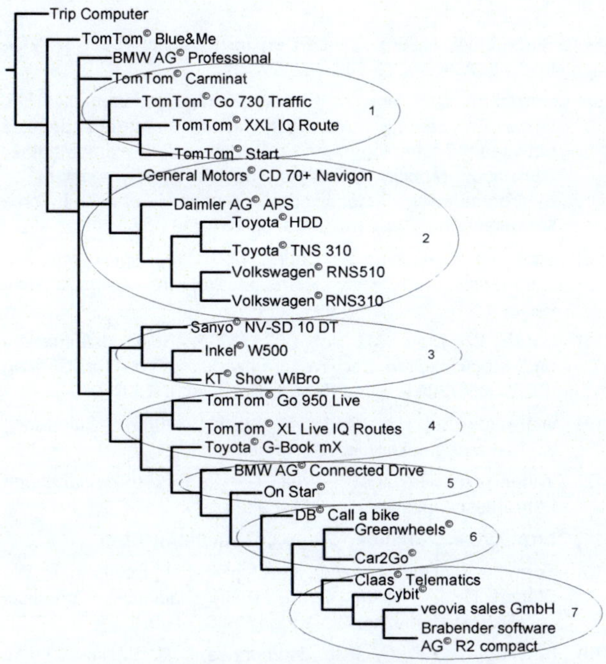

Figure 6 : Cladogram of telematic systems.

In the second sector, fixed embedded OEM systems are located. This group has, in addition to the characteristics of the first sector, various connections such as Bluetooth, CD, DVD or USB. The third sector represents mobile devices that also have characteristics as the independent sending and receiving of changes in road situations, information provision, or TV reception. The branch KT© Show WiBro has also characteristics like live stream street observation and the Sanyo© branch has internet in the vehicle. In the fourth sector the telematic systems do not have the special characteristics of the third group, but have characteristics like vehicle location and IQ Routes. With sector five, again permanently installed systems are represented. These systems also have capabilities to communicate with a control centre, emergency call systems and the transmission and reception of changes in road situations.

In branch six, systems for sharing services are combined. These systems have, compared to the branches previously considered, no emergency call function, navigation and information provisioning. But these branches are equipped with driver identification, monitoring of the usage time or the transmission of the reservation status to the vehicle. The Car2Go©-system also sends information about the tank level to the control centre. All systems have the characteristic of opening the vehicles by a PIN or an access card.

In the sector in the lower right area, there are telematic systems used in commercial vehicles or fleets.

These branches additionally have functions for navigation, job control and status monitoring, for the continuous route observation and monitoring of vehicle data from the CAN. These functions are combined with most of the characteristics available in sector four, five and six. Brabender©, as the most pronounced branch, for example has Bluetooth connectivity, hands-free sets, internet and TV reception in vehicles, but also warnings and information provision opportunities.

5 CONSOLIDATED ANALYSIS OF CLADOGRAMS

When analyzing the cladogram of telematic systems (Figure 6) it turns out that the characteristics are concentrated on different classes of systems. There is no system which combines all characteristics. Instead, special systems are available in different areas. The trend however veers towards offering a system which combines characteristics from the different areas.

The telematic systems in sector one to five are depicted as product-oriented services. Mobile systems (in sector one, three and four) permit a usage independent from the mobility enabler and therefore a service-oriented improvement of the mobility enabler performance. Fixed embedded systems, as in sector two and five, are limited in the application area to the OEMs offering the systems. They are only available in convenient cars, respectively in cars for car sharing or car rental that are using these cars.

The cladogram of telematic systems identifies the systems for sharing services as highly developed. But the telematic systems for commercial vehicles have an even higher quantity of functions. These systems accessorily provide navigation, monitoring of vehicle data or continuous route observation.

In the cladogram mobility enabler (Figure 5) it becomes obvious that rental car, car sharing and car2go© are the most enhanced mobility enablers for passenger transportation today. Those enablers allow travelling with high velocity, long distances, flexible in time and without an initial investment. However, the environmental impact is only at a fair level. Rental car and car sharing need a booking and the points of departure and of arrival are fixed. Car2go© contrarily has flexible stations for departure and is completely flexible in arriving. The service complexity is very high particularly in car sharing and car2go© concepts. To countervail the high service complexity, the described telematic systems for sharing solutions are integrated into the car sharing concepts and into the car2go© concept. The integration of these systems into the cars is a precondition for the implementation of sharing systems in a large scale and with an exactable effort for service organization. The current developments of the mobility enabler in passenger transportation are therefore depending on the development in telematic systems.

This co-evolution points out the changes in mobility due to the upcoming possibilities of integrating service systems into the vehicle. As the cladogram of telematics has clarified that the commercially oriented telematic systems come with further functions than the systems used in sharing concepts today, it can be assumed, that the development of new mobility systems is not yet terminated. Prospectively, the disadvantages of sharing concepts, like the fixation to stations, can be adjusted with further developments in telematics. The monitoring of vehicle data, which is available in telematics for commercial vehicles, e.g. allows reducing the service related complexity for car sharing and for car2go©. Enhanced navigation and route observation furthermore enable the optimization of route planning, the used driveways and the reduction of standby times due to traffic jams, which directly reduces the environmental impacts.

But also completely new concepts are realistic with the increasing usage of telematic systems. Car sharing concepts for example can be integrated with ride sharing concepts. Instead of booking the whole car, the user only books a seat in the car, when the integrated software tool recognizes that different drivers are about to use the same route. The telematic device directs the driver to pick up the passengers and coordinates the ride share. Those services can also be extended to an individualization of the routes and schedules of public transportation or to revised opportunities for the personal adoption of different transportation means.

The prospective developments in passenger transportation therefore can reach an increasing rate of servitisation without losing the advantages of individualization.

For further developments in passenger mobility it is essential to change the view on the product "vehicle", leaving the idea of a single product, coming to the idea of a combined system with vehicle, telematic system, and services.It is expected that the integration of further telematics in the vehicles will change the mobility again and may lead to a third transport revolution. Individual and public transportations could become connected and the advantages of both systems, the individual mobility on the one hand and the economic and ecological advantages on the other hand get interlinked.

6 CONCLUSIONS AND OUTLOOK

At first, as a finding regarding the methodology the application of cladistics points out challenges for getting valid results. Each revision regarding the character selection and coding can have a significant impact to the resulting cladogram. Therefore, much effort has to be spent for these tasks. Additionally, the determination of ancestral and derived states for each character of the street mobility clade shows that the derived state is not necessarily the desirable state. The walker as common ancestor has ancestral states like "low environmental impact" that are worthwhile characteristics for modern mobility systems, too.

With analysing the cladograms of mobility enabler and telematic systems, it becomes obvious that new trends and developments in mobility today are more and more relying on the development in information technology and telematics. The level of service integration is rising significantly due to new potentials in countervailing the high service complexity. In passenger transportation prospective possibilities to pursue trends to a product service orientation and to connect technical developments with service offers can be extended.

Furthermore, the developments in telematics depict that today's mobility concepts are only using a part of the available functions in telematic systems. A great potential for progress in new mobility approaches exists as well for the improvement of the design and for processes in existing concepts. Therewith, future work is required in the field of the symbiotic design of mobility PSS and related telematic solutions.

7 REFERENCES

[1] Braess, H., Seiffert, U. (Ed.) (2007): Handbuch Kraftfahrzeugtechnik, Wiesbaden.

[2] Bundesministerium für Verkehr, Bau und Stadtentwicklung (Ed.), 2009, Verkehr in Zahlen 2009/2010, Hamburg.

[3] Vandermerwe, S., Rada, J. (1988) Servitization of business: Adding value by adding services, in: European Management Journal, vol. 6, no. 4, pp. 314-324.

[4] Communication Networks 14.0, FOCUS Magazin Verlag GmbH, München, 2010.

[5] Heinze, G. W., Kill, H.H. (1991): Evolutionsgerechter Stadtverkehr, VDA-Schrift No. 66., Frankfurt M.

[6] Mikulski, J. (2010): Using Telematics in Transport, in: Mikulski, J. (Ed.) (2010): Transport System Telematics, 10th Conference, TST 2010, Katowice, pp. 175-182.

[7] Winterhoff, M., Kahner, C., Ulrich, C., Sayler, P., Wenzel, E. (2009): Zukunft der Mobilität 2020 - Die Automobilindustrie im Umbruch?

[8] Burgert, W. (1996): Tendenzen im Karosseriebau, in: VDI-Bericht 1256, pp. 29-50.

[9] Herrmann, C., Stehr, J., Kuntzky, K. (2009): Automotive Life Cycle Engineering: Understanding the Interdependencies between Technology and Market of Environmentally Conscious Mobility Concepts, in: 6th Int. Symposium on Environmentally Conscious Design and Inverse Manufacturing, Sapporo, 2009, pp 231-236.

[10] Shaheen, S., Cohen, A., Chung, M. (2009): North American Carsharing - 10-Year Retrospective, in: Transportation Resaerch Record 2110, pp. 35 -44.

[11] Loose, W., Mohr, M., Nobis, C. (2004): Bestandsaufnahme und Möglichkeiten der Weiterentwicklung von Car-Sharing (FE 77.461/2001), Final Report, Öko Institut, DLR.

[12] Wallentowitz, H., Reif, K. (Ed.) (2006): Handbuch Kraftfahrzeugelektronik, Wiesbaden.

[13] Winner, H., Hakuli, S.; Wolf, G. (Ed.) (2009): Handbuch Fahrerassistenzsysteme, Wiesbaden.

[14] **http://www.brabender-group.com/de/home.html**

[15] Forey, P. L., Humphries, C. J., Kitching, I. J., Scotland, R. W., Siebert, D. J., Williams, D. M. (1992): Cladistics: A practical course in systematics, Clarendon Press, Oxford.

[16] Kitching, I. J., Forey, P. L., Humphries, C. J., Williams, D. M. (1998): Cladisitcs, Oxford university press.

[17] Ridley, M. (1985): The Problems of evolution. Oxford University Press, Oxford.

[18] McCarthy, I.(2006): Toward a Phylogenetic Reconstruction of Organizational Life, in: Journal of Bioeconomics, 7:271–307.

[19] McCarthy, I.P., Ridgway, K., Leseure, M., Fieller, N. (2000a): Organisational diversity, evolution and cladistic classifications, in: Omega - The Int. Journal of Management Science, 28, 77–95.

[20] AlGeddawy, T., ElMaraghy, H. (2009): Symbiotic Design of Products and Manufacturing Systems Using Biological Analysis, in: 19th CIRP Design Conference – Competitive Design, Cranfield University, 30-31 March 2009, 469-472.

[21] ElMaraghy, H., AlGeddawy, T., Azab, A. (2008): Modelling evolution in manufacturing: A biological analogy, in: CIRP Annals - Manufacturing Technology 57: 467–472.

[22] AlGeddawy, T., ElMaraghy, H. (2010): Co-evolution hypotheses and model for manufacturing planning, in: CIRP Annals - Manufacturing Technology 59: 445–448.

[23] Rose-Anderssen, C. (2009): A cladistic classification of commercial aerospace supply chain evolution, in: Journal of Manufacturing Technology Management, 20/2:235-257.

[24] Leseure, M.J. (2000): Manufacturing strategies in the hand tool industry, in: International Journal of Operations & Production Management, 20/12:1475-1487.

[25] Baldwin, J.S., Allen, P.M., Winder, B., Ridgway, K. (2004): Modelling manufacturing evolution: thoughts of sustainable industrial development, in: Journal of Cleaner Production, 13/9:887-902.

[26] McCarthy, I. (1995): Manufacturing classification: lessons from organizational systematics and biological taxonomy, in: Integrated Manufacturing Systems, 6/6:37-48.

[27] McCarthy, I., Ridgway, K. (2000b): Cladistics: a taxonomy for manufacturing organizations, in: Integrated Manufacturing Systems, 11/1:16-29.

[28] Nixon, K. C., 1999, Winclada ver. 0.9.9, Ithaka, NY.

Development of an Evaluation Scheme for Product-Service Systems

Kwang-Jae Kim[1], Dong-Hee Lee[1], Chie-Hyeon Lim[1], Jun-Yeon Heo[1], Yoo-Suk Hong[2], and Kwang-Tae Park[3]

[1]Department of Industrial and Management Engineering, Pohang University of Science and Technology, Republic of Korea

[2]Department of Industrial Engineering, Seoul National University, Republic of Korea

[3]Business School, Korea University, Republic of Korea

Abstract

This paper proposes an evaluation scheme for PSS with a focus on constructing evaluation criteria. The proposed scheme has a four-layered hierarchical structure. The four layers refer to perspectives, dimensions, categories, and aspects. The sustainability and customer value perspectives are considered. The sustainability perspective consists of 3P dimensions, while the customer-value perspective consists of quality and cost dimensions. Each dimension is further classified into categories, and finally into more detailed aspects. The scheme has 5 dimensions, 24 categories, and 87 aspects in total. The PSS evaluation scheme can serve as an effective aid in designing as well as evaluating a PSS.

Keywords:

PSS; Evaluation Scheme; Evaluation Criteria

1 INTRODUCTION

Today, governments and industries across the globe are looking for ways to achieve sustainable growth. Thus, they may consider the introduction of product-service systems (PSSs) that have a potential to enhance both competitiveness and sustainability. A PSS is defined as a specific type of value proposition that a business (network) offers to (or co-produces with) its clients, and consists of a mix of tangible products and intangible services so that they jointly are capable of fulfilling customer needs [1]. Representative examples of PSS include chemical management services (CMS), car-sharing, and retailer-consulting services [2, 3].

Conceptual research on PSS such as [2] and [4], had been conducted for about 10 years after its advent in [5]. Various studies on designing PSS have been conducted afterwards. In recent CIRP IPS[2] conferences, issues related to PSS design were discussed by many authors. However, the issues related to PSS evaluation have not received as much attention as those related to PSS design.

The evaluation issues as well as the design issues should be carefully considered. The process of PSS development generally consists of several phases. The outcome of each phase should be evaluated to ensure the effectiveness of succeeding phases. Thus, the process of PSS design should run parallel with its evaluation for the purpose of validation of design. For example, the process of PSS development proposed in the methodology for PSS (MEPSS) [6] consists of design and evaluation modules. In each phase, the outcome of the design module is validated by the evaluation module.

Despite the importance of PSS evaluation, works related to it are rare. In this regard, this paper proposes an evaluation scheme for evaluating PSS. The evaluation scheme should include evaluation criteria, evaluation methods, and an evaluation procedure. This research focuses on the development of evaluation criteria. The evaluation criteria cover not only the provider perspective but also the customer perspective. The results of this research can support the development of PSS.

2 LITERATURE REVIEW

For this research, 33 existing works were collected. These are listed in Table 1. Works in the first and second columns of Table 1 are related to evaluation of PSS and service/product, respectively. Although works in the second column of the table were not conducted for the purpose of PSS evaluation, their approaches can be applied to evaluate PSS. This is because PSS and service/product are essentially similar. The works in Table 1 are assigned identification codes, where PS and SP denote PSS and service/product evaluation, respectively.

The collected works are classified into two categories (i.e., evaluation perspective and research type) as shown in Table 2. Evaluation perspective consists of provider and customer perspectives. Under the provider perspective, economic, environmental, and social values serve as sub-categories. This set of values is used to evaluate sustainability generally referred to as 3Ps (profit, planet, and people). Profit, planet, and people represent economic, environmental, and social values of PSS, respectively. These values are important to a PSS-providing company. The research type consists of methodology developments or case studies.

It is notable that most existing works listed in Table 2 focus on evaluation from the provider perspective. For example, [34] studies the evaluation of activities of companies under all 3P perspectives. [35] also studies the evaluation of activities of nations under all 3P perspectives. However, few works have dealt with evaluation from the customer perspective. For example, [9] suggests a method for measuring customer value based on the concept of net present value (NPV). [10] develops and reinterprets some tools used in product evaluation for evaluating PSS.

J. Hesselbach and C. Herrmann (eds.), *Functional Thinking for Value Creation: Proceedings of the 3rd CIRP International Conference on Industrial Product Service Systems, Technische Universität Braunschweig, Braunschweig, Germany, May 5th - 6th, 2011*,
DOI 10.1007/978-3-642-19689-8_45,

PSS evaluation	Service/Product evaluation	
PS-1 Mont and Plepys (2003) [10]	SP-1 Giirzenich et al. (1999) [33]	SP-15 United Nations (2007) [35]
PS-2 Tukker and Tischner (2005) [7]	SP-2 Hanssen (1999) [30]	SP-16 Park and Tahara (2008) [16]
PS-3 Omann (2006) [8]	SP-3 Vogtlander et al. (2001) [32]	SP-17 Shimomura et al. (2008) [22]
PS-4 Kimita et al. (2009) [11]	SP-4 Khan et al. (2002) [21]	SP-18 Collado-Ruiz et al. (2009) [27]
PS-5 Rese et al. (2009) [9]	SP-5 Global reporting initiative (2002) [34]	SP-19 Liu et al. (2009) [18]
PS-6 Paci and Chiacchio (2009) [13]	SP-6 Vogtlander et al. (2002) [28]	SP-20 Pecas et al. (2009) [20]
PS-7 Steven et al. (2009) [12]	SP-7 Labuschagne (2003) [23]	SP-21 Sugiyama et al. (2009) [25]
PS-8 Mannweilek et al. (2010) [14]	SP-8 Zapf (2004) [24]	SP-22 Alix et al. (2009) [39]
PS-9 Sadek and Steven (2010) [15]	SP-9 Kobayashi (2005) [17]	SP-23 UNEP (2009) [36]
	SP-10 Park et al. (2005) [31]	SP-24 Parent et al. (2010) [37]
	SP-11 Huisman and Stevels (2006) [19]	SP-25 Garvin (1984) [40]
	SP-12 Dreyer et al. (2006) [38]	SP-26 PZB (1985) [41]
	SP-13 Tasaki et al. (2006) [26]	SP-27 Guine (2001) [45]
	SP-14 Ulgiati et al. (2007) [29]	

Table 1: Existing works on evaluation of PSS and service/product.

The research type consists of methodology and case study. Some existing works are classified into methodology research than case study among studied works. For example, [16] suggests an evaluation method by applying data envelopment analysis. [17] suggests an evaluation method that applies quality function deployment. The other works are classified into the case study. [18] evaluates the recycling system of an electronics company in China through a modeling system. [26] evaluates a lease service by developing a mathematical model. It is notable that most works under the case study type evaluate service/product and not PSS. This is because PSSs are not yet widely used in real practice.

Based on the literature, the following three observations are made. First, few works have considered the evaluation criteria on both provider and customer perspectives. Most works have considered the evaluation criteria only from provider perspective. Considering that the main purpose of PSS is to satisfy needs of customers, evaluating PSS on the customer perspective is critical. Therefore, the evaluation scheme should include the evaluation criteria for both provider and customer perspectives.

Second, few works have considered all the criteria for 3Ps. For sustainable growth of a provider company, the three types of values should be considered simultaneously.

Third, most existing works have not considered the life cycle of PSS. Only environmental value is measured and evaluated considering the entire life cycle of PSS (regarding works are related to life cycle assessment research area). The other values, such as economic value, social value, and customer value, are not measured considering the entire life cycle of PSS. Since the other values are also created throughout the entire life cycle, they should be measured and evaluated in the same manner.

3 PROPOSED PSS EVALUATION SCHEME

The PSS evaluation scheme consists of evaluation criteria, evaluation methods, and an evaluation procedure. This research focuses on the development of evaluation criteria. The proposed evaluation scheme pursues a general evaluation model. It attempts to evaluate PSS in a comprehensive manner. First, it considers both customer and provider (company) perspectives. Second, 3P (profit, planet, and people) dimensions are considered in the provider perspective. Third, the evaluation scheme considers all phases of PSS life cycle by including the PSS life cycle-dependent criteria. Section 3.1 describes the structure and detailed items of the evaluation criteria and section 3.2 briefly presents the methods that can be used for the evaluation.

3.1 PSS Evaluation Criteria

Figure 1 shows the structure of the evaluation criteria. The framework of evaluation criteria has a four-layered hierarchical structure. The uppermost level of the structure is the perspective level. The perspective refers to the evaluator's point of view [46]. In the perspective level, two items exist. One is sustainability and the other is customer value. Sustainability and customer value represent the provider (company) and the customer perspectives, respectively. Each perspective has several dimensions, and each dimension has categories as shown in Figure 1. A dimension is a particular part of perspective. Categories refer to "the broad areas or groupings of social, environmental or economic issues of concern to the company stakeholders [6]." A brief explanation of each category is also given in Figure 1. Each category has several aspects. Aspects refer to "the general types of information related to a given category [6]." A category and an aspect need specific adaptations since they depend on a specific context considered.

Many of the evaluation criteria were adapted from some existing works (such as [6], [8], [40], and [45]), while some others were newly added by the authors. Numbers given ahead of each dimension, category, and aspect represent codes for identification. Numbers in the parenthesis in Figure 1 denote the number of aspects included in the corresponding category or dimension. Two sets of aspects associated with the fixed-cost category (code: 10100) and service-related quality category (code: 40200) are given in Figure 1. They are indicated with shadings. The other sets of aspects are not shown in Figure 1 due to the space limitation.

		Evaluation perspective			
		Provider			Customer
		Economic value	Environmental value	Social value	
Research type	Methodology	PS- 1, 2, 4, 6, 7, 8, 9 SP- 3, 5, 6, 7, 9, 10, 15, 22	PS- 1, 2 SP- 3, 4, 5, 6, 7, 9, 10, 14, 15, 16, 27	PS-1,2 SP- 5, 7, 12, 15, 23, 24	PS-1, 5 SP-16, 17, 25, 26
	Case study	SP- 2, 11, 19, 20, 21	SP- 1, 2, 11, 13, 19, 20, 21		SP- 8, 13

Table 2: Classification of the collected works.

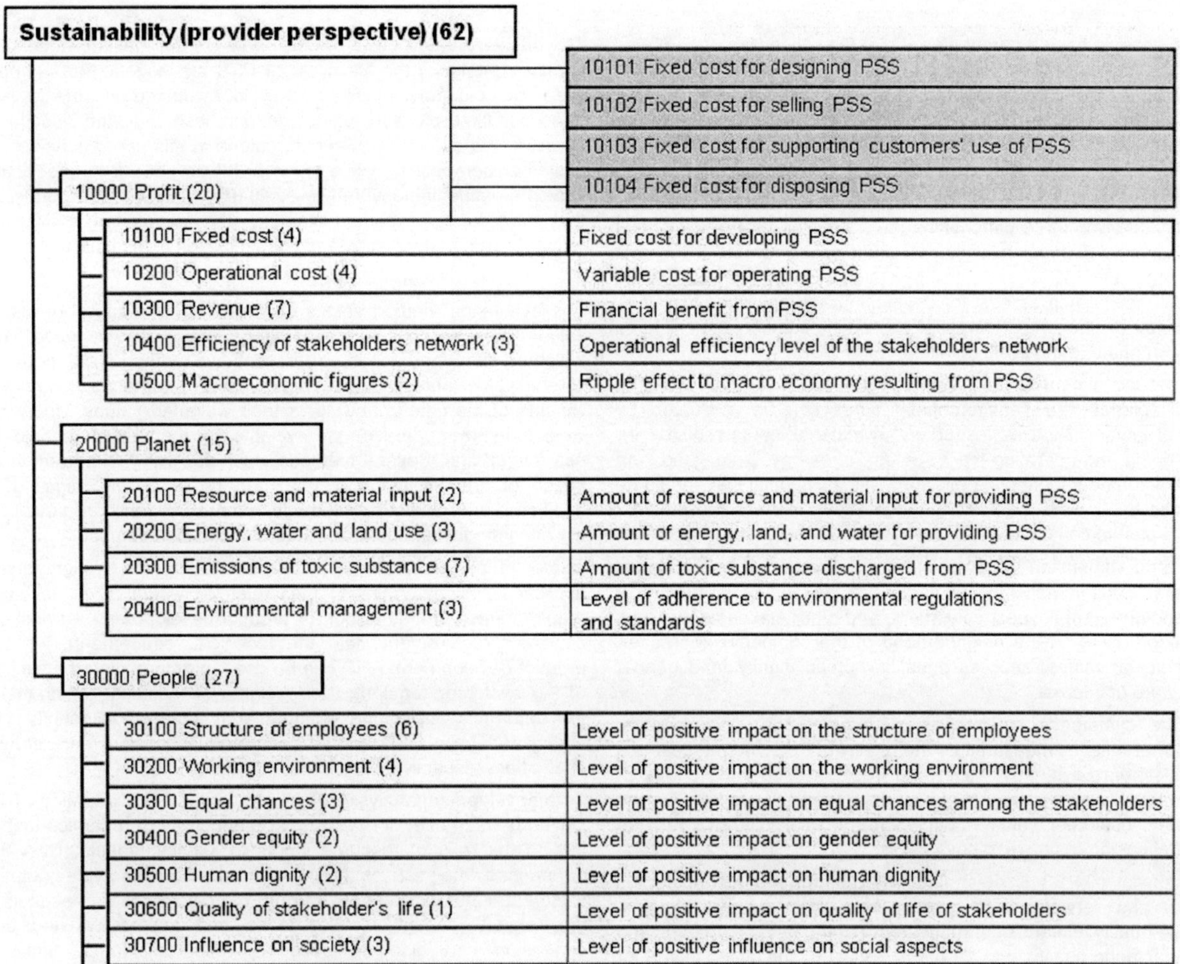

Figure 1: Structure of PSS evaluation criteria.

Some aspects have different degrees of relevance with each phase of PSS life cycle. Such aspects need to be evaluated for each phase. For example, the degree of customization (which is under the "Service related quality" category (40200)) is likely to be different in each phase. Hence, it is evaluated separately for purchase, use, and disposal phases (i.e., 40210, 40211, and 40212).

In sustainability, three dimensions (i.e., 3Ps) exist. The first one is an economic dimension of the provider, called "profit." The profit resulting from PSS is evaluated by several factors (i.e., categories and aspects), such as cost for developing and operating the PSS. Efficiency of network of stakeholders and macroeconomic figures also contribute to the profit resulting from PSS.

The second dimension of sustainability is environmental, which is called "planet." The environmental value can be evaluated by several factors, such as resource and material inputs required for developing and operating the PSS. Use of energy, water, and land has environmental effects. Emissions of toxic substances should be reduced to alleviate the environmental effects. Finally, the provider should plan to construct an environmental management system.

The third dimension of sustainability is social, which is called "people." This dimension evaluates the life and work quality of stakeholders (employees, customers, and other stakeholders, such as people living in the neighborhood of the company). Moreover, other social values, such as equality, human dignity, and gender equity are of interest.

For the customer's perspective, customer value is considered. Customer value is measured by the quality of PSS and the cost that the customer pays. Customer value would be high if the quality of PSS is high and the amount customers have to pay is low. On the contrary, customer value would be low if quality is low and the amount customers have to pay is high.

The quality of PSS is determined by product and manufacturing quality and service-related quality. This paper adopts the well-known quality dimensions suggested in [40] for the product and manufacturing quality, and adopts SERVQUAL suggested in [41] for the service-related quality. While SERVQUAL is applicable for evaluating PSS, it is not comprehensive since it does not sufficiently measure the full benefits of PSS. Therefore, in this paper, some factors critical to the success of PSS, such as convenience, customization, flexible use, and interaction with the provider, are added in the criteria for service-related quality.

It sould be noted that the costs to customers can be divided into those from each phase of the PSS life cycle. In many PSSs, customers and providers have a sustained relationship throughout the entire life cycle. For example, NIKE launced NIKEiD to produce highly customized shoes [44]. A customer designs his/her own shoes with a systematic support (service) via nikeid.com. In this case, there are some costs (either monetary or non-monetary) to the customer in every phase of the PSS life cycle (i.e. design, purchase, use, and dispose).

The suggested evaluation criteria are comprehensive since these include both provider and customer perspectives. Serving as evaluation criteria on the provider perspective, all 3Ps are considered. Moreover, all phases of the life cycle of PSS are considered. In practice, some of the criteria can be selected and used for evaluating a specific PSS. The criteria would be selected by considering several factors, such as the mission of the PSS development, policy of the company, and market information.

This paper adopts the 3Ps criteria of [8] as the criteria on the provider perspective. Compared with the criteria of [8], however, criteria in this paper have two advantages. First, the suggested criteria consider the life cycle of PSS, while the criteria of [8] do not. To consider the life cycle of PSS, new criteria that are dependent on each phase of the life cycle of PSS are added. For example, fixed-cost category (code: 10100) is evaluated for the several phases of life cycle, such as design, purchase, use, and disposal, to measure the exact cost. Second, criteria in this paper consider the customer perspective, while those of [8] do not. More specifically, customer value is measured by evaluating quality of PSS and costs that customers pay. For example, the degree of satisfaction resulting from convenience and customization is measured.

3.2 PSS Evaluation Method

The evaluation method entails scoring and aggregation of criteria values. For scoring, a five-category scale can be used. The weighted-sum approach is a representative method for aggregating criteria values into a single value. It first determines the relative weights of the criteria and sums their weighted values. Under the weighted-sum approach, the weights of the criteria should be determined first. Several methods, such as AHP [42] and methods based on the utility theory [43], can be used for weighting. In addition, some statistical methods, such as regression, structural-equation modeling, and factor analysis, can also be used.

Quality function deployment (QFD) takes the weighted-sum approach in determining the relative weights of technical requirements. QFD is a tool for product development. It translates customer requirements into the technical requirements for new product development. AHP can be used in determining the weights of the customer requirements. Weights of technical requirements are determined using the weighted-sum approach. Recently, [17] and [22] apply QFD for an eco-product design and service evaluation, respectively.

Data envelopment analysis (DEA) can be used for evaluating PSSs. DEA has been used for measuring productivity of a service unit by getting the ratio of resource inputs to service outputs. DEA first determines the list of resource inputs and service outputs. Resource input can be derived from the categories and aspects of the "planet" dimension. Categories and aspects of the cost dimension can also be used as resource inputs. Similarly, categories and aspects of profit, people, and quality dimensions can be used as outputs. Once items are determined, the relative efficiency levels of PSSs are determined by solving linear programming problems. Compared with the weighted-sum approach, the DEA is useful for comparing efficiency of several alternative PSSs. [16] suggests a method for measuring eco-efficiency by using DEA.

4 CONCLUDING REMARKS

Notwithstanding the rapid growth of the PSS literature, systematic evaluation scheme for PSS has not been fully developed. This paper has proposed an evaluation scheme for PSS. The proposed PSS evaluation scheme consists of evaluation criteria, evaluation methods, and evaluation procedure. The focus of this paper is on the construction of the evaluation criteria.

The proposed framework of evaluation criteria has a four-layered hierarchical structure. The four layers refer to perspectives, dimensions, categories, and aspects. Two perspectives are considered, namely, sustainability and customer value. The sustainability perspective consists of 3Ps (i.e., profit, planet, people) dimensions, while the customer-value perspective consists of quality and cost dimensions. The sustainability and customer-value perspectives represent the provider (company) and customer perspectives, respectively. Each dimension is further classified into categories, and finally into more detailed aspects. The framework has five dimensions, 24 categories, and 87 aspects in total.

This research is an attempt to systematically support the process of PSS evaluation. There are three main advantages of using the proposed evaluation scheme. First, the evaluator can utilize the wisdom from rich experience in PSS evaluation in the past, rather than starting from scratch. It would increase the efficiency and effectiveness of PSS evaluation. Second, the proposed evaluation scheme can be used in various stages of PSS development. By selecting and using appropriate criteria for the object to be evaluated, the proposed scheme can be used for evaluating PSS concepts, models, or completed PSSs. Third, the proposed scheme is comprehensive enough to cover the provider and customer perspectives as well as all phases of the PSS life cycle. Although the proposed scheme is not complete at this moment, it has a potential to grow as a generic platform for PSS evaluation in the future.

There are several issues for future research to improve the evaluation scheme. First, the comprehensiveness of the criteria should be continually checked and updated over time. Real PSS cases should be collected and evaluated using the evaluation scheme. While the developed evaluation scheme can be applied to a PSS in general, it would be especially interesting to conduct a case study of a B2C-type PSS. This is because the evaluation scheme explicitly considers the customer perspective, which is a unique characteristic. Moreover, considering that the evaluation scheme encompasses the lifecycle of a PSS, a product-oriented or use-oriented PSS can be effectively evaluated. It will become more reliable and realistic as more cases are tested. Second, the accompanying evaluation method and procedure need to be developed. The evaluation method should include various decision analysis and support tools. The evaluation procedure should provide a step-by-step guide to conducting an evaluation. Third, the feedback mechanism after an evaluation needs to be developed. The evaluation result should be reflected upon and utilized in devising a PSS improvement plan. A detailed guideline on which characteristic of the PSS should be refined or revised when evaluation scores are unsatisfactory should be created.

5 REFERENCES

[1] Tukker, A.; Tischner, U. (2008): Product-services as a research field: past, present and future. Reflection from a decade of research, in:Journal of Cleaner Production, vol. 14, pp. 1152-1156.

[2] Mont, O. (2000): Product Service-Systems. Final report, IIIEE, Lund University.

[3] United States (2007): Technology Review 205/2007, Seizing the White Space: Innovative Service Concepts in the United States, Tekes.

[4] Baines, T. S.; Lightfoot, H.W.; Evans, S.; Neely, A.; Greenough, R.; Peppard, J,; Roy, R.; Shehab, E.;Braganza, A.; Tiwari, A.; Alcock, J.R.; Angus, J.P.; Bastl, M.; Cousens, A.; Irving, P.; Johnson, M.; Kingston, J.; Lockett, H.; Martinez, J.; Michele, P.; Tranfield, D.; Walton, I.M.; Wilson, H. (2007): State-of-the-art in product-service systems, in: Journal of Engineering Manufacture, vol. 221, pp. 1543-1552.

[5] Goedkoop, M.; van Haler, C.; te Riele, H.; Rommers, P. (1999): Product Service-Systems, ecological and economic basics, in: Report for Dutch Ministries of Environment (VROM) and Economic Affairs (EZ).

[6] Halen, C.; Vezzoli, C.; Wimmer, R. (2005): MEPSS Innovation, Van Gorcum.

[7] Tukker, A.; Tischner, U. (2005): New Business for Old Europe, Suspronet.

[8] Omann, A. (2006): Multicriteria tool for Evaluating the impacts of Product Service Systems on Sustainable Development: An application in Austrian companies, Sustainable Europe research institute.

[9] Rese, M.; Karger, M.; Strotmann, M. (2009): The dynamics of Industrial Product Service Systems (IPS^2) – using the Net Present Value Approach and Real Options Approach to improve life cycle management, in: CIRP Journal of Manufacturing Science and Technology, Vol. 1, No. 4, pp. 279-286.

[10] Mont, O.; Plepys, A. (2003): Customer satisfaction: review of literature and application to the product-service systems, in: Final report to the Society for Non-Traditional Technology.

[11] Kimita K.; Hara, T.; Shimomura, Y.; and Arai, T. (2009): Cost Evaluation Method for Service Design Based on Activity Based Costing, in: Proceedings of the 1st CIRP Industrial Product-Service Systems (IPS^2) Conference, pp. 224-228.

[12] Steven, M.; Rese, M.; Soth, T.; Strotmann, W.-C.; Karger, M. (2009): Profitability of Industrial Product Service Systems (IPS²) – Estimating Price Floor and Price Ceiling of Innovative Problem Solutions, in: Proceedings of the 1st CIRP Industrial Product-Service Systems (IPS^2) Conference, pp. 243-248.

[13] Paci, A. M.; Chiacchio M. S. (2009): Sensitivity Cost-Benefit Analysis to Support Knowledge Capture of Industrial Interests, in: Proceedings of the 1st CIRP Industrial Product-Service Systems (IPS^2) Conference, pp. 261- 267.

[14] Mannweileк, C.; Siener, M.; Aurich, J.C. (2010): Lifecycle Cost oriented Evaluation and Selection of Product-Service System Variants, in: Proceedings of the 2nd CIRP Industrial Product-Service Systems (IPS^2) Conference, pp. 21-26.

[15] Sadek T.; Steven M. (2010): Design of PSS Revenue Models, in: Proceedings of the 2nd CIRP Industrial Product-Service Systems (IPS^2) Conference, pp. 245-252.

[16] Park P.; Tahara K. (2008): Quantifying producer and consumer-based eco-efficiencies for the identification of key ecodesign issues, in: Journal of Cleaner Production, Vol. 16, No.1, pp. 95-104.

[17] Kobayashi (2005): Strategic evolution of eco-products: a product life cycle planning methodology, in: Research in Engineering Design, Vol. 16, No. 1-2, pp. 1-16.

[18] Liu X.; Tanaka M.; Matsui Y. (2009): Economic evaluation of optional recycling processes for waste electronic home appliances, in: Journal of Cleaner Production, Vol. 17, No. 1, pp. 53-60.

[19] Huisman, J.; Stevels L. M. (2006): Eco-Efficiency of Take-Back and recycling, a comprehensive approach, in: IEEE Transactions on electronics packaging manufacturing, Vol. 29, No. 2, pp. 83-90.

[20] Pecas, P.; Ribeiro, I.; Folgado, R.; Henriques, E. (2009): A Life Cycle Engineering model for technology selection: a case study on plastic injection moulds for low production volumes, in: Journal of Cleaner Production, Vol. 17, No. 9, pp. 846-856.

[21] Khan, F.I.; Sadiq, R.; Veitch, B. (2002): Life cycle iNdeX (LInX): a new indexing procedure for process and product design and decision-making, in: Journal of Cleaner Production, Vol. 12, No. 1, pp. 59-76.

[22] Shimomura, Y.; Hara, T.; Arai, T. (2008): A service evaluation method using mathematical methodologies, in: CIRP Annals-Manufacturing Technology, Vol. 57, No. 1, pp. 437-440.

[23] Labuschagne, C.; Brent, A. C.; van Erck, R. P. G. (2003): Assessing the sustainability performances of industries, in: Journal of Cleaner Production, Vol. 13, No. 4, pp. 373-385.

[24] Zapf, M. (2004): From the customer to the firm: evaluating generic service process designs for incoming customer requests, in: Computers in Industry, Vol. 55, No. 1, pp. 53-71.

[25] Sugiyama, H.; Fischer, U.; Antonijuan, E.; Hoffmann, V. H.; Horao, M.; Hungerbuhler, K. (2009): How do different process options and evaluation settings affect economic and environmental assessments A case study on methyl methacrylate (MMA) production processes, in: Process Safety and Environmental Protection, Vol. 87, No. 6, pp. 361-370.

[26] Tasaki, T.; Hashimoto, S.; Moriguchi, Y. (2006): A quantitative method to evaluate the level of material use in lease reuse systems of electrical and electronic equipment, in: Journal of Cleaner Production, Vol. 14, No. 17, pp. 1519-1528.

[27] Collado-Ruiz, D.; Ostad-Ahmad-Ghorabi. H. (2009): Comparing LCA results out of competing products: developing reference ranges from a product family approach, in: Journal of Cleaner Production, Vol. 18, No. 4, pp. 355-364.

[28] Vogtlander, J. G.; Bijma, A.; Brezet, H. C. (2002): Communicating the eco-efficiency of products and services by means of the eco-costs value model, in: Journal of Cleaner Production, Vol. 10, No. 1, pp. 57-67.

[29] Ulgiati, S.; Bargigi, S.; Raugei, M. (2007): An emergy evaluation of complexity, information and technology, towards maximum power and zero emissions, in: Journal of Cleaner Production, Vol. 15, No. 13-14, pp. 1359-1372.

[30] Hanssen, O. J. (1999): Sustainable product systems-experiences based on case projects in sustainable product development, in: Journal of Cleaner Production, Vol. 7, No. 1, pp. 27-41.

[31] Park, P.; Tahara, K.; Inaba, A. (2005): Product quality-based eco-efficiency applied to digital cameras, in: Journal of Environmental Management, Vol. 83, No. 2, pp. 158-170.

[32] Vogtlander, J. G.; Brezet, H. C.; Hendriks, C. F. (2001): The virtual eco-costs '99 A single LCA-based indicator for sustainability and the eco-costs-value ratio (EVR) model for economic allocation, A new LCA-based calculation model to determine the sustainability of products and services, in: International Journal of LCA, Vol. 6, No. 3, pp. 157-166.

[33] Giirzenich, D.; Mathur, J.; Bansal, N. K.; Wagner, H. J. (1999): Cumulative Energy Demand for Selected Renewable Energy Technologies, in: International Journal of LCA, Vol. 4, No. 3, pp. 143-149.

[34] Global reporting initiative (2002): Sustainability reporting guidelines.

[35] United Nations (2007): Indicators of sustainable development: guidelines and methodologies.

[36] UNEP (2009): Guidelines for Social Life Cycle Assessment of Products

[37] Parent, J.; Curcuzzella, C.; Reveret, J. P. (2010): Impact assessment in SLCA: sorting the sLCIA methods according to their outcomes, in: International Journal of LCA, Vol. 15, No. 2, pp. 164-171.

[38] Dreyer, L.; Hauschild, M.; Schierbeck, J. (2006): A framework for Social Life Cycle Impact Assessment, in: International Journal of LCA, Vol. 11, No. 2, pp. 88-97.

[39] Alix, T.; Ducq, Y.; Vallespir, B. (2009): Product service value analysis: two complementary points of view, in: Proceedings of the 1st CIRP IPS^2 conference, pp. 157-164, Cranfield, UK.

[40] Garvin, D. A. (1984): What does "Product Quality" really mean?, in: MIT Sloan Management Review, Vol. 26, pp. 25-43.

[41] Parasuraman, A.; Zeithaml, V. A.; Berry, L. (1985): A conceptual Model of Service Quality and Its Implications for Future Research, in: Journal of Marketing, Vol. 49, pp. 41-50.

[42] Saaty, T. (1977): A Scaling Method for Priorities in Hierarchical Structures, in: Journal of Mathematical Psychology, Vol. 15, pp. 234-281.

[43] Keeney, R.; Raiffa, H. (1993): Decision with Multiple Objectives: Preferences and Value Tradeoffs, John Wiley&Sons, New York.

[44] **http://nikeid.nike.com/nikeid/index.jsp#home**

[45] Guine, J. (2001): Handbook on Life Cycle Assessment: Operational Guide to the ISO Standards, Springer.

Economic Relevance of IPS^2 Flexibility

Marion Steven[1], Solmaz Alevifard[1], Johannes Keine genannt Schulte[1]

[1] Faculty of Management and Economics, Ruhr-Universität Bochum, Bochum, Germany

Abstract

IPS^2 flexibility as the ability to adapt a manufacturing system to changing requirements is an important tool for assuring the profitability of the long-term relationship between the supplier and the customer of an IPS^2. In order to reveal the economic relevance of IPS^2 flexibility, the paper first identifies the most relevant drivers of IPS^2 flexibility, then discusses its various dimensions and economic impact and finally concentrates on the calculation of the profitability of IPS^2 flexibility from different perspectives.

Keywords:

Flexibility; IPS^2 Costs; IPS^2 Life Cycle

1 BUSINESS MODELS FOR IPS^2

Industrial product-service systems (IPS^2) are an innovative approach for mechanical engineering companies allowing them to operate successfully in markets with increasing competition and decreasing margins. Instead of selling a standard machine and offering services separately and successively, IPS^2 are an individualized, customer-oriented combination of these items leading to new business models [1].

- The **function-oriented business model** is the closest to the traditional business arrangements of engineering companies. An example is the sale of a machine which is accompanied by training lessons for the customer's employees or by a maintenance contract that guarantees the functionality of the facility for a specified period of time. As customers tend to require additional services for free, it is often difficult to charge them an adequate amount.
- In the **availability-oriented business model**, the supplier's responsibility is extended to guaranteeing the availability of the facility during the time contracted. Full payment can only be achieved if a specified level of availability is kept. Thus the supplier becomes partly responsible for the business processes of the customer and consequently bears a share of the production risk. On the other hand, it is left to him how to achieve the requested level of availability, i.e. by regular maintenance or by preventive servicing.
- The supplier's responsibility for the processes is further extended in the **result-oriented business model**. In this setting, he does not sell the machine, but is paid for the production of a specified number of faultless parts per period. So he has to decide autonomously on the organization and the execution of production and service processes that take place on the grounds of the customer. In this business model, the supplier additionally carries part of the general business risk.

The development of business models depicted above corresponds to the stages resp. types of industrial service providers that were found empirically by Oliva and Kallenberg [2]. The focus of innovative business models is not on production processes or product quantities but on creating a surplus that may be split among contractors according to their involvement. The subsequent analysis is carried out from the IPS^2 supplier's perspective and concentrates on the availability-oriented business model.

Nevertheless, the investment into new machinery for future production processes is characterized by risks related to quantitative and qualitative demand as well as to uncertain developments of the economic environment. **Flexibility**, i.e. the ability to adapt a system to changing requirements, is a means to cope with such uncertainties. The focus of this paper is on the economic aspects of flexibility of an IPS^2. It discusses the question how a sufficient amount of IPS^2 flexibility may be achieved and who is going to pay for it. Section 2 starts with a discussion of the various drivers that necessitate IPS^2 flexibility. Section 3 deals with different dimensions of IPS^2 flexibility and their economic impacts, and section 4 analyzes economic consequences for different IPS^2 configurations. Finally, a survey of future areas of research is given.

2 DRIVERS OF IPS^2 FLEXIBILITY

The necessity to incorporate a certain amount of flexibility into an IPS^2 may result from different areas. Relevant drivers of IPS^2 flexibility are to be found in the environment of the supplier's enterprise (external drivers) as well as within the enterprise itself (internal drivers). These drivers are discussed and allocated to the phases of the IPS^2 life cycle.

2.1 External Drivers

IPS^2 flexibility enables the supplier to react quickly and cost-efficiently to changes in his enterprise's environment. The most important external drivers forcing such a change are [3] [4]:

- **Competition**: Other enterprises in the market offering a similar or even better performance are a strong incentive to change the properties of an IPS^2.
- **Governmental framework**: New regulations i.e. in the fields of industrial safety or environmental policy require an adaptation

J. Hesselbach and C. Herrmann (eds.), *Functional Thinking for Value Creation: Proceedings of the 3rd CIRP International Conference on Industrial Product Service Systems, Technische Universität Braunschweig, Braunschweig, Germany, May 5th - 6th, 2011*,
DOI 10.1007/978-3-642-19689-8_46,

of the operation of the IPS².

- **Changing demand**: Driven by changing market needs, the customer modifies his requirements specified in the contract on the initial IPS² at some point of time during the long-term relationship that is typical for an IPS².
- **Technical progress**: New machines, new technologies or new materials become available that enable the supplier to deliver better service or to operate at lower costs.

2.2 Internal Drivers

There are various requirements concerning the infrastructure of the supplier's enterprise that enable it to provide and facilitate IPS² flexibility at the moment it is needed [4] [5]. A bottleneck in any of these internal drivers may lead to serious problems.

- **Human resources**: The workers concerned with the operation of the IPS² need to have the skills that are required for implementing flexibility.
- **Technology**: Machines and production processes representing the newest technologies that are available tend to be more flexible than old ones.
- **Organization**: IPS² flexibility is enhanced if the organization of production and services allows different ways to carry out a specific task.
- **Financial facilities**: As the provision of IPS² flexibility requires capital investment, sufficient financial resources have to be provided.
- **Machinery**: Multi-purpose machines with short change-over times allow a greater variety of possible products and thus enhance flexibility.
- **Sourcing facilities**: New specifications of products and services may require different kinds of materials that have to be procured by the IPS² supplier. So he has to assure delivery of adequate quantities and qualities by his sourcing network.
- **Business culture**: Offering an IPS² requires a transition from a manufacturing enterprise to an integrated product and service supplier which has to be reflected in business culture.
- **Innovation**: Product and process innovations within the enterprise can lead to new ideas for IPS².

2.3 IPS² Life Cycle

The provision of IPS² follows a well-defined life cycle [6]:

- Starting from customer requirements, an IPS² and its functions are configured in the **design and development phase**. In order to support this development process, the supplier has to provide an appropriate quality of the internal drivers of IPS² flexibility.
- During the subsequent **delivery and operation phase**, the IPS² is implemented and products and services contracted are delivered to the customer according to the chosen business model. As soon as one or more of the external drivers listed above become relevant, the supplier has to react appropriately taking advantage of the IPS² flexibility provided.

If supplier and customer of an IPS² shared a perfect foresight into future developments, the initial IPS² could already be configured in an optimal way and remain unchanged throughout the complete period of the contract. Otherwise, external influences and new requirements will occur and corresponding adaptations of the initial IPS² have to be made. The longer the life cycle, the more important IPS² flexibility becomes because the probability of substantial changes of requirements increases.

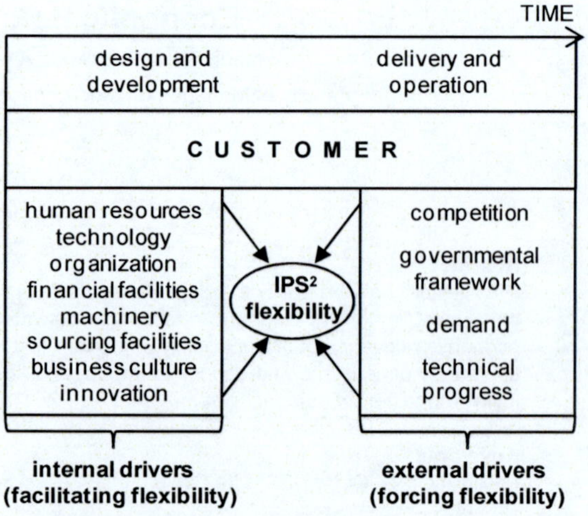

Figure 1: Drivers of IPS² flexibility in the life cycle.

Figure 1 shows the influence of the external and internal drivers on IPS² flexibility during the life cycle.

3 IPS² FLEXIBILITY

Generally, flexibility is the ability of a production system to adapt its performance to changing requirements. This section deals with the relevant dimensions of IPS² flexibility and its economic consequences.

3.1 Dimensions of IPS² Flexibility

IPS² flexibility can be classified according to the following dimensions:

1. **Decision level:**

   From a strategic point of view, **ex ante flexibility** can be incorporated into the initial IPS² as an option allowing to make specific alterations of its future operation – such as the change of plant layout or the use of additional capacity – at relatively low adaptation costs but involving higher development costs. Depending on the contract, ex ante flexibility may be partly paid by the customer.

   Alternatively, on the operative level an ex post reconfiguration can take place at the point of time when requirements change, i.e. for the manufacturing of new product variants. This **ex post flexibility** implies zero additional cost in the development phase but high adaptation costs in the delivery phase which are normally paid for by the customer.

2. **Production characteristics** [7]:

   A production process shows **quantitative flexibility** if the enterprise is able to change the production quantity quickly (output side). This requires that there are sufficient numbers of workers and machines on the input side of the IPS².

   Qualitative flexibility allows a change in the product mix or the substitution of hardware solutions by services i.e. in maintenance processes. It requires sufficient market know-how as well as the knowledge of possible technical solutions.

   Temporal flexibility means the ability to react quickly to changes in production requirements. It is realized by a short reaction time and an abundant availability of production resources and know-how.

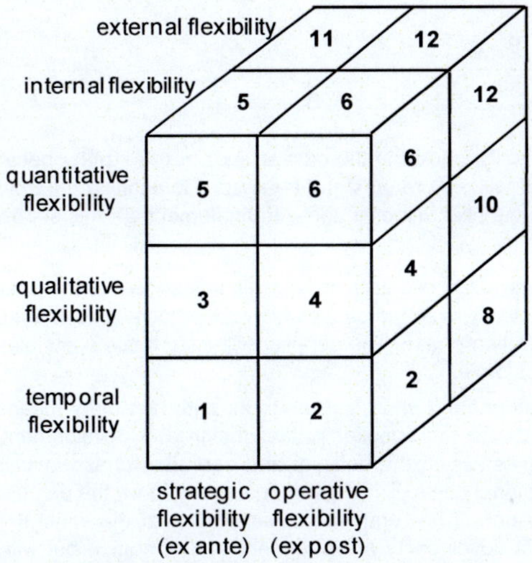

Figure 2: Dimensions of IPS² Flexibility.

3. **Customer interaction:**

   Subdivided by the line of visibility known from service blueprinting [8], processes generating internal resp. external flexibility can be distinguished. While **internal flexibility** allows performing the respective processes more efficiently, **external flexibility** of an IPS² enhances its effectiveness in solving the customer's requirements and thus generates additional profit.

The relations between these dimensions of IPS² flexibility are shown in Figure 2. Every box represents a particular specification of IPS² flexibility which implies specific strategies. In order to illustrate this issue, selected boxes are explained as examples.

- An example for **box 1** is the supplier's competence to configure an IPS² project. This requires flexibility concerning the allocation of resources (internal flexibility) when designing a contract in close contact with the customer. Furthermore the supplier has to react quickly (temporal flexibility) to the customer's requirements in the design and development phase (ex ante flexibility).
- **Box 4** represents the supplier's ability to purchase and provide new machines (internal flexibility) during the delivery and operation phase (ex post flexibility). This allows the supplier to profit from higher productivity of technologically advanced equipment (qualitative flexibility). Lower life cycle costs of advanced machines are very important in the availability-oriented business model.
- Research and development (R&D) processes serve as an example for **box 10**. Successful R&D activities enable the supplier to improve the IPS² (qualitative flexibility) with regard to changing customer requirements (ex post flexibility). Intensive exchange of information is the basis for matching supplier competencies and customer needs (external flexibility).
- Adapting the production volume (quantitative flexibility) finally exemplifies **box 12**. If the customer faces a demand shift (ex post flexibility), the supplier reacts (external flexibility) with an adjusted output volume. This kind of flexibility is most important in availability-oriented business models.

3.2 Economic Impact of IPS² Flexibility

Whereas from the engineering perspective, the main challenge is designing and implementing IPS² flexibility, from the economic point of view, the provision of IPS² flexibility has to be profitable. The most important economic benefits and costs of IPS² flexibility that have a considerable impact on the supplier's economic performance are listed below:

Benefits:

- Flexibility allows the supplier to prolong the **contract** if the business relationship promises further success.
- Flexible capacities facilitate the **adaptation** to changing requirements.
- A high degree of flexibility allows a **quick reaction** to new requirements.
- In case of rising demand total **sales figures** may increase due to possibilities offered by IPS² flexibility.

As a consequence, IPS² flexibility may generate **additional income** in the delivery and operation phase.

Costs:

- Providing IPS² flexibility causes **investment costs** in terms of e. g. asset cost or costs for recruitment of additional staff etc.
- Costs for **training** the workers occur if new competencies are needed to satisfy the changing customer's requirements.
- In order to react quickly and adequately to fluctuating demand, multiple sourcing of materials is recommended. This implies small order quantities and rising **procurement costs**.
- As the supplier needs particular information concerning the market development and the customer's probable requirements, **information costs** are a relevant cost driver.

Generally, a flexible IPS² is profitable from the supplier's point of view if the benefits of flexibility exceed its costs. This will be elaborated in the following section.

4 ECONOMIC ASPECTS OF IPS² FLEXIBILITY

4.1 IPS² Architecture

The IPS² architecture has a major influence on the flexibility of an IPS². In this context, the focus is on the material components of the IPS² because flexibility is much easier to achieve for service components. There are two general design alternatives that are shown in Figure 3:

- **Integrated design** means that the IPS² is built monolithically. If an incident occurs that necessitates an adaptation, the initial IPS² A has to be replaced completely by IPS² B.
- In contrast, a **modular design** makes it possible to replace single modules that are no more compatible with changing requirements. So now the inner structure of the initial IPS² A becomes relevant, in this case it is composed of three modules. In case of an incident, IPS² A can be easily transformed into IPS² A' by replacing module 2 with the technologically advanced module 4. So the modular design enables the supplier to react flexibly to changes without investing in a completely new IPS².

Apart from these structural aspects, there are some economically important characteristics of the design alternatives:

- **Planning effort**: In order to ensure IPS² flexibility, the integrated design requires a lower level of planning activities because replacing the whole IPS² does not have to be planned in detail ex ante. In contrary, IPS² elements have to be grouped in logical modules for the modular design to ensure the required IPS² functionality. The connections between these modules must be taken into account ex ante to ensure flexibility. So structuring a modular design is a task of greater planning complexity.

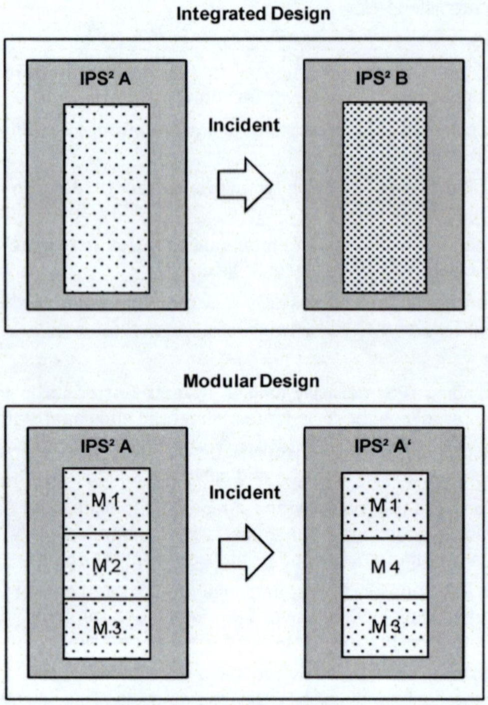

Figure 3: Design alternatives.

- **Costs of repair and maintenance**: Due to the dependencies and close interrelations of IPS² components, an integrated design structure causes higher costs of repair and maintenance. On the other hand, a modular design makes it possible to change modules faster and at lower costs.
- **Flexibility costs**: Ex post flexibility is always paid by the customer who asks e.g. for a higher capacity or new variants. The costs of ex ante flexibility may be carried by the supplier or the customer alone or be shared among them. Installing ex ante flexibility is an investment the risk of which depends on uncertain developments. If the customer does not ask for an adaptation of the IPS², the corresponding costs are sunk costs. Within the frameworks of contract theory, additional costs have to be allocated e.g. with the help of decision rights [9].

4.2 General Economic Conditions

As it is shown in Figure 4, investment in IPS² flexibility is a 2-period problem. In order to make use of flexibility in the delivery and operation phase, the supplier has to invest under uncertainty in the design and development phase.

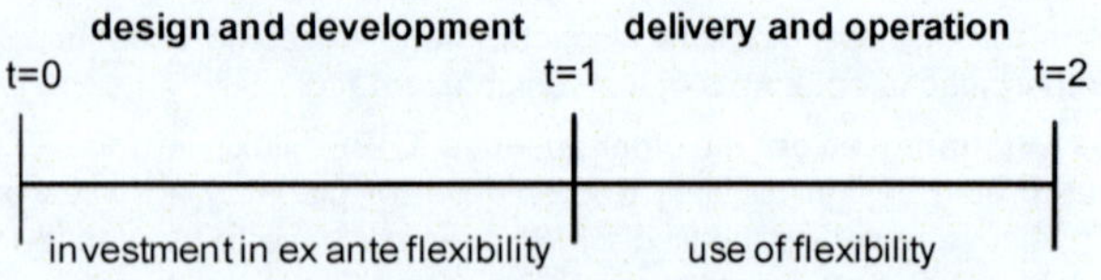

Figure 4: Temporal structure of decisions [9].

Obviously, IPS² flexibility will only be implemented if its benefits (B) exceed the costs (C) involved as well on the collective level of the business model as for each individual participant. So the following **conditions** have to hold:

$$C_{sup} < B_{sup} \tag{1}$$

$$C_{cust} < B_{cust} \tag{2}$$

$$C_{coll} = C_{sup} + C_{cust} < B_{coll} = B_{sup} + B_{cust} \tag{3}$$

This is easily fulfilled in the case of **ex post flexibility** because the customer will only request it if he is sure to achieve a surplus whereas the supplier will only agree to implement it if this seems profitable to him.

Ex post flexibility requires no specific investment in the design and development phase but causes high adaptation costs in the delivery and operation phase. The supplier will pass these costs completely to the customer.

On the other hand, the provision of **ex ante flexibility** means rather high costs for the supplier in the design and development phase whereas its use in the delivery and operation phase involves only low additional costs. The customer is charged for the ex post use of flexibility only if he requests an adaptation of the initial IPS². The rate must cover costs of investment and operation but will still be lower than the costs of using ex ante flexibility. So for ex ante flexibility, condition (1) does not hold if the supplier invests into IPS² flexibility that is not used later in the delivery and operation phase. Figure 5 shows the cost position of the supplier for both types of flexibility.

Depending on his risk averseness and the probability of this case, the supplier might not invest although the investment would generate a surplus on the collective level. In this case, individual and collective rationality do not coincide. A solution to this dilemma is that the customer agrees to increase the rate of the initial IPS² by an additional sum that serves as an **incentive** (I) to the supplier to invest into IPS² flexibility.

	costs of design and development	costs of delivery and operation
ex ante flexibility	**high**	**low**
ex post flexibility	**–**	**high**

Figure 5: Costs of flexibility.

From the customer's point of view, he pays for the **option** of using flexibility if needed later. This leads to the following changes in conditions (1) and (2) while condition (3) remains unchanged:

$$C_{sup} < B_{sup} + I \tag{1a}$$

$$C_{cust} + I < B_{cust} \tag{2a}$$

Obviously, the incentive I must be higher than the difference of the supplier's costs and benefit and it must be lower than the difference of the customer's benefit and costs. Depending on the data of the specific problem, the space of feasible solution may be very small or even empty. In the following, we will assume that there exist feasible solutions and present various bargaining arrangements.

4.3 Negotiations on IPS² Flexibility

The fundamental problem is that the supplier has to convince the customer about the advantages of a modular resp. flexible design. It is obvious that the supplier has an incentive to provide a more optimistic view about the achievable IPS² contribution to the customer´s benefit. Expecting that, the customer has an incentive to offer a lower willingness to pay than his estimation of the real value of the provisioned IPS² configuration. This lower willingness to pay is supported by an **inadequate communication** of generated benefit to the customer. According to that lack of communication, the customer´s risk rises and his willingness to pay for IPS² flexibility further decreases [11]. In order to convince the customer about the advantage of the modular IPS² configuration, the supplier´s credibility is of high importance [12].

So the question arises how the supplier can convince the customer to pay for flexibility that the supplier promises him for some date in an uncertain future. From the customer's point of view, not only the future developments are uncertain, but he faces also uncertainty whether the supplier implements the amount of flexibility he pays for. Possible solutions to this problem arise from **signaling theory** [13]: If the supplier has already carried out similar transactions and other market participants are satisfied with his performance, he can build up reputation and use it for customer acquisition.

Another interesting situation is the following: If the supplier alone has invested into IPS² flexibility, he has a substantial interest that the customer makes use of it in the delivery and operation phase so that he is able to retrieve his money by means of increased rates. In case of a moderate alteration in the environmental conditions, the customer might hesitate to ask for an adaptation of the initial IPS² because he is still doubtful whether his additional earnings will cover the increased rates. On the other hand, the supplier faces **sunk costs** amounting to the money invested if the customer does not request the flexibility. So it may be attractive for him to offer a **discount** on the rate to the customer thus getting at least some of his money paid back. From the point of view of managerial accounting, this decision is just a question of the ratio of (discounted) future costs and future revenues [14].

4.4 Change of Business Model

IPS² flexibility is not only important for the exchange of modules within a given business model. It may also support the changeover from the availability-oriented to the **result-oriented business model**. The customer is typically interested in offering an innovative product on a competitive market. If he lacks experience with this market, he will tend to avoid taking the risk of entering it because this requires an adaptation of the IPS² with higher payments to the supplier. So the supplier may – instead of offering an adaptation of the availability-oriented business model – use the IPS² flexibility provided before in order to propose a different business model. If he adapts the IPS² by exchanging modules to the requirements of the new market, takes over the production risk and payment is modified to a pay-on-production base, the business relation changes to a result-oriented business model.

5 SUMMARY AND FURTHER RESEARCH

It has been shown that IPS² flexibility is a major factor of success for the supplier because it allows him to adapt the performance of his system to changing customer requirements. Still the calculation of the present value of an investment into IPS² flexibility is a multifacet problem. Future investigations will concentrate on modelling a decision model for the determination of the optimal degree of flexibility.

6 ACKNOWLEDGMENTS

This research is financially supported by the German Science Foundation (DFG) through SFB/TR 29 on Industrial Product-Service Systems – Dynamic Interdependencies of Products and Services in the Production Area.

7 REFERENCES

[1] Meier, H., Uhlmann, E., Kortmann, D. (2005): Hybride Leistungsbündel. Nutzenorientiertes Produktverständnis durch interferierende Sach- und Dienstleistungen, in: wt online, Vol. 95, pp. 528-532.

[2] Oliva, R., Kallenberg, R. (2003): Managing the Transition from Products to Services, in: International Journal of Service Industry Management, Vol. 14, No. 2, pp. 160-172.

[3] Schreyögg, G. (2008): Organisation. Grundlagen moderner Organisationsgestaltung, Gabler, Wiesbaden.

[4] Zhang, Q., Vonderembse, M.A., Lim, J.-S. (2002): Value chain flexibility: a dichotomy of competence and capability, in: International Journal of Production Research, Vol. 40, No. 3, pp. 561-583.

[5] Wegerich, C. (2007): Strategische Personalentwicklung in der Praxis, Wiley-VCH, Weinheim.

[6] Müller, P., Schmidt-Kretschmer, M. (2008): Challenges in PSS development processes - New Paradigms, New Development Methodology, in: Transregio 29 (ed.), Transregio 29. Product-Service Systems. Dynamic Interdependency of Products and Service in the Production Area - International Seminar on PSS 21-22 January 2008, Shaker, Aachen, pp. 1-6.

[7] Kistner, K.-P., Steven, M. (2001): Produktionsplanung, Physica, Heidelberg, 3rd Ed.

[8] Kleinaltenkamp, M. (1999): Service Blueprinting. Ein Instrument zur Steigerung der Effektivität und Effizienz von Dienstleistungsprozessen, in: Technischer Vertrieb, pp. 33-39.

[9] Grossman, S. J., Hart, O. D. (1986): The Costs and Benefits of Ownership: A Theory of Vertical and Lateral Integration, in: The Journal of Political Economy, Vol. 94, No. 4, pp. 691-719

[10] Richter, A. (2010): Industrielle Produkt-Service-Systeme: Eine vertragstheoretische Analyse, Arbeitsberichte des Lehrstuhls für Produktionswirtschaft, Ruhr-Universität Bochum, No. 9.

[11] Backhaus, K., Becker, J., Beverungen, D., Frohs, M., Knackstedt, R., Müller, O., Steiner, M.; Weddeling, M. (2010): Vermarktung hybrider Leistungsbündel. Das ServPay-Konzept, Springer, Berlin-Heidelberg.

[12] French, K.R., McCormick, R.E. (1984): Sealed Bids, Sunk Costs, and the Process of Competition, in: The Journal of Business, Vol. 57, No. 4, pp. 417-441.

[13] Spence, A. M. (1974): Market Signaling: Informational Transfer in Hiring and Related Screening Processes, Harvard University Press, Cambridge et al.

[14] Northcraft, G.B., Wolf, G. (1984): Dollars, Sense and Sunk Costs: A Life Cycle Model of Resource Allocation Decisions, in: Academy of Management Review, Vol. 9, No. 2, pp. 225-234.

Customers' Satisfaction on Estimates of Queue Waiting Time in Service Delivery

Satoshi Shimada[1], Kei Taira[1], Tatsunori Hara[1], Tamio Arai[1].

[1] Dept. of Precision Engineering, School of Engineering, The University of Tokyo, Tokyo, Japan

Abstract

Evaluating customer satisfaction is a key factor in service design. Therefore, some studies of customer satisfaction have been carried out in the field of service engineering. In these studies, customers' expectations of service have generally been regarded as being shaped before service delivery. However, expectations of some components of service tend to be formed in the process of service receiving. To analyze service from the view of custoemer satisfaction, expectation that changes while customers are receiving service need to be considered. Therefore, we aim to develop the methodology to support designing services in a way that relates to the formation of customer expectations. In this paper, a model of shaping expectations in service receiving is examined and its effect on customer satisfaction is analyzed.

Keywords:

Service; Customer Satisfaction; Expectation

1 INTRODUCTION

In recent years, the economies of developed countries have been moving from manufacturing industries to service industries. Thus, studies of service are becoming ever more important. The study of the service industry has been carried out in the field of economics, marketing and management. At first, studies were concerned with the characteristics of service that are different from those of products in the manufacturing industry. Because service was seen as being more important in the manufacturing industry, the studies about how products and services were being delivered together attract the interest of engineers. The concept of service engineering [1] has developed from this background. In service engineering, we try to produce methodologies for designing services and to develop the service CAD [2] that supports a computer design.

Studies of services regarding effectiveness and costs are carried out in the same way as those about manufacturing. These studies try to improve services from the viewpoint of the provider. Analyzing and designing service from the viewpoint of the receiver is also important. This is because the difference between the viewpoint of the offer side and that of the receipt side regarding service is larger than that in manufacturing. Therefore customer satisfaction is an important index of evaluation of the service.

Customer expectation of service was introduced to as an element affecting customer satisfaction in preceding studies[3,4]. It is assumed that expectation affects decisions about customer satisfaction in these studies, and it is necessary to understand the expectations that customers have about services when the delivery of services is analyzed from the viewpoint of the receiver.

In this study, we aim to develop a methodology to support designing services in a way that relates to the formation of customer expectations. In order to achieve this, the forming of expectations about service is considered part of the evaluation of customer satisfaction. With regard to fundamental research, the functions of service are simulated in this paper by using a service model that includes the formation of customer expectations about services. Customer satisfaction is evaluated by methods based on the model. In this way, the influence of expectation formation on customer satisfaction can be demonstrated. From the result, suggestions about the methodology to support designing service are obtained.

2 PRECEDING STUDIES ABOUT CUSTOMER SATISFACTION

2.1 The method to estimate customer satisfaction by questionnaire

As described in the previous chapter, various studies of customer satisfaction with services have been carried out. For instance, the American Customer Satisfaction Index (ACSI) [3] is very useful in the field of assessing the effectiveness of service marketing. In the ACSI, customer satisfaction is estimated using questionnaires that are given to customers. ACSI has six indexes of service; "perceived quality," "customer expectations," "perceived value," "overall customer satisfaction," customer complaints," and "customer loyalty." These index have relationship each other.

2.2 The model of service quality expressing gaps in service

The gap model of service quality [4] shows gaps that exist in service. The gaps cause customers' expectation different from the actual service. Some of gaps appear while providers design service and improving organization of providers can make these gaps small. However, other gaps appear on the interaction between customers and service. Those gaps depend on what customers know about service. Thus, to improve them, providers need to support customers getting appropriate information of service.

2.3 The method to estimate customer satisfaction from contents of service

Studies of customer satisfaction have also been carried out in service engineering. To design service, providers have to determine factors of service. Thus, the methodology to evaluate each factor of service is necessary. The concept of satisfaction-attribute value function [5] was introduced to estimate customer satisfaction for each element of the service from a functional viewpoint. The function is expressed in the following equation (1):

J. Hesselbach and C. Herrmann (eds.), *Functional Thinking for Value Creation: Proceedings of the 3rd CIRP International Conference on Industrial Product Service Systems, Technische Universität Braunschweig, Braunschweig, Germany, May 5th - 6th, 2011*,
DOI 10.1007/978-3-642-19689-8_47, © Springer-Verlag Berlin Heidelberg 2011

$$S_r = a\left\{1 - e^{-\frac{b}{a}(v-c)}\right\} \quad (1)$$

The explanation of each parameter is as follows: the parameter "*v*," which expresses the value of a factor of service, represents the input of this function. "*a*" expresses maximum or minimum customer satisfaction regarding the factor, "*b*" expresses customers' sensitivity to the factor, and "*c*" expresses the value of expectation of the factor. Customer satisfaction "*Sr*" with a factor can be calculated when the value of three parameters and a factor are set. Figure 1 shows an example of a graph of that function. Parameter "*a*" has been changed to plus or minus according to whether or not the value of *v* is larger than of *c*. The graph in Figure 1 shows a case in which customer satisfaction decreases with the increase in the value of "*v*." Levels of customer satisfaction about waiting time are demonstrated by the graph. Prospect theory [6] can be applied in deciding the value of "*a*," and in this case, the loss of customer satisfaction is regarded to be larger than the gain.

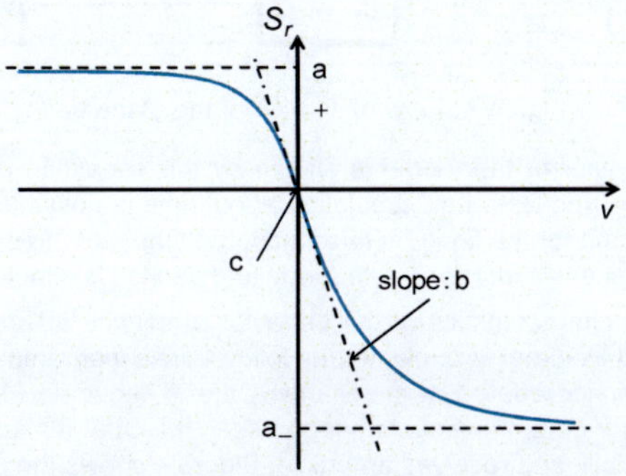

Figure 1: Graph of satisfaction-attribute value function.

As seen from the function, customer satisfaction is decided on by the difference between expectation and the actual quality of service, and customer satisfaction is rated at +-0 when the value of the factor matches the expectation. Thus, customer satisfaction cannot be estimated effectively unless expectations are decided upon

2.4 Customer satisfaction on waiting time

It had been established in previous studies that the expectation that customers hold about service affects customer satisfaction. With regard to the relationship between customer satisfaction and expectations, a study was carried out to assess expectations of waiting time for services [7]. This study dealt with the psychology of customers who are waiting for service and with how providers should manage waiting time. In this study, it is assumed that the expectations of waiting time that customers estimate after beginning to receive service has a more direct influence on customer satisfaction about waiting time than the expectation that customers estimate before they receive services. However, in preceding studies such as [3][4][5], it was shown that the expectations of service tended to be decided before customers received services and the influence of the services was reflected after they received service. The changes of expectation in the process of service delivery had not been modeled. Therefore, it was concluded that it was necessary to construct a framework for expectation formation about service delivery.

3 APPROACH OF THIS STUDY

In this paper, the formation of expectations about waiting time in service delivery is modeled on an actual example, and the influence of expectation formation on customer satisfaction is analyzed to build a frame. The structure of the relationship between expectation and customer, and the structure of the formation has been modeled with reference to [7]. Figure 2 shows this model.

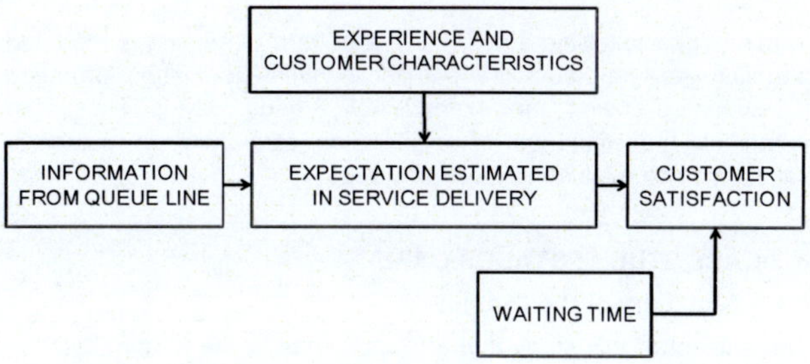

Figure 2: Structure of expectation and customer satisfaction about waiting time.

As shown in Figure 2, the customer obtains information from the length of a queue line during service and forms expectation of the waiting time as "expectations of waiting time estimated after service" on the basis of this information. The formation of expectations is also affected by a customer's experience and personal characteristics. Customer satisfaction is calculated by using the difference between the actual waiting time and the results of expectation formation. For calculation of customer satisfaction, a satisfaction-attribute value function is used, and the influence of expectation formation on customer satisfaction is analyzed numerically.

As explained in the previous chapter, the formation of expectations about service delivery has not been studied in detail until recently, and to enable service analysis from a more practical customer viewpoint, it is necessary to create a framework to model this. Therefore, this study aims to design a supporting framework for the service that considers the expectation formation related to service delivery in four steps:

- Step 1) Modeling service delivery from the viewpoint of both customer activities and provider activities
- Step 2) Incorporating a measurement of the formation of expectations after specifying parameters about expectation formation such as customer characteristic and environmental information, and constituting a way of evaluating types of customer satisfaction
- Step 3) Carrying out simulation and evaluating customer satisfaction
- Step 4) Suggesting a methodology for designing services bearing in mind the formation of expectations in service delivery on the basis of the results of the previous steps

Each step will now be explained in detail. In step1, service is modeled so that the service delivery can be analyzed in the model. For this purpose, the performance of service delivery needs to be expressed. The performance of service delivery includes time needed to deliver service, the interaction between customers and employees, and the interaction among customers. To realize these expressions, the service is modeled with Petri-net. The functions of Petri-net and the model of service expressed by it are explained in the next chapter. For step2, the parameters that affect the expectations formed in service delivery are specified and incorporated in the service model established in step1. In this way, a means of evaluation customer satisfaction, including the expectation formation, is constituted. In step3, service is simulated through the service model by which customer satisfaction is calculated. Regarding the result of this calculation, the influence of expectation formation on customer satisfaction is evaluated. In step4, expectation formation is considered from the standpoint of the design of service and the methodology to support designing service are developed based on the result of step3.

In this paper, following steps1 to 3, a pattern of service is modeled and simulated, and customer satisfaction with expectation formation is evaluated, using the example of Stands Café. This paper examines the influence of expectation formation on customer satisfaction by evaluating those results.

4 CASE STUDY : STANDS CAFE

4.1 Summary of the example

The subject of this study is the Stands Cafe in the terminal station yard in Tokyo. Service of the café is modeled and, using the model, the way in which services are delivered is simulated. The café has three employees, and each is in charge of different duties. Two employees prepare the drink ordered by the customers, and the third handles payment. The products are organized into five categories. The time needed to prepare them and the employee who makes them are different in each product. Table1 shows some detailed information about these products:

Article	Ratio of order [%]	Average Time to prepare it [s]	Employee who makes it
hand drip coffee hdc	45	60	prepare: B finish: A
milk coffee mk	25	40	B
ice coffee ico	20	20	B
caramel milk coffee cmc	5	70	B
ice caramel milk coffee ica	5	50	B

Table 1: Data of each article.

One of the articles, hdc, is made by two employees, and the others are made by only employee B. Of articles of this Stand Café, hdc has the highest ratio of order and it takes rather long time to prepare hdc. This fact makes some customers wait long till they receive articles. However, the employees of this café don't know the influence of hdc to the other articles and the influence of the others to hdc clearly. Thus, the employees would like to understand what causes the long waiting time.

4.2 Modeling of service by Petri-net

In this paragraph, the system of Coloured Petri Nets [8], which is used for modeling the service, is explained. Coloured Petri-net is the tool with which state transitions in a system are modeled and analyzed. As shown in Figure 3, Coloured Petri-net is made up of token, arc, place, and transition.

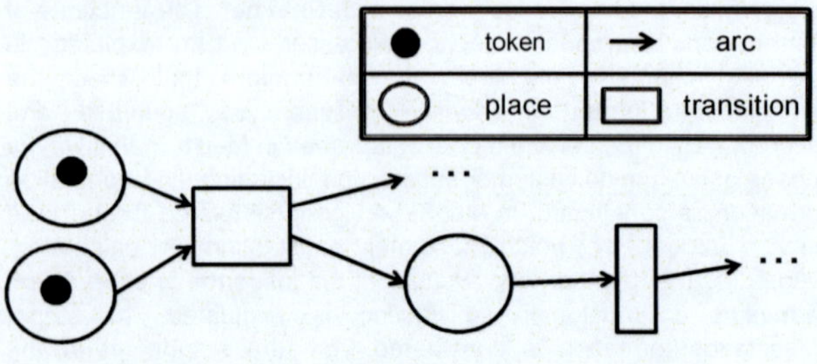

Figure 3: Elements and basic structure of Petri-net.

Place is tied to transition by an arc, and there is a token in place. The behavior of the model advances in the following order:

1. Transitions that all the places just before have tokens catch fire.

2. The tokens move to the former places of transition that caught fire.

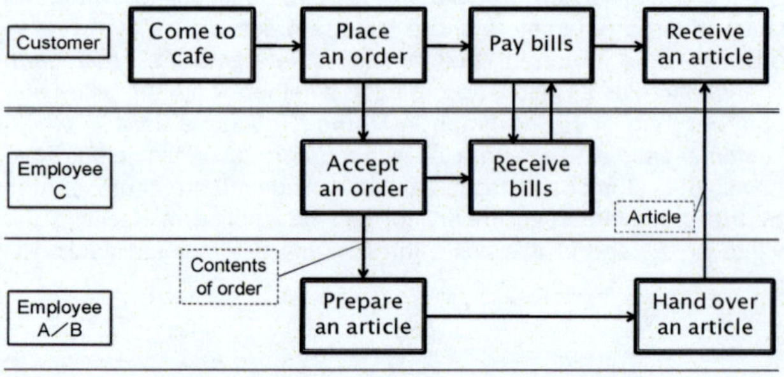

Figure 4: Flow of service of the stand café.

The behavior of the model is shown by this repetition of 1 and 2. With Coloured Petri-net, simulation about time is possible by setting a delay time for the firing. In this study, the time that the duties of the service take is expressed in this way and service is simulated.

How the characteristics of the behavior of service are modeled by Coloured Petri-net is explained as follows. The diagrammatical view of the service offered at the Stands Cafe is shown in Figure 4. As shown in Figure 4, the customer comes to the café, makes an order, pays the bill, and receives an article. Figure 4 shows the process for one customer in the cafe. However, there are usually several customers in the cafe at the same time, and they follow similar actions using the same process. Therefore, in cases where another customer comes to the cafe while one is still ordering, he or she lines up to place an order. The customer who is at the front of the line interacts with employee C for his or her order. The order content is conveyed from employee C to employees A and B, while the customers waits for their order to be delivered. Employees A and B make the required articles in their ordered turns.

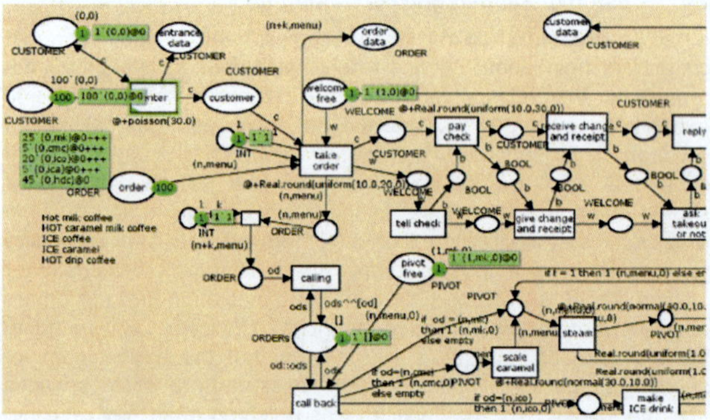

Figure 5: Model of stand café modeled with Petri-net.

Figure 5 shows the flow of the service offer with Petri-net. In modeling with the Petri-net, each employee is expressed as a token, and each takes a part of the service. The condition of divergence with the order contents is incorporated, and, depending on the order content, the process for an offer of an article may change. As described above, the time needed to supply the service is expressed by a delay line in the fire of transition. In addition to this, mistakes in the work are stochastically incorporated and the reshaping of articles is expressed. The time needed to carry out each duty is expressed by the function of dispersion, and includeds unevenness. The fixing of the amount of time needed for each duty and the ratio of each article being ordered is based on real data.

4.3 Setting of expectation formation

As described in the previous chapter, the purpose of this simulation is to analyze the service with regard to expectation formation, and to show the influence of expectation formation on customer satisfaction. The parameters of expectation formation are specified for the model that expresses the behavior of the service, and the level of expectation formed by those parameters is then calculated. The satisfaction-attribute value function is decided upon and the expectation and customer satisfaction is calculated.

In this report, customer satisfaction regarding two kinds of waiting time is evaluated by considering levels of expectation formation using a simulation setting in one service model. For this setting, the letters "*a*" and "*b*" in the satisfaction-attribute value function are determined as follows.

$$a = \begin{cases} 0.5, & v < c \\ -1.0, & v \geq c \end{cases} \quad (2)$$

$$b = 0.00134 \quad (3)$$

"*a*" is set based on prospect theory. One factor for which satisfaction is evaluated is the waiting time for service. Therefore, the smaller (shorter) the factor (time) is, the better the service will be for a customer. When "*v*,"; the value of the factor, is smaller than "*c*,"; the expectation of service, "*a*" is plus, while when "*v*" is larger, "*a*" is minus. "*c*" sets its value from expectation formation. The expectation formation for each factor of service can be explained in detail as follows :

4.3.1 Expectation formation for order waiting time

The order waiting time is the time from entering a shop to placing to order. In order to evaluate expectation formation about order waiting time, queue theory [9] was referred. Here, it is assumed that the customers order in turn same as the turn of coming into the store, and there is only one employee settling the account. These conditions are similar to the M/M/1 model (one window catching a line and the service) in queue theory. In the M/M/1 model of queue theory, "congestion" ρ is calculated by the equation (4) from "average service rate" μ; number of the customers whom the shop side can deal with in one hour, and "average arrival rate" λ; the number of the customers coming to the store in one hour.

$$\rho = \frac{\lambda}{\mu} \quad (4)$$

Furthermore, from ρ, the number of people one waits for until he or she takes service is calculated in equation (5).

$$N = \frac{\rho}{1-\rho} \quad (5)$$

By calculating the product of the *N* and the offer time of service per customer (a reciprocal number of average service rate μ), an estimate can be made of how much time has passed by the time service is received.

In the queue theory mentioned above, the number of people one customer waits for till he or she takes an order is considered so as to estimate waiting time. However, customers can know the number of people in front of them in a queue when they receive the service. Therefore, *N* is provided as a concrete numerical value for each customer, and, from the value of *N*, expectations may be changed in each case. Regarding the offer time for each customer to be served, this is actually fixed by the interaction of exchanges between the employee and the customer. However, the estimation of time will differ according to each customer's experience of similar service and the intention of the individual, so the expectations can be changed by each customer even in same conditions of service. In consideration of this, the expectation formation for order waiting time can be expressed by equation (6):

$$c = t \times n \quad (6)$$

"*c*" is the expectation of waiting time for an order that is expressed by this equation, "*t*" is the offer time for service for one customer that may change in terms of how one estimates it, and "*n*" expresses the number of people lining up in front of a customer. "*n*" is decided by the situation at the time of the service offer and, in simulation, a value of "*n*" for each customer is decided from a simulation result. In addition to "*n*," the difference of the estimate for "*t*" influences expectations of the time needed to offer service to one customer. This expectation formation, as explained above, affects customer satisfaction, and allows the influence of different situations of service to be analyzed.

4.3.2 Expectation formation for article waiting time

The article waiting time is the time from the account being taken to the receipt of the article. The factors which decide article waiting time are the work speed and work content of the employees. The work content of employees depends on the contents of articles that not only the particular customer but also others may order. However, it is hard to imagine that a customer will know how many and what kind of orders have been made by other customers. Furthermore, grasping of the number of people who are waiting for articles is difficult because in fact the customers waiting for their articles rarely line up. Therefore, the only information that a customer can be sure of is the content of his or her order and the expectation of article waiting time is formed from this information. The expectation formation on article waiting time can be expressed as equation (7):

$$c = T_{(order)} \quad (7)$$

$T_{(order)}$ expresses the number to be fixed by the contents of the order. The value of T is fixed from the mean duration needed to make each article. For example, if one customer orders milk coffee (mk), T for him or her is 40 seconds, as shown in Table1.

The Simulation was performed using the setting that 100 customers arrive at the café at the arrival rate of 120 customers per hour, and assuming that one customer comes to the café in 30 seconds on average. The data shown in the next chapter are the results of the simulation.

5 RESULT OF SIMULATION AND DISCUSSION

5.1 Customer satisfaction on order waiting time

Figure 6 expresses as a histogram how many customers were lined up in front of each customer. The horizontal axis represents the number of people in line in front of the customer, and the vertical axis the frequency (100 in total). The number of people in line in front of customer is decided by the average arrival rate of customers and the time needed to order and to fulfill the account.

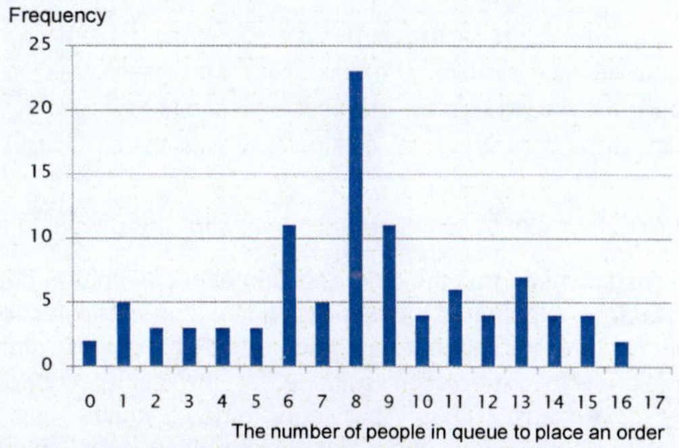

Figure 6: Histogram for people in queue.

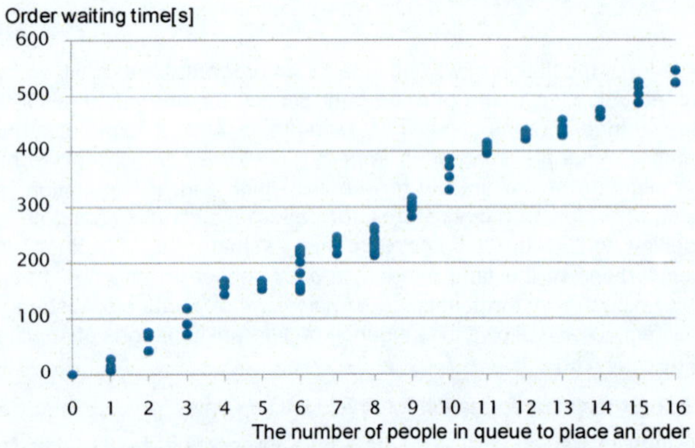

Figure 7: Order waiting time and the number of people in queue.

Figure 7 expresses the relationship between the number of people in the line and order waiting time. The horizontal axis represents the number of people in the line, and the vertical axis represents order waiting time. The time needed to order and create the account has an unevenness, of which the difference is small for each customer. Therefore, as a result, the order waiting time is related to the number of people in line. As mentioned above, the results of the expectations formed from the number of people in line, and of customer satisfaction regarding order waiting time as calculated from the expectations, are shown below.

Figure 8 is a box plot that explains customer satisfaction in relation to order waiting time. The vertical axis represents customer satisfaction, and the horizontal axis expresses an estimate of "t," service offer time in relation to estimation per customer. As is clear from the figure, there is a large difference in expressed as "t." A difference in the unevenness of customer satisfaction by "t" is characteristic. The larger the difference of "t" from 35 (related to both larger and smaller sides), the greater the unevenness becomes. As for the setting of the simulation, the mean of service offer time per one becomes 35, and the variation has a smaller expectation formation that is nearer to a real offer condition.

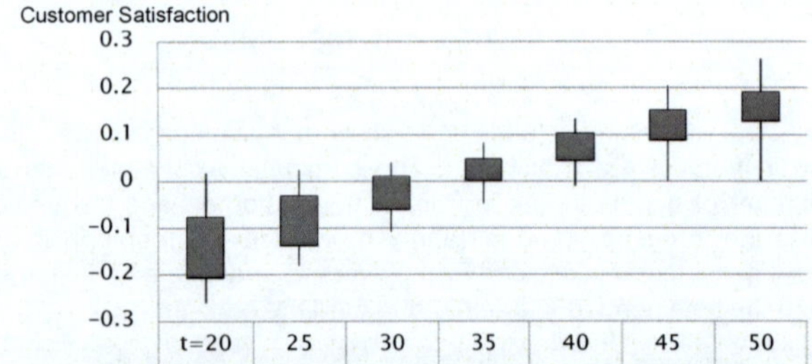

	t=20	25	30	35	40	45	50
3rd quartile	-0.076	-0.0303	0.01257	0.0472	0.10425	0.15365	0.19135
Maximum	0.01841	0.02482	0.03986	0.08218	0.13458	0.2051	0.26201
Minimum	-0.2613	-0.181	-0.1088	-0.0471	-0.0093	0	0
1st quartile	-0.2077	-0.1385	-0.0601	-0.0003	0.04355	0.08774	0.12764

Figure 8: Box plot of customer satisfaction for order waiting time.

These results mean that the differences of expectation that depends on customers' experience have effects on customer satisfaction and the effects are too large to neglect. Addition to that, customer satisfaction with expectation suitable for the actual service has smaller evenness. Thus, providers should give customers information about service and make customers' expectation suitable for service.

5.2 Customer satisfaction on article waiting time

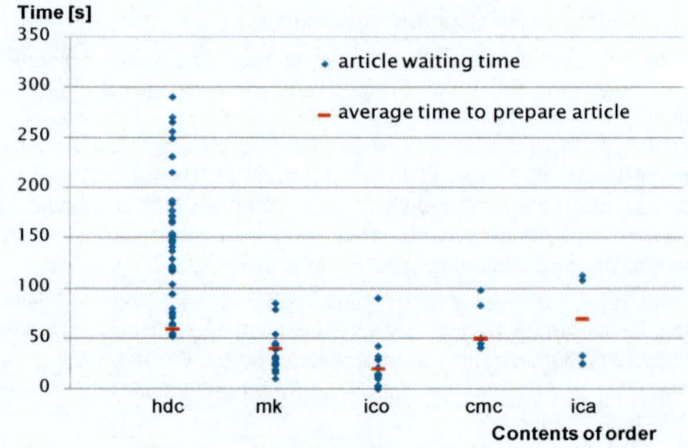

Figure 9: article waiting time and order contents.

Figure 9 showed the relations of order contents and article waiting time. The horizontal axis represents the contents of order and the length of time from the end of ordering to receiving the article is represented by the vertical axis. As described in the previous chapter, an expectation of article waiting time is formed from the order contents and, from this, the degree of customer satisfaction is calculated. The results are shown in Figure 10.

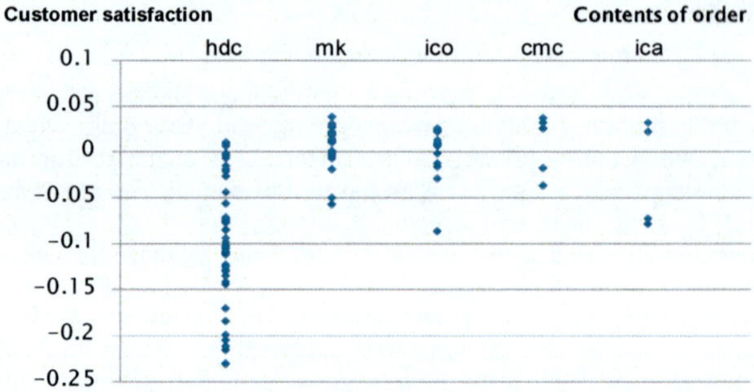

Figure 10: Customer satisfaction for article waiting time for each order content.

Figure 10 shows the distribution of customer satisfaction regarding article waiting for each item. Customer satisfaction is represented by the vertical axis, and the horizontal axis shows the concent of order. As for hdc (hand drip coffee), customer satisfaction levels tend greatly towards dissatisfaction. As shown in Table1, hdc represents a larger ratio for ordering than other articles and, takes longer time to serve. Furthermore, the customer has to wait, even in the case of only one employee A or B is making one article and the other employee is free, because both employee A and B are engaged in making hdc. On the contrary, as for the other articles, only employee B is engaged in making articles and the time to make hdc hardly affects waiting time. Furthermore, the articles that takes a longer time to make than hdc have a much lower order ratio than hdc. Therefore, it is unlikely that the waiting time for these articles will be longer than the time for making hdc.

This paper has demonstrated the results of the simulation using the employees of the Stand Café as a model. One of the employees commented that "It demonstrates the true facts about the Stands Café." The results of this simulation do not mean that sale of hdc should be stopped because of its lower customer satisfaction than other items. Considering the high ratio of hdc order, hdc must be attractive for customers even though waiting time needed to receive hdc tends to be longer. From the result, it can be said that orders of hdc have little influence on the waiting time of the other articles and orders of the others also have small influence on that of hdc. Therefore, if the provider would invest in the improvement of service delivery process for better customer satisfaction, they had better start from the review of waiting time of hdc.

6 CONCLUSION AND FUTURE WORK

In general, it has been assumed that the expectations of customers were formed before services were received. In this research, the formation of expectations in service delivery about waiting time has been modeled using the example of Stands Café. Customer satisfaction is analyzed through the model and the influence of expectation formation on customer satisfaction is shown. The conclusions can be summarized as follows:

To make the unevenness of customer satisfaction less, it is necessary to let customers carry out a process of the expectation formation suitable for the actual service they will get

The example of service that was modeled and simulated in this paper is the case of business-to-consumer service. It means that receivers are not usually experts of services and they tend to have expectation of waiting time that is not based on the actual service. In case of business-to-business service, customers know the service and what they receive well, and waiting time is often included in contracts as the date of delivery. Thus, to satisfy customers, providers should consider contracts from the viewpoint of expectation, instead of considering the date as the span in which they have to finish their jobs and trying only to make lead time short.

Future work includes the following:

Step 1: The differences in the service process, e.g., condition divergence, considered as the differences in expectation formation, will be expressed in the service model. This makes the heterogeneity of the customer clear and the behavior of the actual service possible to be modeled in detail.

Step 2: To express the expectation formation in more detail

A) To list environmental information about expectation formation

B) To manage the expectations from the point of view of customer education

Step 3: Along with the waiting time, customer satisfaction regarding several elements of the whole service is evaluated.

Services depend on the environment more heavily than products and the objective quality brought to the customer as a result of the activity of the provider is even more dynamic. This is why the expectations of the service by the customer can be changed by how the service is received. The customer develops expectations through personal experience, and the expectations that a customer has of the service will be affected by the conditions of the environment at the time of the service receipt. Therefore, the heavier the dependence on the environment of the service industry and of service factor, the more necessary it is to consider the process of expectation formation during the receipt of the service. In future works, these factors will be inspected and considered more deeply, and a methodology for designing services related to expectation formation will be developed.

7 REFERENCES

[1] Arai, T., Shimomura, Y. (2004): Proposal of Service CAD System –A tool for Service Engineering-, Annals of the CIRP, 53(1), pp. 397-400.

[2] Arai, T., Shimomura, Y. (2005): Service CAD System –Evaluation and Quantification-, Annals of the CIRP, 54(1), (ISSN 1660-2773), pp. 463-466.

[3] Fornell, C., Jhonson, M. D., Anderson, E. W., Cha, J., Bryant, B. E. (1996): The American Customer Satisfaction Index: Nature, Purpose, and Findings, Journal of Marketing, 60(Oct), pp. 7-18.

[4] Parasuraman, A., Zeithaml, A. V., Berry, L. L. (1985):A Conceptual Model of Service Quality and Its Implications for Future Research, Journal of Marketing, Vol.49 (Fall), pp. 41-50.

[5] Yoshimitsu, Y., Kimita, K., Hara, T., Shimomura, Y., Arai, T. (2007): Proposal of a Measuring Method of Customer's Attention and Satisfaction on Services, In Proceedings of the 14th CIRP International Conference on Life Cycle Engineering -LCE2007-, pp.417-422.

[6] Kahneman, D., Tversky, A. (1979): Prospect Theory: An Analysis of Decision under Risk, Econometrica, XVLII, pp. 263-291.

[7] Looy, B. V., Gemmel, P., Dierdonck, R. V. (2003): Services Management: An Integrated Approach, Financial Times, Prenticce Hall, Harlow, England.

[8] Jensen, K. (1990): Coloured Petri Nets: A High Level Language for System Design and Analysis, in: G. Rozenberg (Ed.), Advances in Petri Nets 1990, Lecture Notes in Computer Science, Vol. 483, Springer, Berlin, pp. 342-416.

[9] Kendall, D. G. (1951): Some Problems in the Theory of Queues, J. Roy. Statist. Soc., Ser. B 13, pp. 151-185.

Integrating Environmental Assessment Methods: Development of a New Approach

Valentine Moreau[1], Natacha Gondran[1], Valérie Laforest[1]

[1] Institut Henri Fayol, UR SEPIT, Ecole Nationale Supérieure des Mines de Saint-Etienne, Saint-Etienne, France

Abstract

Human activities generate local environmental impacts (on the quality of water, air and soil), as well as global impacts affecting the ecological carrying capacities at world level.

The proposed methodology aims at representing both local and global impacts of a given activity, taking into account both direct and indirect impacts. The final objective is to establish an environmental dashboard based on a set of relevant indicators, in order to assess local and global environmental impacts caused by a given activity.

This paper will present the details of this methodology.

Keywords:

Environmental Assessment; Local and Global Impacts; Environmental Dashboard

1 INTRODUCTION

In 1972, "The Limits to Growth" [1] was one of the first major studies highlighting the ecological dangers of the economic and population growth. It was the first sign of the potential impacts of human beings on the global carrying capacities of the environment. From a legislative perspective, the European Commission has set up two directives: in 1985, the "Environmental Impact Assessments" - EIA Directive [2], and in 2001, the "Strategic Environmental Assessment" – SEA Directive [3]. They both mandate an impact study prior to completion of projects, defined by the importance of their size or their future impacts on the natural environment. For example, in France, the legislation on "installations classées pour la protection de l'environnement" (industrial installations classified for pollution prevention) requires an environmental impact study for the delivering, by the administration, of a permit to exploit. However, up to recent years, the compulsory environmental assessment studies mainly concerned local environmental impacts.

More recently, and mainly within the last ten years, many studies have concerned more global environmental impacts, and, with them, methods of quantifying these impacts.

The purpose of this article is to expose a methodology which retrieves both local and global impacts of a given activity, taking into account both direct and indirect impacts. Direct and indirect impacts are closed and linked. A direct impact is the first consequence of an activity: its results could be an indirect impact. So, there exists a cause-consequence relation and a relation of time between these two categories. For example, transports create emissions of VOC (direct impact), which could participate to the ozone pollution (indirect impact).

There is a relation of cause and effect between the concepts of environmental aspects and impacts. The international standards organization defines an environmental impact as "*any environmental change, negative or beneficial, completely or partially resulting of an environmental aspect of an organization*" [4]. The environmental impact is thus considered as the effect of an environmental aspect, which is a part of the activity that causes this impact [5].

The originality of this methodology is to consider a service approach. Indeed, it evaluates part of an activity of an organization, by inventorying and taking into account both the use of the various products and services that are necessary for a defined service and their life cycle.

2 CONTEXT STUDY

2.1 Environmental assessment

For a better understanding of the environmental impacts that are generated by a given service, it is necessary to conduct studies to quantify them, based on the data and scientific knowledge available. That is the objective of an environmental assessment method.

Environmental assessment is a systematic process to document and evaluate the environmental impacts caused by a given system (product or service). It is then possible to evaluate the performance of a system in a planning and decision making mechanisms in view of predicting and managing environmental impacts [6].

2.2 Existing methodologies

Various environmental assessment methodologies have different objectives, as displayed in figure 1.

The choice of a tool depends on the scale (local or global), but also on the objectives and the type of environmental impacts to be assessed. The final objective of these evaluations is to identify the environmental weak points of the system under study, in order to identify options for action.

These tools are more or less technical and complex and do not consider the same spatial or temporal scales and the same types of impacts. Moreover, they do not have the same limits for a study: it seems to be difficult to use only one tool to assess all the impacts and their interactions.

J. Hesselbach and C. Herrmann (eds.), *Functional Thinking for Value Creation: Proceedings of the 3rd CIRP International Conference on Industrial Product Service Systems, Technische Universität Braunschweig, Braunschweig, Germany, May 5th - 6th, 2011*,
DOI 10.1007/978-3-642-19689-8_48,

Methods	Objectives
Environmental Impacts Study [7]	**Detail all the local environmental impacts of an activity**
Environmental Performance Evaluation [8]	**Follow the impacts of an activity as an environmental management tool**
Life Cycle Assessment [9]	**Characterize the impacts along a product manufacturing chain**
GHG protocol [10] Bilan Carbone® [11]	**Estimate the greenhouse gas that are generated, both directly and indirectly, by an activity or a territory**
Ecological Footprint [12] MASIT [13] MIPS [14] MFA [15]	**Estimate material and energy flows that are generated by a population, an activity or a service**

Figure 1: Environmental assessment methods and their objectives.

Based on these existing methods, the new methodology proposed in this paper aims at building an environmental dashboard.

3 METHODOLOGY APPLIED

The methodology consists of five main parts. First, an identification of stakes and scope of the study is essential to identify the indicators that should be considered. The two next steps consist of data collection and their analysis. The fourth part is the creation of the environmental dashboard. Finally, this environmental dashboard is considered as a basis for the assessment of the arising benefits.

The Information Technology (IT) department of a company already involved in a sustainable development strategy wanted to reduce the environmental impacts of the activities of their centralized and diffused computers. Moreover, results obtained by this study will support their decision between the extension of their data-center and the creation of a more eco-efficient one. In this mindset, a new method was proposed and applied in order to take into account direct and indirect consequences. This paper presents only the methodology and not the results of this case study.

The originality of this methodology is seen through the assessment of a service produced by part of an activity of an organization. In opposition to the LCA methodology, which considers all the cycles of life of a product, the methodology proposed here aims to assess the cycles of life of a set of products used to provide a given service and compare the contribution of the various categories of products, and their phase of life (fabrication, use, waste treatment).

3.1 STEP 1: Identification of scope and stakes

In any environmental assessment method, the scope of the study must be defined first. For the life cycle analysis, it is the functional unit that will allow comparison of two products for the same service.

Defining the scope of the study makes it possible to limit the boundaries of the inventory of data that will be necessary. The objective of our approach is to amass elements to conduct an environmental assessment of the service provided by part of a company.

In this context, the subject of the study is neither the product manufactured by the company studied, nor its geographic scope, but the service provided by some of its activity (activity that is necessary for its internal operation, for example).

It is therefore necessary to define precisely the service provided by this activity and the boundaries of the inventory.

Environmental stakes are defined as *"legitimate operating factors, with implications for operations strategy [and as] the potential for operations to plan for, influence and leverage environmental issues for competitive advantage both internally and externally"* [16].

The identification of environmental stakes relevant for the system under study is essential to identify environmental issues that must be studied in priority, and consequently, the assessment method that has to be privileged. For example, is the activity generating locally worrying water pollution that must be assessed more precisely or is it an activity that seems "cleaner" locally, but that generates important delocalized energy and material flows?

This step involves an analysis of the nature of these issues and the vulnerability of the system under study. Vulnerability is defined as *"the state of susceptibility to harm exposure to stresses associated with environmental and social change and from the absence of capacity to adapt"* [17].

The main environmental stakes identified and the sensitivity of the site to different impacts makes it possible to prioritize them and subsequently, derive the data that must be collected.

3.2 STEP 2: Data collection

The collection of data is often a fastidious, but essential, step in any environmental assessment. This collection can be achieved through three steps:

- An **inventory**, which aims to collect the necessary information, such as a list of incoming and outgoing flow.
- **Surveys** (with operators, for example) to validate the information gathered for the inventory. They help to know the behavior of operators, to assess whether the service has been delivered and finally to assist in developing the protocol for the following phase. Questions asked can concern the value, nature or origin of device, explaining the flows identified in the inventory, etc.
- **Measures** which will aim at validating and/or completing all information previously collected, like the real value of the flow in m^3 of water consumed, kWh of electricity, tons of waste collected, etc.

These three phases can be implemented one after the other or in parallel. A feedback is needed to validate and supplement the information gathered at each level.

3.3 STEP 3: Analysis and Data Processing

Once the data is collected, it is possible to have information related to the different impacts under study. The quantity of devices (given by the inventory (like desktop computers, laptop computers, printers, servers, etc., for example) is a revealing value. Indeed, the amount of raw materials, as well as the depletion of natural resources (direct impact), will be estimated from the number of devices. Origins of these devices give information on the necessary transport to ship to the company. These transports have indirect impacts like, for example, contribution to photochemical pollution. By quantifying transport, contribution this phenomenon can be estimated.

Then, local impacts can be estimated, even if they are not directly generated on the site where the activity under study takes place. For example, the elimination of waste of electric and electronic equipments (environmental aspect directly linked to the activity of the system under study) involves the use of chemical substances generating acidification (indirect local environmental impact). Global environmental impacts can be estimated through, for example, the estimation of the energy necessary during the phase of use of the device that was inventoried (direct environmental aspect). Using conventional factors of emission for the national or European mix of energy, consumption can be translated into greenhouse gases emissions in order to evaluate the contribution of this environmental aspect to climate change.

To make these estimations, coefficients factors have to be retrieved from the literature to convert environmental aspects into environmental impacts.

These coefficients rely on assumptions and conventions published in different environmental assessment methods and databases, such as Ecoinvent (Ecoinvent Database v2.2:2010) [18]. Knowing that any new method published needs an exhaustive list of factors, any new specific case under study requires the search of new factors in the literature or estimation of these factors by assumptions and back-of-the-envelope calculations.

The outcome of environmental assessment is a set of indicators that represent the environmental impacts and performance of a given activity. ISO standard on the Environmental Management, such as Environmental Performance Evaluation Guidelines (EPE), ISO 14031 [8] distinguished two categories of indicators: environmental performance indicators (EPI), in which a distinction is realized between the operational indicators (consumption, for example) and management indicators (for example, issued from surveys of employees of the organization). The second class concerns the environmental condition indicators (ECI) which give information on the environment sensitivity in which the activity takes place.

In this study, EPI were mainly used. Afterwards, information must be synthesized into an environmental dashboard that aims at identifying weak and strong points of the activity under study (major impacts, largest share, potential savings, etc.).

3.4 STEP 4: Environmental Dashboard

Objectives

An environmental dashboard is expressed through a set of environmental indicators intended to represent the state of the environmental pressures (natural resource consumptions, emissions, for example) or the responses to mitigate those impacts. It can describe the evolution of a given phenomenon.

The objectives of a dashboard are various: facilitating the understanding of environmental issues, comparing different sites or activities, following the evolution overtime of an environmental impact, etc.

A dashboard is a management tool, at the crossroads of an information system and a decision-support system. It allows aggregating and synthesizing different information, while following the results of decisions made previously or estimating the various potential consequences of several possible decisions by simulating various scenarios. It can be used to achieve an overview of an initial situation, to compare with another site or activity and then, if maintained regularly, to help monitoring the situation.

Content

On the one hand, a dashboard consists of a series of elements, which represent different activities, sites or years. On the other hand, it contains the expression in terms of environmental impacts of each position studied: the indicators, usually quantitative [19]. They are intended to simplify the information, highlighting the sometimes complex phenomena. They often rely on observable or calculable quantities of materials, activities or services that may impact the environment. The data concerning the activity of the entity studied (consumption, device, releases, etc.) can be transformed (multiplied by conversion factors available in the scientific literature or databases, for example) to represent the importance of the activity with respect to the impact considered [20]. For example, emissions of greenhouse gases related to electricity consumption of a given entity, expressed in $tCO_2eq.$, are estimated by multiplying the annual electricity consumption by a conversion factor defined by a convention (for example, Bilan Carbone® ADEME) [11], to represent the average greenhouse gas emissions per kWh. In this study, measures of consumption realized and European electricity mix conversion factor allow to express greenhouse gas emissions.

Moreover, when the indicators conventionally used in the literature [21] seem difficult to understand, a new representation mode of the impact can be considered. It may use alternative indicators, representing the same impact, but in different units, which can improve the resonance of indicators to future users of the dashboard. Resonance of an indicator is defined as "the ability of an indicator to spark off immediate interest and understanding of the user" [22].

3.5 STEP 5: Assessing benefits

The last step of the methodology lies in identifying options for actions and assessing their benefits in environmental and economic terms. This is divided into three parts:

- **Identifying opportunities for improvement**: the environmental dashboard may highlight weak and strong points that will be the priority lines of thought, such as a process consuming a lot of water or electricity.

- **Scenarios of improvement**: after a feasibility study, the parameters relating to the implementation of improvement solutions will be incorporated into the dashboard to evaluate the new performance. For example, one can calculate the value of the indicators of the dashboard if the process identified as a critical point is replaced by another consuming less water.
  For this example, *in environmental terms*, the benefit is the decrease of the amount of water that is used. For quantification in economic terms, the return on investment can be estimated by dividing the sum of the gains made over the period by the cost of investment. *In economic terms*, the benefit is the cost associated with the substitution of the old process by the new one and the benefits related to smaller water consumption.

- **Comparison of improvements**: the proposed improvements will be analyzed by the decision-makers according to their environmental policy [23].

The ultimate goal is to prioritize the improvement actions achieving the objectives. This aims at helping decision makers in their choices to optimize their organization [24].

The whole methodology is shown in Figure 2.

4 DISCUSSION

4.1 Environmental impacts

Local and global impacts

Most of the environmental impacts associated with human activity have noticeable effects locally. It is possible to classify these impacts on soil, air and water and with which ecosystem the interaction is the most important. The term "global impact" is used when one of the mechanisms ensuring the global balance of the planet is disrupted by human activity. The impacts of the Information Technology activity can be local, such as release of heavy metals on the production sites or eutrophication, but also global with implications on large-scale phenomena, such as climate change. Note that a difficulty with the environmental impacts related to IT hardware is that these impacts are invisible to end users.

As seen above, much of the impacts result from the manufacturing of the devices. Yet, this production does not generally take into account where end users are located. Indeed, some of these impacts can be considered as local, but relocated as they take place away from the end user's location. Similarly, using a computer consumes electricity. This energy has been produced elsewhere, so the use of this energy also has a delocalized impact on the environment that end users often do not perceive, but that has to be highlighted.

Direct and indirect impacts

The concepts of direct and indirect impacts are discussed in literature. Direct impacts are defined as "*impacts resulting from a causal relationship between a project component and a component of the environment*" and the indirect impacts as "*impacts arising from changes on the environment having suffered from a direct impact*" [6].

Indirect impacts are sometimes called second generation impacts because they may result from direct impacts (called first generation impacts). The causes of such an impact as a discharge is often clearly identified by the population, but the consequences are still more difficult to perceive. Environmental knowledge is necessary to understand the causal relationship existing. Today, people are beginning to be aware of these phenomena. Thus, impacts caused by daily activities are often invisible from an end user point of view, which makes awareness-raising difficult.

To summarize, a given activity (providing a service, for example) has direct and indirect impacts on the environment, but also local and global impacts along the fluxes and the lifecycle of devices. For example, the transport that is necessary to deliver raw materials to the manufacturing plant will have direct impacts (like emissions of volatile organic compounds (VOCs) in air). These VOCs are then involved in the phenomenon of tropospheric ozone pollution (indirect impact), particularly irritating to eyes and respiratory mucous membranes and possible cause of reactions such as asthma attacks. This phenomenon takes place locally, even if the VOCs were emitted at a distance more or less important to the place where it occurs. It is therefore a local impact (even if it is relocated) and not a global one. However, the increase in emissions of greenhouse gases resulting from the same transport will lead to a global phenomenon: climate change.

4.2 Conversion factors

Using a conversion factor should be done in a rational way. The availability of this factor is not a sufficient parameter if its characteristics are not known. Scope, methodology, scale (spatial and / or temporal) and objectives are parameters that vary from one method to another. Only an extensive study of these methods allows to spot these differences and the assessment method's user must be careful with their use.

Considering this situation, environmental impacts can be represented by a range of values.

4.3 Complementarities with other methodologies

Originality

This new methodology might be considered as a new approach as it is designed to carry out the collected data and use the environmental dashboard as a basis for the assessment of the arising benefits.

As described in section 3, the assessment of a service produced by part of an activity of an organization represents the original scope of this environmental assessment methodology.

Integration of other methodologies

The new approach used principles, indicators and conversion factors from the existing methods described in figure 1.

First, the expression of local impacts of an activity is based on the principle of the regulatory Environmental Impacts Study [7]. To obtain the detail of these impacts, the same calculation of quantification were used. Second, for the global impacts like greenhouse gas emissions, conversion factors are used from different existing methods and databases. In this way, the French methodology of the Bilan Carbone® [11] and values of the Ecoinvent Database [18] allow to have environmental impacts.

Data, environmental impacts and monitoring indicators are organized in an environmental dashboard. It represents a real environmental management tool like the Environmental Performance Evaluation [8].

Finally, LCA methodology [9] allows the characterization of the impacts along a product manufacturing chain.

Comparison with LCA methodology

In spite of the evident use of different existing methods, this section aims to explain the differences between the proposed methodology and the LCA methods (described in the ISO 14040 [9]).

The first principle of an environmental assessment is the definition of the scope of the study. In the case of LCA, the complete life cycle of a product is considered, ranging from raw material extraction, use, end of life treatment and disposal of waste [9]. The object of our study is not a single product, but a set of interrelated products. This allows the identification of the role of part of an activity in the functioning of the organization studied. Thus, **the study scope differs between the two different methodologies**.

The second principle stated in the standard reference of the LCA is the environmental interest. For the methodology developed, even if the evaluation of environmental impacts is a key objective, **economic and social aspects are taken into account** and are an important part of the methodology. The social aspect is considered during the data collection by taking into account the users/operators working on the part of the activity under evaluation. As for the economic part, it occurs particularly during the last phase in the evaluation of potential benefits. These gains can be expressed in environmental and/or financial terms.

The next principle is an essential element of the LCA methodology: the relative approach and functional unit. Probably the best known step of this method, the functional unit, is the basis of calculation for all analyses and comparisons in the LCA. The methodology developed is not intended to be used as a comparative tool (unlike the LCA), but as a tool for improvement within a single structure. Consequently, the approach is specific to a given entity and a given activity. The concept of functional unit is not used for our methodology.

In contrast, the fourth principle of the LCA, the iterative approach, is a key element of this methodology. These multiple contributing feedbacks (as in LCA) are essential to validate the previous step and also to the completeness and consistency of the study [9].

5 CONCLUSIONS

The concept of environmental assessment methods is not new, but their various potential objectives and applications have resulted in their recent emergence. The methodology proposed is not an additional method, but aims, through the use of indicators already present in existing methods, at synthesizing different approaches. This can lead to an assessment of direct and indirect impacts, both locally and globally.

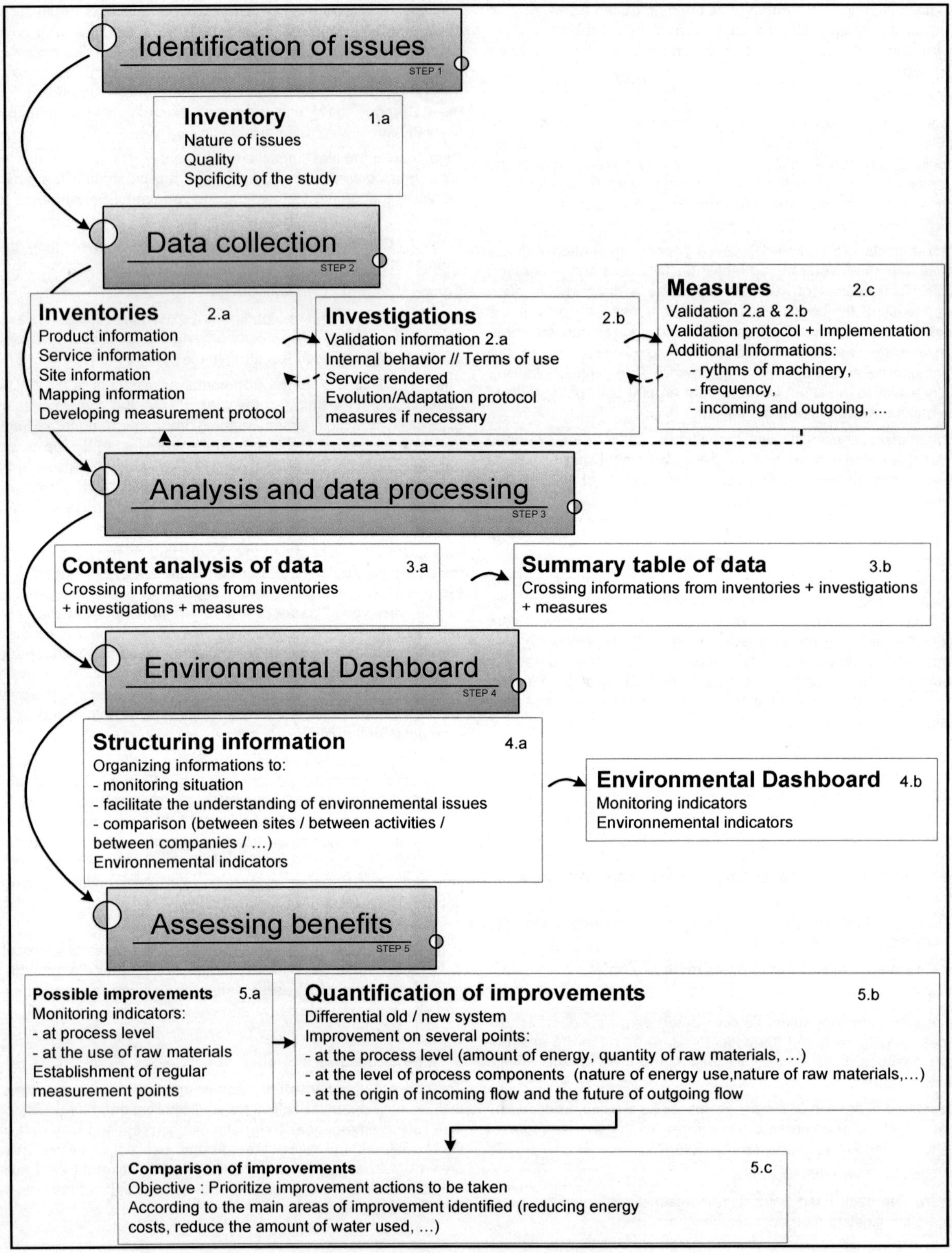

Figure 2: Proposed methodological approach.

6 SUMMARY

Human activity leads to local environmental impacts (pollution problems that affect the local quality of water, air and land), as well as global ones that affect the ecological carrying capacities. Direct or indirect environmental impacts are added to this notion, depending on whether the environmental change generated is a direct or indirect consequence of a phenomenon.

The proposed methodology aims to represent both local and global effects of a given activity, taking into account both direct and indirect impacts. The final objective is to establish an environmental balance dashboard, which is based on a set of indicators, for the assessment of environmental pressures generated by services rendered by part of an activity of an organization at different scales. The originality of the scope of this study is to assess the service provided by part of a business activity (activities necessary to its internal operations, for example).

The details of the methodology were presented: selection of scope and indicators, including the participation of employees and decision-makers within the steering committee of this study, identification of the main strengths and weaknesses, choice of scenarios of improvement and possible expression of environmental benefits through their implementation.

7 ACKNOWLEDGMENTS

We extend our sincere thanks to the company which financed and allowed the application of this methodology. We also thank all the employees who filled out the questionnaire, facilitating the implementation of the project within their walls. Finally, we would like to thank our language reviewer, Charles David Mathieu-Poulin.

8 REFERENCES

[1] Meadows D., (1972) The Limits to Growth, A Global Challenge; a Report for the Club of Rome Project on the Predicament of Mankind, Universe Books, New York, 205p.

[2] European commission (1985) Council Directive 85/337/EEC on the assessment of the effects of certain public and private projects on the environment, in: Official Journal of the European Communities, No. 175. pp. 0040 – 0048.

[3] European Parliament (2001) European Parliament and Council Directive 2001/42/EC on the assessment of the effects of certain plans and programs on the environment, in: Official Journal of the European Communities, No. 197. pp 30 – 37.

[4] ISO (2004) International standard 14001:2004 Environmental management systems – Requirements with guidance for use. 25p.

[5] Moreau V., Gondran N., Laforest V. (2009). Impacts locaux ou planétaires, aspects directs ou indirects: Définitions, Limites et prise en compte dans l'évaluation environnementale, In : Proceedings of international symposium and doctoral consortium of the ISEOR in partnership with the AOM, 16p, Lyon, France.

[6] Andre P. Delisle C.E., Reverét J.-P., Sène A.. (1999). L'évaluation des impacts sur l'environnement – Processus, acteurs et pratique. Presses Internationales Polytechnique. 416 p.

[7] Code de l'environnement (2010), Art. L122-1, Code de l'environnement Articles L122-1 à L122-3 relatifs aux études d'impacts des travaux et projets d'aménagement. Available on www.legifrance.gouv.fr.

[8] ISO (2000) International standard 14031:1999 Environmental management – Environmental performance evaluation. 35p.

[9] ISO (2006) Management environnemental. Analyse de cycle de vie. Exigences et Lignes directrices. Norme européenne – norme française NF EN ISO 14044. AFNOR. 59p.

[10] World Resources Institute (2007). The Greenhouse Gas Protocol Measuring to Manage: A guide to designing GHG accounting and reporting programs. 52p.

[11] ADEME (2009) Bilan Carbone® « entreprises » et « collectivités » Guide des facteurs d'émissions Version 6.

[12] Wackernagel Mathis (1994). Ecological footprint and appropriate carrying capacity : A tool for planning toward sustainability. Thèse de doctorat: Faculty of graduate studies, the University of British Columbia. 219p.

[13] Rafenberg C. (2001) MASIT Nouvel outil d'évaluation comparative pour technologies durables, In : Déchets Revue francophone d'écologie industrielle, n°23, pp.3-8.

[14] Ritthoff M., Rohn H., Liedtke (2002) Calculating MIPS Resource productivity of products and services. Wuppertal Spezial 27è. Wuppertal Institute for Climate, Environment and energy at the science Centre North Rhine-Westphalia. 52p.

[15] Moll, S., Bringezu S. (2003), 'Material flows', in Europe's environment: the third assessment, European Environmental Agency, Copenhagen, 15−23.

[16] Angell, L.C., Klassen, R.D. (1999): Integrating environmental issues into the mainstream: an agenda for research in operations management, in: Journal of Operations Management, Vol. 17, pp. 575-598.

[17] Adger W.N. (2006) Vulnerability. Global Environmental Change, Vol. 16, Issue 3, pp.268-281.

[18] Swiss Centre for Life Cycle Inventories. Ecoinvent Database v2.2 2010. Available with registration on www.ecoinvent.ch.

[19] Donnelly, A., Jones, M., O'Mahony, T. and Byrne, G. (2007). Selecting environmental indicator for use in strategic environmental assessment, in: Environmental Impact Assessment Review, Vol. 27, No. 2, pp.161-175.

[20] Jasch, C. (2000): Environmental performance evaluation and indicators, in: Journal of Cleaner Production, Vol. 8. No. 1, pp. 79-88.

[21] Olsthoorn, X., Tyteca, D., Wehrmeyer, W. and Wagner, M. (2001): Environmental indicators for business: a review of the literature and standardisation methods in Journal of Cleaner Production, Vol. 9, No 5, pp. 453-463.

[22] Breuil F. et Brun E. (2007). Responsabilité sociétale : Terminologie en Français de la R.S. pour le développement durable in Les publications de l'IEPF. 105p.

[23] ISO (2004) International standard 14001:2004 *Environmental management systems - Requirements with guidance for use.* 25p.

[24] Hepting, D.H. (2007): Decision support for local environmental impact assessment in *Environmental Modelling & Software,* Vol. 22, No. 4, pp. 436-441.

Application of Eco-Services in Automotive Industry for Commercial Customers

Katrin Kuntzky[1], Christoph Herrmann[1]

[1] Institute of Machine Tools and Production Technology, Product- and Life-Cycle Management Research Group, Technische Universität Braunschweig, Braunschweig, Germany

Abstract

The contributions of automotive manufacturers to decrease environmental impacts of vehicles can be pursued through technical development. Additionally, eco-services in the phases of the life cycle can make a major contribution to increase resource- and energy efficiency in vehicles.

For automotive manufacturers it is essential to estimate customer requirements and acceptance of eco-services. In this paper a market research is presented, in which the acceptance and the application of eco-services in the group of commercial customers has been analysed in Germany. Selected industrial sectors with a relevant use of vehicles have been consulted and analysed regarding their attitude to eco-services.

Keywords:

Automotive Service System; Market Research; Eco-Services

1 INTRODUCTION

Due to the public discussions on CO_2-emissions and its consequences on the environment, more and more people see the problems of automotive transportation today [1]. At the same time the rising costs, especially for gasoline and diesel, have a deterrent effect. Automotive manufactures are therefore increasingly forced to reduce the environmental impacts of their vehicles in order to fulfil the changing customer requirements.

On the one hand the contributions of automotive manufacturers to decrease the fuel consumption can be pursued through changes on the physical product like new drive concepts, improved aerodynamics and lighter bodies. On the other hand, eco-services as variations of innovative approaches from economy and sociology tend to reduce the ecological impacts in connection with a certain product [2]. The development of eco-services pursues the goal of minimising ecological impacts and rising economic values [3]. Three factors are affected by the eco-service [4]:

- changes on the physical product
- changes at the infrastructure
- changes in the handling of the user

Therefore, the additional introduction of eco-services in the various phases of the automotive life cycle can contribute to increase resource- and energy efficiency in vehicles. Examples include mobility services in the vehicle (eco-services in telematics) as well as services for the repair and maintenance of vehicles (eco-services in workshops).

With the combination of products, services and infrastructure, the customer requirements on mobility can be satisfied and simultaneously the environmental impacts can be reduced. Likewise for product service systems in general, the development of eco-services in automotive industry focuses on supplying the customer with specific results or functions [5]. Changes in the ownership structure can be affected from this [6], [7]. This allows creating synergies between profits, competitiveness and environmental benefits [8].

In order to place eco-services on the market and to succeed in comparison with the competitors, it is essential for automotive manufacturers to know and to estimate the current customer requirements and customer acceptance of offered and planned eco-services. Therefore, this paper presents a market research in which the acceptance and the application of eco-services has been analysed in Germany. The group of commercial customers was selected as target group because significantly more than half of the new cars in Germany are registered as commercial vehicles [1]. Three industrial sectors with a relevant use of vehicles have been consulted and analysed regarding their attitude to eco-services.

2 ECO-SERVICES IN THE AUTOMOTIVE SECTOR

Due to rising motorisation in society and increasing traffic density [1] the emission of e.g. greenhouse gases by street traffic is a major problem. In Germany, the CO_2 emission by street traffic amounts to 144 million tons per year (about 19% of the overall CO_2-emissions of Germany, see Figure 1) [9].

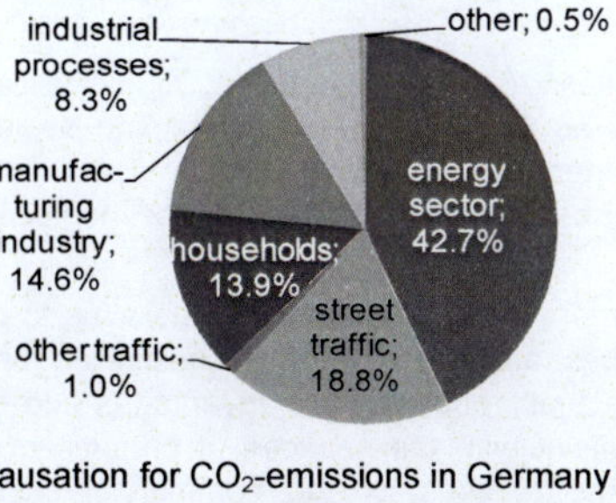

Figure 1 : Causation for CO_2-emissions in Germany in 2009 [9].

In automotive industries a movement towards a servitisation, the adding of value through services to the core product [10] can be observed. This is due to a rising amount of customers who are no longer interested in owning cars but in the fulfilment of their needs for mobility [11]. First steps were made with product-oriented

J. Hesselbach and C. Herrmann (eds.), *Functional Thinking for Value Creation: Proceedings of the 3rd CIRP International Conference on Industrial Product Service Systems, Technische Universität Braunschweig, Braunschweig, Germany, May 5th - 6th, 2011*,
DOI 10.1007/978-3-642-19689-8_49,

services like maintenance or financial services, however, today more and more user-oriented services and result-oriented services are on the market:

- The car sharing market for example is not only rising in Europe [12]. First original equipment manufacturers start to develop special cars optimised for car sharing concepts [13].
- In the area of after sales services, the offered maintenance and repair services concentrate more and more on the reduction of fuel consumption [14].
- Telematic devices are created to enable services like remote diagnostics, reporting of the driving behaviour or influencing the driving routes towards a more economic or ecological way [15].

Thus, the offered automotive services can lead to a reduction of environmental impacts of mobility at a progressive rate.

3 MARKET RESEARCH ON AUTOMOTIVE ECO-SERVICES

The market research on automotive eco-services was structured, conceived, accomplished and analysed following the five phases of market research: definition-, design-, data acquisition-, analyses- and evaluation and recommendation phase [16].

3.1 Definition Phase

The main objective when conducting the market research was the determination of the awareness level, the acceptance and the effective rate of utilisation of eco-services. In particular, the target group of commercial customers (fleets) was of major interest, because significantly more than half of the new cars in Germany are registered as commercial vehicles [1]. This group of customers thus represents a high turnover of additional services. Within this target group, three branches of trade where selected for the questioning, following the strategy to interview three branches with different commercial background and with a high rate of utilisation for their fleet.

- Passenger transport operators (PTOs) (such as cab operators): Each of the cars in the fleets travels wide. Mobility is an original purpose of the activities of PTOs. Thus, particularly strong attachments to the vehicles exist. Therefore, being a tester for new products and services could be well suited for that branch.
- Car sharing providers: Car sharing itself is a product service system. The car sharing concept is comparatively new and therefore the branch is expected to be innovative. Customers are expected to have exceedingly price- and/ or environmental awareness. A high interest in the usage of eco-services is estimated.
- Municipalities: As a public service, municipalities have other characters as the previous branches. The decision structures and the pattern of use vary significantly. The amount of vehicles in the fleet can be estimated as high for municipalities in major cities.

The survey among these customer groups aims to clarify how far the eco-services are known in the considered branches, what services are actually adopted, the advantages and disadvantages which are conjoint with the services, if prejudices exist in some sectors and the reasons for the commercial customers for or against the offers. The answers to these questions are the prerequisite for providers to develop and adopt their service offers. Only in this way, a significant contribution to reducing environmental impacts from passenger transport can be reached by mobile services.

3.2 Design Phase

The survey was designed as a telephone interview. This offers the advantages of a relatively quick survey. A comparatively small influence of the interviewee by the interviewer is possible. The response rate is generally higher than for written questions [17]. Since it is expected that not all eco-services are known to all interviewees, the questionnaire is designed in such a way that it allows the answers some leeway. Due to the selected survey form, the complexity of the questionnaire needs to be adjusted to purely verbal communication. The use of e.g. visual aids to explain issues is not possible.

The survey is divided into four blocks of questions. In the first block, general questions about the vehicle fleet and staff using the fleet are posed. These are e.g. questions on the number of vehicles in the fleet, the number of persons permitted to use them or the annual rate of usage.

In the second block questions about the choice of vehicles in the fleet are presented. It is queried whether fuel-efficient models are in operation which e.g. are aerodynamically optimised, have aligned gear ratio or are equipped with rolling drag optimised tyres, (like Blue Motion line of VW [18] or the eco2 from Renault [19]) or alternative drives like electric drives or natural-gas engines are used in the fleet. Therewith the emissions of the fleets could be reduced. The municipalities are furthermore asked whether product service systems such as car-sharing are used by the employees.

The third section asks for the usage rate of mobility ensuring services that contribute to a reduction of the fleets' environmental impact. Particularly questions on the use of re-repaired components for wear parts in repair and maintenance and emission-saving equipment such as fuel-efficient engine oil or tyres are posed. Furthermore, questions about offered driver training for a fuel saving driving operation, as well as about the implementation of fleet management programs for an optimised deployment of the vehicles are asked.

In the fourth block the use of telematics is queried. The questions in this block refer to the use of telematic systems for the monitoring of travel data and driving violations to allow the controlling of the drivers by the fleet manager. The utilisation of fuel consumption measurements and gear recommendations supports the driver and the fleet manager in controlling the driving manner. Furthermore, remote diagnostic systems foster an early diagnostics of abrasions and defects. Particularly, these systems are aligned for an identification of increased CO_2 emissions [20]. Therefore, the utilisation rate of these systems is questioned as well.

In the four question blocks the customers were interviewed at their recent user behaviour with respect to the particular offer. In addition, questions were asked regarding basic interests in using, as well as reasons for using or not using.

3.3 Data Acquisition Phase

A total of 150 different companies and municipalities were addressed in the survey. A success rate of 28.7% was achieved. The data was created based on 43 companies and municipalities which represent 29,385 cars. Interviewed were the fleet managers responsible for the vehicle pool. Figure 2 shows the allocation of the surveyed companies and the vehicles represented in the target groups.

In the case of PTOs only companies were surveyed with 3 or more vehicles. 13.7 % of the Germany-based passenger transport operators have more than 3 vehicles [21]. 19 companies in this sector participated in the survey. These 19 companies represent 0.8% of the total number of passenger transport vehicles in Germany and 1.4% of the stock of vehicles in companies with more than 3 vehicles.

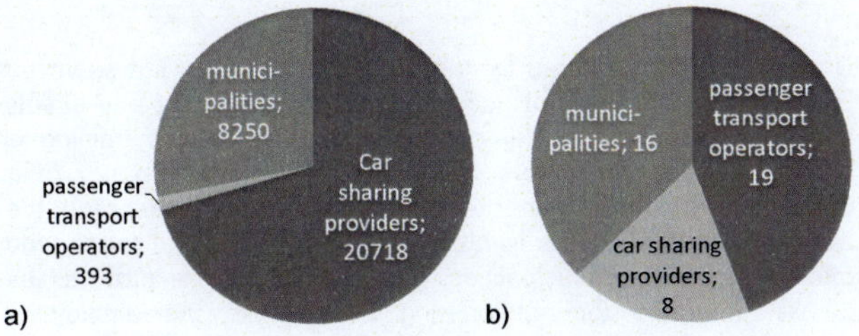

Figure 2 a): Allocation of vehicles to the target groups
b) Allocation of the surveyed companies to the target groups.

In the field of car sharing providers 8 of the 30 car-sharing companies, organized in the Federal Association of Germany e.V. [22], took part in the survey. A trade representativeness of 26.7% was achieved.

Among the 80 major cities in Germany with more than 100,000 inhabitants, 16 were enlisted for participating in the survey. A representativeness of 20% was realized in that sector. The representativeness is presented in Figure 3.

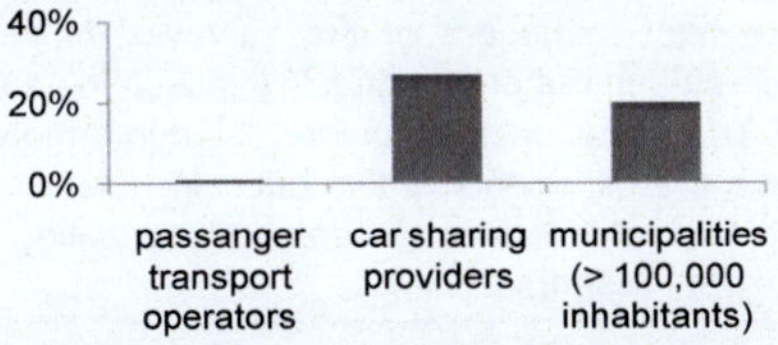

Figure 3: Realized representativeness in the target groups.

4 ANALYSIS

4.1 Surveyed vehicle fleets

Passenger transport operators

The fleet size of the surveyed PTOs varies from 3 to 150 vehicles. 52.6% have between 3 to 9 vehicles, 31.6% have between 10 to 50 vehicles and 10.5% have more than 50 vehicles (Figure 4).

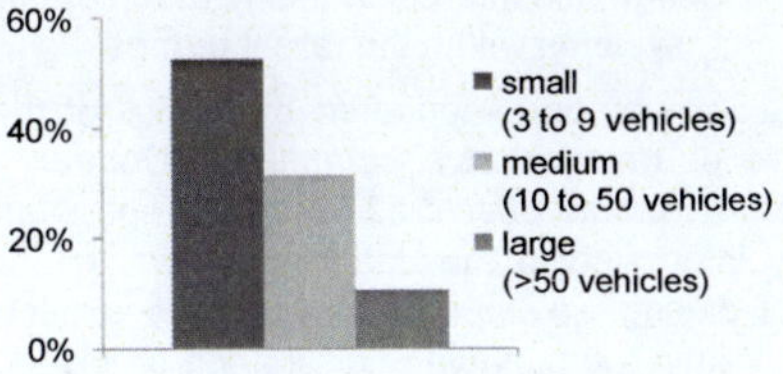

Figure 4 : Fleet sizes of surveyed PTOs.

The annual kilometrage driven by these vehicles is in the average about 30,000 km per year, with variations between 20,000 km and 50,000 km. Overall, PTOs drive significantly more than the German average of 14,200 km/year [23]. The vehicles are maintained, on average, 4.5 years by the responses. In Figure 5, the annual kilometrage of the fleets in the surveyed target groups is depicted.

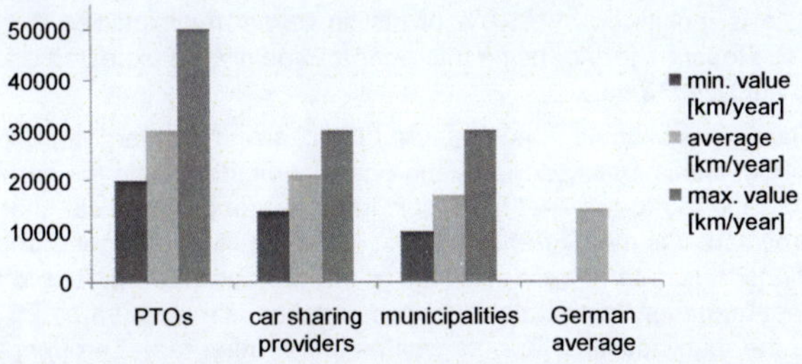

Figure 5 : Annual kilometrage of the surveyed fleets.

Car sharing providers

Within the car sharing providers the fleet size varies extremely. The largest fleet is 20,000 vehicles whereas the smallest is 8 vehicles. 50.0% have less than 100 vehicles, 37.5% have 100 to 300 vehicles and one has 20,000 vehicles (Figure 6). Without the largest provider, the average of vehicles in the fleet is 103 vehicles. The amount of customers per vehicle differs between 7 to 80 customers, with an average of 26 customers per vehicle. With rising fleet size, the annual kilometrage is rising. In average it is with 21,000 km as well distinctively above the German average. In this target group, variability between 14,000 and 30,000 km per year was specified.

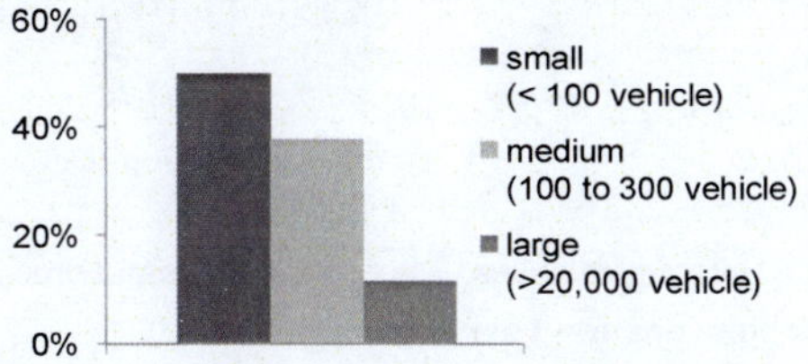

Figure 6 : Fleet sizes in the target group of car sharing providers.

Municipalities

The fleet size within the municipalities is relatively homogeneous. 81.2% of the surveyed municipalities of German major cities with more than 100,000 inhabitants have a fleet between 300 to 800 vehicles. Only 6.2 % have a fleet with less than 30 vehicles and 6.2% have a fleet size above 1,500 vehicles (Figure 7). 6.2% of the municipalities do not have a fleet at all. Employees of these municipalities rather use the offer of a car sharing provider.

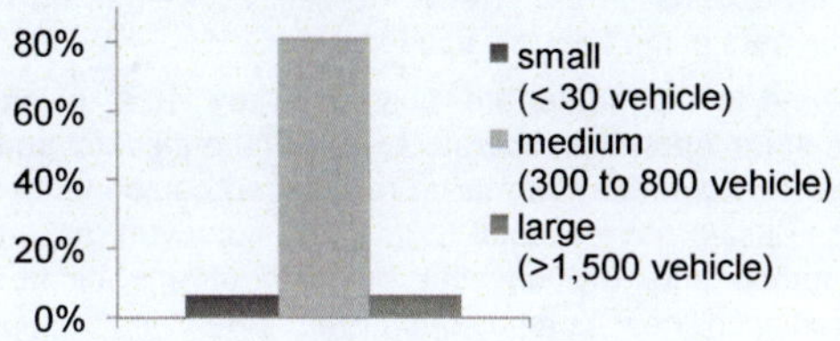

Figure 7 : Distribution of fleet size of surveyed municipalities.

The annual kilometrage in average is 17,000 km per year and varies between 10,000 km to 30,000 km. Therefore, it is slightly above the average kilometrage in Germany.

4.2 Vehicles in the fleets

31.6% of the PTOs use fuel-efficient car models in their fleet. The reasons mentioned for using those models are mostly due to environmental responsibility. Reasons for not using such models are the costs for those models and the not yet sophisticated technology. Among the interviewed municipalities, 31.3% use fuel-efficient car models, and 37.5% of the car sharing providers. Thus, the usage is approximately the same among the target groups.

Alternative drives are used by 26.3% of the fleets in passenger transport. Reasons for not using are costs, poor fuel availability and absent infrastructure, a not totally sophisticated technology and also bad experiences with the technology. The car sharing providers operate in nearly double of the percentage (37.5%) alternative drives. Beside the costs and infrastructural complications, the car sharing providers are also facing the problem that the permanent change in user requires a solid and simple solution for the fuel supply. This is not yet realized in alternative drives. In contrast to the other two target groups, municipalities are using this technology exceedingly. 75.0% have cars with alternative drives in their fleet. Here, the reasons are mainly the expected economic standing for the use of fuel-efficient cars and the cost reduction with alternative drives. Complications with the refilling are not accounted in the fleets of the municipalities (Figure 8).

In total 43.8% of the municipalities use car sharing offers to enlarge their fleet. One municipality uses car sharing of one operator exclusively. 56.2% are basically interested in the usage of car sharing, even though those who are not using it until today are expecting a higher organizational effort when using car sharing.

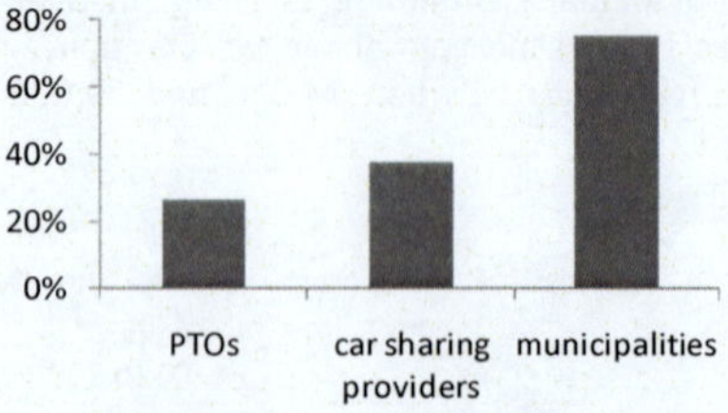

Figure 8: Alternative drives usage rate in the target groups in [%].

4.3 Mobility ensuring services

Fleet management programs are preferentially used by car sharing providers. 87.5% of the surveyed car sharing providers are using these programs, 42.9% of the users have developed their own management program. In contrast to this, none of the small PTOs uses a fleet management program. But 50% of the medium and large size operators are using a program to control their fleet. The small operators declare that the cost-benefit ratio is not positive for such a program. Also in municipalities fleet management programs are already used or the usage is in planning by 25.0%. 37.5% are basically interested in a program. The usage of a program therefore depends on the size of the fleet as well as on the main target the cars in the fleet are used for.

Re-repaired components are only used by 10% of the small passenger transportation operators. 33% of the medium and 50% of the large size operators are using re-repaired components. A basic interest in using have in total 63% of the surveyed, assumed that those components are cheaper but without any disadvantages compared with new components. Emission-saving equipment is used by 26% of the operators. In this purpose the basic interest is at 79%, assuming a cost reduction with the usage. Similarly to this, 12.5% of the target group of car sharing providers uses re-repaired components and another 12.5% is basically interested in the usage. Main reasons are expected limitations compared with new parts and effects on the warranty of the manufacturer. Emission-saving equipment, in contrast, is used in the fleets by 62.5% of the providers. The usage rate of re-repaired components and of emission-saving equipment is shown in.

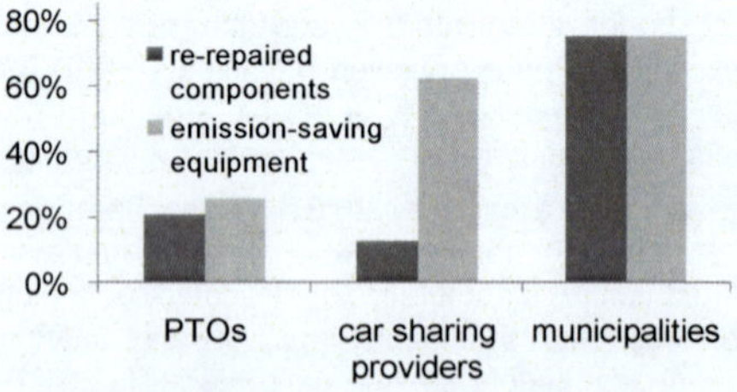

Figure 9 : Usage rate of re-repaired components and of emission-saving equipment in the target groups [%].

Compared to the other two target groups, the usage of re-repaired components is very pronounced among the municipalities. 75.0% are already using re-repaired components. If these components are not used, it is due to influences on the warranty when using other components than from the OEM. Emission-saving equipment is as well used by 75.0% of the municipalities.

Driver's training is offered for the drivers of 52.6% of the surveyed PTOs. Those who do not use trainings do not expect any effects from it. In case of the car sharing providers, driver's training is provided to the customers in 37.5%. Those who do not offer trainings explicate it with the huge amount of drivers and the absence of effects. This is also attested by 33.3% of those who offer trainings. In municipalities, driving trainings are provided by 62.5%. Reasons for not offering trainings to the employees basically are the expectation that the effects are not measurable.

4.4 Usage of telematic systems

Remote diagnostics are only used by 15.8% of the PTOs. This kind of offer was unknown by 52.6% of the surveyed. Likewise, the usage of remote diagnostics is with 12.5% of the car sharing providers and with only 6.3% of the municipalities marginal. Within the car sharing providers the task of remote diagnostics is today realized by the customers, who are urged to report damages and the need for maintenance immediately. Therefore, 87.5% of the car sharing providers have a basic interest in using this function, requiring the equipment can be used OEM comprehensive. Also 43.8% of the fleet managers of the surveyed municipalities are basically interested in remote diagnostics, assuming a positive cost-benefit ratio. The municipalities not interested in remote diagnostics justify this with the high costs of this offer. In Figure 10, the usage rate and the rate of interests in remote diagnostics of the three target groups are presented.

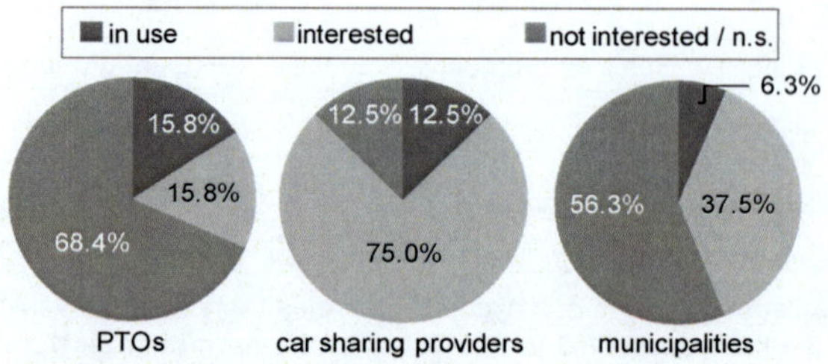

Figure 10 : Usage and interest in usage of remote diagnostic systems within the target groups.

Driving violations are only monitored by 10.5% of the PTOs and only by large or medium size companies. Contrary to this, the monitoring of travel data effects 52.6% of the operators and 73.7% are basically interested in the usage. Reasons for not using the monitoring of driving violations are mainly the expected costs for that service. Within the surveyed car sharing providers, 25.0% use telematic systems for the monitoring of driving violations. In total 50.0% are basically interested in that offer. Main reasons for not using this offer are cost-benefit ratios and doubts due to data privacy. Travel data is not monitored by any provider due to the same reason, even though 37.5% are basically interested in the monitoring. None of the municipalities use telematic systems to monitor driving violations, analogous to the car sharing providers due to data safety and co-determination rights, whereas the travel data is monitored by 25.0% of the surveyed municipalities (Figure 11). Reasons for not using this offer are mainly the expected efforts to analyze the data.

The fuel consumption is measured by 42.1% in that target group of PTOs; this is mainly done by the board computer inside the car and not reported to the fleet manager. In the target group of car sharing providers it is measured by 62.5%. As well as in the target group of PTOs, this data is only available in the car and not reported to the fleet manager. In contrast, fuel consumption is measured by 75.0% of the municipalities. But, as well as in the other target groups, this data is provided by the board computer. A transmission to the fleet manager is not taking place.

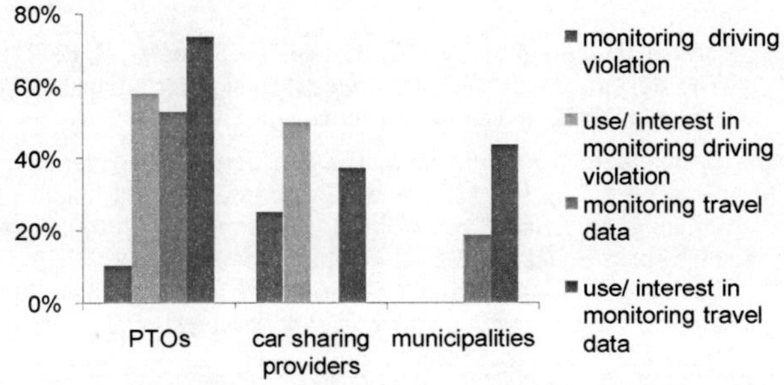

Figure 11 : Monitoring of driving violations and of travel data in the three target groups [%].

Gear recommendations are not used by 26.3% of the PTOs due to the usage of automatic gearboxes. Only 7.1% of the fleets with gears use these recommendations due to absence of interest or high costs. Similarly, these recommendations are only available in cars in 12.5% of the fleets of municipalities and they are used like the fuel consumption inside the car only. A basic interest in this offer does only exist by 18.8% of the municipalities fleet managers.

But 50.0% of the car sharing providers have cars with gear recommendation. They are used as the measurement of fuel consumption only inside the car and not reported to the fleet manager.

In the target group of car sharing providers requests especially for open technologies (OEM independent) in telematics were named as well as on board communication systems for two-way communication.

5 EVALUATION AND RECOMMENDATIONS

5.1 General evaluation of the surveyed services

With analyzing the usage rate of fuel efficient car models and alternative drives, it can be deduced for the three target groups that a concentration on ecological aspects is not sufficient for a high usage rate. Especially the cost factor is of a major interest for the three target groups. With the usage rate of alternative drives in the target group of municipalities, the influence of a positive public image becomes apparent. The lack in infrastructure and the comparingly complex or unknown process of refuel deter the PTOs as well as the car sharing providers. Precisely the car sharing providers were expected to have a high usage rate for alternative drives. The usage rate of this technology of 43%, even though the depicted problems with infrastructure and unknown usage are known, points out the high interest in ecological oriented offers with positive influences on the public image.

The usage rate of re-repaired components shows that the municipalities are primarily interested in cost reduction whereas the other target groups are focusing on a simple handling and do not want to interact in unsecure areas and inexperienced offers. The different usage rate of emission saving equipment is demonstrating the influence of the fleet size on the usage of offered services. The usage of fleet management programs is mainly considered with respect to process optimization. For the large passenger transport fleets and the car sharing fleets, the advantages of the fleet management programs have become obvious. For small fleets, this offer means a higher effort than a benefit.

In the area of telematics it becomes clear that the factor of data privacy and data safety is of major importance to all surveyed target groups. Before using these services, the questions of data safety have to be solved.

The three target groups have a basic interest in the usage of telematic services. They expect positive influences on the driving behaviour and therefore on the upcoming costs when controlling the drivers and the vehicle. Contrariwise, all three target groups doubt if a positive cost-benefit ratio exists. Therefore, it is particularly essential for these services to display the effectiveness of these services with test runs and to convince the customer from the positive economic and ecological impacts. In the usage of telematic services, the three target groups also attach importance to the reduction of complexity in internal processes. A service only for the reduction of environmental impacts will hardly be accepted and established.

5.2 Evaluation of the three target groups

Car sharing providers

Service or service oriented solutions for car sharing providers particularly have to be simple and intuitional in use. The analysis of the usage of eco-service within the car sharing providers group indicates that they are generally open for the usage, but economic benefits stand in the foreground. Especially services accounting to a simplification in internal processes or to significant cost compensation with consistent or reduced efforts are attractive to the car sharing providers. This arises particularly in the high interest of car sharing providers in the remote diagnostics. Here, one of the main challenges in the usage of different services for car sharing providers is becoming apparent. The different standards of the OEM exacerbate the usage of specific services in a large fleet. Services for those fleets only make sense if they can be used for all vehicles in the fleet with one offer and one technology.

Furthermore, the offers are more attractive for car sharing providers if they are focusing on the vehicle itself or on the controlling of the vehicles in the fleet. Services concentrating on changing the single driver's behaviour like driver's training, fail due to the large amount of driver registries at the car sharing provider.

Municipalities

The large usage rate of car sharing offers among the municipalities and of alternative drives clarify the importance municipalities are attaching to a positive public image on the one hand and to a steady overview and reduction on the subsistence costs on the other hand. This arises particularly in the usage rate of re-repaired components and emission saving equipment as well as on the amount of offered driver trainings.

Services offered to this target group therefore need to concentrate on both sides: creating a positive public image due to the ecological aspects and presenting a positive influence on the long term subsistence costs.

Passenger transport operators

The surveyed PTOs have the lowest usage rate for nearly all services questioned in the survey. Alternative drives and fuel efficient car models are used as little as remote diagnostics, emission saving equipment and re-repaired components. Driver trainings are in contrary used more than in the other target groups. Also the monitoring of driving violations and of travel data is used or of a basic interest for a majority of PTOs. A significant difference in the usage rate of the services exists among the fleet sizes. Larger fleets are using services more often than smaller fleets. Service solutions for PTOs have to focus mainly on solutions which are simple to handle, i.e. which do not implicate higher complexity for the driver or the manager. The cost factor is in focus for the decisions in this target group. High initial investments are not tolerated, particularly in small fleets. The services offered to this target group need to depict a clearly economic benefit

5.3 Recommendations

The target group of car sharing providers is primarily interested in services to reduce the effort in detecting failures in the vehicle and in optimizing the internal processes, due to constantly changing users. Environmental orientation is relevant, but can only be in focus if it is not leading to higher complexity and is easy to realize with the changing drivers. Providers therefore have to create solutions with a simple and intuitional handling. Particularly OEM enclosing solutions for the whole fleet are demanded. Services that offer a better overview on the vehicle status and, if applicable, lead to a positive impact on the driving behaviour without addressing every driver personally are of interest.

In contrast, municipalities are focusing on the cost reduction but also on the positive public image. Providers therefore need to focus on the savings that the services offer to the municipalities. At the same time, effectively emphasizing the ecological benefits of the services in advertising can influence the decision for the services.

PTOs are particularly interested in services simplifying the inner processes. Eco-services are only of interest if they are not causing any complexity, and at the same time reduce the costs. Here as well, a focus on the cost effects of the services and on the possibilities to amortize the initial investments is needed. Furthermore, services need to be adapted to the different fleet sizes and the comparatively small fleet sizes in general. The relatively small amount of drivers allows to offer services that directly address the single driver. This can be driver's trainings but as well provisions for identifying and rewarding or punishing extremely positive or negative driving behaviour.

6 CONCLUSION

The survey showed that there is a basic interest in eco-services in all three target groups. The interest is deeply affected by the size of the fleets, the original task of the target group and the customers of the surveyed companies. Therefore, the usage varies in the three target groups, but also in each target group itself. Eco-services are frequently associated with high efforts and costs. Thus, service providers need to know the specific requirements of the individual customers. Neither a single solution nor a single marketing strategy can be attractive for all customers. Rather, individual offers have to be framed out in collaboration between the commercial customers and the providers.

Service providers rely on specialising their offers to single branches to get the required know-how in customer specific interests in mobility eco-services in the branch, or to design the offer in an adoptable way which allows a specification to the single customer of different branches.

Comparably important for all branches are services which not only focus on environmental impacts but also on the improvement of internal processes and reduction of mobility costs. When this is a constitutive part of the advertisement of eco-services, a significant usage rate can be reached. Then, eco-services contribute effectively to the decrease of environmental impacts from the automotive traffic. Especially commercial and fleet customers put the providers of mobility eco-services in the position to implement customized services due to the accessible fleet sizes.

7 REFERENCES

[1] Bundesministerium für Verkehr, Bau und Stadtentwicklung (Ed.), 2009, Verkehr in Zahlen 2009/2010, Hamburg.

[2] Behrendt, S. / Jasch, C. / Kortmann, J. / Hrauda, G. / Pitzner, R. / Velte, D. (2003): Eco-Service Development – Reinventing supply and demand in the European Union, Norton.

[3] Brezet, J.C. / Bijma, A.S. / Ehrenfeld, J. / Silvester, S. (2001): The design of eco-efficient services, Design for sustainability Program, Delft University of Technology, Delft.

[4] Bartolomeo, M. / dal Maso, D. / de Jong, P. / Eder, P. / Groenewegen,, P. / Hopkinson, P. / James, P. / Nijhuis, L. / Örninge, M. / Scholl, G. / Slob, A. / Zaring, O. (2003): Eco-efficient producer services – what are they, how do they benefit customers and the environment and how likely are they to develop and be extensively utilised? in: Journal of Cleaner Production 11(2003), pp. 829-837.

[5] Goedkoop, M. / van Halen, C. / te Riele, H. / Rommens, P. (1999): Product service systems, ecological and economic basis. PricewaterhouseCoopers N.V. / Pi!MC, Storrm C.S., Pre-consultants,1999.

[6] Mont, O. K. (2002): Clarifying the concept of product-service system, in: Journal of Cleaner Production 10 (2002), pp. 237–245.

[7] Roy, R. (2000): Sustainable product-service systems, in: Futures 32 (2000).

[8] Manzini, E. / Vezzoli, C. (2003): A strategic design approach to develop sustainable product service systems: examples taken from the environmentally friendly innovation Italian prize, in: Journal of Cleaner Production 11 (2003), pp.851-857.

[9] www.umweltbundesamt.de.

[10] Vandermerwe, S., Rada, J. (1988) Servitization of business: Adding value by adding services, in: European Management Journal, vol. 6, no. 4, pp. 314-324.

[11] Communication Networks 14.0, FOCUS Magazin Verlag GmbH, München, 2010.

[12] The world's carshare consortium, http://ecoplan.org/carshare/cs_index.htm.

[13] Car2Go (2010): https://www.car2go.com.

[14] ipp: integrierte produktpolitik - instrumente aus der praxis am beispiel automobil, http://www.ipp-bayern.de/archiv/IPPLeitfaden_kfz.pdf.

[15] Gruden, D. (2008): Umweltschutz in der Automobilindustrie - Motor, Kraftstoffe, Recycling, Wiesbaden.

[16] Fritz, W. / von der Oelsnitz, D. (2006): Marketing – Elemente marktorientierter Unternehmensführung, 4. Auflage, Stuttgart.

[17] Hüttner, M. / Schwarting, U. (2002): Grundzüge der Marktforschung, 7. Auflage, Oldenbourg-Verlag, München.

[18] www.volkswagen.de.

[19] www.renault-eco2.com.

[20] www.onstar.com.

[21] Deutscher Taxi- und Mietwagenverband (2009): BZP Geschäftsbericht 2008, http://www.bzp.org.

[22] Bundesverband CarSharing e.V. (2010): http://www.carsharing.de.

[23] Autohaus Extra DAT Report 2009, Autohaus 9/2010, München.

Scenario Management for Sustainable Strategy Development in the Automotive Aftermarket

Uwe Dombrowski[1], Christian Engel[1], Sven Schulze[1]

[1] Institute for Production Management and Enterprise Research, Technische Universität Braunschweig, Braunschweig, Germany

Abstract

The spare parts business is especially in the automotive industry of high relevance because of its high profit margin and its stability. Yet changing conditions pose new challenges to the spare parts management. Examples are the deregulation of the spare parts market or the increasing use of electronic components. To cope with these challenges, the protagonists need to find a convenient strategy and to pursue it consequently. For this purpose different scenarios for the spare parts market in 2030 have been created. In the paper the scenarios, the most important influence factors and possible approaches are described.

Keywords:

Spare Parts Management; Scenario Management; Automotive After Sales

1 INTRODUCTION

The automotive after sales business gains continuously importance for automotive manufacturer and their suppliers. This trend has several main reasons. Firstly, the after sales business is relatively independent of economic cycles and achieves stable, high returns. Within the German automotive market in 2007 the aftermarket is responsible for only 24 % of revenues, but 54.4 % of profit [1]. In this area much higher returns can be achieved compared to the new car or used car business. Even during the current crisis, this segment remained stable while new car sales slumped or could only be kept by state subsidies.

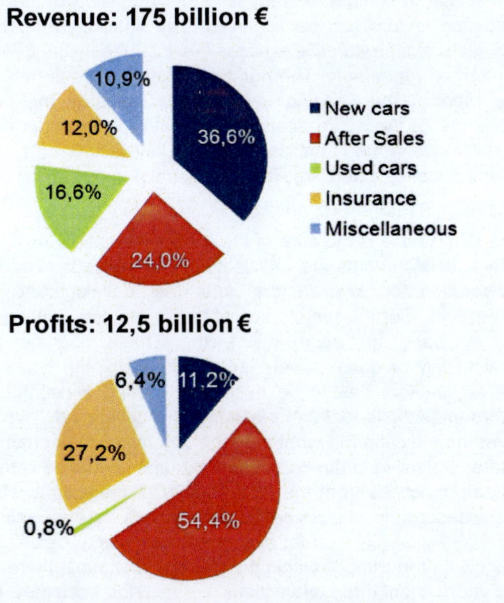

Figure 1: Relations of revenue and profit in the aftermarket [1].

Secondly an efficient after sales business is a way to long-term customer loyalty for the company [2]. The high customer proximity through this service results in high customer satisfaction. Also the requirements for new products can be detected more easily through the customer contact. Thirdly provides the after sales the possibility to set the company apart from the competitors. The development of a good product-related service needs time and high qualified employees and cannot be imitated by new competitors in a short time [3].

However, the aftermarket is also exposed to new conditions that require constant adjustment and optimization of the underlying processes. The rising share of electronics and electrics in the car requires new strategies because the predictability of the demands and characteristics of the parts with respect to storage, availability or reproduction are significant different from mechanical components [4]. The declining depth of added value at the original equipment manufacturer (OEM) also enhances the dependence on the suppliers. New legal rules promote a liberalization of competition. In the European Union for instance the changes in the block exemption regulations or the labeling as original parts can be named. This liberalization results in greater competition and falling prices, which threatens the current high margins of the OEM. Thus, in the grey market spare parts are often 30 to 40% cheaper than the original parts [5]. A rising average age of the vehicle fleet in many markets also causes a higher price sensitivity of customers. Furthermore the development of high-quality components leads in the automotive industry to a decreasing number of required maintenance and repair cases. In Germany in the year 1990 the number of maintenance per car was on average at 1.3 and the number of replacement parts at 1.05. Until 2009 the numbers dropped to 0.85 or 0.6 incidents per year [6].

To meet these new requirements and to maintain the high profitability in the after sales business continues to operate profitably, companies need to adjust their business strategies and prepare for future changes. For this purpose it is necessary to develop systematic and structured scenarios for the future development.

J. Hesselbach and C. Herrmann (eds.), *Functional Thinking for Value Creation: Proceedings of the 3rd CIRP International Conference on Industrial Product Service Systems, Technische Universität Braunschweig, Braunschweig, Germany, May 5th - 6th, 2011*,
DOI 10.1007/978-3-642-19689-8_50,

2 BASICS OF SCENARIO MANAGEMENT

A scenario can be defined as a description of a possible set of events that might reasonably take place. Main purpose is to stimulate thinking about opportunities, risks and the deriving of measures [7].

2.1 Methods for Scenario Management

To create scenarios various methods are described in the literature. These methods are characterized primarily by the criterion of a hard or soft approach and depend on the type of scenarios to be created [8]. The hard methods are based on quantitative methods of calculation, while the soft methods are based on a mixture of quantitative and qualitative approaches. In addition, the soft methods are differentiated by further criteria, such as intuitive or systematic approaches. As intuitive approaches cannot be comprehended completely by others, these approaches are not adequate for use in business or science. However the systematic procedures are characterized by a high accountability and are therefore particularly suitable for supporting business decisions [8].

A systematic approach for the development of scenarios is the scenario management described by Gausemeier. This method of scenario creation is based on a deductive logic and is based on subjective expert opinions as well as mathematical algorithms. The method is particularly suitable for the creation of scenarios that are characterized by a high degree of clarity, thoroughness of content and relevance as well as a harmonious overall appearance and proportions of the scenarios to each other [8].

2.2 Phases of the Scenario Management

The scenario management as described by Gausemeier is divided into five phases. Figure 2 gives an overview about these five stages. Each phase has special methods and tools, which are used for the scenario creation.

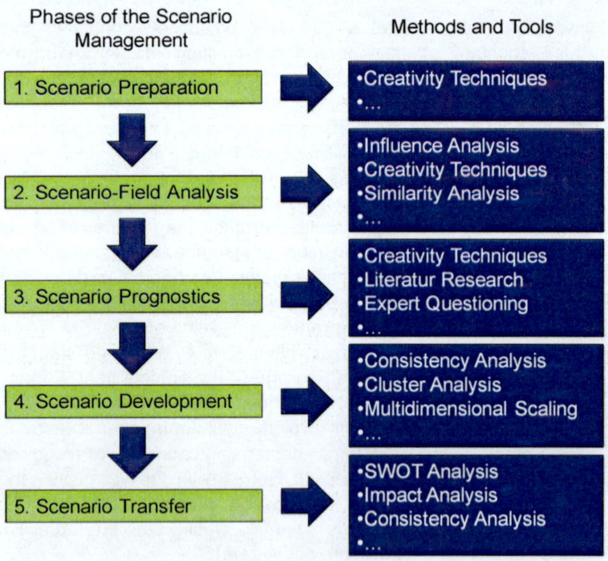

Figure 2: Phases for the scenario creation.

The method is suitable for creating different scenarios such as global scenarios or industrial scenarios but the basic approach remains the same. Only the premises in applying the instruments differ depending on the scenario type to be created [9].

Phase 1: Scenario Preparation

The first phase within the process of the scenario creation is the scenario preparation. At this stage of the project it is necessary to define the specific scope, the so-called decision-field precisely. Depending on the type of scenario to be created, the decision-field might be an industry, a company or a technology [10]. For this, the objectives to be met and tasks, which should be solved in the project, are specified. During the next step the decision-field analysis is accomplished to identify the current state of the considered decision-field. For such an analysis it is useful to divide the decision-field into smaller components that can be considered separately. The result of the first phase is a clear description of the decision-field, which serves as a starting point for the further scenario creation [10].

Phase 2: Scenario-Field Analysis

In the second phase, the scenario-field analysis, the scenario field is determined. An important step is the identification of influence-fields. These influence-fields divide the scenario field in sections that can be regarded individually. They have to be chosen in a way that the scenario field is described completely. To increase transparency, it is helpful to structure the scenario field into hierarchical levels, in which the influenceable and non-influenceable elements could be classified. Based on the segmentation into spheres of influence, the relevant factors can be identified, which describe the state and the development of the influence-fields [11]. To identify the relevant factors creativity techniques such as brainstorming should be applied. Afterwards a detailed description of the factors is necessary. Since the further consideration of all identified factors would end in a very complex procedure, in a further step the key factors from the catalog of the influence-factors have to be determined. Key factors are those factors that have the strongest influence on other factors and thus the scenario field [12]. To identify the key factors the method influence analysis is used, which determines for each factor the direct impact on any other factor. A following similarity analysis identifies those factors which are similar in content, thereby a further reduction of the number of influence factors can be achieved. The final determination of the key factors is an iterative process in which the influence analysis is performed repeatedly. The number of key factors is determined by the ratio of the imaging level, which indicates the percentage coverage of the entire scenario-field with the selected key factors [14]. To assist the identification of the key factors an IT support is useful because extensive calculations have to be carried out.

Phase 3: Scenario Prognostics

The third phase is the core of the scenario creation, as the possible future developments are anticipated. The phase is divided into the preparation of key factors and the development of future projections. For the preparation of the key factors, key figures have to be found for each key factor. These can be measured qualitatively or quantitatively [14]. Additionally, the current state of the key factors has to be described. During a second step, the future projections for each key factor are developed. The projection gives an indication in what way the key factor may changes in the future. Objective in the creation of extreme scenarios is to separate those projections from the total number of projections which cover the widest range of possibilities. But in each case an optimistic and a pessimistic projection for each key factor has to be used. In order to inhibit interferences within the created scenarios, there should be not more than three projections created for each key factor [9]. Finally, the projections have to be described in detail to beware of confusion in the further phases of the scenario management.

Phase 4: Scenario Development

Goal for this phase is the preparation and formulation of scenarios. So-called projection bundles of the key factors serve as base for this step. Projection bundles are a combination of different projections of the key factors which result in a scenario. The decisive criterion for a good scenario is the consistency of the united projections [12]. To ensure that the projections fit within the scenario to each other and not contradict, the consistency analysis provides a check for inconsistencies. As a result of this method the number of combinations of projections for a scenario can be reduced.

A further step in this phase is to create the raw scenarios. Raw scenarios are preliminary descriptions of the scenarios through parameters and are the starting point for the actual scenarios. The cluster analysis which aggregates consistent projections gives information about how many different scenarios optimally should be generated without suffering too much loss of information [9].

After the raw scenarios were created, the visualization of the raw scenarios and projections can be realized with the method multidimensional scaling (MDS). This gives the scenario creator a better understanding of the future development and creates a better basis for communication [15]. At the end the identified raw scenarios are interpreted and the final scenarios formulated.

Phase 5: Scenario Transfer

The fifth phase of the scenario management is used to derive strategic business decisions. For this purpose, the possible effects for the company of the scenarios are simulated. The possible incoming events have to be accomplished as if they might have already occurred. Then strategies can be developed that can counteract negative effects and help to use future opportunities for the company, if the regarded incident occurs in the future. Since the derivation of strategic decisions is performed by the corporate management and during this phase essentially specific corporate requirements and conditions has to be considered, this phase is not in the focus of this paper.

3 SCENARIO MANAGEMENT FOR THE AUTOMOBILE SPARE PARTS MARKET

In a project the method of scenario management was applied in order to provide transparency about the possible developments in the automobile aftermarket. For the creation of the extreme scenarios the first four phases were applied. The scenario transfer has to be done individually by the actors in the aftermarket because of their enterprise specific conditions.

3.1 Phase 1: Scenario Preparation

At the beginning of the project, first the basic conditions for creating the scenario were set and thus the basis of the scenario created. The considered decision field is the automobile spare parts market in Western Europe in the year 2030. The target groups for the application of the scenarios are mainly the OEM and the Original Equipment Suppliers (OES). Nevertheless, the developed scenarios are of interest to customers and all other stakeholders operating in the environment of the automobile aftermarket. The goal of the scenario management is to provide a basis for decisions about the future strategy; therefore the scenarios are so-called orientation scenarios [13].

Since the regarded period of 20 years is quite long, it is not appropriate to create trend scenarios that have the property to identify realistic developments with a high occurrence probability. Rather, in this project extreme scenarios were created that show the stakeholders extreme developments in the spare parts market, that may not occur, but may still be important for the strategy development and strategy review of the stakeholders [13].

3.2 Phase 2: Scenario-Field Analysis

Before the relevant influence factors for the spare parts market in 2030 could be identified in the scenario-field analysis, the scenario-field was divided into four different influence fields. This segmentation of the scenario-field allows identifying the relevant areas of influence. There were four main influence fields that can be described by environment and control systems (Figure 3).

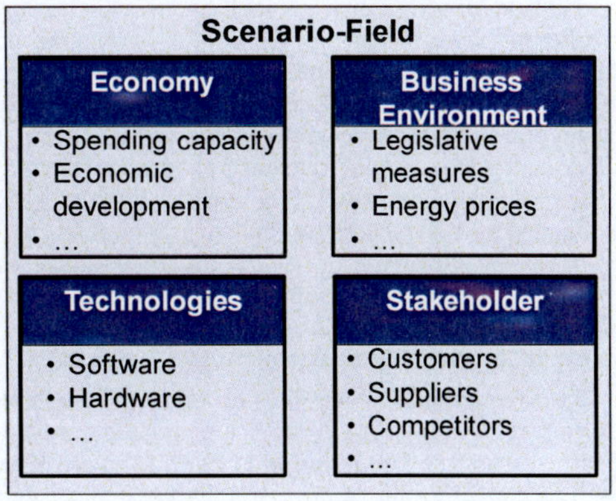

Figure 3: Influence fields.

Economy

The influence field economy covers all influences on the spare parts management, which are driven by financial or economic developments.

Business Environment

The business environment influences the spare parts market, particularly by legal regulations and directives, and non-financial developments of the spare parts business.

Technologies

This influence field includes all factors that change the spare parts market from a technological point of view. The Technologies are divided into hardware and software. Hardware includes those developments in the field of new materials and technical systems, whereas the area of software includes application software and the software to control the technical systems.

Stakeholder

The spare parts market is driven primarily by the stakeholder groups of customers, suppliers and competitors. Each of these stakeholder groups is subject to permanent change.

This segmentation of the scenario field makes it easier to identify the key factors that affect the spare parts market in 2030. In several brainstorming sessions with automotive experts 36 different influence factors were identified. The consideration of all factors would lead to a very complex and expensive procedure. Moreover, the knowledge gained through such a complex approach due to the very long observation period, may not necessarily be helpful. An expert group was therefore invited to carry out an impact analysis of the influence factors. The aim was to identify the key factors and thereby minimize the number of influence factors to those which have a decisive impact on the aftermarket. A total of 15 key factors were identified, that describe the four influence fields. In the following the identified key factors are described briefly.

In the influence field "economy" two key factors were identified. The "economic environment" is of particular importance for cyclical industries such as the automotive industry. As seen in the financial crisis in the last two years the sales of new cars declined rapidly. The resulting state subsidies ("cash for clunkers") had a high impact on the spare parts market since old cars were scrapped in high numbers [16]. The "purchasing power" describes the disposable income in households for consumption purposes. Also new forms of usage like car-sharing or car-pooling are developed and have in the future a high impact on the automobile market [17], the private customers will be an important customer group.

The influence field "business environment" had five key factors. The "cost of energy" were in the past of high relevance for the automobile industry, e.g. in the oil crisis 1973 and 1979/80. The future development is uncertain. Perhaps the renewable energy like solar or wind power may help to stabilize or even lower the price of energy. The energy prize is put together by the price for oil and for energy. "Legislative measures" can have great influence on decisions regarding the automotive industry. Examples are state subsidies or tax incentives for fuel-efficient and low emission cars. Furthermore the legislative may aim to liberalize the spare parts market as it can be seen in the European Union to lower the prizes and increase the quality for the customers [18].

The "customer requirements" can develop to be more related to a price or a performance orientation. An actual trend is the increasing sales of "cheap cars" as can be seen in many Western European countries. The impact for the spare parts market is high, since the lower the purchase price of a car is, the less the repair costs.

The "innovation power" of the OEM and OES is related to the development of customer requirements. Furthermore innovation needs a sustainable business strategy. Expenses for research and development can be more easily made in a good economic environment. A high rate of innovation causes a wide range of parts in the spare parts management.

The influence fields "technology" consists of five key factors. "Hybrid Car and electro mobility" is one main topic in the automobile industry. The effects on the spare parts management are tremendous: Instead of mechanical parts like combustion engine new parts like accumulators and electric motors have to be supplied. So far, natural gas and hydrogen propulsion systems have not yet been able to prevail against the internal combustion engine. The future development will be crucial for the spare parts management. A further propulsion technology is based on "gas or hydrogen". Today the technology is just a niche product, even though it is patronized by the legislative in several countries.

The "share of electronic components" in the automobile is increasing continuously and has a high impact on the spare parts market [2]. Possible developments may be standardization or the usage of central computers to achieve a reduction of the parts variety in the future. The use of "new materials" (especially carbon fiber reinforced plastic) is a common trend in the automobile industry to reduce the weight of vehicles. "Standardization" allows a substantial simplification of the spare parts management by a reduction of the considered components. For electronic components the project AUTOSAR (Automotive Open System ARchitecture) has started, in which a variety of OEM and OES try to drive forward standardization [19].

The last influence field "Stakeholder" contains three key factors. In the past, a sales number "shift between the segments" could be observed. As an example, with the sport utility vehicles (SUV) a completely new segment was established in Europe. Overall, the model range of the OEM has increased considerably.

The "guarantee and warranty deed" is given by law in many countries. Particularly in the European Union a prolongation can be observed in the past. The longer and more comprehensive the warranty commitments of the OEM, the less private customers are active on the aftermarket. "Sustainability" is an increasing important factor for the customer, the society and the company. In economic terms, not only capital but also the long-term costs and benefits generated by the product are considered. But the sustainability of environmental and social perspective will increase more and more as a selling argument for the customers.

3.3 Phase 3: Scenario Prognostics

In the phase of the scenario prognostics future projections for each of the identified 15 key factors have been created. As a basis for these future projections qualitative and quantitative criteria for the key factors were defined to measure the future development In addition, the status quo has been described for the key factors thus later in the scenario creation a unique projection platform is build [14]. With the brainstorming method two extreme projections could be found for each key factor. For the generated future projections no probabilities are specified since the regarded period is too large for reliable probability information and the inclusion of these would have created no value to the credibility. Finally, the projections of the key factors in the spare parts market were briefly described. As an example the key factor "energy price" has the projections decreasing and increasing. Increasing means that the price for energy will rise faster than the inflation, a decreasing energy price declines or rises under the rate of the inflation.

3.4 Phase 4: Scenario Development

The aim of the fourth phase is the creation and description of the extreme scenarios. A correspondence analysis was used to represent consistent projection bundles and future projections. With support by a software tool the millions of different combinations of the future projections were compared and those with projection bundles inconsistencies eliminated. Thus, the number of usable bundle combinations was reduced to a few thousand. A cluster analysis was performed with the goal to get an appropriate number of scenarios. The goal is to reduce the complexity, but to lose as little information as possible. For the automotive spare parts market in 2030, the cluster analysis showed that the three different scenarios can be differentiated. Finally, in this phase a visual representation of the raw scenarios by a mapping of the future place was performed (Figure 4). The mapping of the future place uses the statistical method of Multi Dimensional Scaling (MDS) and makes it possible to identify the corresponding projection bundles for the raw scenarios. The MDS shows the similarities of the key factor projections and projection bundles in a plot without using any dimensions. Likewise, the future projections can be assigned to the raw scenarios.

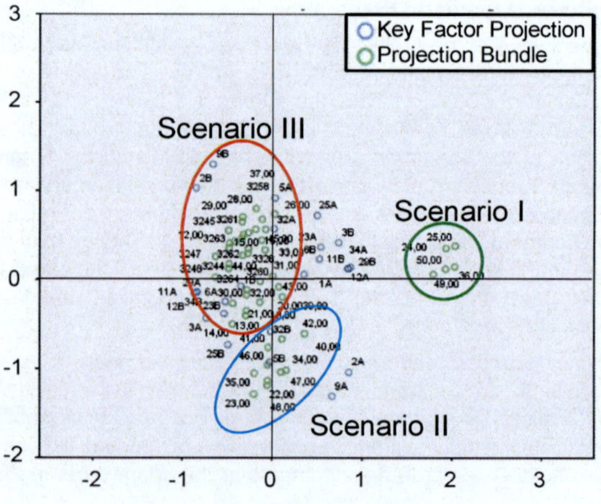

Figure 4: Mapping of the future space.

Key Factor	Projections of the Scenario I	Projections of the Scenario II	Projections of the Scenario III
Economic Environment	Recession	Boom	Boom
Purchasing Power	Decreasing	Increasing	Increasing
Cost of Energy	Decreasing	Decreasing	Increasing
Legislative Measures	Low Influence	High Influence	Low Influence
Customer Requirements	Constant	Increasing	Increasing
Trend to cheap cars	Increasing	Decreasing	Decreasing
Innovation Power	Decreasing	Increasing	Increasing
Hybrid Car/Electro Mobility	Niche product	Niche product	High Market Share
Gas-, Hydrogen-cars	Niche product	High Market Share	Niche product
Share of electronic Parts	Decreasing	Increasing	Increasing
Use of new Materials (like carbon fiber reinforced plastic)	Increasing	Increasing	Increasing
Standardization	Increasing	Decreasing	Increasing
Shifting of market segments	High Diversity	Mass Production	Mass Production
Guarantee, Warranty deed	Not relevant	Selling Argument	Selling Argument
Sustainability	Not relevant	In the focus of customers and enterprises	In the focus of customers and enterprises

Table 1: Key Factors and Projections of the Scenarios.

The final step of the Scenario Development is a verbal description of the raw scenarios. Before the actual formulation of the scenarios took place, initially the used future projections were identified. The selection could be ascribed to the graphical representation and by the selection of the projections from the projection catalog based on the distribution percentage of the projections within the created raw scenarios. One problem in this case is the assignment of projections which cannot be classified clearly into a raw scenario (fuzzy projections). The assignment of the projections to the raw scenarios is the framework for the formulation of the scenarios. Table 1 describes the three extreme scenarios for the automotive spare parts market in 2030.

4 DESCRIPTION OF EXTREME SCENARIOS

Three different extreme scenarios were formulated, that will serve as the basis for strategy development of the participants. Since the observation period with about 20 years is very far, it is not appropriate to create trend scenarios. Therefore, extreme scenarios with two extreme projections for each key factor were found. These extreme developments of the key factors should give an indication of the actual development but it is not realistic that this development will occur in the described way. Thus, in future possibly the key factor "sustainability" will be neither completely irrelevant nor the only relevant factor for the customers and enterprises.

4.1 Scenario I: The spare parts market in times of crisis

In the first scenario, the economy is in recession. Positive effects of the bad economic situation are falling energy prices. The electro mobility will remain a niche product. The trend towards hybrid and electric cars is attenuated by the weak economy. Also the gas-powered cars cannot drive out the traditional internal combustion engine as the power unit in the automotive industry and will remain a niche product in the automotive market.

In the car business longer warranty periods are not used as a selling argument, as the OEM try in the first place to achieve a differentiation from competitors by price leadership. Customers must pay for the spare parts themselves and will more often take cheaper parts from the independent aftermarket instead of the expensive original parts. In the segment of low-cost cars a significant increase of car sales in 2030 will be recognized. The customers do not place great emphasis on innovation in automotive, as this would mean an increased price for the cars.

To reduce the weight of automobiles and to be able to reduce fuel consumption the car manufacturers use advanced materials like carbon fiber reinforced plastic. Due to the increasing use of these new materials the manufacturing costs of components made of these lightweight materials will decrease. This results in a new range of parts for the spare parts management. CFRP components are only partially repairable and must be replaced in many cases of damage. Because of these, the OEMs profit due to higher spare parts sales for these components. A measure to reduce costs is to standardize the components. In particular, electrical control devices will be cheaper to produce and in larger numbers in the future through standardized interfaces and greater integration of functions.

Nevertheless, in both ways automobile manufacturers and the customer do not follow a trend towards a sustainable action. For the car manufacturer, it is primarily important to achieve better quarterly results in the short term and to weather the crisis. Long-term economic and sustainable management decisions are of little importance for the OEMs.

4.2 Scenario II: Spare parts market in economic boom

In the second scenario the automotive market in Western Europe in 2030 benefits from a booming economic development. As a result, the automobile and the spare parts market are in a good situation. The good economic situation is supported by a significant increase in purchasing power in the countries of Western Europe. At the same time, however, the wishes of customers rise concerning products and services offered by the manufacturer. Former optional extras are in the future model already mandatory. It is therefore essential for the car manufacturers to incorporate innovations in the automobiles. Efforts on the part of manufacturers and suppliers to establish standards for the development and construction and to reduce the proportion of the electronic components are due to the amount of innovation not successful. This results in a high number of different spare parts.

Car manufacturers will benefit from the sales of mass models. In the area of new propulsion techniques, the traditional internal combustion engines will be kicked out of the market. The demand for cars with gas engines will increase as a result of the fact that fuel costs can be saved. In the segment of hybrid and electric cars, however, a negative trend can be observed. The significantly higher cost and ongoing low range, limit the usefulness in everyday life.

In the automotive industry more new materials are used to reduce weight and therefore fuel consumption. This results in particular, in a strong increase of lightweight materials. The concept of sustainability is not only in terms of resource conservation and environmental friendliness on the daily focus of the customers and OEMs. That OEMs place great emphasis on sustainability and long-term strategies in their management decisions. Short-term profit-taking at the expense of long term bad operations are avoided by the OEMs.

4.3 Scenario III: Economic boom and electro mobility

In the third scenario the economic trend is positive. Because of the significant increase in purchasing power and the good economic situation the demands of customers increase. Even though models can be produced and sold in high numbers. For the spare parts management another key influence is visible through the standardization of in-car installed control units. To save development costs and achieve economies of scale due to the procurement of large quantities of control devices, suppliers and automobile manufacturers have agreed on standards in the control equipment manufacturing. That does defuse the issue of spare parts management despite a rising share of electronics. The OEMs can reduce the number of spares inventory significantly this results in a reduction of complexity in the spare parts business. The current trend of continuous increase in energy prices will continue in the year 2030. This is forcing car manufacturers to reduce the total weight of automobiles to reduce the fuel consumption. To implement the reduction in weight the car manufacturers use new lightweight materials like carbon fiber.

In the area of new propulsion technologies the cars with electric motors win market share and drive out the cars with conventional internal combustion engines. On the other hand cars that run on gas are still sold in very small quantities and thus cannot gain market share. In addition to the increased demands, in the future customers put a high value on environmentally friendly and sustainable management. This trend will be followed by the OEMs and they make their strategic decisions in terms of sustainability.

5 SUMMARY

This paper stated that the automotive spare parts market is exposed to major changes in 2030. In the face of these changes the actors in the aftermarket have to develop actively strategies that will enable the minimization of possible risks or the use of opportunities for the benefit of the enterprise. For the development of a sustainable strategy, the identified scenarios have to be transferred to the specific conditions of the enterprise. The scenario transfer can be implemented by means of a SWOT analysis, impact analysis and other methods. It is also important to include disturbance events in the planning.

Similarly the extreme scenarios created are a good basis for the review of existing strategies. These can be examined to determine whether the actual strategies can be applied and objectives reached if one of the scenarios becomes reality.

6 REFERENCES

[1] Little, A. D. (2008): Automotive After Sales 2015, in: Automotive Insight.

[2] Dombrowski, U.; Schulze, S. (2008): Repair of Automotive Electronics as a Module of Life Cycle Orientated Spare Parts Management, in: Proceedings of the 15th CIRP International Conference on Life Cycle Engineering, (LCE 2008), pp. 464-469, Sydney, Australia.

[3] Impuls Management Consulting (2008): Global Spare Parts Management 2010.

[4] Dombrowski, U.; Weckenborg, S.; Schulze, S. (2010): Method-supported product development for post-series supply, in: 20th CIRP Design, Nantes, France.

[5] ADAC (2009): Ersatzteile im Preisvergleich, in: Informationen aus der Fahrzeugtechnik.

[6] Deutsche Automobil Treuhand (2010): DAT-Report 2009.

[7] Jarke, M.; Buib, X.T.; Carrollc, J.M (1998).: Scenario Management: An Interdisciplinary Approach, in: Requirements Engineering Vol. 3, No. 3-4, pp.155-173.

[8] Herzhoff, M. (2004): Szenario-Technik in der chemischen Industrie, Dissertation, Berlin.

[9] Fink, A.; Schlake, O. (2000): Scenario Management-An Approach for Strategic Foresight, in: Competitive Intelligence Review, Vol. 11, No. 1, pp. 37–45.

[10] Gausemeier, J.; Stollt, G.; Wenzelmann, C. (2006): Future Scenarios, describing consistent networks of influences from markets, technologies and business environment with a time horizon of 2020, in: Reconfigurable Manufacturing Systems, Springer, Berlin.

[11] Grienitz, V.; Blume, V. (2008): Business Engineering with Scenario Technique, in: Proceedings of the 2008 IEEE IEEM, pp. 1547-1552.

[12] Kosow, H.; Gaßner, R. (2008): Methoden der Zukunfts- und Szenarioanalyse Überblick, Bewertung und Auswahlkriterien, in: WerkstattBericht, No. 103, Berlin.

[13] Bui, T.; Kersten, G.; Ma P. (1996): Supporting Negotiation with Scenario Management, in: Proceedings of the 29th Annual Hawaii International Conference on System Sciences, (IEEE 1996), pp. 209-218, Hawaii, USA.

[14] Gausemeier, J.; Fink, A.; Schlake, O. (1996): Szenario-Management – Planen und Führen mit Szenarien, 2. bearbeitete Auflage, Hanser, München, Wien.

[15] Grienitz, V., Ley, S. (2007): Scenarios for the Strategic Planning of Technologies, in: Journal of Technology Management & Innovation, pp. 21 – 37.

[16] Läufer, N.(2009): Mikro- und makroökonomische Effekte der Abwrackprämie, in: Wirtschaftsdienst, Vol. 89, No. 6, pp.410-441.

[17] Williams, A. (2007): Product service systems in the automobile industry: contribution to system innovation?, in: Journal of Cleaner Production, No. 15, pp.1093-1103.

[18] Baake, P. (2010): Accidents, Liability Obligations and Monopolized Markets for Spare Parts, in: The B.E. Journal of Economic Analysis & Policy, Vol. 10, Iss. 1 (Contributions), Article 36.

[19] Lakshmanan, K.; Bhatia, G.; Rajkumar, R. (2010): Integrated End-to-End Timing Analysis of Networked AUTOSAR-Compliant Systems, in: Design, Automation & Test in Europe Conference & Exhibition (DATE).

PSS Contracts for Rail and Road Infrastructure

Sofia Lingegård[1], Mattias Lindahl[1] and Niclas Svensson[1]
[1]Department of Management and Engineering, Linköping University, Linköping
Sweden

Abstract
The productivity development for rail and road infrastructure has been weak a long time; and explanation can be found in the traditional contracts used, with little room for incentives for innovation. This literature study investigates the use of the few realized PSS contracts within the rail and road infrastructure. The descriptions and the scientific reports are on a synoptic level and a majority of the reports are funded by the involved actors, showing that there is an interest for PSS contracts in the industry and indicating significant potential for further research in the area.

Keywords:
Rail Infrastructure; Road Infrastructure; Literature Review

1 INTRODUCTION

Productivity development in the construction industry, such as road and rail infrastructure, has been weak for a long period of time, possibly due to the traditional form of contracting used. In the recent years, these traditional forms for operation and maintenance have caused increased costs, and thus resulted in an increased interest for new contracting types [1-2]. Construction contracts are currently used to a large extent in Sweden, but this type of contract has shortcomings concerning weak incentives for the development of procedures [3]. Interviews with managers at the Swedish Transport Administration, formerly the Swedish Rail Administration (see more detailed information in Section 3), are in line with this argument, which claims that the major obstacles for technical development, as well as the limited room for innovation, are due to over-detailed specifications of how to do things [4]. This gives the procurer a reason to design the contracts to produce more incentives for cost efficiency [3]. It is also in the strategy of the Swedish Transport Administration to improve the cost efficiency of maintenance by improving the conditions already in the development phase, and to perform maintenance as efficiently as possible considering the entire life cycle of the products [5] . One way of doing this could be with another type of contracting. Performance contracting gives better incentives for contractors to develop the product by e.g. finding a better balance between building and maintaining costs [6]. The contractor is responsible for delivering an agreed upon function and can decide how this should be done, as well as for weighing building costs against future maintenance cost which should provide a cheaper solution for both procurer and contractor [6]. This type of contracting, when the customer only pays when outcomes have been delivered instead of the traditional payment for tasks and activities, is also called Product Service Systems, or PSS [7]. A more extensive review of this type of contracting can be found in Section 4.

This is the starting point of the DORIS (Development of (Integrated product service) Offerings for Rail Infrastructure Systems) project which was launched in cooperation with the Swedish Transport Administration to investigate the possibility for PSS contracts within the rail infrastructure in Sweden. This paper is based on parts of the initial literature study [8] and has the following objective.

1.1 Objective

The objective of this paper is to, through a literature study, investigate what has been published in the area of PSS contracts for rail and road infrastructure, as well as to look into the current state of rail infrastructure procurement. This gives the following research questions:

- RQ1: What types of contracts are currently used when procuring rail and road infrastructure?
- RQ2: To what extent are PSS contracts used for rail and road infrastructure?
- RQ3: In what way are the PSS contracts for rail and road infrastructure documented?

1.2 Limitations

The examples of PSS contracts presented are mostly implemented in Scandinavia, with the exception of one example from the UK. Furthermore, issues concerning performance requirements and measures will not be addressed in this paper, even though these issues represent major challenges for these contracts.

1.3 Outline of the paper

This paper starts with a description of the method used, and subsequently presents the PSS concept. Next, examples of PSS contracts for rail and road infrastructure are presented, followed by the discussion, conclusion and suggestions for further research.

2 METHOD

The starting point for the literature study was an introductory interview with an employee at the Swedish Transport Administration, one well informed with the working process in the organization. The goal was to gain sufficient knowledge within the area to start searching for literature for this study.

Literature reviewed includes several different kinds of sources: scientific articles, reports, homepages, master theses as well as

J. Hesselbach and C. Herrmann (eds.), *Functional Thinking for Value Creation: Proceedings of the 3rd CIRP International Conference on Industrial Product Service Systems, Technische Universität Braunschweig, Braunschweig, Germany, May 5th - 6th, 2011*,
DOI 10.1007/978-3-642-19689-8_51,

doctoral and licentiate theses. The information has, when it was possible, been triangulated using different sources. However, the information concerning the use of PSS contracts for rail and road infrastructure was limited to just a few sources. This could imply that the information was biased, but most of the information was retrieved from the Swedish National Road and Transport Research Institute. No geographical limits were used when searching for literature; instead, the search included literature from several continents, even though some of it is not included in this paper. This is partly due to the difficulty in judging the quality, and partly since the conditions for the contracts were far from the ones in Sweden, and therefore not relevant in this study.

3 THE CURRENT SITUATION FOR PROCUREMENTS OF RAIL AND ROAD INFRASTRUCTURE IN SWEDEN

The Swedish Rail Administration is responsible for 80% of the total rail system in Sweden [9]. Since 2001, the Swedish Rail Administration has exposed its maintenance contracts to competition, and by the end of 2009, 88% (measured in track distance) of the maintenance had been exposed to competitive procurement [10]. Furthermore, competitive procurement has resulted in reduction in costs for maintenance, but despite this, maintenance costs are increasing, and in 2008 maintenance costs and reinvestment cost made up over 50% of the cost for new investments.

It was concluded in a doctorial thesis [11] from 2006 concerning the former Swedish Rail Administration that there has not been significant pressure on the organization, internally or externally, to use life-cycle environmental management, and the work has been focused on environmental issues found locally, and not on life-cycle perspectives. Furthermore, the author states that the railway needs to adopt new perspectives to start working with environmental management of the products, and to set environmental requirements already when designing new products, i.e.before introducing them in the material supply chain.

3.1 Construction contracts

Construction contracts are the most common contracts within the infrastructure construction industry in Sweden [12]. The concept is based on the procurer specifying what, how and how much for the project [6]. This includes the choice of technology, materials and functions, and the tenders are made in unit prices with the choice of tender mainly based on the lowest price [2]. Construction contracts imply that the procurer carries all the risk, and a maximum roof for the price is set which does not create any incentives for the contractors to make the processes more efficient; instead, they benefit from reaching the maximum sum [6]. Furthermore, the author states that it also increases the incentives for the contractors to make additional orders to increase their profitability, or even to present a low tender and then make the money on the additional orders.

3.2 Service contracts

The Swedish Transport Administration has since 2005 used performance contracts for its maintenance [5], meaning that "...the Swedish Rail Administration procures a functionality of the track that has been set in advance. The contractor subsequently decides what measures to take with respect to the performed reviews and regulations for maintenance." [13] The length of the contracts is 5 years, with an additional 2 years option with bonuses for improvements, such as a lower number of errors, and with penalties when the contractor has not reached the levels of e.g. delays [5].

4 PRODUCT SERVICE SYSTEM: ONE CONCEPT, SEVERAL NAMES

The focus for this paper is PSS contracts where the procurer requests a function instead of a specific execution [12]. This kind of functional buying/selling has many different names, and during this literature study several different names for the concept of buying a function have been revealed. The most commonly occurring are presented in Table 1. The concepts in the tables have no intergroup order.

Name	Definition/description	Reference (Name, year, page)
Outcome-based contracting	"...a contracting mechanism that allows the customer to pay only when the firm has delivered outcomes, rather than merely activities and tasks."	Ng et al., 2009, p. 1 [14]
Performance-contracting	"The contract terms are based on that future users are given access to some specific services, not on the contractor fulfilling technical specifications: it is the performance of the asset over the contracting period that matters."	Nilsson et al., 2006, p. 7 [3]
Performance-based contracts	"...are about contracting on performance, rather than tasks or outputs by the service provider."	Ng and Yip, 2009, p. 207 [15]
Functional sales	"The customer purchases a function and the hardware plus service includes the totality of activities that enable the customer to benefit from a total functional provision."	Alonso-Rasgado et al., 2004, p. 515 [16]
Solutions projects	"...solutions projects usually include the responsibility for the provider to manage, resource, support and improve the delivery of the solution through the life of the product or system in use."	Brady et al., 2005, p. 364 [17]
Performance contracts	"Performance Contracts are defining a product and it is up to the contractor how to achieve this. Therefore, work selection, design and delivery are all his responsibility."	Zietlow, 2005, p. 3 [18]
Product Service System, PSS	"a marketable set of products and services capable of jointly fulfilling a user's need"	Goedkoop et al., 1999, p. 18 [7]
Integrated Product Service Offerings, IPSO	"...from a lifecycle perspective, to offer and optimise a solution with a combination of products and services that satisfies an identified customer need, and at the same time increases the suppliers' competitiveness. "	Lindahl et al., 2006, p. 1-2 [19]

Table 1: Different names for performance-based contracts.

From this table it can be concluded that even though the names of the concepts differ they still include the same content, namely that the focus is on the output, and not on how the output is achieved. In this paper, the concept will be called Product Service System, or PSS, since this is the most common name of the concept. The PSS for the infrastructure includes planning, construction and subsequently maintenance of the construction when the usage phase has started [12].

4.1 PSS characteristics

PSS offerings have a life-cycle perspective, and the combination of products and services can be combined into an optimized solution for the customer, as well as to give the manufacturing company the possibility to have control over the product throughout its whole life-cycle [17, 19]. PSS provides the supplier with a possibility to increase the value of the solution for the customer by integrating components in new ways [17], and is thereby a driver for the development of a technical solution [19]. This provides incentives for the supplier to realize a more economical and environmental development when considering the whole life-cycle [19]. Companies acting in a mature industry can use the PSS as a growth strategy and compete with their core competence rather than with physical assets [20]. The author also states that PSS requires a closer and improved relationship with the customer, and the customer no longer has to be the owner of the product.

5 PSS FOR RAIL AND ROAD INFRASTRUCTURE

This section provides examples of projects fulfilling all or parts of a PSS contract for road or rail infrastructure.

5.1 PSS for rail infrastructure

The two rail projects reviewed, as seen in Table 2, are collected from Sweden and the UK, and differ considerably in their content.

Arlandabanan, Sweden

Arlandabanan is a 20 km long railroad section with double tracks, a 7 km tunnel and several underground stations between the city of Stockholm and Arlanda airport [21]. The winning tender for this Public-Private Partnership, PPP,contract was a consortium constructed by six companies from several different countries [22]. The consortium later formed the company A-train AB to finance, build and subsequently run the train traffic. Construction began in 1995 and was finished in 1999, when the ownership of the facility was transferred back to Arlandabanan Infrastructure AB, former A-banan AB. At the same time, A-train got the right to use the tracks for traffic until 2040, as well as the responsibility to run and maintain the facility during the same time period. [21] The ticket revenues from the end customers serve as revenues for A-train in this contract [12]. Arlandabanan Infrastructure AB has the possibility to cancel this agreement in 2010 if A-train is not fulfilling the requirements of the contract [21] .

Two thirds of the construction costs were funded by the consortium, while the last third was a conditional loan from the government. All the income risk was carried by the consortium, as well as most of the cost risk in the project. [23] A-train had the right to a penalty from Arlandabanan Infrastructure AB in case it would not be possible to construct a double track within the time that was agreed, since this would delay the traffic start, and also for other issues such as discovery of archeological findings or incorrect technical information [24]. A-train got the freedom to balance the cost for the initial investment and the cost for maintenance, and subsequently chose to use another solution for the tracks than the original idea of the procurer [25].

The Northern Line, UK

In 2003, three long-term contracts including maintenance and upgrading of infrastructure of the London Underground network were signed dividing the network into three parts e.g. the part including the Jubilee, the Northern and the PiccadillyLines. These PPP contracts spanned over 30 years, with opportunities to review the contracts and requirements every 7.5 years [26]. For the Northern Line, the contract included the leasing of the trains and an area of 50 stations, and full responsibility for the design, manufacturing and cleaning of trains and related equipment [27]. In addition, the contract was formulated so that 80% consisted of capital works and 20% of service elements the first years, while the service made up 100% of the total in the fifth year and beyond.

The contracts were output-specified, and the required performance levels were measured with the following three factors [26]:

- Availability, counting delays and disruptions lasting longer than two minutes.
- Capability of the line.
- Ambience, measuring the quality of the customers' travelling environment.

These measurements were all set in 2003, before the contracts started, and a level of around 5% worse than the historic data of the London Underground was decided to be the benchmark for the first five years, and then subsequently become more challenging with time [26]. The report also states that bonuses and penalties are used as additional adjustment for the performance, and liabilities for environment and safety are included. Furthermore, the same report claims that the availability for the JNP Lines have improved on the whole since the start in 2003/2004. This has, on the other hand, not been consistent for all the lines, e.g. the availability measure for the busy Northern line was 48% worse than the benchmark in 2005/2006 [28]. Furthermore, the article states that the parties then agreed upon a change in the contract to solve the problem, i.e. the program for renewable the tracks was accelerated and initiated two years earlier than stated in the initial contract.

Project	Content	Length	Payment
Arlanda-banan, 20 km (SWE)	Finance, build, operate and maintain.	45 years	Contractor built the tracks and is now leasing them. Ticket revenues.
The Northern Line, 50 stations (UK)	Design, manufacture and service.	30 years	Leasing contract based on a guaranteed number of trains and performance improvement.

Table 2: Reviewed contracts for rail infrastructure.

5.2 PSS for road infrastructure

Three Scandinavian road contracts are reviewed in this paper. Table 3 illustrates their differing financial solutions.

Norrortsleden, Sweden

An ongoing contract for the Swedish Transport Administration includes design, construction and then maintenance for 15 years. The project comprises 7 km of highway, including a tunnel and several road bridges. The public procurement process started in 2003, and three companies sent in their tenders. The differences in price for the contracts concerning the design and construction were minor, while there were significant differences in the pricing concerning the maintenance phase. The contractor built the construction during a period of 3.5 years (2005-2008), and is currently responsible for maintaining it. [29]

The technical requirements were mostly formulated as functional requirements; examples follow below, translated directly from Swedish [29]:

- "The road and the tunnel should be designed for a dimensioned speed of 90 km/h."
- "The road body,..., should by the transfer have a remaining lifespan (residual value) of at least 25 years."
- "The design of the bridges should follow an overall coherent formation concept for the whole of Norrortsleden."

During the construction phase the role of the procurer was one of advisor, since the operation was led by the contractor [30]. The report also refers to the fact that the two organizations were built to mirror each other to ease the cooperation. A fixed annual payment for the construction with a special arrangement for the tunnels, where the contractor classifies the quality of the mountain as well as what activities are needed, serves as guidance for the pricing, and any sums above or below this price are shared by the two parties [29]. Furthermore, the reports states that the compensation to the contractor for the maintenance phase is paid as a fixed sum every year; this sum is adjusted by an index based on nine weights specified in the contract, e.g. amount of traffic [29]. The risk of the project was decided to be assign to the one that was in the best position to manage it, in this case the contractor [30].

E18, Finland

In Finland in 2005, a performance contract was signed for a 51 km road section, including tunnels and bridges, concerning design, construction, operation and finance[3, 31]. The section, Muurla-Lohjanharju, is a cooperation between the public and private sectors, with the Finnish Road Authority buying the entire project from a consortium [32] .

The tenders of the contract were, apart from the price, judged on e.g. experience of similar projects, preliminary plan for building and maintenance, plan for reassuring quality, plan for handling technical disturbances and the contractors' financial situation [3]. The authors claim these parameters influenced the pricing by plus or minus 10%. The contract lasts until 2029, equal to 21 years, and as from the opening of the road in 2008 the Finnish Road Authority pays a service fee depending on the availability and service level of the road including several quality criteria [31-32]. The fee is connected to a yearly index regulation, and the procurer carries the risk for an increase in price for different components used [3]. The authors further state that the maintenance costs are based on a calculated value for the extent of the traffic, and the contractor does not carry the risk for an increase in traffic. Furthermore, they mean that there is a deduction in fee e.g. if the availability is not satisfactory, and the traffic safety is measured every year to give the contractor incentives to work for accident reduction. The report also states that innovations that bring a decreased number of accidents are rewarded, while an increase in the yearly accident count results in a deduction from the yearly payment. Furthermore, the contractor is also refunded for the initial investment when the road is ready for traffic, providing incentives for fast opening. At the end of the contract period, the road section will be transferred and Finnra will attain control [31]. This will be done without any additional fees, and the road section must be in the condition agreed upon [32].

E39, Norway

Three road projects have been procured by PPP contracts [23]. The three projects were 27 km, 38 km and 38 km respectively, and they have similar characteristics so the description will be limited to one of them, namely the first section, E39 Klett- Bårdshaug [33]. Four potential tenders were competing for the contract [23]; however, the potential contractors had to apply for a pre-qualification questionnaire to determine that the contractors fulfilled requirements such as necessary technical and professional knowledge, financial strength to complete the project, documented experience and a list of reference projects [33]. The construction started in 2003 and two years later the road opened, two months before the scheduled time [23]. The authors also state that the contractor is responsible for financing, building and maintenance over a period of 25 years, and after that the Norwegian public roads administration take over the maintenance.

The responsibility of the contractor is the safety and the availability of the road and for standards such as environmental standards [33]. Furthermore, the report points out that the contract includes more detailed requirements, such as quality and performance standards and quality routines for which the NPRA can survey to make sure they are being followed. The public part carries risk for events that the project company has no or minor possibility to influence, such as changes in legislation, while risks related to financing, planning, construction, operation and maintenance are allocated to the project company according to the report.

The project company receives an annual payment depending on the availability, performance, safety and amount of traffic on the road, but out of these, availability and performance are the major contributions to the payment, while the rest are seen as additional payments [33]. These traffic payments increase with increased traffic, as well as the safety payments, which are seen as a bonus related to the number of and the character of the accidents that occur compared to an equivalent average road [23]. Furthermore, the authors state that the Norwegian public roads administration has introduced a toll system for the roads for the first 15 years, from which the income will cover part of the costs for the project.

Project	Content	Length	Payment
Norrorts-leden, 7 km (SWE)	Design, construct and maintain.	15 years	Fixed yearly sum based on index Special payment model for tunnels.
E18, 51 km (FIN)	Finance, design, construct, and maintain.	21 years	Initial investment and service fee to contractor depending on availability and service level of the road. Adjusted for change in traffic.
E39, 27 km (NOR)	Finance, design, construct, and maintain.	25 years	Payment depending on availability and standard of the road. Toll system for procurer.

Table 3: Reviewed contracts for road infrastructure.

6 DISCUSSION

This concluding discussion has the ambition of answering the research questions posed in the introduction by contrasting reviewed contracts.

6.1 What types of contracts are currently used when procuring rail and road infrastructure?

Currently, construction contracts are the most common way to procure rail and road infrastructure in Sweden [12], while maintenance contracts are mostly a type of performance contract spanning over five years with specified functional requirements [5]. This is a short time period compared to the life-cycle of rail tracks, which is why PSS contracts would be of interest instead.

6.2 To what extent are PSS contracts used for rail and road infrastructure?

The two examples reviewed for rail infrastructure have both similarities and differences. The Arlandabanan project included both building the tracks and later maintaining them, as well as operating the train traffic [21]. The Northern Line project, on the other hand, focused on the actual trains and reinvestment, maintenance and operation of them [27]. Both of the contracts involve the operating of the traffic, but only in the case of Arlandabanan is the construction of the tracks included. The Northern Line contract has an output-specified performance level where availability is emphasized when it comes to payment for the contractor [26]. For Arlandabanan, it is a totally different case since that consortium receives its revenues from the end customers [25] and not the procurer. While the Northern Line contract was formulated as a PSS contract, it is not known how specified the contract for Arlandabanan was, even thought there seems to have been margins for flexibility for A-train when formulating their solution [25]. Other similarities are the long time period for the contracts, 45 years for Arlandabanan [21] and 30 years for the Northern Line [26]. This differs a lot from the time periods currently used for maintenance contracts in Sweden.

The three road contracts reviewed differ significantly in size when measured in kilometers as well as the payment mechanism. The E18 and the E39 both have an output-specified payment where the fee depends on availability and performance or service level [32-33]. The E18 contract also includes the fulfillment of some quality criteria in the payment model, and this is also the case for the E39 [32-33], but this model is far from the fixed yearly sum that the payment for the Norrortsleden consists of, where only an adjustment is made by an index [29]. All three contracts do however make adjustments for changes in traffic, since this is something the contractor cannot affect, meaning there has to be some flexibility in the contracts making it possible for the contractor to carry the risk.

All contracts reviewed in this paper seem to have the life-cycle perspective that is included in a PSS contract [19] as well as the combination of products and services [19] that is required. The type of products and services differs from different contracts, even though they are implemented in the same industry. In four of the five cases the procurer owns the product, the rail tracks, while in the fifth case it is the contractor who owns the product, the trains, which is in line with the characteristics of a PSS saying that it is not necessarily the customer who gets the ownership of the product [34]. However, even though most of the contracts are output-specified there are still more detailed requirements from the buyer.

6.3 In what way are the PSS contracts for rail and road infrastructure documented?

Little has been done and consequently documented in the area of PSS contracts for road and rail infrastructure. This is especially true for the rail infrastructure, where few projects have been realized. For the infrastructure that has in fact been procured with a PSS type of contract, the information concerning the implementation includes reports and not articles published in scientific journals. These reports consist of information concerning the planning and building of the infrastructure, but in general the contracts have not been in place long enough to be evaluated. The maintenance phase of the contracts can span over decades, implying a complete evaluation will not be realized in the near future since most of the contracts reviewed in this study have recently transitioned from the building to the maintenance phase. Something worth mentioning is that the reports are mainly financed by the governmental organizations, in the case of Sweden by the Swedish Transport Administration, showing that there is in fact an interest in these types of contracts from the procurer side.

7 CONCLUSION AND FURTHER RESEARCH

Contracting for rail and road infrastructure in Sweden is done using traditional construction contracts, and with maintenance contracts using a small degree of performance dimension. A general conclusion from this literature study is that the PSS contracts used for road infrastructure are more developed than the ones for rail infrastructure. This could be due to the complexity of the rail structure, implying there are more issues to handle than when building roads. It is therefore likely that the rail infrastructure can benefit from the experience from the road infrastructure. On the other hand, the road contracts are output-specified, but still have elements of more detailed requirements.

The lack of publications, in combination with interest from the industry in this field, implies that there is a gap in the area where research is needed both to facilitate the use of the contracts but also to move the research forward. The next step in the DORIS project is to conduct an interview study focusing on possibilities and challenges concerning PSS contracts for rail and road infrastructure in Sweden. The respondents will consist of the Swedish Transport Administration, contractors and track design actors. This will be followed by a quantitative study focusing on the environmental and economical advantages and disadvantages with PSS contracts for rail and road infrastructure.

8 ACKNOWLEDGMENTS

The authors would like to thank the Swedish Transport Administration for financing the study.

9 REFERENCES

[1] Nilsson J.-E., (2009). Nya vägar för infrastruktur. Stockholm: SNS Förlag.

[2] Hedström R., Ihs A., Sjögren L., (2005). Funktionsupphandling av väg- och banhållning, Problem och möjligheter. Linköping: Swedish National Road and Transport Research Institute.

[3] Nilsson J.-E., Ihs A., Leif S., Wiman L.G., Wågberg L.-G., (2006). Funktionsupphandling. Sammanfattning av kunskapsläget och rekommendationer för fortsatt forskning. Linköping: Swedish National Road and Transport Research Institute.

[4] Stenbeck T., (2004). Incentives to Innovations in Roas and Rail Maintenance and Operations. Stockholm: Royal Institute of Technology.

[5] Riksrevisonen, (2010). Underhåll av järnväg.

[6] Nilsson J.-E., Bergman M., Pyddoke R., (2005). Den svåra beställlarrollen. Stockholm: SNS Förlag.

[7] Goedkoop M.J., Halen C.J.G.v., Riele H.R.M.t., Rammens P.J.M., (1999). Product Service systems, ecological and economical benefits. PricewaterhouseCoopers N.V./Pi!MC/Storrm C.S./Pre Consultants, Netherlands.

[8] Lingegård S., (2010). PSS contracts for rail and road infrastructure - a literature study, LIU-IEI-R-- 10/0112--SE. Linköping: Department of Management and Engineering.

[9] Banverket, (2008). Banverkets årsredovisning 2008. Borlänge: The Swedish Rail Administration.

[10] Banverket Produktion, (2009). Banverket Produktion Årsrapport 2009. Banverket Produktion.

[11] Svensson N., (2006). Life-Cycle Considerations for Environmental Management of the Swedish Railway Infrastructure [Doctorial]. Linköping: Linköping University.

[12] Nilsson J.-E., Pyddoke R., (2007). Offentlig och privat samverkan kring infrastruktur- en forskningsöversikt. Linköping: Swedish National Road and Transport Research Institute.

[13] Banverket, (2009). Banverket årsredovisning 2009. The Swedish Rail Administration.

[14] Ng I.C.L., Maull R., Yip N.,(2009). Outcome-based contracts as a driver for systems thinking and service-dominant logic in service science: Evidence from the defence industry. European Management Journal.27(6):377-87.

[15] Ng I., Yip N., (2009). Identifying Risk and its Impact on Constracting Through a Benefit Based-Model Framework in Business to Busniess constracting: Case of the denfence industry. 1st CIRP Industrial Product-Service Systems (IPS^2) Conference. Cranfield University. p. 230.

[16] Alonso-Rasgado T., Thompson G., Elfström B.-O.,(2004). The design of functional (total care) products. Journal of engineering design.

[17] Brady T., Davies A., Gann D.M.,(2005). Creating value by delivering integrated solutions. International Journal of Project Management.23(5):360-5.

[18] Zietlow G., (2004). Implementing performance-based road management and maintenance contracts in developing countries-an instrument of German technical cooperation. German development cooperation.

[19] Lindahl M., E. Sundin, A. Öhrwall Rönnbäck, G. Ölundh, J. Östlin,, (2006). Integrated Product and Service Engineering – the IPSE project. Changes to Sustainable Consumption, Workshop of the Sustainable Consumption Research Exchange (SCORE!) Network (wwwscore-networkorg), supported by the EU's 6th Framework Programme. Copenhagen, Denmark.

[20] Mont O.K.,(2002). Clarifying the concept of product-service system. Journal of Cleaner Production.10(3):237-45.

[21] Arlandabanan Infrastructure AB, (2010). Arlandabanan Infrastructure.

[22] Arlandaexpress AB/A-train, (2010). Arlanda express- About us.

[23] Arnek A., Hellsvik L., Trollius M., (2007). En svensk modell för offentlig-privat samverkan vid infrstrukturinvesteringar. Linköping: Swedish National Road and Transport Research Institute.

[24] Enberg N., Hultkranz L., Nilsson J.-E., (2004). Arlandabanan, en uppföljning av samhällsekonomiska aspekter på en okonventionell projektfinansiering några år efter trafikstart. Swedish National Road and Transport Research Institute.

[25] Nilsson J.-E., (2008). Upphandling, avtalsutformning och innovationer. Borlänge: VTI.

[26] London Underground, (2010). Transport of London, PPP & Performance Report 2009/2010, Report from the financial year ending March 31 2010. London.

[27] Harding A., Watts P.,(2000). The Northern Line Train Service Contract. Proceedings of the Institution of Mechanical Engineers -- Part F -- Journal of Rail & Rapid Transit.214(1):55-60.

[28] (2006). Three years on, and the PPP presents a MIXED picture. Railway Gazette International.162(10):669-74.

[29] Förnyelse i anläggningsbranschen (FIA), (2005). Slutrapport kortversion: Funktionsentreprenad Täby Kyrkby-Rosenkälla. Förnyelse i anläggningsbranschen.

[30] Förnyelse i anläggningsbranschen (FIA), (2008). Funktionsentreprenad i Täby Kyrkby-Rosenkälla, byggskedet. p. 22.

[31] Lehtinen P., Tuomisto T., Mikkola I.,(2006). Finland paves roads for future PPPs. International Financial Law Review.25:2:53-5.

[32] Vägförvaltningen (Finnra), (2005). Riksettan-till födel för alla, Motorvägen E18 Muurla-Lojo. The Finnish Road Administration.

[33] Norwegian Public Roads Administration, (2001). PPP-project E39 Klett-Bårdshaug.

[34] Mont O.K.,(2002). Claryfying the concept of product-service system. Journal of Cleaner Production.10:237-45.

A Study on Determination of Upgradability of Laptop PC Components

Kuniko Mishima[1], Nozomu Mishima[2]

[1] Graduate School of Systems Design and Management, Keio University, Kanagawa, Japan

[2] Advanced Manufacturing Institute, AIST, Ibaraki, Japan

Abstract

Laptop computers are a popular electrical product. Since the technological progress of the product is rapid, after several years, consumers may face the decision whether they should purchase newer products. However, offering an upgrade service may help to reduce environmental burden of production. In this paper, the environmental impact of the components' production stage and other lifecycle stages are estimated based on a disassembly. The paper tries to tell which components are suitable for upgrade. As a result, it is concluded that components with a relatively small environmental impact of the production and a fast technical progress are suitable for upgrade.

Keywords:

Laptop Computer; Reuse; Upgradability

1 INTRODUCTION

According with the increase of environmental consciousness in consumers, it has basically been said 'reuse' is a better end-of-life option than 'recycling.' However, in the actual product, it often happens that no official option of reuse is provided. The laptop computer is a popular electric product widely spread in society. However, the official end-of-life option of laptop computers is material recycling. In implementing the reuse of laptop computers, there are many obstacles such as difficulty in assuring the reliability, etc. As well as other electric and electrical products, laptop computers consist of many parts such as screws, plastic panels, etc. And a certain group of parts corresponds to a certain function of a laptop computer. In this paper, 'component' is defined as a physically connected group of parts having a certain function. One of the difficult problems is the difference of the speed of technical progress in each component. Although some components can still be used, some components don't have updated performance. In such case, consumers might not want to use the product with insufficient performance.

In this case, component reuse instead of product reuse can be the key. Modular design [1], [2] is a well-known technological option for ensuring reuse of components in the product. However, ensuring component reuse technically is not enough to motivate consumers to reuse components, because of the above-mentioned reason. In addition to that, a replacement service, so-called upgrade, of low-performance components is necessary. Without ensuring that the product is always up-to-date, no one would like to use the products for a long time. Therefore, although it might sound strange, upgrade service of components would be helpful in giving incentive to consumers for reuse of other components. In implementing the upgrade-service, the first step is to determine component upgradability properly.

The goal of this paper is to propose a simple and proper method to determine which components of laptop computers are suitable for upgrade. Based on the disassembly experiment of a laptop computer, the paper lists components which can be recognized as components. Then, the paper tries to quantify lifecycle environmental impact and cost for consumers of each component. Using these data and survey of speed of technical progress in typical specification of components, an index to identify the component upgradability will be proposed.

2 DISASSMEBLY OF A LAPTOP COMPUTER

2.1 Disassembly and components identification

Firstly, the authors actually disassembled a laptop computer and investigated which parts can be grouped as components. As a result of disassembly, seven components have been defined. Those are hard disk drive (HDD), AC adaptor, liquid crystal display (LCD), CPU board, battery, CD/DVD drive and main body. Table 1 shows the defined components and corresponding functions. Figure 1 (a), (b) and (c) show some examples of disassembled components of the laptop computer. As it is shown in the figure, some components such as HDD have been modular-designed already, but some of the components such as CPU board and main body cannot be called modules. However, these groups of parts correspond to certain functions of a laptop computer almost one by one. Thus, the groups of parts can be recognized as 'components' with particular functions. There are some other options of components categorization. However, the purpose of this paper is to consider the end-of-life strategies such as reuse or upgrade corresponding to the components. Therefore, components should have one on one relations between performances (specifications). As a result of performance-component matching, seven components shown in Table 1 were listed up.

J. Hesselbach and C. Herrmann (eds.), *Functional Thinking for Value Creation: Proceedings of the 3rd CIRP International Conference on Industrial Product Service Systems, Technische Universität Braunschweig, Braunschweig, Germany, May 5th - 6th, 2011,*
DOI 10.1007/978-3-642-19689-8_52, © Springer-Verlag Berlin Heidelberg 2011

Component name	Function description
Hard disk drive	Storage of data
AC adaptor	Supply of electricity by cable
Liquid crystal display	Display of image
CPU board	Calculate
Battery	Supply of electricity without cable
CD/DVD drive	Reading or writing of data
Main body	Physical support of other components

Table 1: Components and corresponding functions.

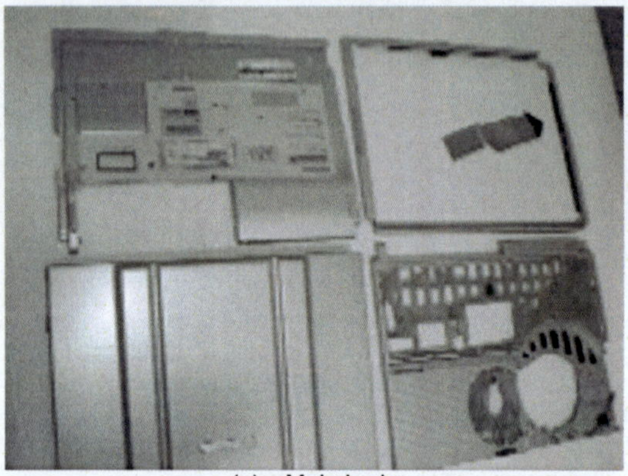

(a) Main body

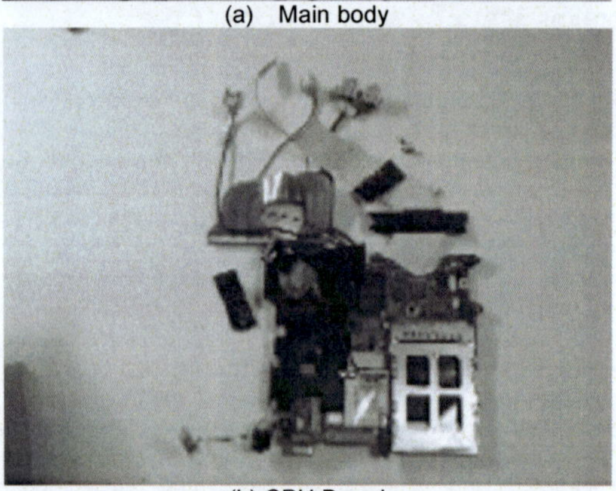

(b) CPU Board

(c) HDD

Figure 1: Examples of disassembled parts.

2.2 CO_2 emissions of components production

Then, the next step is to calculate the CO_2 emission during the manufacturing stage of each component based on existing surveys. For the components that can be disassembled to material level, equation (1) and database [3] on CO_2 intensity of materials were applied to calculate the CO_2 emission. However, the printed circuit board is a common part in electric products and it is difficult to further disassemble the printed circuit board. So, we treated the printed circuit board as a single material. The CO_2 emission in fabricating a printed circuit board was determined based on an existing survey [4]

$$Eci = \sum_i Cmi \times Wpi + Wb \times C_p \quad (1)$$

Eci: CO_2 emission for component production, *Cmi* : CO_2 emission intensity of material *i*, *Wpi* : weight of material *i*, *Cpi* : CO_2 emission intensity of printed circuit board, *Wb* : weight of printed circuit board.

As the result of the disassembly, CD/DVD, main body and CPU board were the three components which could be disassembled to material level. Table 2 shows the results of the measurement of each material in each component. Table 3 indicates the result of the calculation.

Materials (g)		Wpi/ Wb	Wpi/Wb	Wpi/Wb
Name	Cpi	(CD/DVD Drive)	(CPU Board)	(Main Body)
Stainless	1.2	1	—	6
Aluminum	10.3	—	—	9
Magnesium	55	—	—	64
Plastics	1.2	1	—	124
Circuit board	16	67	162	6

Table 2 : Material composition of components.

Component name	CD/DVD Drive	CPU Board	Main Body
Total GHG (kg-CO2, Eci)	1.08	2.59	3.51

Table 3 : CO_2 emissions of component production.

On the other hand, for components that cannot be disassembled to material level, another strategy is necessary. To calculate CO_2 emissions of HDD, LCD, battery and AC adapter, equation (2) was used. Table 4 indicates the weight of the components, weight of the corresponding reference components, CO_2 emissions of the reference components [4, 5] and calculated result of CO_2 emissions of the components production.

$$Eci = Wpi / W0i \times E0i \quad (2)$$

Eci : CO_2 emissions of component production, *W0i* : weight of the component *i*, *W*0i :weight of the reference component, *E0i* : CO_2 emission of the reference component

Components		*W0i* (g)	*Eoi* (kg-CO2)	*Eci* (kg-CO2)
Name	*Wpi*(g)			
AC adaptor	230	175	3.1	4.07
Battery	313	23	0.28	3.81
HDD unit	94	93	2.71	2.73

Table 4: CO_2 emissions of hard-to-disassemble components.

The remaining component to be calculated is the LCD unit. To calculate the CO_2 emission of LCD production, a mixture of the above-mentioned two strategies were used. Based on the existing survey [5] and material composition result, the CO_2 emission of the LCD production was calculated as 12.41[kg-CO_2]. Finally, as the result of the calculation of the CO_2 emission of production of all seven components, Figure 2 has been obtained.

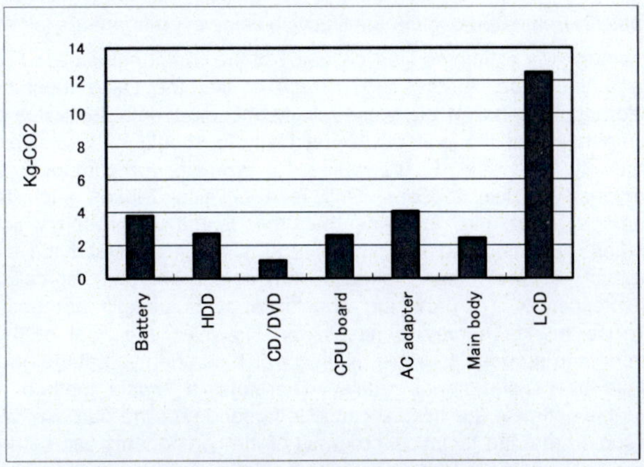

Figure 2: CO_2 emissions of components production.

2.3 CO_2 emissions of other lifecycle stages

The product lifecycle is often divided into production, delivery, use, collection and end-of-life treatment stages. In this paper, we also use this categorization. Since the CO_2 emissions of component production have been calculated in the former section, CO_2 emissions of other life cycle stages are examined here. Basically, the lifecycle CO_2 emission can be calculated by following equation (3).

$$ELi = EPi + ETi + EUi + ECi + ERi \quad (3)$$

EL: lifecycle CO_2 emission of component *i*, *EPi*: CO_2 emission of component *i* production, *ETi*: CO_2 emission of delivery of component *i*, *EUi*: CO_2 emission of usage of component *i*, ECi: CO_2 emission of collection of component *i*, *ERi*: CO_2 emission of end-of-life treatment of component *i*

A) Delivery

A report [6] from a computer maker estimates the CO_2 emission of a laptop computer delivery as 0.5kg-CO_2 per unit. It is distributed to each component based on its weight.

B) Usage

There is a guideline for PC makers to carry out LCA. It says that the average time of use for an average user is about 10 hours a week, 52 weeks per year and 5 years product life. On the other hand, another survey [7] estimates the average power consumptions of electrically driven components. It is estimated 7W, 2W, 41W, 30W for HDD, CD/DVD drive, CPU board and LCD, correspondingly. Based on the guideline and this estimation, it is possible to calculate total electric usage of component through the product lifecycle. By multiplying 0.34kg-CO2/Kwh, it is possible to calculate the CO_2 emissions of components for use.

C) Collection

In this paper, we assumed the distance for collection is same as delivery distance. Again, the CO_2 emission for the whole product is distributed to components based on the weight.

D) End-of-life treatment

The above-mentioned report from a PC maker also estimates the CO_2 emission for material recycling as -3.0kg-CO_2. It means that some material can be recovered and the CO_2 emission which can be reduced by using recycled material is estimated to be that value. In this paper, we also use that value and also distribute to components based on the weight.

Finally, the results of the $LCCO_2$ of components except the production stage are shown in Table 5. And all the sum of $LCCO_2$ including the production stage is indicated in Figure 3.

Component	Delivery	Usage	Collection	EOL
Battery	0.11	0	0.11	-0.69
HDD	0.04	6.19	0.04	-0.22
CD.DVD drive	0.03	1.77	0.03	-0.15
CPU board	0.06	36.24	0.06	-0.36
AC adaptor	0.09	0	0.09	-0.51
Main body	0.08	0	0.08	-0.46
LCD	0.11	26.52	0.11	-0.66

Table 5: Estimated LCCO2 of the components (kg-CO_2/unit).

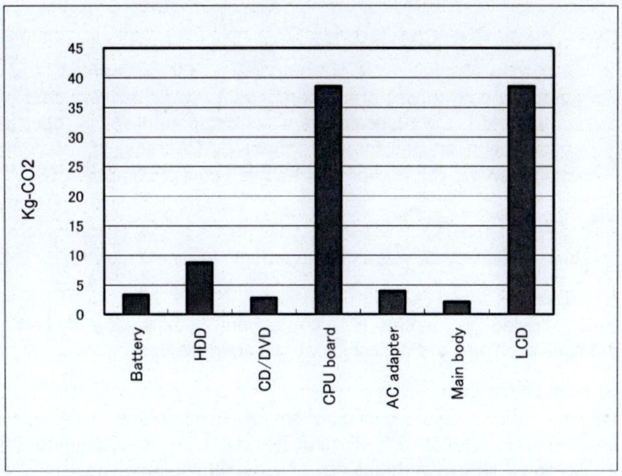

Figure 3: $LCCO_2$ of components.

3 LIFECYCLE COST FOR CONSUMERS

3.1 Purchase cost

In considering the upgradability of the components, it is necessary to take the cost of upgrade service into account. The amount of payment for the upgrade may be critical for consumers in determining whether they use upgrade service or not. In this section, prices of repair parts, price differences between product with CD/DVD drive and without drive, prices of external HDD and auction prices are used to estimate the initial price of components. Table 5 shows the result of the purchase cost estimation for the components of the example laptop PC. The total price of the product was estimated as 300,000JPY as of 2007. In addition, the paper assumes that the prices for the components are equivalent to the 'value' of the components for consumers. Since in an ideal market, prices of products automatically balance with the effects of the products, market prices of the components are equivalent or at least proportional to the component values in this case, too.

Component	Estimated price (JPY)
Battery	25000
HDD	18000
CD/DVD drive	45000
CPU board	66000
AC adaptor	10000
Main body	10000
LCD	60000

Table 6: Estimated price of components.

3.2 Other cost for consumers

Usually, lifecycle cost of products for consumers can be expressed by equation (4). However, in this case these costs affect consumers differently. For example, large expected cost for repair may decrease the component values for consumers. Therefore, a new equation (5) is used in this paper to estimate the component values for consumers. We assumed the cost of consumption articles will be equivalent to the value (effect) of the articles themselves. Then, the cost can be neglected. The other costs, delivery cost, electricity cost, repairing cost, EOL cost, are costs for consumers, that do not add to the product value. Thus, these costs should be subtracted from the component value.

$$Cui = Cpi + Cdi + Cei + Cci + Cfi + Cri \quad (4)$$

Cui : total lifecycle cost of component *i* for consumers, *Cpi* : purchase cost (price) of the component *i*, *Cdi* : delivery cost of component i, *Cei* : electricity cost of component *i*, *Cc* : cost of consumption articles for component *i*, *Cfi* : cost of repairing component *i*, , *Cr* : end-of-life treatment cost of component *i*

$$Vui = Cpi - Cdi - Cei - Cfi - Cri \quad (5)$$

Vui : total lifecycle value of component *i* for consumers

A) Delivery
When buying a laptop PC, consumers usually take it home themselves. Thus, the delivery cost can be neglected.

B) Electricity cost.
Lifetime total power consumption of components has been calculated in section 2.3. Electricity cost can be calculated by multiplying 20JPY/Kwh to the power consumption amounts.

C) Repair cost
There is a report [8] showing practical percentages of repair. It says the percentage is less than 3% of total number of the production. Practically, components that can be repaired are HDD, CD/DVD drive and CPU board, and average repair costs are approximately 30,000 JPY, 20,000 JPY, 50,000 JPY, correspondingly. The paper assumes that the percentages of repair of 3 components are the same.

D) End-of-life treatment
End-of-life treatment cost for consumers is basically the recycling fee. The recycling fee of a laptop PC is fixed to 3,150 JPY. Of course, under the current social system, recycling fee corresponds to the whole product only. Component recycling is not allowed. However, in this paper, the recycling fee is distributed to each component based on the weight in order to calculate the component costs.

3.3 Lifecycle value of components

Based on the calculation shown in the former section and equation (5), lifecycle value of the components can be calculated as Table 7.

Component	Purch-ase	Electri-city	Repair	EOL	Total
Battery	25000	0	0	725	24275
HDD	18000	364	300	227	17109
CD/DVD drive	45000	104	200	160	44536
CPU board	66000	2132	500	375	62993
AC adaptor	10000	0	0	533	9467
Main body	10000	0	0	486	9514
LCD	60000	1560	0	690	57750

Table 7: Estimated component lifecycle value of components (JPY).

However, the estimated lifecycle value of the components based on the initial value. At the end of product life, the value must be different. To consider the upgradability of components, the value of components at the end of its first lifecycle should be evaluated. Basically, it is said [9] that values of products (or components) decrease for two reasons. One is a physical reason such as damage, wear, dirt, etc. And the other is market reason which means more updated products spread in the market and the old product is not that valuable anymore. This is so-called obsolescence. To know the obsolescence is usually not easy. Precise market survey is necessary. However, the goal of this paper is to propose a simple method which can be used at the end-of-life by consumers or retailer. Therefore, a simple method to roughly estimate the obsolescence is needed. For the purpose, we assumed that the technical progress of the components can be the key. Table 8 shows the change of typical performance indexes of laptop PC, from 2004 to 2008.

Date	Battery capacity and life	HDD storage capacity	Drive (speed)	CPU board (clock)	AC adapter capacity	LCD (pixels)
May 2008	5.7A h, 8 h	120GB	Max *8 DVD	2GHz*2	16V 3.75A	1400 *1050
May 2004	7.0 Ah, 7 .5h	40GB	Max *8 DVD	1.3GHz	16V 2.5A	1400 *1050

Table 8: Change of performance indexes of components [10].

Even without any performance obsolescence it is expected that the value of products for users may decrease. For example, when the products become very popular (so-called commodity products), the prices of the products often decrease. In this case, the average price of the example laptop PC has decreased from about 300,000 JPY to 200,000 JPY for the same class high-end model. [11] Considering these factors, the paper proposes equation (6) to calculate substantial value change of laptop computer components through the lifecycle. Table 9 indicates the calculated substantial value of the components.

$$Vsi = Vui * Ct_{2008} / Ct_{2004} * Pi_{2008} / Pi_{2004} \quad (6)$$

Vsi: substantial value of component *i*, Ct_{2008}: price of the total product in 2008, Ct_{2004}: price of the total product in 2004, Pi_{2008}: performance of the component *i* in 2008, Pi_{2004}: performance of component *i* in 2004

Component	Original component price	Substantial component value
Battery	25000	17262
HDD	18000	34218
CD/DVD drive	45000	29691
CPU board	66000	129216
AC adaptor	10000	9340
Main body	10000	6343
LCD	60000	38500

Table 9: Substantial value of components (JPY).

On the other hand, it is assumed that a certain upgrade service fee is necessary, of course. The paper assumes the upgrade service fees of the components are the same as the component purchase prices. If the substantial value of the new component is higher enough comparing to the corresponding upgrade service fee, consumers are expected to choose upgrade service instead of purchasing the whole product. In addition, if the lifecycle environmental impacts of the component are low, upgrade service will not cause a high environmental impact for the new component. So, it is acceptable under environmental aspects . Therefore, $LCCO_2$ estimated in section 2.3, the lifecycle value of components calculated in 3.3, and the substantial component value will be the elements to determine the upgradability of components in the next section.

4 DETERMINATION OF COMPONENT UPGRADABILITY

4.1 Visualization of analysis

In order to clarify and visualize the characteristics of components, a view graph will be helpful. Next Figure 4 is the view graph showing the relations between $LCCO_2$ of components and the substantial component value normalized by the initial component price. Then, Figure 5 shows the CO_2 emissions of component productions versus the normalized substantial value.

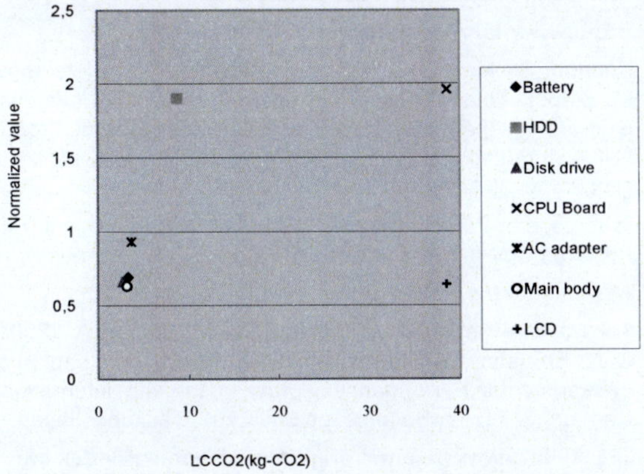

Figure 4: $LCCO_2$ and substantial value of components.

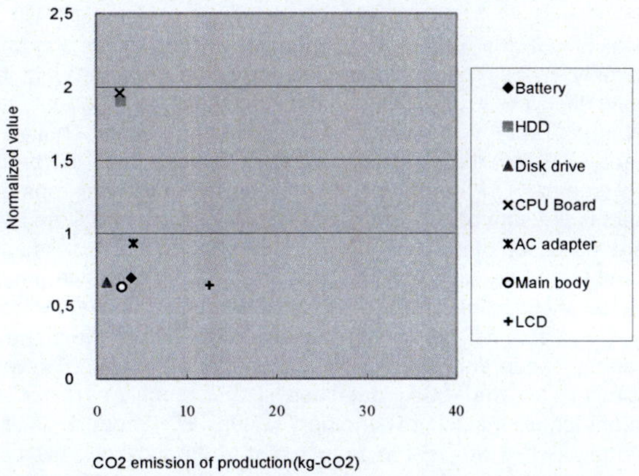

Figure 5: CO_2 emission for production and substantial value of components.

4.2 Proposal of upgradability index

Looking at the two graphs and the estimated data in the former sections, these are the discussed criteria. And the order of these criteria also shows the priority.

a. If the substantial component value is lower than that of the initial component price, which is assumed to be the upgrade service fee, consumers do not want to upgrade.

b. If the CO_2 emission of component production is too large, the component is not suitable for upgrade, because it causes large environmental impacts to produce the components for replacement.

c. If there is a big difference between $LCCO_2$ and CO_2 emission of production, it is an incentive to upgrade the component, because producing new components does not play a big role in increasing the lifecycle environmental impact.

Based on these criteria, we propose a procedure and a simple index to evaluate upgradability of laptop PC components.

(If the initial price of the component is larger than the substantial component value)

The component is not upgradable.

(If the initial price of the component is smaller or almost the same as the substantial component value)

Equation (7) is applied to determine the strategy. If UPi is larger than 1, the components are upgradable. In the example laptop PC, HDD and CPU board are selected as upgradable components in any aspects.

$$UPi = \frac{Vsi}{Cpi} \times \frac{Eci}{\sum Eci} \times \frac{ELi}{Eci} \quad (7)$$

UPi: upgradability index of component *i*

5 EVALUATION OF UPGRADE SERVICE

In the former section, HDD and CPU board have been selected as the upgradable components. In this section, the paper tries to discuss whether the upgrade service of the two components is acceptable for consumers and at the same time good for the environment. In order to evaluate the satisfaction of the two criteria, environmental efficiency [12] can be a simple and a useful index. Without upgrade, the lifecycle value of the product is equal to the sum of original prices of components shown in Table 9, at the beginning of product life. The total CO_2 emission is the sum of CO_2

emissions of the components production shown in Figure 2. Contrarily, by using upgrade service, the value of the product in its second lifecycle is the sum of the substantial component value shown in Table 9. However, the component value should not exceed 2/3 of the initial component value due to product obsolescence. On the other hand, the newly added environmental impact is the sum of CO_2 emissions of production of HDD and CPU board, collection of the products to the renovation site, disassembly, assembly and redelivery of the renovated products. Environmental impacts of collection and redelivery of products can be estimated based on Table 5. This is not an exact calculation of the share the assembly has in the environmental impact of production. However, according to the LCA database [13], roughly 1/3 of the environmental impact of another laptop PC product is from assembly, while the rest is from material production. This paper also assumes 30% of the total environmental impact of production is necessary for disassembly and assembly. In addition, another 100% of CO_2 emission of components production is caused by production of the components for upgrade. Therefore, the eco-efficiency of the component upgrade service can be calculated and is shown below. This calculation suggests that the eco-efficiency of upgrade service of the two components is higher than that of the original product. Table 10 shows a simple comparison of eco-efficiencies.

	Sum of value (JPY)	Sum of environmental impact of upgrade (kg- CO_2)	Eco-efficiency (JPY/kg-CO_2)
Original product	300,000	29.67	10,111
Upgrade service	153,867	13.66	11,264

Table 10: Comparison of eco-efficiencies of original product and the upgrade service.

6 CONCLUSIONS

Although modular design is a hopeful option to promote reuse of parts and prolong the product life, consumers' acceptance is the key to implement the idea in the practical market. To promote reuse of products or parts, it should be assured that the products is always up-to-date and shows sufficient performance. Without that, consumers will never have a motivation to reuse. To give a solution to the problem, upgrade service has been focused on.

In this paper, firstly through the disassembly experiment, parts of the laptop computer were grouped into seven major components. Those are battery, HDD, CD/DVD drive, CPU board, AC adapter, main body and liquid crystal display (LCD). The paper estimates the $LCCO_2$ of seven components and component costs for consumers. Through these analyses characters of seven components have been identified.

Secondly, by investigating the speed of performance progress of the components indicated by specifications, another aspect to characterize components has been discussed. The result of the calculation indicates substantial component value as an important index.

Thirdly, a simple equation to determine the upgradability of components has been proposed. However, effectiveness of the index is totally unclear as, in the example, upgradable components are only HDD and CPU board in any aspects.

Fourthly, a simple calculation has suggested that the environmental efficiency of upgrade service of the suggested components will be higher than that of the original product.

Analysis of this study is for a high-end model laptop computer which costs 300 thousand yen. As for the future work, it is necessary to consider whether the same approach can be applied to more inexpensive models. It is also necessary to investigate how the consumers' preferences affect whether the upgrade-service is really acceptable and affordable for consumers. It is also necessary to test whether the proposed index is useful in other examples. Of course, strategic combination of component reuse and upgrade-service should be discussed.

REFERENCES

[1] Umeda, Y. (2000); Modular Design - Current state and view of future production-, Journal of JSPE, Vol. 66, No.7, pp1009-1010. (In Japanese)

[2] Ulrich, K.; Tung, K. (1991): Fundamentals of Product Modularity, DE-Vol. 39, Issues in Design Manufacture / Integration, ASME.

[3] CREST Yasui team: CO2 intensity data for LCA, http://www.yasuienv.net/CREST/lca-thinking/useful/gentanni_menu.htm. (In Japanese)

[4] JEMAI: Manual for implementing product LCA, LCA of LCD embedded personal computer. (In Japanese)

[5] Nakmura, R.; Ouchi, K.; Yudaki, R.; Istubo, T, (2007): LCA of Lithium-ion battery for cell phones, Proceeding of 2nd conference on LCA, Tokyo, Japan, pp.286-287 . (In Japanese)

[6] Ando, T.: Eco-leaf –LCA communication activity through eco-efficiency, Web page of JEMAI, http://www.jemai.or.jp/lcaforum/pdf/seminar041210_4.pdf

[7] Okumura, H. (2007): Energy saving driving technology of LCD of mobile PC, Toshiba Review, Vol.62, No.11, pp.60-61.

[8] Web page of Multimedia Laboratory Corp., http://www.m2ri.co.jp/newsreleases/1101.htm.

[9] Kondoh, S.; Masui, K.; Hattori, M.; Mishima, N.; Matsumoto, M. (2008): Total Performance Analysis of Product Life-cycle Considering the Deterioration and Obsolescence of Product Value, International Journal of Product Development, Vol.6, Nos.3/4, pp334-352.

[10] Web page of Panasonic Corp: List of abolished product (PC), http://panasonic.jp/pc/support/products/ (In Japanese)

[11] http://kakaku.com/pc/

[12] Tahara, K.; Yamaguch, K.; Sagisaka, M.; Inaba, A. (2004): CO2 Efficiency: A New Evaluation Method of Companies' Environmental Performances, proc. of the 6th International Conference on EcoBalance, pp.269-270, . Tsukuba, Japan.

[13] JEMAI :http://www.jemai.or.jp/english/ecoleaf/index.cfm

A Study on Comparison of Different PSS Concepts Based on Value Creation Efficiency

Yoon-Young Chun[1], Shinsuke Kondoh[2], Nozomu Mishima[2] and Kun-Mo Lee[1]

[1] Department of Environmental Engineering, Ajou University, Suwon, Korea

[2] Advanced Manufacturing Research Institute, AIST, Ibaraki, Japan

Abstract

Proper combinations of products and services are recognized to the keys in establishing sustainable society. In this paper, we try to analyze efficiency to offer value to customers under a certain payment. As a case study we choose water purifier and compare different business options in the water purifier sales. We applied our value quantification method based on weighting of different functional requirements, and integral of value throughout the lifecycle. The paper tries to clarify the suitable combination of product and service that is effective in enhancing the value creation efficiency of business, in the aspect of cost-efficiency and eco-efficiency.

Keywords:

Value Creation; Total Performance Analysis; Business Option

1 INTRODUCTION

Eco-efficiency [1] has been recognized as an important index in satisfying high product quality and low environmental impact simultaneously and has been applied to a number of products on the markets. It is understood that not only reducing the environmental impact of industry but also enhancing eco-efficiency is the key issue to ensure the sustainable manufacturing. One of the simplest expressions of eco-efficiency is added value of products over environmental impact of products. It is named environmental efficiency [2]. The authors' research group has proposed another method to evaluate so-called eco-efficiency of products. [3] We named it Total performance Analysis (TPA). However, it is more appropriate to recognize the method as a method to evaluate 'value creation efficiency.' Because the evaluation formulation has 'product utility value' in the numerator, while it has square root of environmental impact multiplied by lifecycle cost as the denominator. The key to enhance this index is to increase product value efficiently. In addition, it is necessary not only to design new efficient products but also to create the best combination of services and products based on the products characteristics and market conditions. For example, if the product consumes large energy in its usage, such as air-conditioners or automobiles, it is necessary to control usage patterns in addition to the product development. Because, only by the product development, major portion of the product lifecycle will be out of control. Therefore, services have been recognized as a key in industries to decrease environmental impacts and increase attractiveness of the products. In former studies [4] the authors have also investigated and listed many examples of PSS [5].

In the paper, a product in actual industry is focused on and analyzed. In the target product, water server, the maker offers some different business options for consumers. It is not easy for a consumer or the society to determine which option is the most preferable. It is not also easy for a business provider to predict which business option will be most acceptable for consumers and to determine business strategies. A proper method to evaluate value creation efficiencies of different options in PSS will be helpful to both consumers and business providers. Our method, TPA, recognizes the utility value of the product as the integral of product occasional value through the lifecycle. Comparing with other method recognizing product value to be constant, TPA has an apparent advantage. The method can quantify value deviation during the product lifecycle theoretically. Because some of typical service activities in product lifecycle such as upgrade, repair, maintenance, and so on, must affect product value greatly. We try to apply TPA in evaluating the efficiencies of different business options and prove the method is effective in extracting sustainable business strategy.

2 EVALUATION OF VALUE CREATION EFFICIENCY

2.1 Environmental efficiency and cost per value

To enhance social sustainability, it is said that maintaining the quality of life (QOL) and reducing the environmental impact simultaneously is the key. Original formulation of eco-efficiency is the simplest and useful index to evaluate how efficient the product value can be created. It is expressed by (1). On the other hand, it is also necessary to consider how efficiently the product value can be created, or whether the product with the created value is affordable for consumers. This index indicating the balance of product value versus production cost or purchase cost, is often called 'cost per value [6].' It is shown in (2). By considering two indexes at the same time, it is possible to evaluate whether the product is efficient enough plus whether the product is affordable enough.

$$Ef = \frac{V}{LCE} \tag{1}$$

$$CV = \frac{V_c}{LCC} \tag{2}$$

Ef: environmental efficiency

V: added value

J. Hesselbach and C. Herrmann (eds.), *Functional Thinking for Value Creation: Proceedings of the 3rd CIRP International Conference on Industrial Product Service Systems, Technische Universität Braunschweig, Braunschweig, Germany, May 5th - 6th, 2011*,
DOI 10.1007/978-3-642-19689-8_53,

LCE: lifecycle environmental impact

CV: cost per value

V_C:value for consumer

LCC: lifecycle cost

2.2 Quantification of lifecycle value

Definitions of environmental efficiency and cost per value have been shown in the former section. However, the key issue is how it is possible to quantify the numerator. In simple indexes, numerator V is constant, such as product price or added value, etc. However, the value of the product is, of course, not constant during the product lifecycle. And especially for PSS, to assume product value is constant cannot carry out any useful result. Because, if the product value is constant, no service such as cleaning, repair, maintenance, upgrade, etc. are necessary. Consumers just need to use the product as the present form. Therefore, in the former research [3], we proposed an index to evaluate real eco-efficiency of products, by considering product's utility value, cost and environmental impact. New value creation efficiency index was defined by (3) and was named total performance indicator (TPI). The proposed index was the simplest combination of environmental and economical efficiencies. In the calculation, total value of the product is defined as a weighed sum of so-called functional requirements. 'Functional requirement [7]' is a popular concept in software engineering and it is defined as a function of a software system or its component. A function is described as a set of inputs, the behavior, and outputs. As well as the original definition, we defined functional requirements as the set of consumers' requirements for the target product and service activities concerning the product. Then, the total utility value of the product service system is expressed by sum of value of functional requirements as shown in the equation. In addition, since in existing evaluation indexes the value is usually a fixed value, it cannot consider change of the value throughout the product lifecycle. Contrarily, in our proposal, because the utility value of the product can be expressed by integration of occasional values throughout the lifecycle (equation (4)), it can simulate value decrease due to obsolescence and physical factor as shown in Figure 1. We modeled that the product value decreases with the product usage, because of the two reasons. One is physical reason such as damage, wear, dirt, etc. And the other is market reason, based on the appearance of new product in the market. The second reason is often called obsolescence. In analyzing product service systems, to simulate the value change through product lifecycle is an irreplaceable feature. Since the purpose of service activities combined with product is to enhance or to renovate product value, every service activity affects the occasional value of the product. Therefore, analysis along the time line shown in Figure 1 is necessary. TPA, in that sense, is very suitable for analyzing PSS.

$$TPI = \frac{\sum UVi}{\sqrt{LCC}\sqrt{LCE}} \tag{3}$$

TPI: total performance indicator

UVi: utility value of the functional requirement *i*

LCC: life-cycle cost of the product

LCE: life-cycle environmental impact of the product

$$UV_i = \int_{TS}^{TE} V_i dt \tag{4}$$

Vi: occasional value of functional requirement i

TS: start of product use

TE: end of product use

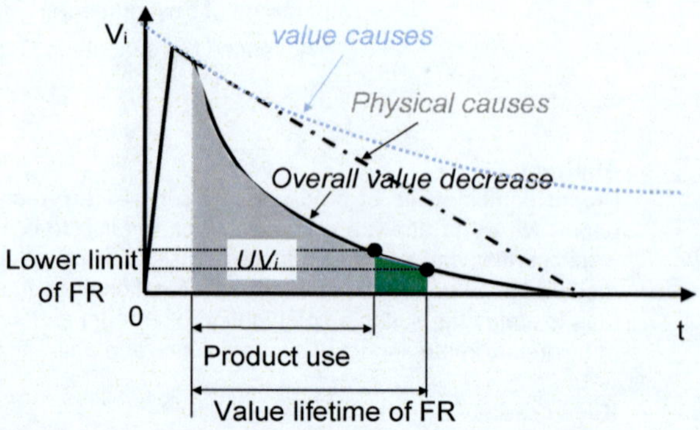

Figure 1: Utility value and its decrease through product lifecycle.

2.3 Value change of PSS through lifecycle

If the product lifecycle is a simple purchase-use scenario, the value decrease curve is monotonous as it is shown in Figure 1. However, once service processes such as upgrade, maintenance, repair, etc. are considered, the curve will be different. Figure 2 is a schematic view of a product value change curve through the product lifecycle with existence of a service activity. Same equation (3) and (4) can be used to evaluate efficiency of a product service system as shown in the Figure 2. However, the value change curve cannot be expressed by a simple mathematical function.

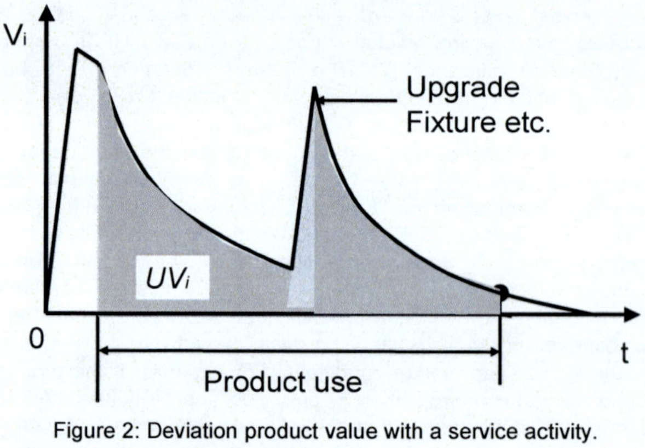

Figure 2: Deviation product value with a service activity.

3 INDUSTRIAL EXAMPLE

3.1 Target; water server

As a first example of a case study, we choose a Korean water server example [8] shown in Figure 3. The product is usually set in a house or in an office and provides purified drinking water and hot water, immediately. There are many types corresponding to the water supplying capability, size, and so on. Basically, the most basic functional requirement of a water server is to supply hot and cold water. But, in detail, water supply capability can be divided to two sub capabilities. Theoretically, those are quantity and temperature. Thus, the first two functional requirements are 'supply appropriate temperature heated/chilled water' and 'supply sufficient amount of hot/cold water.' These functions of water server might cause malfunction under a certain probability. An additional functional requirement of this type of water server is 'purify water.' Value of this function may decrease without exchange of a filter in proper interval. Aforementioned three functional requirements are all very important for this product. Additionally, since the product is a home appliance which is set in a kitchen (office), a good external view is also required. Thus, the forth functional requirement is 'look good as a kitchen furniture.' Because layout change might be necessary in some occasions in a kitchen or in an office, 'easy-to-move' is an extra requirement. However, this functional requirement is not so important, because layout change is not always necessary. Result of above-mentioned consideration is shown in Table 1, as a list of functional requirements and its relative importance. The relative importance was scored according to the three level scoring often used in QFD [9].

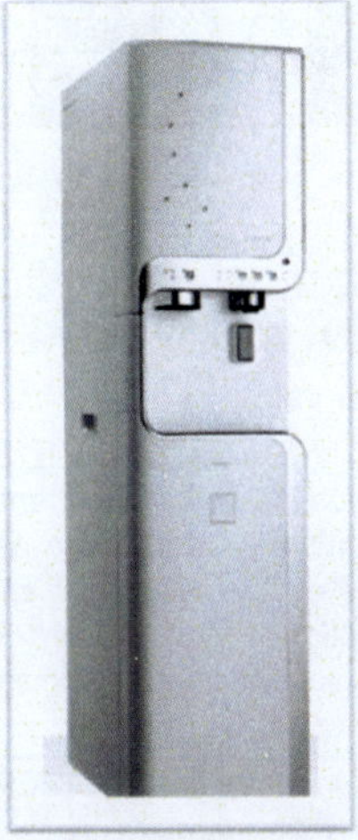

Figure 3: Case study product.

	Functional requirements	Relative importance
FR1	Supplies appropriate temperature of hot/cold water	9
FR2	Supplies enough ammount of hot/cold water	9
FR3	Purifies water appopriately	9
FR4	Looks good as a kitchen furniture	3
FR5	Easy movable	1

Table 1: Functional requirements of the water server.

3.2 Different business options

The supplier provides not only sales, but also provides some rental services of the water server and take care of the maintenance, filter exchange, and so on, during the rental contract. Thus, basically there are some different scenarios for the consumption patterns of the product. These are the brief explanations of each business option.

Type A (sales): Sales of a new product with 8 years service contract

Type B rental and renewal): Rental of the product for 5 years and renewal of the contract (replacement by a new model)

Type C (ownership transfer): Rental of the product for 5 years and transferring of the ownership after the first 5 years (no replacement)

Type A and B include maintenance service including filter exchange, cleansing, and so on. And in these scenarios, we assumed value changes linearly. It is assumed that each scenario has each value change curve shown in Figure 4 A-4C. And the discontinuity in 4C corresponds to the end of the service contract.

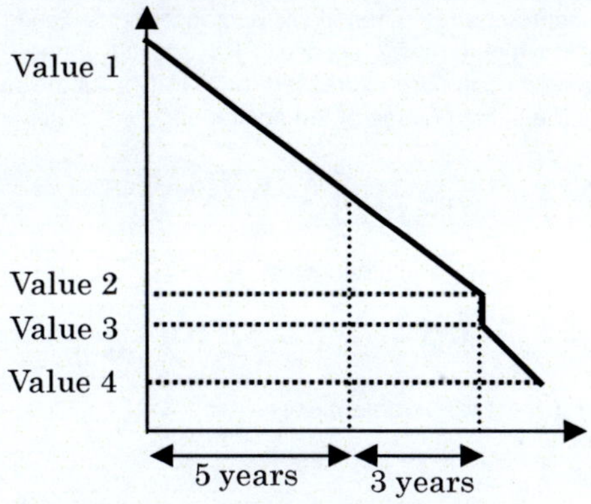

A) Sales

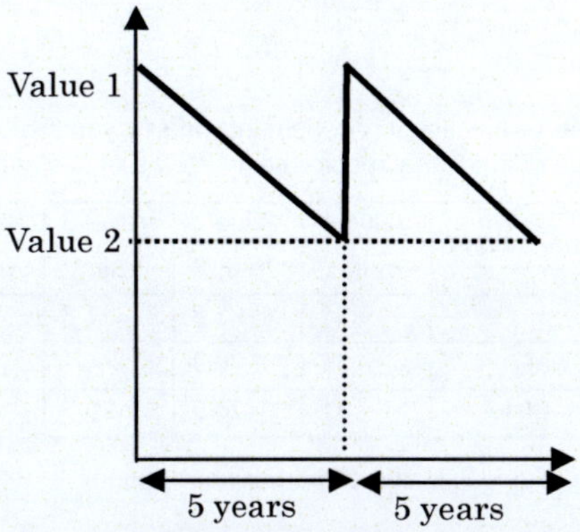

B) Rental and renewal

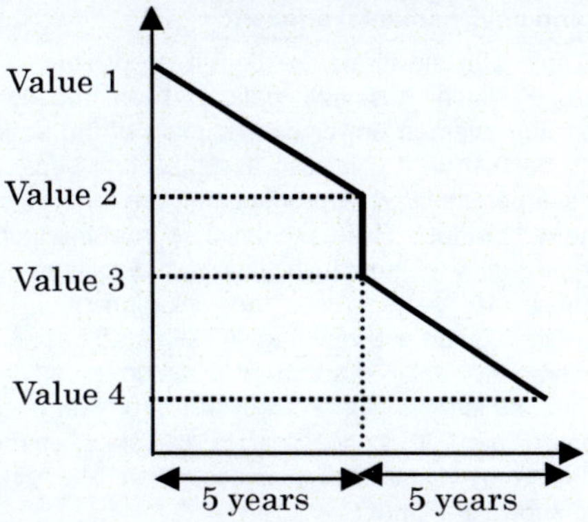

C) Rental and ownership transfer

Figure 4: Schematic images of 3 business options.

3.3 Value quantification

In order to calculate total performance of aforementioned 3 business options, it is necessary to decide some parameters. First of all it is necessary to determine value change curve of the functional requirements of a water server. Functional requirements have been decided as Table 1. We assumed that among the functional requirements, FR1 and FR2 will linearly decrease to 1/2 during 10 years, because of the value cause. And, also it was assumed FR4 decreases to 1/2 during 10 years, because of the physical cause. FR3 do not change, if there is a maintenance service. But, it will decrease to 1/2 immediately, when there the service contract ends. Then, it will decrease to 1/2 during 5 years, if there is no maintenance service. FR5 doesn't change over time. These assumptions are expressed in (4) to (7). As the result, Table 2 shows the value change of the scenarios

For FR1, FR2

$$Vi(t) = V_{0i} \times (1 - t/20) \quad (4)$$

For FR3, FR4 (w/o maintenance)

$$Vi(t) = V_{0i} \times 0.5 \times (1 - (t-5)/10) \quad (t>5) \quad (5)$$

For FR3, FR4 (with maintenance)

$$Vi(t) = V_{0i} \quad (6)$$

For FR5

$$Vi(t) = V_{0i} \quad (7)$$

Vi(t): value of functional requirement *i* after *t* years of use, *Voi*: initial value of functional requirement *i*, *t*: time of use (years)

Name of the option	Value 1 (ratio)	Value 2 (ratio)	Value 3 (ratio)	Value 4 (ratio)
Sales	31	23.8	17.8	14.8
Rental renewal	31	26.5	-	-
Ownership transfer	31	26.5	20.5	11.5

Table 2: Value of the business options.

4 ANALYSIS OF THE BUSINESS OPTIONS

4.1 LCE and environmental efficiency

Based on Table 3 in the former section, it is possible to calculate utility value of each business option. Then, based on the investigation, the average power consumption of the equipment is estimated to be 0.1Kw. It was also assumed that because of the technological progress the new product after the renewal uses 70% power of the old product. This assumption is from the fact [10] that the energy efficiency of home appliances has become almost 2 times in these 10 years. Also, the environmental impact of production of the equipment was roughly estimated based on that of similar equipment. Table 3 shows the assumed environmental impacts (CO_2 emissions) for 2 lifecycles (10 years) based on investigations. Then, it is possible to calculate environmental efficiency defined by (1) for each business option. The results of the analysis are shown in Figure 5.

Name of the option	Impact of using 10 years (kg-CO2)	Impact of production (kg-CO2)
Sales	2400	100
Rental renewal	First 5years: 1200 Next 5years: 840	100*2
Owner-ship transfer	2400	100

Table 3: Environmental impacts of the business options.

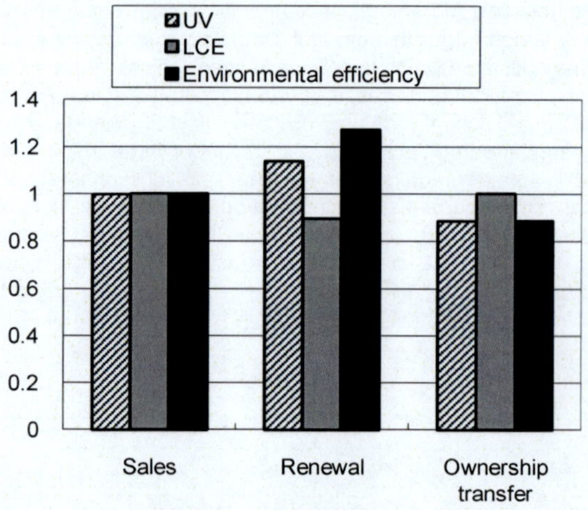

Figure 5: Environmental efficiency analysis of 3 business options.

4.2 LCC and cost per value

As for the next step, the cost of each business option for a consumer which is equivalent to the payment amount, is expressed by ratio by setting the initial product price as 100. These payment amounts were investigated by asking the maker actually. Table 5 indicated the investigation result. Figure 6 is the view graph visualizing the cost per value. Table 4 shows the consumer's costs of the business options.

Name of the option	Rental fee (ratio /month)	Electricity cost (ratio)	Ownershi p cost (ratio)
Sales	-	48	132
Rental renewal	First 5years: 2.4 Next 5years: 2.2	40	-
Ownership transfer	First 5years: 2.4 Next 5 years: 0	48	0

Table 4: Consumer's cost of the business options.

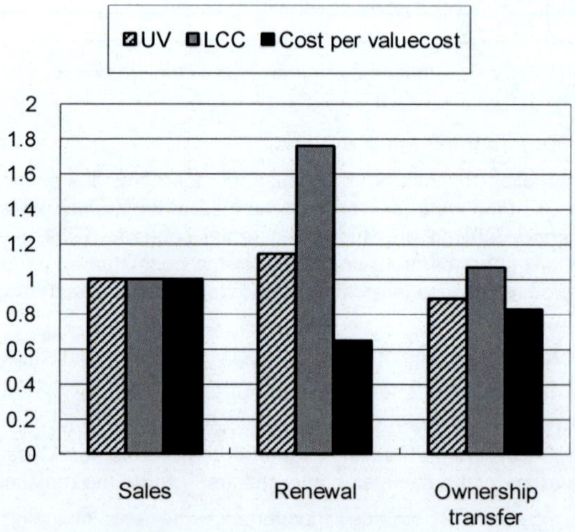

Figure 6: Cost per value analysis of 3 business options.

4.3 Calculation of TPI

Figure 5 indicates the rental and renewal option has the highest environmental efficiency, and Figure 6 shows the sales option is the best in the aspect of cost per value. As an integrated index of these two aspects, TPI has been proposed as (3). Figure 7 shows the calculated result of total performance indicator (TPI) of each business option.

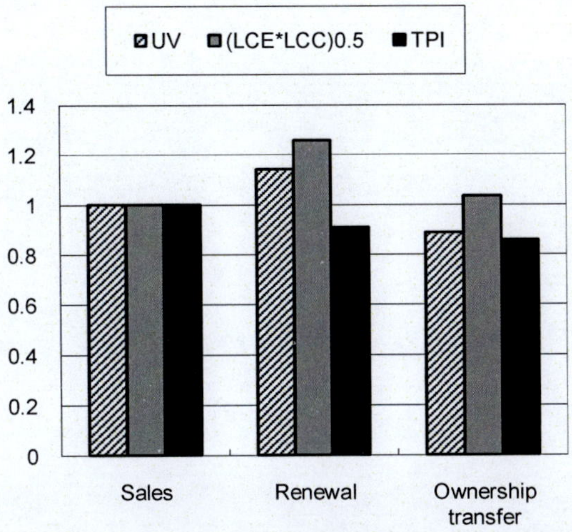

Figure 7: Comparison of TPI of the business options

4.4 Discussions and extraction of sustainable strategy

Under current assumptions, it was estimated that 'rental and renewal' option is the best option to reduce environmental impact. However, in the aspect of affordability, sales option is the best. The business strategies that can be extracted from these facts are depending on the stake holders. For the maker, the more the consumers' payment is, the more the business option is profitable. Therefore, 'rental and renewal' option can be the best strategy for the company. It is also good for environment. However, the analysis result tells that the option is not very affordable for consumers. For consumers, sales option has the advantage. It might result that the most eco-efficient business option will not be selected by the consumers. To match the environmentally conscious option with the affordable option, modification will be necessary. Speaking concretely, it is to discount the rental fee. To persuade consumers and give an incentive to choose eco-friendly option, it is necessary to say cost per value is the same, at least. To satisfy this condition, rental fee should be reduced to about 60% of the current price. In addition, if the company discounts the rental fee averagely, it will motivate consumers to choose 'ownership transfer' rather than 'rental and renewal.' The 'ownership transfer' is not valuable for consumers and not profitable for the company, either. Therefore, the strategy is to focus on the second rental term and reduce the rental fee. Now, we assume that the rental fee of the first 5 years have been reduced to 80% and the second term to be 40% of the old business option. Although it might not be an easy option for a company, this kind of drastic change will be necessary to motivate consumers for environmental conscious consumption behavior. By modifying the business option, it is possible to maintain the relatively high environmental efficiency and make the option more affordable for consumers. Since our proposing index TPI is basically suitable to evaluate the balance of value, cost and environmental impact, it can be used to determine well-balanced business strategy, as well. By applying this modified strategy, the new TPI is shown in Figure 8. By using TPI and reviewing business options, it will be possible to design business options in order to lead consumers to 'win-win' options.

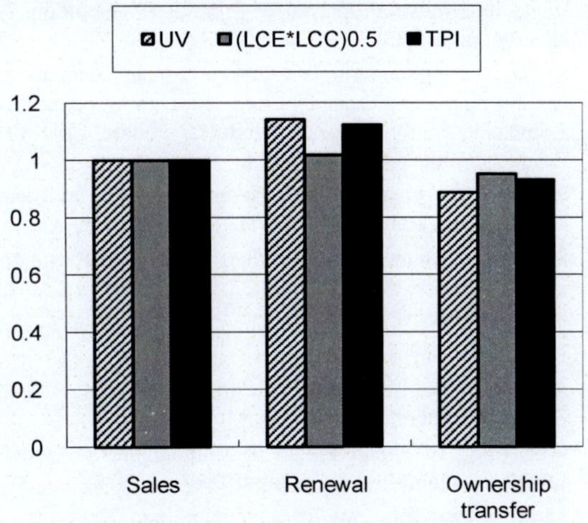

Figure 8: TPI of the new business options.

5 CONCLUSIONS

In the former studies, the authors have proposed a new efficiency index and evaluation method of so-called eco-efficiency of products. However, the index was not very understandable intuitively. Therefore, in this paper, using an industrial example from Korean company, the index was compared with more familiar index, original eco-efficiency and cost per value. As the different options to be compared, there were three business options in the example products. Those were, 'sales,' 'rental and renewal,' and 'ownership transfer.' Eco-efficiency, cost per value and TPI of the business options were estimated and compared. As the result, it was clarified there might be a trade-off between eco-efficiency and 'cost per value.' As for the eco-efficiency, if we consider technological progress, rental and renewal option has the best value. On the other hand, sales option has the best cost per value. To override this trade-off and modify the business options more affordable for consumers, it has been suggested TPI can be useful.

Of course, the calculation in this paper is based on many assumptions. Therefore, more practical and detailed analysis must be carried out in order to prove that TPI is really helpful in evaluating different PSS options and extracting better business strategies, as future work.

6 REFERENCES

[1] DeSimone, L.; Popoff, F. (1997): Eco-Efficiency - the Business Link to Sustainable Development -. Cambridge, MA: MIT Press.

[2] Tahara, K.; Yamaguch, K.; Sagisaka, M.; Inaba, A. (2004): CO2 Efficiency: A New Evaluation Method of Companies' Environmental Performances, proc. of the 6th International Conference on EcoBalance, pp.269-270, Tsukuba, Japan.

[3] Kondoh, S.; Mishima, N. (2009): A Strategic Idea Generation Method for Eco-business Planning, Proc. of ASME IDETC/CIE 2009, Sept. 2009, CA, USA, DETC2009-86961.

[4] Kondoh, S.; Masui, K.; Hattori, M.; Mishima, N.; Matsumoto, M. (2008): Total Performance Analysis of Product Life-cycle Considering the Deterioration and Obsolescence of Product Value, International Journal of Product Development, Vol.6, Nos.3/4, pp334-352.

[5] Sakao, T.; Shimomura, Y. (2007): Service Engineering: A Novel Engineering Discipline for Producers to Increase Value Combining Service and Product. Journal of Cleaner Production 2007, 15(6), 590-604.

[6] Younker, D. L. (2003): Value engineering: analysis and methodology, Marcel Dekker Inc., New York, NY.

[7] Malan, R.; Bredemeyer, D. (2001): Functional Requirements and Use Cases, White Paper 8/3/01, Bredemeyer Consulting.

[8] Web Page of Woongjin COWAY Corp, **http://www.coway.com.my/**

[9] Akao, K. (1990): Quality Function Deployment, Productivity Process, Cambridge, M.A.

[10] Web Page of Japan Electrical Manufacturer's Association, **http://www.jema-net.or.jp/Japanese/kaden/reizou_en/index.html** (In Japanese)..

An Overview on Degradation Modelling for Service Cost Estimation

Pedro Fernandes[1], Rajkumar Roy[1], Jörn Mehnen[1], Andrew Harrison[2]

[1] School of Applied Sciences, Cranfield University, Cranfield, United Kingdom

[2] Rolls-Royce Plc, Derby, United Kingdom

Abstract

The problem of component degradation in aero-engines has become a matter of great interest in the context of Life Cycle Cost (LCC) for Original Equipment Manufacturers (OEMs). On root causes of cost incurred in the operational phase is the uncertainty around component degradation, leading to service support inefficiencies such as under or over capacity in the maintenance network. Different life prediction approaches are available; each has varying applicability to different degradation mechanisms, data quality and availability. This paper considers the necessity of degradation modelling capability for aero-engines LCC and reviews the different approaches found in literature and their underlying concepts.

Keywords:

Life Cycle; Cost Estimation; Component Degradation

1 INTRODUCTION

In the aeronautical industry, the traditional procurement methods in which the airline purchase an engine and subsequently contracts individual maintenance overhauls, has been changed by service agreements such as Rolls-Royce TotalCare^{TM}. Under this type of agreement, the operator pays an agreed support cost per hour of operation and the Original Equipment Manufacturer (OEM) provides the service support, including management of the inherent uncertainty. This has led to a shape change in the product LCC profile for an OEM, increasing the impact of aftermarket services. Therefore, accurate estimation and cost forecasting capabilities for aero-engine maintenance cost have a particular interest for OEMs and services providers.

At the root of the service support costs is the issue of product wear out leading to reduced or lost functionality. Due to the stochastic nature of degradation involved in product wear out; component life is an uncertain quantity. In the time-varying working environment of an aero-engine this uncertainty is greatly enhanced. As stated in [1], operational and environmental conditions have a significant effect on accelerating or decelerating the rate of degradation processes, which occur prior to failure.

The cost incurred in the operational phase of an aero-engine is dominant in the whole product LCC [2]. In order to predict and minimize service cost exposure it is of increasing importance that the cost driving variables are understood. Component unit cost and life are determining factors in the whole engine LCC. Hence, the understanding of the degradation process at a component level for component life prediction is fundamental for an improved cost modelling capability.

2 AERO-ENGINE LIFE CYCLE COST

Life cycle cost is the cumulative cost of a product over its life cycle. A typical LCC cost estimate, consists in a summation of cost throughout the product life cycle, i.e. LCC include cradle to grave cost [3].

The lifetime of an aero-engine normally extends to 30 years in civil applications whereas in military applications this period can exceed 50 years. During this lifetime, the product LCC is structured in different levels accommodating distinct phases of product maturity, utilization and cost drivers. For the different stakeholders involved during a product life cycle, distinct LCC structures and definitions can be found, representing their different perspectives on the cost. Consequently, manufacturers, supply chains and customers tend to have different definitions and LCC structure concepts representing their specific involvement in the product lifecycle and their cost exposure.

In order to bring a common LCC language to the aeronautical engine community, WP2.2 of the VIVACE project [4] proposes an European LCC standard presenting a generic LCC nomenclature and structure by surveying the LCC community of the different industrial partners involved in the project. This LCC structure comprises 3 levels, the first representing the different engine life cycle phases and the remaining providing a more detailed breakdown LCC structure and cost elements identification.

The process of LCC analysis throughout the propulsion system life cycle follows identical, established and fundamental principles. This is supported by a clear LCC structure and cost calculation methodology adaptable to the required level of detail, which increases with the product life cycle. As the product life cycle matures, the amount of available information and data increases in quantity and accuracy. Hence different LCC analysis techniques are employed as a function of data availability and detail requirements of LCC studies.

Each life cycle phase has distinct contributions to the product LCC. With the advent of long term service agreements and increasing market demand for such contracts, the cost expenditure during the product service phase has escalated in relation to past values for OEMs.

2.1 Aero-Engine Life Cycle Cost - Service Phase

The cost incurred in the operational phase of a gas turbine engine is dominant in the whole product life cycle cost. Consequently, this

J. Hesselbach and C. Herrmann (eds.), *Functional Thinking for Value Creation: Proceedings of the 3rd CIRP International Conference on Industrial Product Service Systems, Technische Universität Braunschweig, Braunschweig, Germany, May 5th - 6th, 2011*,
DOI 10.1007/978-3-642-19689-8_54, © Springer-Verlag Berlin Heidelberg 2011

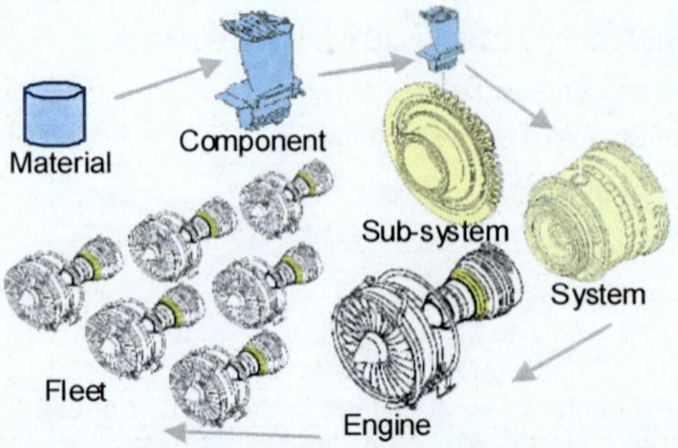

Figure 1: Modelling level hierarchy for life prediction.

phase is of major importance for a LCC analysis. Different operators of one same engine type will see different rates of engine deterioration and consequently different time on wing values (TOW) between major overhaul service events. These differences are a reflection of different engine usage scenarios. Impacting the variability of these scenarios are different environmental variables, such as:

- temperature
- pressure
- relative humidity
- contaminants (e.g. sand, salt, sulphurs, etc.)

and also different operational variables that can be represented by:

- stage length
- derated thrust
- loads
- flight cycle profile

The reliability of a turbine engine and its constituent components together with its service cost, is a complex function of several parameters, which includes the engine operational environment and type as described above, but also:

- component life characteristics (e.g.: failure mode probability, secondary damage, life limitations)
- maintenance strategies adopted
- non-inherent failure causes (e.g. human error, rate of foreign object damage (FOD))

Several models for turbine engines maintenance cost estimation and optimization have been presented [5-7]. Even thought different methodologies for service cost estimation can be found in these models, they all rely on failure time probability distribution functions, obtained from service data normally at a fleet level, for component, module or whole engine lifetime estimation. They consequently predict maintenance shop visit events and the maintenance workscopes (work content) at these events. However, in highly reliable systems such as an aero-engine, few or no failures are expected during its service life especially for critical components due to safety reasons. This leaves limited data for the analyst to infer an accurate failure time probability distribution function based on fleet statistics. Analysis of sub-fleets (smaller engine populations) operating in different environments or operational profiles has proportionately less data available for study. Therefore, the problem of maintenance cost estimation tailored to different operation scenarios remains throughout the product life cycle.

3 MODELLING LEVEL

Left unattended, component failures may result from material degradation (generally causing a shift in some critical device parameter). Such failures would result in engine shop visits where maintenance cost is incurred. In reality maintenance shop visits are generally pre-emptively triggered by engine monitoring signalling the need for maintenance to reduce the risk of developing failure modes. Shop visits can also result from general engine performance indicators deteriorating to pre-defined limit levels, such as Turbine Gas Temperature (TGT). Therefore distinct degradation modelling levels are required to capture the different nature of shop visit drivers and consequent engine maintenance LCC. Figure 1 illustrates this concept by providing a hierarchy level of life prediction modelling objects. These can go from direct material modelling representing the hierarchy lowest level to the highest level by directly modelling the population of systems in service. Note however that although system requirements are made at the highest level, failure is initiated at the lowest level.

A LCC analysis is an activity that tackles the whole product life cycle and consequently all the business processes involved in delivering cost effective products and services solutions. Hence, within an organization an LCC model will also require different levels of modelling in order to represent the different views and requirements of the different stakeholders involved (directly or indirectly making use of its outputs) in a business process.

Note – in the following sections of this paper the term *Failure* should be interpreted as meaning: *the point at which deterioration of a component or system leads to the initiation of maintenance activity (either reacting to loss of functionality or pro-actively prior to loss of functionality).*

4 LIFE PREDICTION APPROACHES

Component failure can be traced to an underlying degradation mechanism that acts and evolves over time. Therefore the identification and understanding of the different potential failure mechanisms present in a gas turbine is essential for accurate life predictions. According to [8] those failure mechanisms are: low cycle fatigue, high cycle fatigue, thermo-mechanical fatigue, creep, overstress, corrosion, erosion, fretting and wear.

Several classification criteria have been used to expose the different techniques for life prediction [9-13]. This classification tends to be made based not only on the distinct characteristics of the different models but also on the model input type (e.g. type of data) representing the system in study and associated degradation mechanism, operational environment and usage conditions. A common classification is given by the following four groups:

- *Model-based methods*: based on the mathematical representation of the component (or any other hierarchy level) and failure mode in study;
- *Knowledge-based methods*: based on expert judgment of the component and failure mode in study;
- *Experience-based methods*: based on failure time data of a component population events record;
- *Data-driven methods*: based on extensive data (normally time-series) of the component and failure mode in study.

Figure 2 presents an extended map from [10] on these different approaches together with some examples of the different techniques.

4.1 Model-base methods

A model-base method consists of a mathematical representation of the dynamic behaviour of a component and associated degradation mechanisms. This modelling approach can be either based on a pure physical representation of the component (together with associated degradation mechanisms, operating environment and usage – physics-of-failure method) or by a statistical / probabilistic method. An example of the latter, is the modelling of defects initiation and growth of gradual deterioration monotonically accumulating over time [14].

Typically, a model-based approach for life prediction is applied by directly modelling the degradation mechanisms and then linking the explanatory variables of the degradation process to the life of the system. In this approach the model input parameters can be either deterministic or distributed random variables [15-17], providing a more robust modelling approach given that their variability impact is accounted for.

Various components and systems physical models have been presented for degradation / failure time estimation and consequent life prediction. These have been applied, for example, in turbine blades [18], gearboxes [19], clutches [17], bearings [20-27] and batteries [28; 29]. In the context of maintenance modelling, [18] tests and compares a preventive maintenance concept based on component usage and applied loads. The presented case study is on a turbine blade where simple physical models are used to estimate component loads and damage accumulation. The only failure mode considered is creep and the parameters describing the usage are the engine average number of operating hours per year and power settings.

In [17] is presented a technique for remaining useful life prediction of a high-power dry clutch system. This technique comprises the physical simulation of the clutch dynamic behaviour in the different operational modes together with the simulation of the associated degradation mechanisms - wear (including its hybrids). The wear coefficients of the different wear mechanisms considered are obtained via a stochastic approach where the dependent parameters (surface roughness, debris size and contacting surfaces temperature) are evaluated probabilistically.

Rolling element bearings have received a considerable amount of attention in the context of condition prognosis and life prediction. In rotating machinery these are seen as one of the root causes of system malfunction. Several physical-based prognostic approaches have been proposed for bearing life prediction [20-23] and special attention has been given in bearing life prediction due to fatigue [24; 25; 30; 31].

An adaptive scheme for bearing remaining life prediction based on a modified Paris's rule for defect growth estimation and indirect measurements of bearing condition is presented in [20]. The residual obtained by the defect growth deterministic model output and the indirect-measured condition (through vibration analysis) is assumed to represent the stochastic nature of the defect growth. This residual is then used to continuously tune the deterministic model parameters allowing a representation of the time-varying defect growth behaviour. This approach is extended in [21] by introducing a lognormal random variable in the defect growth deterministic model, representing the uncertainty in the material properties and environmental factors. The parameters of this stochastic model are then estimated via measured data. [22; 23] presents engine bearing prognostic approaches linking information obtained from sensed data and model based outputs. The model-based approach used consists in fatigue-based accumulation models.

A model for fatigue life prediction in rolling bearings is introduced in [25]. This model assumes a statistical relation between the bearing probability of survival, the fatigue life and a stress-related fatigue criterion. In this model the fatigue life can be estimated for different loads, material and operating conditions. An evaluation between two commonly used methods for bearing life prediction, Lundberg-Palmgren and Ioannides-Harris method, is provided in [24], showing Ioannides-Harris method to be more accurate in predicting bearing fatigue life endurance. This conclusion is corroborated in [26; 27] also in the application context of aero-engines. Also in this context, [22; 23] proposes a hybrid approach for bearing condition and life assessment, combining model-base techniques and sensed data

4.2 Knowledge-base methods

When the accurate mathematical representation of a degradation process, though an understanding of the inherent physical or chemical principles is a difficult task a knowledge-base approach can be a practical alternative. Given the existence of human expert knowledge on the problem, the ability to computationally represent this knowledge and then validate it using knowledge-based methods can be developed. Expert and fuzzy logic systems are two commonly used methods.

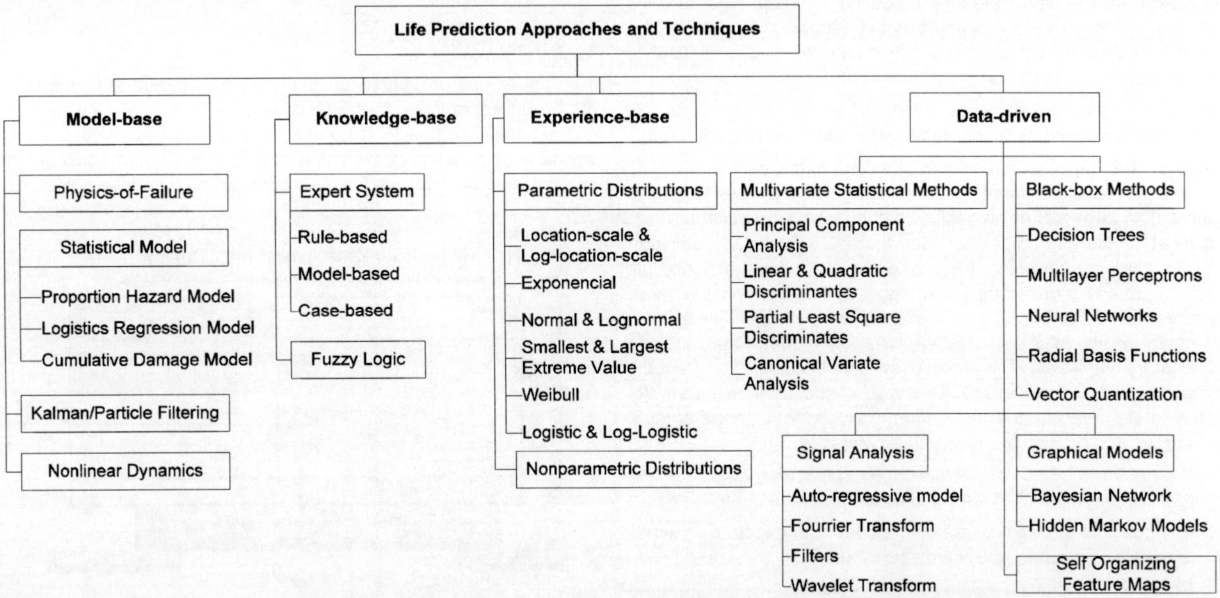

Figure 2: Life prediction approaches and techniques. Extended from [10].

Expert systems have been used for fault diagnostics and prognostics. Different types of expert systems have been developed, from rule-based to model-based and case-based systems. In a rule-based expert system, knowledge is represented in a rule form by *IF (condition) THEN (consequence)* statements forming a knowledge base. These rules can represent different types of knowledge and can be linked using logic operators. To calculate inspections intervals for gas turbine core components, [32] develops a knowledge based prognostics at a component level based on shared information between maintenance provider and operator. In this model the gas turbine usage history is used to estimate component load histories and then calculate component inspection intervals. A review on the application of expert systems for gas turbine diagnostics is given by [33].

In a fuzzy logic system, a balance is offered between a qualitative based simulation (as offered by an expert system) and an analytical simulation. This is achieved by modelling system behaviour in fuzzy sets providing a robust mathematical framework to cope with imprecision and uncertainty. Fuzzy logic modelling is often coupled with other modelling techniques, such as expert systems or neural networks [12]. Examples of the integration of fuzzy logic with expert systems for rotating machinery fault diagnostics can be found in [34], where causes of typical vibration problems are identified and ranked with the aid of a weighting component to reflect the relative significance and in [35] for an industrial gas turbine fault isolation. In [35] the fault isolation is achieved by gas path measurements (exhaust gas temperature, rotor speed and fuel flow) and rules developed from a model of performance obtained from automating the cognitive reasoning of an experience power plant engineer based upon the quantities measured.

4.3 Experience-base methods

Experience-based methods are based on probability distribution functions of failure time histories records from a component population (ideally an identical population). This approach has been extensively used in reliability analysis and a large portfolio of tested probability distributions, from nonparametric to parametric, are available.

Nonparametric estimates are particularly useful when the knowledge on the situation being modelled is limited since it does not require strong model assumptions to be made in the form of the chosen parametric distribution (for example, when it is difficult to determine which distribution best fits the data).

These methods allow the data to be interpreted without distortion that might be caused by using inadequate model assumptions [36].

However the use of parametric distributions might be either useful or even essential in some types of analysis (e.g. reliability data analysis). A large set of probability distributions are available, from the most simple and commonly used that typically include between one to four parameters (e.g. exponential, lognormal, Weibull, logistic, etc.) to more complicated models that can include more parameters. These can involve mixtures, competing failure modes, or other combinations of distributions or models that include explanatory variables. The most used distribution is the Weibull, given its ability to model the full range of the bathtub curve. As stated in [36], parametric distributions complement nonparametric techniques and provide the following advantages:

- Parametric models can be described concisely with just a few parameters, instead of having to report an entire curve.
- It is possible to use a parametric model to extrapolate (in time) to the lower or upper tail of a distribution.
- Parametric models provide smooth estimates of failure time distributions.

4.4 Data-driven Methods

Data-driven approaches use statistical or artificial intelligence techniques on relatively larges sets of measured component degradation / performance to forecast future component state. For the purpose of fault detection, prediction and diagnostics of gas turbines a large number of techniques has been used. These methods range from multivariate statistical methods, to back-box, graphical models, self organising feature maps and also signal analysis. The application of these techniques in the context of prognostics is provided in [13] and also in [37] for gas turbine diagnostics based on performance analysis.

5 ANALYSIS

Based on the methods exposed in this section and on the various comments provided in the references analysed, Table 1 synthesises the advantages / disadvantages of the different approaches for life prediction presented. The requirements of each approach, in terms of model input (data) and model construction are also provided.

It is the authors' view that the different approaches for life prediction can find distinct application suitability for LCC estimation along a product life cycle as represented in Figure 3. At the initial stage, while the product concept is being defined and past service data of the new design is reduced or inexistence, the use of expert knowledge thought appropriate knowledge-based systems containing the trade-off relations between design attributes and consequent impact on LCC, might be an effective solution. As the concept detail is refined and potential failure modes identified, moving to a model-based approach will bring a higher level of modelling accuracy by making use of mathematical models representing the underlying degradation mechanisms triggering the expected component failures. At the product service phase, when rates of in-service events become satisfactory for estimation of probabilistic measures within an acceptable confidence interval, experience-based techniques can be used and past LCC estimates validated. Data-driven may find particular application usefulness for the analysis of events of high rate of occurrence, such as analysis of overhaul workscopes causes and costs.

6 SUMMARY

This paper has presented an overview of different component life prediction approaches and techniques, justifying their usefulness in a gas turbine engine LCC context. When better understanding the degradation process determining component life and consequently influencing cost, improved service cost estimates will be obtained.

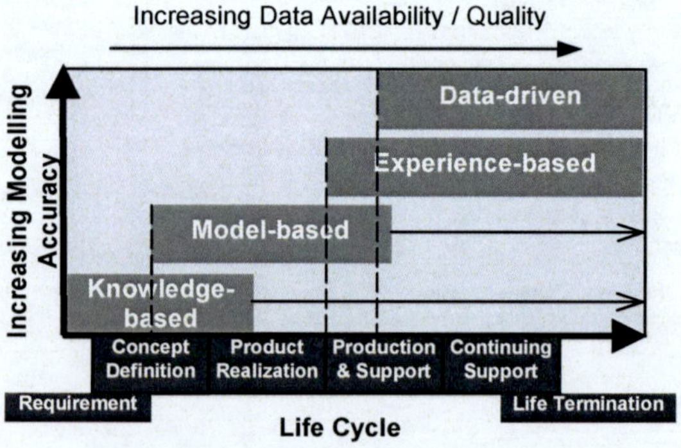

Figure 3: Life prediction approaches suitability along product life cycle.

	REQUIREMENTS	ADVANTAGES	DISADVANTAGES
MODEL-BASED	Ability to mathematically represent the degradation process (component / degradation mechanism / operating environment / usage). Knowledge on component parameters (e.g. materials properties).	Flexibility in the modelling logic and approach (e.g. by modelling all or a combination of: system dynamic behaviour / degradation mechanisms / operating environment / usage). Allows a direct relation to be established between the explanatory variables of the degradation process and the predicted life. Its not highly data dependent. Allows models outputs to represent the uncertainties in the inputs and modelling process by the use of probabilistic methods.	Requires an understanding on the component and degradation process in order to represent component behaviour and associated physical or chemical phenomena's. In some cases, due to the complexity of the problem or modelling limitations, assumptions have to be made reducing the degree of representation of the degradation process and consequent accuracy. Dependent on component parameters, e.g. materials properties.
KNOWLEDGE-BASED	Existence of expert knowledge on the problem.	Not dependent upon a mathematical representation of the problem. Can incorporate the uncertainty, incompleteness and inaccuracy in the expert knowledge using fuzzy logic. By the use of fuzzy logic, a balance between an analytical and qualitative representation of the problem can be obtained.	The acquisition and representation of expert knowledge in abstract rules can be a complex task. Only covers known situations. Since is based on human transmitted knowledge, it will include the inherent errors and misconceptions existing around the problem.
EXPERIENCE-BASED	Existence of failure time data of a population of (near) identical components in (near) identical conditions of operation and usage.	Simple to use and implement. Standard approach in different fields, e.g. reliability. A large set of techniques tested and validated in different domains.	When insufficient failure data is available some of the following compromises may have to be made to meet minimum data levels (generally expansion of the population of study beyond the truly identical: • Failure mode is not explicitly considered in the prediction of component life (multiple modes grouped). • Lack of an explicit relation between component usage and lifetime. • It's assumed that the prevailing environmental conditions either do not change in time, or have no effect on degradation and failure process. • Only quantifies the failure point for a population of components and not the component degradation over time.
DATA-DRIVEN	Extensive data on the problem and conditions that are intended to be covered in the model.	Available techniques to easily cope with static and dynamic modelling requirements. Few assumptions are required in the modelling process, for example in relation to model-based approaches. Able to cope with stochastic and complex degradation processes.	Lack of transparency on the way that model outputs are obtained, therefore making the results difficult to explain theoretically. Models will only be able to simulate similar conditions to those present in the data by which they where obtained / trained.

Table 1: Synthesis of life prediction approaches.

7 REFERENCES

[1] Gebraeel, N. and Pan, J. (2008), "Prognostic degradation models for computing and updating residual life distributions in a time-varying environment", *IEEE Transactions on Reliability,* vol. 57, no. 4, pp. 539-550.

[2] Harrison, A. (2006), "Design for service - Harmonising product design with a services strategy", *2006 ASME 51st Turbo Expo,* Vol. 2, 6-11 May 2006, Barcelona, pp. 135-143.

[3] Barringer, H. P. and Barringer, P. E. (1998), "Life Cycle Cost And Good Practices", *NPRA Maintenance Conference,* 1998, San Antonio, Texas, USA, pp. 1-26.

[4] Pitot, M. (2004), Generic European Life Cycle Cost Nomenclature and Structure, D2.2.1_1, VIVACE.

[5] Burkett, M. A. (2006), "DMTrade - A Rolls-Royce tool to model the impact of design changes and maintenance strategies on lifetime reliability and maintenance cost", *2006 ASME 51st Turbo Expo,* Vol. 2, 6-11 May 2006, Barcelona, pp. 11.

[6] Eres, M. H. and Scanlan, J. P. (2007), "A hierarchical life cycle cost model for a set of aero-engine components", *7th AIAA Aviation Technology, Integration, and Operations Conference,* Vol. 1, 18-20 September 2007, Belfast, Northern Ireland, pp. 41-46.

[7] El Hayek, M., Van Voorthuysen, E. and Kelly, D. W. (2005), "Optimizing life cycle cost of complex machinery with rotable modules using simulation", *Journal of Quality in Maintenance Engineering,* vol. 11, no. 4, pp. 333-347.

[8] Eady, C. (2000), Modes of Gas Turbine Component Life Consumption, RTO-TR-028, Research and Technology Organization of NATO.

[9] Byington, C. S., Roemer, M. J. and Galie, T. (2002), "Prognostic enhancements to diagnostic systems for improved condition-based maintenance", *IEEE Aerospace Conference,* Vol. 6, Big Sky, Montana, USA, pp. 2815-2824.

[10] Bagul, Y. G., Zeid, I. and Kamarthi, S. V. (2009), "Overview of remaining useful life methodologies", *ASME DETC 2008,* Vol. 3, 3-6 August 2008, NY, USA, pp. 1391-1400.

[11] Dragomir, O. E., Gouriveau, R., Dragomir, F., Minca, E. and Zerhouni, N. (2009), "Review of Prognostic Problem in Condition-Based Maintenance", *European Control Conference,* 2009, Budapest, Hungary, pp. 6.

[12] Gorjian, N., Ma, L., Mittinty, M., Yarlagadda, P. and Sun, Y. (2009), "A review on degradation models in reliability analysis", *4th World Congress on Engineering Asset Management,* 28-30 September 2009, Athens, Greece, pp. 16.

[13] Schwabacher, M. A. (2005), "A survey of data-driven prognostics", *InfoTech at Aerospace: Advancing Contemporary Aerospace Technologies and Their Integration,* Vol. 2, 26-29 September 2005, Arlington, VA, pp. 887-891.

[14] Nelson, W. B. (2010), "Defect Initiation, Growth, and Failure – A General Statistical Model and Data Analyses", in Nikulin, M. S., Balakrishnan, N., Huber-Carol, C., et al (eds.) *Advances in Degradation Modeling,* 1st ed, Birkhäuser Boston, pp. 27-37.

[15] Hall, P. L. and Strutt, J. E. (2003), "Probabilistic physics-of-failure models for component reliabilities using Monte Carlo simulation and Weibull analysis: a parametric study", *Reliability Engineering & System Safety,* vol. 80, no. 3, pp. 233-242.

[16] Grooteman, F. (2008), "A stochastic approach to determine lifetimes and inspection schemes for aircraft components", *International Journal of Fatigue,* vol. 30, no. 1, pp. 138-149.

[17] Watson, M., Byington, C., Edwards, D. and Amina, S. (2005), "Dynamic Modeling and Wear-Based Remaining Useful Life Prediction of High Power Clutch Systems", *Tribology Transactions,* vol. 48, no. 2, pp. 208-217.

[18] Tinga, T. (2010), "Application of physical failure models to enable usage and load based maintenance", *Reliability Engineering and System Safety,* vol. 95, no. 10, pp. 1061-1075.

[19] Bartelmus, W. and Zimroz, R. (2009), "Vibration condition monitoring of planetary gearbox under varying external load", *Mechanical Systems and Signal Processing,* vol. 23, no. 1, pp. 246-257.

[20] Li, Y., Billington, S., Zhang, C., Kurfess, T., Danyluk, S. and Liang, S. (1999), "Adaptive prognostics for rolling element bearing condition", *Mechanical Systems and Signal Processing,* vol. 13, no. 1, pp. 103-113.

[21] Li, Y., Kurfess, T. R. and Liang, S. Y. (2000), "Stochastic prognostics for rolling element bearings", *Mechanical Systems and Signal Processing,* vol. 14, no. 5, pp. 747-762.

[22] Orsagh, R., Roemer, M., Sheldon, J. and Klenke, C. J. (2004), "A comprehensive prognostics approach for predicting gas turbine engine bearing life", *2004 ASME Turbo Expo,* Vol. 2, 14-17 June 2004, Vienna, pp. 777.

[23] Orsagh, R. F., Sheldon, J. and Klenke, C. J. (2003), "Prognostics/diagnostics for gas turbine engine bearings", *2003 ASME Turbo Expo,* Vol. 1, 16-19 June 2003, Atlanta, GA, pp. 159.

[24] Harris, T. A. and McCool, J. I. (1996), "On the accuracy of rolling bearing fatigue life prediction", *Journal of Tribology,* vol. 118, no. 2, pp. 297-307.

[25] Ioannides, E. and Harris, T. A. (1985), "New Fatigue Life Model for Rolling Bearings", *Journal of Tribology,* vol. 107, no. 3, pp. 367-378.

[26] Loannides, E., Harris, T. A. and Ragen, M. (1990), "Endurance of aircraft gas turbine mainshaft ball bearings-analysis using improved fatigue life theory. Part 1. Application to a long-life bearing", *Journal of Tribology,* vol. 112, no. 2, pp. 304-308.

[27] Harris, T., Loannides, E., Ragen, M. and Tam, H. (1990), "Endurance of aircraft gas turbine mainshaft ball bearings-analysis using improved fatigue life theory. Part 2. Application to a bearing operating under difficult lubrication conditions", *Journal of Tribology,* vol. 112, no. 2, pp. 309-316.

[28] Chelidze, D. and Cusumano, J. P. (2004), "A dynamical systems approach to failure prognosis", *Journal of Vibration and Acoustics, Transactions of the ASME,* vol. 126, no. 1, pp. 2-8.

[29] Chelidze, D. (2002), "Multimode damage tracking and failure prognosis in electromechanical system", in Willett P.K. and Kirubarajan T. (eds.), *Components and Systems Diagnostics, Progrnostics, and Health Management II,* Vol. 4733, 3-4 April 2002, Orlando, FL, pp. 1-12.

[30] Yu, W. K. and Harris, T. A. (2001), "A new stress-based fatigue life model for ball bearings", *Tribology Transactions,* vol. 44, no. 1, pp. 11-18.

[31] Kotzalas, M. N. and Harris, T. A. (2001), "Fatigue failure progression in ball bearings", *Journal of Tribology,* vol. 123, no. 2, pp. 238-242.

[32] Slottner, P. and Wärja, M. (2008), "Knowledge based prognostics models for gas turbine core components", *2008 ASME Turbo Expo,* Vol. 1, 9-13 June 2008, Berlin, pp. 515-521.

[33] Doel, D. L. (1989), "Diagnostic expert systems for gas turbine engines - Status and prospects", *ASME, SAE, and ASEE, Joint Propulsion Conference,* 10-13 July 1989, Monterey, CA, USA, AIAA-89-2585.

[34] Yang, M. and Shen, Q. (2008), "Fuzzy diagnosis of turbomachines", *International Journal of Knowledge-Based and Intelligent Engineering Systems,* vol. 12, no. 2, pp. 137-146.

[35] Ganguli, R. (2002), "Fuzzy logic intelligent system for gas turbine module and system fault isolation", *Journal of Propulsion and Power,* vol. 18, no. 2, pp. 440-447.

[36] Meeker, W. Q. and Escobar, L. A. (1998), *Statistical Methods for Reliability Data,* 1st ed, Wiley-Blackwell.

[37] Li, Y. G. (2002), "Performance-analysis-based gas turbine diagnostics: A review", *Proceedings of the Institution of Mechanical Engineers, Part A: Journal of Power and Energy,* vol. 216, no. 5, pp. 363-377.

Complexity and Flexibility of IPS² across various Planning Levels

Horst Meier[1], Marion Steven[2], Birgit Funke[1], Mario Boßlau[1], Johannes Keine gen. Schulte[2]

[1] Institute Product and Service Engineering, Chair of Production Systems, Ruhr-University Bochum, Bochum, Germany

[2] Faculty of Management and Economics, Chair of Production Economics, Ruhr-University Bochum, Bochum, Germany

Abstract

The integrated consideration of product and service within Industrial Product-Service Systems (IPS²) leads to new chances and challenges across various planning levels: All services, sold with the machine, have to be planned during the entire life cycle.

The article first describes the strategic positioning of IPS² business model dynamics in the course of the IPS² life cycle.In an economic point of view, also strategic aspects of complexity and uncertainty have to be considered. Furthermore the scheduling approach for IPS² is described handling complexity and flexibility in short term tasks.

Keywords:

IPS² Life Cycle; Flexibility; Complexity

1 INTRODUCTION

The global competition is putting pressure on industrial enterprises to steadily hold ones ground in the B2B market. At the same time, differentiation in product quality only becomes more and more difficult. Already at the end of the 1960s Theodore LEVITT [1] made a statement that is still valid for current industrial applications: „People don't buy products; they buy the expectation of benefits". Furthermore, high technology products require a well skilled customer and for comprehensive customer support. Thus, customer demands are rising and an upward trend towards customized solutions can be recorded. These customer preferences are met by consequently designing Industrial Product-Service Systems (IPS²).

IPS² are specified by delivering value in use to the customer [2] [3]; they imply the integrated development of product and service shares. This integrated consideration of product and service also adds complexity to the delivery phase of this product bundle due to the long-term relationship to the customer; thus the planning of service processes along the entire life cycle gains importance.

Today's rapidly changing and increasingly uncertain economic environment makes business decisions complex and difficult. Industrial enterprises are confronted with new technologies, shorter life cycles, global markets and tougher competition. In this challenging business environment companies should be able to manage their strategic partnerships and still stay flexible to react on market dynamics. Hence, central aspects of the business model have to be defined within the IPS² contract with respect to the factor uncertainty.

As approved by a survey within the project Ogemo.net [4] the main focus for the customer and at the same time the widest scope of offered industrial services exist during the delivery phase; in this phase the customers have to be put back to high plant availability. Services like maintenance, repair, teleservice or overhaul are used to reduce machine downtimes on customer's site to a minimum and thus provide a high efficient plant usage. This kind of efficiency at the customer though presupposes an efficient service design and organization at the machine or service supplier.

The value provision via IPS² comprises several degrees of freedom, such as the partial substitution of product and service shares, alternative resources or processes and the integration of customers' resources. While maintaining the fulfillment of customer's demands, these specific degrees of freedom can be applied with the planning and scheduling. Furthermore, the scheduling has to be optimized during the delivery phase regarding several aims, e.g. costs, delivery time, and constant work load. Due to increasing challenges on competitive markets, innovative business models like IPS² require additional planning effort to incorporate the rising complexity. Furthermore, long term oriented business models are faced with difficulties arising from uncertain environmental conditions. Following, on the way to a successful IPS² obstacles are described, focusing on strategic aspects of IPS² complexity.

2 STRATEGIC POSITIONING OF IPS² BUSINESS MODEL DYNAMICS DURING THE LIFE CYCLE

Industrial Product-Service Systems (IPS²) imply the integrated development of product and service shares across the IPS² life cycle with focus on delivering value to the IPS² customer. Moreover, the possibility of the adaptation to changing customer needs is one of the central characteristics of IPS². It is necessary that IPS² business models can change over time and adapt to the respective requirements flexibly during the IPS² life cycle. The IPS² strategy must therefore imply the continuous adaptation of the respective IPS² business models to the specific market requirements. Particularly the motives for the change of business models as a reaction to changed market conditions have to be considered in the context of the IPS² life cycle. In this section the terms business concept, business model and contract are therefore differentiated across the IPS² life cycle from a strategic perspective.

The IPS² life cycle phases planning, development, implementation, operation and closure are defined by the allocation of the respective phases of the established life cycles of products [5] and services [6] (Figure 1). These IPS² life cycle phases describe the process of generating a customer benefit from a customer need.

J. Hesselbach and C. Herrmann (eds.), *Functional Thinking for Value Creation: Proceedings of the 3rd CIRP International Conference on Industrial Product Service Systems, Technische Universität Braunschweig, Braunschweig, Germany, May 5th - 6th, 2011*,
DOI 10.1007/978-3-642-19689-8_55,

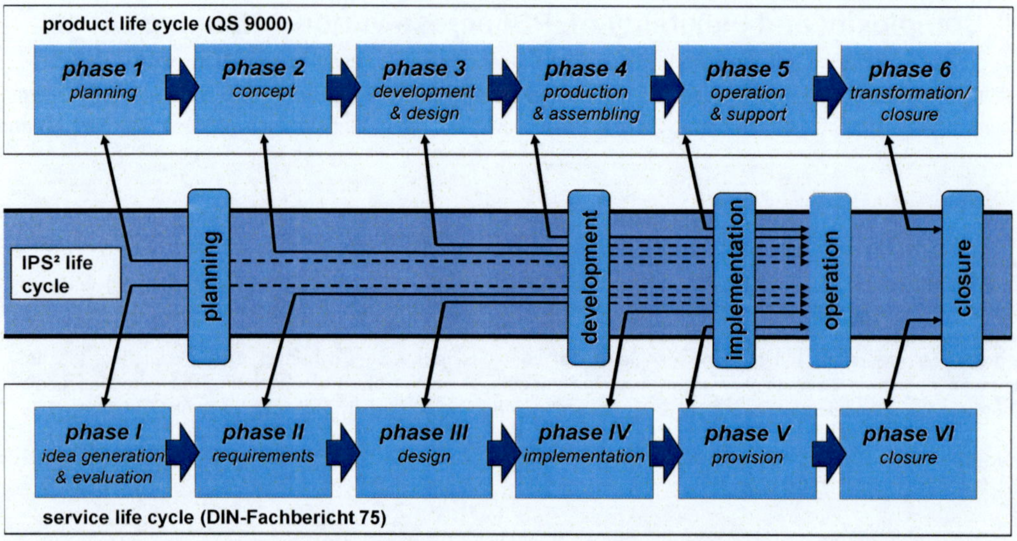

Figure 1: Derivation of the IPS² life cycle [5] [6].

The IPS² strategy expresses itself in the IPS² business concept [7] and describes a consciously intended course of action to deal with the special requirements of IPS² during the IPS² life cycle (Figure 2). HAMEL describes a business concept with four main components: The central strategy with its objective system, the interface to the customer, strategic resources and the value creation network [8]. STÄHLER outlines a business concept as the base of a future business model which is still not converted in practice [7]. In regard to the special requirements of IPS² a business concept represents the base of a future, not yet implemented IPS² business model from the view of the supplier. Thus, it represents a reference framework for the design of innovative IPS²-specific business models.

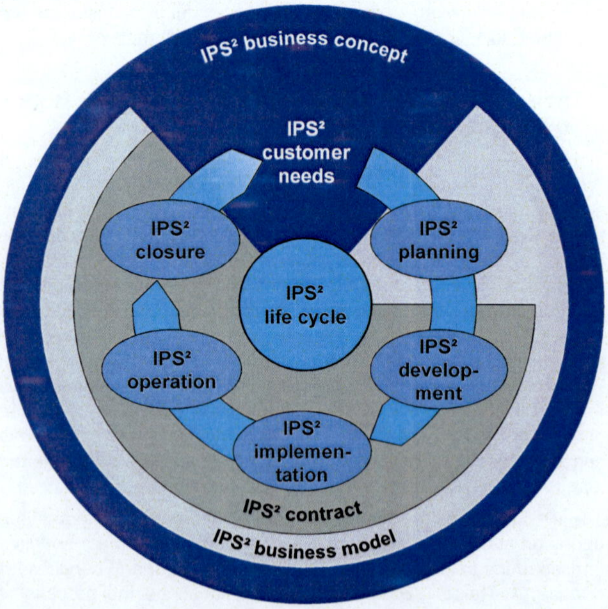

Figure 2: IPS² business concept in the context of the IPS² life cycle.

The IPS² planning is initialized by the first customer contact and covers the identification of the customer needs, the derivation of the specific, resulting customer requirements as well as the contract form. To offer the combination of products and services advantageous to all partners, dynamic business models are necessary considering the special properties of IPS² across the life cycle. An IPS² business model is concretized by the IPS² business concept in the IPS² planning phase; It can be product, use or result-oriented. It covers a value proposition, the value architecture and a revenue model. The value proposition describes the benefits and therefore the value a customer or a value partner gains from the business model. Aim of the value architecture is to create the promised benefit to the customer in an efficient way. The revenue model contains a description from what sources in what ways the firm generates its revenues. [7]

Furthermore, the contractual regulation necessity resulting from the IPS² business model is stated during the IPS² planning phase. In the line of contract negotiations the contents of the IPS² contract are specified. The potential contracting parties grant one another the possibility to influence their objects of legal protections, rights and interests.

The IPS² development is based on the customer needs and requirements identified in the IPS² planning phase. Central aspects are the development of IPS² concepts as well as the design of product and service shares [9]. The IPS² business model is instantiated by an IPS² contract [10]. The IPS² contract is a legal transaction which operationalizes the defined partial models and their dynamic relations within an IPS² business model. It consists of corresponding declarations of intention of at least two parties (IPS² supplier, IPS² customer). Central aspects of the business model are defined within the IPS² contract with respect to the factor uncertainty, for instance the diversification of risks, rights of ownership, modalities of payment, contract period, exit clauses, flexibility etc. The IPS² contract and the underlying IPS² business model are the framework for the phases IPS² development and IPS² delivery (IPS² implementation, IPS² operation and IPS² closure). The IPS² closure phase comprises the termination of the contractual relations between IPS² supplier, IPS² customer and potential other contractual partners.

The variety of many socio-technical systems is so wide that there is in principle uncertainty regarding their behavior. Variety can be defined as "the total number of possible states of a system, or of an element of a system" [11]. When analyzing a dynamic IPS² business model across the life cycle with variable correlations the variety increases disproportionately so that high uncertainty prevails regarding its system behavior. Right exposure with uncertainty according to high dynamics and distinctive complexity must be taken into account mandatorily both in the strategic and in the operative planning. Subsequently, these aspects are examined in more detail with regard to the respective planning level.

3 STRATEGIC ASPECTS OF IPS² COMPLEXITY

The rising planning complexity affects both strategic and short term aspects of IPS². Section 3 concentrates on strategic aspects of IPS² complexity.

Complexity will be categorized into static and dynamic complexity. Static complexity is related to the underlying system and can be found in the development phase of IPS². The system consists of different elements, interdependencies between them as well as the systems novelty as characteristics of static complexity. Compared to that, dynamic complexity appears in the delivery phase of IPS² and cannot be influenced by supplier´s decisions, because it is driven by external factors related to uncertainty and described as set of possible environmental conditions. The longer a business relation is lasting, the higher the amount of dynamic complexity and with it the significance of uncertainty.

Regarding the supplier side, a possible way for reducing IPS² complexity and thereby increasing flexibility, is given by using a modular design of the IPS² [12]. Modular design can be understood as an enabler of the required flexibility [13] [14]. Flexibility itself is the ability to react quickly to changing requirements and is needed to cope with uncertainty. So, different factors have to be regarded, which affect the level of information about uncertain incidents.

On the one hand, uncertain incidents can be measured by probabilities of occurrence and risks. In this case, the relationship between supplier and customer is built on **foreseen uncertainty**. Foreseen uncertainty means that uncertainties can be identified, but it is not sure that these uncertainties will occur. So, supplier and customer are aware of modifications during the delivery phase. The two parties have the choice of signing a contract depending on the foreseen uncertainty or to relinquish on the business relation ex ante.

On the other hand, **unforeseen uncertainty** might occur during the business relation. Unforeseen uncertainty includes incidents that cannot be identified ex ante. So, planning has to be iterative [15]. Further, the relevance of contract design increases considerably compared to foreseen uncertainty. In case of occurrence of unexpected situations, rights and duties have to be determined ex ante. By contracts, the general conditions of the future collaboration are determined, including rights and duties. This can be done with the help of contract theory [16] by allocation of decision rights [17].

The strategic aspects discussed above will be extended by decisions made in the subsequent delivery phase. A possible instrument is resource planning. In this context the focus lies on short term oriented resource planning with e.g. maintenance decisions for ensuring the IPS² availability in the delivery phase. In the following, resource planning will be discussed as a strategy for short term flexibility by means of a heuristic approach.

4 COMPLEXITY AND FLEXIBILITY IN SHORT TERM TASKS

4.1 Complexity and flexibility during IPS² resource planning

The long-term relationship based on IPS² business models enables comprehensive statistics on machine load, operating hours, etc. These statistics can be used to reduce uncertainty during the delivery phase by making predictions as to the service processes in future. This leads to a more certain, and thus less complex, planning situation, compared to classical production planning.

Nevertheless, the before mentioned modular design also induces complexity and flexibility to the delivery phase. This flexibility comprises manifold planning solutions, as e.g. alternative processes are already designed and may be chosen situational.

Inherent with Industrial Product-Service Systems, the following potentials for optimization [18] appear with this flexibility:

- variance in time;
- variance of resources;
- variance of processes;
- variance of allocation time;
- service distribution;
- partial substitution of product and service shares;
- integration of customers' resources;

At the same time, this flexibility requires for a comprehensive analysis of possible planning solutions regarding the current objectives. The heuristic approach for the IPS² resource planning uses these potentials for optimization to adapt the scheduling with unpredictable events.

The Heuristic Resource Planning Approach for IPS² [19] consists of a hybrid metaheuristic structuring the application of the potentials for optimization and defining the exit conditions. Within this metaheuristic the potentials for optimization are applied independent from each other.

The hybrid metaheuristic mainly ranks all the applicable potentials for optimization regarding their purposeful effect. In addition it defines how extensive one specificis used: At first it characterizes up to which degree of relationship an optimization potential is applied on the resources and processes. Furthermore the break condition is set to a certain number of generations, which limits how many applications of this potential for optimization are done consecutively. Additionally, it considers up to which level of complexity (range of alterations) the application of a potential for optimization is purposeful.

The hybrid metaheuristic for the IPS² resource planning's optimization ranks the potentials for optimization regarding to the direct costs they cause [20]. It is structured as follows:

- Use of variance in time
- Use of alternative resources
- Use of alternative processes
- Request to the network (service distribution and/or integration of customer's resources)
- Adjustment of allocation speed
- User intervention

All these steps are done consecutively giving interim solutions for the further optimization in the following steps (iterative optimization). The search within each step is realized by individual metaheuristics again, mainly a combination of genetic algorithms and simulated annealing. Direct search will not be effective with such large scale problems.

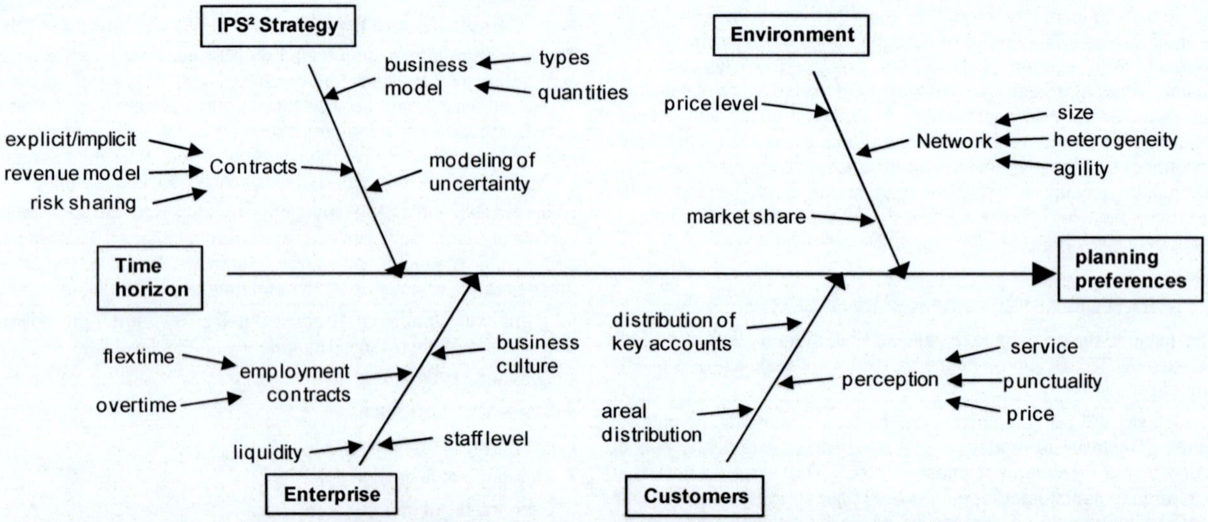

Figure 3: Ishikawa diagram of factors that result in the planning preferences.

4.2 Complexity reduction by setting planning preferences

The preliminary capacity planning has a relatively long term planning horizon and the degrees of freedom are wide. But this variety of possibilities also implies the effective optimization regarding several aims. On the other hand, the operational planning especially has to deal with unscheduled jobs (such as repairing machine break-downs). In this case the optimization aims are switched over more to make the execution of all required processes possible. According to that, planning preferences are set to switch over along the planning horizon.

Planning solutions have to be compared by calculating key performance indicators (KPI) referring to these objectives. The three main objectives of the IPS² resource planning's optimization are: costs, work load of the resources, and delivery reliability (towards the customer).

Costs are the easiest factor in this context: Only direct costs occur during the operational phase; thus planning solutions only differ in direct costs. Therefore, just the ratio of direct costs caused by the planning solutions is calculated. The KPI on the work load of all resources combines several figures, such as the absolute work load of each resource, the minimum and maximum values and the evenness of the work load's distribution as well. The KPI for the delivery reliability from the planner's point of view considers how many jobs will be in time or how much too late. Additionally the delays are weighted by time variance and process' length.

After calculating these KPIs they are combined to an overall KPI for the optimization result. This situational combination of the three main KPIs allows reducing the optimization complexity as there is now only one objective: the optimization of the overall KPI. The calculation of this overall KPI has to be adjusted according to the planning preferences, including strategic aspects.

The planning preference mainly depends on the position within the planning horizon: When the planning problem to be optimized is situated far in the future, so e.g. building up capacities is possible, costs are the main objective of the optimization. The absolute work load on the other hand has secondary importance, beyond costs, for the valuation of the optimization result, because the capacities can be adjusted to the workload. When the planning horizon is near in future the main focus is the reaction on short-term tasks, e.g. due to a machine breakdown. In this case, the delivery reliability for all jobs including the short-term tasks is much more important than costs or workloads.

Moreover, many factors influence the planning preferences (Figure 3). To set up these preferences, strategic as well as objective matters have to be included. Hence, defining planning preferences is an opportunity to reduce complexity, as many aspects of the current technical and economic situation can be covered thereby.

5 SUMMARYAND FUTURE PROSPECTS

As a result of the high combinatorial complexity of IPS² across various planning levels, not all problems can be solved with mathematical standards methods. Hence, these classic solution technologies have to be completed by metaheuristics and simulation methods with regard to dynamics and flexibility. On the one hand, the described optimization heuristic considers the variances and it rates possible solutions with focus on a short term oriented resource planning; on the other hand, it is necessary to provide simulation tools and decision support systems for the strategic aspects of IPS².

Due to the characteristics of IPS² and the integrated analysis of product and service shares, it can be assumed, that a high complexity prevails within business models. Many authors use the partial modeling approach as a structuring attempt to simplify business models and handle complexity. By such a structured approach it is possible to focus the relevant aspects of IPS², the essential partners involved as well as the basic business management relations. Static concepts and most existing approaches dealing with complexity don't cover all relevant aspects regarding the analysis of interdependencies within business models and its periphery. Consequently, these approaches can lead to erroneous decisions with direct effects on the business model success. Dealing with uncertainty - according to high dynamics and distinctive complexity - must be seen as a central challenge in the context of the continuous analysis of IPS² specific business models and requires the development of suitable rules, restrictions and relations for the configuration of IPS² business models across the IPS² life cycle.

6 ACKNOWLEDGMENTS

We express our sincere thanks to the Deutsche Forschungsgemeinschaft (DFG) for financing this research within the Collaborative Research Project SFB / TR 29 on Industrial Product-Service Systems – dynamic inter-dependencies between products and services in the production area.

7 REFERENCES

[1] Levitt, T. (1969): The Marketing Mode, Pathways to Corporate Growth.

[2] Meier, H., Roy, R., Seliger, G. (2010): Industrial Product-Service Systems - IPS², In: CIRP Annals - Manufacturing Technology 59, pp. 607-627.

[3] Meier, H., Völker, O., Funke, B. (2010): Industrial Product-Service Systems (IPS²) - Paradigm Shift by Mutually Determined Products and Services. In: The International Journal of Advanced Manufacturing Technology, Special Issue Product-Service Systems, Springer Verlag.

[4] Meier, H., Lanza, G. (Ed.) (2009): Kooperative Geschäftsmodelle zur Integration von Sach- und Dienstleistungen.

[5] Quality Systems Requirements (2002): QS 9000, Chrysler Corporation, Ford Motor Corporation, General Motors Corporation.

[6] DIN Deutsches Institut für Normung e.V. (1998): Service Engineering: Entwicklungsbegleitende Normung (EBN) für Dienstleistungen, Beuth, Berlin.

[7] Stähler, P. (2002): Geschäftsmodelle in der digitalen Ökonomie: Merkmale, Strategien und Auswirkungen, Josef Eul Verlag, Köln-Lohmar.

[8] Hamel, G. (2000): Leading the Revolution, Harvard Business School Press, Boston, MA.

[9] Sadek, T,, Köster, M. (2010): Eine modellorientierte Methodik zur Unterstützung der Konzeptentwicklung industrieller Produkt-Service Systeme, Proceedings of Dienstleistungsmodellierung 2010, Klagenfurt, Österreich.

[10] Sadek, T. (2008): Ein modellorientierter Ansatz zur Konzeptentwicklung industrieller Produkt-Service Systeme, Dissertation, Fakultät Maschinenbau, Ruhr-Universität Bochum.

[11] Beer, S. (1994): Decision and Control: the meaning of operational research and management cybernetics, Wiley, Chichester.

[12] Richter, A. (2010): Industrielle Produkt-Service-Systeme: Eine vertragstheoretische Analyse, Arbeitsberichte des Lehrstuhls für Produktionswirtschaft, Ruhr-Universität Bochum, No. 9.

[13] Wiendahl, H.-P. et al. (2007): Çhangeable Manufacturing - Classification, Design and Operation, in: CIRP Annals - Manufacturing Technology, Vol.56, No. 2, pp. 783-809.

[14] Richter, A.; Sadek, T.; Steven, M.: Flexibility in Industrial Product-Service Systems and Use- oriented Business Models, in: CIRP Journal of Manufacturing Science and Technology, in print.

[15] De Meyer, A.; Loch, C.H.; Pich, M.T. (2002), Managing Project Uncertainty: From Variation to Chaos, in: MIT Sloan Management Review, Vol. 43, No. 2, pp. 60-67.

[16] Bolton, P.; Dewatripont, M. (2005): Contract Theory, MIT Press, Cambridge, Massachusetts, London, England.

[17] Grossman, S.J.; Hart, O.D. (1986): The Costs and Benefits of Ownership: A Theory of Vertical and Lateral Integration, in: The Journal of Political Economy, Vol. 94, No. 4, pp. 691-719.

[18] Meier, H., Krug, C. M. (2006): Strategische Kapazitätsplanung hybrider Leistungsbündel, PPS Management 11, pp. 48-51

[19] Meier, H., Funke, B.(2010): Heuristic Resource Planning Approach for Industrial Product-Service Systems (IPS²), In: Proceedings of the 7th CIRP International Conference on Intelligent Computation in Manufacturing Engineering (ICME), Naples, in press.

[20] Meier, H., Funke, B. (2010): Resource Planning of Industrial Product-Service Systems (IPS²) by a Heuristic Resource Planning Algorithm, In: 2nd CIRP IPS² Conference, Linköping, Sweden, Linköping University.

Present Situation of Customer Participation in Service Design and Production - Interviewing Tour Agent, Airline and Elevator Maintenance Company -

Tsuyoshi Koga[1], Tatsunori Hara, Yoshinori Taniguchi, Kazuhiro Aoyama[1], Tamio Arai

[1] Department of Systems Innovation, School of Engineering, The University of Tokyo, Tokyo, Japan

Abstract

This paper validates the assumption that customer participation in mass customization, based on modularization, inspection, and configuration, is effective in the design and production of a service. This validation is based on interviews conducted with a number of industrial companies. These companies included a tour agent, an airline company, and an elevator maintenance company. The interview results showed that the current level of customer participation is insufficient. This study concludes that there are significant opportunities to apply modular and mass-customization methodologies to support customer participation and to realize customer-oriented service design and production.

Keywords:

Customer Participation; Service Design; Module; Mass-Customization; Tour Agent; Airline; Elevator Maintenance

1 INTRODUCTION

1.1 Tourism Industry Service and Customer Participation

Japanese tourism industries deal with fewer foreign tourists than most other developed countries. Authors have assumed that one of the major reasons why Japan has fewer tourists is because Japanese tourism resources are difficult to discover, access, and reserve. Hence, removing the barrier of ignorance of Japanese tourism resources, making them accessible, and allowing the reservation of various tourism resources will have a positive effect on increasing the number of foreign tourists. In this paper, customer participation is defined as the customer-participatory design and production of services. Customer-participatory design is defined as the mass-customization of services. Customer participation in the planning and experience stage of his or her travel helps to satisfy various customer needs and remove barriers to the same.

Today, because of the rapid evolution of the global Internet and ICT (Information and Communications Technology), we can reserve flights or hotels more easily than in the past. However, highly valuable services, such as domestic hospitality, are difficult to ascertain and also difficult to reserve.

Japanese manufacturers have developed highly sophisticated product designs and manufacturing methodologies. These include customer participation in the marketing stage, initial design stage, and production stage. Our research area focuses on industrial design and manufacturing. In order to provide greater integration of services in the domain of tourism, we assumed that customer participation methodologies in the industrial domain can be effective applied to the tourism sector. At least they will be suggestive of optimal approaches. Thus, we propose a new field, termed customer participation in service engineering [1]. The value of customer participation is confirmed during service processes [2] [3]. This paper reports the current situation in the service industry, especially in the field of tourism.

1.2 Customer Participation in Industrial Product Design and Manufacturing

In the field of design and manufacturing of industrial products, a computational configuration and ordering system for customer participation is effectively used from the marketing stage through the design and manufacturing stages. Customer participation in the customization stage not only provides the company with the ability to achieve highly lean production and delivery, but also helps the customer to recognize their own requirements. Because the customer can select the functional module, they are able to determine what they want and to find the ideal product.

The authors assumed that customer participation will also work effectively in the service engineering field. In order to qualify the possibility of customer participation, the authors also assumed that functional modules could be decomposed and that they may be used to represent complex service systems. Production, assembly, and inspection methodologies for functional modules have been discussed in the industrial production field [4] [5]. The focus point of this research is the semantic transfer of modularization methodologies from the industrial field to the service-engineering field.

2 FOCUS POINT AND RESEARCH PURPOSE

2.1 Tourism Industry Service and Customer Participation

This research assumed that services such as tourism and maintenance are also the target of design, production, provision, and inspection. Historically, the fields of design and production of industrial products have sought to satisfy the more sophisticated, diversified needs of customers through customer participation, which is qualified by use of modularization and customization methodologies. The authors got the idea from the concept of industrial customer participation, and focused on the possibility of employing customer participation in the design and production of services.

J. Hesselbach and C. Herrmann (eds.), *Functional Thinking for Value Creation: Proceedings of the 3rd CIRP International Conference on Industrial Product Service Systems, Technische Universität Braunschweig, Braunschweig, Germany, May 5th - 6th, 2011*,
DOI 10.1007/978-3-642-19689-8_56, © Springer-Verlag Berlin Heidelberg 2011

This research considered traveling and sightseeing services in the tourism industry [6]. The possibility of applying design, production, and inspection methodologies to services is studied. This research reports the interview results for tourism and maintenance service companies. The purpose of the interview is to understand the current situation and future trends of customer participation in services. In particular, modularization and platformization are considered as approaches to improving the quality of services. Based on the interview result, the improvement of customer satisfaction and service quality are discussed.

2.2 Assumption and Validation

The assembly process of an industrial product is composed of selecting the functional modules and integrating them. If the connectivity of the modular interface is appropriate, all a customer has to do is select the function that he or she wants. Hence, the concept of the functional module and open interface contribute to improving customization to satisfy variable customer needs.

In the field of service engineering, assembly and customization methodologies have not been practically discussed yet. This is because of the primary characteristics of services: they are intangible, inseparable, and perishable. Hence, this research tries to understand service products via module assembly and inspection, and studies the following four assumptions:

(A) Services can be produced by assembling functional components, just as with an industrial product.

A service product is composed of functional modules and their interactions. The value of a complex service can be partially understood by reducing it to the interactions of the functional modules.

(B) Customization of a service product contributes to satisfying various customer requirements.

The flexibility of a service product is derived from a customer's assembly of functional service modules based on their requirements. The configuration via customer participation can provide a highly sophisticated and customized service product.

(C) Customer participation requires a highly independent functional module and standardized modular interface.

The assembly of a service product can be decomposed into selection of functional modules and the integrating of modules.

(D) The modularization and assembly methodologies of industrial production can be applied to service products.

The functional decomposition, structural decomposition, and mapping from functional space to the entity space, in industrial product design, can be used in service product design.

In order to validate hypotheses (A) through (D), the authors interviewed a tourism company, an airline company, and an elevator maintenance company. Based on the interview result, this paper discusses the possibilities of customer participation.

The key enabler of the customer participation is the configurator based on modularization. It is clear that the customer is difficult to design and manufacture the product by themselves, because the customer has less technical knowledge. The standardized interface enables the customer to select and assemble the product easily as their own requirement. A set-theoretical definition of the four aspects of industrial product design and manufacturing (assembly, customization, customer participation, and modularization) by the standardized interface is described using an example of the vehicle platform in referred paper [5].

Based on conclusion of this paper, the authors proposed new research program [7]. In order to discuss major concern in growing customer participation field, this paper also describes hot topics, trends, and concrete problems of the customer participation in the service industry.

3 INTERVIEW METHOD AND TARGET

3.1 Interviewed Companies

The basis of company selection was a desire to survey two different kinds of service products: tourism services and maintenance services. We have selected two companies from the first domain, and one company from the second domain.

For the first service domain, a travel service company and an international airline company are interviewed. The travel service company is the largest travel agency in Japan and one of the largest travel agencies in the world. It specializes in tourism, and has branches all over the world to help Japanese visitors. The airline company is the second largest airline in Japan. It operates services to almost one hundred destinations all over the world.

For the second service domain, an elevator maintenance service company is interviewed. The chosen company is one of the major worldwide producers of elevators.

The main focus of this paper is to assess whether principals of industrial product design and manufacturing such as modularization, customization and customer participation, can be transferred to the design and creation of service. The tourism is just one of the examples. For fair evaluation, we have selected not only pure service industry such as travel agency, but also product-service systems industry such as elevator maintenance company. This is why we should analyze the elevator maintenance company.

3.2 Background information about the interviewed persons and interviewing steps

We have interviewed two to three persons from each company. The interviewed persons work in the research departments of their respective organizations, focused on improving company services. The respondents are selected based on their responsibilities, which meet our purpose. Overall, the interview questions cover the following topics:

1. The dynamics of the service provision,
2. How the service module is identified.
3. The variety of customer requirements.
4. How to meet the variable customer requirements.
5. Whether production and assembly of a module can help with customer participation.

4 INTERVIEW RESULT: TRAVEL SERVICE / PACKAGE TOURING COMPANY

Today's Japanese traveling / touring company provides two plans: (1) package tour, and (2) free plan. The package tour refers to a group tour, where the group is escorted by a chaperone. The package tour includes air flights, sightseeing guidance, and a travel itinerary. In contrast, the free plan has no chaperone and the customer is responsible for organizing all hotels and sightseeing schedules. In recent days, comparing the package tour and the free plan, the free plan is becoming the more common approach to traveling. One of the reasons for this is the variety of customer needs, which are becoming more and more complex, as the knowledge of customers is increasing, as facilitated by Internet technology.

4.1 Designing Package Tour

In this section, we present the interview results regarding the design of package tours. In the marketing and conceptual planning stage, at first, the company starts to create customer segments based on the destinations of travel (e.g. Tokyo, Hokkaido, Kyushu, etc.). Based on some market research, which includes customer satisfaction and the number of customers per generation, a rough tour concept is first written up by defining the rough program, main sightseeing spots, catch phrases, etc. Independent, third party stakeholders then grade the tour concept. These stakeholders may include the ministry of tourism, a consulting company, or an amusement company. The grading is evaluated based on market supremacy, customer satisfaction, relationship with core competencies, contribution to the local economy, and so on. Based on the evaluation score, the detailed business master plan is designed. Based on the detailed plan, a concrete plan is created by selecting stakeholders, such as hotels, amusements, sightseeing spots, and transit agencies, and then negotiating with them. Based on the detailed concrete plan, a pamphlet is designed and provided to travel agencies all over the world.

4.2 Designing Free Plan

In this section, the interview results regarding the design of the free plan are reported. The development process for the free plan is simpler than that of the package tour. All the travel agency has to do is to provide the transportation and accommodation options, and an optional plan (optional sightseeing spots and daily escorting service). This means that the customers create the concept and select the detailed components themselves.

4.3 Modular Configuration for Customer Participation in Designing Package Tour

The package tour is designed based on a framework and methodologies that are similar to those used in industrial product design. The first step of the development of the package tour is the marketing process. The detailed planning stage is similar to the functional deployment stage of the product design. The process of determining the hotels and transportation is almost the same as the selection and assembly process of product components. The planning of the tour schedule is a process similar to the assembly stage, which integrates the interfaces of the functional modules. Hence, in the development process of the package tour, product design methodologies, such as functional deployment, parts inspection, and modular configuration, can be effectively used. There is less chance of customer participation.

The design of a free plan is composed of the same functional modules as the package tour, except for group activities and chaperone escort. The same functional deployment and modular assembly methods can be applied. The major difference between the package tour and the free plan is the assignment of responsibility. The package tour incorporates a tour schedule developed by the travel agent, while the free plan does not. The customer has to take responsibility for his or her travel plan.

4.4 Trend of the Travel Agency

Figure 1 shows the current distribution of customers' selection of subscription methods for the Japanese travel agency. Today, 40% of tourists reserve an accommodation via direct phone call. Web services for travel booking are demonstrating an upward trend. Almost 70% of tourists do not use a travel agency. In order to overcome this situation, the importance of consulting services to the travel agency is increasing [8]

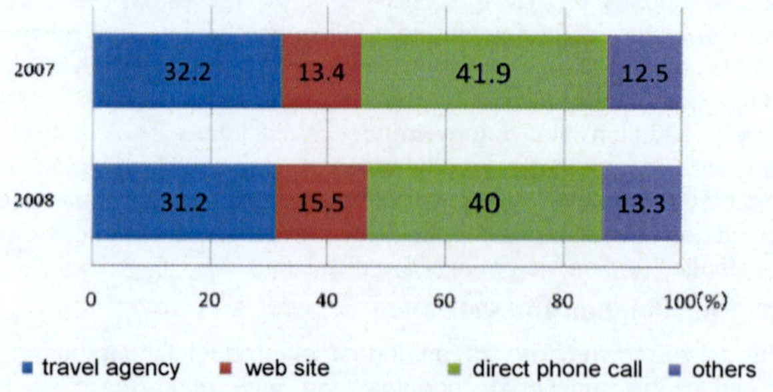

Figure 1: Methods employed for subscribing to travel tours.

5 FLIGHT SERVICE IN AIRLINE COMPANY

The authors interviewed an airline company to research the architecture of flight services and customer participation. The major topics of the interview consist of the following three points:

(A) How can we recognize the functional modules of the flight service?

(B) The variety of customer requirements and opportunities for customer participation in the service production stage.

(C) Whether assembly and inspection of service components is applicable.

5.1 Planning stage of the flight service

Development and sales process of new flight service

Basically, a new flight service is developed based on the recommendation of the marketing / sales division. A unique feature of the development process is that more ticket variation is better in sales stage. Based on marketing analysis, a flight ticket that has potential is developed and added to the lineup.

The customer needs vary with the type of customer. As such, the airline company has various sales channels to reach various customers. Figure 2 shows the combinations of customer types and sales channels. Points 1) – 4) explain the mainstream sales routes.

1) Sales for Company

Large scale manufacturing companies or big factories have a constant demand, thus they require constant, consistent transportation and logistics. For this type of customer, not only flight tickets but also pre and post service, such as limousines or parking, are planned and provided via dedicated channels. This flight ticket is sold directly from the airline company to the customer.

2) Bulk Sales for Travel Agency

Aside from retail sales, the airline company also sells flight tickets in bulk to travel agencies. A volume incentive allows the provider to sell tickets at a discount. In many cases, the airline company restricts the travel agency's sale of these tickets. The travel agency combines other travel services with this ticket, and constructs a travel package.

3) Direct B-to-C Sales via Booking Website

In recent years, flight ticket booking websites are operated directly by the airline companies. Today, travel agency sales are decreasing and airline booking website sales are increasing. By using the booking website, the airline company can determine the price of their product. The airline company studies the sales timing and plans the best price to maximize their income.

Variation in Flight Service to meet Various Customer Requirements

Customers can be categorized into three groups: 1) big companies, 2) business customers, and 3) leisure customers (Figure 2). The customer requirements depend deeply on the customer category. For example, the business customer has a strong time constraint, while the leisure customer does not. Similarly, the leisure customer may have very severe cost constraints. In order to meet customers' varied requirements, the airline company provides a variety of tickets. The ground staff considers the various requirements and differentiates the handling of the customer. For example, a student taking an entrance examination is given higher priority than a first class customer.

Ticket variety improves sales, although too much variety can confuse and complicate broader operations. The confusion increases the cost and decreases adherence to customer requirements. Numerical and scientific design methodologies in the variation of flight tickets are highly desirable. The current situation is such that appropriate ticket variation is often not optimized strategically.

Modularization and Differentiation of Flight Services

The modularization and standardization of flight services can also be a factor in economic obsolescence. The airline company has developed various differentiation measures. A mainstream effort is the up-skilling of the ground operation and paying great care to individual customers. Recent ICT (Internet Communication Technology) devices, such as integrated pass-cards for train or bus travel are one of the major methods of differentiation. A concurrent transporting service and early flight delay or cancellation notification services have been under development.

5.2 Improvement of Flight Service Quality

Evaluation of the Quality

In order to reflect the ground operation and ticket variation, the airline company evaluates the quality of its service via third-party organizations. The total value of the service is studied and compared with that of competitors. The evaluation result is important information for continuous improvement of the service.

Costing of the Flight Service

A bottom up approach is difficult when estimating total cost, because of the various stakeholders involved, such as airport staff, ground staff, maintenance staff, and so on. A major approach, such as that used in IE (Industrial Engineering), estimates the cost of the service module or component independently. Hence, a fair price for one flight is very difficult to estimate. The airline company determines the rough cost of one flight, but it is not clear that this is the appropriate price of any one particular flight.

5.3 Rescheduling Due to Unexpected Incidents

Efforts for High Customer Satisfaction

It is a well-known fact that punctual airline flights result in high customer satisfaction. However, flights are frequently delayed or canceled due to unexpected incidents. These incidents are dealt with by rescheduling, which is intended to maintain a high level of customer satisfaction while adjusting the resources involved.

There is no strict guideline for customer prioritization. The ground staff judges by understanding individual situations. This is a major reason why the airline company cannot reduce the number of ground staff employees.

Rapid and Flexible Rescheduling

A customer's sensitivity and emotional reaction toward delays or cancellations is deeply dependent on the purpose of the customer's travel, such as business or leisure. Recent ICTs allow the ground staff to determine the customer's category and type. However, if the scale of the incident is very large, it is difficult to provide individual attention.

Relieving Customer Frustrations

An induction system of the customer in the airport has been researched. Information on the individual customer allows the airline company to indicate the best way forward, employing real-time projections.

6 ELEVATOR MAINTENANCE COMPANY

Elevators have long product lives. A typical physical lifetime is almost twenty to twenty-five years. The renewal cost includes disassembly and development processes, hence elevators are very expensive. This is the major reason why the elevator is not typically

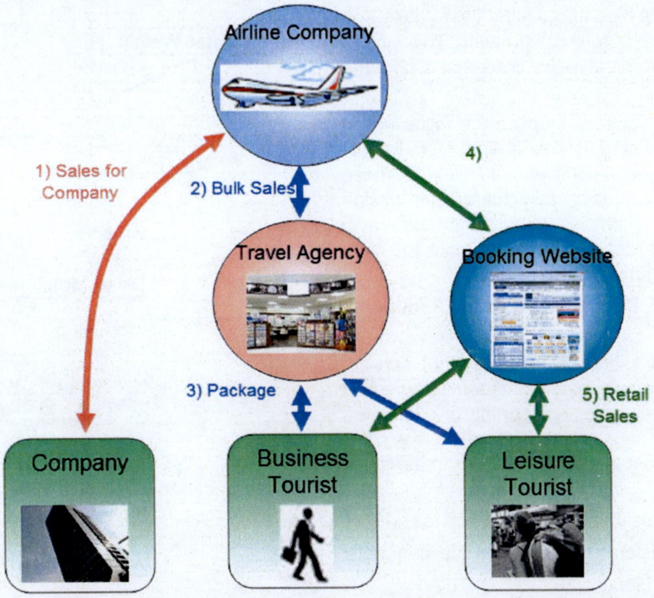

Figure 2: Combination of Customer Type and Sales Channels.

replaced; rather, it is maintained. In the design of the elevator, to the designer must consider future upgrade and maintenance efforts.

Figure 3 shows the functional modules and components of the elevator. The elevator consists of four major modules: machinery, cargo, entrance, and pit. Some modules are rigidly connected to the building, so it is very difficult to replace them.

6.1 Structure and Replacement of the Elevator

The elevator includes modular components and integral components. The modular components include the motor, rope, and cargo, and so on. These modular components do not require expensive adjustment at any point. The control system, however, does require expensive adjustment, so in many cases the control related modules are replaced or upgraded all at once. The easily adjustable modular components have high customization value. The customer can select the functional modules they desire during the construction stage.

6.2 Maintenance Processes

Elevator maintenance requires maintenance parts. The elevator construction company provides the design information to the maintenance company. Based on the designed lifetime, periodic checks and part replacements are performed by the maintenance company.

7 DISCUSSION

The authors confirmed that a package tour is composed of service modules: accommodation, sightseeing spots, and transportation. The following section summarizes our findings from the company interview.

7.1 Package Tour is Composed of Assembled Modules

The package tour is created via detailed plan, by assembling the elemental components, such as accommodation and transportation. The designer creates the tour plan by selecting the sightseeing spots and integrating them. The concept of package tour development is quite similar to that of product development. On the other hand, the use of a free plan does not ensure the success of the tour. Customer participation is high in the latter case, based on the free selection and combination of tour components. This kind of situation is similar to home-built furniture or Personal Computer assembly.

The major characteristics of the package tour are the following: 1) the compatibility of the functional components is very high, and 2) each module has an isolated, complete function. These characteristics allow the customer to easily participate in the design and production of the service. This also means that the travel agency is required to provide additional value in the package tour, to compensate for the lack of customization.

7.2 Flight Service Can Be Decomposed Into Functional Modules

The core service offered by the airline company is flight service. The business' income comes from providing flights to customers. A main function of the flight service is the transporting of customers from a departure airport to their destination. The airline company combines additional pre and post services with this main flight service.

The flight service can be decomposed into three sub-services:

1. Pre service: between the customer's home and the departure airport (until boarding completes)
2. Main service: from boarding to flight landing (flight service)
3. Post service: from landing to the customer's destination

The airline company assembles these three different kinds of functional components to create a flight service. Considering customer characteristics that vary (e.g., travel purpose; business or leisure), a variety of tickets options are developed based on the marketing results. The marketing and product family design practices from product engineering can be applied to the flight ticket development process.

Almost all components of the flight service are functionally independent. The modules of the flight service have temporal relationships. Many different kinds of resources, like airplanes, ground staff and check-in kiosks, are required to fulfill a module's function. Hence, a sequential relationship exists between flight sub-

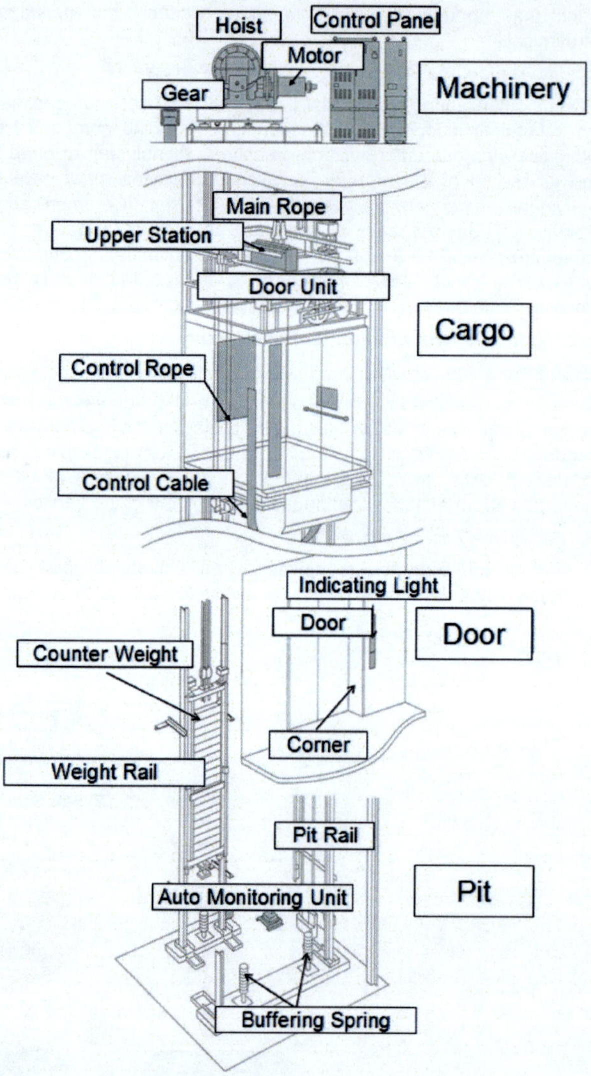

Figure 3: Functional Modules of the Elevator.

services. The resources have time relationship and order relationship. This constraint of order and time is critical when the accidents happen. The punctual flight against the accidents is very important to achieve high customer satisfaction.

8 RESULT OF ASSUMPTION VALIDATION

The following points, (A) to (D), support validation of the prior assumptions, based on the interview results from three companies.

(A) A service can be assembled using functional components, just as an industrial product.

The authors confirmed that service products, such as packaged tours, flight services, and elevator maintenance, consist of functional modules. The design of a service product can be represented as a functional module assembly process.

(B) Customization of the service product contributes to satisfaction of various customer requirements.

The provided service modules can be categorized into two groups: 1) service module selection when the customer purchases the service, and 2) maintenance and upgrades to the service module, over the service lifecycle.

The perishability of a service makes it difficult to replace a service module when an accident takes place.

(C) Customer participation requires highly independent functional modules and a standardized modular interface.

The integration of service modules is relatively easy compared to the integration of industrial product modules. The major reason for this is the flexibility of the interface connection. Customer participation is enabled by the interface connection and functional independency..

A side effect is also confirmed: differentiation is very difficult in this case. Individual characteristics of customers must be considered in order to facilitate customer participation.

(D) The modularization and assembly methodologies, used in the manufacture of industrial products, can be applied to service products.

A service product can also be designed and produced by assembling functional modules. However, the temporal relationship between service modules is stronger than the relationship in an industrial product. In flight service, if an accident takes place, the relationship between service modules and the provision of resources can be a strong constraint. Hence, the modularization and assembly methodologies can be applied almost identically, although we must take care to consider the temporal dependencies between service modules.

9 CONCLUSIONS AND FUTURE WORK

Our findings, obtained from interviewing three service-oriented companies can be summarized as follows. As with industrial products, service products can be produced by the assembly of functional modules. The customization of a service product contributes to satisfying various customer requirements. Customer participation requires the use of highly independent functional modules and a standardized modular interface. The modularization and assembly methodologies of industrial production can also be applied to service products.

The relationships between functional modules can be categorized into two groups. The first is the interaction between service modules, while the second is the relationship between the service module and the customer. The relationship with the customer is a specific problem in the service-engineering domain. Our findings from interviewing the airline company indicate that customer satisfaction can easily decrease with the occurrence of some inevitable accident. Hence, flexible upgrades of the service module, to account for such accidents, taking various customer requirements into consideration, are very important to service production.

The final conclusion of this research is that a service module can be expressed as the assembly of constituent functional modules. In the future, the authors wish to use the decomposition and assembly methodologies, associated with functional modules, and customer participation, in the design and production stage, thereby increasing its sophistication.

Based on the discussion and conclusion in this paper, authors proposed a new research program which is accepted by JST (Japan Science and Technology Agency). Next three years, the authors collaborate with the travel agency and address the customer participation in the tourism field.

10 ACKNOWLEDGEMENTS

This research is supported by Service Science, Solutions and Foundation Integrated Research Program, Research Institute of Science and Technology for Society, Japan Sciecnce and Technology Agency.

11 REFERENCES

[1] Tamio ARAI, Tatsunori HARA, Kazuhiro AOYAMA, and Tsuyoshi KOGA (2009): Report on Production Methodology of Service Product by Module Creation, Assembling, and Inspection, 2009 Project of Japan Science and Technology Agency, 2009.3.

[2] Chan, K.W., Yim, C.K. and Lam S.K. (2010): Is Customer Participation in Value Creation a Double-Edged Sword?, Journal of Marketing, Volume 74, Number 3.

[3] Kellogg, D.L., Youngdahl, W.E., Bowen, D.E. (1997): On the relationship between customer participation and satisfaction: two frameworks, International Journal of Service Industry Management, Vol.8, Issue 3, pp.206-219.

[4] Tatsunori HARA, Yohei KURATA, Naoto YABE, Tsuyoshi KOGA, Taketomi ASANO, Makoto KATO, Kazuhiro AOYAMA, and Yoshiaki HONPO, Research plan on sophisticating tour services for foreign travelers, -Toward architecting service with customer participation based on the analysis of customer experience and design processes –, The International Hournal of Tourism Science, (in press).

[5] Tsuyoshi KOGA and Kazuhiro AOYAMA, "Design Method of Common Platform and Option Module Based on Proposition of a Product Family Model," Journal of Japan Society for Design Engineering, Vol.45, No.12 (2010.12), (in press).

[6] Tsuyoshi KOGA, Hideshi AOKI, and Kazuhiro AOYAMA (2010): "A Modular Design Method for Scenario Embedded Product," New World Situation: New Directions in Concurrent Engineering, Springer – Verlag London Limited, pp.181-190.

[7] Service Science, Solutions and Foundation Integrated Research Program: **http://www.ristex.jp/servicescience/EN/**

[8] Japan Tourist Bureau: Market Insight 2010 - Trends of the Japanese Overseas Travel Market.

The Smart Real-Time Factory as a Product Service System

Bilal Hameed[1, 2], Jörge Minguez[1], Michael Wörner[1], Philip Hollstein[1], Sema Zor[1], Stefan Silcher[1], Frank Dürr[2], Kurt Rothermel[2]

[1] Graduate School for Advanced Manufacturing Engineering, University of Stuttgart, Stuttgart, Germany

[2] Institute for Parallel and Distributed Systems, University of Stuttgart, Stuttgart, Germany

Abstract

In modern manufacturing landscape, companies are increasingly relying on product service systems i.e. bundling of products and services together in order to gain a competitive edge. In this article we present the Smart Real-Time Factory, a smart digital manufacturing environment that can transform the process of production into an informational service for the customers. The different components of the smart factory are discussed at length along with a discussion of the different services that can be offered by the smart factory.

Keywords:

Product Service Systems; Smart Real-Time Factory; Production as a Service

1 INTRODUCTION

For the past decades a high quality product or lower prices were sufficient to differentiate amongst competitors. However, due to globalization, tear down of import duties, and saturation of markets, the possibility to distinguish on the basis of a product alone has become obsolete. Products have converged remarkably during the past decade in terms of quality, technology and pricing. Due to this reason, companies are now facing ever increasing competitive pressure and decreasing profit margins [1].

One way out of this competitive situation is to integrate products with services and offer them together as Product Service Systems (PSS). The bundling of products and services can result in fulfillment of customer needs and establishment of intensive long term relationships with the customers at the same time. Such a scenario creates competitive advantage and additional barriers for competitors to encroach on a company's customers. In addition to this, research has proved that a PSS offering can extend the product life cycle considerably, resulting in continued revenues and more sustainable operations [2].

The most common example of a PSS is a leasing service that companies offer to their customers. This allows the customers to lease the product (laptop, automobiles, etc.) instead of paying out right cash for the product. The leasing service has become such an integral offering especially in the automobile industry that almost all major automobile manufacturers now have a financial arm that offers leasing service for customers.

However, product services are not limited to financial services alone. Other types of services include informational – as in the case of ingredients mentioned on food items, so that consumers are able to get more information about the food that they are buying and peace of mind – as in the case of warranties provided for products, so that customers are assured about the quality of the product that they are buying.

When it comes to product service systems, much of the attention has been focused on bundling services with the product itself, whereas the entire process of manufacturing has remained a black box inaccessible to the customers. Once a customer orders a product, he has no way of finding out if his product has been planned, and how long it will take for his product to be assembled. If the customer is able to get real-time information about the product that he has ordered, the entire process of manufacturing would be transformed into a production service or a PSS.

We believe that today's state of technology can transform factories into smart environments that can provide real-time production information not just to the producers but to the consumers as well. RFID technology can be used in order to transform factories into smart manufacturing environments.

In this paper, we present the different applications that need to be built on top of RFID devices to create a context sensitive smart manufacturing environment that is able to monitor production processes in real-time. The smart factory is not just able to monitor production in real-time but is also able to provide suitable information to the customer. Furthermore, the factory can react to disruptions and changes in production in real-time. In addition to discussing the RFID applications (such as real-time production monitoring, RFID based complex event processing, etc.) for a Smart Real-Time Factory, we also discuss a number of services that the smart factory can offer for prosumers (producers and consumers alike).

The rest of the paper is structured as follows: in Section 2, we take a look at the ways in which the RFID technology is currently being used in manufacturing. In Section 3, the notion of a Smart Real-Time Factory is presented along with the different components that are needed to create such a smart manufacturing environment. The different services that can be offered by Smart Real-Time Factory are discussed in Section 4. In Section 5 we have summarized the numerous benefits of transforming factories into smart real-time manufacturing environments.

2 RFID IN MANUFACTURING

In this section we will take a brief look at some of the ways in which RFID technology is being used in manufacturing. The purpose of

J. Hesselbach and C. Herrmann (eds.), *Functional Thinking for Value Creation: Proceedings of the 3rd CIRP International Conference on Industrial Product Service Systems, Technische Universität Braunschweig, Braunschweig, Germany, May 5th - 6th, 2011*,
DOI 10.1007/978-3-642-19689-8_57, © Springer-Verlag Berlin Heidelberg 2011

this assessment is to gain insight into the current state of the art and to derive requirements for the Smart Real-Time Factory.

2.1 RFID deployment at Ford facility in Mexico

Ford Motor Company has successfully deployed RFID technology at its automated assembly production lines in Cuautitlan, Mexico in order to improve product quality [3]. The Cuautitlan facility produces 300,000 to 400,000 cars and trucks each year. Each of these vehicles are built using just-in-time assembling where suppliers supply parts on an as-needed basis. Due to this, it is crucial that the inventory and production in the plant are precisely tracked.

Prior to the deployment of RFID technology, Ford used a manual coding system to track auto and truck frames as they went through the final assembly, paint and body shop areas of the production line. However, manual tracking was ineffective due to frequent errors and costs associated with resulting production oversights.

With the RFID technology deployed, an RFID tag is placed on a vehicle skid, and then programmed with a serial number that is used to reference the vehicle in Ford's operating system. As a vehicle passes through the different stages of production, different parts of the 23 digit serial number are referenced, indicating what needs to be done at each station. The manual system required an identification sheet to be manually updated at each step, whereas with the RFID deployment the updates are automatically written to the tag. In this way, the risk of operator errors in updating the sheets to determine what needs to be done is totally eradicated.

2.2 RFID deployment at BMW and Vauxhall plants

BMW and Vauxhall [4], [5] have also made use of the RFID technology to accurately customize the cars according to the customer specifications. A read/write smart RFID tag is programmed with the customer specification and is then placed on the car. The tag moves with the car during the production process and at each step is read to determine exactly what needs to be done and which equipment needs to be fitted into this specific car. This tracking ensures that the car is manufactured with the correct color, model, interior, and other user specifications.

2.3 RFID Applications in Manufacturing

Baudin et al. [6] have documented a lot of other applications of RFID deployments in the manufacturing sector. Toyota (South Africa) for example has tagged its carriers to streamline manufacturing and vehicle tracking. The tags are intended to remain with the vehicle throughout its life and hold its maintenance history. Harley Davidson has implemented process automation by tagging bins carrying parts to provide instructions to employees at each stage of the process. Johnson Controls has started using RFID to track cars and truck seats throughout the assembly process. TrenStar tracks their beer kegs to improve demand forecasts and increase efficiency. International Paper tracks their paper roll to reduce lost or misdirected rolls. Gap tracks its denim apparel to improve customer service through better inventory management. Raxel has started tagging reusable plastic biohazard containers to avoid contamination. Michelin on the other hand tags its tyres to comply with the TREAD act and recall management. All the RFID deployments mentioned above were carried out in order to improve shop floor inventory tracking and automate warehouse operations.

From the applications presented in this section it is obvious that although RFID technology is being used in manufacturing, no solution exists that uses RFID devices to track the production processes in real-time. Companies like Ford, BMW, and Vauxhall are using RFID technology to identify what task needs to be done, but there implementations cannot address questions like, where is a certain part at the moment, is everything on the shop floor executing according to plan, and so on.

3 SMART REAL-TIME FACTORY

The Smart Real-Time Factory initiative [7] at University of Stuttgart envisions to create a context sensitive manufacturing environment that can monitor production processes in real-time so that any disruptions in production can be detected and rectified immediately. Based on the discussion of the existing deployments of RFID in manufacturing, we have drafted the following requirements for a Smart Real-Time Factory:

1. It should be able to track the production in real-time. Tracking should not be limited to tracking a specific part or process, instead all parts and mobile machine tools should be tracked in real-time and at all times.
2. It should be able to present the results of production and asset tracking to higher level management and consumers alike in a meaningful way.
3. It should be able to adapt to process changes and system changes in a graceful manner.
4. It should be able to respond to changes (intended and unintended) in manufacturing in real-time. An intended change can be a product configuration change request from a consumer in the midst of production, whereas an unintended change can be a product identified as not being assembled according to a plan.

In the sub-sections below, we present the applications needed to fulfill the requirements of a Smart Real-Time Factory.

3.1 Real-time production tracking

The first requirement of a smart factory i.e. the real-time production tracking of product parts and assets can be accomplished by using RFID devices. However, RFID devices have an accuracy of 80-90% [8] which is further affected by environmental factors such as the presence of metal objects, interference from multiple readers, or presence of multiple tags. The accuracy of an RFID drops to 70% [9], [10], [11] in case it tries to detect more than 5 tags.

This inherent unreliability of RFID devices have impeded their use in real-time production tracking in modern factories. This problem is especially compounded in variant production, where product parts moving through the production lines vary resulting in different variants of the same product being manufactured. In a variant production scenario it is essential to ensure that the product parts move through the production lines in a desired sequence. A disruption in the planned sequence of product parts will result in undesirable final product.

Since RFID readers have an accuracy of 70%, this would result in different readers detecting different sequences of product parts. As an example, consider that three product parts p1, p2, and p3 pass through the production line with p1 being the first part and p3 being the last part. An RFID reader missing out on p3 will assume that the product part sequence is p1, p2 where as another reader that misses out p2 will assume that the part sequence is p1, p3. Failure to detect product parts is not the only issue. RFID readers can detect parts incorrectly, can overshoot and detect parts passing through neighboring production lines, and can have duplicate readings.

In [12], we have proposed an RFID based real-time production tracking framework to solve the issues discussed above. Multiple RFID readers are deployed along the production lines (Figure 1) so that even if a product part is missed out by one reader, it is detected by others. The entire manufacturing plant is divided into topological regions, with each region assigned to a Virtual RFID reader. The physical readers within a region send their data to the virtual reader responsible for the region.

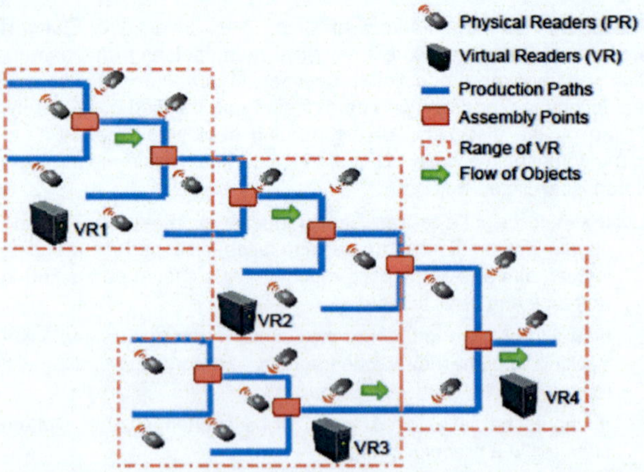

Figure 1: Factory deployment of RFID based real-time production tracking framework [12].

The virtual reader correlates the data sent by all the physical readers and is able to accurately determine the product parts and their respective sequences.

3.2 RFID based complex event processing framework for manufacturing

The deployment of RFID based production tracking component will enable a factory to track the product parts moving across the production lines in real-time. However, raw RFID data poses additional challenges that need to be addressed in order to effectively use this real-time RFID data.

First and foremost, the raw RFID events have implied meaning which needs to be transformed and aggregated with other events to obtain the semantic meaning of those events in the context of manufacturing. Consider for example that we have an RFID event "part p1 at production line 1 at time T" and another RFID event "part p2 at production line 1 at time T". Now from our production database we would know that part p1 needs to be assembled together with part p2 and since our RFID based production tracking framework is telling us that it has read both the parts at the same time at the same place, it means that the two parts have been assembled together. This task of deriving high level complex events from low level raw RFID data is referred to as Complex Event Processing (CEP).

The second problem is that this raw RFID data is temporal in nature and is generated in high volume, so the processing of this data needs to be done in an automatic manner and in real-time due to the temporal nature of the data.

Therefore in order to address the second requirement of the smart factory i.e. to present the results of production tracking to producers and consumers in a meaningful way, we would need to have an RFID based complex event processing framework for the manufacturing domain. Wang et al [13] have proposed a complex event processing framework for RFID data streams. However, their framework provides generic constructs for aggregating and transforming RFID data. Further research needs to be done in order to identify manufacturing specific data needs and to develop the necessary constructs to extract this data from raw RFID events.

3.3 Real-time networked manufacturing

Once a Real-Time production tracking component and the complex event processing component are deployed, we can track the production process in real-time and gain meaningful insight into the production process by viewing high level information such as information about the work in progress products in real-time. However, in case there is a discrepancy the system would merely report the inconsistency.

In order for the factory to react to this newly generated RFID data (which is also the fourth requirement for a smart real time factory), we would have to integrate this data with all the other systems running the factory such as the Manufacturing Execution System (MES), or Materials Control System (MCS). Figure 2 shows the different systems of a factory integrated with one another. This integration will enable the corresponding systems to get RFID data and respond to changes in the production process in real-time.

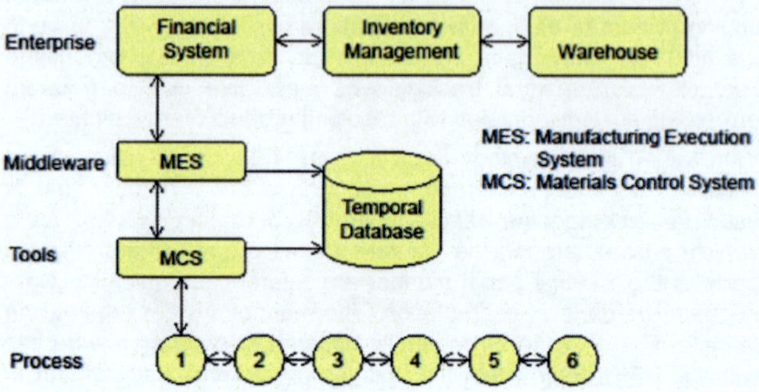

Figure 2: Real-time networked manufacturing system [24].

As an example consider that a product has been fitted together with an incorrect part. Once this information is made available, it makes no sense for the assembly of this product to continue. Therefore, the MES can update its schedule and also let the MCS know that it should not deliver the remaining parts of this product to their respective assembly points. Such a real-time removal of faulty products from the production line would save the factory from complete assembly of a faulty product.

3.4 Service-based integration

One important requirement of the smart real-time factory is to be able to react to process and system changes in a graceful manner. In order to be able to react to unexpected events and adapt the corresponding business processes to new scenarios, it is necessary to know the nature of information flows in event-driven manufacturing environments. The reaction to unexpected events or the detection of an incorrect assembly imply the execution of processes across different domains, such as shop floors, Enterprise Resource Planning (ERP) systems, supply chain, etc. Therefore, there is a need to seamlessly integrate both existing event driven architectures, such as shop-floor systems, business processes, and any future processes and systems.

At the business level, Service-oriented Architecture (SOA) [14] is nowadays the leader paradigm for software design. In manufacturing companies, as well as in many other domains, Enterprise Application Integration (EAI) at the business level is nowadays governed by SOA due to its principles of flexibility and reusability of loosely-coupled services. Based on these premises, the architecture that intends to integrate event-driven systems with enterprise applications at the business level needs to be service-based. Other solutions, usually designed to integrate legacy systems, are based on point-to-point connections or platform-dependent data exchange processes and are much less flexible, components cannot be reused in other processes and lead to high maintenance costs.

In order to achieve the desired flexibility and integration of the Event Driven Architecture (EDA) communication patterns that are typical in manufacturing, we have recently published an architectural reference model, namely the Manufacturing Service Bus (MSB) [15].

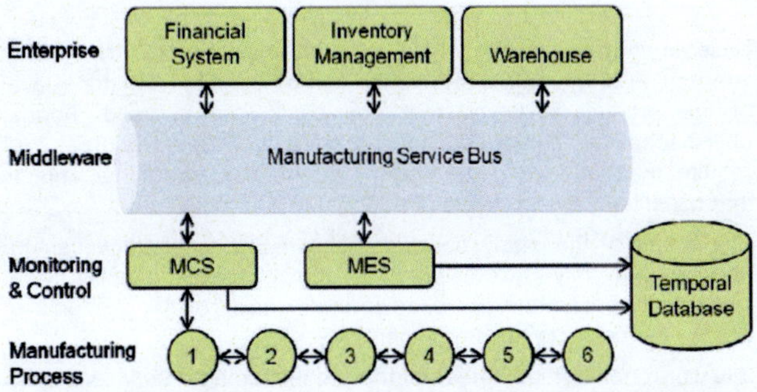

Figure 3: Service-based integration of manufacturing systems.

A prototype implementation of the MSB reference architecture [16] has proven flexibility by introducing service standard-based interfaces. The integration platform itself combines messaging, data transformation and intelligent routing services to connect distributed applications across a real manufacturing environment. These typical characteristics of an Enterprise Service Bus (ESB) along with the usage of an application-independent canonical format for manufacturing event messages support the extensibility and scalability of the architecture when updating services or integrating new services since the impact is limited to the transformation to and from the canonical format. Such an event-driven SOA for manufacturing environments is needed in order to ease the execution of CEP tasks that access data across multiple applications and information systems in the manufacturing company, such as updates of inventory and production orders in case of an RFID-based detection of an incorrect assembly.

Moreover, the loosely-coupling approach of a SOA-based integration increases the reuse of services and degree of flexibility, since services with well defined interfaces can be easily reused in other processes or applications. For example, in the case of a service that provides in real-time the RFID-based locations of all assembly parts moving along the assembly line may be used by an MCS, MES or even a process that feeds a visualization application.

3.5 Real-time visualizations

Production processes need to be flexible and agile. They are expected to use available resources as efficiently as possible and adapt to changes in order specifications, task priorities, and resource availability on short notice, all without being restricted by static productions plans set up weeks in advance. This requires "smart" IT systems that accept general input on what needs to happen from their human supervisor and work out the details of how it could happen based on the current state of the production and general guidelines and constraints.

While computer systems are unrivaled when it comes to quickly finding optimal solutions to clearly defined problems, they are sorely lacking in other aspects, such as making strategic decisions and handling exceptional, unforeseen cases. Additionally, even the most sophisticated computer system is still a machine that cannot take responsibility for its actions. It is therefore necessary to integrate the human controller into all relevant processes. This requires the controller to know and understand the current situation of the production process.

An efficient presentation of the state of the smart real-time factory requires both an automatic limitation to relevant information and a visual representation, as diagrams and images can convey information more efficiently and faster than lists or tables of values. As the state of the real-time smart factory is ever-changing, this visualization needs to be presented in real-time as well. This is not much of a challenge in technical terms, as even modern consumer hardware is capable of generating complex visual image sequences at interactive frame rates. The problems faced in the context of the real-time smart factory are therefore related to how all the necessary data can be gathered in time and how it needs to be visualized to give an efficient overview while still allowing access to relevant details.

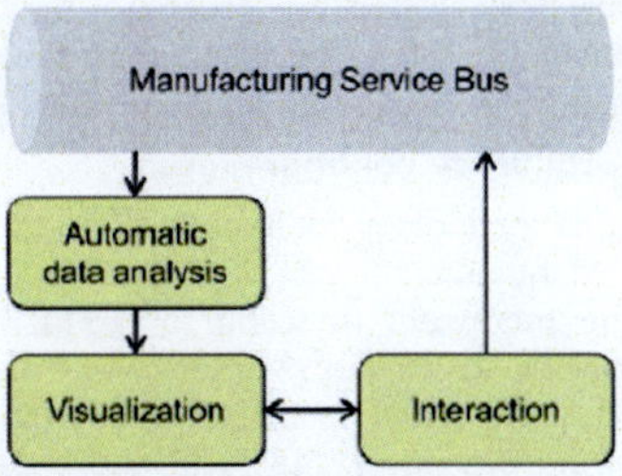

Figure 4: Integrating visualization into the manufacturing process.

These two problems are arguably the reason why visualization is used very sparingly on today's shop floors. Diagrams on sheets of paper posted to a wall show accident rates or compare goals and actual performance. These visualizations, of course, act more like reports on past time periods and are unsuitable for the assessment of the current state of a real-time factory. While many manufacturing machines today collect a wealth of data, this data is not brought together to a central visualization. A Service-oriented Architecture as described Section 3.4 can provide the necessary infrastructure for real-time data provisioning. To fully leverage this data, advanced visualization techniques need to be adapted to the special requirements of real-time smart manufacturing. The operator should not only be able to gain insight into the current state of the production processes, potential problems, and opportunities, but also to act upon these insights. This requires an appropriate interaction concept and a communications channel back across the service bus to the actual machines and devices. Figure 4 shows how visualizations can be integrated into the manufacturing process.

4 SERVICES

In the sub-sections below, we will highlight some of the services that can be offered by a Smart Real-Time Factory to its customers. It is worth noting that since the Smart Factory is able to generate immense amount of additional contextual data as compared to a traditional factory, most of the services that can be offered by establishing a Smart Factory are also informational in nature.

4.1 Production tracking for consumers

In the past, the postal service was a black box process for the customers. Customers used to send post and after doing so have had no information about what happened to their post. Whether the post was delivered successfully was only known to them once they receive a response from the person to whom the post was sent.

This lack of information was such a big market opportunity that saw the rise of companies like FedEx that strived on nothing but speed and informational transparency. Users of FedEx can find out exactly where the post is during each step of the delivery process and can find out at what time the post was delivered and who exactly received it.

Today's factories work in exactly the same way as the postal service used to work in olden days. Once a product is ordered, the user has no way of finding out if the product is planned for production, when it will be produced, and at what stage of production it currently is. This lack of information is especially troublesome for products that have a long manufacturing cycle, such as automobiles or airplanes.

The real-time product tracking component of a smart factory can not only be of use for the producer but can also be offered as an informational service to the customers, thereby allowing them to access real-time information about the state of assembly of their product. We firmly believe that the access of this information will have a similar disruptive affect on the manufacturing segment as the access of postal information had on the postal service.

4.2 Dynamic product re-configurations

Integration of RFID technology for real-time production monitoring with networked manufacturing can significantly reduce response time for both the producers (reaction to production errors) and consumers (changes to previous orders). Let us consider the assembly of a laptop at a computer manufacturer. The customer usually specifies the processor, the RAM, the hard disk, graphics card, screen size and other accessories. With networked manufacturing and production tracking service for consumers (Section 4.1), customers would now have the flexibility to change their order specifications midway through manufacturing. At any moment of time, the customer can track what parts of the laptop has been assembled together and what is still to be done and hence can change the specification of the parts that are still to be assembled together. The factory would respond to these changes in a similar way as it would respond to production errors and will consider the previous customer order to be incorrect and would change the production process such that the new and correct specification is assembled. Since the production environment will be able closely integrated there won't be any paperwork and human interference in order to carry out these changes, which would result in efficiencies obtained through automation. In this way, a company would be able to provide customer satisfaction through tailor-made products supplied reliably and efficiently with exceptionally short response times.

4.3 Production data traceability for producers

The services mentioned above are the ones that can be offered to the customers. However, a Smart Factory can also offer production data traceability to the company itself. Production data should be traced in order to enhance operational efficiency and to infer the quality of different product parts. As a matter of fact, RFID technology has already been used to achieve the above mentioned objectives.

Data traceability for operational efficiency and asset visibility:

Ngai et al [17] have reported a research study of deployment of an RFID-based traceability system in an aircraft engineering company based in Hong Kong. The company is responsible for maintaining and repairing aircrafts produced by both Boeing and Airbus for different airlines.

The company's aim was to enhance operational efficiency and asset visibility during the maintenance cycle. The purpose of the study was to identify, if an RFID-based traceability system would be suitable in tracking both assets (machines and tools required for maintenance) and repairable items (aircraft parts) during the maintenance cycle.

The results of the study concluded that RFID technology can be effectively used to increase operational efficiency and enhance asset visibility.

Data traceability for inference of product quality:

One of the ways in which cost reductions can be achieved in supply chain management is by managing information about a physical object throughout its life cycle [18]. This is one aspect for which RFID technology is being used in supply chain management by retailers like Wal-Mart and Tesco [19] [20]. One side affect of managing the information about a physical object throughout its life-cycle is that this makes it feasible to maintain the traceability information of the object and hence can be used to infer the quality of the product in which that physical object is used. Foods, pharmaceuticals and airline industry are examples of the areas that require maintenance of traceability information about the objects that constitutes the products [21], [22], [23].

The result of the study concluded that RFID technology can be used to effectively track history of an object throughout its life cycle, which in turn can later be used to infer quality of the products of which these objects become a part.

It is worth noting that instead of the product being a service system as is usually the case; the Smart Factory is a service system for both the consumers and the producers. As is obvious from production data traceability service, the smart factory will be able to enhance the customer experience by providing traceability information about the product and its parts throughout the life-cycle of the product.

5 SUMMARY

In this paper, we elaborated on how a Factory can be made "smart" by deploying RFID technology. Furthermore, we also presented the different components/applications that need to be deployed in a smart factory and the services that the factory can offer to both the producers and the consumers. In the end we would like to summarize the benefits that an organization can achieve by deploying a Smart Real-Time Factory:

1. Improved lead time: By deploying a smart factory, a company can keep track of the status of each production order. In case an item is delayed due to scheduling, is not being assembled according to plan, or any other similar mishap, an alert will be sent, and immediate action will be taken by the factory. The smart factory will enable management to solve problems rather than wasting time, trying to find the problems.
2. Competitive differentiation: The Company will gain a competitive advantage over its competitors allowing customers to view the state of production in real-time and to configure their products even during assembly, if they so desire.
3. Breakthrough productivity by automation: The Smart Factory will completely automate the production process. The factory will be able to provide operators with means to keep track of all the production orders. The operators will receive instant warnings in case something goes wrong. The management will be able to see higher level status updates in order to understand the status of production. All these features will result in improved productivity of the production process.
4. Reduce human errors in production: All the product related information in a smart factory will be tracked using RFID technology. Therefore, the smart factory would be able to inform operators in real-time about the errors that they might be committing, and hence can prevent them from doing so.
5. Improved inventory management: As the RFID system provides improved visibility of each item, the inventory levels for expensive product parts can be tracked and kept at a bare minimum, so that the company does not have to invest and tie-up a huge amount of capital on purchasing these expensive parts a long time before use. If the suppliers of the factory can also implement the Smart Factory, the purchase decisions can be automated as well.
6. Reduce manpower and manual data recording: Since most of the operations carried out at the Smart Factory would be automated. This will reduce the need for manual tasks such as

data recording and tracking. This in turn will reduce the manpower needed to perform such tasks.

7. Real time tracking: The Smart Factory will provide real time information on the status of each and every production order. The status information will include full history of a particular product, the parts that have been assembled, the tasks that still needs to be done and so on. This detailed production information can later be further analyzed in order to make the production processes more efficient.
8. Reduction of production parts loss: During production, the production parts are sometimes lost temporarily. However, in a Smart Factory, all the product parts and mobile assets would be tracked by RFID technology. This will eliminate temporary losses of production parts and mobile assets such as machine tools, since the system will be able to notify operators about the location of the part. Even if the part can't be tracked at the moment, the system will be able to tell the last tracked location of the part, which will result in a quick localization of the part.
9. Improved customer relationships: By offering real-time tracking and dynamic product re-configuration services to the customers, the Smart Factory will be able to dramatically improve customer relationships and can attract hordes of new customers.

From the benefits mentioned above it is clear that the Smart Factory not just produces consumer products, but also generates and archives real-time information about these products that enhances the manufacturing process and aids in future maintenance.

6 REFERENCES

[1] Baines, T.S., Lightfoot, H.W., Evans, S., Neely, A., Greenough, R., Peppard, J., Roy, R., Shehab, E., Braganza, A., Tiwari, A., Alcock, J.R., Angus, J.P., Bastl, M., Cousens, A., Irving, P., Johnson, M., Kingston, J., Lockett, H., Martinez, V., Michele, P., Tranfield, D., Walton, I.M., Wilson, H. (2007): State-of-the-art in product service systems, in: Proceedings of IMechE 221: Part B: Journal of Engineering Manufacture.

[2] Spath, D., Demuß, L. (2006): Entwicklung hybrider Produkte – Gestaltung materieller und immaterieller Leistungsbündel, in: Bulling, H.-J., Scheer, A.-W., (ed.), Service Engineering, pp. 463-502, Berling, Heidelberg, New York 2003.

[3] Johnson, D. (2002): RFID tags improve tracking, quality on Ford line in Mexico, in: Control Engineering, 49(11), pp. 16-16.

[4] Brewer, A., Landers, T. (1997): Radio Frequency Identification: A Survey and Assessment of the Technology, in: Technical Report, University of Arkansas, Department of Industrial Engineering.

[5] Zhekun, L., Gadh, R., Prabhu, B.S. (2004): Applications of RFID technology and smart parts in manufacturing, in: Proceedings of DETC' 04: ASME 2004 Design Engineering Technical Conference and Computers and Information in Engineering Conference, Salt Lake City, Texas.

[6] Baudin, M., Rao, A. (2005): RFID applications in manufacturing. http://www.mmt-inst.com/.

[7] Lucke, D., Constantinescu, C., Westkamper, E. (2008): Smart Factory – A Step Towards the Next Generation of Manufacturing, in: Manufacturing Systems and Technologies for the New Frontier, Springer.

[8] Ahmed, N., Kumar, R., French, R.S., Ramachandaran, U. (2007): RF2ID: A reliable middleware framework for RFID deployment, in: IEEE Iinternational Parallel and Distributed Processing Symposium IPDPS, IEEE Computer Society Press.

[9] Prabhu, B. S., Su, X., Ramamurthy, H., Chu, C-c., Gadh, R. (2005): WinRFID A Middleware for the enablement of Radio Frequency Identification (RFID) based Applications, in: Mobile, Wireless and Sensor Networks Technology, Applications and Future. John Wiley and Sons, Inc.

[10] The Basics of RFID Technology, http://www.rfidjournal.com/article/articleview/1337/1/129/.

[11] Hassan, T., Chatterjee, S. (2006): A Taxonomy for RFID, in: HICSS '06: Proceedings of the International Conference on System Sciences, IEEE Computer Society.

[12] Hameed, B., Khan, I., Dürr, F., Rothermel, K. (2010): An RFID Based Consistency Management Framework for Production Monitoring In a Smart Real-Time Factory, in: Proceedings of 2nd International Conference on Internet of Things (IoT-2010), IEEE Computer Society Press.

[13] Wang, F., Liu, S., Liu, P. (2009): Complex RFID event processing, in: The VLDB Journal – The International Journal on Very Large Data Bases, volume 18 issue 4, August 2009.

[14] Erl, T. (2005): Service-Oriented Architecture: Concepts, Technology, and Design, Prentice Hall International

[15] Minguez, J.; Lucke, D.; Jakob, M.; Constantinescu, C.; Mitschang, B.; Westkämper, E. (2010): Introducing SOA into Production Environments: The Manufacturing Service Bus. In: Proc. of the 43rd CIRP International Conference on Manufacturing Systems (ICMS), pp. 1117-1124, Vienna, Austria.

[16] Minguez, J., Ruthardt, F., Riffelmacher, P., Scheibler, T., Mitschang, B. (2010): Service-based Integration in Event-driven Manufacturing Environments. In: Proceedings of the 11th International Conference on Web Information System Engineering (WISE 2010), 1st International Symposium on Web Intelligent Systems & Services (WISS 2010). *To appear.*

[17] Ngai, E. W. T., Cheng, T. C. E., Lai, K. H., Chai, P. Y. F., Choi, Y. S., and Sin, R. K. Y. (2007): Development of an RFID-based Traceability System: Experiences and Lessons Learned from an Aircraft Engineering Company, in: Proceedings of POMS – International Journal on Production and Operations Management, vol. 16 (5), pp. 554–568.

[18] Inaba, T. (2009): Inference of Product Quality by usinig RFID-enabled Traceability Information: A Study on the US Pharmaceutical Supply Chain, in: 9th IEEE International Conference on RFID, pp. 298-305.

[19] Roberti, M. (2005): EPC Reduces Out-of-Stocks at Wal-Mart. http://www.rfidjournal.com/article/view/1927.

[20] Prodromou, A. (2003): Tesco Deploys Class 1 EPC Tags. http://www.rfidjournal.com/article/view/587.

[21] Koh, R., Schuster, E.W., Chackrabarti, I., Bellman, A. (2003): Securing the Pharmaceutical Supply Chain, in: Auto-ID Labs White Paper Series.

[22] Jones, P. (2006): Networked RFID for use in the Food Chain, in: ETFA '06 IEEE Conference on Emerging Technologies and Factory Automation.

[23] Patkai, B., Theodorou, L., McFarlane, D., Schmidt, K. (2007): Requirements for RFID-based Sensor Integration in Landing Gear Monitoring – A Case Study, in: Auto-ID Labs White Paper Series.

[24] http://autoidlabs.org.uk/realtimerfid.html.

Business Model Elements for Product-Service System

Ana Paula B Barquet[1], Vitor P. Cunha[1], Maicon G. Oliveira[1], Henrique Rozenfeld[1]
[1] Escola de Engenharia de São Carlos, Universidade de São Paulo, São Carlos, Brazil

Abstract
Companies are seeking for new ways to do business. The Product-Service System concept can be a potential option since it meets requirements regarding sustainability and also introduces a type of business in which the focus is placed on delivering function instead of providing value thought tangible products. However, the deployment of PSS involves a reassessment of companies' business models. Therefore, this paper identifies and classifies the characteristics of the PSS business model that need to be addressed for its implementation, which was carried out using literature review and considering three types of PSS and nine elements of a business model.

Keywords:
Product-Service System; Business Model; Elements

1 INTRODUCTION

Companies are looking for new ways to do business since the current market scenario is characterized by an increased competition. In this context, the Product-Service System (PSS) concept can be an alternative to companies that want to innovate because it allows selling not only products but also additional services that supplement customers' needs [1].

It is important to consider that the implementation of PSS implies major changes in business operations and thinking, such as: how a company creates value for its customers? How to produce and distribute these values? How to interface with clients and supply chain? [1] Consequently, companies need to adjust the characteristics of their business models to take advantage of PSS benefits entirely [2-4].

Despite of its importance, the literature has little information about shifting toward a PSS business model [5] and one of the first steps required in this sense is the identification and classification of key PSS characteristics. Therefore, this work aims to analyze a PSS business model according to two criteria: the elements of business models [6] and the PSS types.

The PSS business model indicates which characteristics of traditional business models would require modifications and gives a deep understanding about PSS characteristics.

The next session describes the methodology adopted in this work. Following, the literature review section presents the theory concerning PSS and business model used to underpin this study. The subsequent section embraces the identification, analysis and classification of PSS characteristics. Finally, the conclusion section highlights the results achieve by the research, indicating contribution to practice and theory as well as listing future research opportunities.

2 METHODOLOGY

This research applied a systematic literature review to collect and evaluate required information. Three steps adapted from Biolchini et al. [7] were performed: problem formulation, database definition and collection, and data analysis and evaluation.

The topics PSS and business model comprised the focus of the review. Regarding to PSS subject, it was investigated its concepts, characteristcs and types. On the other hand, business model literature were analysed to identify which elements compose a business model.

After identifying PSS characteristics and business model elements in literature, it was carried out a crossing analysis to classify the characteristics of each PSS type that should be considered in each element of a business model.

3 LITERATURE REVIEW

3.1 Product-Service System (PSS)

The product-service system is the result of a strategy that shifts business focus from designing and selling only physical products to deliver a system of products and services that together are capable of meeting specific customers' demands [8].

According to Baines et al. [9], the PSS integrates a product and service into a system which provides the same necessary functionally but reducing the environmental impact.

Most of PSS definitions emphasizes the "sale of use" instead of "selling the product", indeed customer pays for using instead of buying the asset [9]. Furthermore, the focus is not only on the products and services but also in how needs and desires are identified and how organizations define their responsibilities and relationships [8].

In traditional business models, the customer acquires a product, becoming responsible for monitoring its performance, providing assistance and assuring its adequate disposal. On the other hand, in the PSS concept, the producer earns from customers use of a provided function. Thus the ownership of the product is not necessarily transferred to customer and the responsibility to maintain the product along its life cycle and to dispose it can remain with the producer [9].

J. Hesselbach and C. Herrmann (eds.), *Functional Thinking for Value Creation: Proceedings of the 3rd CIRP International Conference on Industrial Product Service Systems, Technische Universität Braunschweig, Braunschweig, Germany, May 5th - 6th, 2011*,
DOI 10.1007/978-3-642-19689-8_58,

The most common categorization available in literature divides the PSS into three types [1] [9] [10] [11] [12]: product-oriented services (POS), use-oriented services (UOS) and result-oriented services (ROS); as shown in Figure 1. These types diverge basically from the relationship between the producer and the customer.

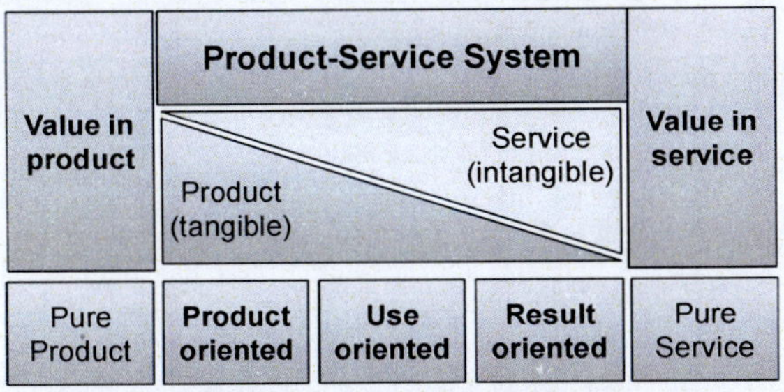

Figure 1: Types of PSS [10].

- Product-oriented services (POS): This involves the sale of a product in the traditional way, in which the customer retains the ownership of the product, although including some associated services. For instance: after-sales service to ensure functionality and durability of the product (maintenance, repair, reuse, recycling, training and consulting). In this case the company is motivated to introduce a PSS to minimize costs of use throughout product lifecycle;
- Use-oriented services (UOS): The product is owned by the producer, which sells the use or functions of a product (e.g., leasing, sharing, etc.). In this case, the company is motivated to create a PSS to maximize the use of the product, to meet demand, extend product life and reuse materials;
- Result-oriented services (ROS): This involves the sale of a result or competence rather than a product. Companies offer a mix of personal service and, when the result is delivered in a product, the PSS provider maintains its ownership and the customer only pays for the results.

The PSS concept has potential to benefit both companies and customers. According to Baines et al. [9], the main advantages for each one are:

Customers:

- More customized offers and higher quality;
- New functionalities and combinations of products and services to better suit customers' needs;
- Responsibility for monitoring and end-of-life tasks transferred to producer;

Companies:

- New market opportunities and competitive advantages;
- Alternative to standardization and mass production;
- Improvement in the total value delivered for the customer through increasing service elements;
- Access to data about product performance during the use phase;

The PSS may also result in a number of environmental benefits. As companies become responsible for the whole life-cycle of their products, they are encouraged to take them back in their end of life, remanufacture them, and later make them available to the market again. Therefore, less waste is discarded or deposited, thereby reducing the consumption of raw materials and energy. Additionally, the inputs required for remanufacturing are often smaller than the inputs used for manufacturing a new product [13].

Despite of the cited benefits, the adoption of a PSS brings great challenges for companies since it requires the development of new relationships with stakeholders [8]. Such requirements should be considered in a company business model. In this sense, the next section presents information about business models.

3.2 Business Model

Regarding its concept, Shafer; Smith and Linder [14] define business model as a representation of a firm's underlying core logic. According to Elbers [15], business models are the way a company creates, delivers and captures value, resulting from strategic choices and in accordance with their network value.

The main contribution of developing a business model is the creation of practices that help companies to capture, understand, design, analyze, and change their business logic [6]. Thus, it has being considered a powerful tool to communicate strategic choices [14].

A structured business model can help decrease the business logic complexity since it represents the company strategy in an understandable form, makes possible to visualize the way a company can make value from its offer, and explain how to implement. Therefore, it allows organization to have a holistic view of the business, clarifying important issues and relationships. The business model can be also considered an abstract representation of the business strategy [16].

Although the term business model is widely used for many authors, there is no common sense for the concept [17] [18] [19]. Osterwalder [16] made an intensive literature review about business model concepts, attributes and their relationships. He, in a posterior research, defined business model as a conceptual tool composed of objects, concepts and their relationships that enable to express simplified description and representation of the business logic of a firm [3]. Another definition comes from Osterwalder and Pigneur [6]. These authors mentioned that a business model describes the rationale of how a company creates, delivers and captures values. They also developed the Canvas business model, which is a tool created for describing, analyzing and designing business models.

3.3 Business Model Canvas

The Canvas business model was applied and tested in many organizations (e.g. IBM and Ericsson), being successfully used to easily describe and manipulate business models to create new strategic alternatives. Moreover, this model presents a clear description of the elements comprising a business model. This approach is used as reference in this paper since the elements that compose the model are described in detail. Defining the elements (also referred as building blocks) which compose a business model is the first step to plan the business and help companies understand and describe the business logic of their firm [16].

The Canvas model comprises nine elements, which cover the four main areas of a business: customers, offer, infrastructure and financial viability. These elements, depicted in Figure 2, are following described.

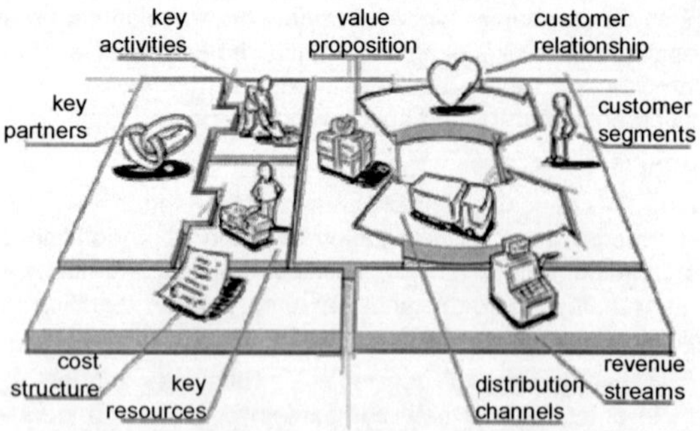

Figure 2: Elements of the canvas business model [6].

I. Customer Segments (distinct segments with common needs, common behaviors, or other attributes): defines the different groups of people or organizations an enterprise aims to reach and serve. Once this decision is made, a business model can be carefully designed around a strong understanding of specific customer needs.

II. Value Propositions: It seeks to solve customer problems, satisfy customer needs and describes the bundle of products and services that create value for a specific Customer Segment.

III. Channels: describes how a company communicates with and reaches its Customer Segments to deliver a Value Proposition, comprising a company's interface with customers.

IV. Customer Relationships: describes the types of relationships a company establishes and maintain with specific Customer Segments.

V. Revenue Streams: result from value propositions successfully offered to customers, this element represents the cash a company generates from each Customer Segment.

VI. Key Resources: are the assets required to offer and deliver the previously described elements.

VII. Key Activities: are performed though the Key Resources to offer and deliver the previously described elements.

VIII. Key Partnerships: describes the network of suppliers and partners that make the business model work. Some activities are outsourced and some resources are acquired outside the enterprise. Companies create alliances to optimize their business models, reduce risk, or acquire resources.

IX. Cost Structure: describes all costs incurred to operate a business model. Costs can be calculated relatively easily after defining Key Resources, Key Activities, and Key Partnerships.

After summarizing the elements that compose the Canvas business model, in the next session it is identified the PSS characteristics related to each of these elements.

4 PSS BUSINESS MODEL

4.1 Identification of PSS characteristics related to the business model elements

It is important to note that Goedkoop et al. [20] have already considered business issues when they defined the PSS. According to them, the PSS concept was defined as a system of products, services, infrastructure and network support that continually strives to be competitive, satisfy customer needs and result in less environmental impact than traditional business models. Baines et al. [9] have also included business issues when they suggested some features for a development of the PSS, for instance:

- Mind change of product to system;
- Increased involvement between the customer and the organization since the beginning of development;
- Design, production and delivery case-by-case, from the perspective of the customer;
- Increase the involvement and responsibility of the company in other phases of the product life-cycle;
- Higher involvement of all stakeholders;

Therefore, one of the main challenges faced by companies which adopt the PSS is to identify specific requirements in the business model. Thus, using results from literature review, the characteristics of the PSS are sorted out across each element of Canvas model.

Concerning to Value Propositions, the value offered in a PSS is based on an integration of a product and service. Then, it can differ basically on the relationship between the producer and the customer, such as: product-oriented service, use-oriented services and result-oriented services [1] [11] [10] [12] [9].

With regards to Customer Segments, it is important to take into account what kind of ownership ideas this specific target group has [11], cultural and regional differences and consumer habits, and also behavior and values [1].

For Channels, sales and retail departments should define how the PSS should be offered and priced to be more attractive than buying a product based option [11]. Training of retail and sales personnel is often necessary as well as changes to pull PSS adoption [5]. It is also important to make careful argumentation in marketing campaigns [11], clarifying customers about the PSS usage [5].

Another key aspect of a PSS is the use of new approaches for Customer Relationship [21]. Therefore, it is necessary to create added value and delivery it through direct relations and intensified contacts with customers [5], which enables the development of long-term relationships instead of short-term and transaction-based relationship found in the traditional "selling products" context [5] [21].

The PSS also gives opportunity to raise revenue through improving the function offered by the PSS provider [22]. However, when it involves more than one company, the revenue distribution needs to be well managed among partners to avoid misunderstandings [13].

Regarding to Key Resources, particularly human resources, PSS providers have to make considerable investments into human asset [23]. New competences about customers' need to be developed, people trained and sometimes additional personnel recruited [5] [24]. It is also required a fundamental shift in corporate culture and market engagement, which requires time and resources [23] [24].

Instead of focusing on the activities related to a physical product, PSS providers should focus on the Key Activities related to their customers. Even when the product provides a core function, more essential activities are involved, which may happen before, during and after the product use [23] [24].

Due to the complex nature of the value proposition and the mix of competencies needed to implement a PSS, a Partnership network is essential. To structure a PSS network, it is necessary to identify what are the required core competences and which actors have the whished competences [13]. In the PSS business model, the relationship between producer and stakeholder is recognized with a wide scope and has a considerable impact on the supply chain structure [25] [5].

A challenge concerning to PSS is how to manage the Cost Structure and how to set a suitable price for the PSS [26]. The financial and accounting functions may need to adapt their practices to new profit centers. The time scale of financial flows changes considerably from an almost immediate return of the capital to an extended use period [5]. That means the PSS provider needs to hold the necessary financial resources or to have the financing partners support to bridge this period [13]. Payments might be based on the availability of the PSS, frequency of use or final result [23]. When the function of products is sold, cost structures may need to be restructured to support different cash-flow requirements. So, in a PSS, the payback period for the value delivered can be longer than in a pure product-selling context.

The identification of PSS characteristics according to business models elements enable to observe some distinctions comparing to the characteristics of traditional business model. Afterwards, the three types of PSS (product-oriented services, use-oriented services and result-oriented services) are also classified based on elements of Canvas business model in order to understand its differences.

4.2 Classification of PSS characteristics

Table 1 sorts the three types of PSS according to each business model element.

	Product-oriented services (POS)	Use-oriented services (UOS)	Result-oriented services (ROS)
Customer segments	Clients who wants or need to have the ownership of the product [1].	Clients who wants to make low initial investments [10] [27] and can accept not to be the owner of the product [1].	Clients who prefer to not own physical products (means assets on their balance sheet) [28]. Clients who need flexibility to change [28].
Value Propositions	Sales of products with extra services added. More tangible value [10]. Product is the core to the system [29] and clients have their ownership [9].	Product's ownership with the provider [10]. The user buys access to a product [11]. Deliver of functional capability to the customer [29] [9].	Deliver of functional result [27] [29] [9], maybe with no product involved. High level of abstraction [10].
Channels	Indirect interface with end client [1]. Marketing has Superior product specifications/qualities [1].	Transitions can occur quickly and are easier to manage, e.g., within a lease contract rather than having to buy and sell a complex and expensive product each time [1].	Direct interface with end client [1]. Marketing need to develop trustworthiness, reliability, knowledge/experience [1].
Customer Relationships	Transactional based [9] [30].There are no major changes in the client base [11].	Blend of transitional and relationship [30]. Enables long term business relation [27].	Relationship based [9]. Longer terms contracts and strong interaction with customers [10] [27] [28] [9].
Revenue Streams	Payment in the form of the sales price [31]. Move from the commoditization of the product [1], e.g., provision of additional services may lead to a more efficient use of products [11].	Payment is based on availability [10]. The same artefact may be shared or used by different users, leading to a lower capital needed in the system [11].	Payment is based on per unit time or unit use [10] and is distributed along the use phase [28]. Need a different cash-flow patterns, since there is different time-scales and amounts [1].
Key Resources	Key resources are material, information and energy. More efficient use of materials and human resources [10] [27] [9]. Technology/machinery and employees viewed as company's critical capital [1].	More use of human resources [10], workers with high level of product knowledge are required [9].	Need of information and kwon how intensive exchange [32]. Workers with high level of communication skills [9] and persuasion are required [1]. Employees as critical capital, unlike equipment [1].
Key Activities	Large production department/machinery [1]. Importance of product and process development [1]. More centralized decision-making [1].	Need for new provider activities [10] since responsibility shifted to him [27]. Need to manage services: which services have to be available at which point and to what extent [27].	New activities related to how to design product and service according to the function required, and how extract customer requirements [29].
Key Partnerships	Outsourcing of services. Incentives and effective control systems are required [30]. Basic information exchange with the partners [30].	Long-term collaboration with service suppliers [30]. Information exchange with the partners includes: demand forecasts; technical training and feedback, complaints and satisfaction of customers [30].	Need to organize outsourced tasks efficiently [10], to reach cooperation [32] and strategic integration between the partners [30]. Information exchange with the partners includes: knowledge, training, integration of capabilities [30].
Cost Structure	Traditional production costs, with additional material and human resource costs [10].	The need for capital is high [10] since various costs are shifted to the provider [11].	Risk premium (responsibility of total cost risks) [28] and/or uncertainty related to the system costs [10] [1]. There is a need to determine of all lifecycle costs [27].

Table 1: Characteristics of business model elements according to the type of PSS.

The PSS business model differs clearly from the traditional model in the way to propose value. In the PSS, the value is no longer based on a tangible product, but in the relationship between producer and customer and how to provide a set of products and services that meet the needs required.

From Table 1 is possible to identify differences between the elements related to each PSS type. In case of Value Propositions, the variations focus on the level of the offer tangibility and for Customer Segments, on the aspects of ownership.

Concerning to Channels, the interface with the client can vary between direct and indirect, which influences the marketing behavior. Customer Relationship mainly varies between transitional and relationship base. In Revenue Streams, the variations of payments (based on product sale, availability, per unit or use) lead to new ways to structure it. It is also true for Cost structure, since PSS provider responsibility for cost and risks alter according to PSS type.

Key Resources differs basically in terms of material and capabilities. It can be observed that Key Activities has the majority of variations between the three types since it encompasses several topics, e.g. product development, management of services, design of the function required. Finally, Key Partnership focuses on the exchange of information and degree of cooperation between partners, which can vary between basic outsourcing and strategic integration.

In this sense, it can be observed that the results show that the elements of PSS business models vary with the type of relationship between producer and consumer. In addition, the classification proposed can be a good practice to help companies do decide which type of PSS fits to their business.

5 CONCLUSION

The analysis of PSS business model characteristics is an important step for companies which want to shift the offer of only product to a combination of product and services. This paper presents the primary characteristics of three types of PSS (product-oriented services, use-oriented services and result-oriented services) classified according to the nine elements that comprises Canvas business model.

This result support companies that aim to implement a PSS to better understand its constraints and to define a business model which fits to its purposes. Furthermore it can also support those which already offer a PSS to make required changes or to improve its current performance.

Considering the characteristics of a PSS and the results of this work, it was noted the necessity to consider two new elements in a PSS business model: the environmental element and the innovation element. The environmental element is addressed in the PSS concept, since this concept has the potential for impact environmental reduction. Also, the ability to innovate is a PSS benefit due to better client contacts and higher network cooperation.

These are going to be addressed in next steps of a research project focused on support PSS implementation, in which empirical studies about each PSS type are also in development.

6 ACKNOWLEDGMENTS

The authors would like to thank the following institutions for their financial support: CAPES (www.capes.gov.br), CNPq (**www.cnpq.br**).

7 REFERENCES

[1] Tischner, U.; Verkuijl, M. (2002): Suspronet Report: First Draft Report Of Pss Review. Suspronet.

[2] Kuo, T. C.; Ma, H.; Huang, S. H.; Hu, H. A.; Huang, C. S. (2009): Barrier Analysis For Product Service System Using Interpretive Structural Model. The International Journal Of Advanced Manufacturing Technology, V. 49, N. 1-4, P. 407-417.

[3] Osterwalder, A.; Pigneur, Y.; Tucci, C. L. (2005): Clarifying Business Models: Origins, Present, And Future Of The Concept. Business, V. 15.

[4] Sakao, T.; Sandström, G. Ö.; Matzen, D. (2009): Framing Research For Service Orientation Of Manufacturers Through Pss Approaches. Journal Of Manufacturing Technology Management, V. 20, N. 5, P. 754-778.

[5] Mont, O. (2004): Product-Service Systems: Panacea Or Myth? Thesis (The International Institute For Industrial Environmental Economics), Lund University, 259p.

[6] Osterwalder, A.; Pigneur, Y. (2009): Business Model Generation. Amsterdam: Self Published.

[7] Biolchini, J.; Mian, P. G.; Natali, A. C. C.; Travassos, G. H. (2005): Systematic Review In Software Engineering. Technical Report Rt – Es 679/05. Rio de Janeiro: Systems Engineering And Computer Science Department, Universidade Federal do Rio de Janeiro.

[8] Manzini, E.; Vezzoli, C. (2002): Product-Service Systems And Sustainability. Opportunities For Sustainable Solutions. Paris: Unep, N. 254.

[9] Baines, T. S.; Lightfood, H.W., Evans, S., Neely, A., Greenough, R., Peppard, J., Roy, R., Shehab, E., Braganza, A., Tiwari, A., Alcock, J.R., Angus, J.P., Bastl, M., Cousens, A., Irving, P., Johnson, M., Kingston, J., Lockett, H., Martinez, V., Michele, P., Tranfield, D., Walton, I.M., Wilson, H. (2007): State-Of-The-Art In Product-Service Systems. Proceedings Of The Institution Of Mechanical Engineers, Part B: Journal Of Engineering Manufacture, V. 221, N. 10, P. 1543-1552.

[10] Tukker, A. (2004): Eight Types Of Product–Service System: Eight Ways To Sustainability? Experiences From Suspronet. Business Strategy And The Environment, V. 13, N. 4, P. 246-260.

[11] Tukker, A.; Tischner, U. (2004): Product-service development as a means to enhance competitiveness and eco-efficiency: final report of SUSPRONET. Network, p. 247.

[12] Tukker, A.; Tischner, U. (2006): Product-services as a research field. Journal of Cleaner Production, v. 14, p. 1552-1556.

[13] Mont, O. (2002): Clarifying The Concept of Product Service-Systems. Journal of Cleaner Production, V. 10, N. 3, P. 237-245.

[14] Shafer, S.; Smith, H.; Linder, J. (2005): The Power Of Business Models. Business Horizons, V. 48, N. 3, P. 199-207.

[15] Elbers, B. F. (2010): Designing Innovative Business Models – A Methodology for full Structured Business Model Innovation. Management Summary.

[16] Osterwalder, A. (2004): The Business Model Ontology - A Proposition in a Design Science Approach. Thesis. Universite de Lausanne, Sciences Politiques de L'université de Lausanne Business, 172 p.

[17] Camponovo, G.; Pigneur, Y. (2003): Business Model Analysis Applied To Mobile Business. Business.

[18] Onetti, A., Zucchella, A.; Jones, M. V.; McDougall-Covinet, P. P. (2010): Internationalization, Innovation And Entrepreneurship: Business Models For New Technology-Based Firms. Journal Of Management And Governance (JMG).

[19] Cortili, M.; Menegotto, M.; Insubria, U. D. (2010): Internationalization And Business Model Decisions: A Business Case In Mobile Telecommunication Industry, V. 6.

[20] Goedkoop, M.; Van Haler, C., Te Riele, H., Rommers, P. (1999): Product Service-Systems, Ecological And Economic Basics.

[21] Williams, A. (2006): Product-service systems in the automotive industry: the case of micro-factory retailing. Journal of Cleaner Production, v. 14, n. 2, p. 172-184.

[22] Mont, O. (2000): Product-Service Systems Final Report. The International Institute Of Industrial Environmental.

[23] Tan, A. R.; McAloone, T. C. (2006): Characteristics of strategies in product / service-system development. Assessment, p. 1-8.

[24] Cook, M.; Bhamra, T.; Lemon, M. (2006): The transfer and application of Product Service Systems: from academia to UK manufacturing firms. Journal of Cleaner Production, v. 14, n. 17, p. 1455-1465,

[25] Kansai, I.; Paper, D. What is Product-Service Systems (PSS)? A Review on PSS Researches and Relevant Policies (2006): Interface, N. 1, P. 0-23.

[26] Sundin, E.; Sandström, G. Ö. ; Lindahl, M.; Rönnbäck, A. O.; Sakao, T.; Larsson, T. C. (2009): Challenges For Industrial Product / Service Systems: Experiences From A Learning Network Of Large Companies. Cycle, April, P. 1-2.

[27] Steven, M.; Soth, T. (2010): Development Accompanying Calculation - How To Calculate IPS² Costs during the Early Development Phase? Proceedings of the 2nd CIRP IPS² Conference.

[28] Lindahl, M.; Sakao, T.; Rönnbäck, A. Ö. (2009): Business Implications of Integrated Product and Service Offerings. Methods, P. 1-2.

[29] Yang, L.; Xing, K.; Lee, S. (2010): A New Conceptual Life Cycle Model for Result-Oriented Product-Service System Development. , P. 23-28.

[30] Rapaccini, M. (2010): Service Sourcing And Supplier Relationships In Product-Centric Servitization : A Preliminary Study.

[31] Lerch, C.; Selinka, G. (2010): Dynamics of Business Models Long-Ranging Impact Assessment of Business Models in the Capital Goods Industry.

[32] Martinez, V.; Bastl, M.; Kingston, J.; Evans, S. (2010): Challenges in Transforming Manufacturing Organisations into Product-Service Providers. Journal of Manufacturing Technology Management, V. 21, N. 4, P. 449-469.

Organizational Buying Behavior in Case of IPS²

Mario Rese[1], Kira Maiwald[1]

[1] Marketing Department, Ruhr-University Bochum, Bochum, Germany

Abstract

Compared to common selling processes of machines, selling processes of IPS² seem to be more complex. The two responsible causes are: (i) the higher degree of newness, and (ii) the risk associated with the IPS² buying situation. Among other things the complexity is reflected in the fact that more people are involved in the buying decision. This characteristic will be the focal point of this paper. The aim is to figure out to what extent the buying behavior in case of an IPS² differs from organizational buying behavior of non-IPS² settings and how customers should be convinced of IPS² advantages.

Keywords:

Buying Center; Resistance to Change; Promoter-Opponent-Model

1 INTRODUCTION

An industrial product service system (IPS²) is an integrated developed system of interrelated product and service components. What is special about this system is that the relation of product and service components can be configured individually and adjusted over time [1]. IPS² solutions only appear in Business-to-Business environments [2], where specific characteristics of buying behavior have to be taken into account.

One main difference that distinguishes Business-to-Business from Business-to-Consumer transactions is the multi-personality of industrial purchasing [3, 4]. In industrial buying processes various people are involved. These people represent the so-called buying center which typically consists of different members occupying in the following five roles: users, buyers, influencers, deciders, and gatekeepers [3]. An expansion of the concept about a sixth role, i.e. initiators, has already been made [5]. Not only must the need of one individual be met, but the interests and goals of several people and the organization as a whole [3]. A supplier within a Business-to-Business context needs to identify all individuals that take part in the decision process. So far, organizational buying behavior has been analyzed and empirically tested for purchase decisions concerning mere products or services [6]. Although the importance of product related services and their effect on industrial purchase decisions is widely known, there is still need for further research [7]. Apart from this an IPS² is a new type of problem solution and therefore implications of previous research cannot just be transferred to IPS² buying situations. On that account, the buying center concept has to be reconsidered and adapted to IPS² buying situations. This article will point out that the buying process of an IPS² is more complex than a common industrial buying situation.

Our first research question concerns the size of the buying center. In case of IPS² there may be other buying center roles apart from those already known. If all buying center members are identified, their needs and interests in this context have to be analyzed. As the IPS² implementation that is indispensably connected to the IPS² purchase probably triggers change processes within the buying organization, some individuals are affected positively by the decision, some negatively. That is why the IPS² buying situation will polarize the buying center members into different interest groups. While a group of promoters will advocate the change process, opponents will try to hinder or delay it [8]. Therefore, the second research question concerns the combination of buying center roles with promoter and opponent roles. In order to successfully sell IPS² solutions it is necessary to know which buying center member is in favor of and which member is against the IPS² purchase.

2 COMPLEXITY OF IPS² BUYING PROCESSES

In accordance with Robinson, Faris, and Wind (1972), buying situations can be distinguished by the newness of the problem, the information requirements, and the consideration of new alternatives. The resulting buyclasses are called new task, modified rebuy, and straight rebuy. The new task situation is the most complex one: while the newness of the problem is high, the information requirements are maximal, and the consideration of alternatives is ranked as important [4]. Empirical validations show no significant difference in the characteristics of new task and modified rebuy [9], so that we only distinguish between new task and rebuy situations.

We characterize complex organizational buying behavior as

- the buying center becoming larger,
- buying center members will be of higher organizational status and authority,
- the potential for conflict between buying center members will increase,
- information search will be active and more information sources will be conducted,
- formal decision rules will be used
- buyer-supplier relationships and communication networks will become more important, and suppliers who offer proven solutions will be favored [10].

However, this characterization does not consider IPS² buying situations which might be even more complex than a new task buying situation of mere product or service purchases. For this reason we will extend this characterization by IPS² in the following.

J. Hesselbach and C. Herrmann (eds.), *Functional Thinking for Value Creation: Proceedings of the 3rd CIRP International Conference on Industrial Product Service Systems, Technische Universität Braunschweig, Braunschweig, Germany, May 5th - 6th, 2011*,
DOI 10.1007/978-3-642-19689-8_59,

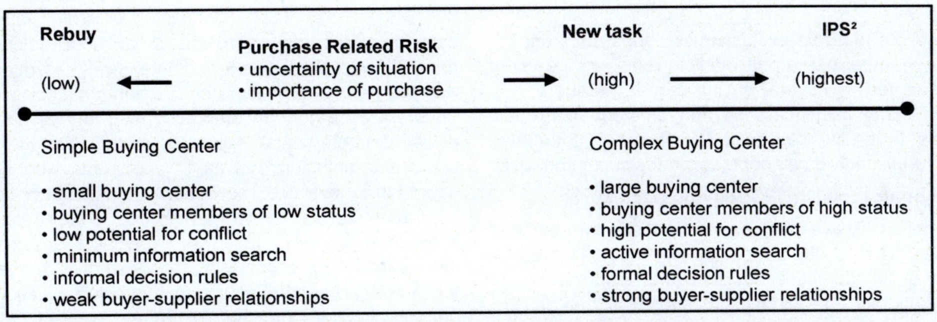

Figure 1: Buying center complexity in case of IPS^2.

In general, the complexity of an industrial buying process increases as higher risk is associated with the purchase [10]. This risk associated with the purchase is based on two dimensions. First, the uncertainty of the situation and, second, the importance of the purchase [11, 12, 13]. Hence, the high complexity of IPS^2 buying situations is based on the risk within this buying situation resulting from the characteristics of an IPS^2 purchase and implementation, mainly from two sources: the flexibility of the IPS^2 solution determining the uncertainty within the situation and the induced change processes that emphasize the importance of the decision.

IPS^2 flexibility: Buyer-supplier relationships in the industrial market are of great importance because these relationships represent resources for developing competitive advantages [14, 15]. In the IPS^2 environment, transactions are always of a relational character. One goal of IPS^2 is to establish a buyer-supplier relationship that cannot be broken up easily by out-suppliers because buyer and supplier work together closely [16]. They do not only have to design a fitting IPS^2 solution collaboratively, but also they have to determine the flexibility and future development of this solution jointly. While a prior relationship or an existing channel of communication between a buying organization and a potential supplier reduces uncertainty [10], a first contract often increases uncertainty. Since there is a risk of being locked-in with a supplier who lacks the needed skills, the buyer has to be sure about choosing the right business partner [17]. Moreover, future behavior of the contractual partner cannot be foreseen, which leads to uncertainty. The difference between an IPS^2 and a common combination of product and services is that the relation between product-based and service-based components is designed integratively [1]. Furthermore, this relation can be adopted flexibly during the life cycle of the IPS^2 by partial substitution of the components [18]. So the buying organization as the external factor has to be highly integrated into the IPS^2 configuration to detect customer needs. In summary, an IPS^2 is a new individual problem solution. This high degree of interaction and the flexible adjustments set up uncertainty concerning IPS^2 purchase decisions.

Induced changes: By the time an IPS^2 is purchased its implementation often induces changes within the buying organization. The implementation of new products, like machines, and the adoption of corresponding services provided by the selling organization, like reparation or maintenance, cause changes in production processes. In general, change processes always imply a movement from the known and established status quo to an unknown status. This unknown status implies a high level of uncertainty [19]. Several employees and the organization as a whole will be affected by the IPS^2 implementation as their field of duty and conditions of employment might change [20]. So the magnitude of negative consequences might be high if a false decision concerning the IPS^2 solution is made. Quite the contrary, the right decision will enable the organization to be ahead with the new technologies resulting in competitive advantages. Therefore, the decision about the IPS^2 purchase is of great importance.

All these aspects emphasize that the buying situation in case of IPS^2 is highly complex and different from common organizational buying processes. Figure 1 emphasizes the previous argumentation by including IPS^2. Thus, we intend to establish a deeper understanding of the decision making process when evaluating IPS^2 solutions. In the following, we will focus on the fact that more participants take part in the buying process. We argue that there are not only more people who take over conventional buying center roles, but there will also be new buying center roles. In addition, we will discuss the buying center members' attitudes regarding the IPS^2 purchase and their corresponding information needs.

3 BUYING CENTER SIZE

3.1 Conventional buying center roles

As already pointed out, the buying center typically consists of five different roles, established by Webster and Wind (1976).

Users are those members who have to work within the new IPS^2, so they have to adjust their work to the new products, technologies, services, and processes that are purchased and implemented.

Buyers are members who have the formal responsibility and authority for contracting with suppliers. They have to pool all purchasing alternatives, the considered suppliers, and their complete offers. Hence, buyers prepare the decision making.

Influencers are involved in the buying process through their direct or indirect provision of information and criteria necessary for the evaluation of alternative buying options. The information provided by influencers either advocates or hinders the purchase. Influencers might be individuals employed in the buying organization, but they can also be outsiders from other organizations, named consultants.

Due to their legitimation, deciders have the authority to choose among different buying alternatives. Even if the majority of buying center members is in favor of the purchase, in the end the decider has the final say and the power to decide against it.

Gatekeepers control the flow of information and materials into and out of the buying center. Thus, they play an important role for the sellers in establishing contact with the potential buying organization. Furthermore, the control of information is a key strategy used to influence the decision process [21]. The selling organization has to persuade gatekeepers to hand over information about their offers and product or service portfolios to the responsible buying center members [3].

Bonoma (1982) characterized a sixth role, the initiators. Initiators are those members who set the buying process off by suggesting the idea. This role can be adopted either by individuals employed within the buying organization or by members of other organizations, mainly the selling organization [5].

It has to be pointed out that role exclusiveness does not exist [3]. One buying center member can act in multiple roles or vice versa one role can be adopted by several buying center members; e.g. there certainly are many influencers so that different individuals adopt the same role, or the buyer often is the initiator of a purchase decision process, so this individual adopts two roles simultaneously.

3.2 New buying center roles in IPS² buying decisions

An IPS² implementation always involves interventions into existent operating processes. That is why people will be highly affected by the consequences of the purchase decision and, therefore, more people will try to influence the situation to their own benefit. For example, unions will not try to influence buying processes of standard machines, but if consequences of a purchase decision are of great importance for employees the unions will take part in the decision making process. However, this is also true in other buying situations when major purchases are contemplated [5], not only because the decision concerns an IPS² purchase and implementation. More interesting is the fact that in the IPS² buying situation there might be specific buying center roles apart from the established ones. These new buying center roles might occur because of the high degree of risk within the situation.

First, the buying organization will seek advice from individuals who are experts in IPS² configurations. In general, these are some special kind of influencers but they exert a much wider influence. These experts are characterized by having specific knowledge and wide experience regarding IPS² implementation and modification because they already helped to develop IPS² solutions for other organizations in the past. Hence, they might be co-developers out of the selling organization or members of another organization. Due to their outstanding knowledge and experience, their influence can be ranked as much higher than the influence of a general influencer. In the following, we will label this role IPS² experts. Their contribution to the decision process is the provision of specific IPS² knowledge and experience regarding the IPS² configuration.

Second, the buying organization needs support by individuals who lead involved employees and the organization as a whole through change processes that occur due to the IPS² implementation. Thus, special change agents will be responsible for this implementation. There are some similarities to traditional change agents as they occur in the change management literature. These change agents have specific knowledge and experience in different change processes and accompany organizations through these [22, 23]. As the change agents' ability to lead the change decides on success or failure of the implementation, they take on a central role in the buying center. However, they will not only influence the buying process but primarily the following implementation and change phase. In the IPS² situation change agents might be individuals of the selling or of a third organization which is specialized in change management. In the following, we will label this role IPS² change agents. Their specific contribution to the decision process is the realistic estimation of the change processes induced and its consequences.

4 ATTITUDES TOWARDS IPS²

4.1 Promoter-Opponent-Model

Every buying center member has a personal attitude regarding the IPS² purchase and implementation. As the implementation induces change processes within the organization, it is likely that some buying center members might be in favor of the purchase and some of them might be against it.

The situation can be analyzed in more detail on the basis of the promoter-opponent-model by Witte (1976). According to the concept of labor division, this model distinguishes between two roles. Promoters actively and intensively advocate an innovation with the help of their hierarchical status or their expert knowledge. Thus, they are differentiated according to their bases of power into power promoters and expert promoters. On the other hand, the same differentiation applies to opponents who actively hinder or delay the innovation. The opponents' resistance to change occurs in terms of administrative barriers due to their hierarchical status or knowledge barriers based on specialist argumentation. Hence, there exists a correspondence between the opponents' barriers and the promoters' contributions to overcome these [8]. While power promoters overcome administrative barriers, expert promoters overcome knowledge barriers. Further investigations of possible barriers detect intra-organizational barriers that hinder innovation processes so that process promoters, who have particular network competencies, are added to this model [24]. A further enhancement of the concept develops the role of relationship promoters, who have network competencies which go beyond the own organizational constraints to overcome inter-organizational barriers [25]. As an IPS² implementation can be seen as an innovation that is advocated or hindered by individuals within the buying organization, the promoter-opponent-model can be easily adapted to the IPS² purchase situation.

Keeping in mind the promoter-opponent-model is of great importance for IPS² supplying organizations. They have to know as exactly as possible which of the identified buying center members might adopt a promoter and which might adopt an opponent role to convince these people of IPS² advantages. For this reason, the promoter-opponent-model will be applied to the buying center concept in the following.

4.2 Buying center roles as promoters or opponents

While employees of the buying organization will be highly concerned by the IPS² indicated change processes, the IPS² buying situation will lead to their polarization into different interest groups, one of promoters and another of opponents. In contrast to common buying situations of mere products or services in which some employees might be indifferent without having a personal interest in the buying decision and its results, in case of IPS² there will probably be no one who is indifferent and therefore abstains from the polarization. Taking a closer look at the buying center members' typical fields of duty in the work place, we can already identify some hints for the combination of their buying center roles to promoter or opponent roles.

The initiators who come up with the purchasing idea can typically be seen as promoters of the IPS² purchase. If they had no intention of advocating the IPS², they would never raise the idea. In the situation considered, the initiation starts outside the own organization, so that the gatekeeper role is a crucial success factor.

As the gatekeepers have the control of information and material flows they can be seen as special influencers. When a supplying organization first informs a gatekeeper about its offers, the gatekeeper is able to decide if the information will be forwarded to persons responsible for the purchase decision. Among other things personal attitudes of the gatekeepers regarding the IPS² solution will influence this information flow. If gatekeepers are promoters, they will forward information, if they are opponents they will not. Hence, we cannot make a general statement about gatekeepers' classification as promoter or opponent. As the gatekeepers usually are in touch with several people outside the own organization their bases of power are their networks and relationships.

The buyers' task is to scan the market for new opportunities and technologies. They prepare a purchasing decision by searching for

Buying Center Role	Indicator		Promoter-Opponent-Model
Initiators	-	⇨	Promoters
Gatekeepers	• forwarding the information (initiator) • not forwarding the information	⇨ ⇨	Relationship promoters Relationship opponents
Buyers	• chance of improving occupational status (initiator) • fear of harming occupational status	⇨ ⇨	Promoters Opponents
Users	• satisfied with their situation • unsatisfied with their situation	⇨ ⇨	Expert opponents Expert promoters
Influencers	• depending on their way of influence	⇨	Promoters or opponents
Deciders	-	⇨	Power promoter or opponent
IPS² experts	• co-developers of the supplying organization • depending on their evaluation of the solution	⇨ ⇨	Expert promoters Expert promoters or opponents
IPS² change agents	-	⇨	Process promoters

Table 1: Buying center roles as promoters and opponents.

and contacting suppliers. Thus, the initiator role is often combined with the buyer role. Even though the supplying organization contacts the buying organization, and therefore a sales person within the supplying organization is deemed as initiator, the buyer presents the purchasing idea to the responsible buying center and therefore initiates a buying process. Buyers initiate the IPS² buying process if they recognize the IPS² advantageousness for the company and a chance of improving their occupational status by coming up with the idea. In this situation, buyers are promoters of the IPS² purchase. Furthermore, buyers might take over gatekeeper roles when the supplying organization contacts them first. If they do not favor the IPS² idea or fear that its implementation might cause problems within the organization so that coming up with the idea might harm their occupational status, they will not initiate the buying process. So both scenarios are possible. Either buyers act as promoters or they are opponents.

Users are typically those individuals who resist forthcoming change processes, as they often fear new situations like modified work conditions. They appear to be concerned with uncertainties based on purchase-related factors within the organization that have an impact on their day-to-day operations [26]. Users have to work within new systems and with new technologies implemented. Due to their experience with and knowledge about the status quo within the organization, they can be characterized as expert opponents. In other situations users might be the initiators of change because they are dissatisfied with the status quo. And even if they do not initiate the buying process by themselves, the users' own conditions of work may improve through the IPS² purchase and implementation. In that case, users are expert promoters.

Influencers within the buying center may be promoters or opponents as well. The IPS² supplier has to identify these buying center members and their way of influence. Some of them might provide information that back up advocacy for the purchase, others might gain and pass over information to the key persons that might hinder the purchase.

Usually, the deciders serve as key persons in a decision process as they have the real decision making power and the final authority to choose among alternative buying actions and suppliers [27]. A decider will be influenced in the decision by all buying center members. Certainly, it depends on the situation if deciders are promoters or opponents, but they might also be primarily indifferent and decide after consulting all other buying center members. Definitely, the IPS² solution has to meet the deciders' requirements which consider the advantageousness of the own company. As the deciders usually have superior hierarchical status in the organization as their basis of power, they are power promoters or opponents.

Whether IPS² experts are promoters or opponents depends on whether they are members of the selling or the buying organization or even of a third organization. If co-developers of the selling organization adopt IPS² expert roles, they are most likely expert promoters, because their aim is to successfully sell the IPS² solution. If the buying organization itself employs IPS² experts, they might be either expert promoters or opponents depending on their personal evaluation of the offered IPS² solution. This is also the case if the IPS² expert is a member of a third organization.

Depending on the assumed extent of change that has to be made and on the IPS² change agents' appraisals which extent of change can be realized in the buying organization, IPS² change agents might be promoters or opponents. Even if induced changes seem to be too complex and problematic, change agents will try to modify the offered IPS² solution, not hinder it. As changes in the organization secure their own jobs in advocating the IPS² purchase, IPS² change agents are rather promoters than opponents. Their typical field of duty is to overcome intra-organizational barriers, thus we label them process promoters.

Table 1 summarizes the findings. Evidently, the combination of promoter and opponent roles to the buying center roles is highly dependent on the personal situation of the buying center member under examination, like current conditions of the workplace, especially the perceived satisfaction with one's performance, or the degree of meeting own objectives.

5 INFORMATION NEEDS OF BUYING CENTER MEMBERS

In view of the fact that specific individuals are targets for marketing effort, not the organization as a whole [3], the selling organization has to convince every single buying center member of the IPS² advantages, because all these individuals contribute to the final decision and try to influence it to their own benefit. Thus, information addressing each buying center member has to be delivered by the selling organization [28]. As we now have categorized the buying center roles corresponding to the promoter-opponent-model, suitable information strategies can be identified.

Buying center roles	Information needs			
	Principles-knowledge	How-to-knowledge	Technical information	Economic information
Users		X	X	
Deciders				X
Influencers	X			X
Gatekeepers	X			X
Buyers				X
Initiators	X	X	X	X
IPS² experts	X	X	X	X
IPS² change agents	X		X	

Table 2: Information needs in case of IPS².

As direct and indirect information search are the most obvious tactics to reduce uncertainty [29], the supplying organization should provide suitable information to buying center members' information needs. For this reason, different knowledge and information have to be defined that can be transferred to buying center members for the purpose of convincing them of IPS² advantages. To create different information strategies we categorize knowledge and information into how-to-knowledge, principles-knowledge, technical information and economic information. How-to-knowledge consists of information necessary in order to correctly use the IPS², answering the question "How does it work?". Information contained in principles-knowledge underlines how the IPS² functions, answering the question "Why does it work?" [30]. Technical information rather consists of technical data on the IPS², like set-up or machining time. Economic information includes, for example, prices or potential cost savings or increases. Hence, we define information strategies as combinations of these different kinds of information. The goal is to figure out which buying center role demands which type of information.

To handle the new situation, users have to know how the IPS² works. In order to overcome users' resistance to change they need to be informed about the change processes. Hence, the information strategy to convince them of the IPS² advantages has to provide how-to-knowledge and technical information.

Deciders very often are capable of being influenced by economic information. As they are indifferent at first and influenced by every other buying center member, the decisive criteria might be some argument like the price, as the major key to the product's success [27], or potential cost savings in the end.

To convince influencers of IPS² advantages, the selling organization has to know about the kind of influence they aim to exert. As influencers are neither working within the new system nor deciding about the purchase, they should obtain principles-knowledge and economic information to understand conditions and consequences of an IPS² purchase and its subsequent implementation.

As they are special influencers, gatekeepers are to be convinced by the same information strategy and should receive principles-knowledge and economic information as well. To make sure that they forward information they have to understand the IPS² idea and how it will work in the buying organization.

Buyers are typically interested in economic information like price or possible cost reductions and savings. Furthermore, to correctly identify and evaluate buying alternatives they have to be informed about the complete product and service portfolio of the supplying organization. This should be the primer information handed out by the supplying organization when offering the IPS² solution.

Because initiators should be prepared to answer all questions concerning the IPS², they should get as much information as possible, independent of the kind of information.

In case of the newly developed buying center roles we have to distinguish between situations in which they are employed by the selling organization or not. If the selling organization employs IPS² experts and IPS² change agents they should obviously know everything about the offered IPS² solution. As the buying or even another organization employs these buying center roles, they should be informed as well.

IPS² experts should have as much information as possible, but as they are experts already, they do not have to be informed about how-to-knowledge or technical knowledge in general but maybe concerning the specific IPS² offered. Furthermore, economic and principles knowledge should be provided.

IPS² change agents need to know everything about the change processes induced by the IPS² purchase or rather its implementation, so they should be given how-to-knowledge and technical information to prepare users to adjust to new processes. Table 2 shows the information requirements of buying center members according to the adopted buying center role.

6 SUMMARY

Our theoretical approach reveals the importance of rethinking and adjusting the existing knowledge of industrial buying behavior for IPS² situations instead of adopting it. Otherwise, several influential and important aspects of the buying process would be ignored. So far, literature has neglected an examination of buying situations that are characterized by implementing consequences of the magnitude as in IPS² buying situations. The risk associated with that kind of buying decision is more distinctive so that the complexity is higher than in other buying situations. That is why not only more people try to influence the decision, but also new buying center roles occur. These new buying center roles contribute to the decision making process by providing their knowledge of and experience with the problem solution and change processes. First, IPS² experts are competent in the field of configuring the right IPS² solution. Second, IPS² change agents' skills back up the IPS² implementation while they lead the organization through induced change processes. To correctly inform buying center members, in accordance with their buying center role, it is necessary to know about their attitudes towards the IPS² purchase. Therefore, we combined the promoter-opponent-model with the buying center concept and then discussed the information needs of the buying center members.

Future empirical work has to investigate the assumed buying center structure in IPS² environments within long-term studies. The previous combination of buying center roles and promoter and opponent roles emerged from a static view. We considered the situation before the IPS² purchase. Given the final purchase decision is in favor of the IPS² implementation, the attitudes of the buying center members might change again. Some of the opponents might be satisfied with the new situation after the change or former promoters might be dissatisfied. Hence, it is necessary to examine the attitudes of the individuals involved during the IPS² implementation and working phase in a dynamic view. The empirical work will also contribute to the change management literature, while the induced change situation is different from usual change processes, because the initiation starts outside of the own organization. Some communication tactics and strategies from the traditional change management literature might be adopted, but it seems to be more complicated to convince the organization as an outsider.

7 ACKNOWLEDGMENTS

We extend our sincere thanks to the German Association of Research (DFG) for funding this research project and to all who contributed to preparing this paper.

8 REFERENCES

[1] Meier, H.; Uhlmann, E.; Kortmann, D. (2005): Hybride Leistungsbündel. Nutzenorientiertes Produktverständnis durch interferierende Sach- und Dienstleistungen, in: wt Werkstattstechnik online, Vol. 95, No. 7/8, pp. 528-532.

[2] Meier, H.; Roy, R.; Seliger, G. (2010): Industrial Product Service Systems – IPS², in: CIRP Annals - Manufacturing Technology, Vol. 59, No. 2, pp. 607-627.

[3] Webster, F. E., Wind, Y. (1972): A General Model for Understanding Organizational Buying Behavior, in: Journal of Marketing, Vol. 36, No. 2, pp. 12-19.

[4] Robinson, P. J., Faris, C. W., Wind, Y. (1967): Industrial Buying and Creative Marketing, Allyn & Bacon, Inc., Boston.

[5] Bonoma, T. V. (1982): Major sales: Who really does the buying?, in: Harvard Business Review, Vol. 60, No. 3, pp. 111-119.

[6] Kauffman, R. G. (1996) Influences on organizational buying choice processes: future research directions, in: Journal of Business & Industrial Marketing, Vol. 11, No. 3/4, pp.94-107.

[7] Sheth, J. N. (1996): Organizational buying behavior: past performance and future, in: Journal of Business & Industrial Marketing, Vol. 11, No. 3/4, pp.7-24.

[8] Witte, E. (1976): Kraft und Gegenkraft, in: Zeitschrift für Betriebswirtschaft, Vol. 46, No. 4/5, pp. 319-326.

[9] Doyle, P.; Woodside, A. G.; Michell, P. (1979): Organizations Buying in New Task and Rebuy Situations, in: Industrial Marketing Management, Vol. 8, No. 1, pp. 7-11.

[10] Johnston, W. J.; Lewin, J. E. (1996): Organizational Buying Behavior: Toward an Integrative Framework, in: Journal of Business Research, Vol. 35, No. 1, pp. 1-15.

[11] Bunn, M. D.; Liu, B. S. -C (1996): Situational Risk in Organizational Buying: A Basis for Adaptive Selling, in: Industrial Marketing Management, Vol. 25, No. 5, pp. 439-452.

[12] Henthrone, T. L.; LaTour, M. S.; Williams, A. J. (1993): How Organizational Buyers Reduce Risk, in: Industrial Marketing Management, Vol. 22, No. 1, pp. 41-48.

[13] Arndt, J. (1967): Role of Product-Related Conversations in the Diffusion of a New Product, in: Journal of Marketing Research, Vol. 4, No. 3, pp. 291-295.

[14] Whipple, J. M.; Lynch, D. F.; Nyaga, G. N. (2010): A buyer's perspective on collaborative versus transactional relationships, in: Industrial Marketing Management, Vol. 39, No. 3, pp. 507-518.

[15] Hald, K. S.; Cordón, C.; Vollmann, T. E. (2009): Towards an understanding of attraction in buyer–supplier relationships, in: Industrial Marketing Management, Vol. 38, No. 8, pp. 960-970.

[16] Rese, M., Karger, M., Strotmann, W. (2009): The dynamics of Industrial Product Service Systems (IPS²) – using the Net Present Value Approach and Real Options Approach to improve life cycle management, in: CIRP Journal of Manufacturing Science and Technology, Vol. 1, No. 4, pp. 279-286.

[17] Stump, R. L.; Heide, J. B. (1996): Controlling Supplier Opportunism in Industrial Relationships, in: Journal of Marketing Research, Vol. 33, No. 4, pp. 431-441.

[18] Rese, M. (2006): IPS² Cost Decisions and Price-Decisions in time of Value based management, Plötner, O.; Spekman, R. (eds), Bringing Technology to Market, Wiley, New York, pp. 61-76.

[19] Leana C.; Barry, B. (2000): Stability and Change as Simultaneous Experiences in Organizational Life, in: Academy of Management Review, Vol. 25, No. 4, pp. 753-759.

[20] Oreg, S. (2003): Resistance to Change: Developing an Individual Differences Measure, in: Journal of Applied Psychology, Vol. 88, No. 4, pp. 680-693.

[21] Dawes, P. L.; Lee, D. Y.; Dowling, G. R. (1998): Information Control and Influence in Emergent Buying Centers, in: Journal of Marketing, Vol. 62, No. 3, pp. 55-68.

[22] Ottaway, R. N. (1983): The Change Agent: A Taxonomy in Relation to the Change Process, in: Human Relations, Vol. 36, No. 4, pp. 361-392.

[23] Lippitt, R.; Watson, J.; Westley B. (1958): The dynamics of planned change, Harcourt, Brace and World, New York.

[24] Hauschildt, J. (2004): Innovationsmanagement, 3rd edition, Vahlen, Munich.

[25] Gemünden, H. G., Walter, A. (1995): Der Beziehungspromotor – Schlüsselperson für inter-organisationale Innovationsprozesse, in: Zeitschrift für Betriebswirtschaft, Vol. 65, No. 9, pp. 971-986.

[26] Spekman, R. E.; Ford, G. T. (1977): Perceptions of Uncertainty within a Buying Group, in: Industrial Marketing Management, Vol. 6, No. 6, pp. 395-403.

[27] Berkowitz, M. (1986): New Product Adoption by the Buying Organization: Who Are the Real Influencers?, in: Industrial Marketing Management, Vol. 15, No. 1, pp. 33-43.

[28] Ghingold, M.; Wilson, D. T. (1998): Buying center research and business marketing practice: meeting the challenge of dynamic marketing, in: Journal of Business & Industrial Marketing, Vol. 13, No. 2, pp. 96-108.

[29] Gemünden, H. G. (1985): Perceived risk and information search. A systematic meta-analysis of the empirical evidence, in: International Journal of Research in Marketing, Vol. 2, No. 2, pp. 79-100.

[30] Rogers, E. M. (2003): Diffusion of innovations, 5th edition, The Free Press, New York.

Building a Network of SME for a Global PSS Infrastructure in Complex High-Tech Systems: Example of Urban Applications

Alan Lelah[1], Fabrice Mathieux[1], Daniel Brissaud[1]

[1] G-SCOP Laboratory, Grenoble University, Grenoble, France

Abstract

While small dynamic firms propose PSS applications based on new technologies quite easily, large infrastructures are often required. SMEs don't have the capacity to do this alone. For complex systems, different SMEs combine efforts to build an efficient infrastructure. Other SMEs can join and propose other PSS, creating a network. As networks grow the SMEs must understand their place within the structure and how they interact with other members. A tool is proposed to visualize relations between partners and illustrate the construction of the service chains composing the PSS. The tool is then applied to the case of urban PSS.

Keywords:

Network of SME; PSS Organization; Machine-to-Machine Application

1 INTRODUCTION

The search for more sustainable models of society has lead to the introduction of Product-Service Systems into Industry (PSS or IPS²). PSS "consists of a mix of tangible products and intangible services designed and combined so that they jointly are capable of fulfilling final customer needs" [1]. Product-based services are an alternative to simply selling products. The responsibility of the producer is extended into anything between simply adding services to the sale of functionalities [1]. The movement towards dematerialization is amplified with the opportunities opened out by Information and Communication Technologies (ICT) [2] capable of providing monitoring systems allowing real-time adaptation of services and facilitating maintenance. Dynamic high-technology Small and Medium Enterprises (SME) propose innovative services and adapt to evolving demands very quickly. However, their small size sometimes makes it difficult to penetrate public or other big markets. At the same time, with growing numbers of complex services, there is a move towards service infrastructures [3] with multiple actors involved, leading to complex organisations of supply chains. One solution is that a larger firm, an operator-integrator, proposes and operates this kind of infrastructure integrating products and services provided by a network of SME partners. This enables the combination of strong well-established operator-integrators together with the flexibility and competitiveness of the SMEs. However, it is necessary to understand the relations between the actors if the SMEs want to have their say and not be relegated to simple subcontractors of the operator-integrators. The partnership should benefit from the diversities of the SMEs and their dynamics without being tied down to the red tape of big operator-integrators pushing for slowly matured replies to market change.

This article seeks to clarify the relative positions of the actors: amongst the SMEs themselves, and with the operator-integrator. After this introduction the transformation from vertical services to a horizontal service infrastructure will be described in more detail along with the organisational problems that are posed. Then a representation of the service-oriented architecture showing how the SMEs interact with one another and the operator-integrator will be described. In the next section the model will be applied to an example of an urban service showing how the representation would penetrate the complexity of the situation and the links between the actors will be described.

2 ORGANISATION FOR GLOBAL INFRASTRUCTURE

2.1 From Vertical Services to a Global Infrastructure

The transformation from direct vertical services provided by different operators or SMEs to a global infrastructure calling on a network of actors is described in Figures 1 and 3. Figure 1a shows an *initial service provider (ISP)* that provides vertical services to a *population (P)*. The services are vertical as they support just one specific service.

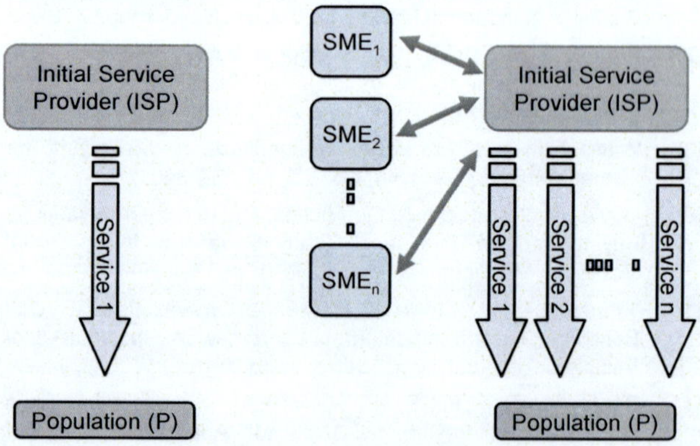

(a) Vertical service (b) Service Provider - SME relations

Figure 1: Vertical Services.

Examples of the couple initial service provider-population could be a local town council providing services to its citizens; a hospital providing health services to its patients; or a truck vendor supplying a fleet of vehicles to industrials. Today, many of these services are

J. Hesselbach and C. Herrmann (eds.), *Functional Thinking for Value Creation: Proceedings of the 3rd CIRP International Conference on Industrial Product Service Systems, Technische Universität Braunschweig, Braunschweig, Germany, May 5th - 6th, 2011*,
DOI 10.1007/978-3-642-19689-8_60, © Springer-Verlag Berlin Heidelberg 2011

high-technology applications and the initial service provider may have neither all the competences internally nor have the intention to acquire these competences. In this case the initial service provider would turn to an SME that can bring skills and know-how to provide the service directly to the population concerned, or through the initial service provider. For example in a truck-vending IPS² the injectors could be supplied by a SME. The life-cycle of the injectors is very much shorter than the life-cycle of the truck and it is environmentally interesting to replace used injectors with remanufactured ones [4]. This could be a good case of PSS as the SME would replace a faulty injector and remanufacture the old one. In the same way, another SME could, for example, deal with the tyres or propose other complementary services. The number of IPS² calling on different SMEs can rapidly become significant. This may lead to the situation shown in Figure 1b. The initial service provider now has to deal with a multitude of SMEs, each proposing a different service.

Another evolution today improves IPS² with ICT [2]. For example in the case of the gear-box the SME providing the IPS² could use Machine-to-Machine (M2M) communication [5] to monitor the gear-box in the trucks wherever the trucks may be. M2M communication here means that some sensors are placed in the gear-box and measure relevant parameters like oil, temperature, wear, etc. The sensors are equipped with a communication system sending the measures regularly to a central managing system in the truck. In turn this system would relay the important information via GSM, through the telecommunication network and finally to a service platform of the SME. This is called M2M because up to the service platform no human intervention is necessary. The machines (sensors, command unit, telecom network and service platform) communicate with one another. At the end of the line, the SME can access real-time knowledge of the functioning of the gear-box PSS, adapting the service or facilitating intervention in case of need. The M2M infrastructure is shown in Figure 2.

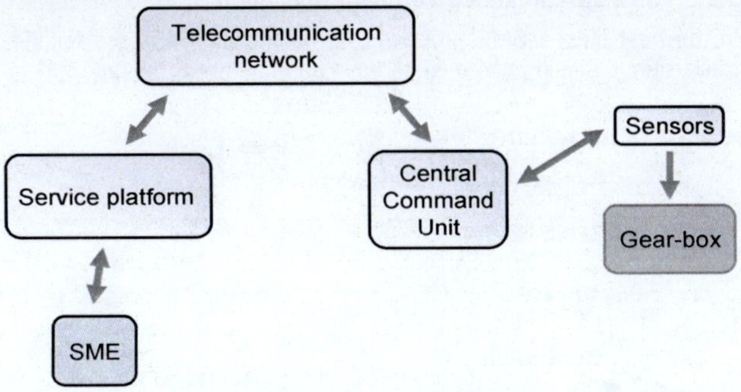

Figure 2: M2M communication.

However, the system can get more complex. The SME that manufactures the gear-box may not be competent in M2M and could turn to another SME who would provide the M2M PSS infrastructure, and so on. Along with the multiplication of services we will soon end up with a complicated network of SMEs and the risk that different infrastructures concurrence each other. Apart from a problem of sustainability that would arise, the infrastructures could start to interfere technically with one another. At one time it would become interesting that the vertical services running in parallel make use of a common service infrastructure (Figure 3).

A new organisation would be set up. The different *SMEs* would form a network, together with a pertinent *operator-integrator (OI)* who would be the natural leader of the group. The initial service provider will only have one interlocutor to speak to and will not have to worry about problems linked to new SMEs joining the group or old ones leaving. All this would be done by the operator-integrator who will ensure the continuity and the coherence of the group. A good example of such a network is SensCity [3, 6], a French project for a global infrastructure of services in urban areas.

Organisational problems are now transferred to the network of SMEs and the operator-integrator. Even though the know-how of the operator-integrator is more adapted to the core question, the organisation of such a network can become quite complicated and it would be important to distinguish the different relations between the actors.

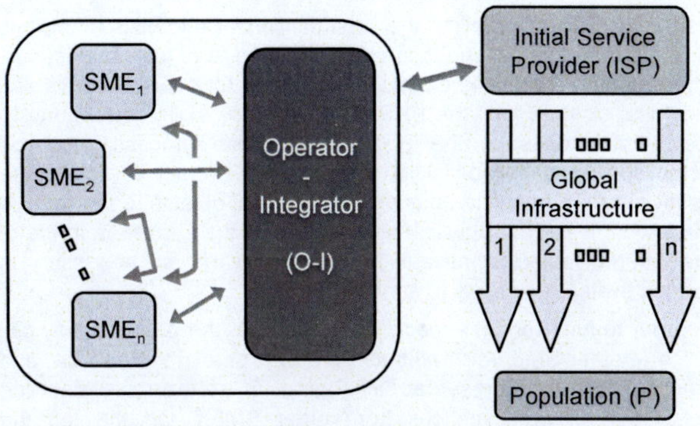

Figure 3: Horizontal services with a global service infrastructure.

2.2 Models for the Organization of PSS

Shostack [7] proposed service blueprints to facilitate the design of services. A line of visibility differentiated what the customer perceived of the more complicated operations necessary to operate the service. Shimomura et al. [8] extended this notion for the case of PSS in order to integrate both the product and the service. Maussang et al. [9] described a PSS as being a set of service units and physical objects, while Tomiyama [10] declared that artefacts were usually service channels or service contents. These representations were very efficient for modelling complex services. However, they did not clearly describe the organisation of the service between the actors. It was not always easy to identify the actor that provided each service and the actor or client that received the service.

We proposed a similar diagram in which the accent was placed on the relations between partners more than the details of the service and its physical support [11]. However, in the case of services providing channels, it is not always clear who commands what service. A modified model is therefore proposed to distinguish services providing channels from services providing contents.

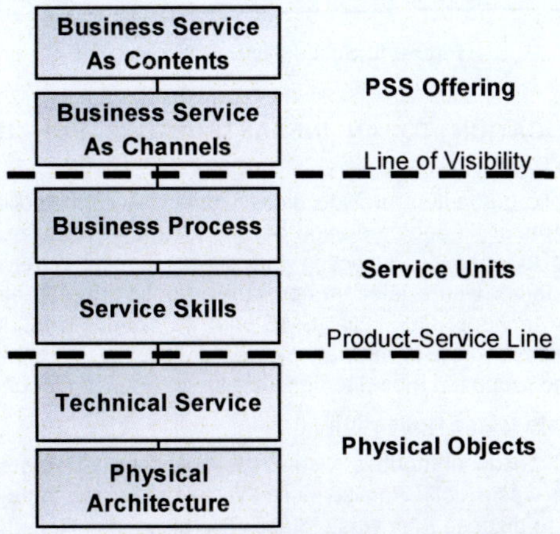

Figure 4: Service offering; service; physical representation.

Basically, in the modified model, above the line of visibility, the PSS offering is proposed by a provider to a receiver. It describes the business service as experienced by the client or partner. The receiver requires a service and turns to the provider who supplies the service. The PSS offering must satisfy the needs of the receiver. Services providing channels and services providing contents are distinguished in two different levels, as shown in Figure 4.

Below the line of visibility, the organisation of the PSS offering is structured and service units are distinguished from physical objects. The product-service line expresses the need for business skills and technical skills to fulfil the PSS. The business skills correspond to the service units of the PSS. A business process must be established and service skills needed to perform the service are associated. The service units rely on physical objects to operate. At the lowest level, the physical architecture of the system is made of the different objects that form it. Technical skills are necessary to run the system and reply to service demands.

In order to fulfil certain aspects of the service, the provider may call on to another SME who will take care of a part of the service, and so on. This is represented in Figure 5 with arrows showing solicitation of new services from other SMEs. On the left the population (P) solicits the final service. The initial service provider (ISP) solicits two SMEs. SME1 provides a channel and SME2 provides contents through the channel. SME2, in turn, solicits another service (not shown in the figure).

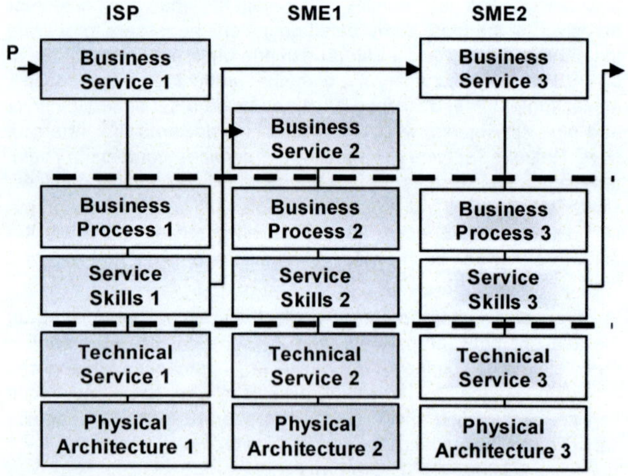

Figure 5: Successive providers.

3 APPLICATION TO AN INFRASTRUCTURE FOR URBAN SERVICES

Local public authorities provide glass waste collection facilities for the residents of an agglomeration of towns. An SME then proposes a service to optimise collecting the cullet. Finally a network of SMEs, together with a telecom operator proposes an infrastructure for urban services. Our initial SME with its optimisation of glass waste collection now forms part of the network. In this section we will use the proposed model to describe the different situations.

3.1 Waste Glass Collection

The local public authorities insure the collection of waste glass through a waste collection service. Waste-banks are installed for residents to dispose their waste glass. The glass is collected in bins that are emptied by trucks going on regular rounds. In the classical service the trucks systematically run through the waste-bank park looking for bins that are sufficiently full.

Service-Oriented Organization Diagram

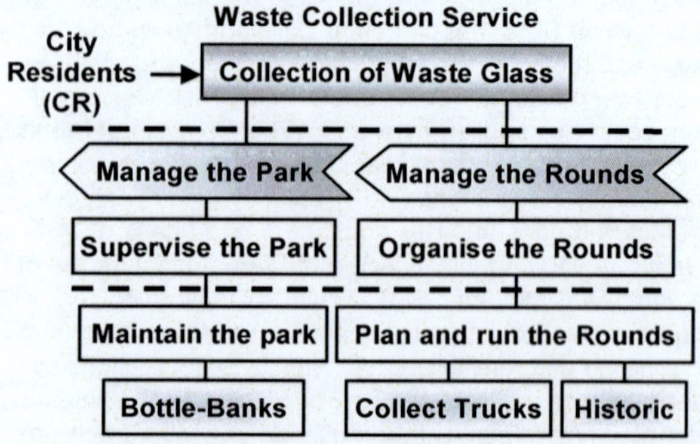

Figure 6: Urban Waste Glass Collection.

The service is represented in Figure 6. As the service is basic and requires no service channel, the top level does not have to be split in two for channel and content services. The population concerned by the service is the *city residents (CR)* and the initial service provider is the *waste collection service* run by the public authorities. There are basically two major business processes in the service: managing the park and managing the rounds. Arrow-shaped boxes have been used to show successive stages of the overall business process provided by the waste collection service. In this case they indicate that in order to manage the park, the rounds have to be managed. Managing the park signifies supervising the park and the corresponding technical service concerns the maintenance of the bins in the bottle-banks. Managing the rounds implies the organization of collection rounds and technically running the trucks and stocking the history of the rounds. The public authorities provide the service alone.

3.2 The M2M-Enhanced Vertical Application

In the next step a SME proposes an optimization process for the truck rounds making use of sensors in an M2M network [12, 13].

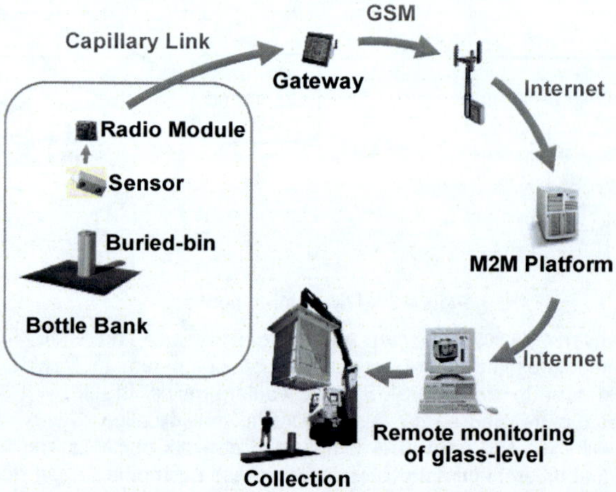

Figure 7: M2M-enhanced waste glass collection.

Figure 7 resumes the service. In the bottle-banks, the bins have embedded ultrasonic sensors that measure the glass levels every day. The sensors are coupled to communication modules that send the information to gateways in the proximity of the banks. In this way the whole park is covered and the data collected is relayed via the telecom network to an urban service platform. Here, the information is treated to optimize the collection of the glass and the results are then sent on to the waste collection service run by the local authorities. The complete service is provided for the local authorities by a single SME. Organization is straightforward: the data is treated within the SME.

Service-Oriented Organization Diagram

Figure 8 depicts the situation. It shows how the SME is solicited by the waste collection service run by the local authorities in order to satisfy the needs of the city residents. Communication between the SME and the waste collection service is done by internet connexions and the two entities must solicit a telecom operator. The SME organizes the waste collection. This means optimizing the rounds using a service platform server. The optimization requires daily information on the glass-levels. The SME must monitor the glass-levels. Technically it is necessary to plan the measures and collect the data. Additionally, server operation and support have to be insured. The SME needs to use classical telecom services of GSM and IP connection to collect the glass level data measured in the bottle-banks. On the field the SME would manage the sensor network with the gateways and insure correct functioning of the park. The ICT services are solicited only as service channels that transfer data between the gateways and the service platform server and then between the SME and the local authorities. The diagram shows the solicitation of ICT services by the SME as well as by the local authorities.

3.3 A Global Urban M2M Infrastructure

An Urban M2M Infrastructure

Vertical applications are relatively easy to build by innovative SMEs, but they are not so efficient in terms of resources. Other than the problem of piling up sensor network structures coming from different applications, the SME must take care of the whole service and installs its own sensor network. It has to use the telecom network as a simple client with no possibility to guarantee QoS in the case of network failure. In addition, the SME has to support maintenance in different towns and even in different countries. It has been proposed [3] to set up an infrastructure to support M2M sensor-enhanced services in urban areas. The infrastructure is built by a consortium of SMEs together with a telecom operator. The telecom operator brings his know-how and reputation, making the consortium a natural partner to work with municipalities or other local authorities. The size and the stature of the telecom operator is a guarantee of the sustainability of the consortium. The local authority will only need to deal with the telecom operator representing the consortium and the consortium as a whole will take on the role of insuring that new members proposing new or better applications can join and that leaving members will be replaced when necessary.

Within the consortium, some SMEs will provide sensors to collect data in the perimeter covered by the infrastructure. The variety of applications proposed to the local authorities will come from their dynamism. Some of these SMEs will provide their own applications while others would just transfer the data collected to specialized SMEs providing a panoply of services. Yet others will deal with the local M2M wireless communication, including the protocols and ensuring the evolution of these technologies in open state-of-the-art standards. Specialized SMEs will provide gateway coverage for all the networks, and when necessary, complete the gateways with repeaters to collect data when it is more convenient.

Other SMEs will be involved with the functioning of an urban service platform. The platform will collect and control the data coming from the sensor network; stock and manage the data; deal with brokerage and other legal questions concerning the access to the information and finally sending information to the end-users or other servers dealing with specific applications.

Service-Oriented Organization Diagram

The organization has now become quite complex to manage and the service-oriented organization diagram can help clarify the relations between the different partners. The case of waste glass collection is shown in figure 9. The infrastructure can be considered as imbrications of successive B2B services. As before, the waste collection service asks the service providing SME to help optimize the collection rounds for the park of bottle-banks in order to satisfy the needs of the city residents. The telecom IP network serves as a channel to exchange data, while the service provider needs to monitor the glass levels in order to run the optimization. This time the service provider will turn to the urban platform host, shown in the diagram just as one entity for simplification. In reality there will be many SMEs working on different parts of the platform (data control, data management, brokerage, etc.) but they will not be considered here. Other service providers can also address the platform. They may offer other applications like providing information on pollution. They are not shown in the figure either. The telecom operator is always seen as a channel for data transfer. However, in the organizational choice described in the figure, the gateway operator is considered as a content provider. In addition to providing a simple gateway, it also manages the sensors and plays a role in stocking information. It can collect data at different moments to be combined and sent together. The gateway operator is thus solicited by the urban platform host to manage the sensors and in turn solicits the different sensor operators in order to collect measured data.

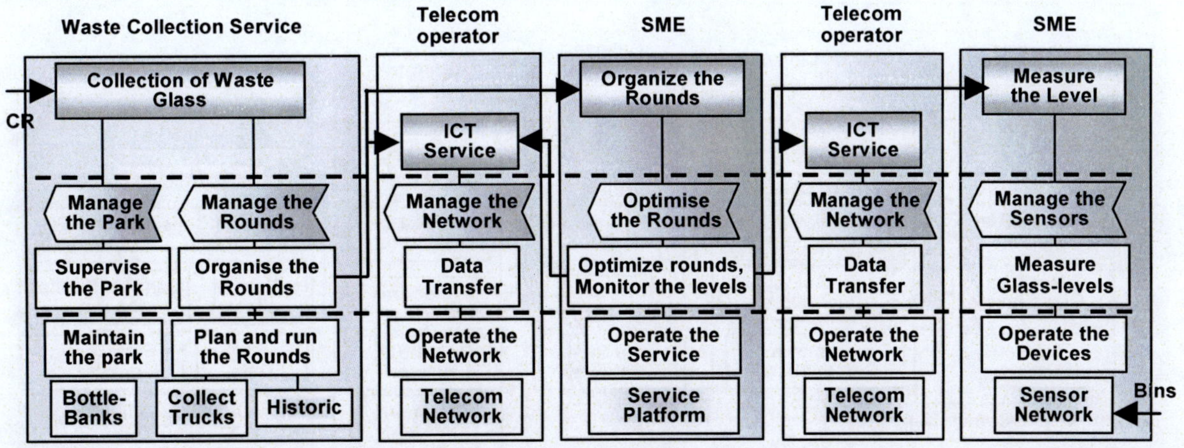

Figure 8: SME service-provider for glass waste collection.

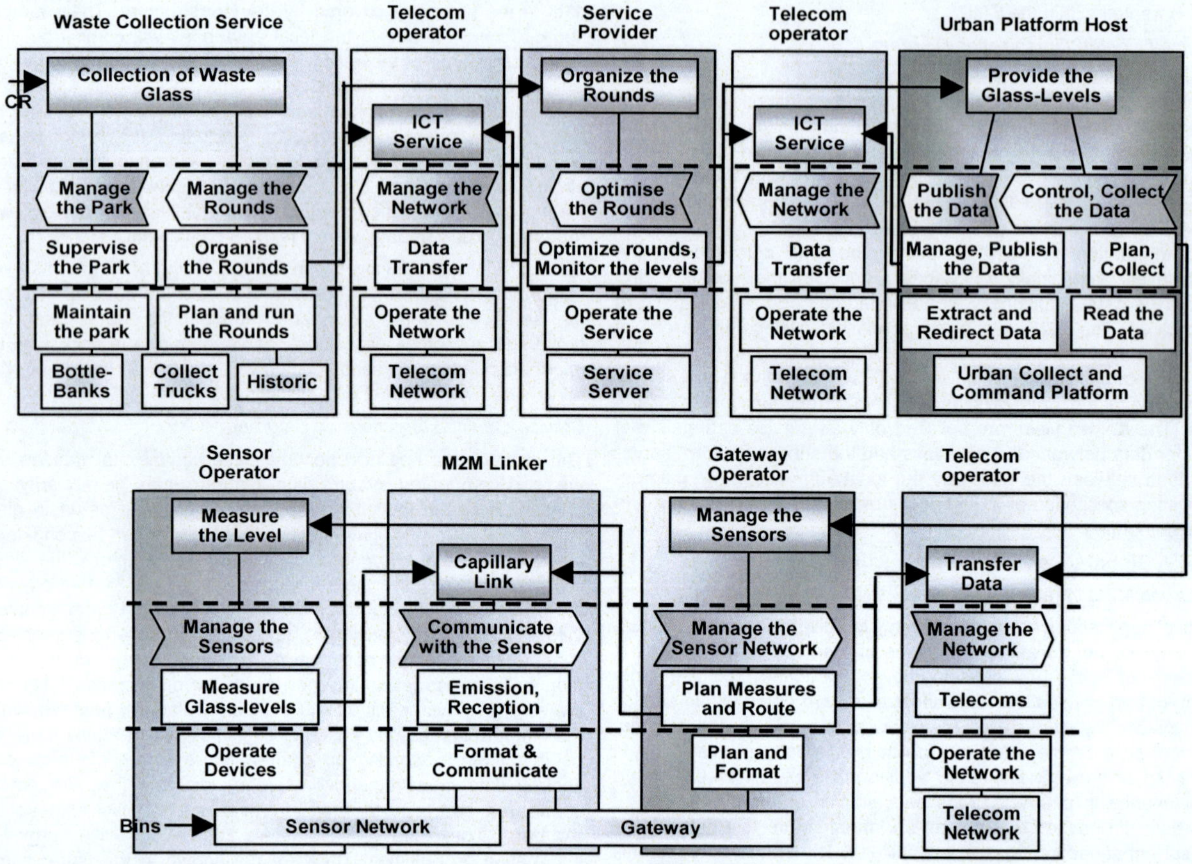

Figure 9: M2M-enhanced PSS network.

Of course the full process is more complicated and the schema has been simplified in order to give an overall view.

Alternative Scenario

However, this is not the only organization possible. Another scenario of organization that can be proposed is shown in figure 10. Here the gateway is considered as a service channel and does not take part in managing the sensors. In this case the gateway operator is solicited by both the SME sensor operator and the telecom operator to relay information. In this scenario, the urban platform directly manages the different sensor networks and takes charge of the collection. The relations between the different SMEs and the telecom operator have changed and the repartition of work done by each actor is modified even though the overall workload is identical.

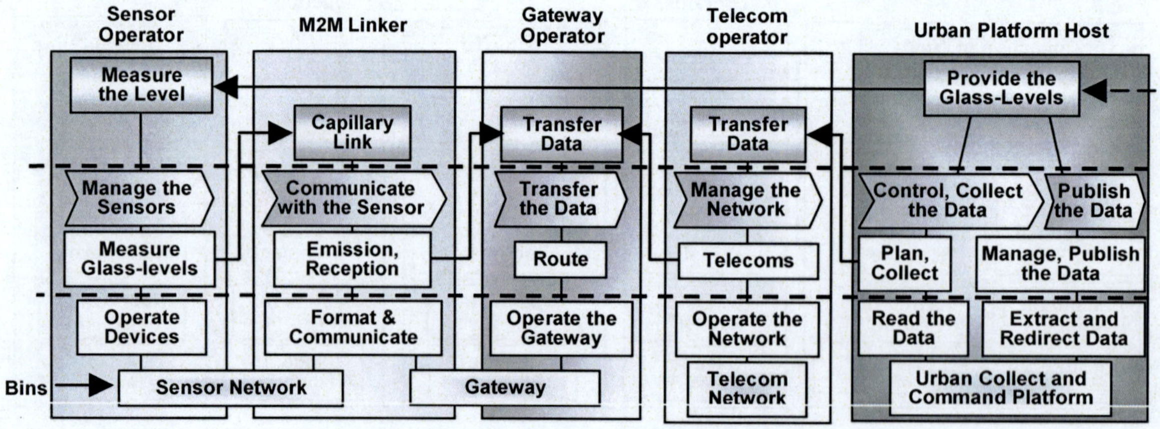

Figure 10: Alternative Scenario.

4 CONCLUSION

A service-oriented organization diagram for a network of SMEs together with an operator-integrator has been proposed. It helps clarifying the passage from vertical services to an infrastructure for PSS in different environments. The diagram was applied to the French project SensCity proposing an infrastructure for M2M-enhanced urban services. The case of glass waste collection showed the different organizational transformations involved in the passage from the basic service to a vertical M2M service and finally to a mutualised infrastructure for services in the city. The diagram was also used to discuss the implications of a possible alternative scenario within the consortium.

5 ACKNOWLEDGMENTS

The authors would like to thank Lucien Vincent and Xavier Boucher, LSTI, as well as Claude Pellegrin, COACTIS, and other participants of the GOSPI cluster in the Rhone-Alps region for the interesting debates that led to the elaboration of the diagrams representing service relations used in this article. We have presented a network of complementary SMEs lead by an operator integrator to provide a global infrastructure for urban services.

6 REFERENCES

[1] Tukker, A., Tischner, U., (2006): Product-services as a research field: past, present and future. Reflections from a decade of research, in Journal of Cleaner Production, 14(17), pp. 1552–1556.

[2] Heiskanen, E., Halme, M., Jalas, M., Kärnä, A., Lovio, R., (2001): Dematerialization: The Potential of ICT and Services, Ministry of Environment, Helsinki.

[3] Lelah, A., Mathieux, F., Brissaud, D., Gimeno, V. (2010): SensCity: a new project opening the way for sustainable services in the city based on a mutualised M2M infrastructure, in Proceedings of the 2nd CIRP IPS² Conference, pp. 99-106, Linköping, Sweden.

[4] Amaya, J., Zwolinski, P., Brissaud, D., (2010): Environmental benefits of parts remanufacturing: the truck injector case, in Proceedings of the 17th CIRP International Conference on Life Cycle Engineering, Hefei, China

[5] Orange, FING, Syntec Informatique (2006): White paper Machine To Machine, stakes and prospects, France.

[6] SensCity, **http://senscity.minalogic.net**

[7] Shostack, G. L., (1987): Service Positioning Through Structural Change, in Journal of Marketing, 51, pp. 34-43.

[8] Shimomura, Y., Hara, T., Arai, T., (2009): A unified representation scheme for effective PSS development, in CIRP Annals - Manufacturing Technology, 58(1), pp. 379-382.

[9] Maussang, N., Zwolinski, P., Brissaud, D., (2009): Product-service system design methodology: from the PSS - architecture design to the products specifications, in Journal of Engineering Design, 20(4), p. 349.

[10] Tomiyama, T. (2001): Service engineering to intensify service contents in product life cycles, in Proceedings of EcoDesign 2001 pp. 613-618, Tokyo, Japan.

[11] Lelah, A., Mathieux, F., Brissaud, D., Vincent, L., (2010): A Network of Complementary SMEs for a Global Infrastructure for Services: the Example of Environmental Urban Services, in Proceedings of Pro-ve 2010, pp. 73-80, St. Etienne, France.

[12] BH Environnement, **http://www.bhenvironnement.com/**

[13] Houang, T., Lelah, A., Mathieux, F., Brissaud, D., Gimeno, V. (2010): Environmental Evaluation of Machine-to-Machine Services: the case of Glass Waste Collection, in Proceedings of the 2nd CIRP IPS² Conference, pp. 75-82, Linköping, Sweden.

Index of Authors

J. Hesselbach and C. Herrmann (eds.), *Functional Thinking for Value Creation: Proceedings of the 3rd CIRP International Conference on Industrial Product Service Systems, Technische Universität Braunschweig, Braunschweig, Germany, May 5th - 6th, 2011*,
DOI 10.1007/978-3-642-19689-8, © Springer-Verlag Berlin Heidelberg 2011